Lecture Notes in Computational Vision and Biomechanics

Volume 36

Research related to the analysis of living structures (Biomechanics) has been carried out extensively in several distinct areas of science, such as, for example, mathematics, mechanical, physics, informatics, medicine and sports. However, for its successful achievement, numerous research topics should be considered, such as image processing and analysis, geometric and numerical modelling, biomechanics, experimental analysis, mechanobiology and enhanced visualization, and their application on real cases must be developed and more investigation is needed. Additionally, enhanced hardware solutions and less invasive devices are demanded. On the other hand, Image Analysis (Computational Vision) aims to extract a high level of information from static images or dynamical image sequences. An example of applications involving Image Analysis can be found in the study of the motion of structures from image sequences, shape reconstruction from images and medical diagnosis. As a multidisciplinary area, Computational Vision considers techniques and methods from other disciplines, like from Artificial Intelligence, Signal Processing, mathematics, physics and informatics. Despite the work that has been done in this area, more robust and efficient methods of Computational Imaging are still demanded in many application domains, such as in medicine, and their validation in real scenarios needs to be examined urgently. Recently, these two branches of science have been increasingly seen as being strongly connected and related, but no book series or journal has contemplated this increasingly strong association. Hence, the main goal of this book series in Computational Vision and Biomechanics (LNCV&B) consists in the provision of a comprehensive forum for discussion on the current state-of-the-art in these fields by emphasizing their connection. The book series covers (but is not limited to):

- Applications of Computational Vision and Biomechanics
- Biometrics and Biomedical Pattern Analysis
- Cellular Imaging and Cellular Mechanics
- Clinical Biomechanics
- Computational Bioimaging and Visualization
- Computational Biology in Biomedical Imaging
- Development of Biomechanical Devices
- Device and Technique Development for Biomedical Imaging
- Experimental Biomechanics
- Gait & Posture Mechanics
- Grid and High Performance Computing on Computational Vision and Biomechanics
- Image Processing and Analysis
- Image processing and visualization in Biofluids
- Image Understanding
- Material Models
- Mechanobiology
- Medical Image Analysis
- Molecular Mechanics
- Multi-modal Image Systems
- Multiscale Biosensors in Biomedical Imaging
- Multiscale Devices and BioMEMS for Biomedical Imaging
- Musculoskeletal Biomechanics
- Multiscale Analysis in Biomechanics
- Neuromuscular Biomechanics
- Numerical Methods for Living Tissues
- Numerical Simulation
- Software Development on Computational Vision and Biomechanics
- Sport Biomechanics
- Virtual Reality in Biomechanics
- Vision Systems
- Image-based Geometric Modeling and Mesh Generation
- Digital Geometry Algorithms for Computational Vision and Visualization

In order to match the scope of the Book Series, each book has to include contents relating, or combining both Image Analysis and mechanics.

Indexed in SCOPUS, Google Scholar and SpringerLink.

More information about this series at http://www.springer.com/series/8910

Gerard A. Ateshian · Kristin M. Myers ·
João Manuel R. S. Tavares
Editors

Computer Methods, Imaging and Visualization in Biomechanics and Biomedical Engineering

Selected Papers from the 16th International Symposium CMBBE and 4th Conference on Imaging and Visualization, August 14–16, 2019, New York City, USA

Springer

Editors
Gerard A. Ateshian
Andrew Walz Professor of Mechanical
Engineering and Professor
of Biomedical Engineering
Columbia University
New York, NY, USA

Kristin M. Myers
Department of Mechanical Engineering
Columbia University
New York, NY, USA

João Manuel R. S. Tavares
Department of Mechanical Engineering
University of Porto
Porto, Portugal

ISSN 2212-9391 ISSN 2212-9413 (electronic)
Lecture Notes in Computational Vision and Biomechanics
ISBN 978-3-030-43194-5 ISBN 978-3-030-43195-2 (eBook)
https://doi.org/10.1007/978-3-030-43195-2

This Springer imprint is published by the registered company Springer Nature Switzerland AG
The registered company address is: Gewerbestrasse 11, 6330 Cham, Switzerland

Preface

The 16th International Symposium on Computer Methods in Biomechanics and Biomedical Engineering and the 4th Conference on Imaging and Visualization (CMBBE2019) were run together at Columbia University, New York City, USA, from August 14–16, 2019. We believe that CMBBE2019 had a strong impact on the development of computational biomechanics and biomedical imaging and visualization, particularly by identifying emerging areas of research and promoting the collaboration and networking between participants. Actually, CMBBE2019 included a record number of 384 oral presentations and 88 poster presentations which more than doubled the number of presentations of the previous CMBBE. In addition, renowned researchers delivered very interesting plenary keynotes, addressing current challenges in computational biomechanics and biomedical imaging. This book includes the extended versions of selected works presented in CMBBE2019.

Briefly, the included 55 chapters address important topics in sports biomechanics, cardiovascular mechanics, brain biomechanics and imaging, spine biomechanics, growth and remodeling, soft tissue mechanics, bone and dental biomechanics, cellular and molecular biomechanics, computer-aided surgery, artificial intelligence in biomechanics, in vivo imaging and visualization, and many others.

The editors would like to take this opportunity to thank the authors of the 55 selected contributions for sharing their works, experiences, and knowledge, making possible their dissemination through this book.

Gerard A. Ateshian
Kristin M. Myers
João Manuel R. S. Tavares
Co-editor and Co-chair of CMBBE2019

Contents

About the Editors

Gerard A. Ateshian, Ph.D. is the Andrew Walz Professor of Mechanical Engineering and director of the Musculoskeletal Research Laboratory at Columbia University in New York City. His primary research is in the field of soft tissue mechanics, with an emphasis on cartilage mechanics, lubrication, and tissue engineering, and the formulation of growth theories for biological tissues. In collaboration with Prof. Clark T. Hung at Columbia, he has translated his findings on cartilage mechanics to the field of functional cartilage tissue engineering, with a particular focus on the role of mechanical loading in tissue growth. Together with Dr. Jeffrey A. Weiss at the University of Utah, he has developed open-source computational tools that facilitate the modeling of tissue mechanics, transport, and growth processes (febio.org). He has applied and validated these tools with the modeling of engineered cartilage growth.

Kristin M. Myers, Ph.D. is an associate professor in the Department of Mechanical Engineering at Columbia University in the City of New York. In 2010, she founded the Myers Soft Tissue Laboratory at Columbia, which uses experimental, theoretical, and computational mechanics to solve problems in women's health and reproductive biomechanics. With the clinical translation in mind, her laboratory is uncovering the structural antecedents of preterm birth. Her current obstetrics research is done in collaboration with the Department of Obstetrics and Gynecology at the Columbia University

Irving Medical Center. She received her Mechanical Engineering doctorate and master's degree from Massachusetts Institute of Technology and her bachelor's degree from the University of Michigan. In 2017, she was given the American Society of Mechanical Engineers Y. C. Fung Young Investigators Award for her contributions to the field of biomechanics, and in 2019, she was awarded the Presidential Early Career Award for Scientists and Engineers (PECASE) for her work in understanding tissue growth and remodeling in pregnancy.

João Manuel R. S. Tavares is a senior researcher at the Instituto de Ciência e Inovação em Engenharia Mecânica e Engenharia Industrial (INEGI) and associate professor with Habilitation at the Department of Mechanical Engineering (DEMec) of the Faculdade de Engenharia da Universidade do Porto (FEUP). He graduated in Mechanical Engineering at the Universidade do Porto, Portugal, in 1992. He earned his M.Sc. degree and Ph.D. degree in Electrical and Computer Engineering from the Universidade do Porto in 1995 and 2001, respectively, and the Habilitation in Mechanical Engineering in 2015 from the same University. He is a co-editor of more than 50 books and co-author of more than 35 book chapters and 650 articles in international and national journals and conferences. Further, he is author of three international as well as three national patents. He has been a committee member of several international and national journals and conferences and is the co-founding and co-series editor of the book series "Lecture Notes in Computational Vision and Biomechanics" published by Springer. He is the founding editor and editor in chief of the journal "*Computer Methods in Biomechanics and Biomedical Engineering: Imaging & Visualization*" published by Taylor & Francis and co-founder and co-chair of the international conference series CompIMAGE, ECCOMAS VipIMAGE, ICCEBS, and BioDental. Additionally, he has been (co)supervisor of several M.Sc. and Ph.D. theses and supervisor of several postdoc projects. He has participated in many scientific projects both as a researcher and as a scientific coordinator. His main research areas include computational vision, medical imaging, computational mechanics, scientific visualization, human–computer interaction, and new product development. More information can be found at his personal page at www.fe.up.pt/~tavares.

Reduced-Order Models for Blood Pressure Drop Across Arterial Stenoses

Jeanne Ventre[1(✉)], Francesca Raimondi[2], Nathalie Boddaert[2], José Maria Fullana[1], and Pierre-Yves Lagrée[1]

[1] Sorbonne Université, CNRS, Institut Jean Le Rond d'Alembert, UMR 7190, Paris, France
jeanne.ventre@dalembert.upmc.fr

[2] Hôpital Necker Enfants malades, APHP, Centre de référence Malformations Cardiaques Congénitales Complexes M3C, Paris, France

Abstract. Stenosis, defined by a partial or full obstruction of the arteries, is a frequent anomaly in the cardiovascular system. The pressure drop across a stenosis indicates the severity of the pathology. There is currently no non-invasive method to obtain this pressure drop. In this communication, we use four different blood flow models to compute the pressure in an idealized geometry of stenosis: the steady RNSP model, the Multi-Ring model, the 1D model, and algebraic models. We compare these models on a test case under a steady flow. We then developed a gradient-based parameter estimation method to compare the complex models (1D and Multi-Ring) with algebraic formulas. We used the parameter estimation to evaluate the influence of the geometry, wall elasticity and flow parameter on the empirical coefficients of the algebraic formulas.

Keywords: Reduced-order models · Pressure drop · Arterial stenoses

1 Introduction

Aortic CoArctation (CoA) is a congenital heart disease that appears in young children and that accounts for 5 to 8% of all congenital heart diseases [5]. CoA is defined as a partial narrowing of an arterial segment called stenosis and is frequently located either in the area where the ductus arteriosus inserts or in the ascending aorta.

In clinical studies, measurements of the blood pressure drop across a stenosis give significant indications regarding the severity of the pathology. Trained medicals can establish a diagnosis based on these specific data. Despite continuous improvements in the field of medical imaging giving the instantaneous velocity field and topology in the vascular network, pressure data cannot be assessed non-invasively. Modeling is therefore a relevant option for computing blood flow in stenosed vessels, and to extract pressure data. In this communication, we propose to study several blood flow models to evaluate the transstenotic pressure drop in large arteries, such as the ones where CoA appears.

G. A. Ateshian et al. (Eds.): CMBBE 2019, LNCVB 36, pp. 1–15, 2020.
https://doi.org/10.1007/978-3-030-43195-2_1

Complex models for blood flow are based on the Navier-Stokes equations [8,18]. Solving the fluid flow coupled to the displacement of the arterial wall in a three-dimensional (3D) domain is complex and requires high computational resources. For real-time medical applications, some simplifications have been introduced.

In the case of large arteries, we can assume axisymmetry in the blood flow and that the axial wavelength is greater than the typical radius. We thus obtain the Reduced Navier-Stokes Prandtl (RNSP) equations [11,16] that we can solve using elastic [19], hyper-elastic, viscoelastic [1] or rigid walls.

By averaging the RNSP model over the cross-section of the vessel, we can derive the classical one-dimensional (1D) model [2]. It also requires a pressure law at the wall [19], and a hypothesis about the shape of the velocity profile to compute the friction.

Finally, by averaging the 1D equations over the length of the artery, we can obtain zero-dimensional (0D), or algebraic, models [27,28]. By solving these models numerically, we can compute the pressure and velocity field in the given domain and thus calculate the pressure drop across a stenosis.

In this study, we defined an idealized geometry of a stenosed artery, computed the pressure in the stenosis and compared the following models: the steady RNSP model, the Multi-Ring model, the 1D model, and algebraic models.

The paper is organized as follows: in Sect. 2, we present the fluid models; in Sect. 3, we compare the models on a steady test case; in Sect. 4, we develop a parameter estimation method to compare the 1D model to the algebraic model by estimating the dependence of the coefficients of the algebraic models parameters on the stenosis properties; in Sect. 5, we extend this parameter estimation to the Multi-Ring model; finally, in Sect. 6, we discuss our results and give perspectives to improve the present study.

2 Fluid Models for Transstenotic Pressure Drop

2.1 The Navier-Stokes Equations

The motion of blood in arteries is governed by the three-dimensional (3D) Navier-Stokes equations. In large arteries, the average shear rate $\dot{\gamma}$ is high enough to consider that the fluid is homogeneous and Newtonian. We can also consider that blood flow is incompressible. It leads to the following mass and momentum conservation equations

$$\nabla \cdot \mathbf{u} = 0, \quad \rho \left(\frac{\partial \mathbf{u}}{\partial t} + (\mathbf{u} \cdot \nabla)\mathbf{u} \right) = -\nabla p + \mu \nabla^2 \mathbf{u}, \tag{1}$$

where $\mathbf{u}$ is the 3D velocity vector, p is the pressure, ρ is the fluid density and μ the dynamic viscosity. In a cylindrical system, the components of the velocity vector $\mathbf{u}$ are (u_r, u_θ, u_x).

2.2 Long-Wavelength Simplification

Considering the geometry of an artery as a cylindrical tube, we can assume that blood flow is axisymmetric. The wavelength of the pulse wave is much larger than the characteristic radius of a large artery, we can thus simplify the Navier-Stokes equations (1) under the long-wavelength assumption and obtain

$$\begin{cases} \dfrac{1}{r}\dfrac{\partial r u_r}{\partial r} + \dfrac{\partial u_x}{\partial x} = 0, & \text{(2a)} \\ \dfrac{\partial u_x}{\partial t} + u_r \dfrac{\partial u_x}{\partial r} + u_x \dfrac{\partial u_x}{\partial x} = -\dfrac{1}{\rho}\dfrac{\partial p}{\partial x} + \nu \left(\dfrac{1}{r}\dfrac{\partial}{\partial r}\left(r \dfrac{\partial u_x}{\partial r} \right) \right), & \text{(2b)} \\ p(x,r,t) = p(x,t), & \text{(2c)} \end{cases}$$

which are referred to as the Reduced Navier-Stokes Prandtl (RNSP) equations. More details about the derivation can be found in [6,16]. We can solve this model with two approaches. The first consists of removing the unsteady term from (2b) and supposing a rigid wall, which we refer to as the steady RNSP in the following. The second consists of coupling (2) with an elastic law for the arterial wall. The pressure law that characterizes the deformation of the wall is the simple elastic law

$$p(x) = K\left(\sqrt{A(x)} - \sqrt{A_0(x)}\right), \tag{3}$$

where the parameter K characterizes the elastic behavior of the arterial wall and depends on the Young modulus and Poisson coefficient. $A(x)$ is the cross-section and $A_0(x)$ the reference cross-section equal to $\pi R^2(x)$, specified later in Eq. (6).

In this approach, we solve the flow by decomposing the fluid domain in concentric rings. We refer to this method as the Multi-Ring model [11] in the following.

2.3 One-Dimensional Model

We obtain the 1D equations by averaging the system (2) over the cross-section of the tube. The 1D equations can also be interpreted as a particular case of the Multi-Ring model with only one ring and an assumption on the velocity profile and the wall shear stress. The averaged equations are

$$\begin{cases} \dfrac{\partial A}{\partial t} + \dfrac{\partial Q}{\partial x} = 0, & \text{(4a)} \\ \dfrac{\partial Q}{\partial t} + \dfrac{\partial}{\partial x}\left(\dfrac{Q^2}{A}\right) = -\dfrac{A}{\rho}\dfrac{\partial p}{\partial x} - C_f \dfrac{Q}{A}, & \text{(4b)} \end{cases}$$

where Q is the flow rate and A the cross-sectional area. The friction coefficient C_f is set to $2(\xi + 2)\mu\pi$ [22]. The parameter ξ determines the friction depending on the assumption on the velocity profile. In the case of a parabolic velocity profile, $\xi = 2$. We use the same elastic pressure law (3) to couple the flow and the wall.

This model has been previously used in the literature [12,17,18,23] to compute the flow in the systemic circulation to study the presence of a stenosis.

2.4 Algebraic Model

The next level of simplifications leads to algebraic, or zero-dimensional, models. They are obtained from averaging the fluid equations (4) over the longitudinal variable. There exists a variety of these models in the literature [17,20,23,27,28] to compute the pressure drop in constrictions and were each derived for a specific application. All of them treat a stenosis as a constriction of a 1D flow and can be considered as generalizations of Bernoulli's equation, or a balance of mechanical energy. All these expressions involve empirical parameters depending on the configuration and the geometrical properties. For unsteady flows, Young and Tsai [28] proposed the following algebraic model

$$\Delta P_{alg} = \frac{K_v \mu}{D_0} U(t) + K_u \rho L_{st} \frac{\mathrm{d}U(t)}{\mathrm{d}t} + \frac{K_t}{2} \left(\left(\frac{A_0}{A_{st}} \right)^2 - 1 \right) \rho |U(t)| U(t), \tag{5}$$

where ΔP_{alg} is the pressure drop across the length of the stenosis L_{st}, D_0 the diameter and A_0 the cross-section of the unobstructed vessel, A_{st} the cross-section at the throat of the stenosis, K_v, K_u and K_t are empirical coefficients. $U(t)$ is the instantaneous input velocity and $|\cdot|$ represents the absolute value of $\cdot$.

The first term captures the Poiseuille viscous loss depending on the coefficient K_v. The second term represents the inertial effect of blood flow in a constriction with an inertial coefficient K_u. The third term accounts for the non-linear effects depending on the coefficient K_t. These expressions are considered in the literature as the gold standard for model comparison and used by physicians to grossly estimate the pressure drop across an arterial stenosis, in cases where they do not have access to an invasive measurement.

Despite the simplicity of this type of model, the drawback is that the coefficients are empirical and were determined in the literature for specific experiments [20].

3 Comparison of the Models on a Steady Case

3.1 Geometry of the Stenosed Artery

Figure 1 defines the geometry of a stenosed artery of length L, radius R_0, stenosis length L_{st} and stenosis degree α. The shape of the radius of the wall is

$$R(x) = R_0 \left(1 + \alpha \exp\left(-\frac{(x - x_{st})^2}{x_l} \right) \right), \tag{6}$$

where x_{st} is the axial position of the throat of the stenosis and x_l is related to the length of the stenosis L_{st}.

The pressure jump ΔP that we evaluate in this study is the pressure difference between upstream and downstream of the stenosis, *i.e.* over the length of the stenosis.

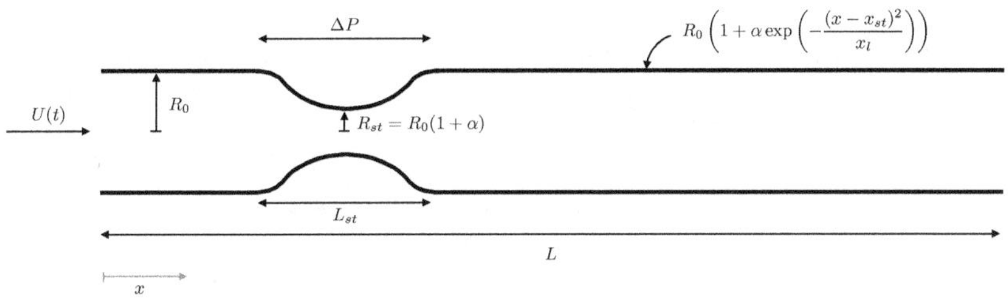

Fig. 1. Geometry of the stenosed artery of length L, radius R_0, stenosis length L_{st} and stenosis degree α ($\alpha < 0$). The shape of the radius of the wall is Eq. (6), with x_{st} the position of the throat of the stenosis, and x_l related to L_{st}. The pressure drop over the length of the stenosis is ΔP.

3.2 Numerical Methods

As all of the models presented in Sect. 2 are non-linear, we do not have access to analytical solutions and thus require numerical schemes to solve them. Different approaches to compute fluid models, such as finite element methods, finite volumes or finite differences, have been introduced in the literature.

For the steady RNSP model, we used an implicit finite difference scheme [16]. For the Multi-Ring model and the 1D model, we used a finite volume approach. We split the system of equations into a convective subproblem that accounts for the transport and a reaction subproblem for the friction source term. We treated the convective subproblem with an explicit method using a kinetic scheme for the flux [3]. We solved the viscous subproblem using an implicit numerical scheme. More details can be found in [11] for the Multi-Ring model, and in [10] and [7] for the 1D model.

We introduced a time discretization with a constant time step $\Delta t = 2 \cdot 10^{-6}$ s for the Multi-Ring model and $\Delta t = 10^{-5}$ s for the 1D model. We used $N_r = 32$ concentric rings and divided the length of the artery into $N_x = 800$ cells for the Multi-Ring model. For the 1D model, we used $N_x = 250$ cells. The RNSP code is steady and thus only requires a spatial discretization, we chose $N_x = 4000$ and $N_y = 1000$. These are typical values for computations with enough precision.

We computed the three models using codes developed in our laboratory written in C or C++.

3.3 Comparison of the Center Pressure

We first compared the steady RNSP, Multi-Ring and 1D models, described in Sect. 2, in a rigid stenosed artery against the Poiseuille pressure in an axisymmetric straight rigid tube $p(x) = -\dfrac{8\rho U_0^2}{R_0 Re_R}(x - L)$.

At the inlet of the vessel, we imposed a steady input flow, *i.e.* at $x = 0$, $U(t) = U_0$. At the outlet of the vessel, we imposed a zero pressure *i.e.* at $x = L$, $p = 0$. The properties of the configuration are reported in Table 1, and the geometry is shown in Fig. 1.

Table 1. Properties of the stenosed artery. R_0: initial radius, α: degree of stenosis, L: artery length, L_{st}: stenosis length, Re_R: Reynolds number based on the radius, U_0: input velocity, ρ: fluid density, μ: dynamic viscosity, K: elasticity. All values are in CGS units.

R_0	α	L	L_{st}	Re_R	U_0	ρ	μ	K
1	−0.4	40	10	100	100	1	$\dfrac{\rho U_0 R_0}{Re_R}$	10^7

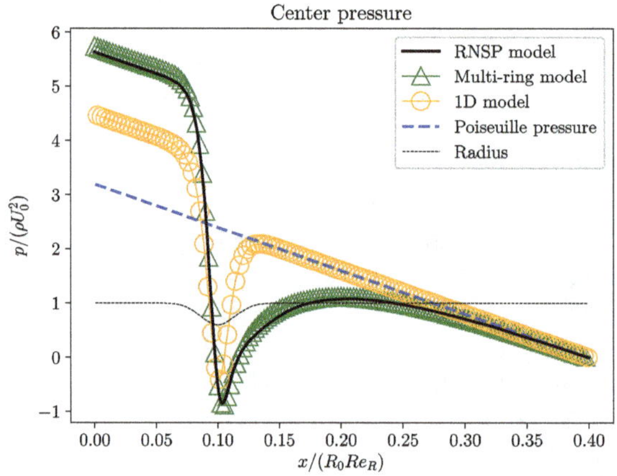

Fig. 2. Dimensionless center pressure along the stenosis represented in Fig. 1 with properties of Table 1. The black solid line (—) corresponds to the steady RNSP model, the green triangles (△) to the Multi-Ring model, the orange circles (◯) to the 1D model, and the dashed blue line (- - -) to the Poiseuille pressure along a straight tube.

We observe that the center pressure drop between the beginning and the throat of the stenosis is similar in all models in the stenosed artery. However, the center pressure downstream of the stenosis is different in the 1D model compared to the steady RNSP and the Multi-Ring. Indeed, the 1D model does not account for the recirculation near the walls and the jet formation in the center after the stenosis, as we impose a constant shape of the velocity profile in each section. The downstream flow is therefore not impacted by the constriction in the 1D model, as opposed to the steady RNSP and Multi-Ring models.

The models presented in this section allow computing the velocity and pressure field in the entire domain. In fact, the relevant indicator for medical diagnosis is the pressure drop ΔP evaluated across the stenosis. Therefore in the following section, we compare the pressure drop using all the models of Sect. 2.

3.4 Comparison of the Pressure Drop

In the Fig. 3, we show a comparison of the dimensionless pressure drop between the algebraic model from [27], the 1D model, the Multi-Ring model, and the steady RNSP model. We show a comparison with in vivo pressure drop measurements from [26]. We excluded one measurement that gave a bad agreement between their experimental and model predictions.

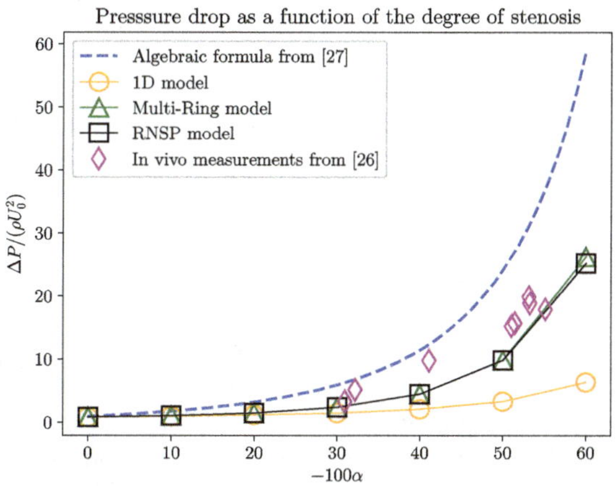

Fig. 3. Comparison of the dimensionless pressure drop as a function of the degree of stenosis in percent -100α for the algebraic model from [27], the 1D model (○), the Multi-Ring model (△), the steady RNSP model (□), and measurements from [26] (◊).

The in vivo measurements from [26] show that the steady RNSP and Multi-Ring are the most accurate models to estimate the pressure drop across the stenosis. However, we can also observe that for mild stenoses, the 1D model gives a reasonable agreement with the in vivo measurements.

For the configuration of Sect. 3.3, we computed the pressure drop across the stenosis with the different models. We obtained $\Delta P = 16$ mmHg for the 1D model and $\Delta P = 33$ mmHg for the steady RNSP and Multi-Ring models, which seem to be reasonable values, *i. e.* within the physiological range. For reference, we compared this pressure drop with the algebraic formula using the values of the empirical coefficients from [27] which gave $\Delta P = 89$ mmHg for the same configuration. The algebraic pressure drop is clearly out of the physical range.

It seems reasonable to think that even if the 1D model might underestimate the pressure drop, it is a better starting point than the current algebraic formula with the empirical coefficients of the literature. Therefore, in the following section, we compare the 1D model to an algebraic formula for which we estimate the empirical coefficients with a parameter estimation process.

4 Comparison Between Algebraic and 1D Models Under Unsteady Flow

For the sake of simplicity, in Sect. 3, we compared the models using a steady input flow. In this section, we investigate the unsteady effects on the pressure drop across a stenosis using the 1D model.

4.1 Description of the 1D Model

We use the same geometry and properties of the stenosed artery as shown in Fig. 1. We study unsteady flows and thus impose at the inlet of the vessel $x = 0$ an oscillating velocity $U(t) = U_0 \sin(2\pi\omega t)$ with U_0 the amplitude of the input velocity and ω the frequency. At the outlet of the tube, we still impose a zero pressure, *i.e.* at $x = L$, $p = 0$.

4.2 Parameter Estimation Method

To compare the 1D model with the algebraic formula (5), we need to estimate the empirical coefficients K_v, K_u, and K_t. The objective is to study the dependence of the parameters on the stenosis properties. We therefore define a cost function J that measures the difference between the pressure drop of the model ΔP_{model}, in this case the 1D model, and the algebraic pressure drop ΔP_{alg} as

$$J(\mathcal{P}) = \left(\int_0^T \left(\Delta P_{model} - \Delta P_{alg}(\mathcal{P}) \right)^2 \mathrm{d}t \right)^{1/2} \tag{7}$$

with $\mathcal{P} = \{K_v, K_u, K_t\}$ the set of parameters to estimate. We minimize the cost function J with respect to $\mathcal{P}$ using a Basin-Hopping algorithm [24] of the SciPy library [15] from Python. The algorithm runs a gradient-based method L-BFGS-B (from the initials of the original authors Broyden [4], Fletcher [9], Goldfarb [13], Shanno [21]) and, to ensure the global minima of the parameters, creates a random perturbation of the parameters $\mathcal{P}$ at each step in the specified parameter space. The method allows finding $\mathcal{P}^{opt}$ that minimizes the cost function J, which we show in the following section.

4.3 Estimation of Empirical Parameters on the 1D Model

We chose to estimate the empirical coefficients K_v, K_u and K_t as a function of several parameters: the geometrical parameters that are the stenosis length L_{st} and the cross-section ratio A_0/A_{st}, the wall rheology parameter K and the flow parameter Re_R.

We first observed that the estimated value of the coefficient K_t was close to zero when using the 1D model.

In Fig. 4(a), we show the estimated value of the coefficient K_v as a function of the cross-section ratio A_0/A_{st}. We compared our estimation with two in vitro experimental measurements from [20] for two configurations that were similar to ours, reported in Table 2. Even though the agreement between the two experimental points and our estimation of K_v is reasonable, we have several explanations for the differences. First, the geometry is different as the constriction in [20] is an abrupt reduction of the radius as opposed to ours that is smooth. Second, they are investigating steady flows. Finally, the aspect ratio of L_{st}/D_0 is smaller in their experiments than in our simulations. Seeley and Young [20] established a theoretical expression for K_v

$$K_v = 32\frac{L_a}{D_0}\left(\frac{A_0}{A_{st}}\right)^2 \tag{8}$$

with $L_a = 0.83L_{st} + 1.64D_{st}$ a correction of the stenosis length L_{st}, which we show in Fig. 4(a). Similarly to our estimation of K_v, Seeley and Young [20] showed that this empirical coefficient only depends on the geometrical characteristic of the constriction.

Figure 4(b) shows that the dependence of K_v on the length of the stenosis L_{st} is linear. As the K_v term in this equation corresponds to the viscous Poiseuille pressure jump, we added for reference the Poiseuille value of K_v in a straight tube that is $32L_{st}/D_0$. The comparison stresses the linear dependence of K_v on the length over which we calculated the pressure drop.

Figure 4(c) shows that the wall elasticity does not influence the value of K_v as the coefficient does not vary significantly with K. When the elasticity of the wall becomes smaller, the coefficient increases a little bit however it does not correspond to the value of the wall elasticity of large arteries [25].

Similarly, we observe that there is no dependence of K_v on the Reynolds number Re_R as shown in Fig. 4(d).

We can therefore conclude that the parameter K_v is only a function of geometric parameters that are the cross-section ratio A_0/A_{st} and the length of the stenosis L_{st}, as predicted by Seeley and Young in [20] and more recently by Heinen *et al.* in [14].

Table 2. Estimation of the empirical coefficient K_v of Eq. (5) from Seeley and Young [20] for comparison with Fig. 4(a).

	α	L_{st}/D_0	K_v
Point 1	−0.375	2	421
Point 2	−0.5	2	1160

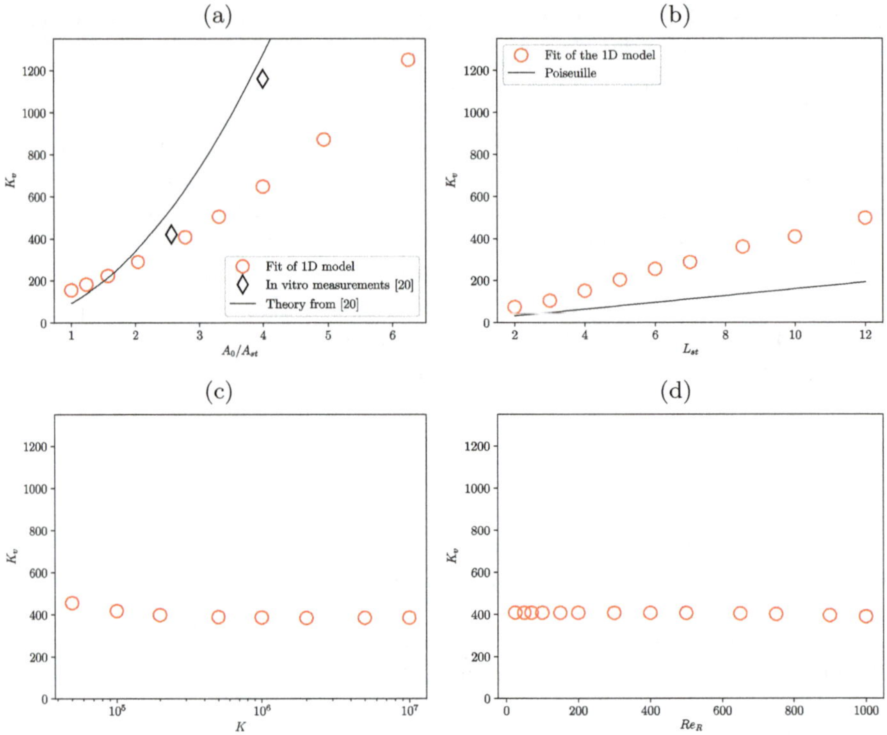

Fig. 4. Values of the coefficient K_v as a function of (a) the cross-section ratio A_0/A_{st}, (b) the length of the stenosis L_{st}, (c) the elasticity K, and (d) the Reynolds number based on the radius Re_R. The coefficient K_v is estimated from the 1D model using the algebraic formula (5). We added on (a) a comparison with in vitro measurements in (◊) and a theory in (—) for K_v from [20]. We added on (b) the Poiseuille value of K_v in a straight tube. L_{st} and K are in CGS units, the other quantities are dimensionless.

Figure 5(a) shows a strong dependence of K_u on the cross-section ratio A_0/A_{st} as Fig. 4(a).

In Fig. 5(b), there is no obvious linear dependence of the coefficient K_u on the length of the stenosis.

Similarly to Figs. 4(c), 5(c) shows that the value of K_u is not significantly affected by a variation in elasticity K.

However, unlike Figs. 4(d), 5(d) shows that K_u depends on the Reynolds number. The coefficient K_u has an asymptotic behavior from $Re_R \approx 200$.

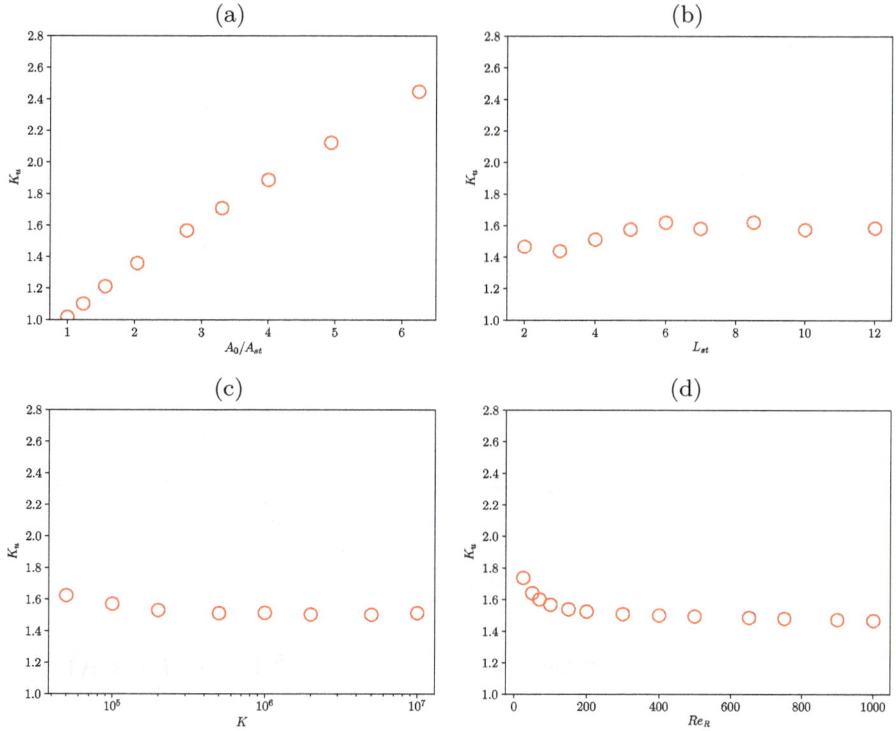

Fig. 5. Values of the coefficient K_u as a function of (a) the cross-section ratio A_0/A_{st}, (b) the length of the stenosis L_{st}, (c) the elasticity K, and (d) the Reynolds number based on the radius Re_R. The coefficient K_u is estimated from the 1D model using the algebraic formula (5). L_{st} and K are in CGS units, the other quantities are dimensionless.

4.4 Comparison Between Algebraic and 1D Models

In Fig. 6, we compare the pressure drop computed with the 1D model and the algebraic pressure drop from Eq. (5) using the estimated optimal parameters K_v, K_u, reported in Table 3 and $K_t = 0$.

With only two terms in the algebraic expression (5), we reproduce exactly the shape of the pressure drop between upstream and downstream of the stenosis under a steady flow.

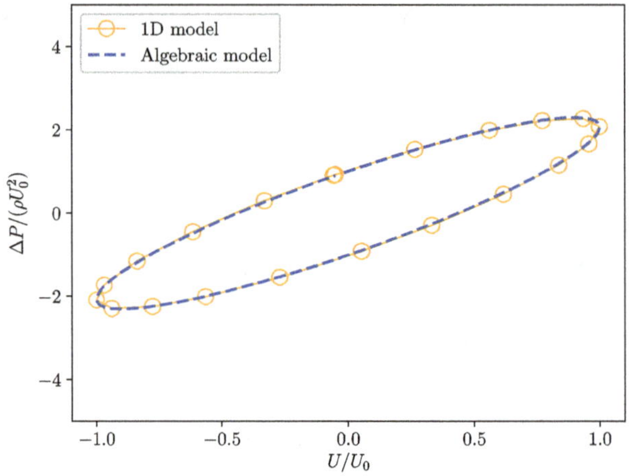

Fig. 6. Comparison of the pressure drop across the stenosis between the 1D model (○) and the algebraic model (5) (- - -) for an unsteady input velocity with pulsation $\omega = 1$ s^{-1}. The properties of the stenosed artery are reported in Table 1 and the parameters K_v and K_u of the algebraic model are reported in Table 3.

5 Comparison Between Algebraic and Multi-Ring Models Under Unsteady Flow

As shown in Fig. 2, the pressure peak at the throat of the stenosis was similar in all the models, however, the upstream to downstream pressure drop was different in the 1D model compared to the steady RNSP and Multi-Ring models. And, in fact, this is the relevant pressure drop for the medical diagnosis. We observed in Fig. 3 that the 1D model gives a reasonable estimation of the pressure drop for mild stenosis. However, for more severe stenoses, the 1D model underestimated the pressure drop and, therefore, we needed to use a more accurate model. We thus applied the same parameter estimation strategy like the one described in Sect. 4.2, using the Multi-Ring model [11]. However, as the non-linear effects are accounted for with more accuracy in this model, we now consider that $K_t \neq 0$ in Eq. (5) and reported the optimal value in Table 3.

In Fig. 7, we compare the pressure drop computed with the Multi-Ring model and the algebraic pressure drop from Eq. (5) using the estimated optimal parameters K_v, K_u and K_t, reported in Table 3. We used the same method to estimate the optimal set of parameters as in Sect. 4.2 where the cost function J (7) now depends on the Multi-Ring pressure drop.

By comparing Figs. 6 and 7, we can see that the amplitude of the maximum pressure drop is higher for the Multi-Ring model than for the 1D, similarly to Figs. 2 and 3. By adding a non-linear term to Eq. (5), we retrieve the exact shape of the unsteady pressure drop across the stenosis of the Multi-Ring model.

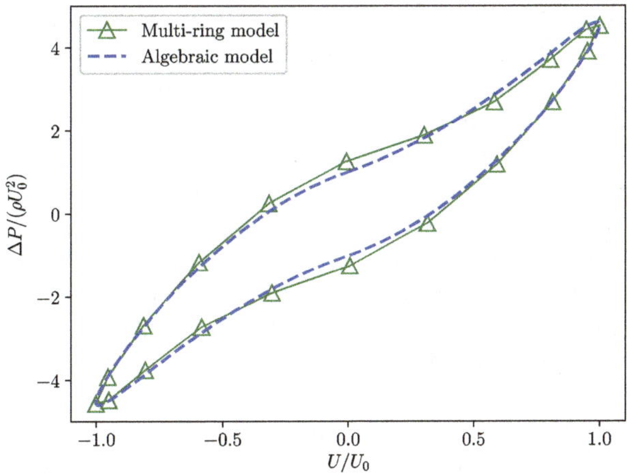

Fig. 7. Comparison of the pressure drop across the stenosis between the Multi-Ring model (△) and the algebraic model (5) (- - -) for an unsteady input velocity with pulsation $\omega = 1\ \mathrm{s}^{-1}$. The properties of the stenosed artery are reported in Table 1 and the parameters K_v, K_u and K_t of the algebraic model are reported in Table 3.

Table 3. Empirical dimensionless coefficients K_v, K_u and K_t of the algebraic models (5).

K_v	K_u	K_t
400	1.5	0.7

6 Conclusion

In this study, we compared four types of reduced-order models (algebraic, 1D, Multi-Ring, steady RNSP) to compute the pressure drop across a stenosis. We showed a comparison of the center pressure in an idealized geometry of stenosis. We highlighted that with a steady input flow, the steady RNSP, Multi-Ring, and 1D model behaved similarly in terms of the shape of center pressure between the beginning and the throat stenosis. Using experimental data from [26], we showed that the Multi-Ring model reproduced the measurements reasonably well. We concluded that, even though the 1D model might underestimate the pressure drop, for mild stenosis it is a good first approximation and is less computationally expensive than the other models.

Therefore we used our models to estimate the empirical coefficients of the algebraic formula using the 1D model in our configuration with a Basin-Hopping optimization method. We first found that the non-linear term is negligible using the 1D model, *i. e.* $K_t = 0$. We found that the coefficient K_v characterizing the viscous loss only depended on the cross-section ratio and the length of the stenosis. The coefficient K_u characterizing the inertial effects depended on the cross-section ratio but also the Reynolds number. An important observation is that the wall elasticity K does not play a significant role in the center pressure drop across the stenosis.

We finally showed that this optimization process can be extended to the Multi-Ring model and give a value for the third coefficient K_t that characterizes the non-linear effects, which is no longer negligible with the Multi-Ring model.

In this study, we computed the models in an idealized geometry of stenosis, which is a limitation. Indeed, to make patient-specific predictions of the pressure drop across a stenosis, we should compute the pressure in real geometries. This would require some high-quality imaging along with a segmentation algorithm to determine the radius at each time and position of the artery. However, the 1D models that treat the arteries as straight axisymmetric elastic tubes have proven reliable in the literature and are a good starting point, if not better, compared to the current algebraic formulas.

Another limitation is that we do not have real invasive measurements to validate the method in the case of a stenosis. Therefore, the main perspective of the present study, to make a patient-specific estimation of the pressure drop, is to obtain invasive pressure measurements to compare our models to. Knowing how the coefficients of the algebraic formula depend on the characteristics of the stenosis would constitute a sort of abacus and thus allow estimating the pressure drop in real-time medical applications. The first results of intra-arterial catheter measurements of the transstenotic pressure drop show that the Multi-Ring model is the most accurate model.

References

1. Alastruey, J., Khir, A.W., Matthys, K.S., Segers, P., Sherwin, S.J., Verdonck, P.R., Parker, K.H., Peiró, J.: Pulse wave propagation in a model human arterial network: assessment of 1-D visco-elastic simulations against in vitro measurements. J. Biomech. **44**, 2250–2258 (2011)
2. Alastruey, J., Parker, K.H., Sherwin, S.J.: Arterial pulse wave hemodynamics. In: 11th International Conference on Pressure Surges, vol. 30, pp. 401–443 (2012)
3. Audebert, C., Bucur, P., Bekheit, M., Vibert, E., Vignon-Clementel, I.E., Gerbeau, J.-F.: Kinetic scheme for arterial and venous blood flow, and application to partial hepatectomy modeling. Comput. Methods Appl. Mech. Eng. **314**, 102–125 (2017)
4. Broyden, C.G.: Quasi-newton methods and their application to function minimisation. Math. Comput. **21**(99), 368–381 (1967)
5. Endorsed by the Association for European Paediatric Cardiology (AEPC), Baumgartner, H., Bonhoeffer, P., De Groot, N.M.S., de Haan, F., Deanfield, J.E., Galie, N., Gatzoulis, M.A., Gohlke-Baerwolf, C., et al.: ESC guidelines for the management of grown-up congenital heart disease (new version 2010) the task force on the management of grown-up congenital heart disease of the European society of cardiology (ESC). Eur. Heart J. **31**(23), 2915–2957 (2010)
6. Chouly, F., Lagrée, P.-Y.: Comparison of computations of asymptotic flow models in a constricted channel. Appl. Math. Model. **36**(12), 6061–6071 (2012)
7. Delestre, O., Ghigo, A.R., Fullana, J.-M., Lagrée, P.-Y.: A shallow water with variable pressure model for blood flow simulation. Netw. Heterogen. Media **11**(1), 69–87 (2016)
8. Figueroa, C.A., Vignon-Clementel, I.E., Jansen, K.E., Hughes, T.J.R., Taylor, C.A.: A coupled momentum method for modeling blood flow in three-dimensional deformable arteries. Comput. Methods Appl. Mech. Eng. **195**(41–43), 5685–5706 (2006)

9. Fletcher, R., Powell, M.J.D.: A rapidly convergent descent method for minimization. Comput. J. **6**(2), 163–168 (1963)
10. Ghigo, A.R., Delestre, O., Fullana, J.-M., Lagrée, P.-Y.: Low-shapiro hydrostatic reconstruction technique for blood flow simulation in large arteries with varying geometrical and mechanical properties. J. Comput. Phys. **331**, 108–136 (2017)
11. Ghigo, A.R., Fullana, J.-M., Lagrée, P.-Y.: A 2D nonlinear multiring model for blood flow in large elastic arteries. J. Comput. Phys. **350**, 136–165 (2017)
12. Ghigo, A.R., Taam, S.A., Wang, X., Lagrée, P.-Y., Fullana, J.-M.: A one-dimensional arterial network model for bypass graft assessment. Med. Eng. Phys. **43**, 39–47 (2017)
13. Goldfarb, D.: A family of variable-metric methods derived by variational means. Math. Comput. **24**(109), 23–26 (1970)
14. Heinen, S.G.H., Van den Heuvel, D.A.F., de Vries, J.P.P.M., Van de Vosse, F.N., Delhaas, T., Huberts, W.: A geometry-based model for non-invasive estimation of pressure gradients over iliac artery stenoses. J. Biomech. **92**, 67–75 (2019)
15. Jones, E., Oliphant, T., Peterson, P.: {SciPy}: open source scientific tools for {Python} (2014)
16. Lagree, P.-Y., Lorthois, S.: The RNS/Prandtl equations and their link with other asymptotic descriptions: application to the wall shear stress scaling in a constricted pipe. Int. J. Eng. Sci. **43**(3), 352–378 (2005)
17. Liang, F., Takagi, S., Himeno, R., Liu, H.: Multi-scale modeling of the human cardiovascular system with applications to aortic valvular and arterial stenoses. Med. Biol. Eng. Comput. **47**(7), 743–755 (2009)
18. Mirramezani, M., Diamond, S.L., Litt, H.I., Shadden, S.C.: Reduced order models for transstenotic pressure drop in the coronary arteries. J. Biomech. Eng. **141**(3), 031005 (2019)
19. Quarteroni, A., Veneziani, A., Vergara, C.: Geometric multiscale modeling of the cardiovascular system, between theory and practice. Comput. Methods Appl. Mech. Eng. **302**, 193–252 (2016)
20. Seeley, B.D., Young, D.F.: Effect of geometry on pressure losses across models of arterial stenoses. J. Biomech. **9**(7), 439–448 (1976)
21. Shanno, D.F.: Conditioning of quasi-newton methods for function minimization. Math. Comput. **24**(111), 647–656 (1970)
22. Sherwin, S.J., Franke, V., Peiró, J., Parker, K.: One-dimensional modelling of a vascular network in space-time variables. J. Eng. Math. **47**(3–4), 217–250 (2003)
23. Stergiopulos, N., Young, D.F., Rogge, T.R.: Computer simulation of arterial flow with applications to arterial and aortic stenoses. J. Biomech. **25**(12), 1477–1488 (1992)
24. Wales, D.J., Doye, J.P.K.: Global optimization by basin-hopping and the lowest energy structures of Lennard-Jones clusters containing up to 110 atoms. J. Phys. Chem. A **101**, 5111–5116 (1997)
25. Wang, J.J., Parker, K.H.: Wave propagation in a model of the arterial circulation. J. Biomech. **37**(4), 457–470 (2004)
26. Young, D.F., Cholvin, N.R., Roth, A.C.: Pressure drop across artificially induced stenoses in the femoral arteries of dogs. Circ. Res. **36**(6), 735–743 (1975)
27. Young, D.F., Tsai, F.Y.: Flow characteristics in models of arterial stenoses–i. Steady flow. J. Biomech. **6**(4), 395–402 (1973)
28. Young, D.F., Tsai, F.Y.: Flow characteristics in models of arterial stenoses–ii. Unsteady flow. J. Biomech. **6**(5), 547–559 (1973)

Impacts of Flow Diverters on Hemodynamics of Intracranial Aneurysms

Trung Bao Le[1(✉)], Elizabeth Eidenschink[2], and Alexander Drofa[2]

[1] Computational Fluid Laboratory, Department Civil and Environmental Engineering, North Dakota State University, Fargo, ND 58105, USA
trung.le@ndsu.edu

[2] Sanford Brain and Spine Center, 700 1st Ave S, Fargo, ND 58103, USA

Abstract. We investigate the hemodynamic impacts of flow diverters in intracranial aneurysms using computer simulations. The geometry of the aneurysm is reconstructed using the open-source software Slicer3D to create a three-dimensional model of the aneurysm. In addition, the geometries of the flow diverters is reconstructed in a mesh-like structure using the commercial software Gridgen and Meshmixer. First, using the provided the blood flow condition at the Internal Carotid Artery as the boundary condition, our in-house code (Virtual Flow Simulator) is applied to simulate the flow dynamics within the aneurysms without flow diverter implantation. The spatial and temporal distribution of wall shear stress (WSS) are computed from the simulation results over the aneurysm dome. At the second step, the virtual implantation of the flow diverter is carried out at the ostium. A second simulation with the implanted flow diverter is carried out to provide hemodynamic conditions after the implantation. Our results show a stark contrast on flow distribution between two cases. The flow diverter not only changes the flow distribution at the ostium level but it also alters the flow distribution across the parent artery. Our results indicate that the flow pulsatility plays a key role in mediating the interaction between the incoming jet and the weave pores. Our simulations suggest that the weave's size correlates well to the small-scale structures of the instantaneous flow in the vicinity of the ostium.

Keywords: Aneurysm hemodynamics · Flow dynamics · High-resolution simulation

1 Introduction

Flow diverter (FD) implantation has become a favorable practice for patients with intracranial aneurysms [1] due to the reduction of cost and efficacy. In the United States, only one type of FD (the Pipeline Embolization Device - PED) has been approved for use by the FDA for wide-necked aneurysms [2]. In spite

G. A. Ateshian et al. (Eds.): CMBBE 2019, LNCVB 36, pp. 16–23, 2020.
https://doi.org/10.1007/978-3-030-43195-2_2

of a promising success rate of 80%, PED implantation does not lead to a complete occlusion (full closure) of aneurysm sac in 20% treated cases [2], which leads to re-treatments and other potential life-threatening complications [3]. Complications of PED implantation involve hemorrhage, stroke, parent artery stenosis, spontaneous rupture and other neurologic complications [1]. There have been a significant number of efforts to understand the root cause of these complications [4].

To date, the sources of complications have not been identified and there is no clear standards how to select patients for PED treatments [6]. Clinical evidence, however, suggest that the treatment might fail if the PED is implanted in aneurysms with complex blood flow dynamics [2,7]. It is thought that the implantation changes the local hemodynamic conditions near aneurysm neck [8] and thus prevents the flow interaction between main parent artery and aneurysmal cavity. Specifically, the healing mechanism of the vessel wall and the occlusion of aneurysm sac is not well understood [8,10]. Such lack of understanding prevents doctors to perform a reliable risk analysis for patients before the actual operation and compare PED implantation to other alternative treatments for an individual. Therefore, there is an urgent need to understand the physiological phenomena of aneurysmal hemodynamics under PED implantation to help establishing the potential links between the success of an implantation with individual conditions [11].

In this work, our goal is to investigate the hemodynamic role of flow-diverter implantation. Our main hypothesis focuses on the role of pore's size in mediating flow structures in the ostium region. We hypothesize that the pore's size correlates to the flow dynamics of inside the aneurysmal cavity.

2 Materials and Methods

2.1 Patient-Specific Data

Magnetic Resonance Angiogram (MRA) data of a 60-year old patient is provided to us by Sanford Health, Fargo, North Dakota as shown in Fig. 1. The patient-specific data has been reviewed, anonymized and approved by the Institutional Review Board (IRB) of Sanford Health. North Dakota State University IRB agreed to rely on the Sanford Health IRB for review and continued oversight of the research. The consent and HIPAA were waived by Sanford Health IRB.

The scanned images are processed by the open-source imaging software - Slicer3D and Osirix. The entire surface is connected, smoothed out, and finally triangulated with 24364 elements using Meshlab and Meshmixer. The geometry and definition of various key geometric parameters of the model are shown in Fig. 1.

Since the waveform of blood flow is not available for this specific patient, the measurement data [5] at the Internal Carotid Artery of a patient is used to prescribe as flow boundary condition.

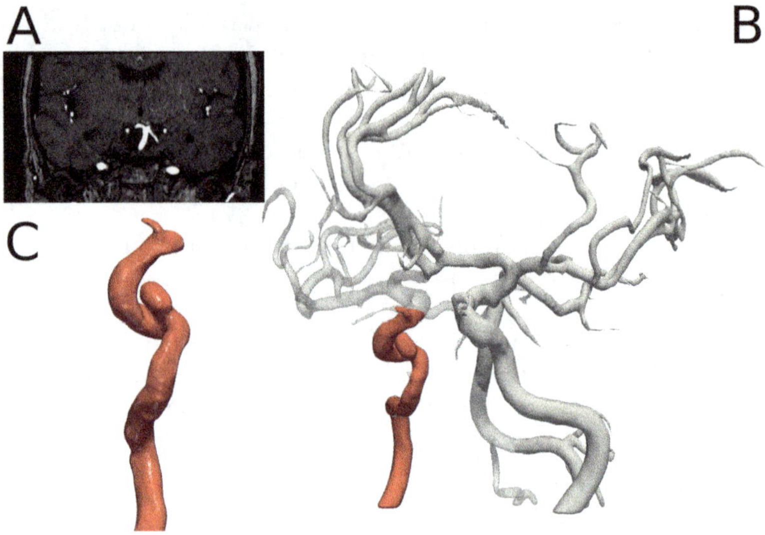

Fig. 1. The reconstruction of the aneurysmal geometries from MRA data. The original DICOM data of MRA is processed using the opensource software Slicer3D (A). The DICOM data is segmented using the commercial software Osirix in (B). The location that harbors the aneurysm is separated using the free software Meshmixer in (C).

2.2 Numerical Methods

The numerical method employed in this work has been extensively described and thoroughly validated against experimental data for aneurysm geometry using three-dimensional volumetric measurements [13]. Therefore, only a brief description of the numerical method is presented in this section. For more details about the method the reader is referred to our previous publications [9].

The governing equations for the fluid (blood) are the three-dimensional, unsteady incompressible continuity and Navier-Stokes equations for velocity vector $\boldsymbol{v}$ and pressure p. The fluid is assumed to be Newtonian, which is considered to be a good assumption for blood flow in large arteries such as the one used herein with approximately 3 mm diameter.

The governing equations are solved in a background curvilinear domain that contains the complex geometry of the intracranial aneurysm model using the sharp-interface curvilinear-immersed boundary (CURVIB) method [9]. The discrete equations are integrated in time using a fractional step method. A Newton-Krylov solver is used to solve the momentum equations in the momentum step and a GMRES solver with multigrid preconditioner is employed for the Poisson equation.

At the inflow boundary, we prescribe uniform velocity profile varying in time in accordance with the prescribed flow waveform. All cases simulated in this work employ the same uniform velocity profile approach at the inlet so that differences in the inflow conditions are excluded as a possible parameter influencing the dome hemodynamics. At the outflow boundary, Neumann-type boundary conditions are specified for the velocity components. No-slip and no-flux conditions are prescribed at all artery wall boundaries, which are considered rigid.

2.3 Simulation Setup

The peak velocity is chosen as $U_0 = 0.67\,\mathrm{m/s}$. The diameter of the parent artery is $D_0 = 4 \times 10^{-3}\,\mathrm{m}$. Blood viscosity is chosen as $\nu = 3.35 \times 10^{-6}\,\mathrm{m}^2/s$. These characteristic parameters define the peak Reynolds number is:

$$Re = \frac{U_0 D_0}{\nu} = \frac{0.67\,\mathrm{m/s} \times 4 \times 10^{-3}\,\mathrm{m}}{3.35 \times 10^{-6}\,\mathrm{m}^2/s} \approx 800 \tag{1}$$

The characteristic time scale τ_0 is defined as:

$$\tau_0 = \frac{D_0}{U_0} = \frac{4 \times 10^{-3}}{0.67} = 6 \times 10^{-3}\,\mathrm{s} = 6\,\mathrm{ms} \tag{2}$$

With a heart rate of 70 bpm, the heart cycle T is:

$$T = \frac{60\,\mathrm{s}}{70} = 0.85\,\mathrm{s} \tag{3}$$

In our computation, the heart cycle T is discretized into 2000 timesteps, so that time step is $\Delta t = \frac{0.85\,\mathrm{s}}{2000} = 0.425\,\mathrm{ms}$. The non-dimensional timestep is therefore: $\tau = \frac{\Delta t}{\tau_0} = 0.071$.

The computational domain is discretized into a structure mesh of size $121 \times 121 \times 361$. At the inlet, uniform flow is prescribed as the boundary condition. The time-dependent flow waveform is used to describe the pulsatile condition at the inlet. A traction-free boundary condition is applied at the outlet of the computational domain.

3 Results

Without the stent implantation, the hemodynamics of this sidewall aneurysm is controlled by the pulsatility of the flow. The formation of the classical vortex ring at the proximal neck of the aneurysm is visible as shown in Fig. 2. This vortex ring is asymmetrical and is inclined to the parent artery at the proximal end. As this vortex ring forms, a separation region is formed simultaneously on the opposite wall of the ostium.

This vortex ring propagates along the neck during late diastole. Finally, it impinges on the distal wall and induces the separation of a secondary vortex on the distal wall. This propagation perturbs the blood flow dynamics at the ostium and the distal end over the entire cardiac cycle. The entire flow dynamics at the neck depends on the trajectory of this vortex ring.

In case the stent implantation, the vortex ring does not form as shown in Fig. 3b. Although the jet is formed at the proximal neck, the presence of the stent

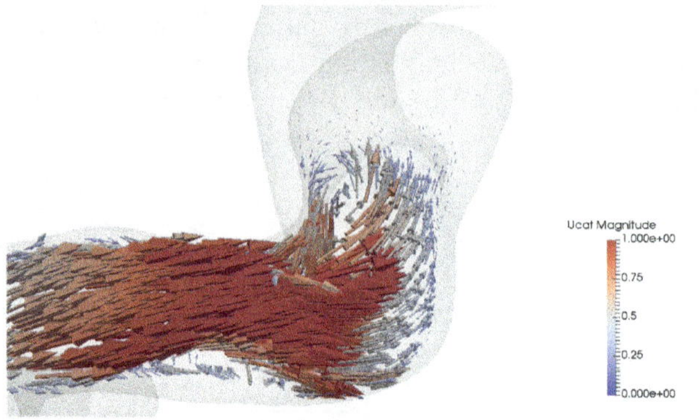

Fig. 2. The formation of the inclined vortex ring at the proximal neck near peak systole before the implantation. Velocity vectors are shown on a central plane of the aneurysm.

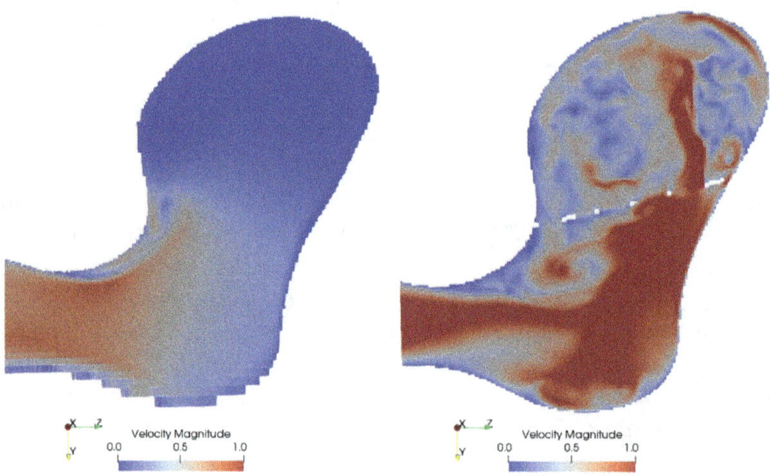

Fig. 3. Differences in hemodynamics at the ostium between the pre-implantation (left) and post-implantation (right) near peak systole. Before implantation, blood flow enters freely to the dome from the parent artery (left). The implantation of the stent (right) induces high speed jet locally in the openings of the weave.

prohibits the rolling-up of the shear layer at the neck. Instead the penetrating jet immediately interacts with the stent strut. Since the weave size is large in this case, the jet still forms albeit is broken up due to smaller scale jet. Such small-scale jets alternatively penetrates into the aneurysm void space as shown in Fig. 3.

Comparison the blood flow dynamics between the pre-implantation and the post-implantation shows the stark contrast between them as shown in Fig. 3. The stent changes the flow dynamics not only at the ostium level but also affects the flow distribution with thin the vessel artery. Because of the stent presence, the jet is now narrow at the aneurysm ostium. Therefore, the velocity of the main jet is increased as shown Fig. 3 (right).

4 Discussions

We have introduced an index to measure the flow unsteadiness namely *the Aneurysm Number (An)* [15]. This number is defined as the product of flow pulsatility (PI) and the ratio between the neckwidth (W) and the arterial diameter (D): $An = PI\frac{W}{D}$. In our case, the value of $PI = 2.31$, $W = 3.2\,\text{mm}$ and $D = 3.3\,\text{mm}$. The resulted *Aneurysm Number* is $An = 2.24$. According to our theory [13,15], this value corresponds to the formation of the vortex ring. Indeed our results in Fig. 2 show the presence of the vortex ring at the ostium level. This result further confirms the applicability of the Aneurysm Number concept in predicting the flow unsteadiness.

The flow unsteadiness plays even more pronounced role when the stent is implanted. There is a significant change of flow distribution at the ostium level as shown in Fig. 3 after the implantation. Comparing the with and without stent scenarios, it is clear that the presence of the stent changes both the flow inside the parent artery as well as in the aneurysmal cavity.

Aneurysm embolization is based on the physiology of thrombus formation [14] in which thrombi fill in the aneurysm cavity. This process can only be successful if the platelet is activated locally at the ostium of the aneurysm. Since platelet activation only occurs at sufficiently high shear stress, it is important to create an appropriate environment for thrombi formation.

The mechanism leads to the thrombi formation might be explained by the existence of small jets through the pores. As shown in Fig. 3, the role of the flow pulsatility is forcing the breakup of the large-scale flow structure via the openings of the stent into smallers jets. These penetrating jets via the openings of the stent has the spatial scale depending on the weave's size. Therefore, the strut density has an important role in the efficacy of the treatment [12].

5 Conclusions

In this work, we show that the impact of flow pulsatility is significant in regulating flow dynamics at the ostium. In addition, the dynamics of the penetrating jet is changed under the implantation of stent significantly, not just at the neck area but also at the proximal end of parent artery. Our results show that the interaction between the jet and the stent is most intense near peak systole. This interaction is dynamically changed during one cardiac cycle. Future works are needed to quantify the correlation between the strut density and the penetrating depth of the break-up jet.

Acknowlegement. This project is partially supported by NSF ND EPSCOR project FAR0030612. We thank graduate student Venkata Kanumuru for preparing the three-dimensional model of the aneurysm. This work is supported by the start-up package of Trung Bao Le from the Department of Civil and Environmental Engineering, North Dakota State University. The computational work has been performed using the computational resources of Center for Computationally Assisted Science and Technology (CCAST) at North Dakota State University.

References

1. Kallmes, D.F., Hanel, R., Lopes, D., Boccardi, E., Bonafé, A., Cekirge, S., Fiorella, D., Jabbour, P., Levy, E., McDougall, C., Siddiqui, A.: International retrospective study of the pipeline embolization device: a multicenter aneurysm treatment study. Am. J. Neuroradiol. **36**(1), 108–115 (2015)
2. Alderazi, Y.J., Shastri, D., Kass-Hout, T., Prestigiacomo, C.J., Gandhi, C.D.: Flow diverters for intracranial aneurysms. Stroke Res. Treat. (2014)
3. Rajah, G., Narayanan, S., Rangel-Castilla, L.: Update on flow diverters for the endovascular management of cerebral aneurysms. Neurosurg. Focus **42**(6), E2 (2017)
4. Cebral, J.R., Mut, F., Raschi, M., Hodis, S., Ding, Y.H., Erickson, B.J., Kadirvel, R., Kallmes, D.F.: Analysis of hemodynamics and aneurysm occlusion after flow-diverting treatment in rabbit models. Am. J. Neuroradiol. **35**(8), 1567–1573 (2014)
5. Cebral, J.R., Castro, M.A., Appanaboyina, S., Putman, C.M., Millan, D., Frangi, A.F.: Efficient pipeline for image-based patient-specific analysis of cerebral aneurysm hemodynamics: technique and sensitivity. IEEE Trans. Med. Imaging **24**(4), 457–467 (2005)
6. Briganti, F., Leone, G., Marseglia, M., Mariniello, G., Caranci, F., Brunetti, A., Maiuri, F.: Endovascular treatment of cerebral aneurysms using flow-diverter devices: a systematic review. Neuroradiol. J. **28**(4), 365–375 (2015)
7. Murthy, S.B., Shah, J., Mangat, H.S., Stieg, P.: Treatment of intracranial aneurysms with pipeline embolization device: newer applications and technical advances. Curr. Treat. Options Neurol. **18**(4), 16 (2016)
8. Mut, F., Raschi, M., Scrivano, E., Bleise, C., Chudyk, J., Ceratto, R., Lylyk, P., Cebral, J.R.: Association between hemodynamic conditions and occlusion times after flow diversion in cerebral aneurysms. J. Neurointerventional Surg. **7**(4), 286–290 (2015)
9. Ge, L., Sotiropoulos, F.: A numerical method for solving the 3D unsteady incompressible Navier-Stokes equations in curvilinear domains with complex immersed boundaries. J. Comput. Phys. **225**(2), 1782–1809 (2007)
10. Becske, T., Potts, M.B., Shapiro, M., Kallmes, D.F., Brinjikji, W., Saatci, I., McDougall, C.G., Szikora, I., Lanzino, G., Moran, C.J., Woo, H.H.: Pipeline for uncoilable or failed aneurysms: 3-year follow-up results. J. Neurosurg. **127**(1), 81–88 (2017)
11. Paliwal, N., Jaiswal, P., Tutino, V.M., Shallwani, H., Davies, J.M., Siddiqui, A.H., Rai, R., Meng, H.: Outcome prediction of intracranial aneurysm treatment by flow diverters using machine learning. Neurosurg. Focus **45**(5), E7 (2018)
12. Bouillot, P., Brina, O., Ouared, R., Yilmaz, H., Lovblad, K.O., Farhat, M., Pereira, V.M.: Computational fluid dynamics with stents: quantitative comparison with particle image velocimetry for three commercial off the shelf intracranial stents. J. Neurointerventional Surg. **8**(3), 309–315 (2016)

13. Le, T.B., Troolin, D.R., Amatya, D., Longmire, E.K., Sotiropoulos, F.: Vortex phenomena in sidewall aneurysm hemodynamics: experiment and numerical simulation. Ann. Biomed. Eng. **41**(10), 2157–2170 (2013)
14. Ngoepe, M.N., Frangi, A.F., Byrne, J.V., Ventikos, Y.: Thrombosis in cerebral aneurysms and the computational modeling thereof: a review. Front. Physiol. **9**, 306 (2018)
15. Le, T.B., Borazjani, I., Sotiropoulos, F.: Pulsatile flow effects on the hemodynamics of intracranial aneurysms. J. Biomech. Eng. **132**(11), 111009 (2010)

Hypertrophic Cardiomyopathy Treatment – A Numerical Study

Asaph Nardi, Guy Bar, Naama Retzabi, Michael Firer, and Idit Avrahami(✉)

Ariel University, 40700 Ariel, Israel
IditAv@ariel.ac.il

Abstract. Considered the leading cause of sudden cardiac death (SCD) in athletes, Hypertrophic Cardiomyopathy (HCM) is diagnosed in up to one of 200 people of the general population, regardless of prior medical condition. A common complication of HCM is Hypertrophic Obstructive Cardiomyopathy (HOCM). This complication is characterized by the obstructive motion of the anterior mitral leaflet causing mitral regurgitation, compromising systolic left ventricular (LV) reduced ejection fraction, and may lead to a significant outflow pressure gradients (>30 mmHg). Common treatment for HCM patients is medication, managing the symptoms. However, this treatment impairs the patient's quality of life restricting their daily activities. Some patients who are unresponsive to medication are prone to highly invasive surgery: myectomy or ablation, exposing them to high mortality and morbidity rates. Our research is aimed at offering a new minimally invasive approach to the treatment of HOCM patients, utilizing a percutaneous device placed in the LV which locally modifies the wall structure and mitral orientation, thus reducing the outflow pressure gradient. In this study we prove the concept using numerical simulations. Models of healthy, pathological and treated LV are used as geometry for computational fluid dynamics (CFD) simulations of the time-dependent flow in the LV. The results analyses show that the suggested procedure may dramatically reduce the pressure gradients during systole and allow better flow during diastole, advising on the improvement of the current treatment and feasibility of the recommended device.

Keywords: HOCM · CFD · SAM · Percutaneous support device · Contact analysis

1 Introduction

Affecting up to one in every 200 people of the general population [1], Hypertrophic Cardiomyopathy (HCM), is the leading cause of sudden cardiac death in young athletes in the United States and is the most common genetic cardiovascular disorder. This disorder causes thickening of the myocardium, directly affecting the heart's blood flow. Hypertrophic Obstructive Cardiomyopathy (HOCM) is a common complication of HCM that develops in up to 70% of these patients. This complication is characterized by the obstruction motion of the anterior mitral leaflet causing mitral regurgitation, compromising systolic left ventricular (LV) reduced ejection fraction, and significant outflow pressure gradients (>30 mmHg) [2].

G. A. Ateshian et al. (Eds.): CMBBE 2019, LNCVB 36, pp. 24–35, 2020.
https://doi.org/10.1007/978-3-030-43195-2_3

The mainstream treatment approach in HCM patients is medication [3], which relieve symptoms by improving diastolic function utilizing beta-blockers [4] and calcium antagonists which may slow down heart rate and decrease the possibility of ectopic heart beats [5]. Patients who are unresponsive to medication, or patients whose symptoms aren't relieved (3% of HCM patients) are subject to invasive procedures. These and HOCM patients, who undergo invasive procedures to either reconstruct the myocardium or neutralize the excessive cells [6], can undergo one of the following treatments: Septal myectomy [7] which is an invasive procedure where parts of the septum are sliced out in order to restore the proper shape of the ventricle. Septal Ablation is a procedure where echocardiography-guided ablation is performed, neutralizing the excessive growth. Recent studies show that utilizing secondary chordal cutting in combination with septal myectomy, can reduce the LV outflow gradient [8]. As gene editing is in its early stages in the field, this solution may be viable following the performance of long and short-term studies, thereby replacing the need for surgery in the future.

Currently, medication is considered the recommended solution for patients. Managing the symptoms and not directly treating the problem has been the main course of action due to high morbidity and mortality rates following invasive treatments. This approach leaves most patients in a compromised state and with limited quality of life. As a last resort, cardiologists advise specific obstructive patients and patients who don't respond to medication to undergo invasive procedures, generating a morbidity rate of up to 40% and 0.9–2.5% mortality. Older patients have a lower chance of recovering from myectomy, making them candidates for septal ablation [3].

We aim to make the invasive procedures redundant by suggesting the implantation of a percutaneous support device in the left ventricle (LV), thus eliminating the Venturi effect caused by the displacement of the martial mitral valve (MV) leaflet and thereby improving the flow regimes in the LV. In this study we propose to investigate the feasibility of the suggested device using numerical simulations.

2 Methods

2.1 Study Overview

We used numerical methods to analyze the dynamics of the fluid domain inside the LV and the impact of the mechanical support on the myocardium and MV. The simulation consisted of three main stages (see Fig. 1):

- CFD simulations of healthy and pathological cases
- Contact (structural) simulation of a pathological LV myocardium with the support device and prediction of a treated LV
- CFD simulations of a treated case

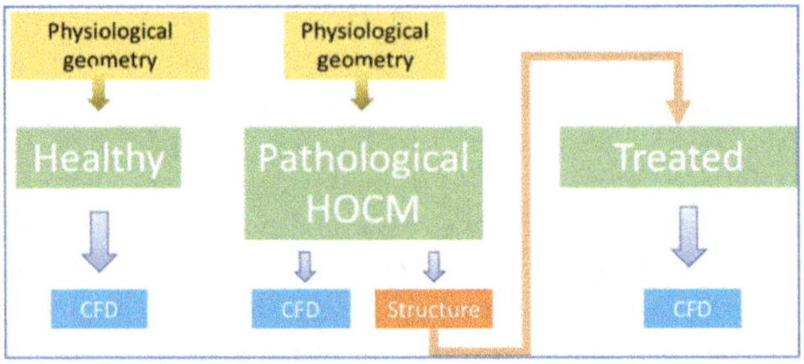

Fig. 1. Schematic description of study work plan

2.2 Geometry and Boundary Conditions

All the models were modeled based on clinical images imported to SolidWorks (Dassault Systems SOLIDWORKS Corp.). Physiological LV images were used to construct the geometry models of healthy and pathological cases. The images were sliced to construct 2D planes on which we sketched closed splines on model edges (see Fig. 2a). These sketches were used to extract the LV lumen from the image (red in Fig. 2b) and construct lofted bodies of the LV lumen, atrium, aorta and leaflets (Fig. 2c) for the CFD simulations. For the pathological case, a model of the LV myocardium was constructed for a structural simulation (Fig. 2d). Using a contact analysis with a model of the support device, a prediction of the treated geometry was achieved, delineating the contours of the treated LV lumen. Based on this prediction, a model of the LV lumen of the treated case was also constructed. The geometric models of the three lumen cases (for CFD) are presented in Fig. 3.

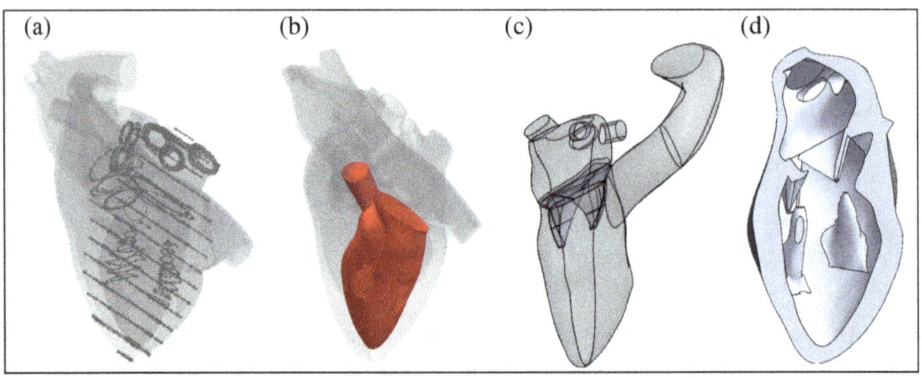

Fig. 2. Stages in 3D LV geometry extraction and construction (a) of LV lumen (b) and myocardium (c)

Boundary conditions for the CFD simulations were based on volumetric change of the LV volume according to a typical time function represents changes in the LV volume (Fig. 4) assigned as wall displacements throughout three successive cardiac cycles heart rate of 80 bpm and cardiac output of 5 l/min [11]. Blood was assumed Newtonian with viscosity of $\mu_b = 0.0035\ \frac{kg}{m \cdot s}$ and density of $\rho_b = 1055\ \frac{kg}{m^3}$ [9].

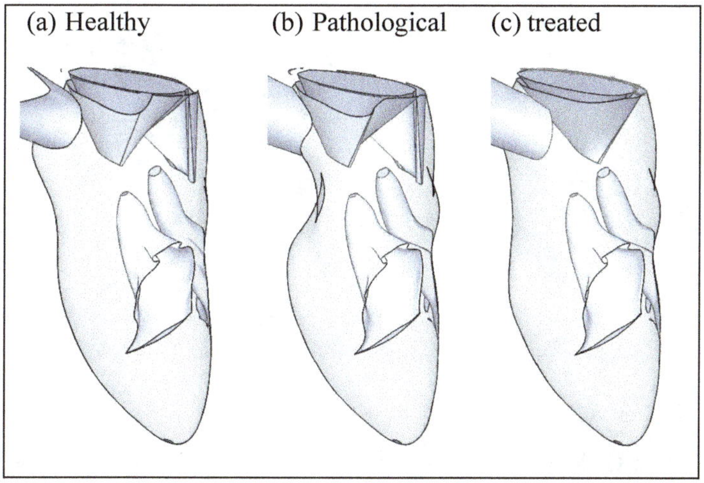

Fig. 3. LV lumen of the three CFD geometric models

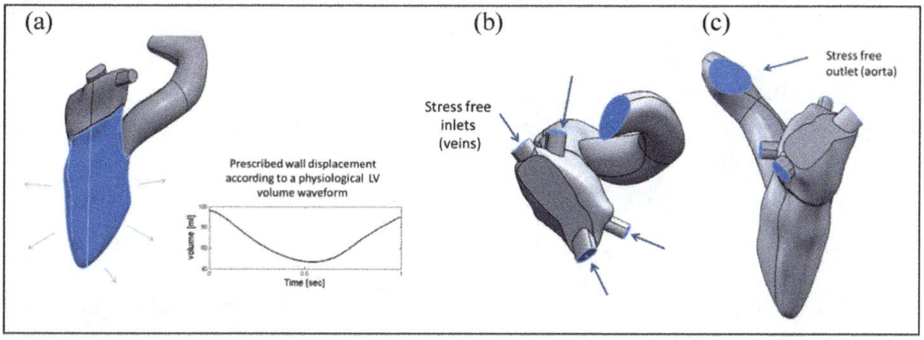

Fig. 4. Boundary conditions for the CFD simulations consisted of (a) prescribed wall displacement according to a typical physiological volume waveform [11], (b) stress free inlets at the veins during diastole and (c) stress free outlet at the aorta during systole.

The structural model assumed effective mechanical properties for both the myocardium and the support device. The effective properties of the myocardium during diastole was assumed to be elastic isotropic with effective Young's modulus of $E_m = 1$ MPa , effective Poisson's ratio of $v_m = 0.3$ and density of $\rho_m = 1060\ \frac{kg}{m^3}$ [9]. The support device was assumed to have effective properties of $E_s = 830$ MPa, $v_s = 0.33$, and $\rho_s = 6250\ \frac{kg}{m^3}$ [10]. The structural analysis simulated the contact between the HOCM myocardium and the support device. Device implantation was simulated by compression of the device following by device inflation (Fig. 5).

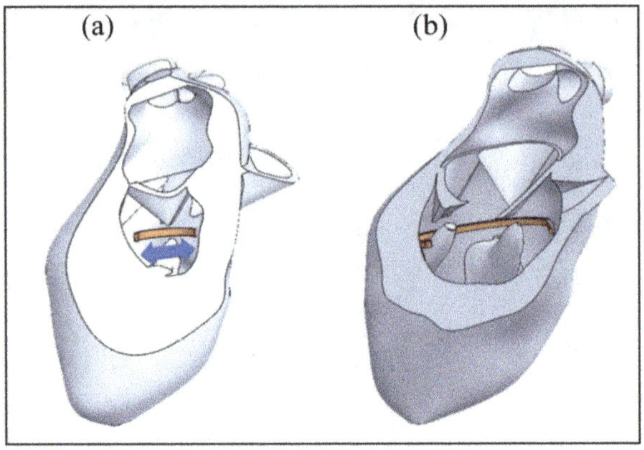

Fig. 5. Boundary conditions for the structural with a contact between the support device and the myocardium. The support device in its pre-inflation position (a) and post-inflation (b)

2.3 The Mathematical Model

The fluid domain Ω_f is defined by the filling and ejection of the blood flow in the LV. This domain is solved by the Navier-Stokes equations for laminar, homogeneous, Newtonian and incompressible fluid.

$$\begin{gathered} \nabla \cdot \mathbf{U}_f = 0 \\ \rho_f \left(\frac{\partial \mathbf{U}_f}{\partial t} + \mathbf{U}_f \cdot \nabla \mathbf{U}_f \right) = -\nabla P + \mu \cdot \nabla^2 \mathbf{U}_f \end{gathered}, \tag{1}$$

where U_f is the velocity vector, t is time and ∇P is pressure. The integral volume change, also known as the flux, is time dependent, and is explicitly shown as the mass conservation equation:

$$\frac{d\forall}{dt} = \frac{d}{dt} \int (\boldsymbol{d}_s \cdot \mathbf{n}) d\mathbf{A} = Q(t), \tag{2}$$

where $\forall$ is volume, Q is flux, $\mathbf{d_s}$ is the vector of displacement, $\mathbf{n}$ is a unit vector normal to the wall and $\mathbf{A}$ is the wall surface area.

The solid domain Ω_s is dominated by the structure's dynamic equation:

$$\rho_s \frac{\partial^2 \boldsymbol{d}_s}{\partial t^2} - \nabla \boldsymbol{\sigma}_s = \boldsymbol{f} \tag{3}$$

Where $\boldsymbol{\sigma}_s$ is the Cauchy stress tensor, $\mathbf{d_s}$ is the vector of structure displacement, ρ_s is the wall density and $\mathbf{f}$ represents the body forces applied on the structure.

2.4 The Numerical Model

The commercial software ADINA (ADINA R&D, Inc.) was used to solve the governing equations in the fluid and structure domains using the Finite element method (FEM). Wall motion of the fluid domain used the Arbitrary Lagrangian-Eulerian (ALE) method to calculate the dynamic mesh displacements with time.

A double-sided contact element group was defined to allow the support device to first compress and then reopen, simulating an interaction with a moving wall. A non-continuous contact segment normal was defined allowing the segment another degree of freedom. Contact birthtime was set to 0.21 [s], allowing contraction and relaxation.

Mesh and time-step independence tests were conducting to ensure a stable calculation process and analysis error within the range of <10% (estimated error in pressure gradients is 7% and in maximal velocity is 0.4).

The structural model consisted of 60,000 3D 1st order tetrahedral elements and the fluid domain consisted of 270,000 3D tetrahedral fluid elements.

2.5 Cases Studied and Investigated Parameters

Our study consisted of three models:

1. CFD simulation of a healthy LV lumen, used as a control case.
2. CFD simulation of a pathological LV lumen
3. Structure model of pathological LV myocardium in contact with a model of the support device
4. CFD simulation of a treated LV lumen (as predicted by the structural simulation, 3)

The critical parameter showing the improvement of treated subjects is the pressure gradients. Therefore, pressure gradients were measured between the center of the LV and the aortic root at three time instances: during peak flow, mean systolic flow and of averaged over the cardiac cycle. Maximal velocity was recorded as well, due to the physical link between the pressure and velocity.

A flux independence test was performed to validate the moving wall boundary conditions by calculating the time-dependent flow rate in a cross section at the aorta outlet according to Eq. (2). Results show (Fig. 6) an average error during the tested period of $Avg|Err| = 0.8\%$. The flux values were compared to expected flux values from literature [11].

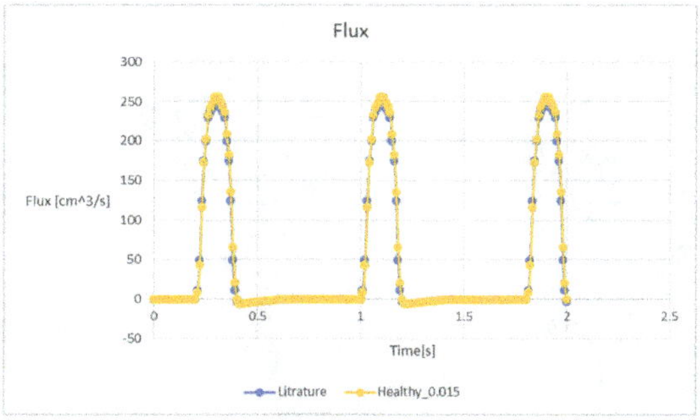

Fig. 6. Calculated outlet flux (through the aortic outlet) in the healthy case (yellow) compared with expected aortic flow [11].

3 Results

3.1 Structure Simulation

First we will present the structure simulation, which presents the interaction between the pathological myocardium and the support device. The myocardium and septum displacements following the implantation of the support device is shown in Fig. 7. Results show a displacement value of $\Delta \cong 1.04$ [cm] at the contact points of the wall with the device. An explicit ring indentation is noticeable at the contact points.

Elevated stress values (up to $\tau = 3$ [MPa]) can be seen in Fig. 8 both on the myocardium and on the support device. High stress value points are seen at the auricle of the left atrium where the veins are fixed. The elevated stress values on the myocardium and septum are consistent with contact points.

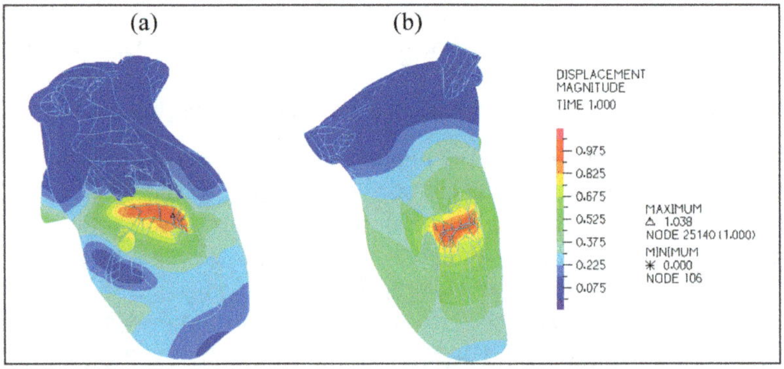

Fig. 7. Displacement magnitudes as calculated in the contact analysis between the pathological myocardium and the support device

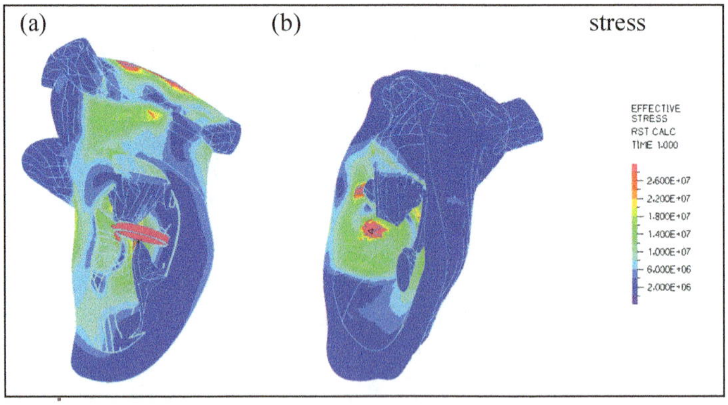

Fig. 8. Effective stress magnitudes as calculated in the contact analysis between the pathological myocardium and the support device

3.2 CFD Simulation

The "treated" geometry was constructed according to the results obtained from the pathological structural simulation, and was used to calculate the flow field in the treated case. Results were compared with the healthy and pathological cases. Results of the flow parameters in three cases are presented in this section.

Figure 9 shows velocity vectors in the LV out-tract during peak systole. Elevated values are found at the Aortic root in all cases. Elevated velocities aren't distinguishable in the pathological case over the healthy model. Lower velocities appear in the treated model during systole following the device's placement (Table 1).

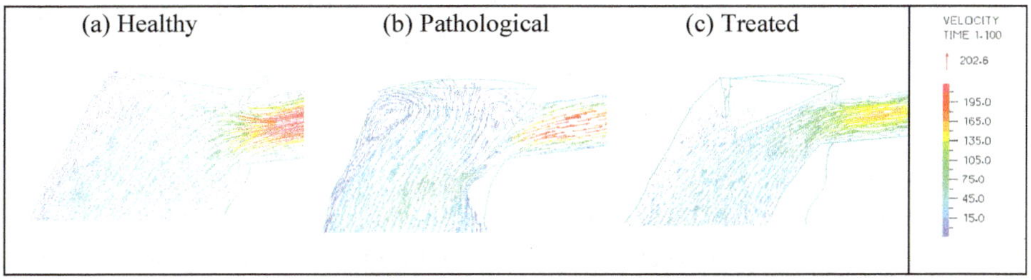

Fig. 9. Velocity vectors in the LV out-tract during peak systole

Table 1. Calculated maximal velocity in the three cases (in cm/s)

Velocity [cm/s]	Healthy	HOCM	Treated
Peak systolic flow	213	327	222
Mean systolic flow	140	217	126
Averaged over the cardiac cycle	76	148	87

Figure 10 presents pressure gradients in the LV out-tract during peak systole. Table 2 presents calculated gradient values for the peak systole, the mean systolic flow and averaged along the entire cardiac cycle. Results show a severe gradient due to the combination of the myocardium growth and reshapement of the LV together with the MV leaflet SAM. Healthy case shows low pressure gradients consistent with the

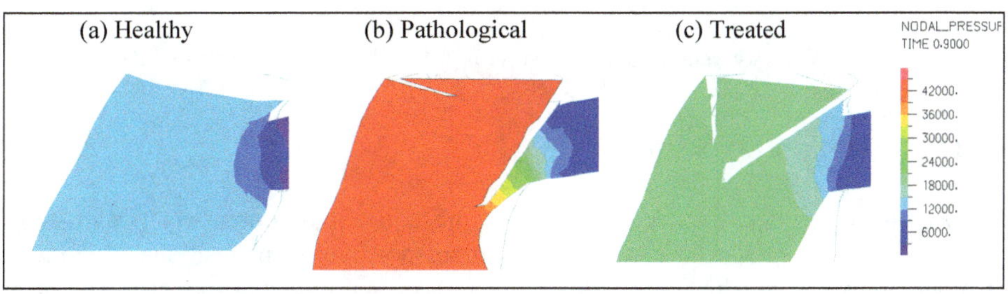

Fig. 10. Pressure gradients in the LV out-tract during peak systole

literature, while a treated LV provides a lowered gradient while repositioning the MV and clearing the outflow tract.

Table 2. Calculated pressure gradients in the three cases (in mmHg)

ΔP [mmHg]	Healthy	HOCM	Treated
Peak systolic flow	20	73	27
Mean systolic flow	7	34	11
Averaged over the cardiac cycle	2.6	10	9

Device implantation not only changed velocities and pressure gradients magnitudes, but also changed flow patterns during systole, as shown in Fig. 11. In the pathological case, velocity vectors appear to bypass the MV by flowing around it, thus maintaining mass conservation despite the obstruction. In the treated case, the velocity vectors are straight forward.

Flow patterns in the healthy model show smooth velocity vectors leading to the aortic root during systole. During diastole we observe vortices in the LV near the septum and myocardium adjacent to the MV circumference. Pathological case study shows a detour in flow pattern to bypass the obstruction. This might cause elongation of the posterior MV leaflet in the angular direction following high velocities, as well as the longitudinal, as described in the literature [12, 13].

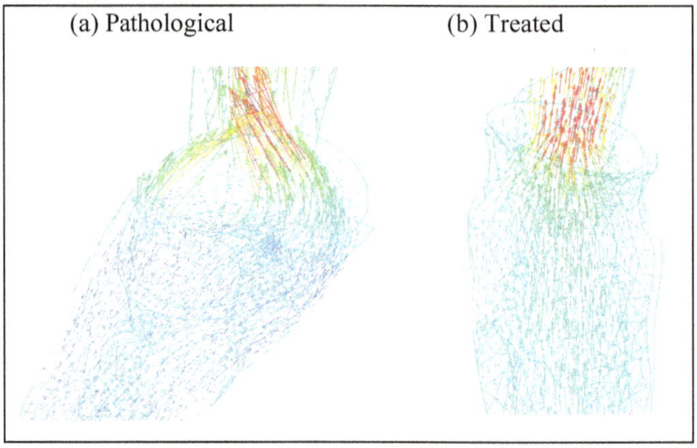

Fig. 11. Velocity vectors during systole in the pathological and treated cases.

4 Discussions and Summary

We present a numerical model that proves the concept of our novel percutaneous support device for HOCM patients who are unresponsive to medications. The model calculates numerically the flow regimes in the LV. We used the models to compare between healthy, pathological and treated cases. In order to predict the geometry of the treated case, we used structural model using contact analysis between the myocardium and the support device.

This study is aimed at comparing the overall behavior of flow in the three cases (of healthy, pathological and treated LV) and therefore represents only a quantitative analysis to compare the different cases. Consequently, the numerical models used several simplifications. The assumptions and simplifications that we made to solve the problem need to be carefully evaluated and their implications are discussed below.

Our main simplification is related to the specific motion of the LV wall during systolic contraction. We assumed a uniform contraction, where the wall contraction at each node should be a function of time and space. This simplification should have a strong effect on the actual flow, since it does not consider the complex 3D geometry motion. A more precise simulation would have used a patient's-specific wall motion and requires a massive effort [e.g. 14, 15]. Yet, we believe that this assumption should provide a good first-order estimation of the systolic flow in the aortic track for quantitative comparison between the cases.

Moreover, the models do not target a specific patient geometry or boundary condition. The models are based on representative prototypes of anatomical geometry constructed according to a typical anatomical database, and the boundary conditions are based on a typical time functions garnered from the literature. Since patient specific anatomies and physiology come in large variations, especially for HOCM patients, whatever models we use will be inaccurate for the specific patient. However, the typical geometry we chose should be feasible for non-specific comparison analyses. Moreover, the calculated results of the healthy cases agree with more accurate, patient-specific simulations [16–18].

The mechanical properties of the myocardium and of the support device were estimated to be linearly elastic. The assumption of linear elastic material properties for the myocardium may be questionable. Physiologic myocardium tissue is known as anisotropic and hyperelastic [19, 20]. Moreover, in HOCM patients, the situation is even more complex, since the abnormal tissue growth increasing the complexity of the tissue properties. In our model, we assumed homogeneous, isotropic and linear elastic material only to show a methodological concept and allow a global prediction of the modified structure, to delineate the lumen geometry for CFD. A more realistic research should be made while improving the device's design. We expect that better choice of the material properties and support design should have some effect on the resulted lumen geometry (and this on the flow results).

The assumption of Newtonian fluid should be reasonable in this case. This is because the shear rates in the aorta are generally greater than 100 s^{-1} [21]. Another simplification is related to the flow assumption.

The average Reynolds number is 700; therefore the simulation uses laminar flow models. This assumption may be problematic due to the complex geometry and induced flow instabilities, especially during the diastolic deceleration phase. In these conditions it was shown that the blood flow can undergo a transition from a well-structured laminar state to a chaotic turbulent state [22]. This transition is a challenge to model. Several studies partially succeeded in developing and using low-Reynolds turbulence models that showed good agreements between measurements and numerical results within an accuracy of up to 10%. However, these models required high computational resources with mesh sizes of $\sim$5 Million elements in order to ensure convergence. Due to the high demands of the FEM solution, the assumption of laminar flow highly simplified the model and made the simulation possible, thus it was inevitable. We will address this issue in future work where more accurate models will be considered.

The use of a fixed-leaflet geometry (of either open or closed valve) is another common simplification [23]. Although the opening and closing phases of the leaflets are very short (of the order of 1–10 ms), they may have some influence on the flow field and it might add some error to the calculated values. However, including a fluid-structure interaction simulation of the leaflets in this case is much beyond the scope of the present study, especially for HOCM patients, where the mechanical properties of the leaflets are not clear.

In summary, the purpose of this study was to depict the major dominant factors of the LV flow (especially at the systolic phase) using a simple and feasible model. Despite the simplifications and assumptions mentioned above, the comparative results of this simple model may provide essential information and shed light on the hemodynamic mechanisms that determine the flow field in the pathological and treated LV.

Using the numerical simulations, we were able to show that there was improvement in the pressure gradient in the treated case using the suggested device. Our results show a dramatic change in gradient pressures between the different cases. The device also Re-positioned the MV leaflet reducing the SAM effect, together with pressing the thickened wall. This allows improved systolic flow, improved flow regimes and reduced pressure gradients in the LV out-track.

Acknowledgements. The research was supported by the Ariel grant for advancing women in academia.

References

1. Kramer, C.M., et al.: Hypertrophic cardiomyopathy registry (HCMR). Am. Heart J. **170**(2), 223–230 (2015)
2. McIlvennan, C.K., Eapen, Z.J., Allen, L.: Hospital readmissions reduction program. Circulation **131**(20), 1796–1803 (2016). HHS Public Access
3. Maron, B.J., et al.: The 50-year history, controversy, and clinical implications of left ventricular outflow tract obstruction in hypertrophic cardiomyopathy: from idiopathic hypertrophic subaortic stenosis to hypertrophic cardiomyopathy. J. Am. Coll. Cardiol. **54**(3), 191–200 (2009)
4. Leviner, D.B., Hochhauser, E., Arad, M.: Inherited cardiomyopathies - novel therapies. Pharmacol. Ther. **155**, 36–48 (2015)
5. Gersh, B.J., et al.: 2011 ACCF/AHA guideline for the diagnosis and treatment of hypertrophic cardiomyopathy: Executive summary. J. Thorac. Cardiovasc. Surg. **142**(6), 1303–1338 (2011)
6. Sherrid, M.V., Balaram, S., Kim, B., Axel, L., Swistel, D.G.: The mitral valve in obstructive hypertrophic cardiomyopathy a test in context. J. Am. Coll. Cardiol. **67**(15), 1846–1858 (2016)

7. Nishimura, R.A., Seggewiss, H., Schaff, H.V.: Hypertrophic obstructive cardiomyopathy: surgical myectomy and septal ablation. Circ. Res. **121**(7), 771–783 (2017)
8. Ferrazzi, P., et al.: Transaortic chordal cutting mitral valve repair for obstructive hypertrophic cardiomyopathy with mild septal hypertrophy. J. Am. Coll. Cardiol. **66**(15), 1687–1696 (2015)
9. Hazan, O., et al.: An innovative appendage invagination procedure to reduce thrombus formation–a numerical model. Comput. Methods Biomech. Biomed. Eng. **21**(4), 370–378 (2018)
10. Brand, M., Avrahami, I., Einav, S., Ryvkin, M.: Numerical models of net-structure stents inserted into arteries. Comput. Biol. Med. **52**, 102–110 (2014)
11. Van Slochteren, F.J., et al.: Left ventricular PV loop estimation using 3D echo. Internal report BMTE 07.42, Eindhoven University of Technology (2007)
12. Hagège, A.A., et al.: The mitral valve in hypertrophic cardiomyopathy. J. Cardiovasc. Transl. Res. **4**(6), 757–766 (2011)
13. Maron, B.J., et al.: Systolic anterior motion of the posterior mitral leaflet: a previously unrecognized cause of dynamic subaortic obstruction in patients with hypertrophic cardiomyopathy. Circulation **68**(2), 282–293 (1983)
14. Su, B., et al.: Cardiac MRI based numerical modeling of left ventricular fluid dynamics with mitral valve incorporated. J. Biomech. **49**(7), 1199–1205 (2016)
15. Tang, D., et al.: Image-based patient-specific ventricle models with fluid–structure interaction for cardiac function assessment and surgical design optimization. Prog. Pediatr. Cardiol. **30**(1–2), 51–62 (2010)
16. Bavo, A.M., et al.: Patient-specific CFD simulation of intraventricular haemodynamics based on 3D ultrasound imaging. Biomed. Eng. Online **15**, 1–15 (2016). Article number. 107
17. Chnafa, C., Mendez, S., Nicoud, F.: Image-based simulations show important flow fluctuations in a normal left ventricle: what could be the implications? Ann. Biomed. Eng. **44**(11), 3346–3358 (2016)
18. Khalafvand, S.S., et al.: Assessment of human left ventricle flow using statistical shape modelling and computational fluid dynamics. J. Biomech. **74**, 116–125 (2018)
19. Humphrey, J.D.: Continuum mechanics. In: Cardiovascular Solid Mechanics, pp. 68–106. Springer, New York (2002)
20. Sacks, M.S., Chuong, C.J.: Biaxial mechanical properties of passive right ventricular free wall myocardium. J. Biomech. Eng. **115**(2), 202–205 (1993)
21. Fung, Y.C.: Biomechanics: Mechanical Properties of Living Tissues. Springer, New York (1993)
22. Bluestein, D., Einav, S.: Techniques in the stability analysis of pulsatile flow through heart valves. In: Leondes, C. (ed.) Cardiovascular Techniques, Biomechanical Systems Techniques and Applications. CRC Press, Boca Raton (2001)
23. Avrahami, I., Rosenfeld, M., Raz, S., Einav, S.: Numerical model of flow in a sac-type ventricular assist device. Artif. Organs **30**(7), 529–538 (2006)

Computational Assessment of Risk of Subdural Hematoma Associated with Ventriculoperitoneal Shunt Placement

Milan Toma[1,2(✉)] and Sheng-Han Kuo[3]

[1] Department of Mechanical Engineering, College of Engineering and Computing Sciences, New York Institute of Technology, Old Westbury, NY 11568-8000, USA
tomamil@tomamil.eu
http://www.tomamil.com

[2] Department of Osteopathic Manipulative Medicine, College of Osteopathic Medicine, New York Institute of Technology, Old Westbury, NY 11568-8000, USA

[3] Department of Neurology, Columbia University Medical Center, Columbia University, New York City, NY, USA

Abstract. Hydrocephalus is an accumulation of cerebrospinal fluid within the brain. The condition yields increased pressure inside the skull. When excess cerebrospinal fluid collects in the ventricles of the brain, ventriculoperitoneal (cerebral) shunt is routinely implanted. A ventriculoperitoneal shunt redirects fluid from the ventricles to the abdominal cavity. However, the ventriculoperitoneal shunt is sometimes associated with several complications, e.g. subdural hematoma. A subdural hematoma is a collection of blood outside the brain caused by the sudden shrinkage of the brain as the cerebrospinal fluid is drained away by the shunt implant. However, the mechanism of the development of subdura hematoma remains not entirely clear due to the dynamic alterations between cerebrospinal fluid, intraparenchyma, and ventricular pressure. Therefore, we aim to establish a model to simulate this interaction to understand the mechanism.

Keywords: Computational geometry · Smooth-particle hydrodynamics · Fluid-structure interaction

1 Introduction

Management of patients with normal-pressure hydrocephalus (NPH) routinely involves placement of a ventriculoperitoneal (VP) shunt to aid cerebrospinal fluid (CSF) drainage. The VP shunts include adjustable valves that enable non-invasive adjustment of the amount of CSF drainage. The valves help to maximize symptoms relief and minimize over-drainage. Without adjustable valves there

G. A. Ateshian et al. (Eds.): CMBBE 2019, LNCVB 36, pp. 36–47, 2020.
https://doi.org/10.1007/978-3-030-43195-2_4

would be frequent need for repeated surgical interventions to manage shunt pressure [1,2].

Patients who have undergone CSF shunting procedures to relieve symptoms related to NPH are commonly at risk of developing symptomatic subdural hematoma (SDH). To reiterate, with the subsequent decrease of their intracranial pressure, these patients are susceptible to SDH formation caused by overdrainage of CSF [1,2].

In essence, the brain can be structurally divided into the cerebrum, cerebellum, and brainstem. The cerebrum is divided into two approximately equal hemispheres connected by the corpus callosum and a shared ventricular system. The brainstem is further divided into the midbrain, pons, and medulla oblongata. The CSF fills a system of cavities at the center of the brain, known as ventricles, and the subarachnoid space surrounding the brain and spinal cord [3]. The CSF cushions the brain within the skull and serves as a shock absorber for the central nervous system [4,5].

The three published constitutive models representing CSF behavior are the solid-like, viscoelastic, and fluid-like CSF. However, they all are solid material models with different material properties. The solid-like constitutive model represents CSF as a linear, nearly incompressible elastic solid with a bulk modulus much larger than the shear modulus [6,7]. The viscoelastic CSF constitutive model is a linear viscoelastic model with shear relaxation behavior [8,9]. The fluid-like CSF is modeled using elastic solid material with fluid-like constitutive behavior using an equation of state constitutive model [10]. In recent years, fluid-structure interaction (FSI) models are being implemented by multiple groups [11–15].

While the significance of including CSF in the numerical simulations has been shown [16], the current finite element studies reported in the literature often lack more detailed anatomical structures [9,17–21]. The brain has a more complex structure than what has been described in prior models. For example, CSF is commonly presented as existing only outside of the brain, i.e. between the skull and a single, solid mass representing the brain. The model used in this study is the first FSI model that uses fluid material properties to represent the CSF, both between the brain parenchyma and skull and also in the ventricular system, and includes all major anatomical features of the brain [11–13]. The model has been validated against cadaveric data [13]. A detailed analysis of the CSF cushioning mechanism is useful for the treatment and prevention of brain injuries. The importance of studying CSF interaction with brain parenchyma, and its implication on human diseases such as traumatic brain injury and hydrocephalus, is axiomatic.

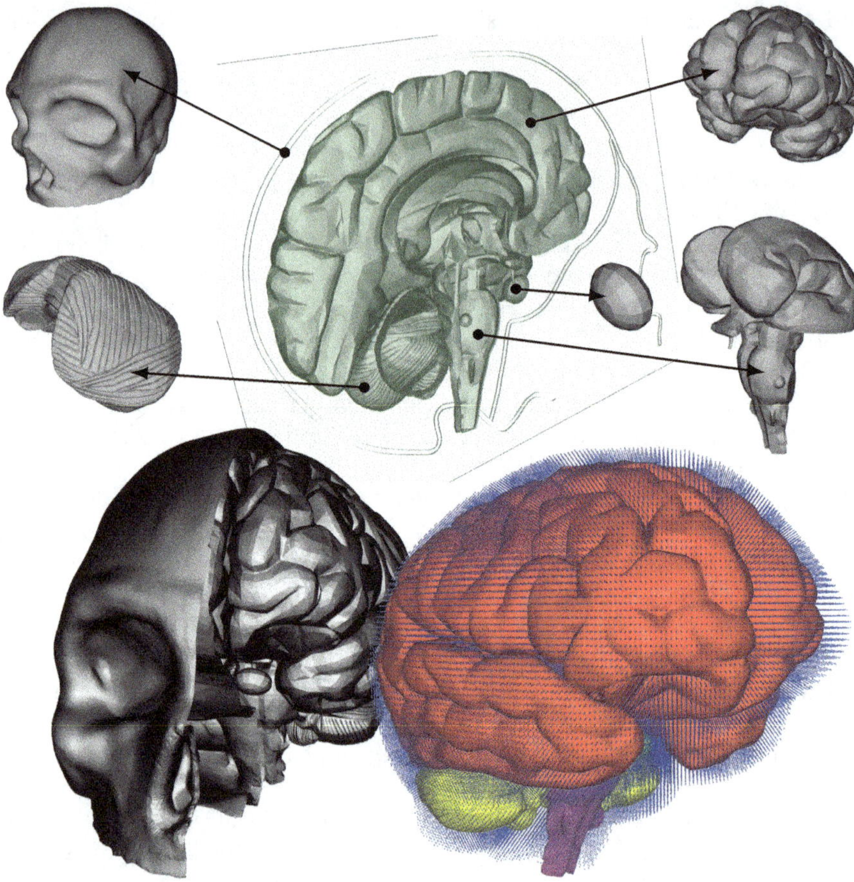

Fig. 1. The depiction of the entire head model with skull, cerebrum, cerebellum, pituitary gland and brain stem, respectively. The subarachnoid space and other cavities are filled with fluid particles (blue dots surrounding the brain model, in the lower right corner). The entire model with half the skull is also shown (lower left).

2 Methods

The geometrical model, numerical method, computational simulations, and initial and boundary conditions are described in the following subsections.

2.1 Head Model

The five distinct anatomical structures used in this model are shown in Fig. 1. The skull, cerebrum, cerebellum, pituitary gland, and brainstem each have unique material properties. The patient-specific model is based on the Digital Imaging and Communications in Medicine (DICOM) images acquired from an online database. Anatomical features missing in this model include the skin, spinal cord, meninges, and the arachnoid granulation. When compared to the very short impact impulse time history used in these simulations, the CSF flow in the head can be neglected, too. The CSF flow is relatively slow, 0.05–0.08 $\mathrm{m \cdot s^{-1}}$, i.e. during the impact impulse time history the CSF flows by 0.2–0.3 mm. These assumptions also make the presence of the granulations negligible.

2.2 Smoothed Particle Hydrodynamics Formulation

The smoothed particle hydrodynamics (SPH) has recently been used in modeling the interaction of the blood flow with the arterial walls, heart ventricles and native and artificial heart valves [22–29]. Simulating the interaction of the cerebrospinal fluid with the brain and skull structures using the SPH has also begun [11–13,30]. In recent years, the SPH has been increasingly used in biomedical applications [31].

In this study, next generation SPH method with high accuracy is used. For completeness, the basic governing equations of the SPH method is summarized in this section. The SPH is governed by a set of ordinary differential equations as follows.

The continuity equation, i.e. conservation of mass, is

$$\frac{d\rho_i}{dt} = -\rho_i \sum_{j=1}^{N_p} (\mathbf{v}_i - \mathbf{v}_j) \frac{m_j}{\rho_j} \cdot \nabla_i W_{ij} + 4^{th} \text{ order diffusive term},$$

where W is the Kernel function and the 4th order diffusive term improves the pressure evolution.

The momentum equation, i.e. Euler equation, is

$$\frac{d\mathbf{v}_i}{dt} = -\sum_{j=1}^{N_p} \left(\frac{P_i + P_j}{\rho_i \rho_j} + \Pi_{ij}\right) m_j \cdot \nabla_i W_{ij} + \mathbf{f}_i,$$

where Π_{ij} is the artificial viscosity.

The energy equation, i.e. first law of thermodynamics, is

$$\frac{d\epsilon_i}{dt} = -\sum_{j=1}^{N_p} \left(\frac{P_i}{\rho_i^2} + \frac{\Pi_{ij}}{2}\right)(\mathbf{v}_i - \mathbf{v}_j) m_j \cdot \nabla_i W_{ij}.$$

More detail on the numerical methods used in the cardiovascular FSI simulations, e.g. the integration method, artificial viscous term, laminar viscous term, speed of sound used, can be found published elsewhere [22,26].

The SPH interacts with the structural finite elements using a penalty-based contact algorithm. In penalty-based contact, when a penetration is found, a force proportional to the penetration depth is applied to resist and eliminate penetration. Linear contact spring stiffness is based on the nodal masses that come into contact and the time step size as follows

$$k = 0.1 \frac{m}{\max(\Delta t_{global}^2, \Delta t_{SPHcontact}^2)}.$$

The resulting contact stiffness is independent of the material constants so it is well suited for treating contact between fluid and structure.

2.3 Computer Simulations

The IMPETUS Afea Finite Element and γSPH solvers® (IMPETUS Afea AS, Norway) were used to model the impact scenario. Fluid motion and boundary interaction calculations were solved with the IMPETUS Afea γSPH Solver, while large deformations in the solid parts were simultaneously solved with the IMPETUS Afea Solver. Both the solvers use a commodity GPU for parallel processing. All solid elements were fully integrated, removing the possibility of hourglass modes and element inversion that plagues the classic under-integrated elements. Both fluid and solid domains and their interaction were solved with an explicit integration scheme. All simulations were solved on a standard workstation. Parallel acceleration was achieved with a Tesla K40 GPU with 12 GB of Graphic DDR memory and 2880 CUDA Cores.

The contact used is relatively simple, i.e. particle to structure contact, which is why γSPH is ideal for these complex applications because it can easily account for movement in any direction, as opposed to finite element fluid solvers which involve more complicated contact and usually require remeshing of the fluid during the simulation. Needless to say, consequently, compared to the SPH method used, parallelization of standard finite element FSI solvers is not as efficient [32]. The fact that IMPETUS can allow for a very high resolution in terms of particle density provides for a very accurate particle to structure contact, especially with such a detailed and complex structure as brain. The other critical part of the simulation is the structural model. The IMPETUS Aset® Element Technology provides high order tetrahedron elements that are accurate for nonlinear dynamic response. This allows for automeshing complicated structures such as those found in biomedical applications. The accuracy of these elements has been demonstrated in many commercial applications when compared with hexahedron elements and is very accurate in bending and plasticity.

2.4 Simulating Normal-Pressure Hydrocephalus

Simulating NPH numerically, additional fluid particles are introduced to enter the skull inside the choroid plexus where CSF production is assumed. A pipe-like structure is incorporated into the model. The pipe is filled with fluid particles of the same properties as the fluid particles already filling the subarachnoid space and other cavities within the skull. At the far end of the pipe an impermeable piston is prescribed to move in the direction of the pipe. Consequently, the piston pushes all the particles from inside the pipe into the skull. Thus, the additional fluid particles are introduced to increase the intracranial pressure inside the skull. These new particles simulate the mechanism that yields excess CSF.

2.5 Simulating Cerebrospinal Fluid Drainage

After pressure increase has been simulated by introducing new particles pushed through the pipe structure into the skull (described in Sect. 2.4), the piston that has been pushing these particles is deleted and, due to an increased gravitational force applied to the fluid particles, the fluid particles are allowed to drain back down the pipe outside of the skull yielding decreased intracranial pressure. This method is used to simulate the drainage of excess CSF, e.g. via lumbar drain where a small flexible tube is placed in the lumbar spine. The tube drains some of the CSF that filled the subarachnoid space. To assess NPH, draining CSF is used to determine if the patient could benefit from a shunt. However, as mentioned above, the patients are then susceptible to SDH formation.

2.6 Bridging Veins

Many bridging veins (BVs) are located along sagittal sinus. Most acute SDHs develop after rupture of one or more BVs. The anatomy of the BVs makes them susceptible to tearing within the border cell layer of the dura mater [33]. In the numerical model, the forces exerted on the veins during the NPH and CSF drainage simulations are measured using a series of nonlinear spring connectors defined in the subarachnoid space between the dura mater and skull. While BVs are located along the sagittal sinus, for comparison, spring connectors are placed all around the brain in the entire subarachnoid space (Fig. 2).

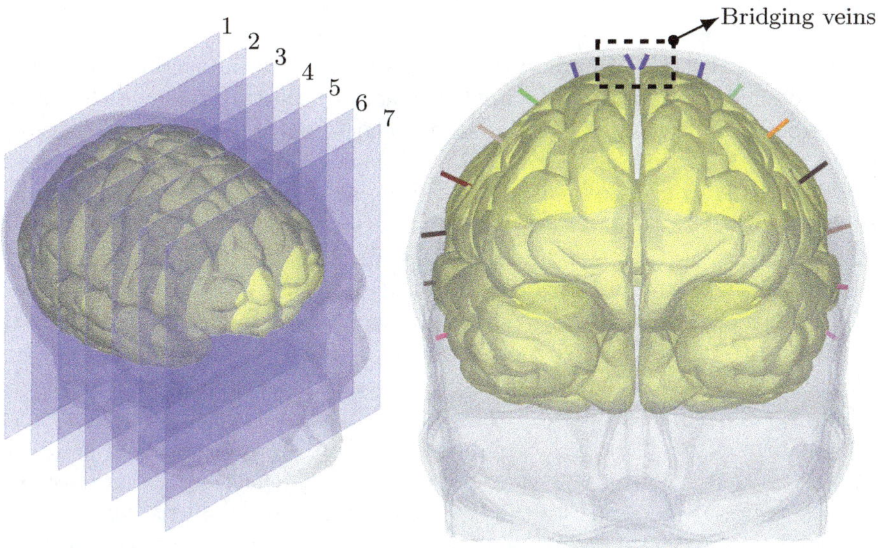

Fig. 2. Spring connectors located in the subarachnoid space between the dura mater and skull (one plane, i.e. coronal plane, out of seven, with the spring connectors is shown). Bridging veins are located along sagittal sinus (dashed rectangle).

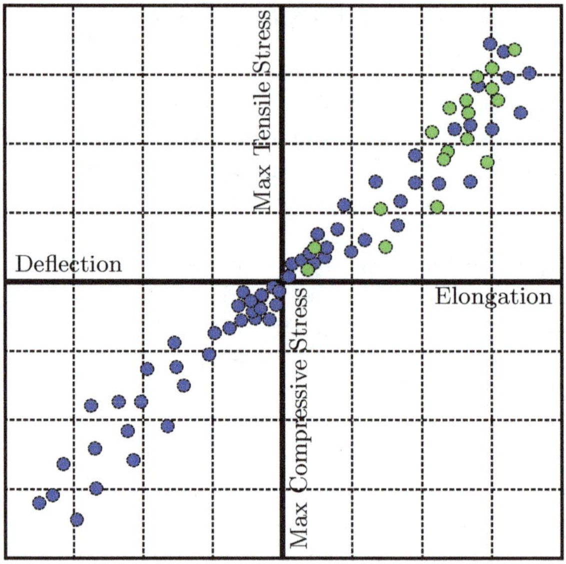

Fig. 3. Over hundred spring connectors are placed all around the brain connecting skull with dura mater. During the CSF drainage, some experienced deflection (compression) and some experienced elongation (extension). However, all springs placed along the sagittal sinus (green) experienced elongation.

3 Results

An experimental study of human tolerance level for BV-induced acute SDH has been conducted by delivering occipital impacts in sagittal plane to the cadaver [34]. Moreover, finite element analysis has been recently conducted to assess BV rupture associated with acute SDH [35]. However, none of the previous studies measured the SDH risk in relation to VP shunt placement and CSF drainage. The maximum BV strains are found at anterior parietal regions (i.e. the BVs depicted in Fig. 2). The maximum BV strain found is 0.61, which is within BV strain range (0.5 ± 0.16) from the study on the effects of strain rate on tensile failure properties of BVs studied in cadavers [36] and similar to another numerical FSI study assessing the SDH risk [14].

More than a hundred nonlinear spring connectors were placed in the subarachnoid space between the brain and skull. The maximum stress values exerted on all the connectors during the drainage phase are summarized in Fig. 3. It can be observed that some of the springs experienced deflection, i.e. compression, and some experienced elongation, i.e. extension. However, all spring connectors placed along the sagittal sinus known as the bridging veins (green) experienced elongation. As mentioned above, the highest values are found in anterior parietal region. Large amount of spring connectors experienced none or very little deflection/elongation.

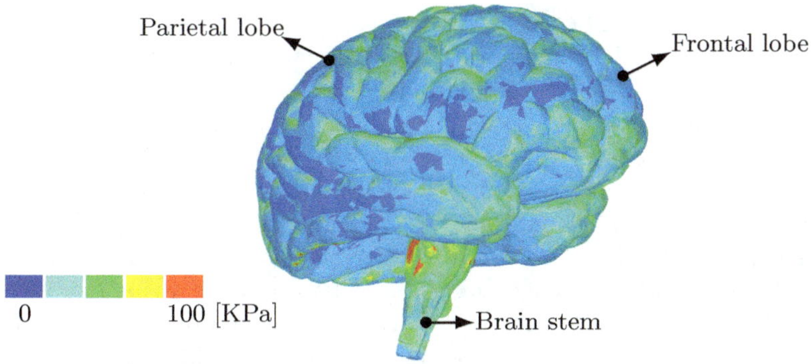

Fig. 4. Effective stress contour plot at the end of the NPH phase. Large value regions can be seen on brainstem, anterior frontal and anterior parietal lobes.

The effective stress values extracted at the end of the NPH phase are shown in Fig. 4. As explained in Subsect. 2.4, the NPH phase is when additional fluid particles are introduced into the model, using impermeable piston and pipe, to increase the intracranial pressure by simulating overproduction of CSF.

4 Conclusion

In engineering mostly, laboratory experiments are increasingly replaced by high fidelity and computationally expensive simulations. In order to decrease the run-time required to calculate the simulated results, the users tend to simplify the geometries used. Simplifying the geometries also decreases the time necessary to process the medical images in order to create geometric 3D models. Thus, understandably, it is very tempting to use simpler geometries instead of computationally expensive fully comprehensive patient-specific models. However, once simplified it can no longer be considered truly patient-specific model as it does not represent the patient anymore. The brain geometry is more complex than most human organs. Moreover, brain is submerged in CSF. Thus, any numerical analysis involving brain, especially when CSF is involved, should be an FSI analysis.

The study presented in this paper is a single-case study. However, the brain geometry is similar to all humans. The purpose of this study was to better understand the mechanism in which a person is under higher risk of developing SDH associated with VP shunt placement and subsequent CSF drainage. It has been demonstrated that this approach has the potential to assess the patient's risk if their subject-specific geometry is used. Other drawbacks of this model include some of the missing brain features, e.g. meninges. However, even with some neglected features this is the most comprehensive model that is used in FSI analysis currently found in the literature. Moreover, it can be argued that these missing features would have negligible effect on the results. The omission of cerebral vasculature is considered to be a major limitation of the model in this study and therefore it is currently being integrated. The cerebral vasculature alters brain stiffness. The network of arteries and veins creates a spring-like suspension system that restricts the brain motion.

Other variable necessary to consider is the speed by which the CSF is drained. If patient-specific geometry is used, it would be possible to find the best use of the adjustable valves in order to minimize the SDH risk due to the CSF drainage. However, it takes many steps and heavy calculations to make this approach truly patient-specific. Often, some of the steps, e.g. creating detailed 3D geometry, are tedious and time consuming. For this approach to be efficiently used in the clinical settings, a lot of research still has to be done to automate many of the steps performed. On the other hand, the calculations with these numerical methods are relatively fast and do not require expensive supercomputers. As all the solvers use a commodity GPU for parallel processing, a single GPU workstation is sufficient to complete all the calculations in reasonable runtime.

As the brain model used is highly comprehensive, i.e. it preserves the folds and depressions in the brain called gyri and sulci, respectively, it enables the identification of not only the brain region but the exact gyrus most affected under given conditions (Fig. 4). Thus, a model like this can be used to predict the symptoms a patient is to experience even before they are to undergo a certain brain trauma or intervention, such as VP shunt placement with subsequent CSF drainage.

Funding Information. This study has been partially supported by a faculty start-up fund provided by the New York Institute of Technology and by a donation from the New York Thoroughbred Horseman's Association.

Data Availability Statement. The datasets generated during and/or analysed during the current study are available from the corresponding author on reasonable request.

Ethics Approval. No formal approval was necessary for this study.

Conflict of Interest. All authors declare no conflict of interest. No benefits in any form have been or will be received from a commercial party related directly or indirectly to the subject of this manuscript.

References

1. Chrissicopoulos, C., Mourgela, S., Kirgiannis, K., Sakellaropoulos, A., Ampertos, N., Petritsis, K., Spanos, A.: What is the appropriate shunt system for normal-pressure hydrocephalus? Acta Neurochir. Suppl. **113**, 119–121 (2012). https://doi.org/10.1007/978-3-7091-0923-6_24
2. Hoya, K., Tanaka, Y., Uchida, T., Takano, I., Nagaishi, M., Kowata, K., Hyodo, A.: Treatment of traumatic acute subdural hematoma in adult hydrocephalus patients with cerebrospinal fluid shunt. Clin. Neurol. Neurosurg. **114**, 211–216 (2011)
3. Tangen, K.M., Hsu, C.Y., Zhu, D.C., Linninger, A.A.: CNS wide simulation of flow resistance and drug transport due to spinal microanatomy. J. Biomech. **48**(10), 2144–2154 (2015). https://doi.org/10.1016/j.jbiomech.2015.02.018

4. Rengachary, S.S., Ellenbogen, R.G.: Principles of Neurosurgery. Elsevier Mosby (2005)
5. Linninger, A.A., Tangen, K.M., Hsu, C.Y., Frim, D.: Cerebrospinal fluid mechanics and its coupling to cerebrovascular dynamics. Annu. Rev. Fluid Mech. **48**, 219–257 (2016). https://doi.org/10.1146/annurev-fluid-122414-034321
6. Ho, J., Kleiven, S.: Dynamic response of the brain with vasculature: a three-dimensional computational study. J. Biomech. **40**(13), 3006–3012 (2007). https://doi.org/10.1016/j.jbiomech.2007.02.011
7. Chen, Y., Ostoja-Starzewski, M.: MRI-based finite element modeling of head trauma: spherically focusing shear waves. Acta Mech. **213**(1–2), 155–167 (2010). https://doi.org/10.1007/s00707-009-0274-0
8. Watanabe, D., Yuge, K., Nishimoto, T., Murakami, S., Takao, H.: Impact injury analysis of the human head. AutoTechnology **7**(6), 34–37 (2007). https://doi.org/10.1007/BF03247021
9. Chafi, M.S., Dirisala, V., Karami, G., Ziejewski, M.: A finite element method parametric study of the dynamic response of the human brain with different cerebrospinal fluid constitutive properties. Proc. Inst. Mech. Eng. Part H: J. Eng. Med. **223**(8), 1003–1019 (2009). https://doi.org/10.1243/09544119JEIM631
10. Madhukar, A., Chen, Y., Ostoja-Starzewski, M.: Effect of cerebrospinal fluid modelling on spherically convergent shear waves during blunt head trauma. Int. J. Numer. Method Biomed. Eng. **33**(12), e2881 (2017). https://doi.org/10.1002/cnm.2881
11. Toma, M., Nguyen, P.D.H.: Fluid–structure interaction analysis of cerebrospinal fluid with a comprehensive head model subject to a rapid acceleration and deceleration. Brain Inj. **32**(12), 1576–1584 (2018). https://doi.org/10.1080/02699052.2018.1502470
12. Toma, M.: Predicting concussion symptoms using computer simulations. In: Advances in Intelligent Systems and Computing, vol. 880, pp. 557–569 (2018). https://doi.org/10.1007/978-3-030-02686-8_42
13. Toma, M., Nguyen, P.D.H.: Coup-contrecoup brain injury: fluid-structure interaction simulations. Int. J. Crashworthiness (2019). https://doi.org/10.1080/13588265.2018.1550910
14. Zhou, Z., Li, X., Kleiven, S.: Fluid–structure interaction simulation of the brain-skull interface for acute subdural haematoma prediction. Biomech. Model. Mechanobiol. **18**(1), 155–173 (2019). https://doi.org/10.1007/s10237-018-1074-z
15. Zhou, Z., Li, X., Kleiven, S.: Biomechanics of acute subdural hematoma in the elderly: a fluid–structure interaction study. J. Neurotrauma **36**(13), 2099–2108 (2019). https://doi.org/10.1089/neu.2018.6143
16. Vorwerk, J., Clerc, M., Burger, M., Wolters, C.H.: Comparison of boundary element and finite element approaches to the EEG forward problem. Biomed. Eng./Biomedizinische Technik. **57**(SI-1 Track-O), 795–798 (2012). https://doi.org/10.1515/bmt-2012-4152
17. Luo, Y., Li, Z., Chen, H.: Finite-element study of cerebrospinal fluid in mitigating closed head injuries. J. Eng. Med. **226**(7), 499–509 (2012). https://doi.org/10.1177/0954411912445729
18. Liang, Z., Luo, Y.: A QCT-based nonsegmentation finite element head model for studying traumatic brain injury. Appl. Bion. Biomech. **2015**, e837585 (2015). https://doi.org/10.1155/2015/837585

19. Bei, L., Shijie, R., Haiyan, L., Shihai, C., Lijuan, H.: The effects of different mesh density of the cerebrospinal fluid on the dynamic responses of a 6 years old child finite element head model. In: Eighth International Conference on Measuring Technology and Mechatronics Automation (ICMTMA), pp. 756–767 (2016). https://doi.org/10.1109/ICMTMA.2016.186
20. Gilchrist, M.D., O'Donoghue, D.: Simulation of the development of the frontal head impact injury. Comput. Mech. **26**(3), 229–235 (2000). https://doi.org/10.1007/s004660000179
21. Toma, M., Njilie, F.E.A., Ghajari, M., Galvanetto, U.: Assessing motorcycle crash-related head injuries using finite element simulations. Int. J. Simul. Model. **9**(3), 143–151 (2010). https://doi.org/10.2507/ijsimm09(3)3.164
22. Durrwachter, J.: Hemodynamics of the left ventricle: validation of a smoothed–particle hydrodynamics fluid-structure interaction model. M.S. thesis, Georgia Institute of Technology (2016)
23. Caballero, A., Mao, W., Liang, L., Oshinski, J., Primianok, C., McKay, R., Kodali, S., Sun, W.: Modeling left ventricular blood flow using smoothed particle hydrodynamics. Cardiovasc. Eng. Technol. **8**(4), 465–479 (2017)
24. Toma, M., Bloodworth IV, C.H., Einstein, D.R., Pierce, E.L., Cochran, R.P., Yoganathan, A.P., Kunzelman, K.S.: High resolution subject-specific mitral valve imaging and modeling: experimental & computational methods. Biomech. Model. Mechanobiol. **15**(6), 1619–1630 (2016). https://doi.org/10.1007/s10237-016-0786-1
25. Mao, W., Li, K., Sun, W.: Fluid-structure interaction study of transcatheter aortic valve dynamics using smoothed particle hydrodynamics. Cardiovasc. Eng. Technol. **7**(4), 374–388 (2016)
26. Toma, M., Einstein, D.R., Bloodworth IV, C.H., Cochran, R.P., Yoganathan, A.P., Kunzelman, K.S.: Fluid–structure interaction and structural analyses using a comprehensive mitral valve model with 3D chordal structure. Int. J. Numer. Methods Biomed. Eng. **33**(4) (2017). https://doi.org/10.1002/cnm.2815
27. Toma, M., Bloodworth IV, C.H., Pierce, E.L., Einstein, D.R., Cochran, R.P., Yoganathan, A.P., Kunzelman, K.S.: Fluid-structure interaction analysis of ruptured mitral chordae tendineae. Ann. Biomed. Eng. **45**(3), 619–631 (2017). https://doi.org/10.1007/s10439-016-1727-y
28. Toma, M., Jensen, M.O., Einstein, D.R., Yoganathan, A.P., Cochran, R.P., Kunzelman, K.S.: Fluid-structure interaction analysis of papillary muscle forces using a comprehensive mitral valve model with 3D chordal structure. Ann. Biomed. Eng. **44**(4), 942–953 (2016). https://doi.org/10.1007/s10439-015-1385-5
29. Singh-Gryzbon, S., Sadri, V., Toma, M., Pierce, E.L., Wei, Z.A., Yoganathan, A.P.: Development of a computational method for simulating tricuspid valve dynamics. Ann. Biomed. Eng. **47**(6), 1422–1434 (2019). https://doi.org/10.1007/s10439-019-02243-y
30. Toma, M., Nguyen, P.D.H.: Fluid-structure interaction analysis of cerebral spinal fluid with a comprehensive head model subject to a car crash-related whiplash. In: 5th International Conference on Computational and Mathematical Biomedical Engineering - CMBE2017 (2017)
31. Toma, M.: The emerging use of SPH in biomedical applications. Significances Bioeng. Biosci. **1**(1), SBB.000502 (2017). https://doi.org/10.31031/SBB.2017.01.000502
32. Toma, M., Oshima, M., Takagi, S.: Decomposition and parallelization of strongly coupled fluid-structure interaction linear subsystems based on the Q1/P0 discretization. J. Comput. Struct. **173**, 84–94 (2016). https://doi.org/10.1016/j.compstruc.2016.06.001

33. Miller, J.D., Nader, R.: Acute subdural hematoma from bridging vein rupture: a potential mechanism for growth. J. Neurosurg. **120**(6), 1259–1502 (2014). https://doi.org/10.3171/2013.10.JNS13272
34. Depreitere, B., Lierde, C.V., Sloten, J.V., Audekercke, R.V., Perre, G.V.D., Plets, C., Goffin, J.: Mechanics of acute subdural hematomas resulting from bridging vein rupture. J. Neurosurg. **104**(6), 950–956 (2006). https://doi.org/10.3171/jns.2006.104.6.950
35. Cui, Z.Y., Famaey, N., Depreitere, B., Ivens, J., Kleiven, S., Sloten, V.J.: On the assessment of bridging vein rupture associated acute subdural hematoma through finite element analysis. Comput. Methods Biomech. Biomed. Eng. **20**(5), 530–539 (2017). https://doi.org/10.1080/10255842.2016.1255942
36. Lee, M.C., Haut, R.C.: Insensitivity of tensile failure properties of human bridging veins to strain rate: implications in biomechanics of subdural hematoma. J. Biomechan. **22**(6–7), 537–542 (1989). https://doi.org/10.1016/0021-9290(89)90005-5

Ligament Shear Wave Speeds Are Sensitive to Tensiometer-Tissue Interactions: A Parametric Modeling Study

Jonathon L. Blank[1], Darryl G. Thelen[1,2], and Joshua D. Roth[1,2(✉)]

[1] Department of Mechanical Engineering, University of Wisconsin-Madison, Madison, WI 53706, USA
roth@ortho.wisc.edu
[2] Department of Orthopedics and Rehabilitation, University of Wisconsin-Madison, Madison, WI 53705, USA

Abstract. Shear wave tensiometry is based on the observation that wave propagation speed varies directly with axial stress in tendons and ligaments. Hence, empirical measures of shear wave speed could be used to estimate axial tissue stress. However, this requires that the sensor measure shear wave speed without altering wave propagation within the tissue. We currently monitor shear wave speed using a piezoelectric tapping device and two accelerometers secured over the tissue. Three important factors in the design and use of shear wave tensiometers (i.e., mass, stiffness, and application force of tensiometer components) might alter the shear wave propagation and thus the measured shear wave speeds. Accordingly, the primary objective of this study was to determine the sensitivity of ligament shear wave speed to these three factors. To do this, we developed a parametric finite element model of a tensiometer and ligament. Our model generally agreed with prior analytical estimates of shear wave speed. The measured shear wave speeds were most sensitive to the application force of the tensiometer onto the ligament surface, which likely alters the local stress environment. Thus, it is important to remember that user factors, in addition to design factors, might alter the stress predicted from measured shear wave speeds. Moving forward, our parametric finite element model provides a platform for investigating robust tensiometer designs for non-uniformly loaded tissues, which extends beyond the current analytical model. This will enable the use of shear wave tensiometry to predict non-uniform tissue stresses that are common in orthopedic and rehabilitation applications.

Keywords: Total knee arthroplasty TKA · Soft tissue balance · FEBio finite element model · Shear wave tensiometry · Parametric sensor design

1 Introduction

Shear wave tensiometers are novel sensors that enable the direct non-invasive measurement of soft tissue loading by measuring shear wave propagation speed [1]. Typically, a shear wave tensiometer consists of one actuator that induces a shear wave in the

G. A. Ateshian et al. (Eds.): CMBBE 2019, LNCVB 36, pp. 48–59, 2020.
https://doi.org/10.1007/978-3-030-43195-2_5

tissue of interest and two sensors that track shear wave propagation. Because the spacing between the two sensors is known, the shear wave propagation speed is computed by dividing the sensor spacing by the time it takes the shear wave to travel from one sensor to the other [1]. Previous studies have shown that the relationship between squared shear wave speed and stress in axially loaded soft tissues is linear [1–3]:

$$c^2 = \frac{\sigma + k'\mu}{\rho} \quad (1)$$

Here, c is the shear wave speed, σ is the axial stress, μ is the shear modulus of the tissue, k' is a shear correction factor, and ρ is the effective density that includes both the density of the tissue and added mass effects due to surrounding tissue, fluid medium, and the tensiometer components themselves.

Current shear wave tensiometers consist of a piezoelectric actuator (i.e., tapper) and two miniature accelerometers. Both the tapper and accelerometers require direct contact with the tendon or ligament of interest. Hence, the interaction between the sensor and the tissue of interest can vary based on the components used and how they are secured over the tissue of interest by an end-user. Conceivably, a tensiometer might change the relationship between the shear wave speed and stress by altering the local tissue stress, tissue stiffness, and/or the effective density. These changes might in turn introduce errors into the computed axial stress based on the measured shear wave speed.

There are three important factors in the design and use of a shear wave tensiometer that might affect the relationship between measured shear wave speed and axial stress. These are the (1) mass and (2) stiffness of the tensiometer components and the (3) application force with which the tensiometer components are pressed against the surface of the tissue. The mass of the tensiometer components might add effective mass to the soft tissue and thus increase the tissue's effective density (ρ). The stiffness of the tensiometer components might change the dynamics of the tensiometer-tissue interaction by altering the application force as the shear wave passes each sensor. The overall application force might give rise to elevated local stresses (σ) in the tissue through small deformations at the contact points and large deformations transverse to the axis of loading. It is likely that all three of these factors vary between tensiometer designs, and the application force likely varies between both tensiometer trials and end-users.

Accordingly, the objectives of this study were to (1) develop a parametric finite element model to investigate tensiometer-tissue interactions, and (2) determine the sensitivity of measured shear wave speeds to tensiometer component mass, stiffness, and application force.

2 Materials and Methods

2.1 Model Geometry

We created a parametric finite element model of the interaction between a tensiometer and a representative ligament (FEBio version 2.8.5.0 [4]). We modeled the ligament as a 60-mm long cylinder with a radius of 3 mm. The actuator (i.e., tapper) and accelerometers (i.e., sensors) of the tensiometer were modeled as rigid cylinders (6 mm long with a radius of 1.5 mm). We modeled the stiffness and damping of the tensiometer components with a linear spring and damper in parallel behind the tapper and each sensor. We modeled the interaction between tensiometer components and the ligament surface with a sliding-elastic contact formulation. The ligament mesh was composed of 19,200 8-node hexahedral elements arranged in a butterfly pattern with fibers oriented along the long axis of the cylinder. This mesh density was determined based on the results of an a priori mesh convergence study (see Appendix A).

2.2 Constitutive Model

We modeled the ligament as an incompressible, transversely isotropic material using a Mooney-Rivlin hyperelastic model. The matrix, collagen fibers, and tissue's volumetric response to loading were characterized using the following uncoupled strain energy density function, W:

$$W = F_1\left(\tilde{I}_1, \tilde{I}_2\right) + F_2\left(\tilde{\lambda}\right) + \frac{K}{2}(\ln J)^2 \tag{2}$$

$\tilde{I}_1$ and $\tilde{I}_2$ represent the first and second invariants of the right Cauchy-Green deformation tensor, $\tilde{\lambda}$ indicates the deviatoric part of the stretch ratio along the fiber direction, K represents the bulk modulus, and finally J represents the Jacobian of the deformation. $F_1\left(\tilde{I}_1, \tilde{I}_2\right)$ represents the matrix contribution of the isotropic material model to the strain energy density function:

$$F_1\left(\tilde{I}_1, \tilde{I}_2\right) = C_1\left(\tilde{I}_1 - 3\right) + C_2\left(\tilde{I}_2 - 3\right) \tag{3}$$

The hyperelastic model was reduced to an uncoupled Neo-Hookean formulation by setting the Mooney-Rivlin coefficient C_2 to zero. $F_2\left(\tilde{\lambda}\right)$ represents the contribution of the collagen fibers to the strain energy density function:

$$\begin{aligned} \tilde{\lambda}\frac{\partial F_2}{\partial \tilde{\lambda}} &= 0 & \tilde{\lambda} &\leq 1 \\ \tilde{\lambda}\frac{\partial F_2}{\partial \tilde{\lambda}} &= C_3 e^{C_4\left(\tilde{\lambda}-1\right)} - 1 & 1 < \tilde{\lambda} &\leq \lambda^* \\ \tilde{\lambda}\frac{\partial F_2}{\partial \tilde{\lambda}} &= C_5\tilde{\lambda} + C_6 & \tilde{\lambda} &> \lambda^* \end{aligned} \tag{4}$$

λ^* is the stretch at which the collagen fibers fully engage and axial stress increases linearly with stretch. The hyperelastic model coefficients, C, were chosen from previous literature [5–7] (Table 1).

Table 1. Hyperelastic (uncoupled Neo-Hookean) model parameters.

Parameter	Value
Density, ρ [kg/m^3]	1600
C_1 [MPa]	0.2
C_3 [MPa]	1
C_4	50
C_5 [MPa]	600
λ^*	1.05
K [MPa]	200

2.3 Code Verification

The shear wave speeds computed using our model were compared against those predicted by an analytical shear wave speed-stress relationship for tensioned beams with similar geometry [1, 2]. The effective density and shear modulus used in the hyperelastic model (Table 1) were used in the analytical shear wave speed equation (Eq. 1).

2.4 Model Boundary Conditions

The simulation was split into two separate analyses: a static loading analysis and a dynamic tapping analysis. In the static loading analysis, the ligament was strained to either 3% or 6% strain by prescribing the axial displacement of a rigid grip located on

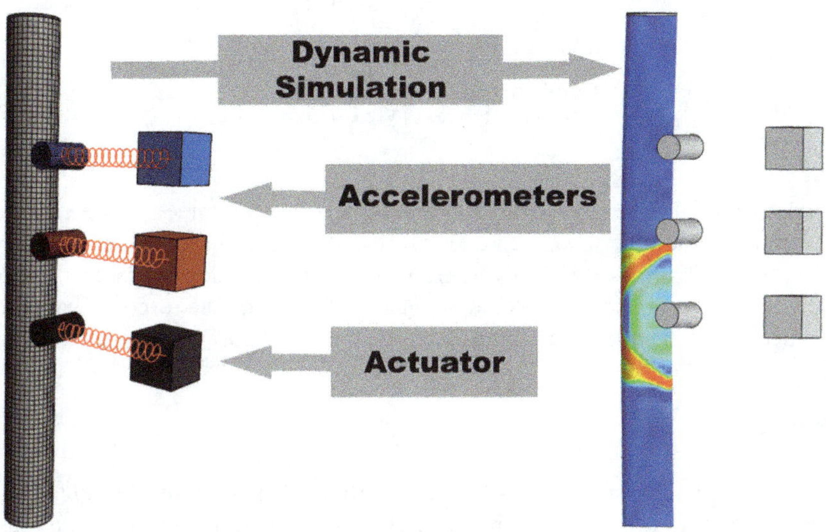

Fig. 1. In our parametric finite element model, two accelerometers and an actuator (i.e., tapper) were pressed against the ligament surface at a fixed initial displacement prescribed to apply a specific application force. Then, a dynamic analysis was performed to determine the propagation speed of a shear wave induced by the actuator. Shear wave propagation was tracked using the transverse accelerations of the two rigid accelerometer bodies.

the top of the cylinder with the bottom of the cylinder held fixed. Fixed transverse displacements were prescribed to the actuator and accelerometers to achieve a specified application force based on a calibrated second-order force-displacement relationship. To begin the dynamic tapping analysis, a 20-micron tap was applied via a ramp input over a duration of 0.62 ms transversely into the ligament to replicate the experimental response time of a typical piezoelectric actuator (Fig. 1). Shear wave speeds were calculated by cross-correlating the rigid body accelerations of the two accelerometer bodies [1].

2.5 Parametric Modeling and Analysis

To demonstrate the utility of our parametric finite element model, we used a three-factor Box-Behnken design to generate a total of 13 models to investigate how shear wave speed was altered by changes in (1) the force with which the actuator and sensors were pressed into the ligament, (2) the stiffness of the springs behind each tensiometer component, and (3) the mass of the tensiometer components (Table 2). All 13 models were run through the simulation workflow described in Sect. 2.4 (Fig. 1) at each of the two strain levels. Simulations were computed sequentially using single compute nodes and were on average completed in under twenty minutes using a standard laptop computer.

Table 2. Parametric model parameters. Ranges of values were selected to span the range of tensiometer components used in our previous studies.

Parameter	Value(s)
Stiffness, $k\left[\frac{N}{mm}\right]$	1.0, 5,5, 10.0
Damping, $\mu\left[\frac{N\cdot s}{m}\right]$	1
Mass, m[g]	0.2, 10.1, 20.0
Application Force, F[N]	1.0, 5,5, 10.0

To determine the sensitivity of shear wave speeds to the three tensiometer factors, we performed a multiple linear regression with interaction on the results from the 13 models at each strain level. In each regression, the independent variables were component stiffness, component mass, and application force; and the response variable was shear wave speed. The level of significance was set to 0.05.

3 Results

The parametric model successfully passed two model quality checks. Element-level volume changes in the ligament were limited to under 4% for the most extreme cases of deformation. The shear wave speeds predicted by the analytical model (46.8 m/s at 3% strain (3.1 MPa) and 89.4 m/s at 6% strain (12.4 MPa)) were within one standard deviation of the mean shear wave speed computed using the parametric models at each strain level (Fig. 2).

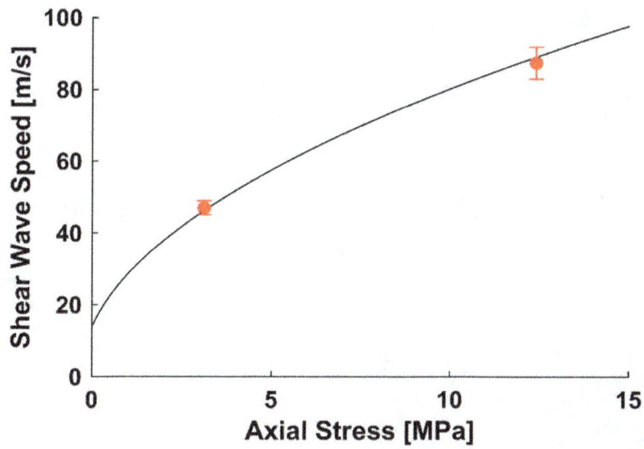

Fig. 2. Plot shows the comparison between shear wave speeds predicted by a previously validated analytical model [1] (solid line) and the mean (red dots) and standard deviation (red error bars) of those computed using our parametric finite element model.

The measured shear wave speeds were most sensitive to tensiometer application force, with increases in application force leading to increases in shear wave speed. However, this effect was only significant at 3% strain (Figs. 3 and 5, Table 3).

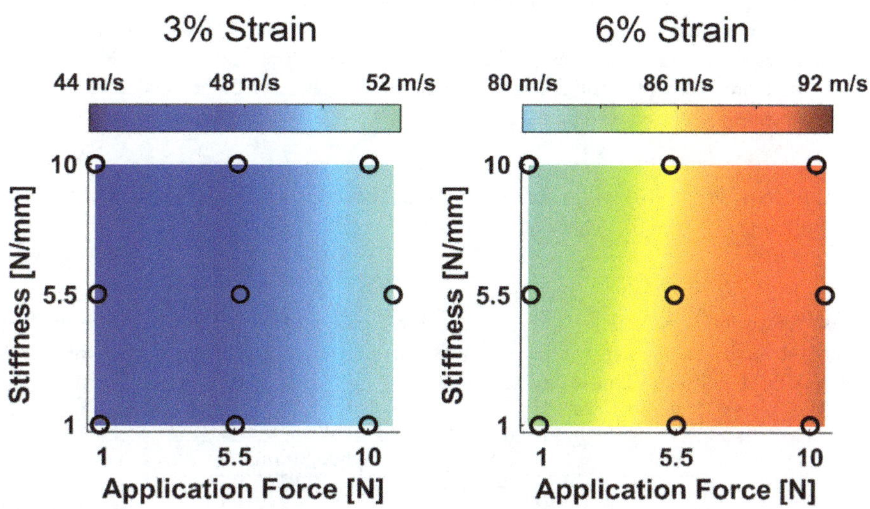

Fig. 3. Contour plots show the response surface fit of the shear wave speeds computed at 3% (left) and 6% (right) strain. Open black circles indicate the specific component stiffness and application force values of the 13 models. Shear wave speed measurements were sensitive to the application force of the tensiometer against the surface of the ligament. The shear wave propagation speed increased with application force.

In addition to application force sensitivity at 3% strain, there was also a significant interaction between application force and component mass at the 3% strain level (Figs. 4 and 5, Table 3). At this strain level, the highest shear wave speeds occurred with the highest application force and the lowest component mass.

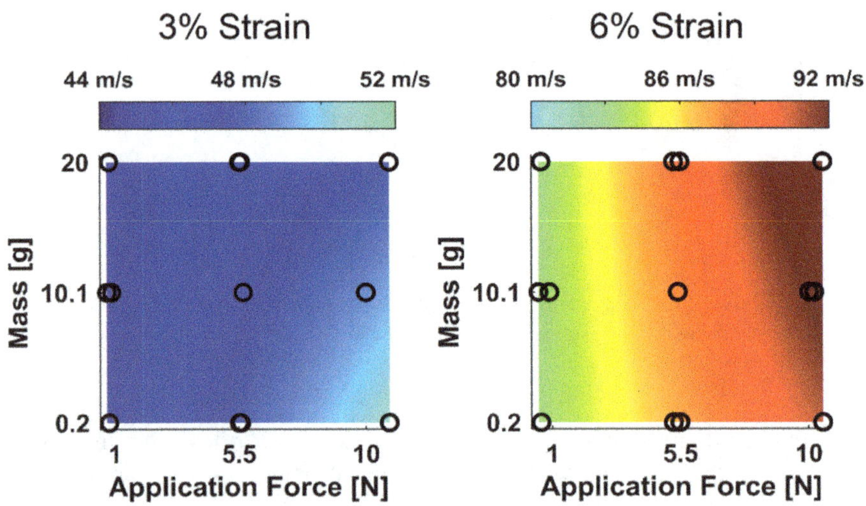

Fig. 4. Contour plots show the response surface fit of the shear wave speeds computed at 3% (left) and 6% (right) strain. Open black circles indicate the specific component mass and application force values of the 13 models. Shear wave speed measurements were sensitive to interactions between component mass and application force at low strain levels. Generally, shear wave speeds increased with application force and were higher for higher masses at 3% strain, with the opposite effect demonstrated at 6% strain.

We did not detect any significant effects of the component stiffness or component mass at either strain level (Table 3).

Table 3. Sensitivity coefficients and p-values for terms in the multiple regression analysis including the main and interaction effects of three tensiometer factors. Asterisks indicate p-values less than 0.05, which were considered significant.

Effect	Regression Coefficient (p-value)	
	3% Strain	6% Strain
Force	0.579 (0.002) *	0.750 (0.121)
Stiffness	−3.9e09 (0.735)	−2.0e−04 (0.662)
Mass	0.028 (0.616)	−0.020 (0.926)
Force: Stiffness	5.0e−06 (0.729)	1.4e−05 (0.808)
Force: Mass	−0.017 (0.024) *	0.019 (0.445)
Stiffness: Mass	1.8e−06 (0.791)	7.6e−06 (0.776)

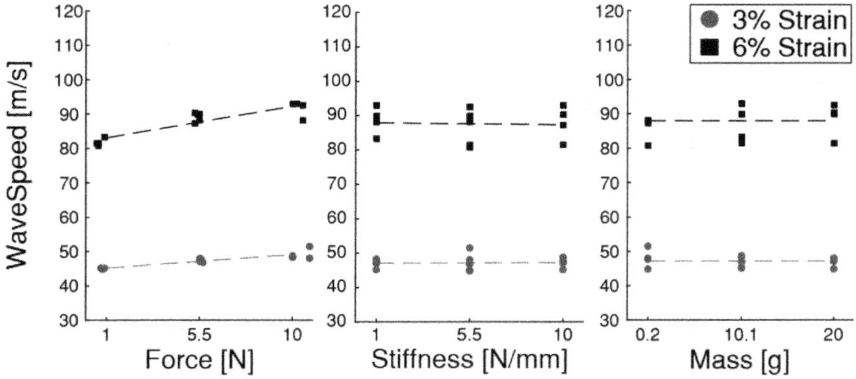

Fig. 5. Scatter plots show the sensitivity of shear wave speed to differences in application force (left), stiffness (middle), and mass (right) at two strain levels. Shear wave speeds were most sensitive to application force.

4 Discussion

4.1 Key Findings

There were two objectives of the present study. The first was to develop a novel parametric finite element model capable of manipulating key factors in the design and use of shear wave tensiometers. The second was to use this model to evaluate the effect of application force, component stiffness, and component mass on shear wave speed measures. The key findings were (1) the parametric finite element model predictions were generally consistent with a validated analytical model of shear wave propagation speed in uniformly loaded tensioned beams, and (2) the shear wave speeds were most sensitive to application force of the tensiometer components against the tissue surface.

The close agreement between the shear wave speeds computed using our parametric finite element model and a previously validated analytical model has important implications for the future applications of our model (Fig. 2). This close agreement justifies the future use of this model (1) as a design tool for evaluating new tensiometer designs and (2) to complement experimental tensiometry studies (e.g., to investigate the local shear wave propagation, which is difficult to characterize experimentally). Further, our parametric finite element model enables the investigation of non-uniform loading cases that are common in the applications of interest (e.g., non-uniform tension in the superficial medial collateral ligament [8]) but are beyond the scope of the current analytical model.

The high sensitivity of measured shear wave speeds to application force also has important implications for the use of shear wave tensiometers. The increase in measured shear wave speeds due to increased application force is likely caused by elevated local stresses due to both tensiometer-ligament contact and transverse displacement of the ligament (Figs. 3, 4, 5 and 6). Based on the analytical model predictions (Eq. 1), the range of shear wave speeds measured at 3% and 6% strain would lead to predicted stresses ranging from 2.8 to 3.8 MPa at 3% strain and 10.0 to 13.4 MPa at 6% strain.

Relative to the expected average tissue stress, the stress predicted using the measured shear wave speeds exhibited errors (bias ± precision [9, 10]) of 0.1 ± 0.3 MPa at 3% strain and −0.5 ± 1.2 MPa at 6% strain. Thus, when using a shear wave tensiometer, the minimum application force necessary to maintain coupled transverse motion of the sensors and the soft tissue of interest should be applied. Excessive application forces, especially in low-load conditions common during measurements of ligament loading, might elevate local stresses leading to errors in the prediction of the tissue stress without the tensiometer in place.

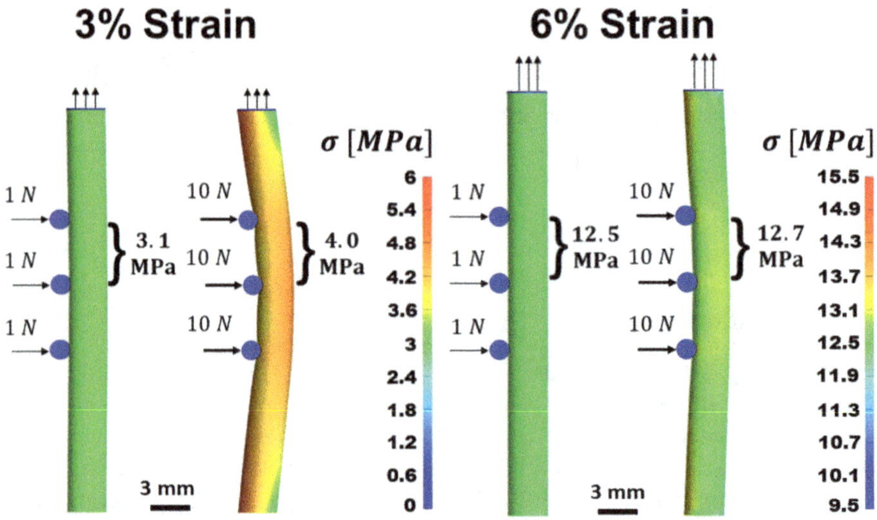

Fig. 6. Representative comparison between models with a 1-N and a 10-N application force under 3% and 6% strain. Before the tensiometer was applied to the model, the average stresses at 3% and 6% strain were 3.1 MPa and 12.4 MPa, respectively. At 3% strain, the 1-N model had an average local stress between the accelerometers of 3.1 MPa and displaced transversely 0.3 mm between the accelerometers, on average; in comparison, the 10-N model had an average local stress between the accelerometers of 4.0 MPa and displaced transversely 2.8 mm between the accelerometers, on average. At 6% strain, the difference between the local stresses and the average stress as well as the average transverse displacement in the 1-N and 10-N models were much less than those at 3% strain.

There are three limitations in this study that should be considered when interpreting the results. First, the geometry of the ligament used in the parametric model was simplified to a cylinder. The sensitivities of our parameters will likely change for an anatomic ligament geometry, which would more accurately depict the propagation of shear waves along its length. Second, the constitutive model used did not include transverse strain stiffening. Thus, the effect of application force might be conservative in our model because stiffer tissue beneath the accelerometers might further increase the shear wave speed. Third, the current study utilized a Box-Behnken experimental design to limit the number of simulations performed on a single local computer.

However, a greater number of simulations, which are possible using a Monte Carlo design on a high-throughput computing grid, will enable a more complete investigation of the interaction between factors in the design and use of shear wave tensiometers.

4.2 Applications and Future Directions

Shear wave tensiometry is a new measurement technique developed at the University of Wisconsin-Madison. Traditional shear wave imaging techniques, such as shear wave elastography, can utilize ultrasonic imaging to track shear waves and estimate tendon elasticity [11–15]. Similarly, shear wave tensiometry uses shear wave propagation speeds to estimate the axial stress in tendons and ligaments. Using a pair of miniature accelerometers and a piezoelectric tapping device embedded in a handheld or mountable tensiometer, shear wave propagation speeds are determined using the delay in shear wave arrival at each accelerometer during clinically relevant motions. These measures enable the objective quantification of in situ stresses, which has implications to improve early diagnosis and track rehabilitation. Additionally, in ligaments, this technology can be used during orthopedic procedures, for example, by enabling surgeons to objectively and noninvasively balance collateral ligaments during total knee arthroplasty.

The utility of such a measurement is prominent in biomechanics; however, the sensor's design is a category of engineering research alone. The parametric model used in this study introduces a broad area of design and optimization research aimed to produce reliable tensiometers. Here, we varied application force, component stiffness, and component mass. Other important design parameters include the sensor's dimensions, component spacing, component damping, and actuator input rate and magnitude, among many others. Moving forward, we plan to utilize high-throughput parallel computing to explore the full design space of this technology using our parametric model. Additionally, this model enables investigation of shear wave propagation patterns arising from non-uniform loading often present in ligaments and tendons, which is important to characterize local tissue stress.

4.3 Conclusion

Because the shear wave speeds computed using our parametric finite element model were generally consistent with a validated analytical model, our parametric finite element model is a useful tool for evaluating novel tensiometer designs, performing complementary analyses to future experiments, and investigating shear wave propagation in non-uniformly loaded ligaments and tendons. Of the three factors studied, shear wave speeds were most sensitive to the application force; hence, it is important to remember that user factors, in addition to design factors, might alter the stress predicted from measured shear wave speeds.

Acknowledgements. We would like to thank Dr. Matthew Allen for his technical insights on the interpretation of the shear wave data. We would also like to thank Dr. Steve Maas for his advice during the early development of this model.

This study was supported by the National Institute of Biomedical Imaging and Bioengineering and the National Institute of Aging of the National Institutes of Health under award numbers [NIBIB: R21EB024957; NIA: T32AG000213].

A Appendix

A.1 Mesh Convergence

A mesh convergence analysis was performed for a low strain, low application force model. Six different mesh densities ranging from 2,400 to 64,800 elements were used in the ligament geometry (Fig. 7). Shear wave speed was measured using the same technique outlined in the methods of this study.

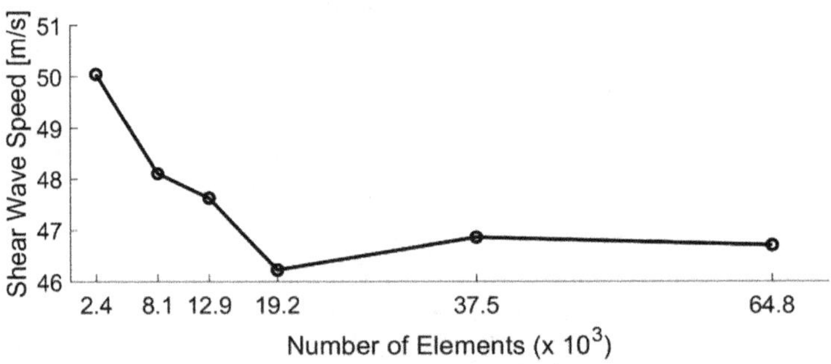

Fig. 7. Scatter plot shows that shear wave speed did not vary much with a change in mesh density (standard deviation = 1.38 m/s). A mesh density of 19,200 elements was chosen because it had a consistent shear wave speed with other generated meshes and met our preferences for simulation computation time.

References

1. Martin, J.A., Brandon, S.C.E., Keuler, E.M., et al.: Gauging force by tapping tendons. Nat. Commun. **9**, 2–10 (2018). https://doi.org/10.1038/s41467-018-03797-6
2. Martin, J.A., Schmitz, D.G., Ehlers, A.C., et al.: Calibration of the shear wave speed-stress relationship in ex vivo tendons. J. Biomech. **90**, 9–15 (2019). https://doi.org/10.1016/j.jbiomech.2019.04.015
3. Blank, J.L., Roth, J.D., Thelen, D.G.: Shear wave speeds track axial stress in porcine collateral ligaments. J. Mech. Behav. Biomed. Mater. 103704 (2020). https://doi.org/10.1016/j.jmbbm.2020.103704
4. Maas, S.A., Ellis, B.J., Ateshian, G.A., Weiss, J.A.: FEBio: finite element for biomechanics. J. Biomech. Eng. **134**(011005), 1–10 (2012)
5. Weiss, J.A., Gardiner, J.C., Bonifasi-Lista, C.: Ligament material behavior is nonlinear, viscoelastic and rate-independent under shear loading. J. Biomech. **35**, 943–950 (2002). https://doi.org/10.1016/S0021-9290(02)00041-6

6. Gardiner, J.C., Weiss, J.A.: Subject-specific finite element models can predict strain in the human medial collateral ligament. J. Orthop. Res. **21**, 1098–1106 (2003). https://doi.org/10.1016/S0736-0266(03)00113-X
7. Cui, L., Maas, H., Perreault, E.J., Sandercock, T.G.: In situ estimation of tendon material properties: differences between muscles of the feline hindlimb. J. Biomech. **42**, 679–685 (2009). https://doi.org/10.1016/j.jbiomech.2009.01.022
8. Robinson, J.R., Sanchez-Ballester, J., Bull, A.M.J., et al.: The posteromedial corner revisited. J. Bone Joint Surg. Br. **86**, 674–681 (2004). https://doi.org/10.1302/0301-620x.86b5.14853
9. Standard Practice for Use of the Terms Precision and Bias in ASTM Test Methods 1 (2010)
10. ISO 5725-1 International Standard
11. Brum, J., Bernal, M., Gennisson, J., Tanter, M.: In vivo evaluation of the elastic anisotropy of the human Achilles tendon using shear wave dispersion analysis. Phys. Med. Biol. **59**, 505–523 (2014). https://doi.org/10.1088/0031-9155/59/3/505
12. Cortes, D.H., Suydam, S.M., Silbernagel, K.G., et al.: Continuous shear wave elastography: a new method to measure viscoelastic properties of tendons in vivo. Ultrasound Med. Biol. **41**, 1518–1529 (2015). https://doi.org/10.1016/j.ultrasmedbio.2015.02.001
13. Bercoff, J., Tanter, M., Fink, M.: Supersonic shear imaging: a new technique for soft tissue elasticity mapping. IEEE Trans. Ultrason. Ferroelectr. Freq. Control **51**, 396–409 (2004). https://doi.org/10.1109/TUFFC.2004.1295425
14. Helfenstein-Didier, C., Andrade, R., Brum, J., et al.: In vivo quantification of the shear modulus of the human Achilles tendon during passive loading using shear wave dispersion analysis. Phys. Med. Biol. **61**, 2485–2496 (2016)
15. Aubry, S., Risson, J.R., Kastler, A., et al.: Biomechanical properties of the calcaneal tendon in vivo assessed by transient shear wave elastography. Skeletal Radiol. **42**, 1143–1150 (2013). https://doi.org/10.1007/s00256-013-1649-9

Computational Parametric Studies for Preclinical Evaluation of Total Knee Replacements

Steven P. Mell(✉), Spencer Fullam, Markus A. Wimmer, and Hannah J. Lundberg

Rush University Medical Center, Chicago, IL, USA
steven_p_mell@rush.edu

Abstract. Aseptic loosening remains a leading cause of long-term failure of total knee replacement (TKR) that can limit the lifetime of the implant past the second decade. To help accelerate and facilitate advances in TKR design, our goal was to develop a computational framework based on previous work, that could accurately and efficiently predict the effect of design, surgical, and patient variability on TKR wear. The framework can accommodate patient-specific, population-specific, or standardized motions and forces as inputs. The wear model that is fully integrated into the finite element model the framework is built around was calibrated from materials testing (wheel-on-flat) experiments. Validation of the model is carried out by comparison to pin-on-disk and mechanical knee simulator studies. We present two applications to test the effectiveness of the framework for performing parametric studies: one to test the sensitivity of TKR wear to the transverse-plane rotational alignment of the tibial component, and the second to test the sensitivity of TKR wear to femoral center of rotation. We demonstrate that wear is highly sensitive to femoral component alignment, consistent with previous studies, and that choice of femoral center of rotation is important during simulator testing of TKR components. The two reported applications represent initial attempts to study variability in component alignment- future work will include studying variability in kinematics and kinetics. In a design setting, the finished framework will allow new innovations to be tested as soon as they are developed, supplementing mechanical testing, and reducing the amount required.

Keywords: Total knee replacements · Polyethylene wear · Finite element analysis · Computational modeling

1 Introduction

A leading cause of disability globally, osteoarthritis of the knee [1], leads to severe pain and/or loss of knee function. Total knee replacements (TKRs) are the gold standard treatment for end stage osteoarthritis [2], and involve replacing the articulating surfaces of the joint with artificial materials. The number of TKR surgeries performed is increasing rapidly [3–5], growing by 134% in the last decade [6, 7]. In 2007, Kurtz et al. predicted that TKR procedures would increase by 601%, from 402,100 to 3.48

G. A. Ateshian et al. (Eds.): CMBBE 2019, LNCVB 36, pp. 60–85, 2020.
https://doi.org/10.1007/978-3-030-43195-2_6

million by 2030 [4]. This is supported by the current data [5, 7, 8] with updated projections placing the number of primary TKR surgeries performed per year at 1.38 million by 2020 [4]. With the increase in operations, and an increasing number of younger patients [3, 7, 9, 10], further study of the performance of TKRs becomes crucial to ensure implants last throughout a patient's lifetime. It is therefore imperative to understand the underlying causes of failure in order to better design implants in the future. With a 15-year survival rate of up to 92.2% [11, 12], analysis of current implant designs suffers a significant time delay from any new design going to market until its long-term clinical outcome can be determined.

One of the long-term causes of TKR failure is aseptic loosening attributable to osteolysis from the biological responses to ultra-high molecular weight polyethylene (UHMWPE) wear debris [13, 14]. Despite reduced wear rates offered by technological advances such as the introduction of highly crosslinked polyethylene, the consequences of polyethylene wear remain a problem limiting the lifetime of TKRs past the second decade. TKR designs constantly change in response to new innovations and technology, and preclinical wear testing using mechanical simulators is the gold standard used to ensure that new designs meet or exceed the standards set by those currently in use. However, due to the time and cost of testing with mechanical simulators, it is only performed late in the design process.

Due to the time and expense of using mechanical simulators, a technological advance that could move the field forward is increased usage of computational modeling of the wear process for TKR. One of the most valuable aspects of computational models is the ability to perform large parametric studies to complement experiments and help guide the design process before experimental testing is performed. Recently, the FDA published guidelines for submitting computational modeling results for preclinical data during medical device approval [15]. Increased use of computational modelling could help broaden understanding of how patient, surgical, and design factors all contribute to TKR wear long term.

In this body of work, we present the design of a computational framework that can efficiently predict the effect of design, surgical, and patient variability on TKR wear. Requirements for the framework included the ability to accept large numbers of input variables, and to be scalable to patient-specific, population-specific, or generic inputs and outputs. Based on previously published work [16–18], the framework combines finite element analysis (FEA), an execution engine, a frictional energy-based wear model, and subject specific inputs if desired. In this work we detail the development of this framework from an analytical wear model based on Wheel-on-Flat experiments [19] to the development of a TKR contact model linked to the wear model [16–18]. Next, we detail the process of validating the framework and finally, verification and application of the framework is performed in two parametric studies looking at the sensitivity of wear to implant positioning and femoral center of rotation [17].

2 Computational Framework Development

The computational framework encompasses an analytical wear model coupled to a FEA model. The analytical wear model is based on frictional energy and is developed specifically for UHMWPE under rolling and sliding contact [19]. The FEA model makes use of implicit integration for model stability and improved accuracy of contact stress. A mesh-deformation subroutine is used to apply material loss (as calculated by the analytical wear model) during the analysis as opposed to periodically updated between simulations. The FEA model of TKR wear is then incorporated into a larger computational framework capable of conducting large scale parametric studies of TKR wear. This combination of modeling techniques, and the analytical model of wear used, has led to a model of wear with high predictive accuracy.

2.1 Analytical Wear Model Development from Experimental Data

A previously proposed analytical wear model was linked to an FEA contact model [16, 19]. The analytical wear model uses the concept of frictional work to predict the energy required to remove a unit volume of material both parallel and perpendicular to the primary molecular orientation of UHMWPE and is based on data from wheel-on-flat experiments [19].

The data from wheel-on-flat experiments [19] were reanalyzed to determine the necessary coefficients for the computational wear model. The wheel-on-flat apparatus (Fig. 1) consisted of a cobalt chromium molybdenum (CoCrMo) wheel articulating on a UHMWPE flat. The wheel rotates about its center, while the flat translates along a path that keeps the area of contact in the centerline of the flat. The flat also rotates to introduce cross-shear, which refers to the sliding direction dependency of polyethylene wear. Cross-shear occurs because polyethylene wear is highly dependent on the direction of predominant fibril orientation, or primary molecular orientation (PMO), in relation to sliding direction [20, 21]. Load is applied via an actuator that pushes the wheel down towards the flat. The CoCrMo wheel was machine polished to a medical grade surface finish. The UHMWPE flats were machined from slab molded GUR1050 and gamma sterilized at a dose of 2.50–3.25 Mrad at a temperature of 45 °C and stored in nitrogen. Surface profiles were measured before and after testing using an optical coordinate measuring machine (Smartscope, Optical Gaging Products Inc., Rochester NY USA) to reconstruct linear wear.

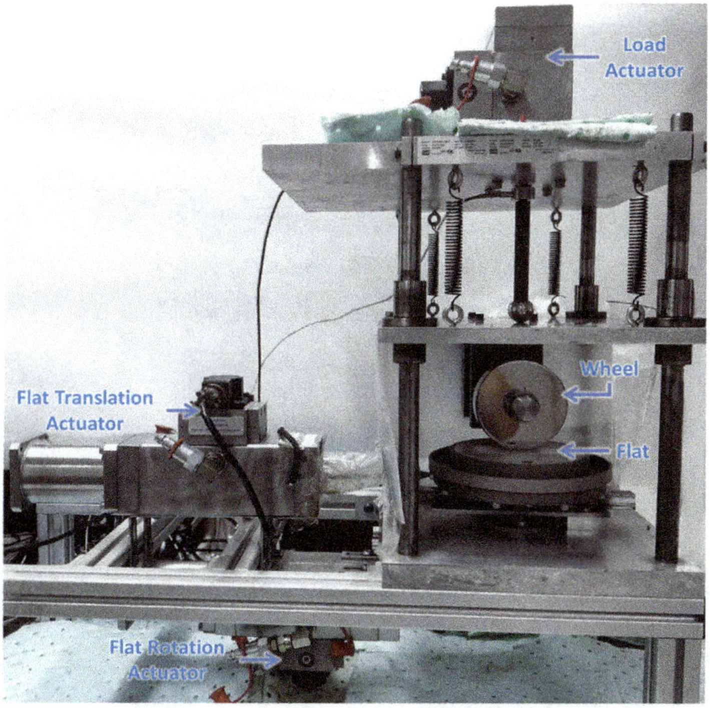

Fig. 1. Wheel-on-flat wear testing machine. A linear actuator applies load to the wheel. Two linear actuators provide translation and rotation of the UHMWPE flat. A rotational actuator (obscured from view) provides rotation to the CoCrMo wheel.

The degree of cross-shear was varied from 0° to ±10.2° along the length of the flat while load and relative sliding velocity were kept constant at 1500 N and 50 mm/s. The direction of flat rotation was switched between each cycle to account for molecular reorientation. The experiment ran for 1 million cycles (MC). Wear depth, measured from surface profiles on the flat, was fit to a mathematical equation based on a velocity-vector description of cross-shear (Eq. 1), where zero cross-shear is experienced at a sliding angle parallel to the primary molecular orientation (0° and 180°), and maximum cross-shear at a sliding angle perpendicular to the primary molecular orientation (90° and 270°).

$$z(\theta) = \frac{(1 + \sin(2 \cdot \theta - 90^{\circ}))}{2} \quad (1)$$

Here z represents wear depth at angle θ. The resulting equation can be seen in Fig. 2.

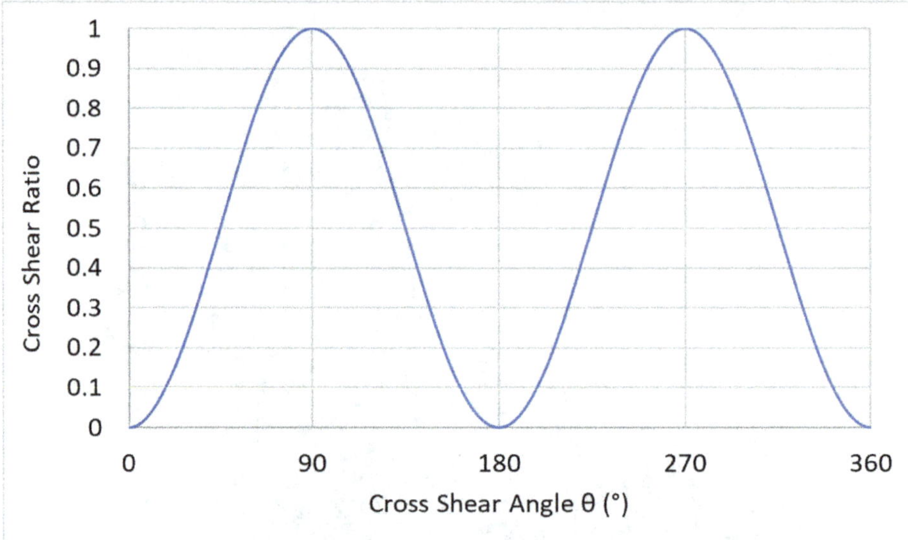

Fig. 2. Cross-shear model where cross-shear is defined as the portion of the velocity vector perpendicular to the primary molecular orientation (PMO) [19]. Here maximum cross-shear occurs at 90° and 270°.

Equation 1 was modified with curve fitting parameters a and b and fit to the linear wear profiles from the experiment (Eq. 2):

$$z(\theta) = \frac{(1 + \sin(2 \cdot \theta - 90°))}{2} \cdot a + b \tag{2}$$

where $a/2$ is 0.13 and b is 0.0178 [19].

The final equation, including a term for the number of cycles n and scaled to the total number of cycles is shown below (Eq. 3):

$$z(\theta, n) = ((1 + \sin(2 \cdot \theta - 90°)) \cdot 0.13 + 0.0178) \cdot \frac{1}{1 \cdot 10^6} \cdot n \tag{3}$$

The wear profile is then separated into portions that are caused by wear parallel to (y-direction) and perpendicular to (x-direction) the primary molecular orientation (PMO). The linear wear depth for each portion is calculated as shown below (Eqs. 4 and 5):

$$z_y(n) = \frac{b}{1 \cdot 10^6} \cdot n \tag{4}$$

$$z_x(\theta, n) = ((1 + \sin(2 \cdot \theta - 90°)) \cdot a) \cdot \frac{1}{1 \cdot 10^6} \cdot n \tag{5}$$

The total volume for each portion is calculated in Eqs. 6 and 7 below:

$$V_y = d_{wheel} \cdot \int_{s_i}^{s_f} z_y ds \left[mm^3\right] \tag{6}$$

$$V_x = d_{wheel} \cdot \int_{s_i}^{s_f} z_x ds \left[mm^3\right] \tag{7}$$

Here, s is the position along the wear path, s_i and s_f are the start and end positions respectively, and d_{wheel} is the width of the wheel (20 mm). Integrals are evaluated numerically using the Scientific Python Computing Environment (SciPy). The resulting wear volumes are:

$$V_y = 1.7800 \cdot 10^{-5}\,\text{mm}^3 \text{ and } V_x = 2.7293 \cdot 10^{-6}\,\text{mm}^3$$

Next, the amount of work done by the force of friction (F_f) in each direction is calculated using the sliding velocity vector (v) and the cross-shear angle (θ). Note the cross-shear angle varies as a function of time (t). (Equations 8 and 9):

$$W_y = \int_{t_i}^{t_f} \overrightarrow{F_f} \cdot \vec{v} \cdot \sin^2\theta(t)\, dt\ [J] \tag{8}$$

$$W_x = \int_{t_i}^{t_f} \overrightarrow{F_f} \cdot \vec{v} \cdot \cos^2\theta(t)\, dt\ [J] \tag{9}$$

Integrals are evaluated numerically using SciPy and the total frictional work done in each direction is:

$$W_y = 6.8618 \cdot 10^3\,\text{J and } W_x = 9.6803 \cdot 10^1\,\text{J}$$

From the values of total volumetric wear and total frictional work done in each direction, it is possible to calculate the work required to remove a unit volume of material in the directions parallel and perpendicular to the PMO (Eqs. 10 and 11):

$$\partial Work_y = W_y/V_y \left[\frac{J}{mm^3}\right] \tag{10}$$

$$\partial Work_x = W_x/V_x \left[\frac{J}{mm^3}\right] \tag{11}$$

Calculated values for unit work are:

$$\delta Work_y = 3.855 \cdot 10^8\,\text{J/mm}^3 \text{ and } \delta Work_x = 3.547 \cdot 10^7\,\text{J/mm}^3$$

Volumetric wear is calculated using the final equation below (Eq. 12):

$$V_{total} = \int_{t_i}^{t_f} \frac{\overrightarrow{F_f}(t) \cdot \vec{v}(t) \cdot \sin^2 \theta(t)}{\delta Work_x} + \frac{\overrightarrow{F_f}(t) \cdot \vec{v}(t) \cdot \cos^2 \theta(t)}{\delta Work_y} dt \; [mm^3] \tag{12}$$

The wear model is well-suited for incorporation into a FEA contact model as it needs frictional energy as input, which can be calculated from sliding velocity and the force of friction.

2.2 TKR FEA Contact and Wear Model

As previously reported, CAD models of a left sided NexGen Cruciate Retaining TKR (Zimmer Biomet, Warsaw, IN) and Abaqus v6.13-2 (Dassault Systèmes, Waltham, MA) were used to generate an FEA mesh using semi-manual meshing techniques [16]. The tibial insert geometry was manually partitioned into segments suitable for automatic meshing using the tools available within the Abaqus CAE software (Dassault Systèmes, Waltham, MA). The UHMWPE tibial component was composed of 78,108 linear hexahedral reduced integration elements (C3D8R). The cobalt chromium molybdenum (CoCrMo) femoral component was modeled as a rigid surface with 4080 2nd order quadrilaterals (SFM3D8) and 547 2nd order tetrahedrons (SFM3D6). The J-2 Plasticity model [22] was used to model the UHMWPE tibial component, with a density of 9.4E−7 kg/mm^3 [23, 24], an elastic modulus of 1,051 MPa [22], and a Poisson's ratio of 0.46 [22]. The J-2 Plasticity model is an elastic-plastic empirical model of UHMWPE yielding and has been shown to have an R^2 of 0.87 when combining small, intermediate, and large monotonic strains, as well as cyclic strain [22]. Table 1 contains the material parameters used in the implementation of the J-2 Plasticity model. The penalty method was used to model contact between the femoral component (master surface) and the tibial insert (slave surface), with a friction coefficient of 0.04 on the articular surface [25–27]. A reference point on the tibial component, located according to International Organization for Standardization (ISO) specifications for knee simulator wear testing [28], was used for the application of velocity boundary conditions for internal/external rotation and anterior/posterior translation. A superior-inferior axis located 5 mm medial from this reference point was used to apply axial force to the tibial component [28]. Consistent with experimental tests and ISO specifications, medial/lateral translation and abduction/adduction were left free. Flexion/extension was applied to the femoral component which had all other degrees of freedom fixed. The boundary conditions applied to the model are summarized in Fig. 3.

Table 1. Parameters for the J-2 Plasticity model for UHMWPE

Stress (MPa)	Strain
12.1	0.00
21.4	0.03
23.8	0.11
44.0	0.55
92.4	0.98
135.0	1.09
515.0	1.34

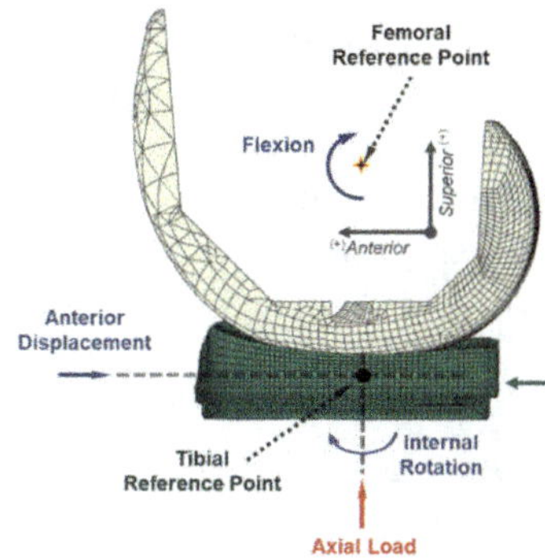

Component	Degree of Freedom	Constraint
Femoral Component Rigid Body Quadratic Surface Elements (4627)	Abduction Internal Rotation Anterior Displacement Medial Displacement Superior Displacement	Fixed
	Flexion	Applied Velocity
UHMWPE Tibial Insert Deformable Body Linear Hexahedral Elements (78108)	Flexion	Fixed
	Abduction Medial Displacement	Free
	Superior Displacement	Loaded
	Anterior Displacement Internal Rotation	Applied Velocity

Fig. 3. FEA model loading and boundary conditions. Reprinted from [17], with permission from Elsevier.

In order to utilize the analytical wear model in FEA, Eq. 12 was discretized with respect to time and space, converted to linear penetration, and applied to each articulating node of the model (Eq. 13) [16]. The symbols used in the wear model are defined in Table 2. The discretized wear model was applied to the FEA mesh using the built in Abaqus Fortran user subroutine UMESHMOTION. Using Eq. 13, linear wear is calculated at every node and for every time increment, then scaled by the number of cycles in the current update. The wear step is repeated until the total number of cycles is reached. The optimal number of cycles per mesh update was determined through a convergence study to be 500,000 for a total number of cycles of 1 million [16].

$$WearDepth_{node} = \sum_{Start\ time}^{End\ time} \frac{F_f(t) * v(t) * cos^2(|\varphi - SD|)}{\delta Work_x} + \frac{F_f(t) * v(t) * sin^2(|\varphi - SD|)}{\delta Work_y} * \frac{\Delta t}{\Delta A} \tag{13}$$

Table 2. Symbols used in the discretized wear model (Eqs. 13 and 14).

SD	Sliding Direction
v	Sliding Velocity
F_f	Tangential Force
φ	Primary Fiber Orientation
A	Area
t	Time
$\delta Work_x$	Unit work in X (perpendicular to PMO)
$\delta Work_y$	Unit work in Y (parallel to PMO)

To account for UHMWPE's tendency to align fibrils to the direction of sliding, each node was initialized to an initial PMO prior to wear simulation. This initial direction is found by calculating total frictional work in the anterior-posterior and medial-lateral directions at every node of the tibial articular surface in an initial FEA run without simulating wear. The initial direction for each node is then assigned as the angle between the total work for direction, or resulting direction of total frictional work (Eq. 14, symbols in Table 2).

$$\varphi_{node} = \tan^{-1}\frac{\sum F_{f,ML} * v_{ML}}{\sum F_{f,AP} * v_{AP}} \tag{14}$$

Mesh convergence was ensured by confirming that local contact pressure errors outputted by Abaqus/Standard were an order of magnitude less than contact pressure at each node as per developer guidelines [29]. Using this technique, a balance was found between computational efficiency and accuracy. The model runs in an acceptable amount of time: running on a 14 core Intel® Xenon® E5-2680 CPU @2.40 GHz, the initialization cycle without wear takes ~1 h and the wear cycle takes ~5 h for a total run time of 6 h per complete cycle.

2.3 Development of a Computational Framework for Parametric Evaluation of TKR Wear

In order to utilize the developed FEA wear and contact model to its full potential, it must have the capability to perform large parametric studies in an automated or semi-automated manner, and execute a study design with little to no input from the user. To do this, a computational framework around Simulia's Isight Execution Engine (Dassault Systèmes, Waltham, MA) was designed and built. The framework is flexible, in that it accepts multiple forms of input data: patient-specific, population-specific, or standardized motions and forces. Multiple parameters can be investigated, including component alignment, isolated parameters within the kinematic and loading profiles, and different material parameters and combinations. Output variables can include volumetric wear and wear scar shape parameters. Isight can automatically generate a design of experiments (DOE) analysis from input and output variable(s) of the user's choice. SciPy and Matlab scripts are used to link the components of the framework together and aid in the calculation of inputs and outputs. Isight controls the execution of the framework and stores input and output variables in a database for easy access and analysis after a parametric study has completed. The results are further using statistical methods such as multiple linear regression.

To efficiently determine the effect of multiple parameters and their interactions in computational analyses, the Latin Hypercube Sampling (LHS) DOE technique is often used [30–35]. The LHS approach uses a stratified sampling scheme which divides each parameter into n-evenly spaced increments, and then randomly combines them to generate n-unique combinations. The LHS approach is useful in computational analysis when investigating many parameters makes traditional DOE techniques impractical. For example, a full factorial study with 9 parameters and 2 levels would require 512 simulations alone. If 3 levels are studied this number increases to 19,683. The LHS technique improves on the coverage of the input space compared to more structured DOE designs, by sampling over the entire range, offers flexibility in the number of simulations performed, and increases that chance that an individual parameter will have significant contribution to the variability [36]. LHS is more efficient than random sampling techniques such as Monte Carlo when running thousands of simulations is impractical. Finally, by partitioning the sample space so each interval has equal probability, LHS ensures even coverage of the sample space [32, 36, 37].

Statistical analysis of the LHS DOE results is performed within Isight and using The Statistics and Machine Learning Toolbox of Matlab (The Mathworks Inc., Natick, MA). Matlab scripts are used to analyze the results using multiple linear regression. Linear regression models are useful to identify which variables are important and predict what effect they have on the output variables of interest. They can also be utilized as surrogate models [32], allowing for the prediction of wear given a set of parameters without running additional FEA simulations.

3 Validation of the Wear and FEA Models

Validation tests were performed for multiple levels of the framework. This includes computational comparisons for experiments that were different than the original wheel-on-flat experiments used to fit model coefficients, and wear simulator tests to ensure that volumetric wear and wear scar appearance was similar for multiple kinematic and load inputs and implant alignment. The below sections describe the experiments and computational simulations used to validate the framework.

3.1 Validation Against Pin-on-Disk Experiments

Validation of the UHMWPE wear model was carried out by comparing predicted outputs from the wear model to experimental pin-on-disk (POD) tests. Predicted wear calculated with three computational implementations of the wear model were compared to the experiment. The three computational implementations, in order of increasing complexity, were (1) an analytical model, (2) a numerically integrated model developed within SciPy, and (3) a FEA model of pin-on-disk wear.

Experiments were carried out using a six station AMTI OrthoPOD multi-directional POD wear testing machine (AMTI, Watertown, MA). GUR 1050 UHMWPE pin samples were provided by Orthoplastics Ltd (Lancashire, England) and irradiated in nitrogen to 30 ± 5 kGy (3.0 Mrad). Diameter of the pins was 9.5 mm. The UHMWPE pins were soaked in deionized water until equilibrium weight was obtained. Pins articulated against CoCrMo disks in a 15 mm × 15 mm square motion path at 1.0 Hz for 1.25 MC. A 200 N load was applied to the pins for a nominal contact pressure of 2.8 MPa. Six pins total were tested, and a seventh pin was used as an unloaded soak control and did not undergo articulation. In each station, 15 mL of lubricant was placed in the chamber. Lubricant was bovine serum (Life Technologies Corporation, Carlsbad, California) diluted with saline to a protein content of 30 g/L and pH of 7.6. Ethylenediaminetetraacetic acid (EDTA) was added at 200 mg/L to avoid the precipitation of metal-phosphates. A heated water bath was used to keep chambers at a constant 37.0 °C. The CoCrMo disks were polished to a medical grade finish. Disk roughness was assessed using an optical interferometer (Zygo 6400, Zygo Corp., Middlefield, CTETC). Average disk roughness pre-test was RMS = 0.008 μm, Ra = 0.006 μm. Pins were cleaned and weighed every 250,000 cycles and wear was assessed gravimetrically.

For the first validation of the wear model, Eq. 12 was modified so that the total volume loss in a POD test could be directly calculated as a function of the number of cycles (Eq. 15).

$$V_{total} = \left(\frac{F_n \cdot \mu \cdot v \cdot \sin^2 \theta}{\delta Work_x} + \frac{F_n \cdot \mu \cdot v \cdot \cos^2 \theta}{\delta Work_y} \right) \cdot n \left[mm^3 \right] \tag{15}$$

Here n is the number of cycles (1 million), the frictional force is encompassed by Fn, the normal force (200 N), and μ, the coefficient of friction (0.46). v is the velocity of the pin (60 mm/s), θ is the cross-shear angle (an average of 45° for a square motion path and $\delta Work_x$, $\delta Work_y$ are the unit work values, derived previously. As an analytical model, this represents the simplest form of the wear model, and can only be used for simple geometries and motion paths.

The second method utilized a model of the POD machine written in SciPy where the pin is modelled as a single node, with an assumed PMO of 45°. Frictional force and velocity are applied to this node according to the relative motion of the pin with respect to the disk. The model is stepped forward in time using a two stage second-order Runge-Kutta method [38]. The time step used was 0.01 s. Wear is calculated using Eq. 13 and scaled to 1 MC. The surface area of the pin is used to scale the linear wear calculation to volumetric wear. This implementation of the model can handle more complex motion paths but is limited again to simple geometries.

The final validation of the POD wear experiment was conducted using an FEA model. Abaqus/Standard v6.13-2 was used to create a model and mesh of the POD wear testing apparatus using the same wear model implementation as the TKR computational wear framework, allowing for complex geometries such as a TKR. The UHMWPE pin was modeled as a deformable body using 11,004 linear hexahedral elements (C3D8). The material model used for the pin was the above described J-2 plasticity model

(Table 1). The CoCrMo disk was modeled as a rigid body utilizing 4282 quadratic surface elements (SFM3D8). Wear was applied using the Fortran subroutine UMESHMOTION as previously described. Models were run using Abaqus/Standard v6.13-2 and the mesh was updated every 250,000 cycles to a total of 1.25 MC.

Wear rate from the experimental POD test was 8.41 ± 0.12 mg/MC (Fig. 4). Wear rate is reported as the slope of the linear regression ± standard error. The chamber containing the lubricant for pin 4 leaked during the period between 1 MC to 1.25 MC and its data was excluded from the analysis.

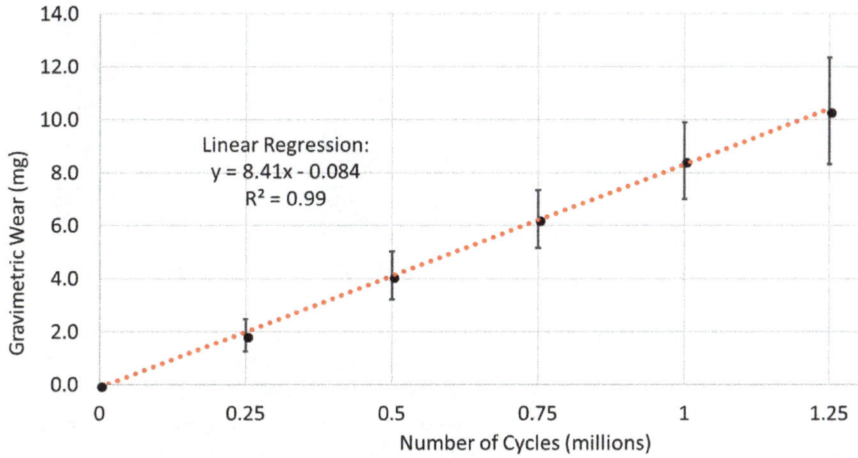

Fig. 4. Average gravimetric wear results from POD test. Error bars are ± standard deviation. Red dotted line is the linear regression fit to the data from the experiment.

The predicted wear rate from the analytical method was 8.50 mg/MC. Wear rate from the numeric SciPy model was 8.47 mg/MC. Wear rate predicted by the POD FEA model was 8.50 mg/MC.

3.2 Comparison of FEA Wear Prediction to Knee Simulator Tests

To test the capability of the model to predict TKR wear, a series of experimental tests were conducted using a mechanical knee simulator. Predicted gravimetric wear was compared to the results of three separate knee simulator tests under differing inputs: the displacement-controlled ISO standard from 2004 (ISO:2004) [39], the newer displacement-controlled ISO standard from 2014 (ISO:2014) [28], and a population average for stair ascent [40]. The distribution of linear wear within wear scars was also compared for ISO:2004. These results have been previously reported [16–18], and are briefly summarized here.

NexGen tibial liners (Zimmer, Inc., Warsaw, IN) were wear tested using a four-station knee wear simulator (Endolab, Rosenheim, Germany) in displacement control. Applied degrees of freedom were identical as those previously described for the FEA model. Each test was run to a maximum of 3 MC at 1.0 Hz. Testing was paused for

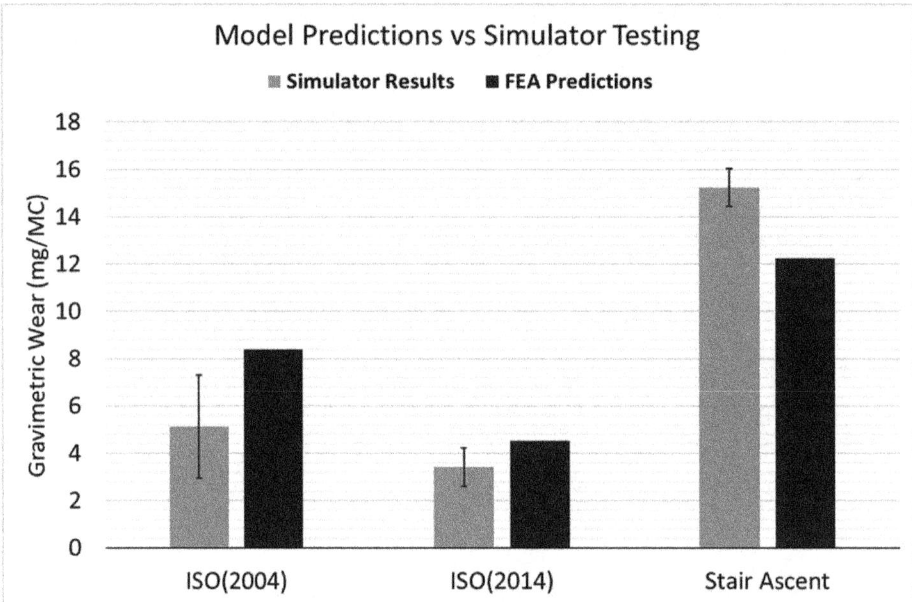

Fig. 5. Comparison of 3 simulator experiments to FEA wear predictions: ISO(2004), ISO (2014), and the average of 20 TKR patients stair ascent data as input (Stair Ascent). The wear model predictions have the same relative trends as the experiments. The wear model had a high coefficient of determination (R^2 = 0.84) when compared to simulator tests. Results are reported in mg/million cycles (MC). Error bars are ± standard deviation. Reprinted from [17], with permission from Elsevier.

cleaning and weighing of the tibial liners every 0.5 MC. Gravimetric wear from simulator tested components was compared to FEA wear by converting FEA computed volumetric wear to gravimetric wear using the density of UHMWPE (9.4E−7 kg/mm^3). For ISO:2004, the articular surfaces of the tibial liners were digitized, and wear contours were reconstructed [41].

Model and mechanical simulator produced wear were similar (Fig. 5), with the model having an R^2 of 0.84 [17]. ISO:2014 simulator produced wear rates were 3.42 ± 0.814 mg/MC versus a model prediction of 4.52 mg/MC. Simulator produced wear rates for ISO:2004 were 5.14 ± 2.18 mg/MC versus a model prediction of 8.40 mg/MC [16]. Patient average stair ascent simulator wear rates were 15.24 ± 0.79 mg/MC versus a model prediction of 12.24 mg/MC [17].

The distribution of wear on the tibial liners predicted by FEA compared well with the reconstructed wear scars from the simulator tested components (Fig. 6) [16]. Wear scars on the simulator-tested components were highly variable, and the FEA predictions fell well within that variability. Highest wear was seen on the medial side of the medial compartment for both FEA-predicted and simulator tested wear scars. The lateral compartment wear scar was smaller and shallower. Both FEA-predicted and simulator-tested wear scars also were located farther anterior on the medial compartment than the lateral compartment, indicating an overall external rotation of the wear scar.

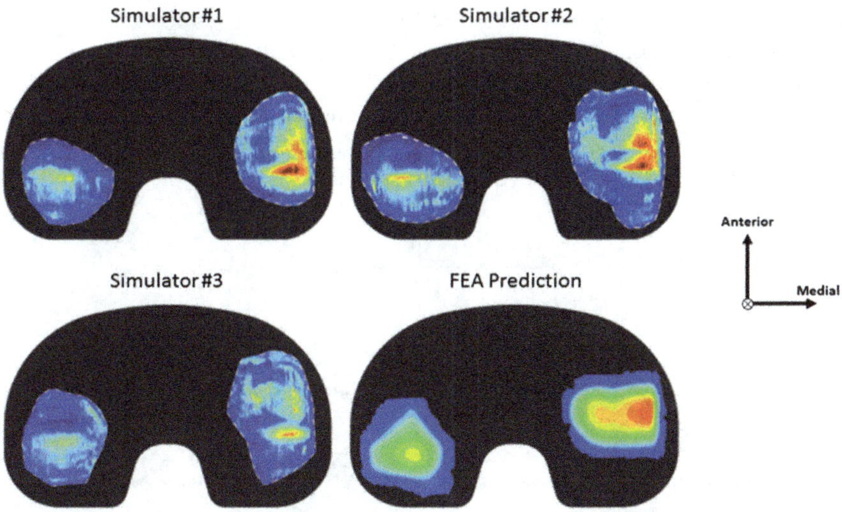

Fig. 6. Comparison of wear distribution on 3 simulator tested components and a FEA prediction for ISO:2004. The computational model predicts wear scar location and distribution of linear penetration depth with a high degree of accuracy. Reprinted from [16], with permission from Sage.

4 Verification and Application of the Framework

With the combined wear and FEA model validated the framework as a whole was verified by conducting a preliminary experiment alongside mechanical testing. This initial study looking at the effect of rotational alignment of the tibial component on wear serves as a first application as well as framework verification. Next a larger parametric study looking at the effect of femoral center of rotation (CoR) on wear was performed [17] using the finished framework and is summarized here to demonstrate the capabilities of the system.

4.1 Parametric Study of Internal/External Rotation of Tibial Component

The transverse plane rotational alignment of the tibial component relative to the femoral component is one variable that could lead to differences in TKA wear simulator results. Transverse plane alignment, particularly excessive internal rotation of the tibial component, has also been shown to lead to poor functional outcomes in TKRs [42, 43]. While it is well established that femoral malrotation can lead to poor outcomes, tibial component malrotation has only recently been identified as a problem [27]. Malrotation of the tibial component can lead to high stresses within the component [44–46] which may lead to higher polyethylene wear. Here, the computational framework is leveraged to study the effect of transverse plane malrotation on TKR wear. We also perform experiments on a knee wear simulator to compare to computational results.

The computational framework is used to perform an 11-point parameter study of tibial component malrotation on TKR wear (Fig. 7). The initial rotation of the tibial component is positioned between 25° of external rotation and 25° of internal rotation in increments of 5°. This adjustment to initial alignment is handled by Isight, which directly modifies the FEA model inputs. Kinematic and loading inputs followed ISO:2014 [28], and one million cycles are simulated.

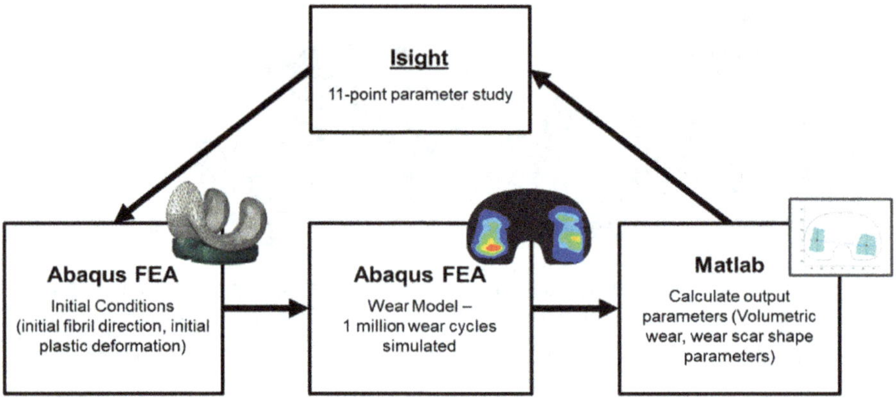

Fig. 7. Computational workflow. An 11-point parameter study is performed to investigate the effect of tibial component internal/external malrotation on wear. First, initial conditions are determined with an FEA model run. Next, a FEA-coupled-wear step is performed and scaled to simulate one million gait cycles. Finally wear scar dimensions are calculated with a custom Matlab script.

To verify the computational framework, short term wear simulator experiments were conducted to generate wear scars under different tibial component rotation angles. In these experiments, the articulating surfaces of NexGen CR UHMWPE inserts were colored with marker, and simulator tests were run in displacement control for 5000 cycles using kinematic and kinetic inputs defined by ISO:2014. Marker that was removed during the test indicated the wear scar for a specific tibial rotation angle [47]. A coordinate measuring machine (Smartscope, OGP, Rochester, NY) was then used to digitize the wear scar. The experimental tests were simulated using FEA using identical motion and load inputs, and transverse plane alignments as the experimental model. Once wear scars were obtained, ImageJ (NIH, Bethesda, MD) was used to calculate area and dimensions of FEA and simulator generated contact areas.

The experimentally generated contact areas and computationally predicted wear scars compared optically very well throughout the range of external (Fig. 8) and internal (Fig. 9) tibial component rotation. As a quantitative comparative assessment, the angle between the medial and lateral centroid (Fig. 10) and the total wear scar areas (Fig. 11) were computed. Wear scar area initially increased then decreased for both internal and external rotation as the component had more contact with the edge of the tibial plateau. The angle between the medial and lateral wear scar centroids was very linear, with an angle of approximately zero degrees with no initial transverse plane tibial rotation. As can be seen from the figures, the FEA model was able to reproduce the wear scar areas and trends generated by the short-term simulator tests with reasonable accuracy.

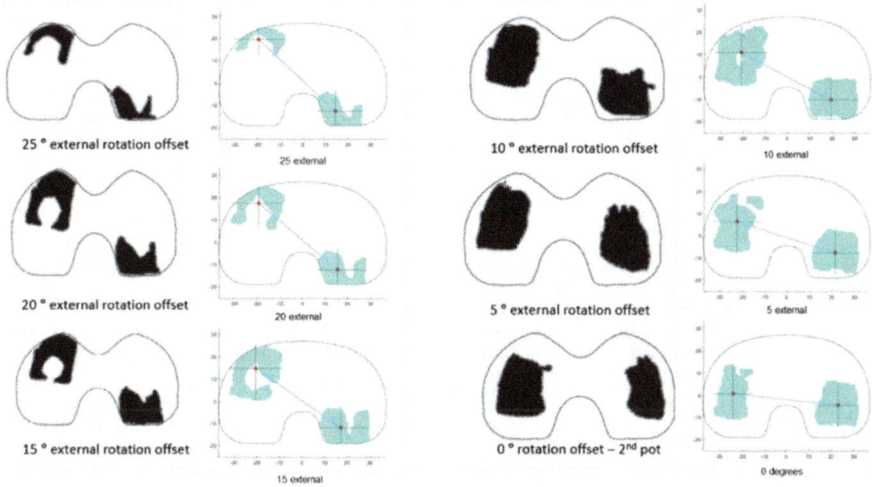

Fig. 8. Comparison of short-term simulator generated contact areas (black) and computationally predicted wear scars (blue) for 0 to 25° of tibial external rotation.

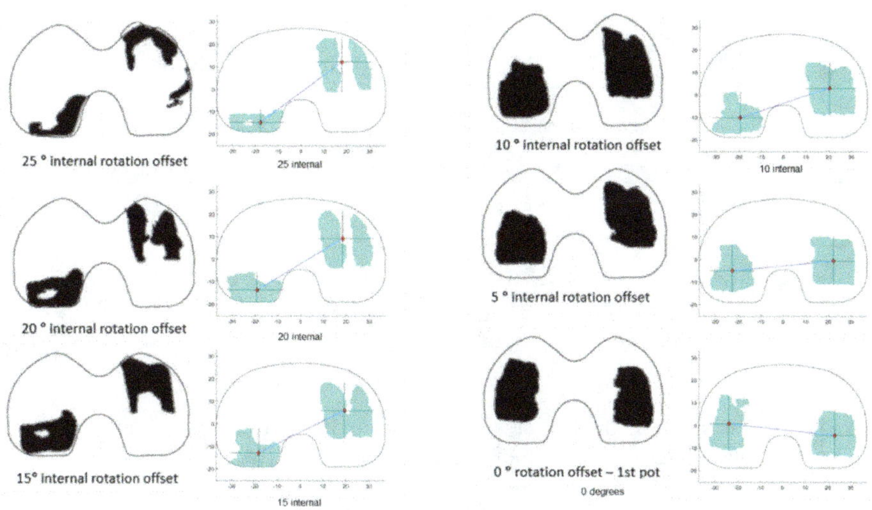

Fig. 9. Comparison of short-term simulator generated contact areas (black) and computationally predicted wear scars (blue) for 0 to 25° of tibial internal rotation.

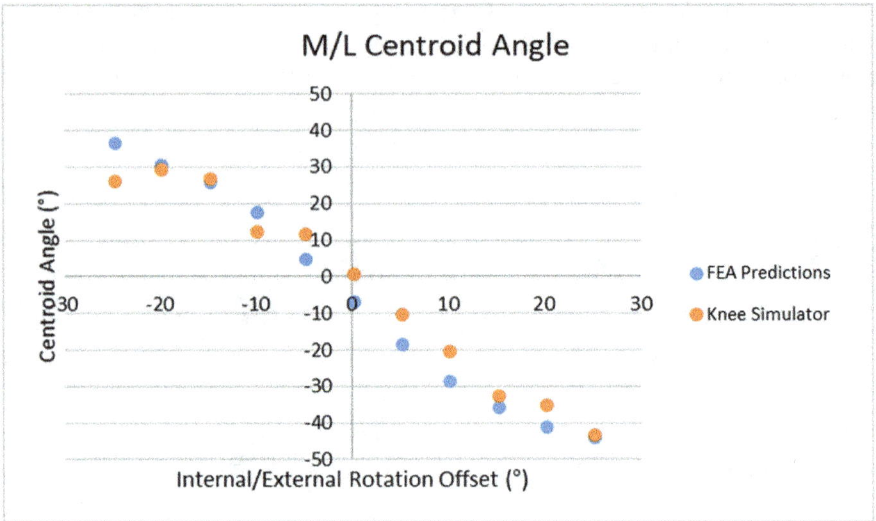

Fig. 10. Comparison of the angle between the medial and lateral computationally predicted wear scar centroids (blue) and experimentally measured contact area centroids (blue). Here external rotation is positive.

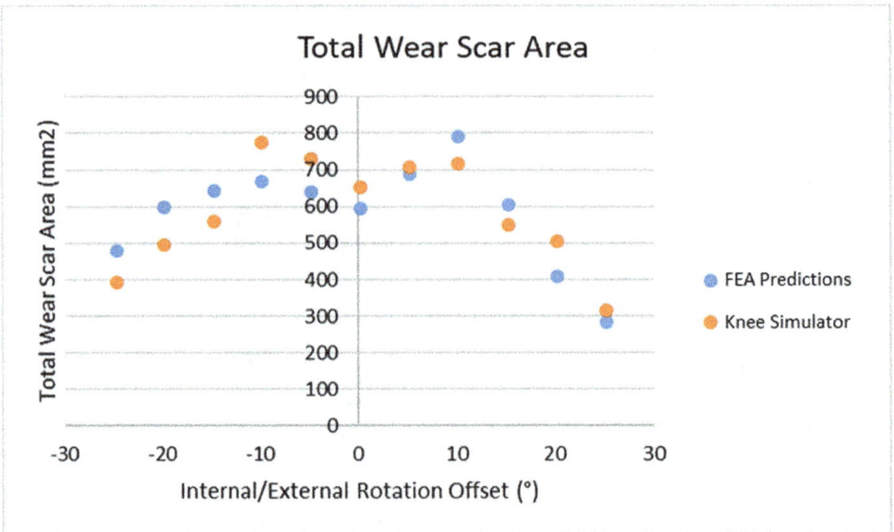

Fig. 11. Comparison of total area of computationally predicted wear scars (orange) and experimental contact areas (blue). Here external rotation is positive.

The FEA model was also used to predict the volumetric wear rates for all 11 rotational conditions of the tibial component. Those are summarized in Fig. 12. A large increase in wear is predicted for all instances of excess internal rotation of the tibial component. With just 5° of internal rotation, wear increased from 4.66 mm^3/MC to 8.95 mm^3/MC, almost double the value for the tibial component neutral position. At 25° of internal rotation, predicted wear was 15.26 mm^3/MC. Interestingly, for external rotation of the tibial component, wear initially increased then decreased at rotations of 20° and 25°.

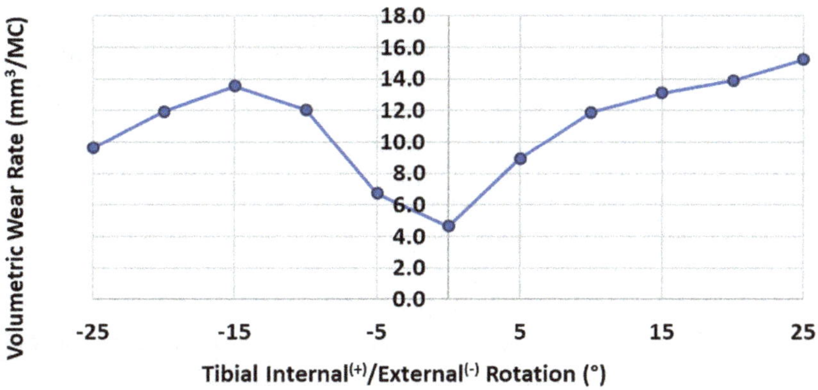

Fig. 12. Volumetric wear rate (mm^3/MC) as a function of tibial component rotation angle. Here internal rotation is positive and external rotation is negative. A large increase in wear is predicted for excess internal rotation. External rotation is initially more forgiving, but then similarly excessive for 10 and 15° of malrotation. Interestingly, any further increase is followed by a decrease in wear.

Consistent with previous findings [44–46], tibial component malrotation led to drastically increased wear predictions. The results of this study indicate that proper positioning of the tibial component relative to the femoral component is important not only clinically, but also when conducting mechanical wear testing. Further refinement of this investigation should be conducted, with more total simulations performed at smaller intervals, particularly at rotations between −5° to 5° internal-external rotations, where the rate of change is highest, to determine how small errors in alignment can impact wear and to determine the rotation angle that produces the least wear. The decrease in wear at higher angles of external rotation is interesting. One explanation is that contact area decreases due to edge contact of the femoral component and tibial insert, as can be seen in Fig. 9. Why this is not a factor in internal rotation is not clear, however examining Fig. 9 and Fig. 10, there is more edge loading in the external rotation case, and there is less contact area overall for corresponding external rotations ($\sim$100 mm^2 difference). In a mechanical test, this edge loading may lead to additional wear modes such as fatigue wear that the wear model is not designed to predict.

4.2 Parametric Study on Sensitivity of Wear to Center of Rotation

One potential source of variability within TKR wear tests is the femoral center of rotation (CoR) [48–50]. In a previously published work, the computational framework described here was leveraged to perform a parametric study on the effect of femoral CoR on TKR wear [17]. A 40-point LHS DOE was performed using the framework described above. Model boundary conditions were identical to those described previously with the exception of the location of the femoral CoR. Kinematics and loading were applied as defined by the displacement-controlled knee wear testing ISO standard [28]. A square spanning −10 mm to 10 mm in the anterior-posterior direction and −10 mm to 10 mm in the superior-inferior direction was used to bound the sample space for the LHS DOE. The CoR from wear simulator studies conducted in our lab was used as the origin and reference point of the sample space. Here we define negative (−) to be posterior and inferior, and positive (+) to be anterior and superior. The location of the sample points investigated are depicted in Fig. 13. A simulation was preformed, and wear was predicted to 1 MC for every sample point. Stepwise linear regression up to and including second order terms was used to fit a model to the results. This regression was used as a surrogate model to predict volumetric wear as a function of femoral CoR location. Terms were eliminated from the model that did not contribute to the explained variance.

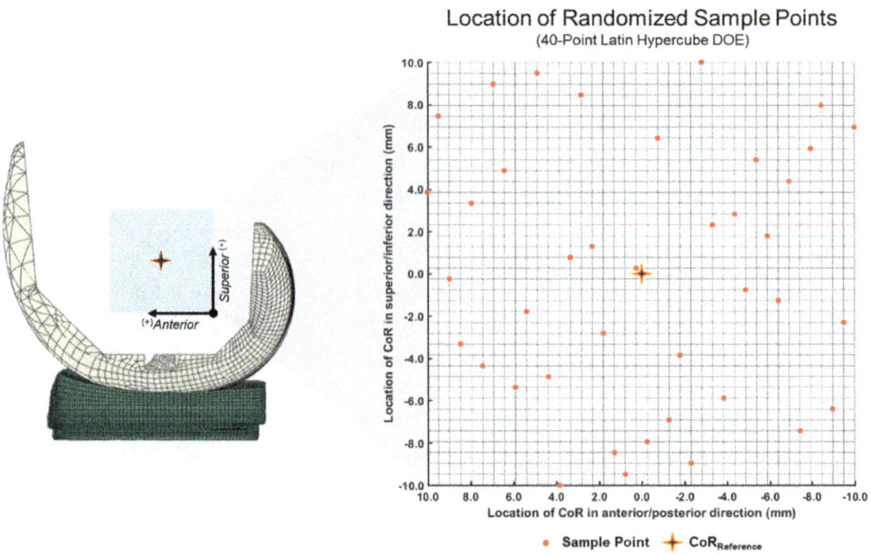

Fig. 13. Span and location of the Reference CoR and sample points generated by the 40-point Latin Hypercube DOE. Reprinted from [17], with permission from Elsevier.

A second order polynomial approximation (Fig. 14) was generated after linear regression with $R^2 = 0.987$, an RMSE error of 0.0685, and $p < 0.001$. All model terms were significant ($p < 0.001$) with the exception of the second order term which corresponds to the location of the femoral CoR in the anterior/posterior direction. This second order term was excluded from the statistical model as the stepwise algorithm determined that it did not contribute to the explained variance. CoR locations at the extreme ends of the sample space also produced wear distributions that varied when compared to both the reference CoR and when compared to each other.

The computational framework predicted that choice of femoral CoR can affect volumetric wear by up to 50% difference within the bounds of the sample space investigated. The results from this study demonstrated that volumetric wear and wear distribution are more sensitive to alterations in the femoral CoR in the superior-inferior direction than in the anterior-posterior direction (Fig. 14). The conclusion of this study was that differences in the location of the femoral CoR represent a source of variability in the preclinical testing and evaluation TKR wear that may not be accounted for.

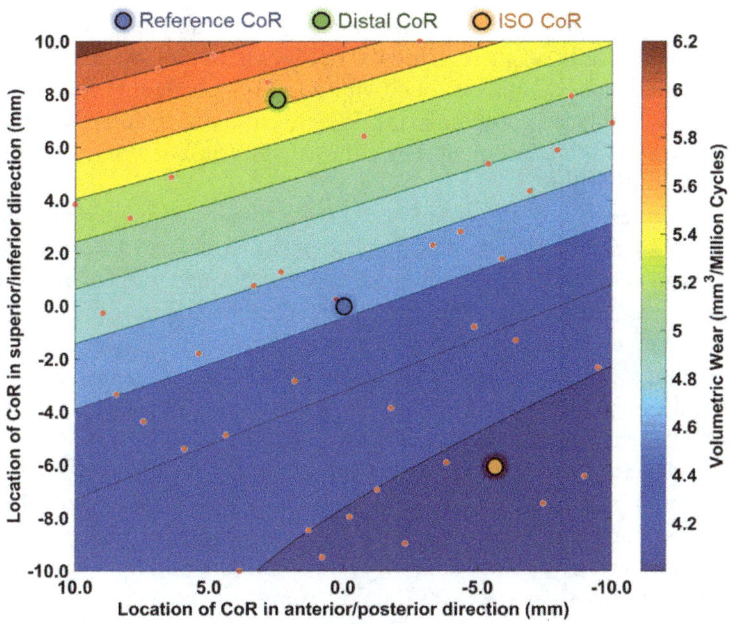

Fig. 14. Polynomial approximation of volumetric wear as a function of femoral center of rotation (CoR) location (contours). Sample points in the Latin Hypercube DOE are shown as red dots. Also shown are CoR locations used in the literature. The reference CoR is located at 0 mm anterior and 0 mm superior. The distal CoR is located at 2.4 mm anterior and 7.8 mm superior [48]. The ISO Standard CoR is located 5.7 mm posterior and 6.0 mm inferior. Reprinted from [17], with permission from Elsevier.

5 Discussion

TKR failure due to polyethylene wear is a problem that persists to this day, despite recent advancements in materials and prosthesis design. With the number of TKR surgeries increasing, and with the demographic of patients becoming younger, TKR longevity is a problem that will likely persist well into the future. Polyethylene wear is one of the primary contributors to late TKR implant failure, and if components are expected to last past the second decade, it is imperative to investigate the factors that contribute to wear and damage.

We describe a novel framework for computational prediction of TKR wear that can efficiently perform large parametric studies. Using this method, we demonstrated (1) that TKR wear is sensitive to the initial internal/external rotation of the tibial component and that (2) the choice of femoral center of rotation matters when considering TKR wear. The first finding corroborates well with previously published work and indicates that proper rotation of the tibial component is important. Internal malrotation of the tibial component of >10° has been identified as a negative prognostic factor [42], and here we predict an almost 3-fold increase in wear at that level. The second finding is important to minimize wear. This is not only true for simulator testing, but also clinically, as there is high variability in the rotational alignment of the tibial component. In our study, we found an up to 50% change in volumetric wear at the ends of the investigated solution space. We also found volumetric wear to be more sensitive to center of rotation changes in the superior-inferior direction than in the anterior-posterior direction.

Through a series of model validation tests, we found that the computational wear model had a high coefficient of determination when compared to experimental data ($R^2 = 0.84$), and replicated the distribution of wear through similar wear scars [16]. There are, however, several opportunities for further development of the framework. First, a long-term mechanical knee simulator test for wear should be conducted to verify the findings of this computational study of tibial component malrotation. This would not only help to further validate the wear model, but also serve to test the robustness of the framework.

There are several sources of error that are currently not captured by the computational framework and could contribute to differences in volumetric wear between the experiments and computational predictions. These error sources include differences in the lubricant, the lack of consideration of UHMWPE chain mobility in the computational wear model, as well as UHMWPE fluid absorption which can mask experimental material loss if wear rates are low [51–53]. Differences in the UHMWPE material used to calibrate the model versus that used in the TKR samples, such as degree of crosslinking may also have been present. The wheel-on-flat experiments, which were used to generate the unit work parameters used in the model [19], were also carried out at room temperature, while our knee simulator tests were run at 37 °C.

There were several limitations related to the FEA model itself, such as that no tibial tray was present in the model, and backside wear was not considered. It has however been shown previously that backside wear represents a small fraction of the total wear [24, 54], and so it was excluded for computational efficiency. An additional limitation is that viscoelastic effects (creep) are not included in the UHMWPE material model. A viscoelastic material model would likely effect linear penetration, not volumetric wear because strain due to the J2 plasticity model would be much larger than strain due to creep. In addition, viscoelasticity would increase processing time. As such, discrete linear penetration values are not included in the comparison of knee simulator wear scars and the predicted FEA wear scars. A comparison of the current plastic model versus a fully viscoelastic model should be done in a future study.

In addition to the studies performed thus far, the framework can perform parametric studies that assess patient, surgical, and design factors that affect wear in TKRs. Many future studies are planned that leverage this capability, including how wear rates are affected by variability within patient gait, surgical alignment of the components beyond just femoral internal-external rotation, and different material properties of the polyethylene.

In conclusion, we have just begun to use the capability of the framework to efficiently perform large variable studies. The two applications represent initial first attempts to study variability in component alignment across large solution spaces. Once more fully developed, this framework could be directly used in the design process to reduce the number of physical prototypes. Wear tests could be conducted without having to build the prosthesis. Material properties and design features could be adjusted to minimize wear debris under full consideration of all other design aspects and constraints. The substantial number of simulations that could be conducted—level walking, stair climbing, running, just to name a few—could boost confidence in the product and ultimately make it safe for individual use. The value of the framework will be in the ability to test new TKR design innovations as soon as they are developed, and identifying which designs should go on for further physical testing.

Acknowledgements. We thank Michel Laurent for useful discussions and Catherine Yuh for help with figures. This work was supported by the National Institutes of Health (R01 AR059843, MAW). Zimmer-Biomet (Warsaw, IN) provided CAD models for FEA analysis.

References

1. Cross, M., et al.: The global burden of hip and knee osteoarthritis: estimates from the global burden of disease 2010 study. Ann. Rheum. Dis. **73**(7), 1323–1330 (2014). https://doi.org/10.1136/annrheumdis-2013-204763
2. Chakravarty, R., Elmallah, R.D.K., Cherian, J.J., Kurtz, S.M., Mont, M.A.: Polyethylene wear in knee arthroplasty. J. Knee Surg. **28**(5), 370–375 (2015). https://doi.org/10.1055/s-0035-1551833
3. Crowninshield, R.D., Rosenberg, A.G., Sporer, S.M.: Changing demographics of patients with total joint replacement. Clin. Orthop. **443**(Journal Article), 266–272 (2006). https://doi.org/10.1097/01.blo.0000188066.01833.4f
4. Kurtz, S., Ong, K., Lau, E., Mowat, F., Halpern, M.: Projections of primary and revision hip and knee arthroplasty in the United States from 2005 to 2030. J. Bone Jt. Surg. - Am. **89**(4), 780–785 (2007)

5. Kurtz, S.M., Ong, K.L., Lau, E., Bozic, K.J.: Impact of the economic downturn on total joint replacement demand in the United States: updated projections to 2021. J. Bone Jt. Surg. - Am. **96**(8), 624–630 (2014). https://doi.org/10.2106/JBJS.M.00285
6. Siqueira, M.B.P., Klika, A.K., Higuera, C.A., Barsoum, W.K.: Modes of failure of total knee arthroplasty: registries and realities. J. Knee Surg. **28**(2), 127–138 (2015). https://doi.org/10.1055/s-0034-1396014
7. Losina, E., Katz, J.N.: Total knee arthroplasty on the rise in younger patients: are we sure that past performance will guarantee future success? Arthritis Rheum. **64**(2), 339–341 (2012). https://doi.org/10.1002/art.33371
8. Wier, L.M., et al.: HCUP Facts and Figures: Statistics on Hospital-based Care in the United States: 2009 (2011)
9. Kurtz, S.M., Lau, E., Ong, K., Zhao, K., Kelly, M., Bozic, K.J.: Future young patient demand for primary and revision joint replacement: national projections from 2010 to 2030. Clin. Orthop. **467**(10), 2606–2612 (2009). https://doi.org/10.1007/s11999-009-0834-6
10. Ravi, B., Croxford, R., Reichmann, W.M., Losina, E., Katz, J.N., Hawker, G.A.: The changing demographics of total joint arthroplasty recipients in the United States and Ontario from 2001 to 2007. Best Pract. Res. Clin. Rheumatol. **26**(5), 637–647 (2012). https://doi.org/10.1016/j.berh.2012.07.014
11. Roberts, V.I., Esler, C.N.A., Harper, W.M.: A 15-year follow-up study of 4606 primary total knee replacements. J. Bone Joint Surg. Br. **89**(11), 1452–1456 (2007). https://doi.org/10.1302/0301-620X.89B11.19783
12. O'Connor, M.I.: Implant survival, knee function, and pain relief after TKA: are there differences between men and women? Clin. Orthop. **469**(7), 1846–1851 (2011). https://doi.org/10.1007/s11999-011-1782-5
13. Gallo, J., Goodman, S.B., Konttinen, Y.T., Wimmer, M.A., Holinka, M.: Osteolysis around total knee arthroplasty: a review of pathogenetic mechanisms. Acta Biomater. **9**(9), 8046–8058 (2013). https://doi.org/10.1016/j.actbio.2013.05.005
14. Fraser, J.F., Werner, S., Jacofsky, D.J.: Wear and loosening in total knee arthroplasty: a quick review. J. Knee Surg. **28**(2), 139–144 (2015). https://doi.org/10.1055/s-0034-1398375
15. Morrison, T.M.: Reporting of Computational Modeling Studies in Medical Device Submissions. U.S. Department of Health and Human Services Food and Drug Administration, 21 September 2016
16. Mell, S.P., Fullam, S., Wimmer, M.A., Lundberg, H.J.: Finite element evaluation of the newest ISO testing standard for polyethylene total knee replacement liners. In: Proceedings of the Institution of Mechanical Engineers, Part H, p. 0954411918770700, April 2018. https://doi.org/10.1177/0954411918770700
17. Mell, S.P., Wimmer, M.A., Lundberg, H.J.: The choice of the femoral center of rotation affects material loss in total knee replacement wear testing - a parametric finite element study of ISO 14243-3. J. Biomech. **88**, 104–112 (2019). https://doi.org/10.1016/j.jbiomech.2019.03.027
18. Lundberg, H.J., Mell, S.P., Fullam, S., Wimmer, M.A.: Development and application of a computational total knee arthroplasty evaluation framework for preclinical parametric studies. Orthop. Proc. **101-B**(Suppl. 5), 109 (2019). https://doi.org/10.1302/1358-992x.2019.5.109
19. Schwenke, T., Wimmer, M.A.: Cross-shear in metal-on-polyethylene articulation of orthopaedic implants and its relationship to wear. Wear **301**(1–2), 168–174 (2013)
20. Strickland, M.A., Taylor, M.: In-silico wear prediction for knee replacements-methodology and corroboration. J. Biomech. **42**(10), 1469–1474 (2009)

21. Willing, R., Kim, I.Y.: A holistic numerical model to predict strain hardening and damage of UHMWPE under multiple total knee replacement kinematics and experimental validation. J. Biomech. **42**(15), 2520–2527 (2009). https://doi.org/10.1016/j.jbiomech.2009.07.008
22. Bergström, J.S., Kurtz, S.M., Rimnac, C.M., Edidin, A.A.: Constitutive modeling of ultra-high molecular weight polyethylene under large-deformation and cyclic loading conditions. Biomaterials **23**(11), 2329–2343 (2002)
23. Kurtz, S.M.: The UHMWPE Handbook: Ultra-High Molecular Weight Polyethylene in Total Joint Replacement. Elsevier (2004)
24. O'Brien, S., Luo, Y., Wu, C., Petrak, M., Bohm, E., Brandt, J.-M.: Computational development of a polyethylene wear model for the articular and backside surfaces in modular total knee replacements. Tribol. Int. **59**, 284–291 (2013). https://doi.org/10.1016/j.triboint.2012.03.020
25. Godest, A.C., Beaugonin, M., Haug, E., Taylor, M., Gregson, P.J.: Simulation of a knee joint replacement during a gait cycle using explicit finite element analysis. J. Biomech. **35** (2), 267–275 (2002)
26. Halloran, J.P., Petrella, A.J., Rullkoetter, P.J.: Explicit finite element modeling of total knee replacement mechanics. J. Biomech. **38**(2), 323–331 (2005). https://doi.org/10.1016/j.jbiomech.2004.02.046
27. O'Brien, S.T., Bohm, E.R., Petrak, M.J., Wyss, U.P., Brandt, J.-M.: An energy dissipation and cross shear time dependent computational wear model for the analysis of polyethylene wear in total knee replacements. J. Biomech. **47**(5), 1127–1133 (2014)
28. ISO 14243-3: Implants for surgery – wear of total knee-joint prostheses – part 3: loading and displacement parameters for wear-testing machines with displacement control and corresponding environmental conditions for test. Int. Organ. Stand., no. Journal Article (2014)
29. Dassault Systems: Modeling Contact with Abaqus/Standard 6.14. Dassault Systemes (2014)
30. Dar, F.H., Meakin, J.R., Aspden, R.M.: Statistical methods in finite element analysis. J. Biomech. **35**(9), 1155–1161 (2002)
31. Dopico-González, C., New, A.M., Browne, M.: Probabilistic finite element analysis of the uncemented hip replacement—effect of femur characteristics and implant design geometry. J. Biomech. **43**(3), 512–520 (2010). https://doi.org/10.1016/j.jbiomech.2009.09.039
32. Fitzpatrick, C.K., Hemelaar, P., Taylor, M.: Computationally efficient prediction of bone–implant interface micromotion of a cementless tibial tray during gait. J. Biomech. **47**(7), 1718–1726 (2014). https://doi.org/10.1016/j.jbiomech.2014.02.018
33. Rohlmann, A., Boustani, H.N., Bergmann, G., Zander, T.: A probabilistic finite element analysis of the stresses in the augmented vertebral body after vertebroplasty. Eur. Spine J. **19** (9), 1585–1595 (2010). https://doi.org/10.1007/s00586-010-1386-x
34. Rohlmann, A., Nabil Boustani, H., Bergmann, G., Zander, T.: Effect of a pedicle-screw-based motion preservation system on lumbar spine biomechanics: a probabilistic finite element study with subsequent sensitivity analysis. J. Biomech. **43**(15), 2963–2969 (2010). https://doi.org/10.1016/j.jbiomech.2010.07.018
35. Fitzpatrick, C.K., Baldwin, M.A., Rullkoetter, P.J., Laz, P.J.: Combined probabilistic and principal component analysis approach for multivariate sensitivity evaluation and application to implanted patellofemoral mechanics. J. Biomech. **44**(1), 13–21 (2011). https://doi.org/10.1016/j.jbiomech.2010.08.016
36. Iman, R.L.: Latin hypercube sampling. In: Wiley StatsRef: Statistics Reference Online. American Cancer Society (2014)

37. Mckay, M.D., Beckman, R.J., Conover, W.J.: A comparison of three methods for selecting values of input variables in the analysis of output from a computer code. Technometrics **42** (1), 55–61 (2000). https://doi.org/10.1080/00401706.2000.10485979
38. Ralston, A.: Runge-Kutta methods with minimum error bounds. Math. Comput. **16**(80), 431–437 (1962). https://doi.org/10.2307/2003133
39. ISO 14243-3: Implants for surgery – wear of total knee-joint prostheses – part 3: loading and displacement parameters for wear-testing machines with displacement control and corresponding environmental conditions for test. Int. Organ. Stand., no. Journal Article (2004)
40. Freed, R.D., Simon, J.C., Knowlton, C.B., Villaseñor, D.A.O., Wimmer, M.A., Lundberg, H.J.: Are instrumented knee forces representative of a larger population of cruciate-retaining total knee arthroplasties? J. Arthroplasty **32**(7), 2268–2273 (2017). https://doi.org/10.1016/j.arth.2017.01.054
41. Knowlton, C.B., Wimmer, M.A.: An autonomous mathematical reconstruction to effectively measure volume loss on retrieved polyethylene tibial inserts. J. Biomed. Mater. Res. B Appl. Biomater. **101**(3), 449–457 (2013). https://doi.org/10.1002/jbm.b.32782
42. Panni, A.S., et al.: Tibial internal rotation negatively affects clinical outcomes in total knee arthroplasty: a systematic review. Knee Surg. Sports Traumatol. Arthrosc. **26**(6), 1636–1644 (2018). https://doi.org/10.1007/s00167-017-4823-0
43. Sternheim, A., et al.: The benefit of revision knee arthroplasty for component malrotation after primary total knee replacement. Int. Orthop. **36**(12), 2473–2478 (2012). https://doi.org/10.1007/s00264-012-1675-6
44. Kuriyama, S., Ishikawa, M., Furu, M., Ito, H., Matsuda, S.: Malrotated tibial component increases medial collateral ligament tension in total knee arthroplasty. J. Orthop. Res. Off. Publ. Orthop. Res. Soc. **32**(12), 1658–1666 (2014). https://doi.org/10.1002/jor.22711
45. Liau, J.J., Cheng, C.K., Huang, C.H., Lo, W.H.: The effect of malalignment on stresses in polyethylene component of total knee prostheses–a finite element analysis. Clin. Biomech. Bristol Avon **17**(2), 140–146 (2002)
46. Osano, K., Nagamine, R., Todo, M., Kawasaki, M.: The effect of malrotation of tibial component of total knee arthroplasty on tibial insert during high flexion using a finite element analysis. Sci. World J. **2014**, 695028 (2014). https://doi.org/10.1155/2014/695028
47. Orozco, D., Schwenke, T., Wimmer, M.A.: Wear scar prediction based on wear simulator input data - a preliminary artificial neural network approach. J. ASTM Int. **3**(9) (2006). https://doi.org/10.1520/jai100249
48. Brockett, C.L., Abdelgaied, A., Haythornthwaite, T., Hardaker, C., Fisher, J., Jennings, L. M.: The influence of simulator input conditions on the wear of total knee replacements: an experimental and computational study. Proc. Inst. Mech. Eng. Part H **230**(5), 429–439 (2016). https://doi.org/10.1177/0954411916645134
49. DesJardins, J., Rusly, R.: Single flexion-axis selection influences femoral component alignment and kinematics during knee simulation. Proc. Inst. Mech. Eng. Part H **225**(8), 762–768 (2011). https://doi.org/10.1177/0954411911400534
50. Zietz, C., et al.: Experimental testing of total knee replacements with UHMW-PE inserts: impact of severe wear test conditions. J. Mater. Sci. Mater. Med. **26**(3), 134 (2015). https://doi.org/10.1007/s10856-015-5470-y
51. Brandt, J.-M., Charron, K.D.J., Zhao, L., MacDonald, S.J., Medley, J.B.: Commissioning of a displacement-controlled knee wear simulator and exploration of some issues related to the lubricant. Proc. Inst. Mech. Eng. Part H **225**(8), 736–752 (2011). https://doi.org/10.1177/0954411911406061

52. Schwenke, T., Wimmer, M.A., Kaddick, C.: Fluid composition influences wear testing of artificial knee implants. In: Medical Device Materials II - Proceedings of the Materials and Processes for Medical Devices Conference 2004, pp. 173–177 (2005)
53. Wang, A., Essner, A., Schmidig, G.: The effects of lubricant composition on in vitro wear testing of polymeric acetabular components. J. Biomed. Mater. Res. B Appl. Biomater. **68B** (1), 45–52 (2004). https://doi.org/10.1002/jbm.b.10077
54. Muratoglu, O.K., Rubash, H.E., Bragdon, C.R., Burroughs, B.R., Huang, A., Harris, W.H.: Simulated normal gait wear testing of a highly cross-linked polyethylene tibial insert. J. Arthroplasty **22**(3), 435–444 (2007). https://doi.org/10.1016/j.arth.2006.07.014

Poromechanical Modeling of Porcine Knee Joint Using Indentation Map of Articular Cartilage

Mojtaba Zare, Daniel Tang, and LePing Li(✉)

University of Calgary, Calgary, AB T2N 1N4, Canada
leping.li@ucalgary.ca

Abstract. The knowledge of site-specific properties of articular cartilage of a knee joint may be important for understanding the onset of cartilage degeneration in the knee. Few earlier studies have focused on the rate-dependent poromechanical response of knee joints to site-specific material properties across the joint. The objective of the present study was to develop a methodology to implement the in-situ cartilage mechanical properties in an anatomically accurate computational model of the porcine knee joint. Fresh porcine knee joints were used to reconstruct the knee geometry using magnetic resonance imaging. An automated indentation test was used to determine the site-specific cartilage properties. The variations of the recorded reaction forces over different sites were not solely due to nonuniform cartilage thickness. The nonfibrillar matrix and fibrillar network of the tibial cartilage had higher stiffness compared to that of the femoral cartilage as determined in the data fitting procedure. Considering the site-specific properties in finite element simulations, the force-compression relationship of the joint was determined by both compression-magnitude and compression-rate. The preliminary results indicated that a realistic implementation of site-specific tissue properties may be necessary for understanding the load distribution in the joint. The methodology will be further refined and tested.

Keywords: Articular cartilage · Fibril-reinforced model · Finite element · Image based modeling · Indentation test

1 Introduction

Articular cartilage injuries and diseases alter the load support capacity of the tissue [1]. As such, the changed mechanical response of knee joints to external loadings may be used to evaluate tissue damage and early osteoarthritis (OA) in vivo. Since cartilage is aneural and avascular, early OA and shallow damage often does not engender pain [2], but can progress to advanced OA over time if left untreated due to a physical and biological domino effect on the surrounding healthy articular cartilage [3]. The onset of OA may ensue from an altered stress and strain state in cartilage due to, for example, an injury of ligament, cartilage, or meniscus [4]. To assess the onset of OA and the possible failure sites in articular cartilage, computational models may be supplemental to current medical diagnosis. As such, a realistic description of joint geometry, joint kinematics, and tissue mechanical properties are required for a potential clinical application.

G. A. Ateshian et al. (Eds.): CMBBE 2019, LNCVB 36, pp. 86–106, 2020.
https://doi.org/10.1007/978-3-030-43195-2_7

The complex structure of cartilage, comprised of a depth-dependent collagen fibril network and proteoglycans (PGs) saturated with an interstitial fluid, supplies its unique function in joint articulation [2, 5, 6]. A material model is required to well describe the mechanical properties and microstructure of the soft tissue. The specific arrangement and hierarchical organization of the fibers have major influences on the mechanical behavior of the tissue [7]. Accordingly, the best material law embodies a fibril-reinforced, fluid-saturated composite and emulates the microstructure of the tissue [8]. A nonlinear fibril-reinforced model has reasonably described the transient behavior of articular cartilage observed in unconfined compression, e.g. the high spike of reaction force in displacement controlled experiments [9]. Experimental data have shown the dependency of the transient stress and stiffness of articular cartilage on strain and strain rate [10, 11] which may be explained by the interplay of fibril reinforcement and fluid pressurization in the nonlinear fibril-reinforced model [12].

Confined compression, unconfined compression, and indentation tests have broadly been employed to study the compressive behavior and lateral expansion of cartilage and curve-fit material model parameters [13, 14]. The in-situ indentation, nevertheless, can simulate a more realistic compression behavior of cartilage inside a joint than the other two in vitro compression tests. Furthermore, it was demonstrated that, depending on the indenter size, Young's modulus obtained from indentation tests was greater than the values from unconfined or confined compression tests [15]. An indentation creep experiment was conducted on young bovine femoral condylar cartilage in-situ to obtain the biphasic model aggregate modulus, Poisson's ratio and permeability with a numerical algorithm [16]. As a significant finding of this study, the Poisson's ratios of cartilage obtained were noticeably smaller than the values reported from the tests performed on excised osteochondral plugs. The asymptotic response of the indentation test, or the equilibrium response in creep and relaxation, is normally required in most curve-fitting procedures, which often requires long durations for experiments. A principal component analysis was thus developed to predict the full indentation creep curve based on the first few minutes' deformation history [17]. Less than 5% difference between the biphasic curve-fitting data and the experimental values was observed for the mechanical properties of bovine knee and porcine temporomandibular joint (TMJ) condylar cartilages. Micro-indentation technique was also applied on a murine articular cartilage for detecting altered cartilage mechanics in small animal models of OA [18]. This indentation technique was also performed on porcine TMJ condylar cartilage at multiple regions in creep tests and showed that the osmotic pressure associated with fixed charge density in PGs significantly increased the tissue's apparent stiffness [19].

Finite element (FE) simulation is a ubiquitous precious tool for investigating functional relationships between structure and tissue properties of articular cartilage and for studying stress and strain distributions within healthy or injured knee joints. At the tissue level, many attempts have been carried out to detail the unique mechanical behavior of articular cartilage stemming from its multiphasic nature, flow-dependent and flow-independent viscoelasticity and material anisotropy [20–22]. Furthermore, the effect of cartilage defect and OA on the altered mechanical response and fluid flow properties of the tissue has been documented [23, 24]. Nevertheless, the in vivo interaction between various tissues, such as contact between cartilages and menisci, introduces exceeding nonlinearities and complexities in numerical modeling, although emulates more realistic kinetics and kinematics of the joint. On this account, 3D anatomically accurate FE knee models have been developed in the past decades from computed tomography or magnetic resonance imaging (MRI) [25–27]. Implementing 'split lines' aligned according to principal stress directions [28] in an FE model of human knee joint demonstrated their biomechanical role of protecting the cartilage by limiting the deformation to the area of higher cartilage thickness [29]. Cadaveric knee models were employed for different degrees of malalignment, tibiofemoral kinematics, and articular cartilage pressure in static, quasi-static, and dynamic loadings with the application of diagnosing potential knee injury mechanisms and subject-specific surgical planning for OA [30, 31]. The development of knee OA due to cartilage defect, proteoglycan loss, and collagen degeneration was simulated by computational modeling with a degeneration algorithm [32]. Excessive levels of tissue stresses predominantly caused the degeneration of the collagen fibril network, while excessive levels of tissue strains mainly resulted in a decrease in PG content and an increase in permeability [32, 33]. Covering vast simulation processes and modeling of knee joint biomechanics in the literature, a standardized modeling and simulation workflow was proposed [34].

Despite the outstanding advances over the last years, further developments in modeling are required for a practical and pragmatic knee joint model of clinical use. In most published simulations and modeling, the material properties of cartilage are uniform or at most depth-dependent for accounting the collagen arrangement. The major tissue changes occurring in OA, such as the variation of proteoglycan content, fluid flow, and collagen fibril network, may be modeled with the site-specific material properties of cartilage determined from the indentation test. Therefore, the objective of this study was to develop a methodology to obtain the site-specific mechanical properties of knee cartilages and implement them in a three-dimensional computational porcine knee joint model. The site-specific properties may give out a more realistic condition of the initiation and development of OA.

2 Methods

Owing to the availability and close anatomy to human knee joints, porcine stifle joints were used for generating a 3D model and performing indentation tests on articular cartilage [35]. MRI scans were used to build the knee geometry thanks to its high contrast for soft tissues. MIMICS was employed to segment all tissues. Automated indentation testing and thickness mapping were performed on intact femoral and tibial cartilages. The measurements from different sites were then used to fit the site-specific material parameters of the fibril-reinforced model. Eventually, a finite element analysis of the reconstructed joint was performed with the site-specific cartilage properties determined from the indentation tests.

2.1 3D Reconstruction of Porcine Knee Joints

Fresh porcine knee (stifle) joints, with preserved femoral heads and partially retained ankle joints to keep the tibia and fibula together, were purchased within 24 h of the slaughter of the animals (Red Deer Lake Meat Processing, Calgary, Canada). These joints were kept in sealed bags, hydrated with Phosphate Buffered Solution (PBS) and stored at 4 °C before imaging in 1–6 days.

MRI was conducted using a GE Optima MR430s 1.5T unit (GE Healthcare, Illinois, United States) at the Centre for Mobility and Joint Health, University of Calgary. With minimal straightening of the joint, a knee specimen was placed in a 123 mm coil then inserted into the magnetic unit (Fig. 1a). The knees were imaged using a couple of sequences during the testing stage, e.g. two-dimensional gradient echo in the sagittal plane (SAG GRE 2D) and sagittal proton density fast-spin echo (SAG PD FSE). The images obtained by SAG GRE 2D proved to be the best for reconstruction, as all other sequences were unable to capture tri-planar images with good pixel resolution (Fig. 1b, c). The images were produced with a slice thickness of 1.4 mm, field of view of 160 mm, spacing of 0 mm, and matrix of 512 × 512 pixels.

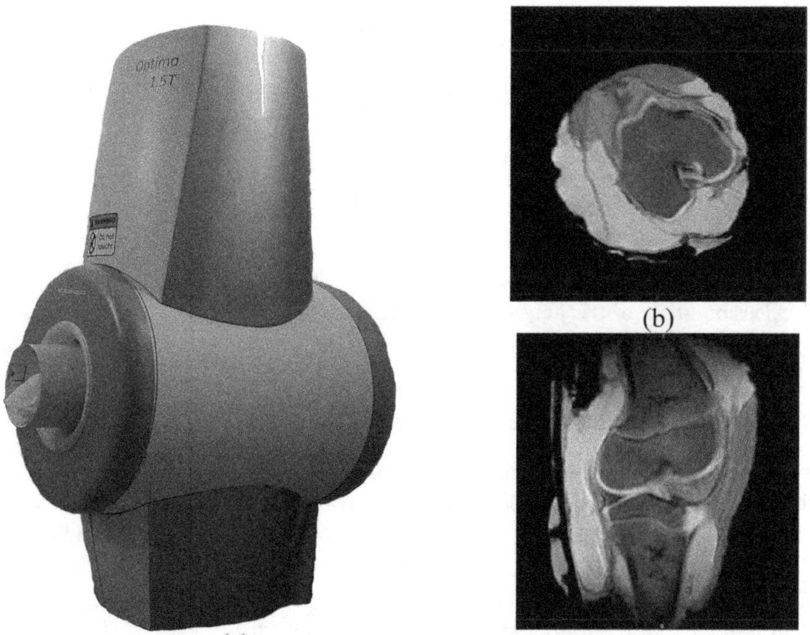

Fig. 1. MRI scan of a porcine knee joint. The MRI was conducted using a GE Optima MR430s 1.5T unit in a 123 mm coil (a). The images in the SAG GRE 2D sequence in the transverse (b) and sagittal (c) planes demonstrated the best resolution for reconstruction.

The MR images were processed and compiled using the Materialize Interactive Medical Image Control System (MIMICS, Materialize NV, Leuven, Belgium). Articular cartilages, menisci, bones, and ligaments were segmented to produce 3D models for meshing and subsequent finite element analyses.

MIMICS provides a tool to create masks which are threshold filters that allow segmentation of specific components based on the selected pixels. To ensure articulation between articular cartilage and subchondral bone, congruence was evaluated and enabled between the two masks associated with the cartilage and bone respectively. Model refinement for producing 3D parts was performed to eliminate any geometric errors. The final parts were observed to be reasonably smooth and ready to be meshed (Fig. 2).

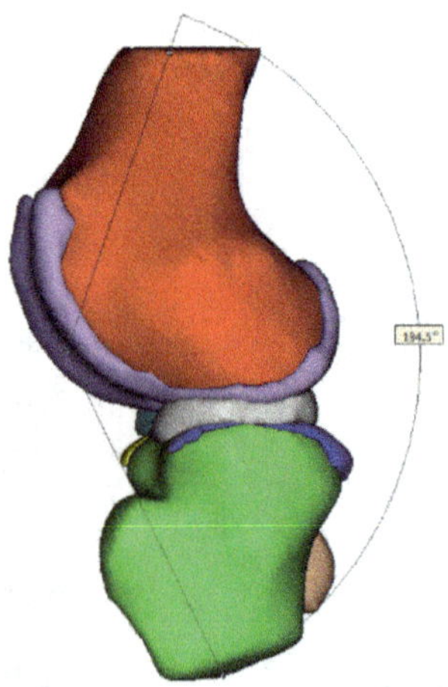

Fig. 2. Reconstructed model of a porcine knee joint with bones and cartilaginous tissues. The natural extension angle of the joint during standing of the animal was being preserved during imaging and reconstruction.

2.2 Indentation Testing on Intact Articular Cartilage

An automated indentation technique was adapted in this study to measure the topographical variation and mechanical properties of intact cartilage on the bone. The technique was developed earlier for measuring cartilage thickness and instantaneous modulus of porcine TMJ, murine and lapin knee joints [36–38]. The Mach-1 V500CSS (Biomomentum, Laval, Canada) was used for testing the mechanical response of articular cartilage using automated normal indentation testing (Fig. 3). A 17-N load cell was used, which has a resolution of 0.85 mN, and a data acquisition rate of 100 Hz.

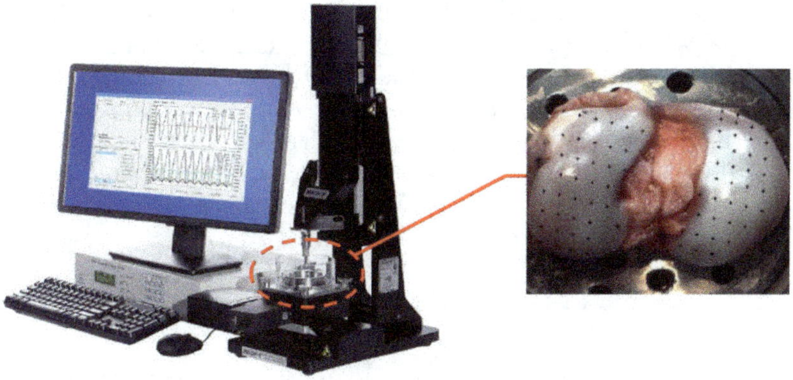

Fig. 3. Automated indentation system for cartilage. Mach-1 V500CSS tester used for the indentation was capable of one vertical and two horizontal stages for compression, tension and shear testing.

Fully intact femoral and tibial cartilages attached on bones were extracted from the joint after MR imaging. The cartilage-bone sample was securely held using a 6-screw holder and surrounded by a 3″ chamber wall. First, the Mapping Toolbox software of the tester was used to map the cartilage boundaries, assign test sites and define reference points. For example, the boundaries were determined from a live image of the sample from the integrated camera. A sequence of functions was then created in the Mach-1 Motion software, where parameters of the test were defined (such as indentation rate and amplitude) and locations of the test sites were imported from the mapping step. Phosphate Buffered Solution (PBS) was added to cover the highest point of the specimen by at least 1 cm and 15 min were given for the sample to equilibrate. The first indentation test was conducted using a 6.35-mm diameter spherical indenter, but a 2-mm indenter was used later for new joint tests. During indentation, the system first mapped the surface geometry of the test point and obtained the surface angle of the test point by locating the geometric positions (x, y, and z coordinates) of 4 vicinity points 0.3 mm from the test point in the +x, −x, +y, and −y directions. The indenter then approached and compressed the surface with a constant rate of 200 μm/s until a 200-μm dent in the tissue was reached. The indenter was held in place for 60 s as the force response was continually recorded at an acquisition rate of 100 Hz, then removed and repositioned to conduct testing of the next point. Indentation occurred in the vertical (z) direction, however by using the surface angle, the vertical force response was automatically converted to the component normal to the surface at the test point.

Cartilage thickness mapping was then conducted using the needle probe method, by attaching a Precision Glide BD 26G 3/8″ bevel needle to the load cell. The thickness at each test point was crucial to the determination of the mechanical properties of articular cartilage through indentation data fit. With the Mach-1 needle probe sequence, the needle pierced the sample at 0.5 mm/s until either a user-defined reaction force was reached (contact criterion), or a set distance of stage movement was reached. The set distance necessarily limits the needle penetration depth to prevent damage to the load cell. The contact criterion was set at 2 N since the reaction force of the needle does not reach this value until the needle has fully pierced the cartilage and reached the subchondral bone. When this occurred, the needle repositioned to return to the point that it first experienced a force response (at 2X load cell resolution) before retracting fully from the sample and moving to the next point. The vertical tissue thickness at each site could then be obtained with the force response against the position of the needle (Fig. 4). By evaluating the surface angle obtained through the normal indentation test, the product of vertical thickness and the cosine of the surface angle would provide the normal thickness at the test point.

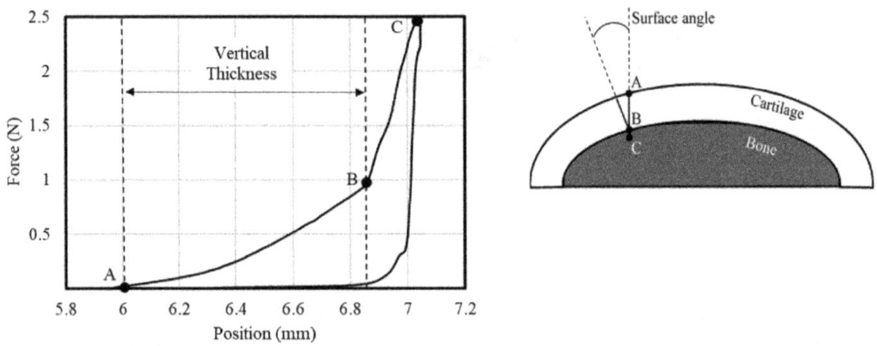

Fig. 4. Schematic representation of force vs. position curve in the needle test performed on articular cartilage in automated thickness mapping. Points along the thickness curve generated by the Mach-1 software indicate when (A) the needle contacts the cartilage surface, (B) the needle contacts the underlying bone, and (C) the reaction force exceeds the contact criteria.

2.3 Acquisition of Site-Specific Mechanical Properties from the Indentation Data

For the simplicity of implementing the mechanical properties in FE modeling, the indentation results were averaged for a small number of regions of cartilages (Fig. 5). The collagen network architecture, such as split line pattern in articular cartilages of porcine stifle joints, was analogous to that of humans [39]. The porcine cartilages were partitioned so that each region had approximately the same split line pattern as in human cartilages [28]. Furthermore, the recorded reaction and thickness at each region were similar.

An FE model was created in ABAQUS to emulate the indentation test on each point (Fig. 6). In order to speed up the simulations for the large number of indentation sites, an axisymmetric cylindrical cartilage disk was assumed to simulate the test and determine the tissue properties at each indentation site. The bottom surface of the cartilage was assumed to be completely fixed to the subchondral bone beneath. The fluid could exude out of the cartilage via the free surfaces. During the indentation, the fluid could either discharge from the top surface of the cartilage or flow across the surrounding tissue. It was assumed that at distances far from the pressurized zone, the tissue was not affected by the indentation. As such, the radius of the cylinder was taken to be 3 mm which was large enough to assume the zero fluid pressure condition on the outer surface of the cartilage (Fig. 6).

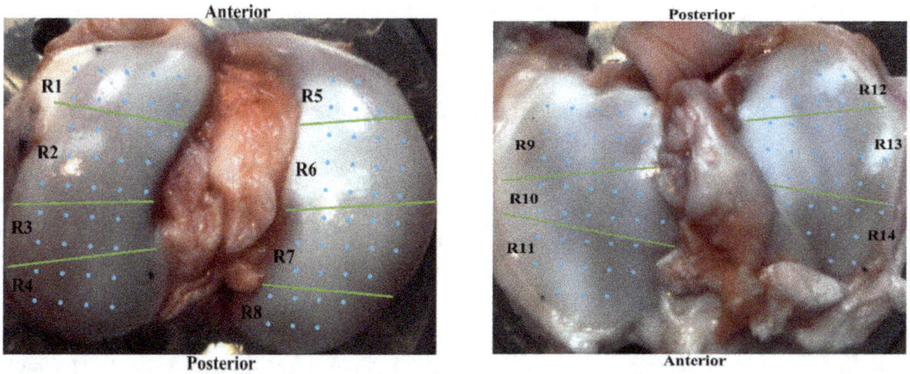

(a) (b)

Fig. 5. Divided regions with the same average of cartilage thickness and reaction force during the indentation test on a right porcine knee cartilage: (a) Femoral cartilage, inferior view; medial condyle on the right. (b) Tibial cartilages, superior view; medial plateau on the right. The divided regions were based on the cartilage split line pattern and the proximity of the present measurement.

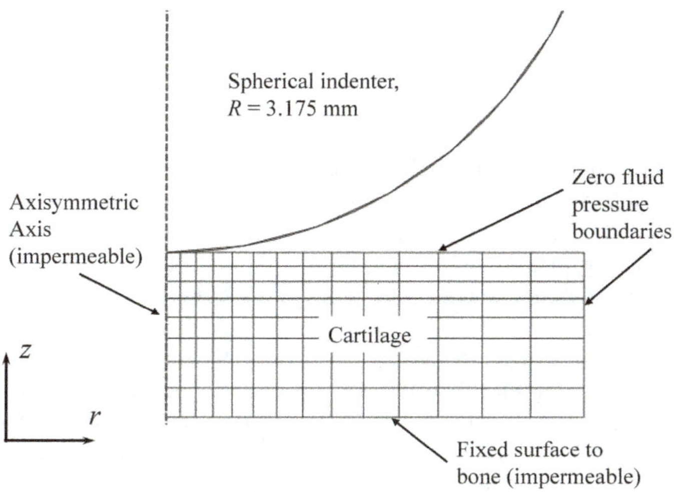

Fig. 6. Finite element model used to extract the mechanical properties at each indentation site. The cartilage tested under the indenter was assumed as a cylindrical plug and hence, an axisymmetric modeling was possible. The spherical indenter was rigid. A biased sizing of the cartilage plug was considered to apply finer mesh size near the indenter. The bottom surface and axisymmetric edge were impermeable, while the fluid could discharge from the other surfaces. The bottom surface was fully restrained to the bone and the horizontal displacement of the axisymmetric edge was zero.

The tissue was modelled with 4-node axisymmetric full integration elements including pore pressure, i.e. the element CAX4P in ABAQUS. The linear element was selected for faster simulations, while no noticeable difference was observed between the results of quadratic and linear elements for the mesh size considered in the model. A non-uniform distribution of elements along all edges was considered by defining the size of the coarsest and finest elements along the edges. The ratios of elements' dimensions with respect to cartilage dimensions varied between 0.08 to 0.12 through the thickness direction and 0.08 to 0.25 along the radius, with the lower ratios being closer to the indenter. As such, the total number of elements for each indentation site varied based on the tissue thickness, and the mesh was convergent for a variety of thicknesses. The indenter was modeled as a rigid part with 2-node linear axisymmetric rigid links in planar geometries which are RAX2 elements in ABAQUS. The same displacement magnitude and rate used in the indentation test were applied in the FE simulation. The fibril-reinforced material model was employed [40]. Three different phases were considered in this model: nonfibrillar PG matrix, collagen fibrillar network and synovial fluid. The isotropic nonfibrillar porous matrix requires Young's modulus, E_{m}, and Poisson's ratio, v_{m}. The hydraulic permeability, $k\left(\mathrm{mm}^4/\mathrm{N}\cdot\mathrm{s}\right)$, on the other hand, was direction dependent for the fitting procedure. So, four parameters were used to model the nonfibrillar matrix. The stress of the fibrillar network in the n direction, which can be either x, y or z (r, θ or z in the axisymmetric model), was

$$\sigma_{fn}(t) = \int_0^t \left(1 + \sum\nolimits_{m=1}^{3} g_{mn} exp\left(-\frac{t-\tau}{\lambda_{mn}}\right)\right) \frac{\partial\left(A_n\varepsilon_n + B_n\varepsilon_n^2\right)}{\partial\varepsilon_n}\frac{\partial\varepsilon_n}{\partial\tau}d\tau \tag{1}$$

Hence, the material model had 8 distinct parameters for each direction of the fibrillar network. For example, in the x-direction, there are two tensile properties, A_x and B_x and six reduced relaxation coefficients, $g_{1x}, \lambda_{1x}, g_{2x}, \lambda_{2x}, g_{3x}$, and λ_{3x}. In order to reduce the number of parameters to simplify the data fit during this preliminary study, the reduced relaxation coefficients λ_{mn} were selected from the literature $(\lambda_1 = 10, \lambda_2 = 100, \lambda_3 = 1000)$ [40]. Furthermore, the elastic properties, A_n and B_n, in the y and z-directions were set to be 0.3 and 0.9 times that of the x-direction, respectively, for the femoral and tibial cartilages (these values will be tested and may be adjusted later). Therefore, the total number of material parameters for the data fit decreased to nine: four for the nonfibrillar porous matrix mentioned above and five for the fibrillar network after simplification. The x-direction corresponded to the fiber orientation in the 3D joint modeling to be introduced in the next subsection.

In order to adopt the variation of the cartilage thickness and parameters, a script was developed in Python to fit the parameters and find the best match between the FE results and experimental data (Fig. 7). A set of test simulations was run to attain the range of variables and the sensitivity of each one on the results. The parameters of both nonfibrillar and fibrillar matrices had a significant effect on the maximum reaction force, whereas the effects of E_m and permeability (k_x) were more noticeable on the relaxation response. Initial values were set for all parameters of the material model, which were considered sufficient to obtain a reaction force greater than what was recorded during the experiment. For example, $E_m = 0.16, v_m = 0.38, A_x = 0.36, B_x = 360, g_1 = 0.6, g_2 = 0.031$, and $g_3 = 0.212$ were selected as initial values for region R1. The increment sizes for different parameters during iterations were selected according to their sensitivity found during the test simulations. A 5% error for the reaction forces was defined for the iteration to stop and report the parameters. Owing to the considerable effect of permeability on the results and convergence of the simulations, the values of permeability were assigned for each region within a reasonable range that was determined during the test simulations.

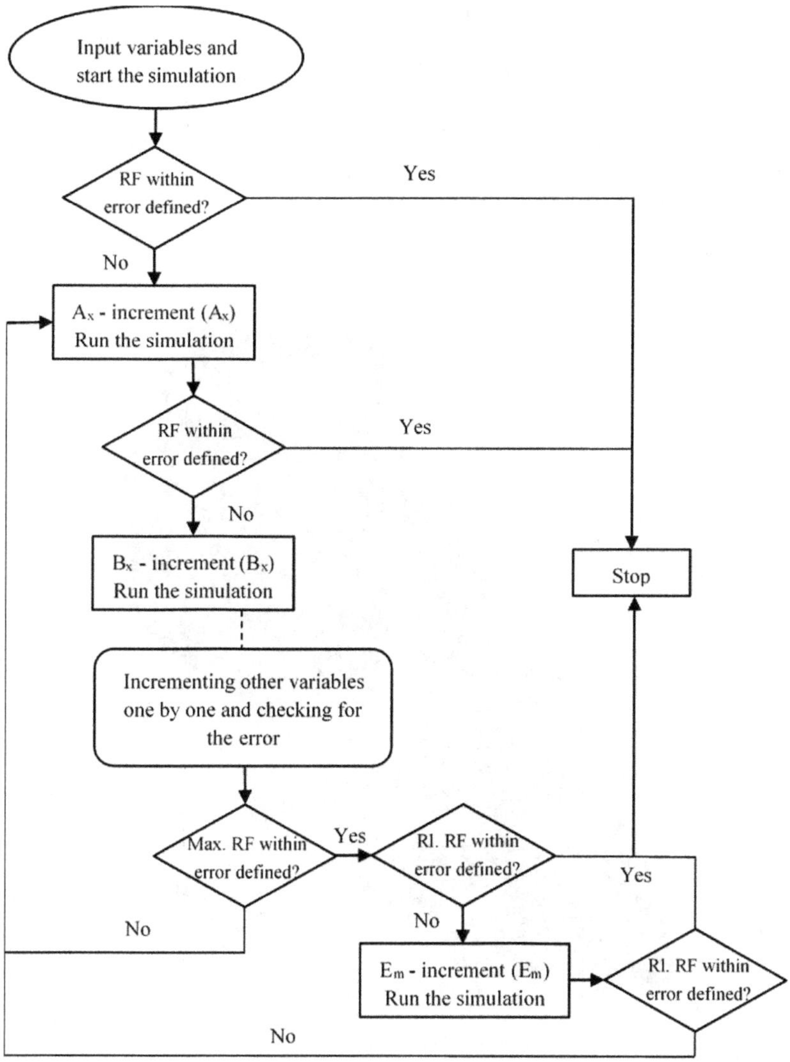

Fig. 7. Flowchart of the script for fitting the material model parameters with the indentation test results. RF stands for reaction force which could be either maximum (Max.) or relaxed (Rl.) at 60 s (equilibrium was not achieved).

2.4 Finite Element Simulation of the Knee Joint

To implement the indentation results in a whole joint model, a total meniscectomized knee joint without ligaments was constructed as the first step of joint modeling. After importing the reconstructed porcine knee joint to ABAQUS (Simulia, Providence, USA), soft tissues were partitioned into smaller pieces for better meshing. Femoral and tibial cartilages were modeled as fibril-reinforced fluid-saturated composites, while the bones were considered as rigid bodies due to their greater stiffness as compared to soft tissues. It has been reported that the rigid body assumption for bones changed the contact stresses by less than 2% [41]. As such, the bones were treated as rigid hollow bodies with their preserved external boundaries so that soft tissues could be accurately attached to them and the computing time could be somewhat reduced. In this study, soft tissues were discretized with a combination of hexahedral and tetrahedral elements to avoid distortion of elements at nonuniform regions. An 8-node hexahedron, trilinear displacement, trilinear pore pressure (C3D8P) and a 4-node linear tetrahedron, coupled displacement-pore pressure (C3D4P) were selected from the ABAQUS element library. Bone surfaces were discretized using 3-node 3D rigid triangular elements of type R3D3. To probe the effect of mesh sensitivity on the convergence of results, two

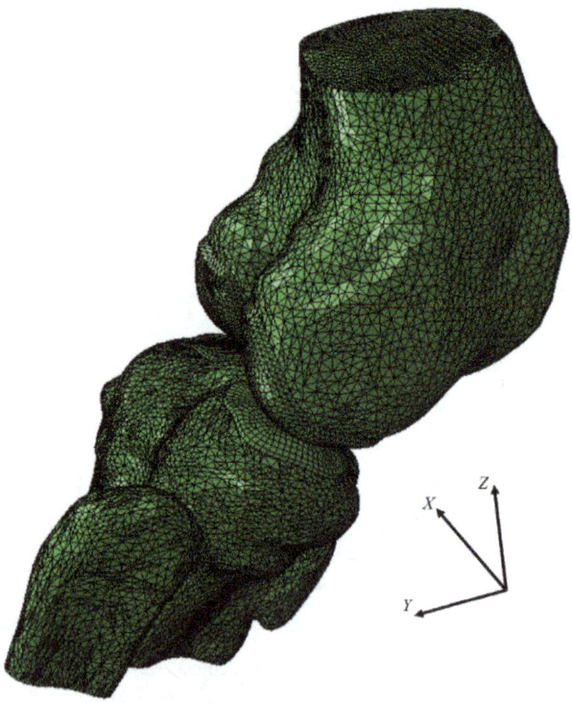

Fig. 8. Finite element model of a right porcine knee joint, including femur, tibia, fibula, and articular cartilages. Tetrahedral and hexahedral pore pressure continuum elements were applied to cartilages whereas the bone surfaces were discretized using 3-node rigid triangular elements. The distal femur was along the z direction and a 45.5° natural extension angle of the joint was assumed, which simulated the standing position of the animal.

additional meshes were tested, which had 1.1 and 1.18 times of the number of elements of the original mesh. It was found that the contact distribution in each tissue was almost unchanged after the mesh refinement and the change in maximum contact pressure was virtually negligible, with less than 2.5% change for the finest mesh. The final mesh chosen (Fig. 8) had 12162, 18393, 10889, 3438 elements for femur, tibia and fibula, femoral cartilage, and tibial cartilage, respectively.

For the meniscectomized knee, only two contact pairs were necessarily defined between femoral and tibial cartilages on medial and lateral sides. The nonlinear surface to surface contact discretization based on a hard contact constraint and finite sliding was chosen. The linear penalty method was used for the contact constraint enforcement. The friction coefficient on the articular surfaces was 0.087 [42]. The material model was implemented by a user-defined material subroutine developed previously [40]. An average 60% porosity consistent with the literature was used for all cartilages [43]. Fluid flow was allowed across the free surfaces of cartilages using zero pore pressure conditions. The global frame of reference was oriented to approximate the anatomical directions: the x-axis was the frontal axis and the z-axis was in the distal-femoral direction. The knee joint was assumed to be in the natural extension of standing and tibia was restrained in all six degrees of freedom.

To study the time-dependent response of the knee joint, a compressive displacement of 800 μm was applied to the femur in the z-direction. Different compression rates, 0 (static), 10, 100, and 200 μm/s, were applied in the loading phase. The soil consolidation procedure in ABAQUS/Standard was used to solve the time-dependent problem with an implicit scheme. Geometrical nonlinearities in large deformations were used.

3 Results

The indentation testing had only been simulated for one joint at the time this manuscript was written. The maximum reaction force was recorded for each location after a 200-μm compression was applied at 200 μm/s. The "relaxed force" refers to the load after holding the indenter at its 200-μm position for 60 s when the tissue was partially relaxed, which was not obtained for the tibial cartilage of the first joint tested. The medial femoral condyle had a larger thickness and slightly greater relaxed reaction force than the lateral one (Table 1); nonetheless, the maximum reaction force was almost equal for both condyles. Overall, greater reaction forces were recorded for the lateral tibial plateau compared to the medial one, with the thickness being quite the same for most regions investigated. The average maximum force in the tibial cartilage was appreciably higher than in the femoral cartilage; the average thickness was lower for the tibial cartilage. No obvious correlation was observed between the reaction forces and cartilage thickness in different regions, nor between the maximum and relaxed forces in the femoral cartilage.

Table 1. Measured thicknesses of cartilage and reaction forces during the indentation test. Regions R1 to R8 are in the femoral cartilage (Fig. 5a); and Regions R9 to R14 in the tibial cartilage (Fig. 5b). The relaxed loads were not measured for the tibial cartilage of the joint tested.

Region	No. of indentation points	Average maximum reaction force (N)	Average relaxed reaction force (N)	Average thickness of cartilage (mm)
R1	10	0.77	0.11	1.39
R2	14	0.98	0.1	1.89
R3	8	1.18	0.09	1.72
R4	8	1.06	0.11	1.55
R5	5	1.27	0.25	1.81
R6	12	0.97	0.13	2.21
R7	13	1.07	0.12	2.03
R8	7	0.91	0.2	2.03
R9	14	8.04	–	1.04
R10	11	12.26	–	0.92
R11	13	11.35	–	0.96
R12	9	7.17	–	0.65
R13	16	7.3	–	1.03
R14	10	10.7	–	0.88

The fitted parameters may not be unique for the material model considered. Nevertheless, the obtained values show the best match with the experimental data (Table 2). The Young's modulus of the nonfibrillar matrix and the two tensile properties of the fibrillar network (A_x, B_x) were considerably lower for the femoral cartilage compared to the tibial cartilage. Femoral cartilage had higher permeability in the x direction as compared to tibial cartilage, which was compatible with the literature [44].

The force-compression relationship of the joint was determined by both compression-magnitude and compression-rate in the FE simulations of the knee joint (Fig. 9). The reaction force was more sensitive to the compression rate when the compression was large. A strong nonlinear and rate-dependent response was also seen: the load response was almost linear at a static compression and became strongly nonlinear at a fast compression. The same trend was recorded in a set of experiments on porcine knee joints for determining the loading-rate dependence, creep and relaxation of normal, dehydrated and meniscectomized joints [45] which partially verified our finite element model (Fig. 9).

The joint response depended on site-specific material properties (Fig. 10). The joint was compressed 800 μm at a constant rate of 200 μm/s and then relaxed for 52 s, the total simulation being 60 s. The maximum and relaxed reaction forces decreased by about 9% and 15%, respectively when uniform material properties were considered for each cartilage (Fig. 10a). The contact pressure, however, slightly decreased when the region-dependent properties were considered for the lateral tibial plateau (Fig. 10b).

Table 2. Site-specific material parameters of the fibril-reinforced model obtained from data fit. The error shows the deviation of the FE results from the indentation results (Table 1). A_y, B_y were 0.3 and 0.9 times of the corresponding parameters in the x direction for the femoral and tibial cartilages, respectively. The units of E_m and k are, MPa and $mm^4/N \cdot s$, respectively

Region	E_m	v_m	k_x	k_y	A_x	B_x	g_1	g_2	g_3	%Err. of Max. RF	%Err. of Rl. RF
R1	0.11	0.31	0.013	0.002	0.21	280	0.54	0.029	0.126	0.52	2.36
R2	0.13	0.25	0.01	0.003	1.53	462	0.92	0.031	0.341	3.06	2.33
R3	0.11	0.26	0.013	0.003	1.95	481	0.93	0.032	0.547	4.24	4.84
R4	0.12	0.28	0.01	0.003	1.48	323	0.89	0.046	0.439	1.89	2.94
R5	0.26	0.29	0.003	0.001	1.86	311	0.77	0.041	0.361	1.57	2.80
R6	0.19	0.33	0.008	0.003	2.09	426	0.86	0.038	0.307	0.82	3.23
R7	0.12	0.26	0.013	0.003	2.35	521	0.9	0.021	0.502	4.67	2.50
R8	0.18	0.29	0.002	0.001	1.03	288	0.77	0.028	0.361	1.10	3.03
R9	0.79	0.41	0.003	0.002	8.97	694	0.98	0.028	0.757	0.75	–
R10	1.06	0.41	0.002	0.001	9.62	783	0.96	0.037	0.913	1.88	–
R11	0.99	0.42	0.002	0.001	9.31	720	0.89	0.043	0.826	1.32	–
R12	0.34	0.39	0.002	0.001	2.93	354	0.71	0.035	0.342	1.39	–
R13	0.63	0.37	0.002	0.001	6.36	413	0.84	0.031	0.585	1.23	–
R14	0.79	0.41	0.002	0.001	7.91	653	0.92	0.031	0.666	1.00	–

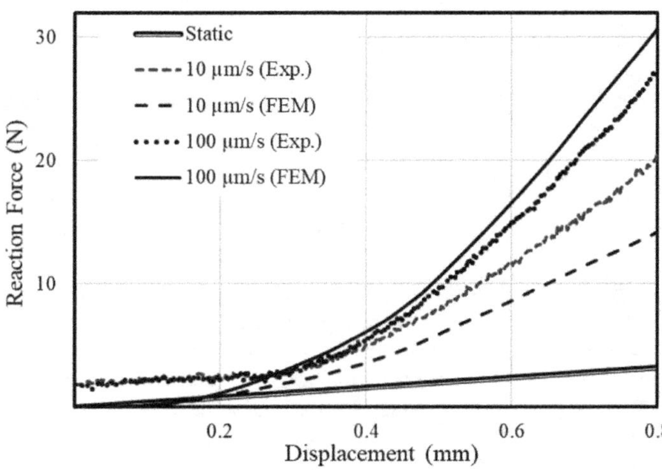

Fig. 9. Predicted and measured reaction force in a meniscectomized porcine knee joint as a function of compression magnitude and compression rate. In the legend, FEM stands for finite element method, while Exp. stands for experimental data. The experiments were performed on a wide assortment of fresh porcine knee joints [45]. The joint for the indentation test was different from the joints used in the whole-joint reaction measurement.

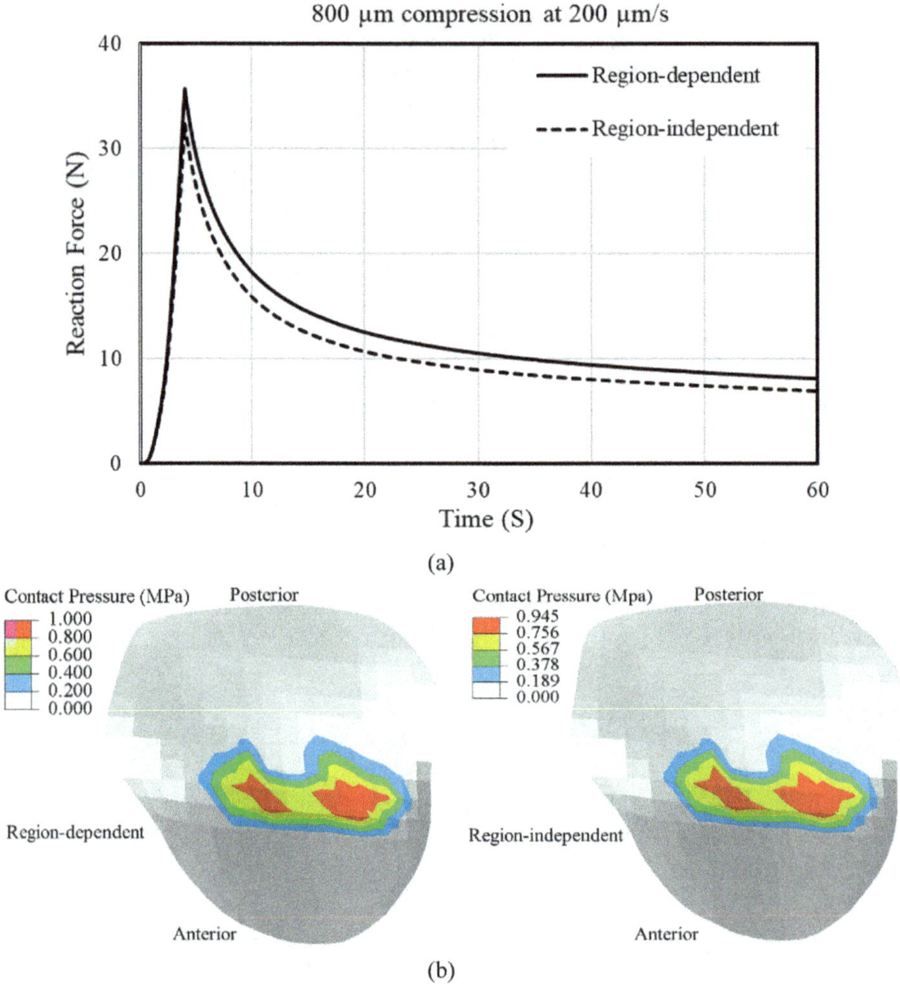

Fig. 10. Sensitivity of FE results to site-specific material properties. The "Region-dependent" result was obtained using different material parameters for different regions of cartilages (Table 2). The joint was compressed for 800 µm and partially relaxed at 60 s of loading. The resultant reaction force was obtained for the whole joint (a). The contact pressure distributions of the lateral tibial plateau at 60 s is shown for the region-dependent and region-independent material properties (b).

4 Discussion

A methodology was developed to implement the site-specific material properties of articular cartilages in a 3D FE model of the porcine knee joint. Indentation tests were carried out on the femoral and tibial cartilages of a right porcine knee to map the tissue properties across the knee joint. MRI images were segmented to reconstruct an FE model of the porcine joint for meshing and modeling. Only limited experimental data and computational results were available for the present analysis. The material properties extracted from a single joint requires validation and the numerical results were mainly presented here for the demonstration of the methodology. Ongoing work includes further testing on the data fitting procedure as well as further experiments and FE analyses.

The indentation results on porcine knee femoral and tibial cartilages revealed extensively varied results for different sites (Table 1). This finding is consistent with creep indentation tests on porcine TMJ at anterior, posterior, central, lateral, and medial regions, which indicated significant differences in the aggregate modulus among the regions [19]. The variations in the reaction forces over regions were not mainly due to the cartilage thickness. For instance, different forces were measured in regions R7 and R8 with essentially the same thickness. One source of difference was because of the microarchitecture of articular cartilage which contributed to the varied material and structural properties. The collagen fiber orientation was shown to have an appreciable effect on the mechanics of articular cartilage [46]. While higher reaction forces could be partially attributed to the smaller thickness of tibial cartilage, the different properties of the constituents could still set the stage for different results. Polarized light microscopy (PLM), digital densitometry and Fourier transform infrared imaging showed the site-specific depth-wise profiles of collagen orientation angle, proteoglycan content and collagen content in a rabbit knee joint [47]: collagen orientation within 1–69% cartilage depth were 12% different between femoral and tibial cartilages; PG content within 1–82% cartilage depth was 10% higher in the femoral compared to the tibial cartilage. Any disease and degeneration of cartilage can alter the microstructure of the tissue and consequently the mechanical response in the indentation test. For instance, microscopic MRI and PLM have verified quantitative changes in collagen fiber architecture in early OA of canine tibial cartilage, since the change of collagen fibrillar orientations in cartilage due to cartilage defect and OA can cause variations in, for example, MRI intensity [48]. Only healthy joints were investigated in the present study. However, it may be interesting to further test degenerated joints to quantify alterations in cartilage properties from degeneration.

The site-specific material properties were obtained for the fibril-reinforced cartilage model with indentation data fit. The fibrillar network has a strong instantaneous response because of great tensile stiffness of collagen fibers. As such, a greater fibrillar modulus was seen for the regions with a greater maximum reaction force (Table 2). The tibial cartilage also had a stiffer PG matrix than the femoral cartilage. Although more indentation data, and possibly an improved fitting procedure, are required to determine typical tissue properties, the present tissue parameters from our first data fit were in agreement with earlier investigations which demonstrated higher modulus of elasticity, tensile and compressive strength for the porcine knee tibial cartilage as compared to the femoral cartilage [44].

The preliminary finite element simulation of the knee joint considering the site-specific material properties showed substantial compression-rate dependence of the load response of the tissue on different compression rates at physiologically reasonable deformation (Fig. 9). Two major sources may explain this phenomenon: flow-dependent [49] and flow-independent properties of cartilage [50]. The contribution of the intrinsic viscoelastic nature of cartilages to augmenting the rate-dependence was shown in an earlier study considering the tensile loading of bovine knee cartilage [51]. During tension, all the burden is carried by the solid matrix, mainly by collagen fibers. During compression through the cartilage thickness direction, the tissue experiences lateral expansion or tension in the direction tangential to the articular surface. Hence, during joint compression, both the rate of fluid exudation and the time-dependent behavior of the solid matrix underlie the compression-rate dependence of cartilaginous tissues. This feature requires an accurate and realistic material model to approximate the rate-dependent response of biological tissues with small and large deformations.

The effect of region-dependent material properties on the predicted joint response was of interest in the further development of patient-specific joint modeling. The poromechanical response obtained so far indicated the necessity of implementing site-specific material properties in a joint model. The significance of implementing site-specific tissue properties in the joint modeling is expected to be demonstrated with ongoing measurements and modeling.

Briefly, a methodology was applied to measure the site-specific mechanical properties of porcine knee articular cartilage with indentation testing and implement the measured properties in a finite element model of a porcine knee joint. The different reaction forces measured at different sites of cartilage were attributed to the microstructure of cartilage. The regions with greater maximum reaction force had higher tensile properties of the collagen fibers. The tibial cartilage had stiffer collagen fibers in all directions and a stronger PG matrix compared to the femoral cartilage. The joint reaction force was sensitive to both compression-magnitude and compression-rate, which is consistent with the observations from mechanical tests on cartilage explants and the ex-vivo testing of porcine knee joints [45, 51]. An accurate implementation of site-specific properties of tissues may possibly impact the predicted poromechanical response of the knee joint. This preliminary study requires further experimental and mathematical development though. For example, the properties of menisci need to be determined with a modified indentation setup that is suitable for the application of indentation on the steep surface of the tissue. A successful simulation with the porcine joints may be extended to human knee joints for considering a variety of physiological loading conditions. The current modeling may be applied to injured and defected cartilage to study the effect of localized cartilage degeneration on the poromechanical response of knee joints. Since site-specific properties were believed to reveal a deeper perception of the regions prone to the initiation of osteoarthritis, a comprehensive measurement and implementation of the properties in a joint model may deliver a better understanding of cartilage degeneration leading to osteoarthritis.

Acknowledgments. The present study was supported by the Natural Sciences and Engineering Research Council of Canada. The first indentation test of a porcine joint was performed at Biomomentum (Quebec, Canada) with a Mach-1 tester (the original photos for Figs. 3 and 5 were taken at Biomomentum and modified for using here with permission). All subsequent indentation tests and joint reconstructions were performed using Dr. Brent Edwards' facility at the Human Performance Lab, where Andrew Sawatsky trained Daniel Tang for the use of Mach-1. The MRI images were obtained at the Centre for Mobility and Joint Health, Dr. Steven Boyd's lab at the University of Calgary.

References

1. Sophia-Fox, A.J., Bedi, A., Rodeo, S.A.: The basic science of articular cartilage: structure, composition, and function. Sport. Health Multidiscip. Approach **1**, 461–468 (2009). https://doi.org/10.1177/1941738109350438
2. Troken, A.J., Mao, J.J., Marion, N.W., Wan, L.Q., Mow, V.C.: Cartilage and meniscus, properties of. In: Encyclopedia of Medical Devices and Instrumentation. Wiley, Hoboken (2006)
3. Wang, Y., Ding, C., Wluka, A.E., Davis, S., Ebeling, P.R., Jones, G., Cicuttini, F.M.: Factors affecting progression of knee cartilage defects in normal subjects over 2 years. Rheumatology **45**, 79–84 (2006). https://doi.org/10.1093/rheumatology/kei108
4. Korhonen, R.K., Julkunen, P., Li, L.P., van Donkelaar, C.C.: Computational models of articular cartilage. Comput. Math. Methods Med. **2013**, 254507 (2013). https://doi.org/10.1155/2013/254507
5. Huber, M., Trattnig, S., Lintner, F.: Anatomy, biochemistry, and physiology of articular cartilage. Invest. Radiol. **35**, 573–580 (2000)
6. Bhosale, A.M., Richardson, J.B.: Articular cartilage: structure, injuries and review of management. Br. Med. Bull. **87**, 77–95 (2008). https://doi.org/10.1093/bmb/ldn025
7. Mow, V.C., Huiskes, R.: Basic Orthopaedic Biomechanics & Mechano-biology. Lippincott Williams & Wilkins, Philadelphia (2005)
8. Taylor, Z.A., Miller, K.: Constitutive modeling of cartilaginous tissues: a review. J. Appl. Biomech. **22**, 212–229 (2006)
9. Li, L.P., Soulhat, J., Buschmann, M.D., Shirazi-Adl, A.: Nonlinear analysis of cartilage in unconfined ramp compression using a fibril reinforced poroelastic model. Clin. Biomech. **14**, 673–682 (1999). https://doi.org/10.1016/S0268-0033(99)00013-3
10. Gao, L.L., Zhang, C.Q., Gao, H., Liu, Z.D., Xiao, P.P.: Depth and rate dependent mechanical behaviors for articular cartilage: experiments and theoretical predictions. Mater. Sci. Eng. C **38**, 244–251 (2014). https://doi.org/10.1016/J.MSEC.2014.02.009
11. Oloyede, A., Broom, N.: Stress-sharing between the fluid and solid components of articular cartilage under varying rates of compression. Connect. Tissue Res. **30**, 127–141 (1993)
12. Li, L.P., Herzog, W.: Strain-rate dependence of cartilage stiffness in unconfined compression: the role of fibril reinforcement versus tissue volume change in fluid pressurization. J. Biomech. **37**, 375–382 (2004)
13. Boschetti, F., Pennati, G., Gervaso, F., Peretti, G.M., Dubini, G.: Biomechanical properties of human articular cartilage under compressive loads. Biorheology **41**, 159–166 (2004)
14. DiSilvestro, M.R., Suh, J.K.F.: A cross-validation of the biphasic poroviscoelastic model of articular cartilage in unconfined compression, indentation, and confined compression. J. Biomech. **34**, 519–525 (2001). https://doi.org/10.1016/S0021-9290(00)00224-4
15. Korhonen, R.K., Laasanen, M.S., Töyräs, J., Rieppo, J., Hirvonen, J., Helminen, H.J., Jurvelin, J.S.: Comparison of the equilibrium response of articular cartilage in unconfined compression, confined compression and indentation. J. Biomech. **35**, 903–909 (2002). https://doi.org/10.1016/S0021-9290(02)00052-0
16. Mow, V.C., Gibbs, M.C., Lai, W.M., Zhu, W.B., Athanasiou, K.A.: Biphasic indentation of articular cartilage—II. A numerical algorithm and an experimental study. J. Biomech. **22**, 853–861 (1989). https://doi.org/10.1016/0021-9290(89)90069-9
17. Chen, X., Zimmerman, B.K., Lu, X.L.: Determine the equilibrium mechanical properties of articular cartilage from the short-term indentation response. J. Biomech. **48**, 176–180 (2015). https://doi.org/10.1016/J.JBIOMECH.2014.10.036

18. Cao, L., Youn, I., Guilak, F., Setton, L.A.: Compressive properties of mouse articular cartilage determined in a novel micro-indentation test method and biphasic finite element model. J. Biomech. Eng. **128**, 766 (2006). https://doi.org/10.1115/1.2246237
19. Lu, X.L., Mow, V.C., Guo, X.E.: Proteoglycans and mechanical behavior of condylar cartilage. J. Dent. Res. **88**, 244–248 (2009). https://doi.org/10.1177/0022034508330432
20. Hosoda, N., Sakai, N., Sawae, Y., Murakami, T.: Finite element analyses of articular cartilage models considering depth-dependent elastic modulus and collagen fiber network. J. Biomech. Sci. Eng. **5**, 437–448 (2010). https://doi.org/10.1299/jbse.5.437
21. Klika, V., Gaffney, E.A., Chen, Y.C., Brown, C.P.: An overview of multiphase cartilage mechanical modelling and its role in understanding function and pathology. J. Mech. Behav. Biomed. Mater. **62**, 139–157 (2016). https://doi.org/10.1016/J.JMBBM.2016.04.032
22. DiSilvestro, M.R., Zhu, Q., Wong, M., Jurvelin, J.S., Suh, J.K.F.: Biphasic poroviscoelastic simulation of the unconfined compression of articular cartilage: I—simultaneous prediction of reaction force and lateral displacement. J. Biomech. Eng. **123**, 191 (2001). https://doi.org/10.1115/1.1351890
23. Richard, F., Villars, M., Thibaud, S.: Viscoelastic modeling and quantitative experimental characterization of normal and osteoarthritic human articular cartilage using indentation. J. Mech. Behav. Biomed. Mater. **24**, 41–52 (2013). https://doi.org/10.1016/J.JMBBM.2013.04.012
24. Mäkelä, J.T.A., Han, S.K., Herzog, W., Korhonen, R.K.: Very early osteoarthritis changes sensitively fluid flow properties of articular cartilage. J. Biomech. **48**, 3369–3376 (2015). https://doi.org/10.1016/J.JBIOMECH.2015.06.010
25. Bendjaballah, M., Shirazi-Adl, A., Zukor, D.: Biomechanics of the human knee joint in compression: reconstruction, mesh generation and finite element analysis. Knee **2**, 69–79 (1995)
26. Li, G., Gil, J., Kanamori, A., Woo, S.L.: A validated three-dimensional computational model of a human knee joint. J. Biomech. Eng. **121**, 657–662 (1999). https://doi.org/10.1115/1.2800871
27. Gu, K.B., Li, L.P.: A human knee joint model considering fluid pressure and fiber orientation in cartilages and menisci. Med. Eng. Phys. **33**, 497–503 (2011)
28. Below, S., Arnoczky, S.P., Dodds, J., Kooima, C., Walter, N.: The split-line pattern of the distal femur: a consideration in the orientation of autologous cartilage grafts. Arthrosc. J. Arthrosc. Relat. Surg. **18**, 613–617 (2002). https://doi.org/10.1053/JARS.2002.29877
29. Shim, V.B., Besier, T.F., Lloyd, D.G., Mithraratne, K., Fernandez, J.F.: The influence and biomechanical role of cartilage split line pattern on tibiofemoral cartilage stress distribution during the stance phase of gait. Biomech. Model. Mechanobiol. **15**, 195–204 (2016). https://doi.org/10.1007/s10237-015-0668-y
30. Kiapour, A., Kiapour, A.M., Kaul, V., Quatman, C.E., Wordeman, S.C., Hewett, T.E., Demetropoulos, C.K., Goel, V.K.: Finite element model of the knee for investigation of injury mechanisms: development and validation. J. Biomech. Eng. **136**, 011002 (2013). https://doi.org/10.1115/1.4025692
31. Mootanah, R., Imhauser, C.W., Reisse, F., Carpanen, D., Walker, R.W., Koff, M.F., Lenhoff, M.W., Rozbruch, S.R., Fragomen, A.T., Dewan, Z., Kirane, Y.M., Cheah, K., Dowell, J.K., Hillstrom, H.J.: Development and validation of a computational model of the knee joint for the evaluation of surgical treatments for osteoarthritis. Comput. Methods Biomech. Biomed. Eng. **17**, 1502–1517 (2014). https://doi.org/10.1080/10255842.2014.899588
32. Mononen, M.E., Tanska, P., Isaksson, H., Korhonen, R.K.: A novel method to simulate the progression of collagen degeneration of cartilage in the knee: data from the osteoarthritis initiative. Sci. Rep. **6**, 21415 (2016). https://doi.org/10.1038/srep21415

33. Mononen, M.E., Tanska, P., Isaksson, H., Korhonen, R.K.: New algorithm for simulation of proteoglycan loss and collagen degeneration in the knee joint: data from the osteoarthritis initiative. J. Orthop. Res. **36**, 1673–1683 (2018). https://doi.org/10.1002/jor.23811
34. Erdemir, A., Besier, T.F., Halloran, J.P., Imhauser, C.W., Laz, P.J., Morrison, T.M., Shelburne, K.B.: Deciphering the "art" in modeling and simulation of the knee joint: overall strategy. J. Biomech. Eng. **141**, 071002 (2019). https://doi.org/10.1115/1.4043346
35. Proffen, B.L., McElfresh, M., Fleming, B.C., Murray, M.M.: A comparative anatomical study of the human knee and six animal species. Knee **19**, 493–499 (2012). https://doi.org/10.1016/J.KNEE.2011.07.005
36. Lavoie, J.F., Sim, S., Quenneville, E., Garon, M., Moreau, A., Buschmann, M.D., Aubin, C. E.: Mapping articular cartilage biomechanical properties of normal and osteoarthritic mice using indentation. In: Osteoarthritis Research Society International, Seattle, WA, USA (2015)
37. Sim, S., Matuska, A., Garon, M., Quenneville, E., McFetridge, P., Buschmann, M.D.: Cartilage stiffness and thickness distributions revealed by an automated indentation technique in the temporomandibular joint. In: TMJ Bioengineering Conference, Barcelona, Spain (2016)
38. Moshtagh, P.R., Pouran, B., Korthagen, N.M., Zadpoor, A.A., Weinans, H.: Guidelines for an optimized indentation protocol for measurement of cartilage stiffness: the effects of spatial variation and indentation parameters. J. Biomech. **49**, 3602–3607 (2016). https://doi.org/10.1016/J.JBIOMECH.2016.09.020
39. Rieppo, J., Hyttinen, M.M., Halmesmaki, E., Ruotsalainen, H., Vasara, A., Kiviranta, I., Jurvelin, J.S., Helminen, H.J.: Changes in spatial collagen content and collagen network architecture in porcine articular cartilage during growth and maturation. Osteoarthr. Cartil. **17**, 448–455 (2009). https://doi.org/10.1016/J.JOCA.2008.09.004
40. Kazemi, M., Li, L.P.: A viscoelastic poromechanical model of the knee joint in large compression. Med. Eng. Phys. **36**, 998–1006 (2014). https://doi.org/10.1016/J.MEDENGPHY.2014.04.004
41. Donahue, T.L., Hull, M.L., Rashid, M.M., Jacobs, C.R.: A finite element model of the human knee joint for the study of tibio-femoral contact. J. Biomech. Eng. **124**, 273 (2002). https://doi.org/10.1115/1.1470171
42. Teeple, E., Fleming, B.C., Mechrefe, A.P., Crisco, J.J., Brady, M.F., Jay, G.D.: Frictional properties of Hartley guinea pig knees with and without proteolytic disruption of the articular surfaces. Osteoarthr. Cartil. **15**, 309–315 (2007). https://doi.org/10.1016/J.JOCA.2006.08.011
43. Lu, X.L., Miller, C., Quo, X.E., Van Mow, C.: A new correspondence principle for triphasic materials: determination of fixed charge density and porosity of articular cartilage by indentation. ASME-BED, Vail, Colorado (2005)
44. Pal, S.: Mechanical properties of biological materials. In: Design of Artificial Human Joints & Organs, pp. 23–40. Springer US, Boston (2014)
45. Rodriguez, M.L., Li, L.P.: Compression-rate-dependent nonlinear mechanics of normal and impaired porcine knee joints. BMC Musculoskelet. Disord. **18**, 1–10 (2017). https://doi.org/10.1186/s12891-017-1805-9
46. Meng, Q., An, S., Damion, R.A., Jin, Z., Wilcox, R., Fisher, J., Jones, A.: The effect of collagen fibril orientation on the biphasic mechanics of articular cartilage. J. Mech. Behav. Biomed. Mater. **65**, 439–453 (2017). https://doi.org/10.1016/J.JMBBM.2016.09.001
47. Ronkainen, A.P., Fick, J.M., Herzog, W., Korhonen, R.K.: Site-specific cell-tissue interactions in rabbit knee joint articular cartilage. J. Biomech. **49**, 2882–2890 (2016). https://doi.org/10.1016/J.JBIOMECH.2016.06.033

48. Alhadlaq, H.A., Xia, Y., Moody, J.B., Matyas, J.R.: Detecting structural changes in early experimental osteoarthritis of tibial cartilage by microscopic magnetic resonance imaging and polarised light microscopy. Ann. Rheum. Dis. **63**, 709–717 (2004). https://doi.org/10.1136/ARD.2003.011783
49. Oloyede, A., Flachsmann, R., Broom, N.: The dramatic influence of loading velocity on the compressive response of articular cartilage. Connect. Tissue Res. **27**, 211–224 (1992)
50. Hayes, W.C., Bodine, A.J.: Flow-independent viscoelastic properties of articular cartilage matrix. J. Biomech. **11**, 407–419 (1978). https://doi.org/10.1016/0021-9290(78)90075-1
51. Ahsanizadeh, S., Li, L.P.: Strain-rate-dependent non-linear tensile properties of the superficial zone of articular cartilage. Connect. Tissue Res. **56**, 469–476 (2015)

Precise Mean Axis of Rotation (MAR) Analysis for Clinical and Research Applications

Mayar Abbasi[1,2,3(✉)], Aslam H. Khan[3], Karim Bayanzay[4], Asmaa Rana[5], and Abdullah Mosabbir[6]

[1] McGill University, Montreal, QC, Canada
[2] Averna Technologies, Montreal, QC, Canada
[3] KKT Orthopedic Spine Center, Toronto, ON, Canada
mayar@kktspinetreatment.com
[4] Jersey Shore University, New Jersey, NJ, USA
[5] Wilfrid Laurier University, Waterloo, ON, Canada
[6] Baycrest Health Sciences, Toronto, ON, Canada

Abstract. The movement of individual cervical vertebrae from extension to flexion can be defined as a rotation about a certain axis point, known as the Mean Axis of Rotation (MAR). Researchers have used the position of the MAR to define normal and abnormal pathologies, making it an important indicator. However, the 2–3 h required to accurately calculate the MAR manually severely limits the usage of MAR in clinical and research environments. Multiple attempts have been made to automate the analysis, but none have been able to achieve the reliability of the manual method. We present here a redesigned semi-automatic MAR Analysis tool, which leverages advanced software architectures and computer vision methods to overcome previous limitations, and faithfully replicate the manual method. Initial testing of this MAR tool shows promise that the desired reliability can be achieved with this tool.

Keywords: Spine · Vertebra · Axis of rotation · Semi-automatic · Computer vision

1 Introduction

Mean Axis of Rotation (MAR) Analysis is a technique to measure the movement of the cervical vertebra as the subject moves from full extension to full flexion position. It has been used in many studies as a measure of spinal pathology, with some early works connecting symptoms such as migraines with abnormally located MAR for a selected set of data. However, in 1991 Amevo determined that the techniques used to determine the MAR were prone to significant interobserver errors (13–38% relative variance) [1], compromising the MAR data. Amevo presented an optimized technique to measure MAR, where the relative variance of interobserver data was 3–11% [2]. With the technical accuracy of the MAR analysis defined, Amevo used this technique to measure MAR for 40 normal subjects, and formalized the normal range of MAR locations, which can be used as a basis in subsequent studies, relating symptoms to abnormal MAR locations.

G. A. Ateshian et al. (Eds.): CMBBE 2019, LNCVB 36, pp. 107–122, 2020.
https://doi.org/10.1007/978-3-030-43195-2_8

In 2012, Desmoulin et al. used Amevo's optimized MAR technique to evaluate the effectiveness of Khan Kinetic Treatment (KKT), a non-invasive treatment which adjusts the alignment of the spine and upregulates the gene in the discs [3]. He determined that the KKT treatment successfully adjusts abnormal MAR to normal, which correlated with a reduction of neck pain.

While Amevo's optimized MAR technique is reliable, it is difficult to perform (2–3 h effort/patient), and therefore has not been generally adopted. Therefore, the challenge of automating Amevo's MAR analysis became pertinent. The existence of a tool to calculate MAR quickly and reliably would allow MAR analysis to be performed in clinical settings and would provide researchers with opportunities to relate various conditions to MAR data.

In 2013, we developed a software which used computer vision techniques to automate Amevo's MAR technique [4]. However, our 2015 validation study of this tool determined that the relative variance of interobserver errors was 6–26%, and therefore less reliable than Amevo's method [5].

In 2016, Pinheiro developed a tool in Matlab to calculate the MAR, after the user marked 8 points on each vertebra (4 on extension x-ray and corresponding 4 on flexion x-ray) [6]. However, this tool also did not achieve the reliability of Amevo's manual technique.

To achieve the goal of developing an MAR tool which can match the reliability of Amevo's manual method, we carefully compared each step of Amevo's procedure with the corresponding step in our 2013 MAR tool to determine the source of variance. We determined that our 2013 MAR tool did not accurately replicate every maneuver and decision of the manual method. Based on these findings, we designed a new computer vision-based tool, with added intelligence and decision-making capabilities, revised algorithms, and advanced software engineering architectures, to fully replicate Amevo's manual method. This new tool represents a major improvement over previous efforts, and initial test results show promise that MAR Analysis performed using this tool achieves similar reliability to Amevo's manual method.

2 Methods

The MAR analysis requires the following major steps to be performed

- **Step 1 - Initial Trace of X-Rays:** Initially trace of C2–C7 vertebra on extension and flexion radiograph
- **Step 2 - Trace Quality Assurance:** compare each vertebra trace from the extension x-ray to the corresponding vertebrae trace in the flexion x-ray. Continue to adjust the extension and flexion traces until a "best fit" is obtained. The average of the two traces shall then be calculated and used for the remaining procedure.
- **Step 3 - Calculate the MAR:** Perform geometrical analysis to determine the movement of C2–C7 from extension to flexion, with respect to the adjacent lower vertebra
- **Step 4 - Normalization:** Impose a coordinate system on the vertebra and calculate the normalized position of the MAR relative to the vertebra.

To present the advances made in the 2019 MAR Tool compared to the 2013 MAR Tool for each step in the MAR analysis, the following subsections are presented for each step of the MAR analysis procedure:

1. Review of original method of Amevo (short: Amevo Manual Method)
2. Review of our 2013 replication of Amevo's methods (short: 2013 Abbasi Tool)
3. Explanation of limitations of our 2013 tool.
4. 2019 MAR tool implementation to overcome limitations (short: 2019 Abbasi Tool).

2.1 Step 1 - Initial Trace of X-Rays

Amevo Manual Method
The original approach for tracing the extension and flexion X-Rays involved superimposing transparent acetate film on top of the physical X-Ray, and then using a 0.25 mm ink pen to trace the cancellous margin of the vertebra bodies. The cancellous margin was defined as the first region where the cancellous cavities touch a continuous line of cortical bone.

2013 Abbasi Tool
The 2013 MAR tool was developed in LabVIEW 2010 programming language. The software first displayed a digitized image of the extension X-Ray on the screen and provided an interface where the user can click on points on the image to trace the cancellous margin of the vertebra. The coordinates of these points were saved into an array, to define the trace of the vertebra. As the user clicked on points on the image, lines were drawn on the X-Ray image connecting the points, to show the full trace of the vertebra. Typically, 50 clicks would be required to complete the trace around a single vertebra. This procedure repeated for each vertebra, for the extension and flexion X-Rays. A zoom feature was provided to assist the user in tracing the vertebra.

Limitations of 2013 Abbasi Tool
The following limitations were identified in the tracing interface, which reduced the accuracy of the traces:

1. Due to memory limitations, the maximum resolution of the digitized X-Rays was 1024*1024 pixels.
2. Clicking a mouse multiple times along the border of the vertebra required more effort than simply tracing the vertebrae with a pen on acetate paper. The extra effort sometimes affected the ability of the user to precisely mark the vertebra landmarks.
3. Graphically, the average of 50 points connected with lines trace appeared to faithfully define the vertebra. However, subsequent investigation revealed that the resolution was not sufficient to achieve the desired accuracy. Note that the original Amevo method used 0.25 mm pen on acetate film to ensure maximum accuracy of the traces.

2019 Abbasi Tool
The 2019 MAR tool was developed in LabVIEW 2018. The author achieved Certified LabVIEW Developer (CLD) certification in 2016 and applied Object-Oriented

Programming methods and Queued Message Handler architecture to design a new MAR Tool. This tool provides an interface similar to the 2013 tool to trace the extension and flexion X-Rays, with the following improvements:

1. Instead of clicking and individual points around the vertebra, the user can simply hold the mouse button down, and then move the mouse along the border of the vertebra, which is comparable to tracing the vertebra with a pen on acetate paper. On touch screen computers such as the Microsoft Surface Book, the Surface pen can be used instead of the mouse, improving the user experience even more.
2. Using the mouse movement as the tracing method, instead of mouse clicks, each pixel that the mouse touches along the vertebra boundary is recorded and added to the trace point array. Using this method, the point coordinate array defining a single traced vertebra consists of 500–600 pixel point coordinates, which represents a ten-fold increase in trace resolution.
3. Using memory optimization methods, there is no limit to the size and resolution of the X-Ray image used. Images with 2056*3072 pixel resolution were used for this study. Future studies will use images with 3456*4608 pixel resolution
4. Parallel processing methods were used to maintain the CPU responsiveness while drawing on high-resolution images (Fig. 1).

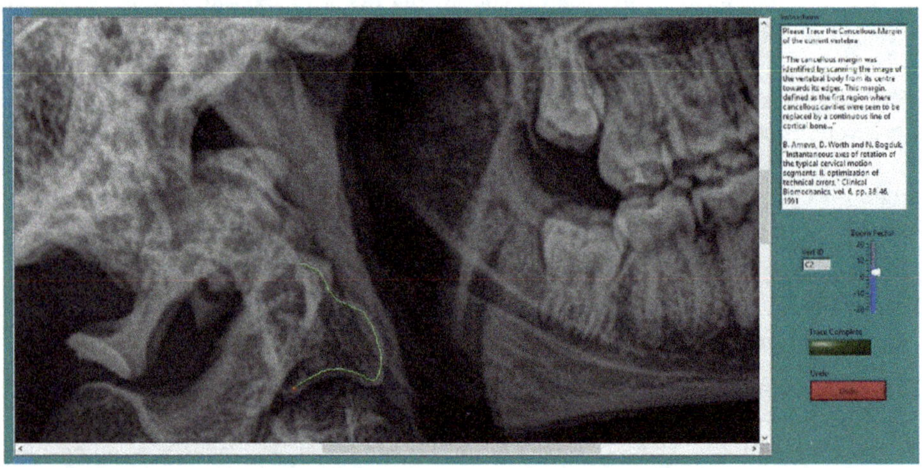

Fig. 1. Vertebra trace interface

2.2 Step 2 - Trace Quality Assurance

Amevo Manual Method

Once the initial trace of each vertebrae on the extension and flexion X-Ray is complete, the traces are compared to ensure they are identical. To achieve this, the acetate with the extension traces is overlaid on the acetate with the flexion traces. With the extension vertebra trace super-imposed on the corresponding flexion trace, and the extension trace is adjusted as needed. The similar procedure is repeated with the flexion acetate overlaid on the extension traces. The extension and flexion traces are continuously adjusted until a best fit is achieved. The following definitions were provided:

- **Ideal Best Fit:** "each of the four borders of the flexion vertebra trace could be perfectly and simultaneously super-imposed on the corresponding borders in the extension tracing for at least 75% of their respective lengths" [2].
- **Adequate fit:** "at least two or more orthogonal margins (preferably the posterior and inferior) could be simultaneously superimposed for at least 75% of their lengths with at least 50% of each of the remaining margins being fully superimposed" [2]

Vertebra traces which did not achieve the above fitting criteria were excluded from the remainder of the study.

2013 Abbasi Tool
The 2013 Tool, a Normalized Cross Correlation (NCC) based Pattern Matching algorithm was used to determine the translation/rotation of corresponding vertebrae from the extension to the flexion. Using this data, the flexion traces were superimposed on the extension image, and similarly the extension traces were superimposed on the flexion image, revealing the differences in the traces. An interface was provided to redrawing any trace if required. Finally, using edge detection algorithms, an "average" trace was drawn, by finding the mid-point between the 2 traces, as shown in Fig. 2. The averaged trace was used for subsequent steps.

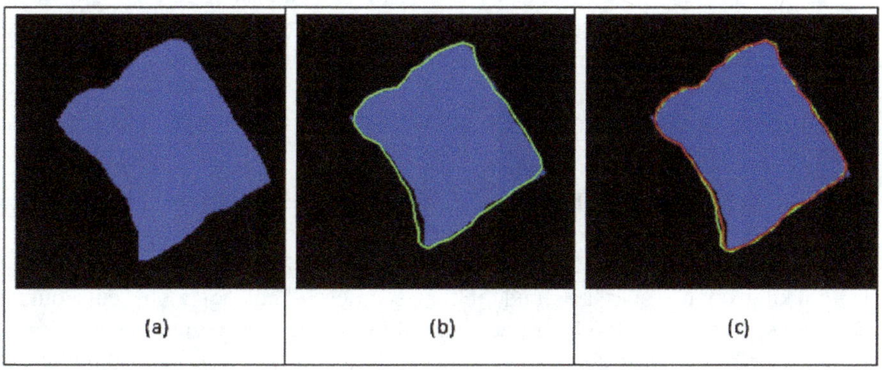

Fig. 2. (a) Flexion trace (b) flexion trace with extension trace overlaid in green (c) flexion trace and extension trace, with average of the two traces overlaid in red

Limitations of 2013 Abbasi Tool
Achieving a "Best Fit" of the extension and corresponding flexion traces is vital to calculating the MAR accurately, and millimeter differences in traces significantly affect the final MAR value. Here, the 2013 MAR Tool did not provide the user enough functionality to replicate Amevo's method.

In Amevo's technique, after superimposing the extension trace on the flexion trace (and vice versa), areas of discrepancies between the traces were identified. Then, the segment of the trace in the identified areas were adjusted as needed. After completing the adjustment, the updated extension trace was again superimposed on the flexion trace, and remaining discrepancies were identified, and then those were corrected as well. This procedure continued until all traces satisfied the best fit criteria of 75% of the borders perfectly super-imposing. After achieving the best fit criteria, the traces were redrawn on a new acetate sheet to prepare for the next step of the analysis. When redrawing the traces, if the extension and flexion traces still didn't correspond to each other for any segment of the trace, then a line passing through the middle of the extension and flexion traces was drawn to represent the final representation of the trace.

By contrast, the 2013 MAR tool did not provide the user a method to fix any erroneous segments of any trace. Rather, after identifying trace errors by viewing the superposition of the extension trace on the flexion trace, the only option provided to the user was to accept the current trace or to redraw the entire trace. Practically, redrawing the entire trace from scratch repeatedly proved time consuming for the user, and even if some trace errors were corrected, new errors were introduced, making it difficult to achieve faithful traces.

Although the 2013 MAR tool did calculate the average of the extension and flexion traces, without the ability to correct segments of the trace, such an average does not ensure the final traces are faithful to the original vertebra on the X-Ray.

2019 Abbasi Tool

The 2019 MAR tool introduced a trace adjustment interface, which displays the extension and flexion X-Rays side by side. On the extension X-Ray, the original extension vertebra trace is drawn in green, and the flexion trace (rotated and translated based on the NCC Pattern matching algorithm) overlaid on the extension image is drawn in a red dotted line.

The following tools are provided to allow the user to adjust the traces to achieve best fit.

1. **Selective Trace Display:** The user can show or hide either trace using the buttons, to easily compare each trace with the original X-Ray.
2. **Adjust Trace Functions:** The user can select a specific segment of a trace to adjust by clicking on the start and end of that segment. The tool then "cuts out" the selected segment from the total trace and allows the user to redraw the specific part of the trace only. Once the new segment is drawn, it is merged into the trace, and then the pattern matching algorithm is again used to recalculate the extension trace and flexion trace rotation and displacement. The updated superposition of the flexion traces on the extension x-ray (and vice versa) is displayed on each X-Ray image.
3. **Trace Averaging:** In addition to the original and rotated extension and flexion traces, the tool also calculates the average traces for each, as well as the averaged extension trace superimposed on the flexion X-Ray, and vice versa. The display of the average traces can also be turned on or off.

4. **Auto-Adjust Trace Segment:** The user can select a specific segment of a trace to auto-adjust using computer vision edge detection. After selecting the trace segment, auto-adjust will scan a 5-pixel wide region along the trace line, and slide each original trace point onto the strongest edge (based on pixel intensity gradient) close to it. This feature is useful to assist the user to draw the trace exactly on the border of the cancellous cavity and the continuous cortical bone. However, the feature is sensitive to noise in the image, and therefore should only be used on trace segments in relatively noise free parts of the trace. **Note:** this feature was not implemented at the time of the validation tests presented in this paper. However, it is ready to use in subsequent validation studies.
5. **Match Score:** after every trace adjustment, the pattern matching algorithm recalculates the rotation/translation required to overlay the flexion trace on the extension trace (and vice versa). The match score is a measure of the match of the 2 traces, with 1000 being a perfect match.
6. **MAR Intersection Errors:** the technique for geometrically calculating the MAR is explained in the following section and involves finding the intersection of multiple bisecting lines. It is possible that the multiple lines don't all intersect at the exact point, in which case Amevo's procedure instructed the user to recheck the traces to improve the fit of each vertebra. The bisector intersection errors are mainly caused by tracing errors, so the user can continue to adjust the traces until the MAR intersect errors are minimal (Figs. 3, 4 and 5).

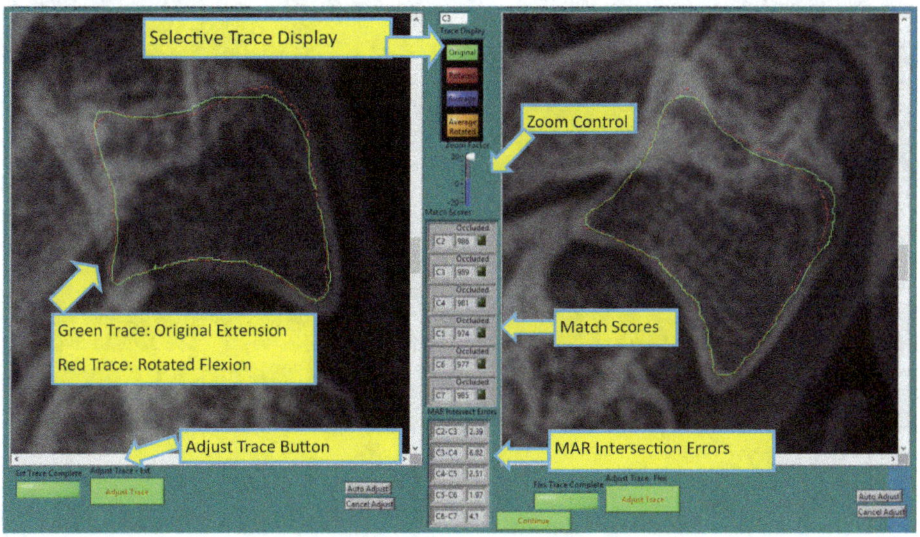

Fig. 3. Adjust trace interface – initial screen

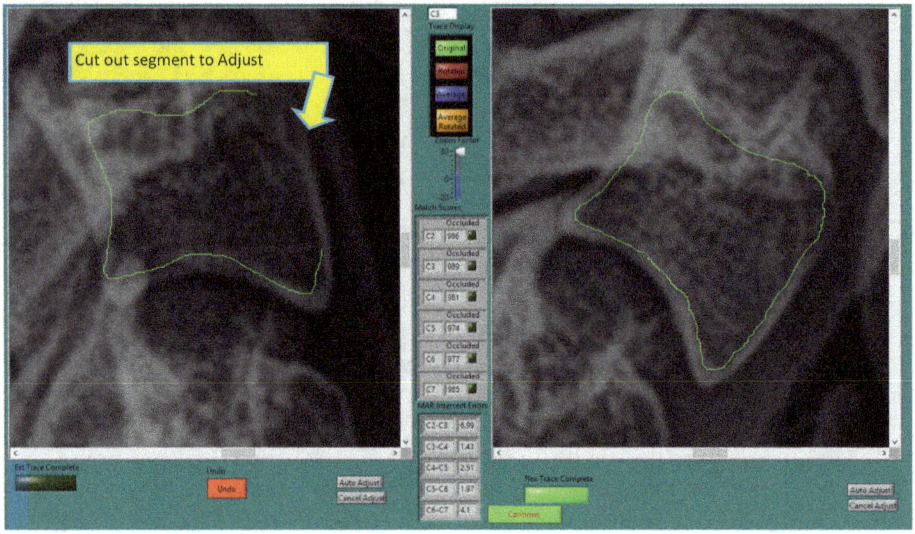

Fig. 4. Adjust trace – cut out segment to redraw

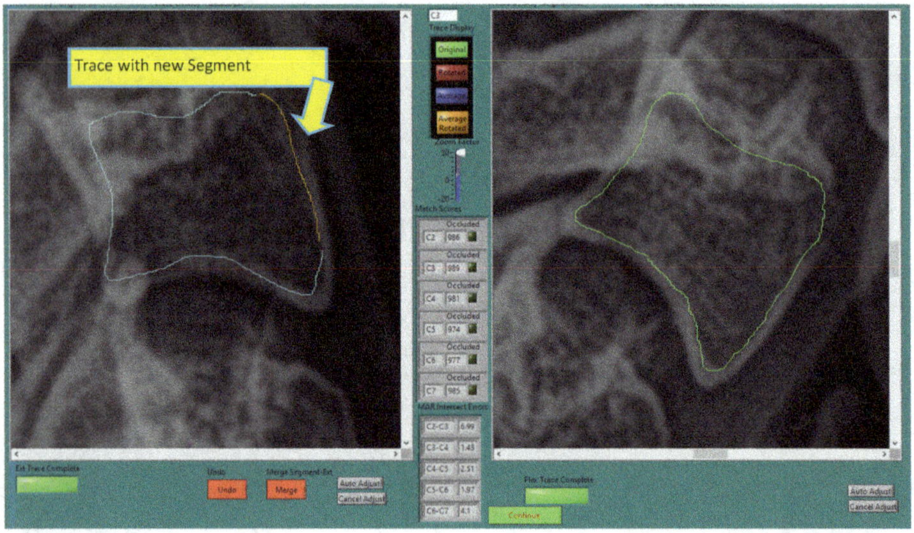

Fig. 5. Adjust trace – new segment drawn to replace old segment

Using the trace adjusting interface, users can now easily replicate Amevo's steps to achieve a best fit of corresponding traces.

The user can continue to adjust the extension and flexion traces until he/she achieves the best fit criteria defined by Amevo.

2.3 Step 3 - Construction of the MAR

Amevo Manual Method

As the original MAR procedure involved drawing lines on acetate paper, calculating the MAR was also performed geometrically, using rulers and protractors to find the intersection point of 4 bisecting lines representing the movement of a vertebra Fig. 6. A key indicator of successful construction of the MAR is that all 4 bisector lines converge perfectly to a single point. If 3 out of 4 bisectors converged, then the 4th was ignored, and the MAR was determined from the 3 lines only. In the case that even 3 bisecting lines would not intersect, then this vertebra was excluded from further study.

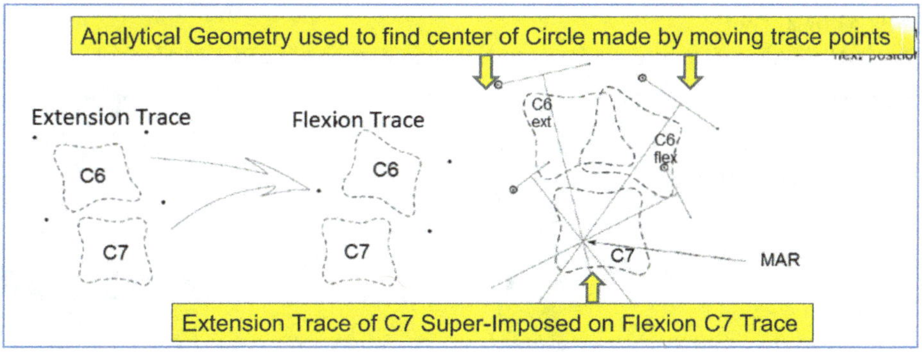

Fig. 6. Intersection of 4 bisecting lines determines MAR location

2013 Abbasi Tool

The 2013 Abbasi tool calculated the intersection point of the bisecting lines mathematically using linear equations.

Limitations of 2013 Abbasi Tool

Using a purely mathematical method to compute the intersection point of 4 lines, the user does not have an indication whether all four lines converged exactly, or if one or more lines would not converge. The final MAR point returned was the mathematical average of the different intersection points of the different lines, which could introduce substantial error in the case of non-converging lines.

2019 Abbasi Tool

To faithfully replicate Amevo's procedure, the 2019 MAR tool mathematically calculates the intersection of every combination of 2 bisecting lines (total 6 combinations), and then checks if any 3 lines converge perfectly. The best 3 lines are selected, and their average intersection point is calculated, along with a measure of the intersection error (distance in pixels). As mentioned previously, this MAR intersection error is displayed to the user as he/she performs the Trace adjustment. This allows the user to continue to adjust the trace until the Best Fit is achieved, and the convergence of the bisector lines is confirmed.

2.4 Step 4 - Normalization

Amevo Manual Method

The final step of the MAR Analysis is to impose a coordinate system on the vertebra, to represent the MAR in (x, y) coordinates. Amevo presents instructions on how to place the X axis on the inferior margin, and the Y axis on the posterior margin, after rotating the acetate such that the vertebra appears flat.

2013 Abbasi Tool

The 2013 MAR tool provides the user with a Normalization interface, A rotation knob allows the user to rotate the trace of the vertebra until it appears flat. Then, the user can use the mouse to place the X and Y axis bars as needed.

2019 Abbasi Tool

The 2019 MAR tool provides the same functionality as the 2013 MAR tool.

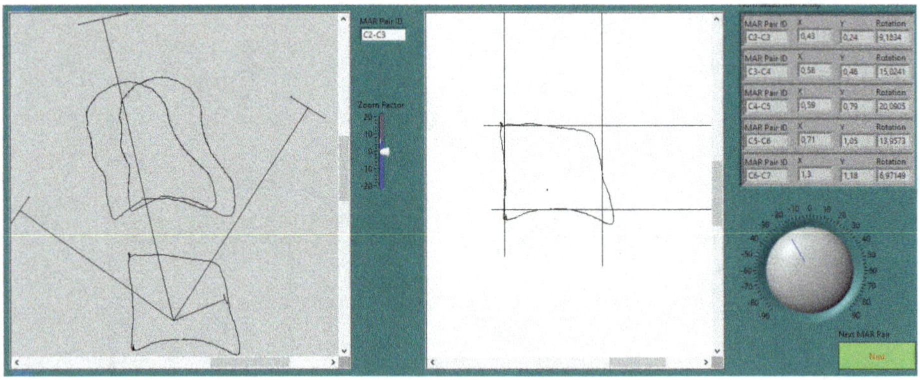

3 Results

The MAR Analysis was performed by 2 different observers on 10 sets of extension/flexion X-Rays using the 2019 MAR Tool.

The following analysis was performed:

1. Compare vertebra traces from each observer.
2. Validate MAR Dataset. If the range of motion of the vertebra from extension to flexion is 5° or less, the MAR is discarded from the dataset due to un-manageable technical errors
3. Calculate Interobserver Differences in valid dataset
4. Calculate Relative Variance of Interobserver differences in valid dataset.

3.1 Comparing User Traces

The traces for each X-Ray were visually compared with each other using a special software designed for this purpose. Vertebrae traces which did not satisfy the best-fit criteria between the 2 observers were discarded from future study. It is important to note that Amevo's definition of a best fit required most of the vertebra traces to superimpose perfectly on each other. A difference of 3–5 pixels between trace lines is enough to disqualify a trace from the best-fit (Table 1).

Table 1. Comparison of user traces of vertebrae landmarks

	Best fit trace	Not best fit trace	Occluded in X-Ray
C2	3	7	0
C3	7	3	0
C4	8	2	0
C5	7	3	0
C6	4	1	5
C7	1	0	9

We observe that substantial interobserver differences occurred in identifying the same exact margin of the vertebra, especially in C2. The reasons for these differences are not related to the MAR tool, but rather they are related to the skillset of the observers. Specifically:

1. In the X-Ray set used for the validation study, many of the vertebra (especially C2) had some indistinguishable margins (Figs. 7 and 8). For the present validation study, only minimal training was provided to the observers. Amevo had defined a strict criterion for identifying the vertebrae landmarks, and all observers received extensive training on understanding the criterion. Any future validation study of the MAR Tool requires the observers to be trained on how to apply these criteria in a common way for noisy images.
2. The second cause of interobserver differences was that while tracing vertebrae margins with a mouse, occasionally the observers traced the margin lines with 2–3 pixels of difference between the lines. These minor trace differences can be overcome with additional practice/training of the observers. Additionally, the new "Auto-adjust trace segment" feature, which identifies the best edge based on pixel intensity gradients, will assist testers to trace margins with optimal precision in future validation studies.

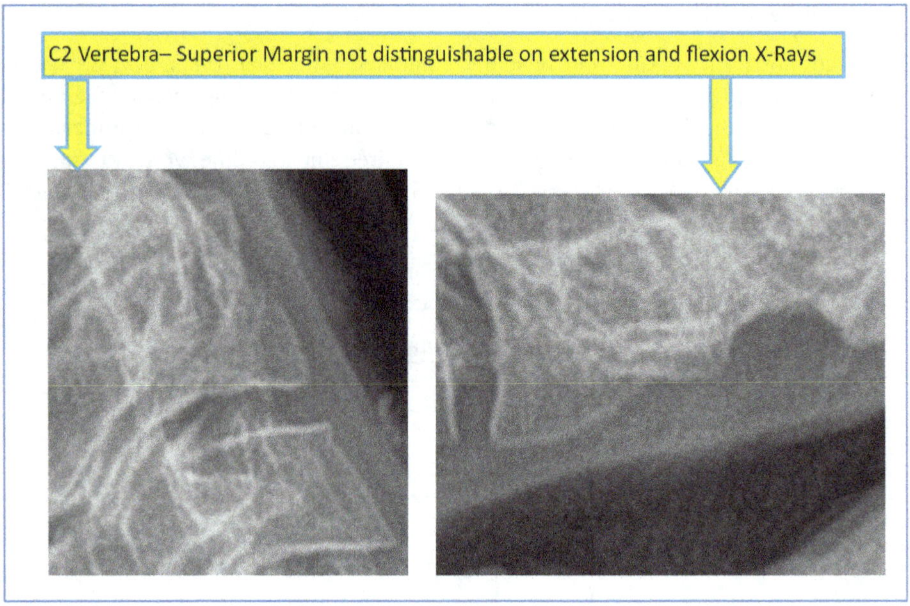

Fig. 7. C2 vertebra sample 1 - Superior and posterior margins difficult to identify

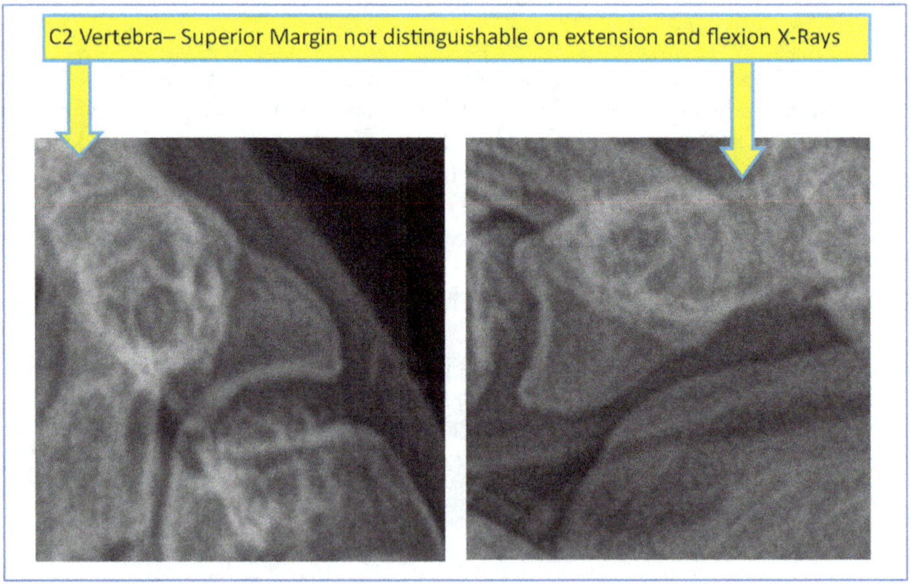

Fig. 8. C2 vertebra sample 2 - Superior and posterior margins difficult to identify

Finally, we observe that in the dataset used in the current study, the majority of C6 and C7 vertebra were partially or completely occluded in the image, depriving the study of data related to these vertebrae Fig. 9.

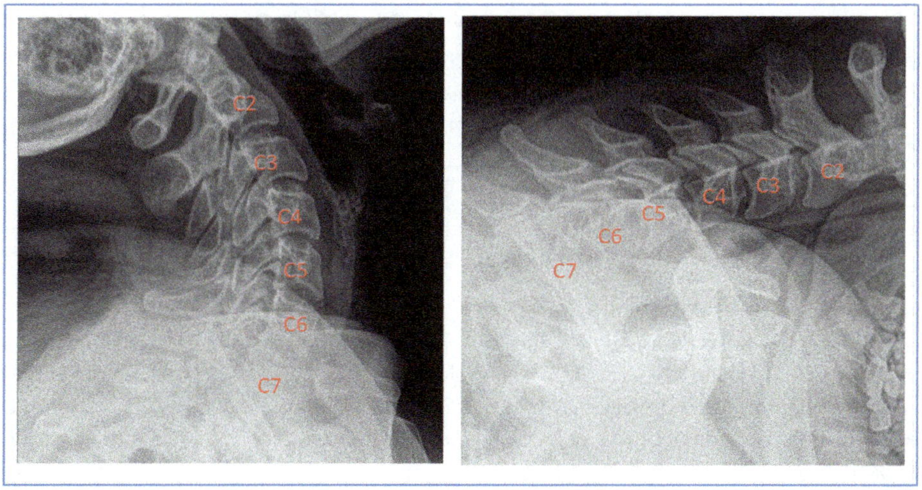

Fig. 9. C6 and C7 vertebra completely occluded in X-Ray

The resulting dataset is admittedly very small, and therefore not adequate to make definite claims about the reliability of the MAR Tool. Additional validation studies are planned in the future, which will use a larger dataset.

3.2 Validating MAR Datasets

After calculating the MAR for each vertebrae pair, we also calculate the total range of motion of the upper vertebra about the lower vertebra. If a particular vertebra rotates 5% or less about its adjacent lower vertebra, then the MAR coordinate calculated based on such small motion will be subject to significant technical errors and cannot be accepted as reliable. Accordingly, vertebra pair MAR based on rotation of less than 5% were excluded from the dataset (Table 2).

Table 2. Validation of MAR dataset

	Best fit pairs	Rejected, small range	Total MAR pairs in dataset
C2–C3	2	1	1
C3–C4	4	1	3
C4–C5	4	0	4
C5–C6	3	1	2
C6–C7	1	0	1

3.3 Interobserver Differences and Relative Variance in MAR Dataset

After removing the invalid datasets, the mean and standard deviation of the interobserver errors was calculated and are shown in Table 3.

Table 3. Interobserver differences in MAR dataset

Segment	# of samples	X Coordinate		Y Coordinate	
		Mean (pixels)	sd	Mean (pixels)	sd
C2–C3	1	6.3	n/a	9.5	n/a
C3–C4	3	−2	5.8	−1.94	5.6
C4–C5	4	3.5	3.3	0.32	4.86
C5–C6	2	7.96	2.0	5.6	4.2
C6–C7	1	10.7	n/a	15.9	n/a

While the dataset is small, we see that the mean interobserver error is less than 4 pixels for C3–C4 and C4–C5 (3 & 4 samples), and less than 10 pixels for C5–C6 (2 samples). There was only one sample of C2–C3 and C6–C7, so the statistics are not relevant. Future studies will present more meaningful statistics, as a larger dataset will be used.

Amevo measured the significance of the interobserver errors by calculating the relative variance, which is determined by taking one standard deviation range of possible interobserver errors and dividing them by the mean value of the quantity being measured. The relative variance for the interobserver differences in using the 2019 Abbasi tool is presented in Table 4.

Table 4. Relative variance of interobserver differences

Segment	Mean coordinate		Relative variance	
	X (pixels)	Y (pixels)	X	Y
C2–C3	10.2	28.3	N/a	N/a
C3–C4	31.26	47.72	0.18	0.11
C4–C5	49.61	50.52	0.06	0.09
C5–C6	58.95	76.82	0.03	0.05
C6–C7	79.45	113.6	N/a	N/a

We can see that the relative variance of all the segments is between 3% and 18%. There was only 1 sample for C2–C3 and C6–C7, so relative variance is not possible to calculate.

The best results were achieved the C4–C5 and C5–C6 segments. It is known that the technicians who take lateral X-Rays generally center the X-Ray beam on C4 or C5, resulting in more distinct image features around this region, and less distinct image features away from this region. This may also explain why the interobserver trace errors in C4 and C5 are less than on the other vertebra.

The statistics presented in Table 4 are based on a vary small dataset, and therefore they cannot be relied upon for any formal conclusion. However, these initial results do give hope that future validation studies will confirm the 2019 Abbasi tool to be scientifically reliable enough to use in clinical and research applications.

4 Conclusion

We have presented a new computer-vision based tool to help perform the MAR Analysis, faithfully replicating every step of Amevo's manual method. The tool leverages advanced software methods and architectures to provide functionality equivalent to the manual procedure.

Initial testing showed that for the C4–C5 and C5–C6 pairs, relative variance was 3%–9%, while the C3–C4 relative variance was 11%–18%. Notwithstanding the small sample size, the discrepancy of the variance between these 2 regions show that interobserver differences are caused more by human error in tracing, rather than errors introduced by the tool.

Human tracing errors can be overcome by additional training, as well as additional functionality to assist the user in tracing the vertebra. Already, an automatic adjust feature has been developed which uses pixel intensity gradients-based edge detection to identify the exact pixels along a vertebra margin. As additional user feedback is received, other features can also be added to the toolbox, including image processing functionality to improve contrast and sharpen edges. An extensive validation study of the MAR tool is planned, which will include a large data set, some new features of the MAR Tool and proper training for the testers.

The current MAR tool is a work in progress, and we intend to continuously improve it with additional features to improve the accuracy and facilitate the procedure for its users. In the future, we are interested in exploring learning techniques and other AI methods to reduce the human effort required in the analysis.

Finally, KKT Orthopedic Spine Center, a global leader in spine treatment with over 26 clinics in 13 countries, has been collecting extension and flexion X-Rays of their patients across the world for many years. Once the reliability of the MAR tool is confirmed, the KKT team will perform MAR analysis in its clinics for their patients as well. With the extensive MAR data collected, and the other clinical data collected from the patients, KKT researchers can explore the connection of MAR positions with other symptoms and conditions. As all the data is collected and stored digitally in a database, the data-mining possibilities are vast. Furthermore, the MAR tool can be integrated directly with the database, so that not only can the MAR tool send data to the database for analysis, but the database can send information back to the MAR tool as well, which may be relevant in machine learning applications. With the close collaboration of multi-disciplinary research teams at KKT, the opportunities will continue to open.

References

1. Amevo, B., Worth, D.: Instantaneous axes of rotation of the typical cervical motion segments: a study of normal volunteers. Clin. Biomech. **6**, 111–117 (1991)
2. Amevo, B.: Instantaneous axes of rotation of the typical cervical motion segments: optimization of technical errors. Clin. Biomech. **6**, 38–46 (1991)
3. Desmoulin, G.T., Hunter, C.J., Hewitt, C.R., Bogduk, N., Al-Ameri, O.S.: Noninvasive intervention corrects biomechanics and upregulates disk genes for long-term spinal health. Glob. Spine J. **2**, s-0032 (2012)
4. Abbasi, M., Desmoulin, G., Khan, A.H.: Semi automation of mean axis of rotation (MAR) analysis using computer vision. In: Proceedings of the 11th International Symposium, Computer Methods in Biomechanics and Biomedical Engineering (CMBBE), Salt Lake City, Utah, p. 515 (2013)
5. Abbasi, M., Khan, A.H., Payat, D.: Reliability of a semi-automated mean axis of rotation (MAR) analysis tool. In: International Symposium on Computer Methods in Biomechanics and Biomedical Engineering (CMBBE), Montreal, p. 54 (2015). https://live-mccaig.ucalgary.ca/sites/default/files/miscellaneous/cmbbe2015-proceedings.pdf
6. Pinheiro, J.: A software for quantitative analysis for intervertebral movement based on medical images, Lisbon, Portugal, November 2016. https://fenix.tecnico.ulisboa.pt/downloadFile/1407770020544895/Dissertation.pdf

A Preliminary Sensitivity Study of Vertebral Tethering Configurations Using a Patient-Specific Finite Element Model of Idiopathic Scoliosis

J. P. Little[1,2,3,4(✉)], R. D. Labrom[1,2,3], and G. N. Askin[1,2,4]

[1] Biomechanics and Spine Research Group, IHBI at Centre for Children's Health Research, Queensland University of Technology, Brisbane, Australia
j2.little@qut.edu.au

[2] Queensland Children's Hospital, Brisbane, Australia

[3] The Wesley Hospital, Brisbane, Australia

[4] Mater Health Services, Brisbane, Australia

Abstract. Vertebral Body Tethering (VBT) surgery for skeletally immature idiopathic scoliosis (IS) patients involves anteriorly placed vertebral screws securing a deformable Polyethylene-Terephthalate (PET) tether. Before securing the tether, compressive force is applied between the screw heads along the axis of the spine. There are no clear guidelines regarding the force magnitude required to optimize deformity correction. In the current study, a validated, patient-specific finite element (FE) model of the thoracolumbar spine/ribcage for a 10-year-old IS patient was analysed, to investigate the effect of four different VBT loading scenarios on spinal alignment and biomechanics.

The patient-specific FEM was previously validated using clinical results for pre-/post-operative deformity. Linear elastic continuum elements (PET material) were used to tether laterally oriented screws at spinal levels T5–T12, with roughened contact simulating the screws when locked onto the tether. Compressive forces measured intra-operatively during VBT and during anterior scoliosis fusion surgery (FS) were the basis for the four loadcases. The inferior L5 endplate was fixed.

After the surgical loadcases, patient-specific, level-wise gravitational loads at all vertebral levels simulated standing. In this preliminary series of VBT analyses, model predictions for corrected Cobb angle/Kyphosis angle/Axial trunk rotation, correction in major curve intervertebral disc wedge angle; and vertebral bone stress, were compared to determine how different surgical tether tension magnitudes affected deformity correction/spinal loading.

Results demonstrated varying degrees of improvement in coronal deformity correction could be achieved with different patterns of tethering loads. However, resultant loads on surrounding anatomy must be considered, with associated high spinal tissue loads and increased propensity for asymmetric growth modulation with increasing tether forces.

Keywords: Idiopathic scoliosis · Vertebral Body Tethering · Finite element model

G. A. Ateshian et al. (Eds.): CMBBE 2019, LNCVB 36, pp. 123–132, 2020.
https://doi.org/10.1007/978-3-030-43195-2_9

1 Introduction

With the development of fusionless spinal tethering as a surgical alternative for skeletally immature patients presenting with idiopathic scoliosis [1–3], there is a growing interest from both the surgical field and the patient population in the use of Vertebral Body Tethering (VBT) systems. These systems provide a dynamic stabilisation of the deformed spinal levels, with anteriorly placed vertebral screws securing a deformable Polyethylene Terephthalate (PET) cord between the screw heads [2, 4].

Such fusionless correction strategies aim to create compressive force along the anterior column of the spine, on the convex side of the deformity, while encouraging distraction on the concave side of the deformity [5]. Drawing on the Hueter-Volkmann principle of mechano-regulated bone growth, this asymmetric loading configuration encourages down-regulation of bone growth on the convexity of the spinal column due to compressive forces from the tether [6]. If growth modulation is applied using vertebral tethering at an appropriate stage in the patient's skeletal growth curve, this mechanical tethering can not only stabilise but reduce the spinal deformity. Stokes [7] proposed a linear growth algorithm describing the attenuation of yearly vertebral bone growth rate as a result of the difference between mean normal stress and the maximum local stress applied to the growth plates. This algorithm was proposed for the AIS spine and has since been applied in a finite element model (FEM) of VBT growth modulation [8].

There are many surgically and biomechanically relevant questions relating to the VBT surgery that have come to light with this relatively new surgical technique [1, 9]. Early biomechanical studies establishing the efficacy of vertebral tethering in modulating both vertebral and intervertebral disc growth were focused on providing evidence for the feasibility of fusionless surgical interventions using animal models [5, 10, 11]. Newton *et al.* [10] conducted biomechanical studies in an *in vivo* porcine model, to establish whether tether pre-tension significantly affected the vertebral growth modulation and intervertebral disc health in the growing porcine spine. This study noted that tether pre-tensioning resulted in greater immediate postoperative coronal deformity creation and greater intervertebral disc wedging than for the un-tensioned tether.

Clinically, the tether is pre-tensioned intra-operatively, either by a tensioning device or via inter-segmental compression using a rod compressor [1–4]. This tensioning can create some immediate deformity correction. However, the magnitude of this tensioning force is currently dependent upon the surgeon's experience, as no definitive biomechanical studies have demonstrated an ideal tension force profile between vertebral anchors [6], nor have clinical studies established the influence of varying magnitudes of pre-tension on the progressive change in deformity correction. As yet, there are limited data describing what magnitude of forces are required to produce an optimum VBT correction for an individual patient, nor is there clear empirical evidence describing over which levels in the deformed region of the spine the tether should extend. Similarly, there are limited studies providing a means to systematically investigate this topic.

To address these questions, a validated, patient-specific finite element model of an idiopathic scoliosis patient treated in Brisbane, Australia was analysed. The predicted coronal plane correction, axial de-rotation and change in sagittal plane kyphosis were assessed to investigate the first of these clinical questions relating to the biomechanical effect of different intra-operative VBT tethering strategies on post-operative spinal alignment and biomechanics. We hypothesised that increasing magnitudes of inter-segmental compression applied over the major curve would result in increasing coronal plane correction and increased propensity for vertebral bone remodeling post-operatively.

2 Methods

2.1 FEM Details

Custom-software (*VirtuSpine* [12]) was utilized to reconstruct the osseo-ligamentous anatomy for the thoracolumbar spine and ribcage of a 10 year old idiopathic scoliosis patient [13] (Fig. 1a). The patient was not directly involved in this study, rather their clinical imaging data was referenced from a historic, clinically indicated, low-dose computed tomography (CT) data series (Institutional Ethics #15000000084) which was previously used for pre-operative surgical planning [14]. FE geometry was individualized to the patient's anatomy using this historical CT dataset. A sensitivity study of patient-specific spinal FE models created using *VirtuSpine* demonstrated low variation in user-selected anatomical landmarks and in model predictions for spinal curvature [15].

FE predictions for the patient-specific model had previously been validated against clinical results for the patient's pre-operative deformity (FE vs Clinical: Standing Cobb angle 40° vs 44°; Fulcrum Flexibility Cobb 23° vs 26°) and post-operative surgical correction (FE vs Clinical: Cobb angle 13° vs 14°) [16]. All predicted results in the prior study [16] were within clinical measurement variability, when analysing a surgical loadcase representing thoracoscopic anterior spinal fusion (TASF) surgery for AIS [17]. These surgical compressive forces were measured intra-operatively [17] and were used as a benchmark for inter-level compressive forces applied over the deformed region of the spine in the current study.

This patient-specific FEM was analysed in the current sensitivity study, to investigate different combinations of tether tension forces applied between vertebral screws on the convex side of the curve. Following clinical advice, the surgically modified anterior tethering of the simulated spine was modelled by including titanium screws oriented laterally at spinal levels from T5 to T12 (Fig. 1b), using a similar method to that detailed in [16]. A PET tether was modelled between screw heads as a linear elastic material ($E = 108$ MPa, $\nu = 0.4$), with an initially frictionless contact relationship between the screw heads and tether elements.

In this preliminary series of VBT analyses, four surgically relevant loading scenarios were investigated, where either tether tension was applied between all levels in the major curve (VBT_01, VBT_02, VBT_03); or no tether tension was applied (VBT_04) (Table 1). The magnitude of tether tension was based on both corrective forces measured intra-operatively during TASF (VBT_02, VBT_03) [17] and on tether tensions measured intra-operatively during VBT surgery, using instrumented surgical tools (Globus Medical, PA, USA) (VBT_01). A zero tension loadcase (VBT_04) was investigated in light of the experimental findings of Newton *et al.* [5], to establish the different potential for growth modulation when a tether tension force is applied. Tether force magnitudes are listed in Table 1.

The inferior L5 endplate was assumed fixed. After each inter-segmental force was applied between adjacent screws, the tangential contact between the screws and rod was changed to bonded contact, to simulate locking the screw onto the tether. Following the VBT surgical loading steps, patient-specific, level-wise gravitational loads were applied at each vertebra to simulate standing. The change in the coronal, axial and sagittal plane rotation of the vertebra T5 to T12 were calculated and used to determine the change in overall spinal deformity and in segmental rotation pre- to post-operative (with a standing loadcase).

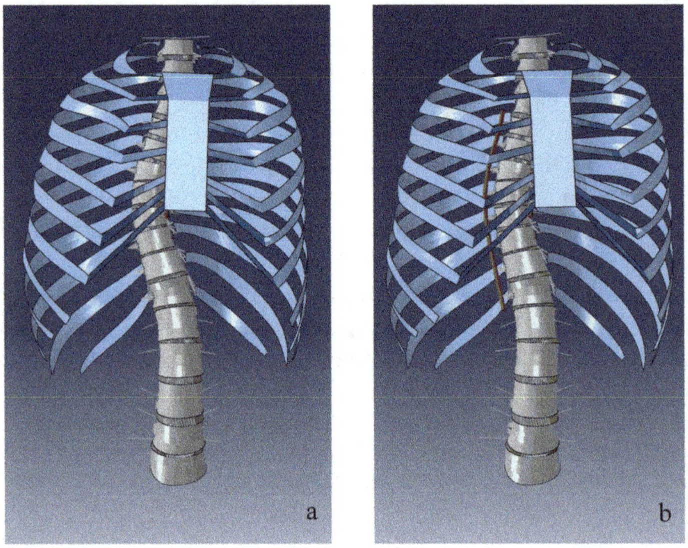

Fig. 1. Patient-specific FEM showing unloaded (a) and surgically simulated (b) spine for a 10 year old female, IS patient.

Table 1. Four tether tension surgical loadcases (Newtons, N), based on both a small series of intra-operatively measured forces from an early series of VBT patients (VBT_01) and on intra-operatively measured forces during TASF (VBT_02, VBT_03, [17]). VBT_05 is a reference case based on the biomechanical testing of Newton e*t al.* [5].

	VBT_01 (N)	VBT_02 (N)	VBT_03 (N)	VBT_04 (N)
T5T6	190	320	240	0
T6T7	200	470	350	0
T7T8	350	550	410	0
T8T9	100	660	495	0
T9T10	350	675	505	0
T10T11	135	580	435	0
T11T12	50	400	300	0
Details	*Average of surgically measured level-wise tensions for two VBT patients*	*Surgically measured level-wise forces for TASF (Fairhurst et al. 17])*	*75% of surgically measured level-wise forces for TASF*	*No tether tension (Newton et al. [10])*

2.2 Predicted Load-Related Modulation of Vertebral Growth Rate

A baseline vertebral growth rate of 1.1 mm/year per vertebra was calculated on the basis of sitting height change for a 10 year old female [7, 18]. Using FE predicted bone stresses, a predicted vertebral bone growth rate was calculated for each loadcase. The growth modulation relationship proposed by Stokes [7] was applied (Eq. 1), where σ was the local maximum/minimum axial stress on the endplates and σ_m was the mean axial stress on the endplate. A scaling factor, β, of 1.5/Mpa was used [7].

$$G = G_m(1 - \beta(\sigma - \sigma_m)) \tag{1}$$

The propensity for growth modulation in the instrumented vertebra in the early post-operative period, was determined for the different surgical loading strategies. Newton *et al.* [10] observed that in the first 6 months post-surgery in a porcine model, there was a greater observed growth differentiation in the porcine vertebra with inter-segmental tether tension applied than for those with no tension. This porcine model was "creating" rather than correcting a deformity. In the current pilot sensitivity study, load cases VBT_01 and VBT_04 were compared to investigate this differential growth potential in the deformed AIS spine.

The local growth rate in regions of peak axial vertebral bone stress were calculated using predicted mean and peak axial stress applied to the inferior endplate in each of the instrumented joints (T5 inferior to T11 inferior) after the standing loadcase. The relative difference in growth rate between the no tension (VBT_04) and VBT surgery (VBT_01) loadcases indicated the propensity for different levels of growth modulation with different spinal loading. The difference between the maximum and mean axial stress on the inferior endplates in the instrumented curve was assumed to indicate up-regulated bone growth. Conversely, the difference between the minimum and mean axial stress on these endplates indicated down-regulated bone growth.

3 Results

Predicted deformity correction following the simulated surgery/standing loadcase are shown in Table 2. With the exception of VBT_04, all loadcases resulted in improved coronal Cobb angle and axial trunk rotation. While there was a minimal increase in kyphosis after the VBT surgical forces were simulated, this was not evident after the standing loadcase was applied (Table 2).

The largest magnitude surgical loadcase result in the largest overall coronal plane deformity correction (Fig. 2, Table 2). Similarly, segmental deformity correction in the intervertebral discs was typically of higher magnitude for VBT_02. Predicted cancellous bone stresses were less than the compressive strength of cancellous bone in all loadcases [19], suggesting there was no risk of screw plough.

In both the VBT_01 and VBT_04, the maximum and minimum stresses were typically located contra-laterally on the endplates in the instrumented vertebra, although not always at the lateral-most margins.

The predicted modified growth rates were compared to the baseline value (1.1 mm/year) and expressed as a percentage of this baseline value (Fig. 3). This comparison demonstrated a higher relative change in growth rate for the VBT_01 loadcase. Compressive stress at some vertebral levels for the VBT_01 loadcase were sufficiently high to indicate there would be no longitudinal growth in this region (Fig. 3).

Table 2. Change in 3D deformity following simulated VBT surgery and standing loadcases. Improvement in coronal Cobb angle, sagittal kyphosis angle and axial rotation between the limits of the major curve (degrees) - negative angles indicate a reduction compared to pre-operative.

	VBT_01	VBT_02	VBT_03	VBT_04
	Change in major curve angle (T5-T12) from pre-operative condition (degrees)			
Cobb angle	−4	−9	−7	0
Sagittal kyphosis	0	0	0	0
Axial trunk rotation	−2	−4	−3	0

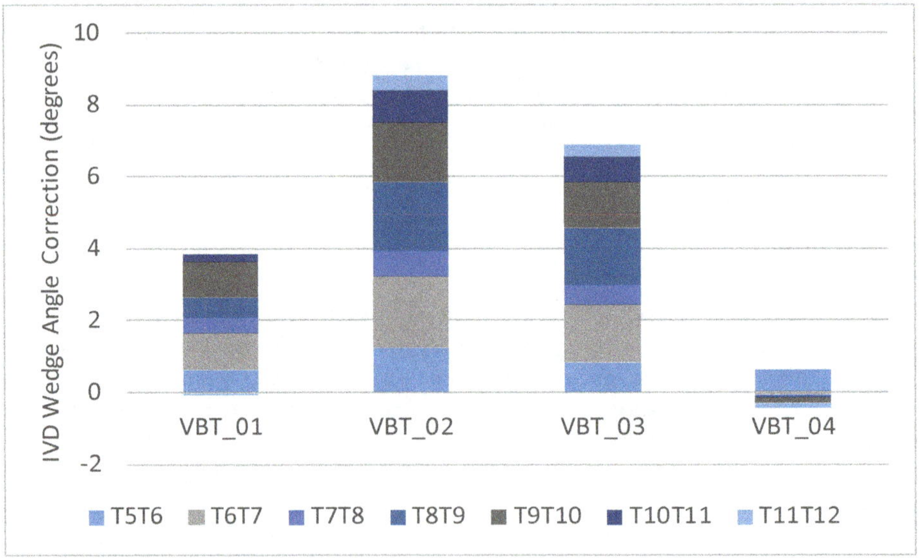

Fig. 2. Coronal plane intervertebral disc correction (degrees) over the major curve

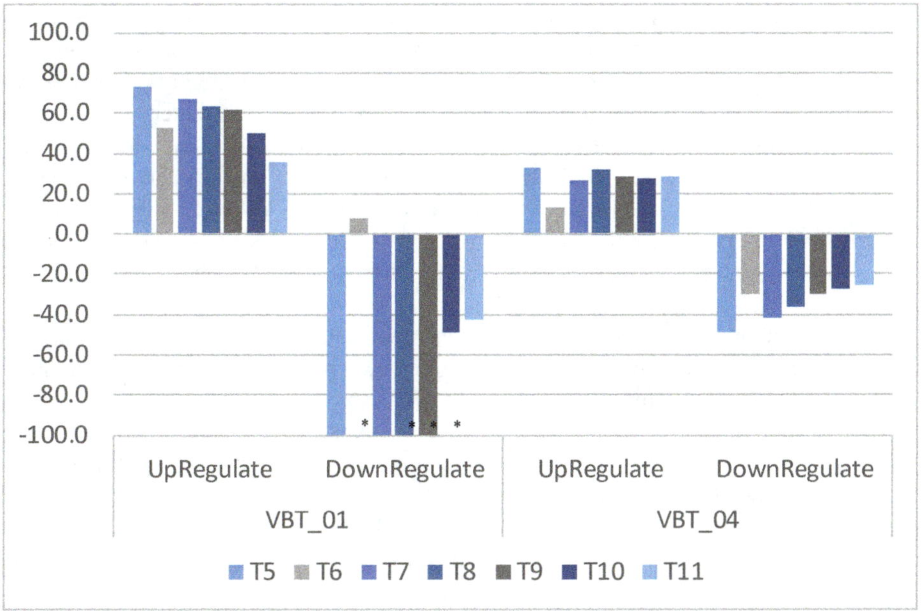

Fig. 3. Mechanically induced change in growth rate, relative to baseline 1.1 mm/year, expressed as a percentage of baseline value. Change in growth rate is for the peak maximum (tensile) and minimum (compressive) stresses at each level. Tensile stress can lead to up-regulation of bone growth, and compressive stress can lead to down-regulation. * growth rate = 0 mm/year

4 Discussion

To provide a tool capable of parametrically investigating clinically driven inquiries relating to the VBT surgery, this study utilised a validated, patient-specific FEM of a patient with idiopathic scoliosis. This patient-specific FEM was analysed to simulate four different clinically relevant, intra-operative loading scenarios. Analysis results demonstrated the utility of such a computational approach in carrying out 'what-if' studies to establish the sensitivity of clinically-derived output parameters to surgical loading.

As was predicted by Cobetto *et al.* [8], the magnitude of tether tension had minimal affect on the sagittal and axial plane correction, but a distinct influence on coronal plane correction. Coronal plane deformity correction increased with increasing magnitudes of tether tension and this coronal correction decreased slightly following simulated standing. The differing coronal and axial plane deformity correction with different tether tension was in keeping with the findings for TASF [16] and demonstrated that discernible differences in early post-operative deformity correction can be achieved with different patterns of tethering loads. However, the resultant loads on the surrounding anatomy must be considered, with the potential for a relationship of diminishing returns between improved correction and risk of tissue overload due to increased forces active on the spinal tissues.

Biomechanical and early clinical studies of VBT have demonstrated the success of these implants both in providing a dynamic, internal brace to halt deformity progression and in perturbing the mechano-biology of the deformed vertebral bodies, in order to bring about a more 'typical' or non-deformed vertebral growth pattern [1, 2, 5, 10]. Predicted results for the change in growth potential in the instrumented vertebrae are in keeping with Newton *et al.* [10], whereby, applying a tether tension between adjacent vertebrae creates a favourable mechanical environment for differential growth at the proximal ends of the vertebral body. At least in the early post-operative period, this will likely lead to great potential for growth modulation in the tensioned compared to the non-tensioned tether.

To date, there are no published biomechanical data describing an ideal range of VBT tether tension magnitudes to achieve both arrest of deformity progression and to encourage differential growth in the young deformed spinal column. As such, the current computational sensitivity study based one loadcase on intra-operative forces measured for two VBT patients in an early clinical series. Additionally published data for intra-operative TASF surgical forces were used as a basis for other simulated VBT loadcases, as both these surgical procedures involve anterior fixation using vertebral screws. Further intra-operative measurements on a larger series of VBT patients will be used in future FEM analyses to provide more definitive data for intra-operative segmental loading. Additionally, further work is necessary to improve the biofidelity of the growth modulation algorithm applied in the FEM, both in terms of the implementation of the algorithm to facilitate automatic re-meshing of the vertebrae and in terms of the form of the growth modulation algorithm. The latter is based on small animal models, which may not be ideally suited to replicate human adolescent vertebral growth.

In the absence of detailed data describing tissue mechanics for paediatric spinal tissue, the current patient specific FEM includes tissue parameters derived using data for adult tissues. This is comparable to similar models in the field [8, 20] and a constant limitation associated with modelling young tissues due to lack of *ex vivo* samples. While the modelling approach and patient-specific FEM used in the current study have previously been validated using clinical and *in vivo* kinematic data for the AIS spine, the FEM was a representative AIS patient. The model geometry was not based on a specific patient who had received VBT surgery, rather provided a general indication of the predicted biomechanics for this surgical loadcase.

Drawing on these preliminary results, further investigations of different tether tension magnitudes and different screw configurations/orientations will be investigated to optimise the post-operative deformity correction. These analyses will be carried out with consideration of the balance required between providing a favourable mechanical environment for mechanically modulated growth asymmetry while not risking bone overload and potential screw-bone interface failure.

References

1. Crawford, C.H., Lenke, L.G.: Growth modulation by means of anterior tethering resulting in progressive correction of juvenile idiopathic scoliosis: a case report. JBJS (Am) **92**(1), 202–209 (2010)
2. Samdani, A.F., Ames, R.J., Kimball, J.S., Pahys, J.M., Grewal, H., Pelletier, G.J., Betz, R.R.: Anterior vertebral body tethering for idiopathic scoliosis. Spine **39**(20), 1688–1693 (2014)
3. Samdani, A.F., Ames, R.J., Kimball, J.S., Pahys, J.M., Grewal, H., Pelletier, G.J., Betz, R.R.: Anterior vertebral body tethering for immature adolescent scoliosis: one-year results on the first 32 patients. Eur. Spine J. **24**, 1533–1539 (2015)
4. Joshi, V., Cassivi, S.D., Milbrandt, T.A., Larson, A.N.: Video-assisted thoracoscopic anterior vertebral body tethering for the correction of adolescent idiopathic scoliosis of the spine. Eur. J. Cardiothorac. Surg. **54**, 1134–1136 (2018)
5. Newton, P., Fricka, K.B., Lee, S.S., Farnsworth, C.L., Cox, T.G., Mahar, A.T.: A symmetrical flexible tethering of spine growth in an immature bovine model. Spine **27**(7), 689–693 (2002)
6. Courvoisier, A., Eid, A., Bourgeois, E., Griffet, J.: Growth tethering devices for idiopathic scoliosis. Exp. Rev. **12**(4), 449 (2015)
7. Stokes, I.A.: Analysis and simulation of progressive adolescent scoliosis by biomechanical growth modulation. Eur. Spine J. **16**, 1621–1628 (2007)
8. Cobetto, N., Parent, S., Aubin, C.E.: 3D correction over 2 years with anterior vertebral body growth modulation: a finite element analysis of screw positioning, cable tensioning and postoperative functional activities. Clin. Biomech. (Bristol, Avon) **51**, 26–33 (2018)
9. Lavelle, W.F., Moldavsky, M., Cai, Y., Ordway, N.R., Bucklen, B.S.: An initial biomechanical investigation o fusionless anteiror tether constructs for controlled scoliosis correction. Spine J. **16**, 408–413 (2016)
10. Newton, P., Farnsworth, C.L., Upasani, V.V., Chambers, R.C., Varley, E., Tsutsui, S.: Effects of intraoperative tensioning of an anterolateral spinal tether on spinal growth modulation in a porcine model. Spine **36**(2), 109–117 (2011)
11. Braun, J.T., Hoffman, M., Akyuz, E., Ogilvie, J.W., Brodke, D.S., Bachus, K.N.: Mechanical modulation of vertebral growth in the fusionless treatment of progressive scoliosis in an experimental model. Spine **31**(12), 1314–1320 (2006)
12. Little, J.P., Adam, C.J.: Patient-specific modelling of scoliosis. In: Gefen, A. (ed.) Patient-Specific Modelling in Tomorrow's Medicine. Springer, Berlin (2012)
13. Little, J.P., Adam, C.J.: The effect of soft tissue properties on spinal flexibility in scoliosis: biomechanical simulation of fulcrum bending. Spine **34**(2), E76–E82 (2009). (Phila Pa 1976)
14. Kamimura, M., Kinoshita, T., Itoh, H., Yuzawa, Y., Takahashi, J., Hirabayashi, H., Nakamura, I.: Preoperative CT examination for accurate and safe anterior spinal instrumentation surgery with endoscopic approach. J. Spinal Disord. Tech. **15**(1), 47–51 (2002). discussion 51–42
15. Little, J.P., Adam, C.J.: Geometric sensitivity of patient-specific finite element models of the spine to variability in user-selected anatomical landmarks. Comput. Methods Biomech. Biomed. Eng. **18**(6), 676–688 (2014)
16. Little, J.P., Izatt, M., Labrom, R.D., Askin, G.N., Adam, C.J.: An FE investigation simulating intra-operative corrective forces applied to correct scoliosis deformity. Scoliosis Spinal Disord. **8**(1), 9 (2013)
17. Fairhurst, H., Little, J.P., Adam, C.J.: Intra-operative measurement of applied forces during anterior scoliosis correction. Clin. Biomech. (Bristol, Avon) **40**, 68–73 (2016)

18. Fredriks, A.M., van Burren, S., van Heel, W.J.M., Kijkman-Neerincx, R.H.M., Verloove-Vanhorick, S.P., Wit, J.M.: Nationwide age references for sitting height, leg length, and sitting height/height ratio, and their diagnostic value for disproportionate growth disorders. Arch. Dis. Child. **90**, 807–812 (2005)
19. Keaveny, T.M., Morgan, E.F., Niebur, G.L., Yeh, O.C.: Biomechanics of trabecular bone. Ann. Rev. Biomed. Eng. **3**, 307–333 (2001)
20. Rohlmann, A., Richter, M., Zander, T., Klockner, C., Bergmann, G.: Effect of different surgical strategies on screw forces after correction of scoliosis with a VDS implant. Eur. Spine J. **15**(4), 457–464 (2006)

Transpositions of Intervertebral Centroids in Adolescents Suffering from Idiopathic Scoliosis Optically Diagnosed

Saša Ćuković[1(✉)], William R. Taylor[2], Christoph Heidt[3], Goran Devedžić[1], Vanja Luković[4], and Tito Bassani[5]

[1] Faculty of Engineering, University of Kragujevac, Sestre Janjić 6, 34000 Kragujevac, Serbia
{cukovic, devedzic}@kg.ac.rs

[2] Institute for Biomechanics, Swiss Federal Institute of Technology – ETH Zurich, Leopold-Ruzicka-Weg 4, 8093 Zurich, Switzerland
bt@ethz.ch

[3] Children's Hospital, Pediatric Orthopedic Department, University of Basel, Spitalstrasse 33, 4056 Basel, Switzerland
christoph.heidt@ukbb.ch

[4] Faculty of Technical Sciences, University of Kragujevac, Svetog Save 65, 32102 Čačak, Serbia
vanja.lukovic@ftn.kg.ac.rs

[5] Laboratory of Biological Structures Mechanics, IRCCS Galeazzi Orthopaedic Institute, R. Galeazzi 4, 20161 Milan, Italy
tito.bassani@grupposandonato.it

Abstract. This paper describes new optical method and software tool developed to extract important intrinsic indicators of adolescent idiopathic scoliosis (AIS) non-invasively. Using optical digitalization, we scanned dorsal surfaces of 372 patients (141 males and 231 females) in a clinical environment from 2014–2017. Patient datasets were processed using our in-house "ScolioSIM" tool to generate key deformity indicators, along with 3D deformity visualizations. This method relies on a digitalized dorsal surface of AIS patient, on a multiscale and registerable 3D generic spinal model, CAD and KAx technologies. In this paper we focused our attention towards transpositions of intervertebral disks and their centroids in anteroposterior plane (AP). By knowing the exact transpositions values of all intervertebral centroids (L5/L4 to T1/C7) from the local spinal axis (DM-C7) it is possible to model 3D middle spinal alignment and to calculate the range of AIS curvature which is a basic for a new 3D classification scheme.

Keywords: Scoliosis (AIS) · Optical diagnosis · Intervertebral transpositions

1 Introduction

Adolescent idiopathic scoliosis (AIS) is the most frequent spinal deformity type, diagnosed in about 80% of all scoliosis cases (e.g. idiopathic) [1]. Scoliosis can affect normal life of patients, so it is of crucial importance to detect this deformity in the early

G. A. Ateshian et al. (Eds.): CMBBE 2019, LNCVB 36, pp. 133–141, 2020.
https://doi.org/10.1007/978-3-030-43195-2_10

stage to monitor progress before bone maturity is completing. There are different methods to evaluate scoliosis in clinical praxis, and many of them are based on x-ray imaging which is not always convenient for adolescents due to cumulative ionizing effect which can cause cancer development [2]. Besides preliminary diagnosis, regular radiological follow-ups are required to monitor the deformity every 3–6 months. On these radiographs in 2 or 3 planes it is possible to estimate or measure important scoliosis indicators, but also position of intervertebral discs centroids and centroids of vertebral bodies. By knowing the exact values of their transpositions from specific spinal axis it is possible to determine the position of the most transposed vertebra or disk e.g. apex vertebra/apex disk in AP plane which is crucial in scoliosis diagnosis. Also, it is possible to conclude which vertebra or disk is "end vertebra" or "end disk" in primary and secondary curve of scoliosis which is of significant importance for positioning of reference lines for Cobb angles measurement. Similar analysis can be conducted in sagittal and axial planes.

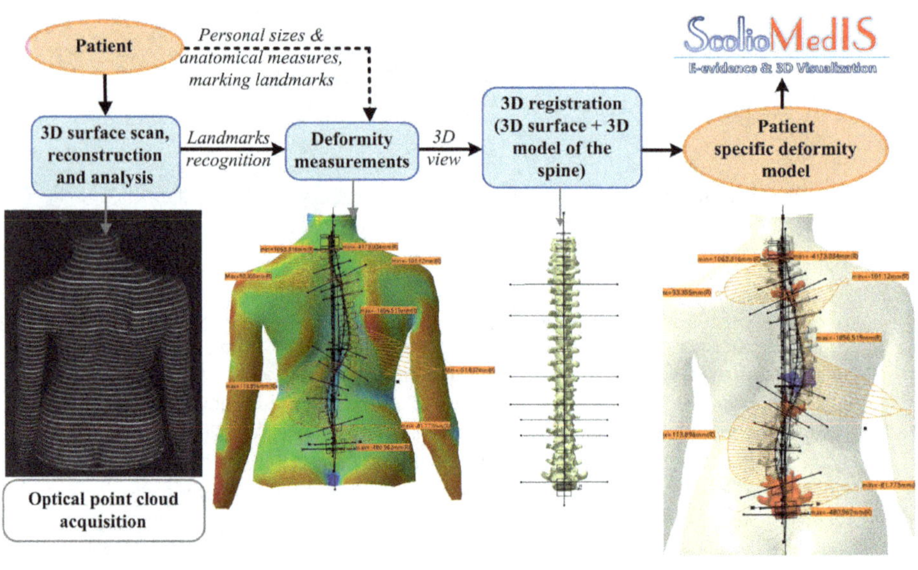

Fig. 1. General scheme – 3D optical non-invasive diagnosis of AIS patients

In order to reduce ionizing radiation of AIS patients, we recently developed a 3D non-invasive method and software tool (ScolioSIM) aiming to calculate a number of deformity indicators and to visualize patient-specific skeletal distortion, based only on a 3D virtual dorsal surface, optically digitalized (Fig. 1) [3]. This tool generates over 100 deformity indicators and also positions of vertebral and intervertebral centroids in 3D coordinates. We focused our attention here only on intervertebral disks transpositions in anteroposterior (AP) plane and their analysis.

2 Materials and Methods

Using 3D optical scanning device, we digitalized dorsal surfaces of 372 patients (141 males and 231 females) in a clinical environment from 2014–2017 [3]. Patients' datasets were processed using our in-house "ScolioSIM" tool to generate over 100 extrinsic and intrinsic deformity parameters, along with 3D deformity visualizations. This tool relies on a multiscale and registerable 3D generic spinal model developed in Materialise Mimics [4] and knowledgeware technologies (KAx), and implemented in CAD software CATIA v5 (Fig. 2) [5]. This tool will be an integral module of the Web-based information system for scoliosis monitoring called ScolioMedIS [6, 7].

ScolioSIM tool performs: surface 3D and asymmetry analysis, anatomical landmarks detection (e.g. C7, DL, DR, etc.), extraction of external and internal alignments (3D spinous processes line (green line in Fig. 2) and 5th degree B-Spline of middle spinal alignment (blue line in Fig. 2)), 3D registration of CAD spinal model to back surface, and calculates the key deformity indicators (Cobb angles, SOSORT-these angles, axial vertebral rotations, etc.).

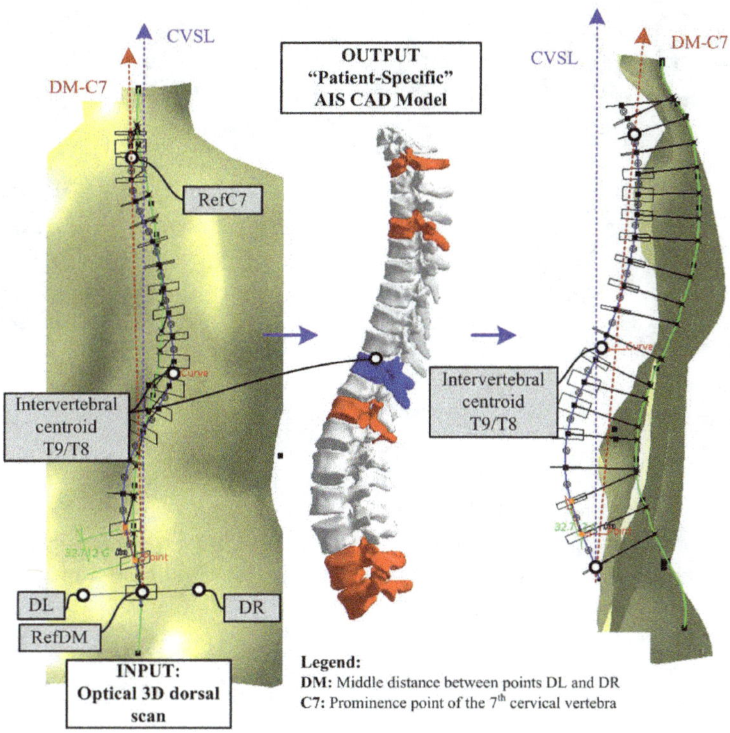

Fig. 2. 3D optical diagnosis of AIS and intervertebral centroids positions from DM-C7 and CVSL axis in frontal plane [5].

In this research, we focused our attention towards the estimated positions of intervertebral disks centroids from the DM-C7 line in the AP plane, located on the 5th degree B-Spline which represents the 3D middle spinal alignment [8]. These measures are important to determine whether AIS has apex disc or apex vertebra. We measured normal distances *a* and *b* of the specific reference points in lumbar, thoracic and cervical vertebral models from DM-C7 and CVSL lines, as it is illustrated on Fig. 3.

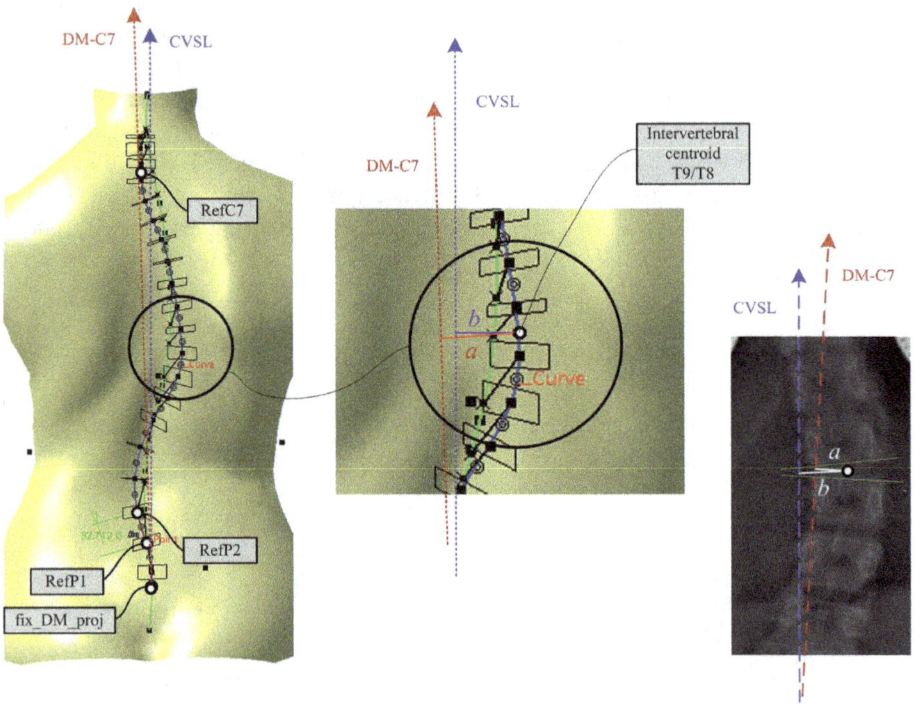

Fig. 3. Intervertebral transpositions from CVSL and DM-C7 spinal axes denoted as *a* and *b* lengths

To determine whether the AIS deformity is left or right oriented (*sinistroconvex* or *dextroconvex*) it is of crucial importance to determine which centroid is the mostly dislocated from the spinal axis (DM-C7): centroid of vertebral body (apex vertebra) or centroid of intervertebral disc (apex disc).

3 Results

We statistically processed all datasets and deformity indicators for male and female separately. The following Table 1 gives a summary of all min, max and average intervertebral transpositions in male and female patients from DM-C7 axis in AP plane, where negative values assign negative transpositions on left side, while positive describe right transpositions [9]. In case of multiple segments of the AIS curve (double, triple, etc.), there will be more than one apex centroid. However, the most important one is apex of the primary AIS curve, where the Cobb angle has the maximum value. Comparative analysis of these values is given in Figs. 4 and 5.

Table 1. Descriptive statistics on min, max and average positions of intervertebral centroids from C7-DM line [mm] in male and female cases; ("−") assigns left and ("+") right oriented scoliosis (frontal plane).

Parameters	Males, n = 141			Females, n = 231		
	Min	Max	Mean	Min	Max	Mean
T1/C7	−3.2	1.9	−0.3 ± 0.8	−5.8	2.8	−0.2 ± 0.9
T2/T1	−10.4	6.6	−0.9 ± 2.8	−10.6	9.3	−0.4 ± 3.0
T3/T2	−20.4	13.4	−0.8 ± 5.1	−17.4	19.4	0.5 ± 5.6
T4/T3	−27.5	24.9	−0.2 ± 7.5	−25.3	28.2	2.5 ± 8.3
T5/T4	−30.0	34.9	0.4 ± 9.2	−33.2	33.9	4.4 ± 10.5
T6/T5	−28.9	42.7	1.1 ± 10.3	−38.4	40.5	6.1 ± 12.2
T7/T6	−24.1	46.6	1.8 ± 10.9	**−43.3**	45.4	7.4 ± 13.4
T8/T7	−20.2	48.2	2.2 ± 11.2	−39.4	47.8	8.2 ± 13.7
T9/T8	−22.4	48.0	2.2 ± 11.1	−41.5	49.0	8.2 ± 13.5
T10/T9	−25.5	46.8	1.9 ± 10.9	−39.6	**49.6**	7.5 ± 12.9
T11/T10	−29.1	51.9	1.2 ± 10.8	−35.4	44.2	6.0 ± 12.1
T12/T11	−32.5	55.1	0.3 ± 11.1	−31.0	42.9	3.9 ± 11.8
L1/T12	**−33.4**	**55.4**	−0.3 ± 11.4	−30.9	38.1	1.7 ± 11.7
L2/L1	−30.3	44.4	−0.8 ± 10.6	−33.3	34.8	0.0 ± 10.9
L3/L2	−24.5	30.2	−0.8 ± 8.6	−25.0	27.4	−0.8 ± 8.8
L4/L3	−17.1	17.3	−0.4 ± 5.5	−12.2	20.4	−0.9 ± 5.8
L5/L4	−9.5	12.0	−0.0 ± 2.5	−6.7	9.3	−0.3 ± 2.2

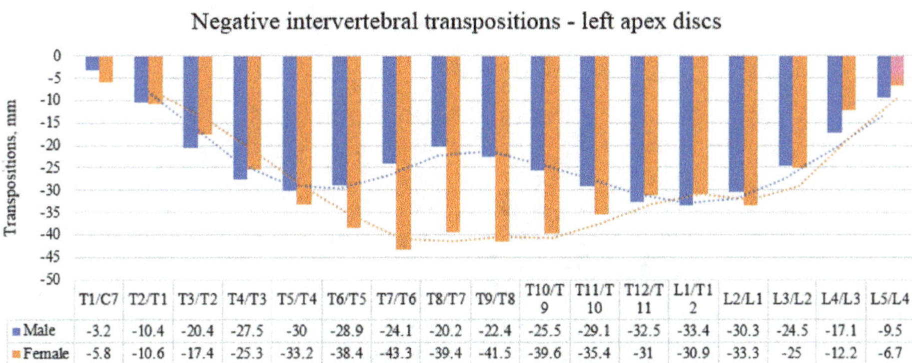

Fig. 4. Maximal negative intervertebral transpositions from DM-C7 spinal axis

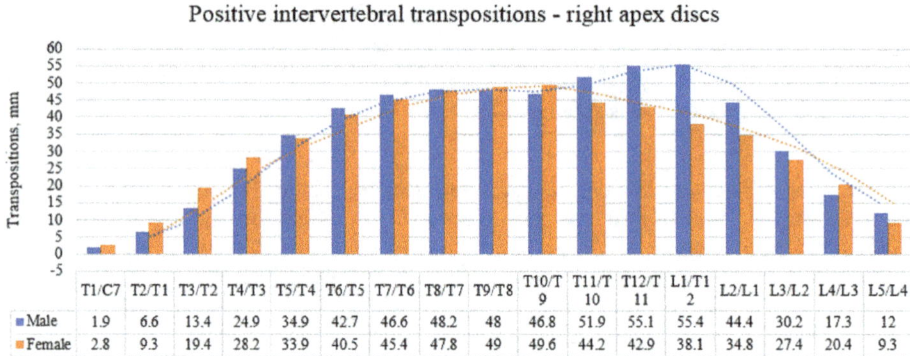

Fig. 5. Maximal positive intervertebral transpositions from DM-C7 spinal axis

Transpositions of all intervertebral centroids (L5–L4 to T1–C7) from the DM-C7 line (Table 1) revealed that the most transposed intervertebral disc (apex disc) in male AIS patients was located between L1–T12 in left thoraco-lumbar curves (−33.4 mm) than the right-transposed disc between the same vertebrae (55.4 mm). In female AIS patients, apex disc was located between T7–T6 in left thoracic curves (−43.3 mm) and the right-transposed disc between T10–T9 (49.6 mm).

According to the frequent analysis, we had the highest number of apex intervertebral discs transposed in male patients (disc L1–T12 in 26/141 cases) and female patients (disc L2–L1 in 38/231 cases).

Descriptive statistics on absolute average positions of intervertebral centroids measured from CVSL line is presented in the Table 2.

Table 2. Descriptive statistics on absolute average positions of intervertebral centroids from CVSL line [mm] in male and female cases (frontal plane).

Parameters	Males, n = 141	Females, n = 231
	Mean	Mean
T1/C7	5.2 ± 3.7	5.5 ± 3.6
T2/T1	5.7 ± 4.2	6.0 ± 4.1
T3/T2	6.2 ± 4.7	6.1 ± 4.7
T4/T3	6.9 ± 5.6	7.0 ± 5.8
T5/T4	7.6 ± 6.6	8.8 ± 7.0
T6/T5	8.2 ± 7.2	10.3 ± 8.2
T7/T6	8.5 ± 7.7	11.2 ± 9.3
T8/T7	8.6 ± 7.9	11.5 ± 9.9
T9/T8	8.3 ± 8.0	11.2 ± 10.0
T10/T9	7.9 ± 7.9	10.7 ± 9.4
T11/T10	8.1 ± 7.5	9.8 ± 8.5
T12/T11	8.5 ± 7.4	9.2 ± 7.8
L1/T12	8.8 ± 7.7	8.7 ± 7.7
L2/L1	8.1 ± 7.2	8.3 ± 6.9
L3/L2	6.5 ± 5.8	7.0 ± 5.4
L4/L3	4.2 ± 3.7	4.8 ± 3.6
L5/L4	2.0 ± 2.1	2.0 ± 2.4

Similar analysis was performed in [10] for vertebral centroids. It revealed that the most frequent apices in AIS patients are located between T1–T11 with predominant thoracic curve (72/141 males, 51.1%; 146/231 females, 63.2%) followed by the thoracolumbar curves with apices between T12–L1 (47/141 males, 33.3%; 55/231 females, 23.8%).

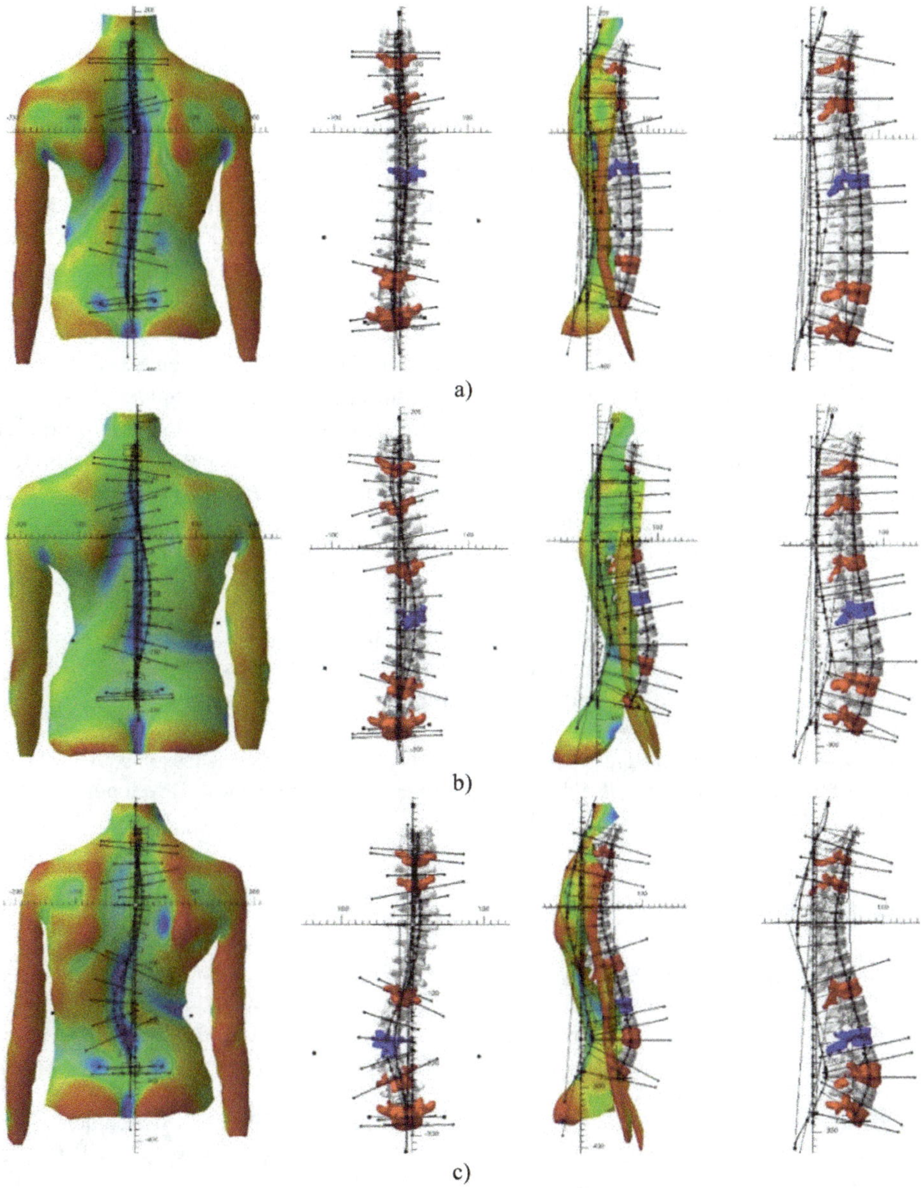

Fig. 6. Case studies – Patient specific CAD models: (a) apex disc in upper thoracic region, (b) apex disc in lower thoracic region, (c) apex disc in lumbar region

3.1 Case Studies

By knowing the exact values of transpositions of centroid of intervertebral, it is possible to model the range of AIS curves in left and right directions from the local spinal axis and to determine whether the AIS is right or left oriented. In our samples we had the most frequent dextroconvex AIS (Fig. 6a and b) in female patients (158 cases, 68.4%), while in males we had the most frequent sinistroconvex AIS (Fig. 6c) (71 cases, 50.4%).

4 Conclusion

The clinical relevance of the presented study is high, since the degree of the deformity and the localization of the apex vertebra or intervertebral disk on AIS middle spinal alignment define a range, as well as the horizontal boundaries of AIS spinal curves, and may be the basis for investigating the range of motion and flexibility of the spine. We demonstrated that our radiation-free imaging method and ScolioSIM tool enable a reliable 3D visualization of deformity and detection of subtle postural changes in a large cohort affected by the most translated disk in the primary AIS curve. Due to its non-ionizing nature, the integrated method of data acquisition and analysis has enormous potential to be safely used in clinical settings, for monitoring AIS, and to predict important indicators of the underlying structures and the associations between the intervertebral transpositions and other parameters of spinal deformity. After the final validation of our 3D visualization module against 200 biplanar EOS radiographs, we expect its implementation in clinical practice.

Ethics Statement

The local ethics committee of San Raffaele Hospital (Milan, Italy) approved this study (no. EOSCC2013). Subject assent and parental permission to participate in the study and use anonymized data were given by signing an informed consent.

Acknowledgments. Presented research is supported by the Swiss-SERI-SGES grant [2017.0024] and by the Serbian Ministry of Education, Science and Technological Development grant [III-41007].

References

1. Akbarnia, B.A., Yazici, M., Thompson, G.H.: The growing spine: management of spinal disorders in young children. Springer, Berlin (2011). ISBN 978-3-540-85206-3
2. Ronckers, C.M., Land, C.E., Miller, J.S., Stovall, M., Lonstein, J.E., Doody, M.M.: Cancer mortality among women frequently exposed to radiographic examinations for spinal disorders. Radiat. Res. **174**(1), 83–90 (2010)
3. Ćuković, S.: Non-rigid registration of sculptured surfaces in internet environment, PhD Thesis, University of Kragujevac, Faculty of Engineering, Kragujevac, Serbia (2015)
4. Mimics 18.0, Materilaise, Belgium. https://www.materialise.com/

5. Devedzic, G., Ristic, B., Stefanovic, M., Cukovic, S., Lukovic, T.: Development of 3D parametric model of human spine and simulator for biomedical engineering education and scoliosis screening. J. Comput. Appl. Eng. Educ. **20**(3), 434–444 (2010)
6. Devedžić, G., Ćuković, S., Luković, V., Milošević, D., Subburaj, K., Luković, T.: ScolioMedIS: web-oriented information system for idiopathic scoliosis visualization and monitoring. J. Comput. Meth. Programs Biomed. **108**(2), 736–749 (2012)
7. Luković, V., Ćuković, S., Milošević, D., Devedžić, G.: An ontology-based module of the information system scoliomedis for 3D digital diagnosis of adolescent scoliosis. J. Comput. Meth. Programs Biomed. **178**, 247–263 (2019)
8. Ćuković, S., Devedžić, G., Luković, V., Anwer, A., Zečević-Luković, T., Subburaj, K.: 3D modeling of spinal deformities shapes using 5th degree B-splines. J. Prod. Eng. **18**(2), 103–106 (2015)
9. Stokes, I.A.F.: Three-dimensional terminology of spinal deformity. A report presented to the scoliosis research society by the scoliosis research society working group on 3D terminology of spinal deformity. Spine **19**(2), 236–248 (1994)
10. Ćuković, S., Heidt, C., Studer, D., Taylor, W.: Apex vertebra transpositions in the 3D optical diagnosis of 372 patients with adolescent idiopathic scoliosis, In: 5th Virtual Physiological Human (VPH) Conference - VPH2018, Zaragoza, Spain, pp. 1 (2018)

Mechano-Physiological Modeling to Probe the Role of Satellite Cells and Fibroblasts in Cerebral Palsy Muscle Degeneration

Stephanie Khuu[1], Kelley M. Virgilio[2], Justin W. Fernandez[1,3], and Geoffrey G. Handsfield[1(✉)]

[1] Auckland Bioengineering Institute, University of Auckland, Auckland, New Zealand
{s.khuu, g.handsfield}@auckland.ac.nz

[2] Department of Biomedical Engineering, University of Virginia, Charlottesville, VA, USA

[3] Department of Engineering Science, University of Auckland, Auckland, New Zealand

Abstract. Cerebral palsy (CP) is a neural disorder that greatly affects the musculoskeletal system, but the causes and progression of muscle degeneration are poorly understood. The CP muscle environment is altered compared to typical, particularly with the presence of fibrosis and fewer satellite cells (SCs). Healthy regeneration of muscle requires both SCs—a progenitor cell population for muscle cells—and fibroblasts—the primary instigators of extracellular matrix (ECM) remodeling. SCs and fibroblasts interact, but their dynamics at the muscle level are complex and nonlinear; nevertheless, detailed knowledge of these dynamics is necessary for a thorough understanding of muscle degeneration in CP and as a precursor to the development of novel clinical interventions. In this work, computational agent-based modeling (ABM) was used to investigate muscle regeneration by representing muscle tissue adaptation at the cellular level following injury. We used an ABM to vary SC levels in simulated muscle regeneration, which showed that muscle fiber recovery was impaired when SC levels were decreased, whereas fibroblast activity was enhanced. Complete recovery of damaged muscle tissue was sensitive to the level of injury. Coupling of this ABM with finite element modeling will contribute to the development of a mechano-physiological model to probe muscle injury and regeneration in CP.

Keywords: Cerebral palsy · Agent based modeling · Muscle regeneration

1 Introduction

Cerebral Palsy (CP) is a neuromusculoskeletal disorder that commonly results in impaired movement, gait, and posture [1]. The cause of CP is a nonprogressive lesion in the central nervous system that occurs at or near birth [2, 3]. Secondary to this are musculoskeletal effects including spasticity, contractures, and impaired growth leading to muscles which are both weak and functionally short [4–7]. Further research is needed

G. A. Ateshian et al. (Eds.): CMBBE 2019, LNCVB 36, pp. 142–157, 2020.
https://doi.org/10.1007/978-3-030-43195-2_11

into the nature of the impaired growth in CP, but evidence suggests that skeletal muscle growth is inhibited in both the along-fiber and cross-fiber directions via inhibition of serial sarcomere addition [1, 8, 9] and cross-sectional hypertrophy [2, 9]. Reduced muscle growth has been shown to precede the development of increased muscle stiffness in children with CP [9, 10], suggesting that an understanding of pathological growth in CP may underpin a greater understanding of the disease as a whole.

1.1 Muscle Growth and Regeneration

Muscle injury is commonly linked to mechanical loading, such as eccentric exercise [11]. The process of muscle repair and regeneration that follows has four interrelated phases: necrosis and degradation of tissue, acute inflammation, satellite cell (SC) activation and repair, and extracellular matrix (ECM) remodeling [12, 13]. Firstly, tissue damage triggers recruitment of leukocytes that control the acute pro- and anti-inflammatory periods of muscle repair, as well as the timing of SC recruitment [13]. The duration of SC activation and proliferation is affected by the presence of secreted factors such as HGF, TNF-α, IGF-1, FGF, and TGF-β from surrounding cells [14, 15]. Muscle regeneration takes place when SCs differentiate into myoblasts and fuse to form nascent myotubes [12, 16, 17]. ECM remodeling by fibroblasts occurs continuously throughout the entire repair process [18].

1.2 Regenerative Potential

Muscle adaptation following injury requires the presence of a myogenic population that differentiates to form muscle tissue [19]. SCs are the mononucleated progenitor cell population pivotal to physiological muscle repair and regeneration [20–22]. In an uninjured state, SCs reside between the plasma membrane of the muscle fiber and the basal lamina. SC content of muscle differs between age groups, activity levels, and location along the muscle fiber [23, 24]. SCs on myofibers proliferate and migrate to the site of injury and initiate muscle regeneration [20]. Aberrant regeneration can occur when too few SCs are present or when fibroblasts become overactive [18]. Fibroblasts make up a small portion of cells in skeletal muscle [25], but they synthesize and assemble the majority of the components required for functional ECM remodeling [18, 26, 27].

The physiological cascade of events that impairs skeletal muscle growth and leads to the development of contractures in CP is largely unknown [1, 2]. Several *in vitro* studies suggest that the muscle milieu in patients with CP is drastically altered at the cellular level [9, 28–30]. Two recurring observations from these studies are the reduced potential for myogenic growth due to SCs, and perturbed ECM remodeling performed by fibroblasts.

In CP muscle, SC content is diminished, and excess collagen deposition results in fibrosis [1, 30, 31]. Smith et al. [32] investigated populations of SCs, inflammatory cells and endothelial cells in biopsies of spastic human muscle. They found significantly fewer SCs (up to 70% decrease) compared to typically developing (TD) individuals. Transcriptional levels of inflammatory and endothelial cells were found to be similar in CP and TD groups. ECM remodeling is also impaired following stretch-induced injury, due in part to the reduced SC content [2]. The progression of both ECM remodeling and muscle regeneration in CP at the cellular level remains poorly understood. Although transcriptional abnormalities have been highlighted, no mechanism has been defined for the decrease in SC population. It is also unclear how SCs interact with other cellular factors in the process of impaired muscle regeneration in CP. Computational simulation approaches may be particularly appropriate here, in order to probe the mechanisms and interactions between SCs, fibroblasts, related growth factors, and their contribution to impaired muscle growth.

1.3 Agent-Based Modeling

Agent-based modeling (ABM) describes a bottom-up computational modeling technique that ascribes sets of rules and behaviors to individual agents (e.g., cells) that define their interactions with other agents and with the environment [33, 34]. The ABM approach has previously been applied to studies of tumor formation, vascular remodeling, bone tissue regeneration, wound healing, disuse atrophy, and disease modifications in muscle [35–39].

One feature of ABM is its ability to demonstrate emergent phenomena—macroscale behaviors that are non-intuitive consequences of individual interactions occurring in large numbers [40, 41]. Cellular and non-cellular components of an ABM can be programmed to perform biologically relevant behaviors such as proliferation, apoptosis, and migration [42, 43]. In physiology, ABMs are a powerful tool in their ability to connect tissue and organ level behaviors to simple cellular interactions. *In silico* experimentation that can repeatedly and efficiently test experimentally driven hypotheses while incorporating cellular and molecular dynamics from the literature, may provide a systematic solution to investigating the impaired growth process that manifests as muscle degeneration in CP [44].

2 Methodology

2.1 Construction of an Agent-Based Model for Healthy Muscle Regeneration

To probe the decrease in regenerative potential in CP muscle compared to typical, we created an ABM of muscle regeneration in both healthy and CP muscle, built in Repast Simphony, a Java-based modeling platform (Argonne National Laboratory, Lemont, IL). Rules were developed based on literature descriptions of physiological interactions. The model was built to simulate the progression of events during muscle regeneration. Agents included fibers, SCs, fibroblasts, macrophages, ECM, and secreted growth factors. Model geometry consisted of 4865 2-D grid elements that represented 17 muscle fiber cross-sections in ECM (see Fig. 1 and section below). We simulated injury-induced muscle regeneration over 28 days with time steps of one hour. The injury was modeled as a user-defined percentage of muscle fibers set to a damaged phenotype, which occurred at model instantiation. End-point fiber count was used as a measure of muscle fiber regeneration.

Agent Interactions

Fibers & ECM

Fiber geometry is based on cross-sections of 17 muscle fibers from the mouse lower limb (Fig. 1) [38]. The histological images were masked in MATLAB (Mathworks, Natick, MA) to extract fiber and ECM pixels for mapping on to the ABM grid. Injury was modeled as a prescribed amount of damage that was stochastically applied to the myofibrils to damage either 20% (normal injury) or 40% (double injury) of the total fibrils. Each of the ECM components in the damaged fibrils' Von Neumann neighborhood was also damaged to allow for ECM remodeling and muscle growth. The extended area of damage simulated necrotic tissue.

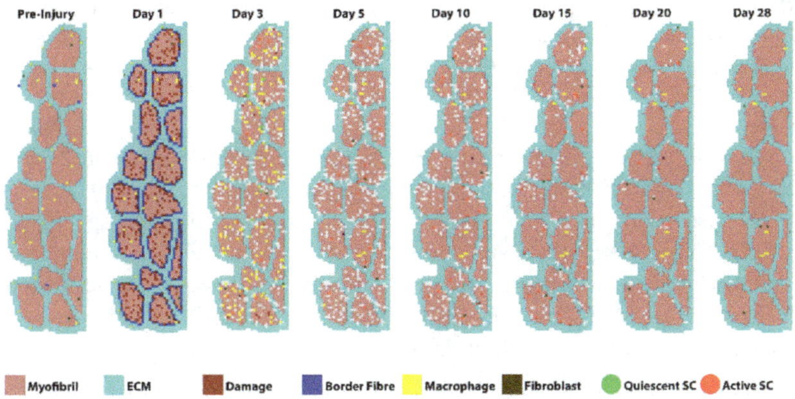

Fig. 1. Agent-based model simulation of muscle regeneration. Injury to healthy muscle is simulated as damaged fibrils randomly distributed throughout the geometry. Over 28 days, activity of macrophages, fibroblasts, and SCs repair the damaged tissue and return it to a state similar to pre-injury.

Macrophages

Macrophages are critical to the repair process as they express several inflammatory mediators that drive the cellular responses following injury [13, 45]. In this model, macrophages search a Moore neighborhood for damaged fibers and ECM components. Once the damage is located, the angle of desired movement is calculated, and macrophages move towards a specific damaged point. This method simulates the behavior of chemical factors that attract macrophages. When a macrophage encounters a damaged fiber, the fiber is set to 'degenerating' and is eligible to be phagocytosed. Phagocytosis allows for macrophage proliferation as well as increases in the levels of TNF-α present in the model.

Satellite Cells

In this simulation, SCs are seeded according to physiological values of approximately 0.24 SCs per fiber for a 20-μm thick section [38]. During initialization, SCs are stochastically placed on border fibers to represent their location in muscle between the sarcolemma and basement membrane of muscle fibers [20]. The activation of quiescent SCs requires the presence of hepatocyte growth factor (HGF) [46]. SC division can take place after 10 h for each agent given that the active SC is in close proximity to a fibroblast [27, 47, 48]. The proliferation of active SCs then occurs to mimic the transit-amplifying cells present during repair [49]. This process takes place in the timeframe when levels of IGF-1 and TGF-β are increased. SCs are protected from apoptosis while pro-inflammatory macrophages are present in the ABM [38]. Activated SCs express Pax7, MyoD, and Myf5 [22]. There was a 10% probability for activated SCs to commit self-renewal rather than differentiation [38, 50]. Three flags signal for differentiation to occur: loss of Pax7, loss of Myf5, and an increase in myogenin over time [20, 50]. Once differentiated, SCs do not express Myf5 and become myoblasts [15]. A myoblast agent is created in the fiber class, and the myoblasts fuse to damaged fiber edges to repair the injured muscle. Simulations were run with varying levels of SCs. SCs per fiber was varied between 0.24 (control) and 0 with a 0.06 decrement. SC levels were tested at baseline and double levels of initial injury.

Fibroblasts

Fibroblast levels are seeded according to Mackey et al. [27]. Fibroblasts proliferate based on a positive TNF-α gradient and the presence of neighboring activated SCs [27, 51]. Fibroblasts become activated myofibroblasts in the presence of a positive TGF-β gradient [52]. Myofibroblasts search the area for empty cells that neighbor ECM components. Myofibroblasts compete with myoblasts to regenerate tissue by depositing collagen near damaged ECM edges. When there were no SCs present in the muscle, fibroblasts deposited collagen within damaged fiber edges.

Secreted Factors

Levels of secreted factors IGF-1 [39], TNF-α [39, 53] and TGF-β [39, 54] per hour were represented by the following equations:

$$\frac{dIGF}{dt} = 8.8e^{-5}Fb + 5.1e^{-5}F \tag{1}$$

$$\frac{dTNF}{dt} = 5.8e^{-2}PM + 4.9e^{-6}Fb \tag{2}$$

$$\frac{dTGF}{dt} = 8.75e^{-3}AM + 1.7e^{-6}Fb \tag{3}$$

where Fb was the number of fibroblasts, F was the number of fibers, PM was the number of pro-inflammatory macrophages, and AM was the number of anti-inflammatory macrophages.

3 Results

We developed an ABM to explore the cellular pathophysiology of CP and probe the regeneration of fibers with differing levels of injury, SCs, and fibroblast activity. To compare our model's predictions with experiments from the literature, we made a comparison of SC counts over a 14-day period following injury to the experimental observations of Murphy et al. [51] (Fig. 2). The ABM followed the timing and amplitude of SC proliferation *in vitro* for the first five days following injury.

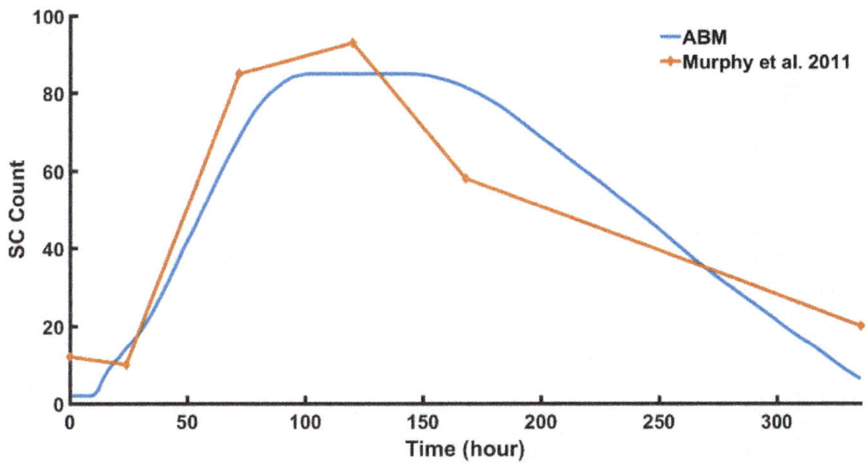

Fig. 2. Satellite cell count comparison between experimental results from Murphy et al. [51] and ABM following 20% injury with control 0.24 initial SC number. Satellite cell proliferation in the first five days was comparable to *in vitro* values.

In an exploration of the effects of SC concentration and injury magnitude, we varied each to examine recovery timing and magnitude. Following a 20% injury, fibers recovered and exceeded the original fiber count when SCs per fiber was set to 0.12, 0.18, and 0.24 (Fig. 3). Peak recovery times occurred in the two to three-week time-frame. Reducing SCs by 75% (0.06 SC per fiber) affected the ability of the muscle cells to regenerate beyond the pre-injury state, resulting in a lack of hypertrophy growth, but nearly returning the muscle to the original fiber count (98%).

The magnitude of injury also affected the overall regeneration and recovery outcomes. Increasing the extent of injury by a factor of two altered recovery profiles by preventing full recovery at all SC levels (Fig. 4). The level of recovery scaled with the number of SCs per fiber following both 20% and 40% injury, but greater injury inhibited the ability of the model to repair the damage fully (Fig. 5). Recovery percentages were increased following 40% injury at 0.12, 0.18, and 0.24 SC per fiber but were not sufficient for total fiber count regeneration.

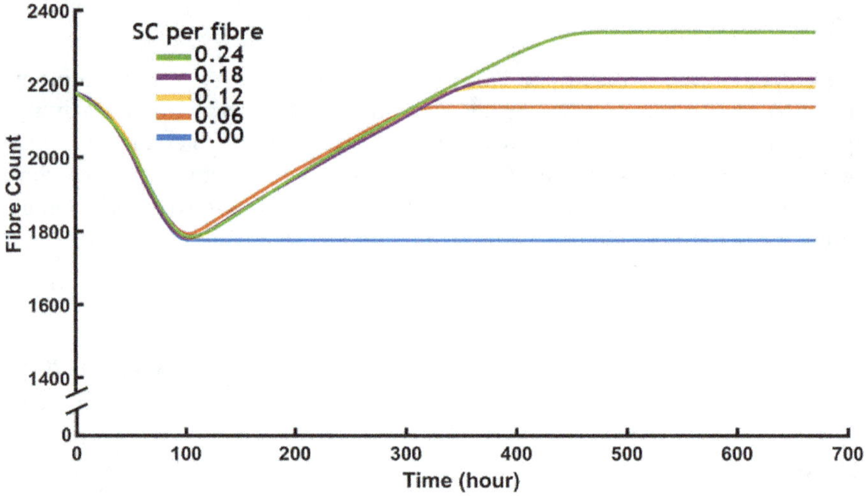

Fig. 3. Regeneration of muscle following 20% injury at different levels of SCs per fiber. Fiber count recovery scaled with the concentration of SCs per fiber. Control levels (0.24 SC per fiber) showed marked fiber count improvement. Mid-levels of SCs (0.18 & 0.12) resulted in fiber count recovery above the original, while a 75% reduction in SC concentration led to sub-optimal regeneration.

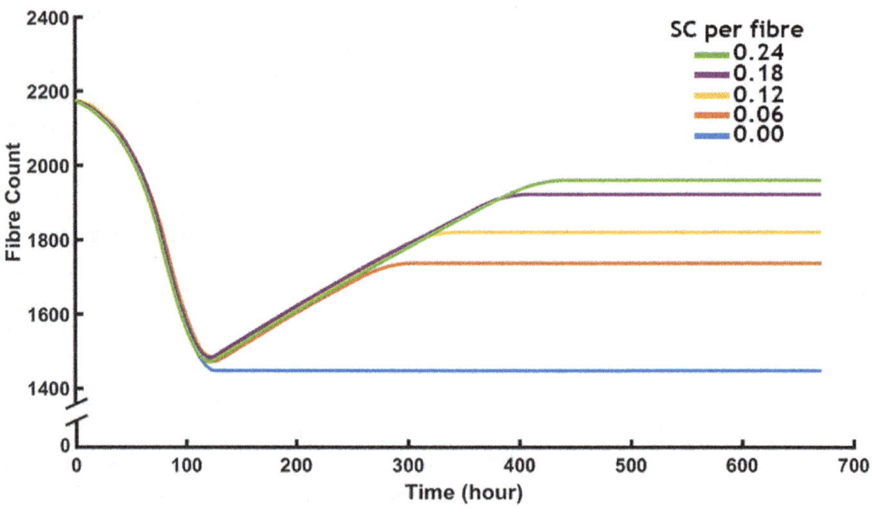

Fig. 4. Regeneration of muscle following 40% injury at different levels of SCs per fiber. Regeneration scaled with SC concentration; however regeneration of fiber count was impaired at every SC level, and no net improvement in fiber count was observed.

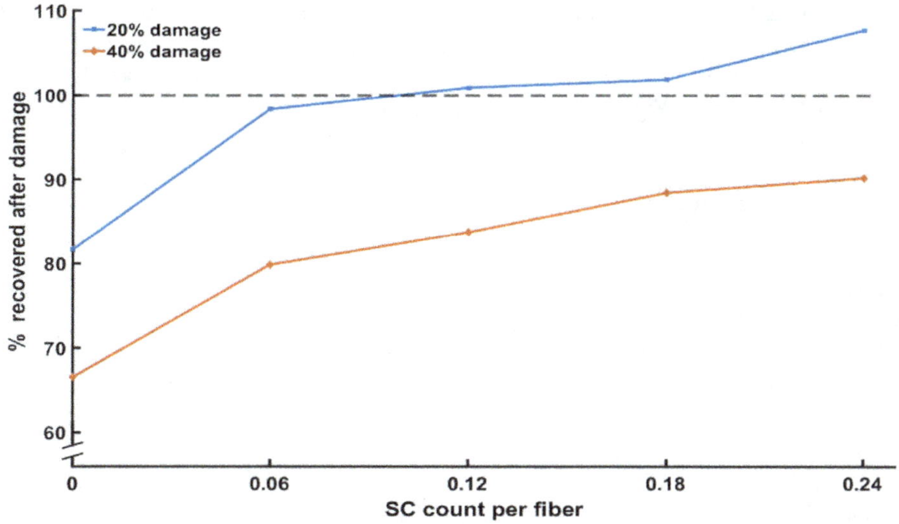

Fig. 5. Percentage of recovery increased with SC count following both 20% and 40% damage. The percentage of recovery was greater at 0.12, 0.18, and 0.24 SC per fiber following 40% injury despite incomplete fiber recovery.

The recovery of the ECM reflects the movement and actions of fibroblasts. Reflective of the competitive nature of fibroblasts and satellite cells, recovery of ECM components was greatest when simulations had no SCs on fibers or with 0.06 SC per

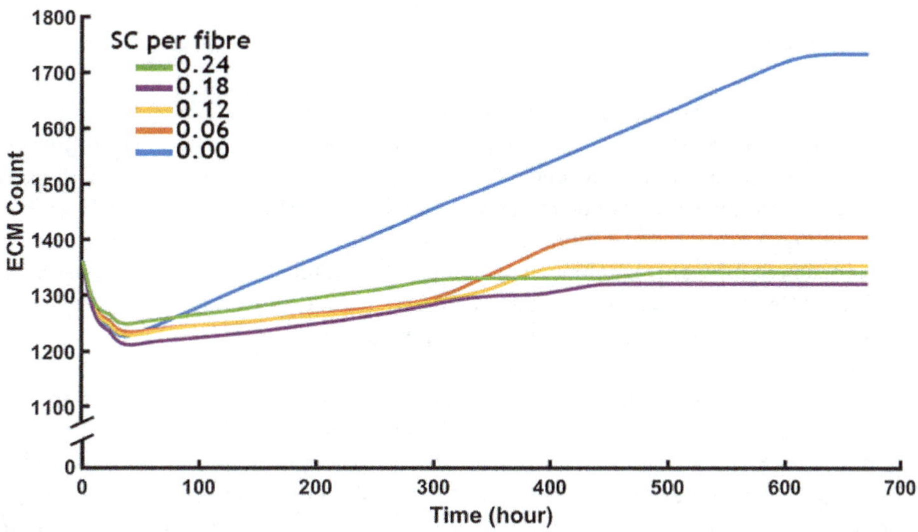

Fig. 6. ECM recovery over 28 days following 20% injury with varying levels of SCs on fibers. ECM Count regenerated faster at lower SC concentrations of 0.00 and 0.06. At all other concentrations, ECM did not recover to original levels.

fiber following 20% and 40% injury (Fig. 6). In the simulations with 0.00 and 0.06 SC per fiber during 20% injury, ECM recovery was above original levels and can be interpreted to represent fibrosis. This outcome was also apparent at all levels of SCs following 40% injury (Fig. 7) where ECM recovery was highest.

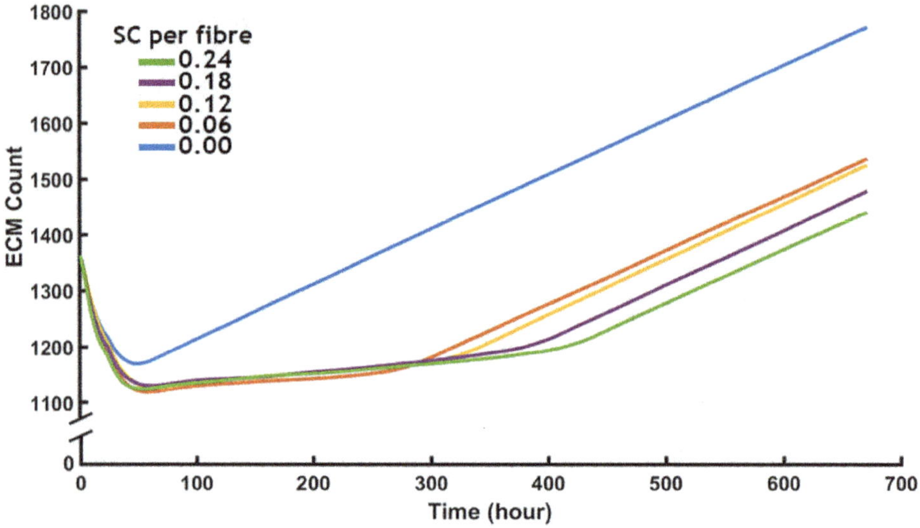

Fig. 7. ECM recovery over 28 days following 40% injury with varying levels of SCs on fibers. ECM count recovery was higher at low SC concentrations, and all recovery profiles exceeded the original ECM count. There was a higher rate of recovery when no SCs were present in the simulation.

4 Discussion

The motivation for this research stems from the hypothesis that the observed pathological muscle degeneration in CP results from an impairment in the muscle regeneration process. There is a need to understand this process in individuals with CP better as they grow. In this study, we developed an *in silico* model of the micro-scale muscle milieu during muscle regeneration. To understand muscle degeneration in CP at the most fundamental level, modeling that incorporates cellular dynamics is an essential step. This ABM was used to investigate decreases in SC concentration and an increase in fibrosis observed in the CP muscle environment.

SCs are not required for hypertrophy in adulthood [55]; however, they are known for their regenerative potential following injury [21, 56]. In our ABM, regeneration of fiber count scaled with the levels of SCs following both normal and double damage. The effect of SC concentration on fiber recovery showed non-linearities following normal injury where mid-levels of SCs did not affect the overall fiber recovery greatly. At 0.06 SC per fiber and below, the fiber regeneration was impaired compared to control. CP muscle is weaker and has a smaller physiological cross-sectional area compared to typically developing muscle [57]. Interestingly, our model showed that a combination of high injury and reduced SC concentration would contribute to deficient muscle regeneration. Continued loading of a CP muscle may act as repeated injury. This, combined with the up to 70% decrease of SCs in CP muscle, would lead to slower and otherwise impaired tissue regeneration over time.

Our model successfully captured the timing of healthy regeneration. Active muscle regeneration peaks at seven to ten days post-injury and reaches a peak at two weeks [12]. Mature healthy muscle completes its regeneration process in three to four weeks post-injury [23, 24]. Our ABM shows that recovery following single magnitude injury was more resistant to changes in SC concentration compared to the double injury simulation. This is inconsistent with work from Virgilio et al. [38], which found that initial damage magnitude at control levels of SCs did not lead to significant changes in the fiber recovery of cross-sectional area after 28 days. It is also important to recognize that SCs can undergo symmetric and asymmetric division to replenish the quiescent SC pool [50]. Implementing this process in the current ABM would provide a detailed picture of myogenic potential based on the number of active and differentiating SCs over time. In order to compare the recovery at different magnitudes of injury, the ABM requires more detailed rules to govern the action of SCs and fibroblasts.

In our model, consistent with literature descriptions of the physiology, scar-tissue formation competes with the regeneration of myofibers [24, 58]. Fibroblasts deposit collagen for ECM remodeling, and under homeostatic conditions, fibrosis does not occur [25]. Muscle fiber growth was able to occur during fibroblast activity throughout the simulation, with the ECM remodeling around the newly fused myoblasts. However, a lack of SCs activity led to recovery of ECM components above the original level around fiber edges and within the fiber boundaries. When SC count was decreased, the ECM count increased despite no change in the speed of fibroblast activity and no change in the number of seeded fibroblasts (Fig. 5). To estimate the amount of fibrosis that is caused by decreased SC count and overactive fibroblasts, more rules regarding the response of fibroblasts to inflammatory cytokines and growth factors need to be implemented.

The increase in initial injury led to incomplete ECM recovery due to the insufficient muscle fiber regeneration. Insufficient regeneration could occur due to delayed debridement and phagocytosis of tissue [59]. Delayed clearance of necrotic tissue has the potential to enhance myofibroblast activation due to the prolonged presence of TNF-α expressed by inflammatory macrophages [60]. Future ABMs should investigate the inflammatory stage of muscle regeneration and how delayed clearance can influence fibroblast dynamics and fibrosis.

The process of muscle regeneration is largely guided by the chemical environment [52, 60–62]. The inclusion of a more detailed sub-cellular environment allows for a more physiologically accurate model of muscle regeneration. The estimation of growth factor activity in our model currently includes TGF-β, HGF, TNF-α, and IGF-1 secretion from a limited number of cell types. Recreating a more detailed growth factor environment in healthy and impaired muscle could reveal important insights into how changes in growth factor levels can help or hinder different stages in the muscle regeneration process. The agent rules do not currently include migration dynamics and restrictions to movement along fiber edges; thus, the practical implications of implementing additional growth factors include the ability to apply spatial gradients of growth factors across the grid environment to guide the speed and direction of various cell types.

5 Conclusions

Changes in macroscale tissue morphology can often be explained by microscale phenomena. This study aimed to characterize the adaptations in the microscale muscle environment during muscle regeneration in CP via an *in silico* modeling approach. The physiological processes of muscle regeneration in CP are not well understood. ABM offers a way of studying populations of cells, which incorporates the non-linearity of physiological interactions. These tools combined with experimental validation can be used to test targeted hypotheses and elucidate potential therapeutic avenues for muscle repair and regeneration in CP.

6 Future Work

Finite element modeling (FEM) can be used to numerically approximate how mechanical loading affects muscle deformation at the whole muscle, fascicle, or fiber level using a continuum approach that homogenizes information below these structural levels [63]. Because biological tissues are subject to large deformations, the stress-strain relationships are non-linear, and the underlying tissue has preferred material property directions, whole muscles and muscle fibers have previously been represented as transversely isotropic incompressible hyperelastic materials [63, 64].

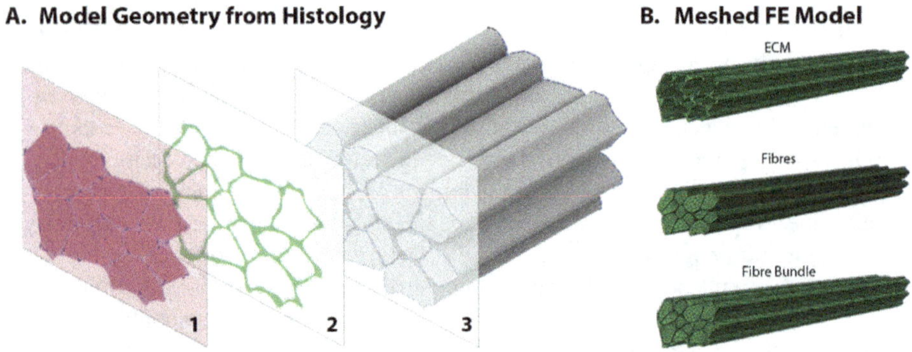

Fig. 8. (A) Construction of muscle fiber volumes by (1) taking coordinates from a histological cross-section in MATLAB, (2) recreating a coordinate map of 2-D cross-section in Autodesk Inventor, and (3) extruding volumes based on the 2-D cross-section to create 3-D muscle fibers and ECM. (B) Meshed ECM and separate fibers are then combined to create a FEM of a fiber bundle.

The pathological differences in muscle geometry and material properties that may lead to spatial changes in deformation in the tissue microstructure can be observed by coupling ABM and FEM. FEM is a numerically efficient approach for determining areas of high strain where tissue injury would be localized (Fig. 8). These areas of damage would inform injury occurrence in the ABM. These damage sites would influence cell migration in the ABM during the cycle of repair following injury. Following the repair time course, the resultant ABM geometry and cell counts would inform the reconstruction of the FE model geometry and material property evolution, thereby providing progressive morphological data (Fig. 9). As the two models provide iterative feedback, the effect of pathological muscle morphology on chronic injury and regeneration can be observed and compared to typical muscle regeneration and muscle morphology. By coupling FEM with ABM in a two-way coupling, the biological processes of tissue adaptation can be explored temporally.

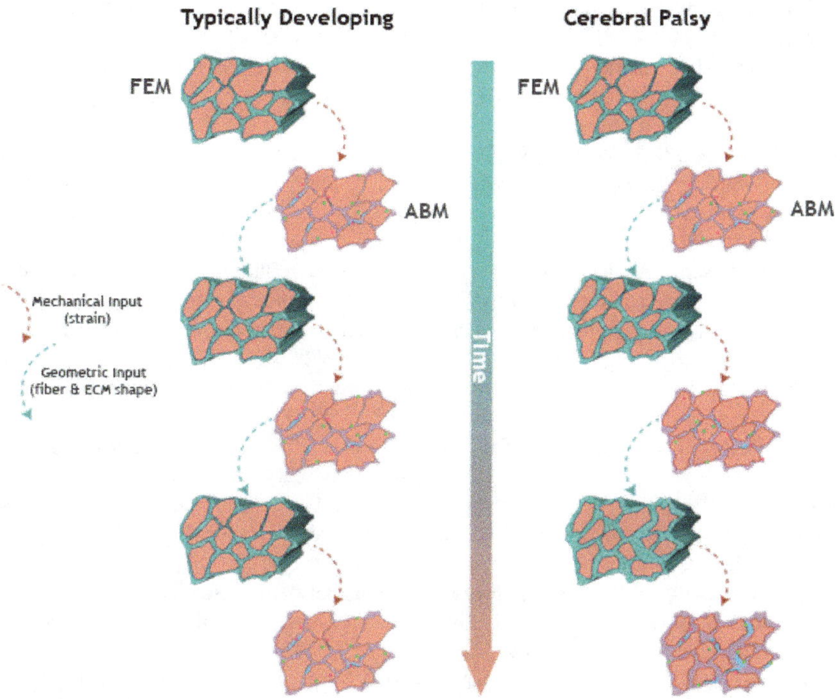

Fig. 9. Feedback between mechanical FEM stimulus and cellular level ABM of muscle regeneration would be used to illustrate the process of healthy muscle regeneration in typically developing muscle and the chronic degeneration process in cerebral palsy muscle that leads to pathological changes.

References

1. Graham, H.K., Rosenbaum, P., Paneth, N., Dan, B., Lin, J.P., Damiano, Di.L., Becher, J.G., Gaebler-Spira, D., Colver, A., Reddihough, Di.S., Crompton, K.E., Lieber, R.L.: Cerebral palsy. Nat. Rev. Dis. Prim. **2**, 19–20 (2016). https://doi.org/10.1038/nrdp.2015.82
2. Kinney, M.C., Dayanidhi, S., Dykstra, P.B., McCarthy, J.J., Peterson, C.A., Lieber, R.L.: Reduced skeletal muscle satellite cell number alters muscle morphology after chronic stretch but allows limited serial sarcomere addition. Muscle Nerve **55**, 384–392 (2017). https://doi.org/10.1002/mus.25227

3. Fahey, M.C., Maclennan, A.H., Kretzschmar, D., Gecz, J., Kruer, M.C.: The genetic basis of cerebral palsy. Dev. Med. Child Neurol. **59**, 462–469 (2017). https://doi.org/10.1111/dmcn.13363
4. Norton, N.S.: Cerebral palsy. xPharm Compr. Pharmacol. Ref. 1–5 (2007). https://doi.org/10.1016/b978-008055232-3.60641-5
5. Lieber, R.L., Runesson, E., Einarsson, F., Fridén, J.: Inferior mechanical properties of spastic muscle bundles due to hypertrophic but compromised extracellular matrix material. Muscle Nerve **28**, 464–471 (2003). https://doi.org/10.1002/mus.10446
6. Kerr Graham, H., Selber, P.: Musculoskeletal aspects of cerebral palsy. J. Bone Jt. Surg. **85**, 157–166 (2003). https://doi.org/10.1302/0301-620X.85B2.14066
7. Friden, J., Lieber, R.L.: Spastic muscle cells are shorter and stiffer than normal cells. Muscle Nerve. **26**, 157–164 (2003). https://doi.org/10.1002/mus.10247
8. Mathewson, M.A., Lieber, R.L.: Pathophysiology of muscle contractures in cerebral palsy. Phys. Med. Rehabil. Clin. N. Am. **26**, 57–67 (2015). https://doi.org/10.1016/j.pmr.2014.09.005
9. Smith, L.R., Lee, K.S., Ward, S.R., Chambers, H.G., Lieber, R.L.: Hamstring contractures in children with spastic cerebral palsy result from a stiffer extracellular matrix and increased in vivo sarcomere length. J. Physiol. **589**, 2625–2639 (2011). https://doi.org/10.1113/jphysiol.2010.203364
10. Willerslev-Olsen, M., Choe Lund, M., Lorentzen, J., Barber, L., Kofoed-Hansen, M., Nielsen, J.B.: Impaired muscle growth precedes development of increased stiffness of the triceps surae musculotendinous unit in children with cerebral palsy. Dev. Med. Child Neurol. **60**, 672–679 (2018). https://doi.org/10.1111/dmcn.13729
11. Proske, U., Morgan, D.L.: Muscle damage from eccentric exercise: mechanism, mechanical signs, adaptation and clinical applications. J. Physiol. **537**, 333–345 (2001)
12. Chargé, S.B.P., Rudnicki, M.A.: Cellular and molecular regulation of muscle regeneration. 209–238 (2009). https://doi.org/10.1152/physrev.00019.2003
13. Sciorati, C., Rigamonti, E., Manfredi, A.A., Rovere-Querini, P.: Cell death, clearance and immunity in the skeletal muscle. Cell Death Differ. **23**, 927–937 (2016). https://doi.org/10.1038/cdd.2015.171
14. Tidball, J.G., Villalta, S.A.: Regulatory interactions between muscle and the immune system during muscle regeneration. Am. J. Physiol. Regul. Integr. Comp. Physiol. **298**, R1173–R1187 (2010). https://doi.org/10.1152/ajpregu.00735.2009
15. Ten Broek, R.W., Grefte, S., Von Den Hoff, J.W.: Regulatory factors and cell populations involved in skeletal muscle regeneration. J. Cell. Physiol. **224**, 7–16 (2010). https://doi.org/10.1002/jcp.22127
16. Arnold, L., Henry, A., Poron, F., Baba-Amer, Y., van Rooijen, N., Plonquet, A., Gherardi, R.K., Chazaud, B.: Inflammatory monocytes recruited after skeletal muscle injury switch into antiinflammatory macrophages to support myogenesis. J. Exp. Med. **204**, 1057–1069 (2007). https://doi.org/10.1084/jem.20070075
17. Järvinen, T.A., Kääriäinen, M., Äärimaa, V., Järvinen, M., Kalimo, H.: Skeletal muscle repair after exercise-induced injury. In: Schiaffino, S., Partridge, T. (eds.) Skeletal Muscle Repair and Regeneration, pp. 217–242. Springer, Heidelberg (2008)
18. Lu, P., Takai, K., Weaver, V.M., Werb, Z.: Extracellular matrix degradation and remodeling in development and disease. Cold Spring Harb. Perspect. Biol. **3**, a005058–a005058 (2011). https://doi.org/10.1101/cshperspect.a005058
19. Lepper, C., Conway, S.J., Fan, C.M.: Adult satellite cells and embryonic muscle progenitors have distinct genetic requirements. Nature **460**, 627–631 (2009). https://doi.org/10.1038/nature08209

20. Yin, H., Price, F., Rudnicki, M.A.: Satellite cells and the muscle stem cell niche. Physiol. Rev. **93**, 23–67 (2013). https://doi.org/10.1152/physrev.00043.2011
21. Wang, Y.X., Rudnicki, M.A.: Satellite cells, the engines of muscle repair. Nat. Rev. Mol. Cell Biol. **13**, 127–133 (2012). https://doi.org/10.1038/nrm3265
22. Snijders, T., Nederveen, J.P., McKay, B.R., Joanisse, S., Verdijk, L.B., van Loon, L.J.C., Parise, G.: Satellite cells in human skeletal muscle plasticity. Front. Physiol. **6**, 283 (2015). https://doi.org/10.3389/fphys.2015.00283
23. Charge, S.B.P., Rudnicki, M.A.: Cellular and molecular regulation of muscle regeneration. Physiol. Rev. **84**, 209 (2004). https://doi.org/10.1152/physrev.00019.2003
24. Ambrosio, F., Kadi, F., Lexell, J., Kelley Fitzgerald, G., Boninger, M.L., Huard, J.: The effect of muscle loading on skeletal muscle regenerative potential: an update of current research findings relating to aging and neuromuscular pathology. Am. J. Phys. Med. Rehabil. **88**, 145–155 (2009). https://doi.org/10.1097/PHM.0b013e3181951fc5
25. Chapman, M.A., Meza, R., Lieber, R.L.: Skeletal muscle fibroblasts in health and disease. Differentiation **92**, 108–115 (2016). https://doi.org/10.1016/j.diff.2016.05.007
26. Mendias, C.L.: Fibroblasts take the centre stage in human skeletal muscle regeneration. J. Physiol. **595**, 5005 (2017). https://doi.org/10.1113/JP274403
27. Mackey, A.L., Magnan, M., Chazaud, B., Kjaer, M.: Human skeletal muscle fibroblasts stimulate in vitro myogenesis and in vivo muscle regeneration. J. Physiol. **595**, 5115–5127 (2017). https://doi.org/10.1113/JP273997
28. Smith, L.R., Pontén, E., Hedström, Y., Ward, S.R., Chambers, H.G., Subramaniam, S., Lieber, R.L.: Novel transcriptional profile in wrist muscles from cerebral palsy patients. BMC Med. Genom. **2**, 44 (2009). https://doi.org/10.1186/1755-8794-2-44
29. De Bruin, M., Smeulders, M.J., Kreulen, M., Huijing, P.A., Jaspers, R.T.: Intramuscular connective tissue differences in spastic and control muscle: a mechanical and histological study. PLoS One. **9** (2014). https://doi.org/10.1371/journal.pone.0101038
30. Booth, C.M., Cortina-Borja, M.J.F., Theologis, T.N.: Collagen accumulation in muscles of children with cerebral palsy and correlation with severity of spasticity. Dev. Med. Child Neurol. **43**, 314–320 (2001). https://doi.org/10.1111/j.1469-8749.2001.tb00211.x
31. Foran, J.R.H., Steinman, S., Barash, I., Chambers, H.G., Lieber, R.L.: Structural and mechanical alterations in spastic skeletal muscle. Dev. Med. Child Neurol. **47**, 713–717 (2005). https://doi.org/10.1017/S0012162205001465
32. Smith, L.R., Chambers, H.G., Lieber, R.L.: Reduced satellite cell population may lead to contractures in children with cerebral palsy. Dev. Med. Child Neurol. **55**, 264–270 (2013). https://doi.org/10.1111/dmcn.12027
33. Wilensky, U., Rand, B.: An Introduction to Agent-based Modelling: Modelling Natural, Social and Engineered Complex Systems with NetLogo. The MIT Press, Cambridge (2015)
34. Abar, S., Theodoropoulos, G.K., Lemarinier, P., O'Hare, G.M.P.: Agent Based Modelling and Simulation tools: a review of the state-of-art software. Comput. Sci. Rev. **24**, 13–33 (2017). https://doi.org/10.1016/j.cosrev.2017.03.001
35. Borgiani, E., Duda, G.N., Checa, S.: Multiscale modeling of bone healing: toward a systems biology approach. Front. Physiol. **8**, 287 (2017). https://doi.org/10.3389/fphys.2017.00287
36. Rouillard, A.D., Holmes, J.W.: Coupled agent-based and finite-element models for predicting scar structure following myocardial infarction. Prog. Biophys. Mol. Biol. **115**, 235–243 (2014). https://doi.org/10.1016/j.pbiomolbio.2014.06.010
37. Zhuan, X., Luo, X., Gao, H., Ogden, R.W.: Coupled agent-based and hyperelastic modelling of the left ventricle post-myocardial infarction. Int. J. Numer. Method. Biomed. Eng. **35**, e3155 (2018). https://doi.org/10.1002/cnm.3155

38. Virgilio, K.M., Martin, K.S., Peirce, S.M., Blemker, S.S.: Agent-based model illustrates the role of the microenvironment in regeneration in healthy and mdx skeletal muscle. J. Appl. Physiol. **125**, 1424–1439 (2018). https://doi.org/10.1152/japplphysiol.00379.2018
39. Martin, K.S., Blemker, S.S., Peirce, S.M.: Agent-based computational model investigates muscle-specific responses to disuse-induced atrophy. J. Appl. Physiol. **118**, 1299–1309 (2015). https://doi.org/10.1152/japplphysiol.01150.2014
40. Thorne, B.C., Bailey, A.M., Peirce, S.M.: Combining experiments with multi-cell agent-based modeling to study biological tissue patterning. Brief. Bioinform. **8**, 245–257 (2007). https://doi.org/10.1093/bib/bbm024
41. Macal, C.M., North, M.J.: Tutorial on agent-based modelling and simulation. J. Simul. **4**, 151–162 (2010). https://doi.org/10.1057/jos.2010.3
42. Gorochowski, T.E.: Agent-based modelling in synthetic biology. Essays Biochem. **60**, 325–336 (2016). https://doi.org/10.1042/EBC20160037
43. An, G.: Integrating physiology across scales and formalizing hypothesis exploration with agent-based modeling. J. Appl. Physiol. **118**, 1191–1192 (2015). https://doi.org/10.1152/japplphysiol.00243.2015
44. Bonabeau, E.: Agent-based modeling: methods and techniques for simulating human systems. Proc. Natl. Acad. Sci. **99**, 7280–7287 (2002). https://doi.org/10.1073/pnas.082080899
45. Mackey, A.L., Kjaer, M.: The breaking and making of healthy adult human skeletal muscle in vivo. Skelet. Muscle **7**, 1–18 (2017). https://doi.org/10.1186/s13395-017-0142-x
46. Miller, K.J., Thaloor, D., Matteson, S., Pavlath, G.K.: Hepatocyte growth factor affects satellite cell activation and differentiation in regenerating skeletal muscle. Am. J. Physiol. Cell Physiol. **278**, C174–C181 (2000). https://doi.org/10.1152/ajpcell.2000.278.1.C174
47. Siegel, A.L., Kuhlmann, P.K., Cornelison, D.D.W.: Muscle satellite cell proliferation and association: new insights from myofiber time-lapse imaging. Skelet. Muscle **1**, 1–7 (2011). https://doi.org/10.1186/2044-5040-1-7
48. Pawlikowski, B., Vogler, T.O., Gadek, K., Olwin, B.B.: Regulation of skeletal muscle stem cells by fibroblast growth factors. Dev. Dyn. **246**, 359–367 (2017). https://doi.org/10.1002/dvdy.24495
49. Rangel-Huerta, E., Maldonado, E.: Transit-amplifying cells in the fast lane from stem cells towards differentiation. Stem Cells Int. **2017**, 1–10 (2017). https://doi.org/10.1155/2017/7602951
50. Kuang, S., Gillespie, M.A., Rudnicki, M.A.: Niche regulation of muscle satellite cell self-renewal and differentiation. Cell Stem Cell **2**, 22–31 (2008). https://doi.org/10.1016/j.stem.2007.12.012
51. Murphy, M.M., Lawson, J.A., Mathew, S.J., Hutcheson, D.A., Kardon, G.: Satellite cells, connective tissue fibroblasts and their interactions are crucial for muscle regeneration. Development **138**, 3625–3637 (2011). https://doi.org/10.1242/dev.064162
52. Delaney, K., Kasprzycka, P., Ciemerych, M.A., Zimowska, M.: The role of TGF-β1 during skeletal muscle regeneration. Cell Biol. Int. **41**, 706–715 (2017). https://doi.org/10.1002/cbin.10725
53. Tipping, P.G., Hancock, W.W.: Production of tumor necrosis factor and interleukin-1 by macrophages from human atheromatous plaques. Am. J. Pathol. **142**, 1721–1728 (1993)
54. Vignola, A.M., Chanez, P., Chiappara, G., Merendino, A., Zinnanti, E., Bousquet, J., Bellia, V., Bonsignore, G.: Release of transforming growth factor-beta (TGF-β) and fibronectin by alveolar macrophages in airway diseases. Clin. Exp. Immunol. **106**, 114–119 (1996). https://doi.org/10.1046/j.1365-2249.1996.d01-811.x

55. McCarthy, J.J., Mula, J., Miyazaki, M., Erfani, R., Garrison, K., Farooqui, A.B., Srikuea, R., Lawson, B.A., Grimes, B., Keller, C., Van Zant, G., Campbell, K.S., Esser, K.A., Dupont-Versteegden, E.E., Peterson, C.A.: Effective fiber hypertrophy in satellite cell-depleted skeletal muscle. Development **138**, 3657–3666 (2011). https://doi.org/10.1242/dev.068858
56. Dayanidhi, S., Lieber, R.L.: Skeletal muscle satellite cells: mediators of muscle growth during development and implications for developmental disorders. Muscle Nerve **48**, 153–154 (2013). https://doi.org/10.1002/mus.23878
57. Mockford, M., Caulton, J.M.: The pathophysiological basis of weakness in children with cerebral palsy. Pediatr. Phys. Ther. **22**, 222–233 (2010). https://doi.org/10.1097/pep.0b013e3181dbaf96
58. Cholok, D., Lee, E., Lisiecki, J., Agarwal, S., Loder, S., Ranganathan, K., Qureshi, A.T., Davis, T.A., Levi, B.: Traumatic muscle fibrosis: from pathway to prevention. J. Trauma Acute Care Surg. **82**, 174–184 (2017). https://doi.org/10.1097/TA.0000000000001290
59. Novak, M.L., Koh, T.J.: Phenotypic transitions of macrophages orchestrate tissue repair. Am. J. Pathol. **183**, 1352–1363 (2013). https://doi.org/10.1016/j.ajpath.2013.06.034
60. Wynn, T.A., Vannella, K.M.: Macrophages in tissue repair, regeneration, and fibrosis. Immunity **44**, 450–462 (2016). https://doi.org/10.1016/j.immuni.2016.02.015
61. Kirk, S., Oldham, J., Kambadur, R., Sharma, M., Dobbie, P., Bass, J.: Myostatin regulation during skeletal muscle regeneration. J. Cell. Physiol. **184**, 356–363 (2000). https://doi.org/10.1002/1097-4652(200009)184:3%3c356:AID-JCP10%3e3.0.CO;2-R
62. Goetsch, S.C., Hawke, T.J., Gallardo, T.D., Richardson, J.A., Garry, D.J.: Transcriptional profiling and regulation of the extracellular matrix during muscle regeneration. Physiol. Genomics **14**, 261–271 (2015). https://doi.org/10.1152/physiolgenomics.00056.2003
63. Blemker, S.S., Pinsky, P.M., Delp, S.L.: A 3D model of muscle reveals the causes of nonuniform strains in the biceps brachii. J. Biomech. **38**, 657–665 (2005). https://doi.org/10.1016/j.jbiomech.2004.04.009
64. Sharafi, B., Blemker, S.S.: A mathematical model of force transmission from intrafascicularly terminating muscle fibers. J. Biomech. **44**, 2031–2039 (2011). https://doi.org/10.1016/j.jbiomech.2011.04.038

Deep Learning-Based Segmentation of Mineralized Cartilage and Bone in High-Resolution Micro-CT Images

Jean Léger[1(✉)], Lisa Leyssens[2], Christophe De Vleeschouwer[1], and Greet Kerckhofs[2,3,4,5]

[1] ICTEAM, UCLouvain, Louvain-la-Neuve, Belgium
jean.leger@uclouvain.be
[2] iMMC, UCLouvain, Louvain-la-Neuve, Belgium
[3] Institute of Experimental and Clinical Research, UCLouvain, Louvain-la-Neuve, Belgium
[4] Department of Materials Science and Engineering, KU Leuven, Leuven, Belgium
[5] Prometheus, Division of Skeletal Tissue Engineering, KU Leuven, Leuven, Belgium

Abstract. High-resolution 3D micro-CT imaging is a powerful tool for the visualization of the mineralized tissues. However, it remains challenging to discriminate automatically between mineralized cartilage and bone as they have similar greyscale values. Currently, manual contouring is still the standard way to segment these two tissues but it is time-consuming and user-biased. In this work, we have optimized a 3D fully convolutional neural network, i.e. U-net, to automatically segment mineralized cartilage from bone in high-resolution micro-CT images of the Achilles tendon-to-bone interface. Using the 3D U-net, we reach an average Dice Similarity Coefficient of 0.85 compared to manual annotations for twelve 3D datasets. The proposed method shows comparable results to a 2D U-net approach while ensuring better 3D segmentation consistency. We also found that reducing the resolution of the 3D micro-CT images for the network training did not importantly impact the performance while considerably reducing the training time.

Keywords: Mineralized tissue segmentation · CNN · Deep learning · Cartilage versus bone · High-resolution micro-CT

1 Introduction

The musculoskeletal system supports and stabilizes the human body and coordinates the movements of the muscles and the skeletal system. It is composed of bone, muscle, cartilage, tendon, ligament, and other connective tissues. In particular, tendons and bone are joined in a specific way in order to facilitate joint motion, forming the insertion site [1]. This is also known as the enthesis or the bone-to-tendon interface. The tissue that makes up this interface is a complex transitional tissue, which is essential for physiologic musculoskeletal motion.

G. A. Ateshian et al. (Eds.): CMBBE 2019, LNCVB 36, pp. 158–170, 2020.
https://doi.org/10.1007/978-3-030-43195-2_12

It serves to integrate and minimize stress concentrations between the bone (a stiff, structural, hard tissue) and the tendon (a compliant structural soft tissue). More precisely, the bone-to-tendon interface can be divided into four specific zones with varying compositional and mechanical properties and functions [2]: the tendon, the non-mineralized fibrocartilage, the mineralized fibrocartilage, and the bone.

Because of its function of mediating load between very dissimilar tissues, the bone-to-tendon interface is a common site of injury. However, in case of such injury, the natural tissue is not regenerated after healing [1,3]. Instead, it is replaced with a so-called scar tissue, which is isotropic and significantly less stiff than either tendon or bone. It has been reported that depending on the severity and the location of the injury, the regenerated tissue will rupture again in 20 to 94% of cases [4]. A solution to this problem would be to use a construct composed of a scaffold (made of a biomaterial) with growth factor and/or cells to obtain a tissue that presents the same properties as the original one.

In this context, characterizing the bone-to-tendon interface properties and its 3D sub-architecture is primordial. Towards this goal, high-resolution 3D contrast-enhanced microfocus computed tomography (CE-CT) is, thanks to its high spatial and contrast resolution, a powerful tool for the visualization of both the unmineralized and mineralized cartilage, along with the bone [5]. However, it remains challenging to discriminate automatically between the mineralized cartilage and the bone in both high-resolution micro-CT and CE-CT images, as they have similar greyscale values. Indeed, when using only greyscale-based segmentation, it is very difficult to accurately detect the boundaries between these two mineralized tissues. Currently, manual delineation of the two tissues is still used to allow for its 3D structural analysis and that of the bone. Nevertheless, this is time consuming and highly user-biased.

Interestingly, mineralized cartilage has a discriminant porous texture in comparison to bone, because of the presence of chondrocyte lacunae. Consequently, our goal is to develop an automatic mineralized cartilage segmentation tool exploiting its texture, shape and 3D consistency. For this purpose, we use a fully convolutional neural network [6], named U-net, which is state-of-the-art for the semantic segmentation in 2D and 3D biomedical images [7,8]. The 3D U-net model is compared to the 2D U-net model and the impact of reducing the 3D image resolution for GPU memory limitations is discussed.

To the best of our knowledge, no fully automatic segmentation algorithms have been proposed for the segmentation of mineralized cartilage from bone. Indeed, all the related works mentioned below focus on the easier unmineralized cartilage segmentation. Most works focus on the unmineralized cartilage versus bone segmentation in MR images. To this end, classical medical image segmentation approaches relying on active shape models [9] and atlas databases [10] have been proposed. However, those methods perform poorly in case of high ROIs shape variability and require a relatively long segmentation times at inference. In contrast, deep learning algorithms are supposed to be robust to shape and appearance variations if those variations are captured in the training database and they are fast at inference. In a first study on deep-learning based knee cartilage segmentation, they used a triplanar CNN [11]. This was followed by 2D encoder-decoder architectures such as Segnet [12], U-net [13] and U-net complemented with conditional random fields to promote 3D consistency [14]. Very recent papers use 3D approaches for knee cartilage segmentation, such as 3D U-net [15] or a variant of it, namely μ-Net [16]. Only few methods have been proposed to automatically segment unmineralized cartilage on CT images. They include an atlas-based segmentation of acetabular cartilage [17] and a registration-based segmentation of knee articular cartilage [18] on CE-CT images. One CNN-based unmineralized cartilage segmentation on high-resolution micro-CT images has been proposed. It uses a 2D U-net model followed by a 3D refinement. However, the method has only been presented in a one-page abstract and is therefore hard to reproduce [19].

2 Materials and Methods

2.1 Data Acquisition, Annotation and Preprocessing

The training and test data for the U-net consist of 3D high-resolution micro-CT images of 12 murine bone-to-Achilles tendon interfaces. After harvest, all the samples were fixed in 4% paraformaldehyde during 16 h and then stored at 4 °C in a phosphate-buffered saline solution (PBS). Next, the samples were dissected in order to isolate the tendon and the bone, and to remove all tissue that was unnecessary for the purpose of the experiments. A Phoenix Nanotom M - Computed Tomography System (GE Sensing & Inspection Technologies GmbH, Wunstorf, Germany) was used with the following scanning parameters: 60 kV, 87 µA, 1.25 µm voxel size, 2400 images, 500 ms exposure time, 20 min scan time.

The mineralized cartilage has been manually annotated on all the slices of the micro-CT images by an expert using the CTAn software from Bruker MicroCT (Kontich, Belgium). Those annotations serve as ground truth in this work and they are stored as 12 3D binary masks with the same size and resolution as the 3D micro-CT images.

The original isotropic voxel size is 1.25 µm. The 12 3D datasets have been cropped to 1024 × 1024 × 256 voxels around the region of interest (i.e. mineralized cartilage and bone). The ROI is fully contained in every 1024 × 1024 slice. The bottom slices contain only mineralized cartilage (and no bone) whereas mineralized cartilage and bone start to be difficult to distinguish visually on the upper slices. This preprocessing is necessary since the memory limitations of the GPUs is a bottleneck for deep learning algorithms with such high-resolution data. Nvidia 1080Ti 11 GB GPUs are used.

2.2 Network Architecture

The 3D U-net fully convolutional neural network considered in this study is the same as in Brion et al. [8]. More precisely, the network follows the same architecture (i.e. number and composition of layers) as in Ronneberger et al. [20], where the 3×3 convolutions, the 2×2 max-pooling and the 2×2 up-conversion operations have been replaced by their $3 \times 3 \times 3$ and $2 \times 2 \times 2$ counterparts, as in Ciçek et al. [6]. The 3D input goes through a contracting path to capture context and an expanding path to enable precise localization. In the last layer, a sigmoid is applied and the network outputs the probability for each voxel to belong to the mineralized cartilage. To obtain the final binary segmentation mask, a threshold of 0.5 is chosen. The main advantage of fully convolutional neural networks is that they output a prediction with the same size as the input. Hence, the thresholded output of the network directly gives the desired prediction of the ROI with the same size and resolution as the input 3D data.

The network is trained with the Dice loss. The optimization algorithm used is Adam with learning rate 10^{-4}. The number of epochs is chosen such that convergence is reached and without using early-stopping. The hyper-parameters mentioned here are the same as in [8] and showed to be satisfactory on the data used in this work. For this reason, no validation set is considered here. We only used two sets: a training set and a test set both containing 6 3D images. The batch size depends on the learning strategy that is used, but has not been tuned.

The results of 3D U-net are compared to those of 2D U-net. The 2D architecture is exactly the same as the 3D one presented above, where the $3 \times 3 \times 3$ convolutions, the $2 \times 2 \times 2$ max-pooling and the $2 \times 2 \times 2$ up-conversion operations have been replaced by their 3×3 and 2×2 counterparts. Online data augmentation (i.e. rotation, shift, shear, flip) is implemented for both 2D and 3D approaches.

2.3 Learning Strategies

Two major interests of the considered datasets are (i) their high-resolution allowing an accurate tissue visualization, in particular for the porous texture characterizing the mineralized cartilage region, and (ii) their ability to capture the 3D architecture of the imaged tissues in a uniformly sampled 3D grid, in opposition to histology where only stacks of 2D slices are available. However, such 3D large high-resolution data are particularly challenging for deep learning models in terms of training time and GPU memory limitations. In order to deal with those constraints, three strategies could be applied.

1. The high-resolution 3D image is uniformly downsampled in every dimension. A 3D model can then be trained and tested on the full 3D image directly. This solves the GPU memory limitation at the expense of a lower resolution.
2. The 3D image is split in 2D slices along a chosen dimension. In this case, a 2D model can be trained and tested at high-resolution, but the 3D consistency is no more ensured.
3. The high-resolution 3D image is split in 3D patches without reducing their resolution. However, this approach requires many training patches in order to capture the variability of the full 3D image. This increases considerably the training time.

In this work, the 12 1024 × 1024 × 256 voxels 3D datasets have been downsampled by a factor 4 in every dimension, resulting in 256 × 256 × 64 voxels 3D datasets. The downsampling factor has been chosen in order to visually preserve the porous texture of the mineralized cartilage. The same operations are performed on the annotation 3D binary masks. A 3D U-net model is then trained on 6 of those low resolution 3D datasets and tested on the 6 remaining ones. In order to determine whether the 3D approach improves the segmentation performances compared to the 2D approach, a 2D U-net model has been trained on 384 low resolution images (6 3D datasets multiplied by 64 slices in each 3D dataset) of 256 × 256 pixels. The 2D test predictions are stacked in order to reconstruct a 3D binary mask and compare it with the 3D binary mask predicted with the 3D approach. In order to evaluate the impact of reducing the spatial image resolution, a 3D U-net model has also been trained on 96 high-resolution patches (16 patches on each of the 6 3D datasets) of 512 × 512 × 16 voxels. The sampling of the patches in the 1024 × 1024 × 256 high-resolution 3D datasets is the following. The patches are sampled uniformly in the third 3D image dimension (i.e. a patch every 16 slices in order to cover the 256 slices) and randomly in the two first dimensions, provided that at least one voxel belongs to the cartilage in the patch. The batch size is set to one for the 3D approaches and 64 for the 2D approach. This is the maximum batch size allowed by the GPU memory limitation. A summary of the three strategies properties is provided in Table 1.

Table 1. Characteristics and training hyper-parameters for the three learning strategies.

Parameters	Strategy 1	Strategy 2	Strategy 3
Model	3D U-net	2D U-net	3D U-net
Input data size	256 × 256 × 64	256 × 256	512 × 512 × 16
Batch size	1	64	1
Number of training samples	6	384	96
Training time per epoch	∼40 s	∼3 s	∼640 s
Number of epochs	∼150	∼150	∼100

2.4 Performance Assessment

We use a two-fold cross-validation (with 6 3D datasets in each fold) in order to obtain the predictions on the 12 3D datasets. The absence of a validation set is motivated in Sect. 2.2. In order to evaluate our results, we use the Dice similarity coefficient (DSC), which measures the overlap between two binary masks. More specifically,

$$\text{DSC} = \frac{2|A \cap B|}{|A| + |B|}, \tag{1}$$

where A and B are the predicted and ground truth segmentation binary masks. The DSC reaches respectively zero and one for no or complete overlap between both binary masks. The low resolution ROIs predictions of the deep learning model data are upsampled by a factor 4 before the DSC computation. Hence, the DSC are computed in the original high-resolution.

3 Results and Discussion

3.1 Quantitative Comparison of the Learning Strategies

Table 2 presents the segmentation performances obtained with the three considered learning strategies. The 3D U-net approach trained on the low resolution datasets reaches an average DSC close to 0.85. This corresponds to visually acceptable segmentation outputs as shown in Figs. 1 and 2, which respectively present the results on Set 2 (DSC = 0.889) and Set 3 (DSC = 0.836). The red and blue regions correspond to the mineralized cartilage ROIs. In particular, in Fig. 1d, the 3D U-net model is able to capture the complex mineralized cartilage structure, while reasonably avoiding the bone. Three types of errors are mostly present on the predicted ROIs.

1. *Errors on the mineralized cartilage extremities.* This can be observed in Figs. 1e and 2e. The extremities of the ROIs are too long compared to the manual annotations. A noteworthly point is that the annotations might also be less accurate in those regions.
2. *Errors in small isolated regions.* This can be observed in Fig. 1f where a small mineralized cartilage region is detected in a bone region.
3. *Large errors on the external border of the bone.* This can be seen on the bottom of the bone region in Fig. 2d. This is explained by the fact that only mineralized cartilage is present on the first micro-CT slices in the 3D image. Bone progressively appears in the imaged structure as we go up in the stack of slices. In the slices where bone starts to appear, the 3D U-net model overestimates the presence of mineralized cartilage. Again, the annotations might have been less accurate in those regions.

Interestingly, the three types of errors appear equally on the three learning strategies presented in this work. The alternative strategies reach an average DSC performance close to 0.85 as well, with one failure case on Set 9 for the third strategy. Since the DSC standard deviation is high compared to the difference of average DSC between the approaches, we cannot identify a significantly better strategy. Hence, a qualitative comparison of the learning strategies is provided in the next sections.

But, before moving to a qualitative comparison, it is worth noting that the DSC only converges to 0.903 (and not 1) on the training set for the first strategy despite the predictions being visually consistent with the ground truth. We might expect that the average DSC measured when comparing annotations provided by different annotators also saturates around 0.9.

Table 2. Comparison of the DSC on the high-resolution test 3D datasets for the different U-net training strategies. Strategy 1: 3D U-net on the low resolution 256 × 256 × 64 3D datasets; Strategy 2: 2D U-net on the low resolution 256 × 256 images; Strategy 3: 3D U-net on the high-resolution 512 × 512 × 16 patches. DSC: Dice Similarity Coefficient. SD: Standard Deviation.

Datasets	Strategy 1	Strategy 2	Strategy 3
Set 1	0.881	0.877	0.858
Set 2	0.889	0.877	0.879
Set 3	0.836	0.850	0.872
Set 4	0.787	0.806	0.812
Set 5	0.868	0.879	0.856
Set 6	0.858	0.857	0.881
Set 7	0.842	0.880	0.816
Set 8	0.878	0.867	0.878
Set 9	0.786	0.860	0.606
Set 10	0.884	0.879	0.899
Set 11	0.861	0.794	0.804
Set 12	0.848	0.781	0.779
Mean ± SD	0.851 ± 0.033	0.851 ± 0.035	0.828 ± 0.076

3.2 Qualitative Impact of the 3D Consistency

Based on Table 2, the 3D U-net model does not show a significant improvement of the DSC performances compared to a 2D U-net approach. The 3D model is however harder to train and test since it is fed with 3D input data, reaching faster the memory bottleneck of the GPU. On the other hand, the 2D approach does not ensure the 3D consistency of the segmentation. This is shown in Fig. 3. Three consecutive slices of Set 2 are shown on it. A discontinuity is present in the segmentation of adjacent slices with the 2D approach, which is not desirable.

3.3 Qualitative Impact of a Reduction in the Spatial Image Resolution

In the first learning strategy, the resolution of the 3D micro-CT datasets is reduced by a factor 4 in every dimension. The ROI prediction is performed on the low resolution datasets as well. In order to report the DSC performance, the low resolution prediction is upsampled to the original resolution. This leads to a rougher edge of the segmented ROI as shown in Fig. 4. Indeed, the low resolution 3D approach is not able to segment accurately the extremity of very irregular regions such as the one shown in Fig. 4c. However, those errors are small compared to the errors presented in Sect. 3.1, as shown in Table 2. Hence, we conclude that reducing the image resolution by a factor 4 preserves most of the information (e.g. the image texture) required to discriminate between the mineralized cartilage and bone.

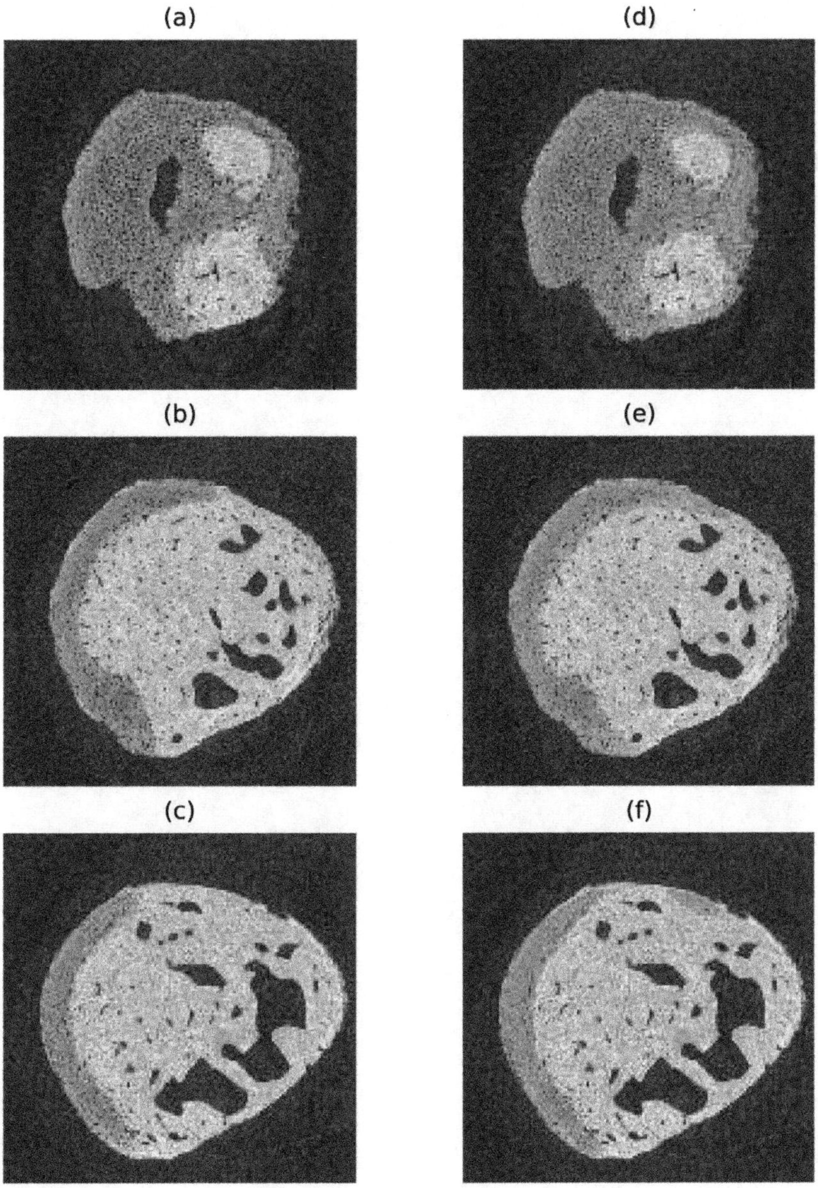

Fig. 1. Comparison of mineralized cartilage segmentations between the ground truth in red (GT) and the prediction of 3D U-net in blue (strategy 1: low resolution) for the slices indices 64, 128, 192 of Set 2.

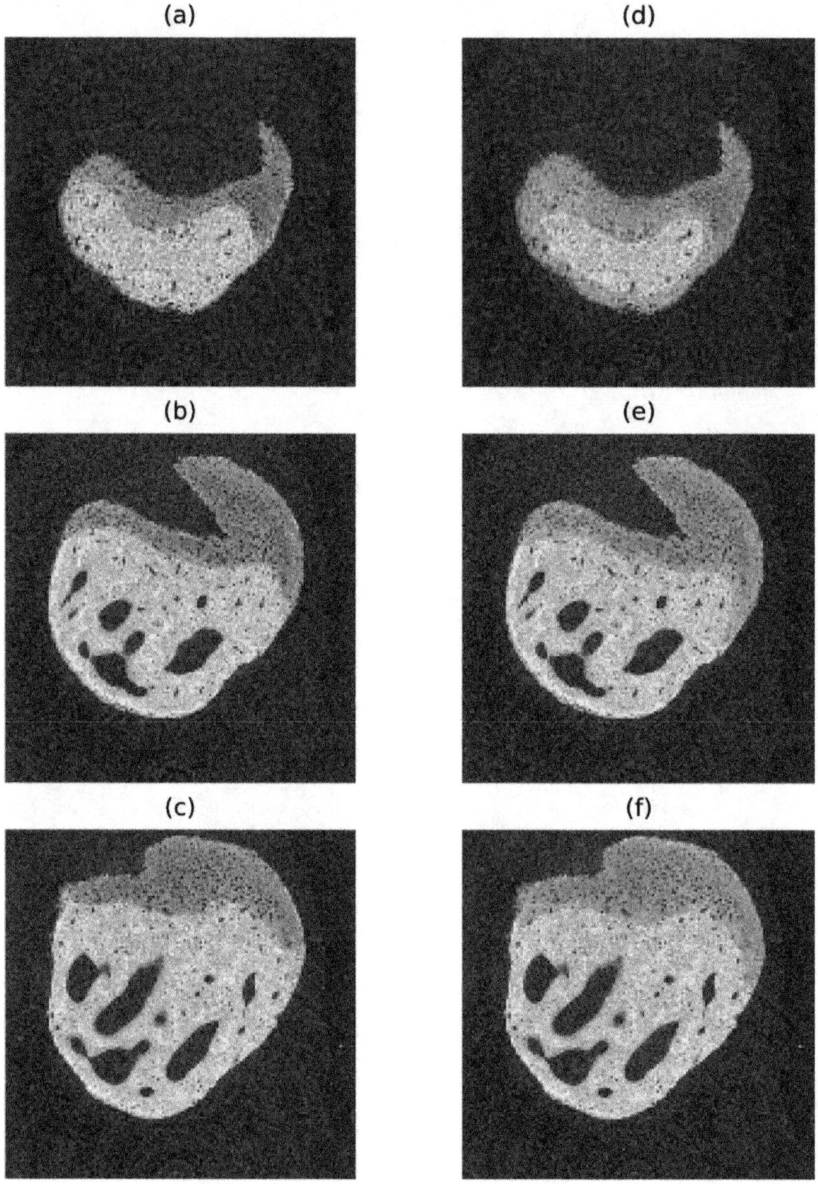

Fig. 2. Comparison of mineralized cartilage segmentations between the ground truth in red (GT) and the prediction of 3D U-net in blue (strategy 1: low resolution) for the slices indices 64, 128, 192 of Set 3.

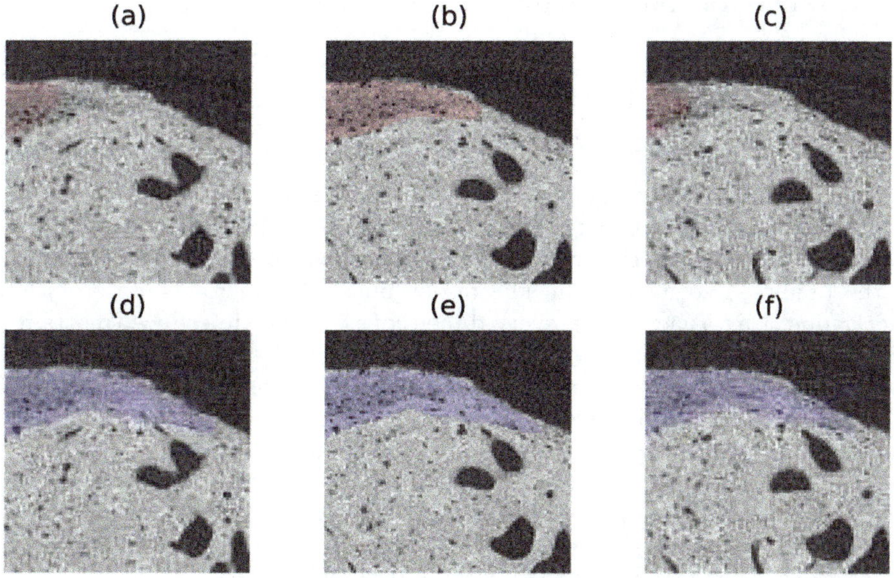

Fig. 3. Comparison of mineralized cartilage segmentations between the 2D U-net in red (strategy 2) and the 3D U-net in blue (strategy 1: low resolution) on a detail of Set 2.

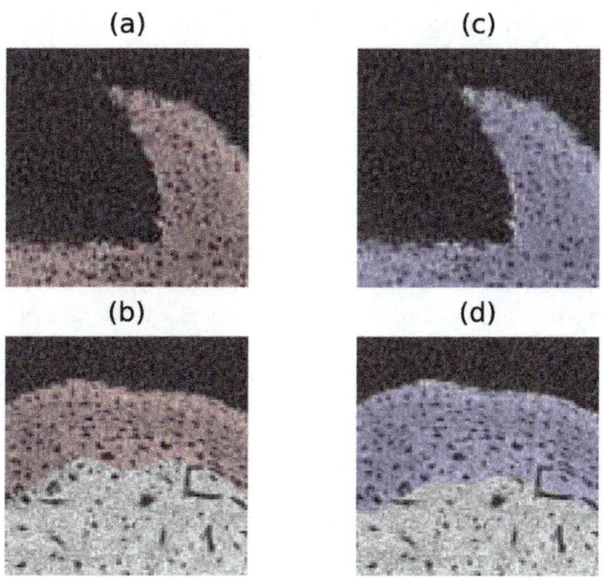

Fig. 4. Comparison of mineralized cartilage segmentations between the patch-based 3D U-net in red (strategy 3) and the low resolution 3D U-net in blue (strategy 1) on a detail of Set 3.

4 Conclusion and Future Perspectives

To our best knowledge, this is the first study presenting a deep learning strategy for automatic segmentation of mineralized cartilage from bone in high-resolution micro-CT images. For this purpose, we use a fully convolutional neural network, i.e. U-net. The average DSC computed on 12 3D micro-CT datasets reaches 0.85 compared to a manually annotated ground truth. The 3D U-net model is compared to the 2D U-net model. The difference in DSC performance is not statistically significant. However, the 3D approach ensures better 3D consistency in the predicted segmentation. The resolution of the micro-CT datasets have been reduced by a factor 4 in every dimensions in order to reduce the training time and to better manage the GPU memory limitations. No major difference in DSC has been observed compared to a 3D U-net trained on full resolution 3D micro-CT patches. Importantly, manual segmentation of mineralized cartilage can be subjected to inter-expert variability. This point has not been considered in the current work and would improve the performance assessment of the proposed methods. Also, the robustness of the deep learning segmentation methods to errors in the annotations should be further investigated since getting high quality annotations is challenging on these kind of datasets.

Acknowledgments. Jean Léger is a Research Fellow of the Fonds de la Recherche Scientifique - FNRS and Christophe De Vleeschouwer is Senior Research Associates with the Belgian F.R.S.-FNRS. Lisa Leyssens acknowledges the Fonds Spéciaux de Recherche (FSR) and the UCLouvain-FSR for funding her PhD. This research project was funded by a research project of the Research Foundation Flanders (FWO; Grant no. G088218N). The high-resolution micro-CT images have been generated at the X-ray computed tomography facilities of the Department of Development and Regeneration of the KU Leuven, financed by the Hercules Foundation (project AKUL 13/47).

References

1. Apostolakos, J., Durant, T.J., Dwyer, C.R., Russell, R.P., Weinreb, J.H., Alaee, F., Beitzel, K., McCarthy, M.B., Cote, M.P., Mazzocca, A.D.: The enthesis: a review of the tendon-to-bone insertion. Muscles Ligaments Tendons J. **4**(3), 333 (2014)
2. Genin, G.M., Kent, A., Birman, V., Wopenka, B., Pasteris, J.D., Marquez, P.J., Thomopoulos, S.: Functional grading of mineral and collagen in the attachment of tendon to bone. Biophys. J. **97**(4), 976–985 (2009)
3. Liu, Y., Thomopoulos, S., Birman, V., Li, J.S., Genin, G.M.: Bi-material attachment through a compliant interfacial system at the tendon-to-bone insertion site. Mech. Mater. **44**, 83–92 (2012)
4. Liu, Y., Birman, V., Chen, C., Thomopoulos, S., Genin, G.M.: Mechanisms of bimaterial attachment at the interface of tendon to bone. J. Eng. Mater. Technol. **133**(1), 011006 (2011)

5. Kerckhofs, G., Sainz, J., Wevers, M., Van de Putte, T., Schrooten, J.: Contrast-enhanced nanofocus computed tomography images the cartilage subtissue architecture in three dimensions. Eur. Cells Mater. **25**, 179–89 (2013)
6. Çiçek, Ö., Abdulkadir, A., Lienkamp, S.S., Brox, T., Ronneberger, O.: 3D U-net: learning dense volumetric segmentation from sparse annotation. In: International Conference on Medical Image Computing and Computer-Assisted Intervention, pp. 424–432. Springer (2016)
7. Litjens, G., Kooi, T., Bejnordi, B.E., Setio, A.A.A., Ciompi, F., Ghafoorian, M., Van Der Laak, J.A., Van Ginneken, B., Sánchez, C.I.: A survey on deep learning in medical image analysis. Med. Image Anal. **42**, 60–88 (2017)
8. Brion, E., Léger, J., Javaid, U., Lee, J., De Vleeschouwer, C., Macq, B.: Using planning CTs to enhance CNN-based bladder segmentation on cone beam CT. In: Medical Imaging 2019: Image-Guided Procedures, Robotic Interventions, and Modeling, vol. 10951, p. 109511M. International Society for Optics and Photonics (2019)
9. Fripp, J., Crozier, S., Warfield, S.K., Ourselin, S.: Automatic segmentation and quantitative analysis of the articular cartilages from magnetic resonance images of the knee. IEEE Trans. Med. Imaging **29**(1), 55 (2010)
10. Tamez-Pena, J.G., Farber, J., Gonzalez, P.C., Schreyer, E., Schneider, E., Totterman, S.: Unsupervised segmentation and quantification of anatomical knee features: data from the osteoarthritis initiative. IEEE Trans. Biomed. Eng. **59**(4), 1177–1186 (2012)
11. Prasoon, A., Petersen, K., Igel, C., Lauze, F., Dam, E., Nielsen, M.: Deep feature learning for knee cartilage segmentation using a triplanar convolutional neural network. In: International Conference on Medical Image Computing and Computer-assisted Intervention, pp. 246–253. Springer (2013)
12. Liu, F., Zhou, Z., Jang, H., Samsonov, A., Zhao, G., Kijowski, R.: Deep convolutional neural network and 3D deformable approach for tissue segmentation in musculoskeletal magnetic resonance imaging. Magn. Reson. Med. **79**(4), 2379–2391 (2018)
13. Norman, B., Pedoia, V., Majumdar, S.: Use of 2D U-net convolutional neural networks for automated cartilage and meniscus segmentation of knee MR imaging data to determine relaxometry and morphometry. Radiology **288**(1), 177–185 (2018)
14. Zhou, Z., Zhao, G., Kijowski, R., Liu, F.: Deep convolutional neural network for segmentation of knee joint anatomy. Magn. Reson. Med. **80**(6), 2759–2770 (2018)
15. Ambellan, F., Tack, A., Ehlke, M., Zachow, S.: Automated segmentation of knee bone and cartilage combining statistical shape knowledge and convolutional neural networks: data from the osteoarthritis initiative. Med. Image Anal. **52**, 109–118 (2019)
16. Raj, A., Vishwanathan, S., Ajani, B., Krishnan, K., Agarwal, H.: Automatic knee cartilage segmentation using fully volumetric convolutional neural networks for evaluation of osteoarthritis. In: 2018 IEEE 15th International Symposium on Biomedical Imaging (ISBI 2018), pp. 851–854. IEEE (2018)
17. Tabrizi, P.R., Zoroofi, R.A., Yokota, F., Tamura, S., Nishii, T., Sato, Y.: Acetabular cartilage segmentation in CT arthrography based on a bone-normalized probabilistic atlas. Int. J. Comput. Assist. Radiol. Surg. **10**(4), 433–446 (2015)
18. Myller, K.A., Honkanen, J.T., Jurvelin, J.S., Saarakkala, S., Töyräs, J., Väänänen, S.P.: Method for segmentation of knee articular cartilages based on contrast-enhanced CT images. Ann. Biomed. Eng. **46**(11), 1756–1767 (2018)

19. Frondelius, T., Tiulpin, A., Lehenkari, P., Nieminen, H., Saarakkala, S.: Fully automatic deep learning based segmentation of bone-cartilage interface from micro-CT images of human osteochondral samples. Osteoarthritis Cartilage **26**, S469 (2018)
20. Ronneberger, O., Fischer, P., Brox, T.: U-net: convolutional networks for biomedical image segmentation. In: International Conference on Medical image computing and computer-assisted intervention, pp. 234–241. Springer (2015)

Statistical Finite Element Analysis of the Mechanical Response of the Intact Human Femur Using a Wide Range of Individual Anatomies

Mamadou T. Bah[1](✉), Reynir Snorrason[2], and Markus O. Heller[2]

[1] University of Winchester, Winchester SO22 4NR, UK
Mamadou.Bah@winchester.ac.uk
[2] University of Southampton, Southampton SO17 1BJ, UK

Abstract. This paper attempts to obtain an improved and more physiological distribution of the applied joint and muscle forces on the intact human femur and to gain an understanding of inter-subject variability on the mechanical response. A set of 109 CT-based femur models of individual anatomies were simulated using the Finite Element method during walking. Heterogeneous material properties, physiological boundary and loading conditions were applied to each femur model to form a reference initial load configuration [1]. To correct the imbalance in the force system, an optimisation scheme was adopted that iteratively updated the locations of both muscle and joint attachments across a 5-mm radius circle centred at the initially defined node in the reference load configuration [2, 3]. Across all patients, a 28–48% reduction in the resultant reaction force magnitude measured at the femoral head was achieved. A clear gender bias was present in terms of reaction forces and strains in both the initial and optimised models. The optimisation scheme mostly affected the medial-lateral component of the reaction force. The change in the average strain was found to be highly dependent upon the percentage reduction achieved in the optimisation process. This reduction was higher for males than females and is most likely due to size differences. Body weight and bone density highly influenced reaction forces and strains. Femoral anteversion linarly increased with reaction forces; other anatomical parameters such as neck length, neck offset, and functional femoral length or CCD angle did not have a clear influence on these forces.

Keywords: Inter-subject variability · Anatomy · Femur · Walking · Finite Element · Reaction force · Strain · Optimisation · Statistical analysis

1 Introduction

With technology to enable subject-specific Finite Element (FE) analysis becoming more widely available in the field of computational biomechanics [4], the issue of variability in patient anatomy is being increasingly tackled [5, 6]. Whilst inter-individual variability also extends to the musculoskeletal loading conditions [7, 8], determination of subject specific loads might not be feasible in larger cohort studies and this variability in these internal forces is only rarely considered [9, 10]. Often, only

G. A. Ateshian et al. (Eds.): CMBBE 2019, LNCVB 36, pp. 171–180, 2020.
https://doi.org/10.1007/978-3-030-43195-2_13

typical or reference loading conditions [11] are applied to the long bones which may differ in their anatomy from those for which the loading conditions were initially derived. As a result of such a geometric mismatch, however, the force system applied to the bone may no longer be in balance. Under such conditions, the specific choice of displacement constraints to remove rigid body motion in the FE analysis can have a substantial influence on the deformation pattern and the strains in the bone [1, 12]. It is further conceivable that artefacts introduced by such unbalanced force systems mask or at least skew the true effect that variations in the anatomy can have on the strain distribution.

Previous work on the loading conditions in the femur has indeed shown that non-physiological boundary conditions can lead to substantial increase of the forces at the hip joint, reaching values of up to 4,107 N (4.7 BW) and 5,271 N (6 BW) during walking and stair climbing, respectively [1], and thus considerably exceeding previously reported in vivo data [13]. Therefore, it is important to maintain both force and moment equilibrium when conducting FE analyses. Although variation in soft tissues anatomy is thought to exceed variability in bone morphology [12], the variability of muscle anatomy could be actively exploited to re-balance the force system when a set of loading and boundary conditions is to be mapped from a reference bone to a femur with different geometry.

The objective of the present work is therefore to compare variations in the reaction forces and strain distribution in the human femur across a larger sample of individual anatomies and bone stiffnesses and using different boundary conditions. To establish reference conditions, FE analyses were first performed using a typical reference load configuration of joint and muscle forces and physiological displacement boundary constraints. Further analyses then used an optimised set of muscle and joint forces where the location of muscle and ligament attachments on the femur were varied within a range consistent with inter-individual soft-tissue anatomy variability to minimise reaction forces. For both analysis conditions, it is sought to explore the influence of key parameters of femoral morphology as well as bone mechanical properties on the hip reaction forces and the strain distribution across the sample.

2 Materials and Methods

A convenience set of 109 segmented 3D femur bone models obtained from CT scans of 75 male and 34 female subjects aged between 43 and 106 years (average age: 64.6 ± 19.7 years) formed the basis for the current study [5]. Each femur model was segmented using ScanIP software (Simpleware Ltd., UK), followed by the application of a mesh registration and morphing technique. This enabled the description of each femur model with the same number of 65031 nodes and 304638 tetrahedral elements and a unique element connectivity matrix using a previously published image-to-mesh generation process and bone density assignment [5].

2.1 Reference Load Configuration

Displacement boundary conditions were applied as described by Speirs et al. [1], see Fig. 1a. The forces (Fx, Fy, Fz) applied at the hip joint contact (P0), abductor and tensor fascia latae (P1) and vastus lateralis (P2) of the reference subject with body weight 110 kg are: (611.54 N, 311.42 N, −2524.17 N), (−713.22, −30.86, 901.89 N) and (−713.22, −159 N, −1028.29 N), respectively. Peak hip joint contact forces and eight muscle forces during walking [11] were scaled to reflect each other patient's estimated body weight. The muscle force components, defined with respect to the coordinate system defined in Fig. 1a were applied to the FE models by distributing the load over a number of nodes. First, the location of each muscle point from the reference model was mapped onto each remaining femur model to identify the corresponding node and its surrounding elements. The force components were then distributed over all of the nodes (approximately 10) associated with these selected elements. This process was employed to emulate physiological conditions where soft tissue attachments are distributed over an area but also to minimise local artefacts in the strain distribution.

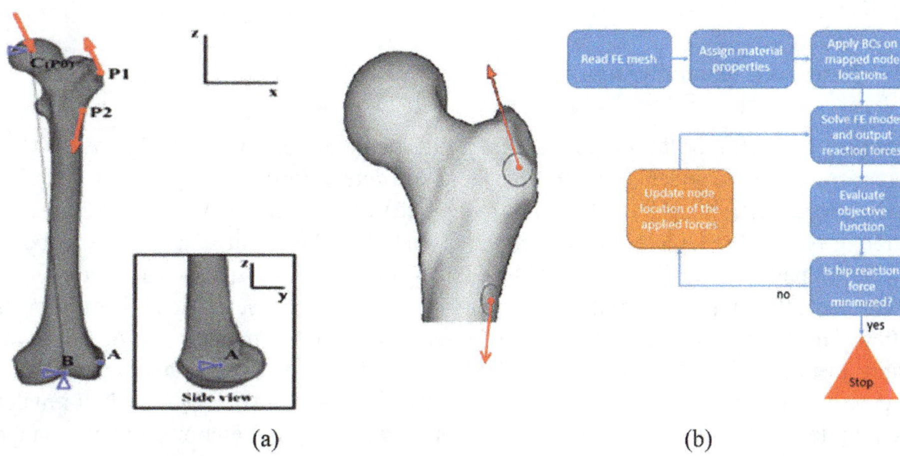

Fig. 1. The boundary conditions used for the FE models (a); the search domain and a flowchart of the adopted optimization scheme; reduced search domain where the abductor force is applied to the femur (b).

2.2 Optimised Load Configuration: Minimising Reaction Forces

A value of 10 mm for the typical variability of the location of muscle and ligament attachments [2, 3] defined the extent of the search domain to optimise soft tissue attachments. Here, a circular area with a radius of 5 mm centered at the initially defined node locations defined the permissible search area (Fig. 1b). For each femur model considered, the optimisation process involved a single objective function, i.e. minimising the reaction force on the constrained node located at the femoral head.

The design variables were the nodes at which two force vectors were applied. Each force vector can be applied to a finite number of nodes within the previously described search area. The first femur model for example led to 672 possible combinations for the two applied muscle forces.

2.3 Comparison Between Reference and Optimised Load Configuration

All FE models were solved using ANSYS Mechanical APDL (Ansys Inc., USA). Post-processing of the strain distributions focussed on the reaction forces at the constrained nodes, as well as the percentage of bone volume within specific strain limits (0–0.3%, 0.3–0.5%, 0.5–0.7% and >0.7%), and the maximum and average strain. To check for statistical differences between females and males in terms of reaction forces, two-tailed t-tests based on independent samples were performed. Finally, the individual effects of patient anatomical parameters and bone densities on reaction forces and strains were assessed and linear regression analyses were performed.

3 Results

3.1 Reference Load Configuration

Reaction Forces. For the initial load configuration, reaction forces as high as 3000 N were recorded at the nodes where the displacement constraints were applied. The median reaction force at the distal lateral epicondyle tended to be higher for males than for females although results for females showed a wider spread of data with a maximum value of 220 N compared to 150 N for males. At the knee-joint centre, the spread of reaction forces was in general wider for males than females; the median values were also higher for males with wider 25–75% percentile ranges. At the femoral head, the median reaction force was higher in males (215 N) than females (190 N); in the medial-lateral axis the 25–75% percentile ranges of reaction forces extended up to values of 340 N and 260 N for males and females, respectively. Regression analysis showed a strong relationship between the magnitude of the reaction forces and subject weight with high coefficients of determination R^2: 1 for the Z-component of reaction force at knee-centre FZ and 0.9 for the X-component of the reaction force at head centre FX, respectively.

The analysis of the relationship between anatomical parameters and reaction forces revealed that the anteversion angle had more substantial effect on their magnitude with coefficients of determination of 0.46 and 0.3 for the posteriorly directed reaction force (FY) at the distal lateral condyle and knee centre, respectively. T-tests revealed a clear difference between males and females in terms of reaction forces ($p < 0.05$). A clear linearity was also found in the data, indicating a normal distribution.

Strain Distribution. Under the reference loading conditions, a clear difference in the strain distribution was found between females and males. In males, approximately 95% of the femur experiences strains lower than 0.3% compared to 90% in the females, who demonstrated increased average and maximum strains. No clear relationship was observed between anatomical parameters and strain distribution. However, a linear relationship was found between bone mean density and computed strains.

3.2 Optimised Load Configuration

Reaction Forces. Across the entire population, the reduction in the resultant reaction force magnitude measured at the head centre varied from 28% to 48% (Fig. 2a). This implies that varying the nodes at which soft tissue forces were applied reduced the

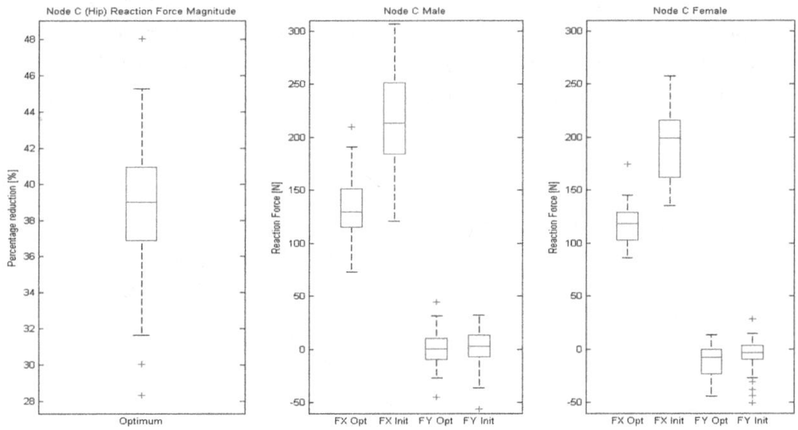

(a)

* Init and Opt refer to initial and optimised load configuration.

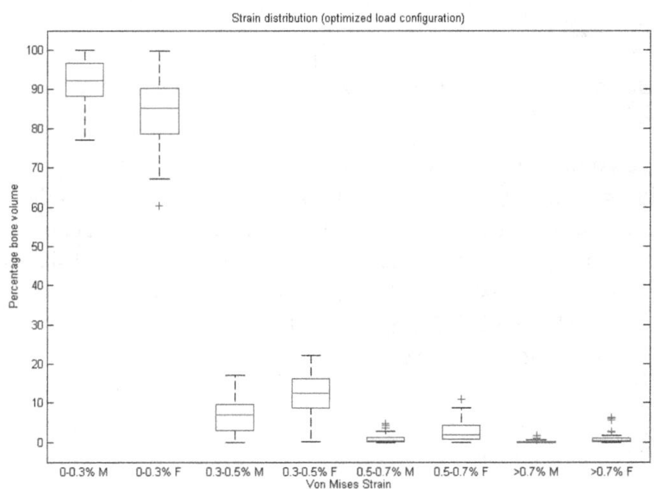

(b)

Fig. 2. Percentage reduction in hip reaction force at the head centre (a) and strain distribution in the optimised load configuration (b).

imbalance in the force system as measured by the reaction forces (Fig. 1), but could not achieve a completely balanced set of loading conditions. This Figure also revealed that the optimisation mostly affected the medial-lateral component of the reaction force FX in both females and males rather than the anterior-posterior component FY.

Strain Distribution. Similar to the reference loading configuration, a clear difference in the strain distribution was found between females and males (Fig. 2b). In males, approximately 92% of the femur experiences strains lower than 0.3% compared to 85% in the females. When the bone average strain is analysed, similar trends are observed for both females and males. Figure 3 shows an interesting relationship between the percentage difference in average strain and the percentage reduction in the hip reaction force.

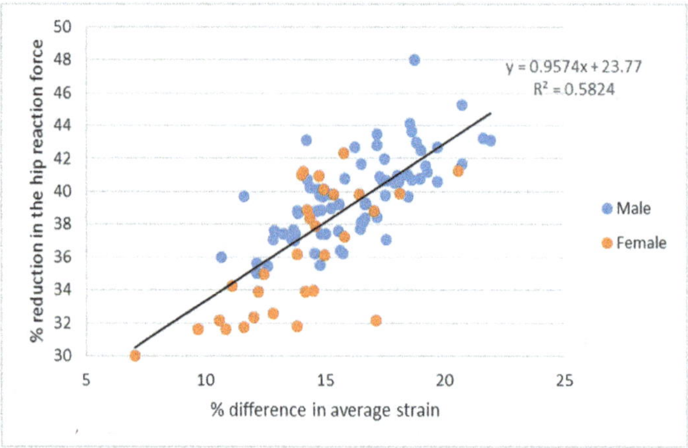

Fig. 3. Relationship between percentage reduction in hip reaction force and bone average strain.

It is demonstrated that the change in the average strain is highly dependent upon the percentage reduction achieved in the optimisation process. A higher percentage reduction in the optimisation results in a higher percentage difference in the average strain. This reduction was higher for males than females and is most likely due to size differences. Finally, when analysing the individual effects of anatomical parameters (anteversion angle, neck length, head medial offset and femoral functional length) on strain distribution, no clear trends were found.

4 Discussions

The objective of the study was to compare variations in reaction forces and strain distribution using a wide range of individual anatomies and different boundary conditions. Note that the same force vectors were applied to the femur models in both the initial and optimised load configurations. When analysing the femur with the highest reduction in reaction force, it was found that the neutral axis along the femoral shaft shifted laterally. The increased bending in the femur imposed larger strains at the surface of the cortical bone in the femoral shaft [12, 14]. This behaviour can be related to force imbalance [1, 15] and reaction forces should be reduced further in order to obtain a physiologically more accurate representation of the femur loading.

In terms of generated strains, the optimal load configuration resulted in an increase in the average strain in both male and female subjects. A decreased amount of bone volume was found within the first strain interval, 0–0.3%, after the optimisation process in both groups (Fig. 2b). A strong relationship was found between the achieved reduction in reaction forces and bone average strain (Fig. 3). Male subjects showed in general a greater reduction in reaction forces and subsequent strain distribution.

No evident effects of anatomical parameters were found except for the anteversion angle. Reaction forces are expected to depend on the geometry of the femur but not on bone elastic modulus or density. Out of the 352 possible node combinations in the optimisation, three subjects were identified (Table 1). Subject040 produced the largest percentage reduction of the reaction force (48%). Subject060 and Subject008 were found to produce the median and smallest percentage reduction, respectively.

Table 1. Anatomy data and reduction of hip reaction force reduction of seven selected subjects.

Femur ID	40	60	2	107	8	31
Gender	M	M	F	M	F	M
Age	53	79	76	50	35	55
Weight [kg]	83	47	52	133	60	72
Anteversion [°]	9.1	21.2	14.2	0.3	17.5	−0.5
CCD angle [°]	130	121	130	133.8	115.4	136.2
Neck length [mm]	55.3	54.4	47.1	47.4	47.1	48.4
Functional femoral length [mm]	411.6	386	357.7	471.1	392.7	448.2
Mean element density [g/cm^3]	0.47	0.53	0.47	0.64	0.62	0.51
Hip reaction force reduction	48%	39%	34%	38%	28%	40%

Figure 4 displays the distribution of both displacement and strain before and after optimisation for these three subjects, the ID being labelled on the undeformed femur shape. In the optimised load configuration, larger deflections can be seen; the femur is clearly under increased bending and torsion which results in even greater strain in the cortical bone on the femoral shaft. This could be expected as the reaction force on the hip was reduced significantly in the optimised load configuration. This also suggests that there is still a free moment present in the mechanical system that affects the balance. The neutral axis has also shifted laterally. Both the deflection and strain values are not far from the physiological standards of [1, 14].

The smallest female (Subject002) and tallest male (Subject107) subjects were scrutinised further. Deflections are larger in Subject107 (who had a weight of 133 kg) due to the larger applied forces. These larger deflections however, lose their physiological relevance according to the literature [14]. The reason for this is that the force and constraints models used in this study are developed for a reference femur geometry of a relatively normal size, i.e. not taking into account the extremities found within the population. Although larger deflections are observed in Subject107, strains appear to be higher in Subject002 due to the lower bone density of the subject.

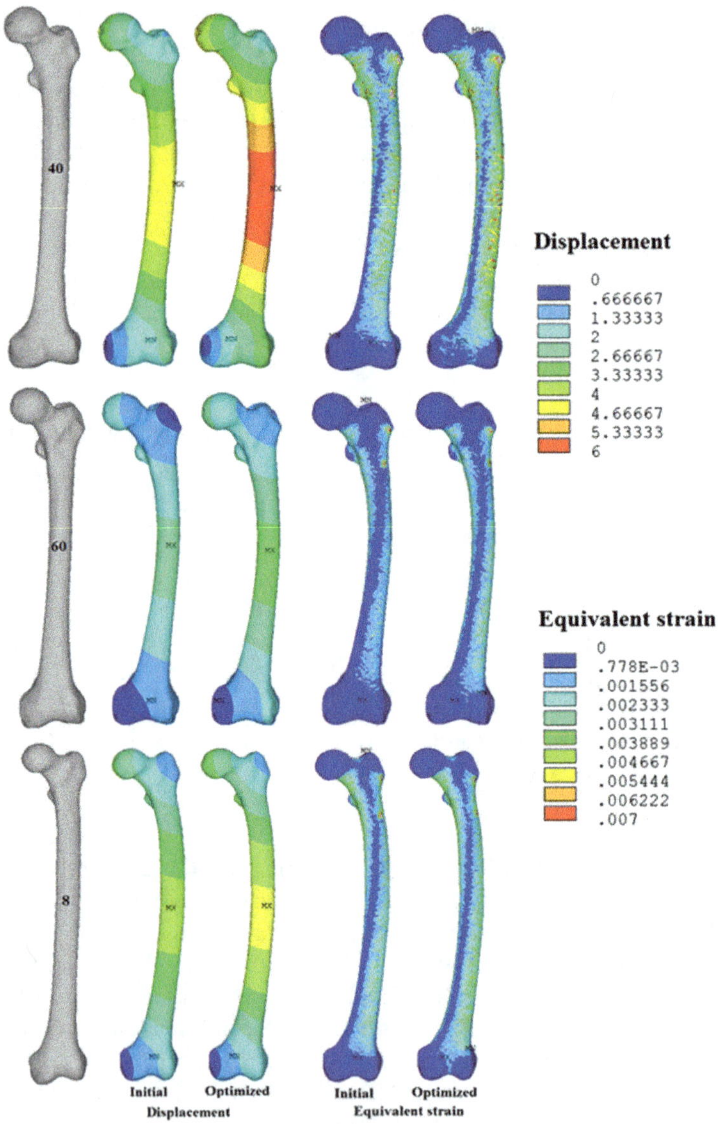

Fig. 4. Displacement (mm) and equivalent strain distribution. in three selected patients. Left grey femur shows the undeformed shape together with the subject ID number.

Subjects with the smallest (Subject008) and largest (Subject031) CCD angle were also compared, see Table 1 for more details. The mean CCD angle in males and females is 126 ± 5° and 125 ± 5°, respectively. FE results showed that Subject031 generated lower strains and smaller deflections than Subject008. This must be explained by the difference in femur anatomy, i.e. he CCD angle. For a smaller CCD angle, the force vector imposes bending in the femoral neck which results in an increased bending and torsion in the overall geometry. Increased bending and torsion leads to higher strain on the surface of the femur and larger displacements. For a larger CCD angle, the femoral neck is under compression instead of bending when the hip contact force is applied to the femoral head. This will decrease the bending and torsion in the overall geometry and result in lower strain and smaller deflections within the femur. For this specific comparison of the extremities in the CCD angle it is clear that the anatomy has an effect on the displacement and strain distribution.

In summary, a further reduction in reaction forces can be achieved by enlarging the search domain but this would lead to a design space that is physiologically not justifiable [2, 3]. The location of the knee-joint constraint could also be modified but this could impose larger bending on the femoral shaft and as a result cause higher strain and larger deflections. The knee-joint constraint modification might have influenced this behaviour but is unlikely to be causing it alone. Therefore, the location of the knee-joint constraint could be added as a design variable in the optimisation, allowing a change along the medial axis or within a spherical volume around the initially assumed knee centre node. Also, due to the uncertainty in the magnitude of applied forces, the design space could be extended further by allowing these forces to vary within specific limits to enable a shift in the balance of the force system and but also a reduced reaction force on the hip.

Conflict of Interest. All authors have no conflict of interest.

References

1. Speirs, A.D., Heller, M.O., Duda, G.N., Taylor, W.R.: Physiologically based boundary conditions in finite element modelling. J. Biomech. **40**(10), 2318–2323 (2007)
2. Kepple, T.M., Arnold, A.S., Stanhope, S.J., Siegel, K.L.: Assessment of a method to estimate muscle attachments from surface landmarks: a 3D computer graphics approach. J. Biomech. **27**(3), 365–371 (1994)
3. Matias, R., Andrade, C., Veloso, A.P.: A transformation method to estimate muscle attachments based on three bony landmarks. J. Biomech. **42**(3), 331–335 (2009)
4. Taylor, M., Prendergast, P.J.: Four decades of finite element analysis of orthopaedic devices: where are we now and what are the opportunities? J. Biomech. **48**(5), 767–778 (2015)
5. Bah, M.T., Shi, J., Browne, M., Suchier, Y., Lefebvre, F., Young, P., Heller, M.O.: Exploring inter-subject anatomic variability using a population of patient-specific femurs and a statistical shape and intensity model. Med. Eng. Phys. **37**(10), 995–1007 (2015)
6. Bah, M.T., Shi, J., Heller, M.O., Suchier, Y., Lefebvre, F., Young, P., King, L., Dunlop, D. G., Boettcher, M., Draper, E., Browne, M.: Inter-subject variability effects on the primary stability of a short cementless femoral stem. J. Biomech. **48**(6), 1032–1042 (2015)

7. Bergmann, G., Deuretzbacher, G., Heller, M., Graichen, F., Rohlmann, A., Strauss, J., Duda, G.N.: Hip contact forces and gait patterns from routine activites. J. Biomech. **34**(7), 859–871 (2001)
8. Trepczynski, A., Kutzner, I., Kornaropoulos, E., Taylor, W.R., Duda, G.N., Bergmann, G., Heller, M.O.: Patellofemoral joint contact forces during activities with high knee flexion. J. Orthop. Res. **30**(3), 408–415 (2012)
9. Pancanti, A., Bernakiewicz, M., Viceconti, M.: The primary stability of a cementless stem varies between subjects as much as between activities. J. Biomech. **36**, 777–785 (2003)
10. Taddei, F., Palmadori, I., Taylor, W.R., Heller, M.O., Bordini, B., Toni, A., Schileo, E.: Safety factor of the proximal femur during gait: a population-based finite element study. J. Biomech. **47**(17), 3433–3440 (2014)
11. Heller, M.O., Bergmann, G., Kassi, J.-P., Claes, L., Haas, N.P., Duda, G.N.: Determination of muscle loading at the hip joint for use in pre-clinical testing. J. Biomech. **38**(5), 1115–1163 (2005)
12. Duda, G.N., Heller, M., Albinger, J., Schulz, O., Schneider, E., Claes, L.: Influence of muscle forces on femoral strain distribution. J. Biomech. **31**(9), 841–846 (1998)
13. Kennedy, J.C., Hawkins, R.J., Willis, R.B., Danylchuck, K.D.: Tension studies of human knee ligaments. Yield point, ultimate failure, and disruption of the cruciate and tibial collateral ligaments. J. Bone Joint Surg. Am. **58**(3), 350–355 (1976)
14. Taylor, M.E., Tanner, K.E., Freeman, M.A., Yettram, A.L.: Stress and strain distribution within the intact femur: compression or bending? Med. Eng. Phys. **18**(2), 122–131 (1996)
15. Carbone, V., van der Krogt, M.M., Koopman, H.F.J.M., Verdonschot, N.: Sensitivity of subject-specific models to errors in musculo-skeletal geometry. J. Biomech. **45**(14), 2476–2480 (2012)

Shrinking Window Optimization Algorithm Applied to Pneumatic Artificial Muscle Position Control

William Scaff(✉), Marcos de Sales Guerra Tsuzuki, and Oswaldo Horikawa

Escola Politécnica of São Paulo University, São Paulo, Brazil
{william.scaff,mtsuzuki,ohorikaw}@usp.br

Abstract. Pneumatic Artificial Muscles (PAMs) are compliant actuators that can be safely used to interact with humans. Additionally, they can be made compatible with functional Magnetic Resonance Imaging (fMRI) for neurorehabilitation procedures and MRI-guided surgery. A major drawback of PAMs, however, is the precise positioning control due to its highly nonlinear dynamics. This control problem can be solved using an optimal control approach, defining an objective function and solving for the controller's parameters that minimizes this function. Typically this is done using a model of the system, and further trial and error tuning is required for better performance in the implementation stage, to compensate for the imperfections of the model. Optimizing the parameters using the real system to evaluate the objective function is harder because of the time, physical limitations and noise of the objective function, but it can provide better quality parameters and eliminate further tuning. For that, we use a novel optimization algorithm called Shrinking Window (SW), which is capable of finding the controller's parameters faster and with similar quality to state-of-the-art global optimization algorithms such as Bayesian Optimization (BO). We describe the SW algorithm and compare its performance on a synthetic noisy function and in tuning the controller for a positioning system powered by a PAM, against random search and BO. On learning the controller's parameters, we report a 48,15% shorter learning time and a 8% improvement on parameter quality compared to BO.

Keywords: Optimization algorithm · Pneumatic artificial muscle · Position control · Optimal control

1 Introduction

The usage of robotics in the rehabilitation field is becoming more popular. In Brazil, the Lucy Montoro Rehabilitation Network offers robotic rehabilitation since 2011 [6]. There are also many other applications of robotics in the health area, such as robotic surgeries, robotic health care, orthoses and prostheses.

G. A. Ateshian et al. (Eds.): CMBBE 2019, LNCVB 36, pp. 181–187, 2020.
https://doi.org/10.1007/978-3-030-43195-2_14

But whenever a robot interacts with a human, a safety concern takes place. This is because standard robotic systems have rigid parts, which can cause serious injuries in case of malfunction. This, however, can be solved by using soft robotics.

An interesting soft robotic actuator is the Pneumatic Artificial Muscle (PAM). This actuator have similar characteristics to the human skeletal muscle in terms of force and contraction ranges. Another interesting fact is that it can be build without ferromagnetic or conductive materials, which may be used in Magnetic Resonance Imaging-guided (MRI-guided) and functional MRI-guided (fMRI-guided) surgeries and rehabilitation.

PAMs have passive compliance which minimizes the risk of injury when the control or the system fails. This compliance, however, in conjunction with other effects such as hysteresis, makes this actuator difficult to model and control precisely [1,3,5,7,10,12–14].

There are several papers on PAM's control using different techniques. PID controllers are used for their simplicity and high adaptability [8]. However, many of the proposed PID controllers are tuned using the Ziegler-Nichols methods and trial and error, producing low quality gains [2,8,9]. Sliding-mode controllers were proposed, but they are known to have chattering problems [4]. Neural network (NN) based control was reported to suffer from time shift in response due to the iterations of the NN [12].

In general, because of the high nonlinear behavior of the muscle, control techniques that requires linearized models may suffer from unmodeled effects and nonlinear behavior. Sofisticated hybrid control techniques, such as Adaptative Fuzzy Non-singular Terminal Sliding Mode Control (AFNTSMC) may achieve robust, stable, smooth and fast control actions, but lack on positioning accuracy and are complex to implement [12].

Instead of using a model to design the controller, which at the end will need further trial and error tunning to improve the control performance, this paper proposes an optimal control approach directly at the bench. This avoids modeling and related problems, further trial and error tunning at the implementation stage, flexible parameter training which can be set with an objective function, with fewer iterations than a NN and without the time shift in response.

For that, it is used the Shrinking Window optimization algorithm with a simple PID controller. The algorithm is compared against random search and Baysian Optimization, which has been shown to outperform other state-of-the-art global optimization algorithms [11].

2 The Shrinking Window Algorithm

The Shrinking Window Algorithm is a novel sequential black-box optimization algorithm proposed by the authors. It was designed to optimize objective functions subjected to noise. Even though its sequential, it is easy to improve the learning time through parallelization on the function evaluation.

The algorithm is mainly consisted of a convergence factor, a movable search area (Window), a sampler and a chooser. The convergence factor is responsible for tuning the algorithm for exploration or exploitation, by shrinking the search window or changing the probability of choosing a point. The search window is the domain in which the sampler will acquire data. The chooser attributes a probability to each data point and randomly selects one point.

The algorithm works by repeating the process of calculating the convergence factor, adjusting the window size, sampling and choosing a data point until the stop criterion is reached.

By attributing a non-zero probability to the sampled data points, the algorithm has the ability to escape from local minima.

2.1 Implementation

Let J be a real-value cost function in $\Re^n$, ξ be the convergence vector, ε be the search domain size vector, Ω be the complete search domain set and N the number of samples per iteration. The objective of the algorithm is to find an element x in the domain Ω which minimizes, or nearly minimizes, the value of $J(x)$.

Initially, N samples are taken from Ω, which can be made in parallel if possible. Then, a probability is attributed to each sample according to their value. Lower values are given higher probability of being chosen than lower values. The convergence of the algorithm is critically dependent on this probability attribution. An example is given below:

$$P(x_i|x_{best}, x_{max}) = \left|\frac{x_i - x_{max}}{x_i}\right| \frac{1}{e^{x_i - x_{best}}} \tag{1}$$

Where $P(x_i|x_{best}, x_{max})$ represents the probability of the solution x_i being chosen given x_{best} and x_{max}. The variable x_{max} is the value of the worst solution from the N samples and the variable x_{best} is the current best solution.

To normalize the sample's probabilities such that $\sum_{i=1}^{N} P_i = 1$, a range from $0 \rightarrow 1$ is filled with sub-regions for each sampled data point. The size of each region is proportional to the probability of that point being chosen. Then a random number from $0 \rightarrow 1$ is generated, which determines the selection of the point.

In every following iteration, the window size ε is updated according to the convergence vector ξ:

$$\varepsilon_{i+1} = \varepsilon_i \odot \xi_i \tag{2}$$

Where $\odot$ is the Hadamard product, or element wise product. The convergence vector can also be updated according to some criterion, e.g. considering the variance of the samples or, for simplicity, each element can be constant for a geometrical window shrinking.

The next window is centered at the current solution x_n, and the process of sampling, probability attribution and choosing a point is repeated until the stop criterion.

3 Optimization on Synthetic Benchmark Functions

To verify the performance of the SW algorithm, some benchmark functions were used. They simulate noisy functions with local minima. The SW is compared against pure random search and BO. The pure random search samples random points inside Ω and the BO is the scikit-optimize Python library implementation of the algorithm.

The following benchmark function is used in the first test

$$J_1(a, b) = a^2 + b^2 + 4\sin(a) + 10\sin(b) + \chi(0, 5) \tag{3}$$

where $\chi(0, 5)$ represents a Gaussian noise with mean value of 0 and standard deviation of 5.

The complete domain Ω is $-5.12 \rightarrow 5.12$ for both a and b and the optimization is limited to 100 and 500 function evaluations.

Another function J_2 was used to test the optimization with four parameters:

$$J_2(a, b, c, d) = a^2 + b^2 + c^2 + d^2 + 4(\sin(a) + \sin(c)) + 10(\sin(b) + \sin(d)) + 2\chi(0, 5) \tag{4}$$

Table 1 summarizes the optimization results for J_1 and Table 2 for J_2.

Table 1. Optimization results for the J_1 benchmark function.

Algorithm	$\bar{min}(J_1)$	σ	Evaluations	J_{1best}	Time of execution
Random	–12.918	3.149	100	–21.790	0.2 s
BO	–21.819	2.192	100	–30.034	8671.8 s
SW	–19.650	2.971	100	–31.622	0.2 s
Random	–13.791	4.225	500	–20.948	0.01 s
BO	–25.763	1.834	500	–28.896	31279.0 s
SW	–24.584	2.073	500	–29.200	0.11 s

Table 2. Optimization results for the J_2 benchmark function.

Algorithm	$\bar{min}(J_2)$	σ	Evaluations	J_{2best}	Time of execution
Random	–21.929	3.469	500	–28.972	0.12 s
BO	–15.640	8.523	500	–25.200	41118.97 s
SW	–43.235	8.112	500	–62.344	0.41 s

Note that the best solution for the 500 evaluations tests are worst than for the 100 evaluation tests. This is because the number of repetitions for the 100 and 500 evaluations tests are 100 and 10 respectively. Also note that in terms of comparison, the time of execution for the BO on J_1 would be 10 times longer, reaching 87 h approximately.

Although BO is better on average for J_1, SW found the best solutions on all tests. For J_2, BO performed worse than random search, with SW almost 3 times better on average solution and best solution.

4 PAM Optimal Control

SW and BO were used to optimize the parameters of a PID controller on a nonlinear system. The system is a PAM positioning system with on-off valves illustrated in Fig. 1.

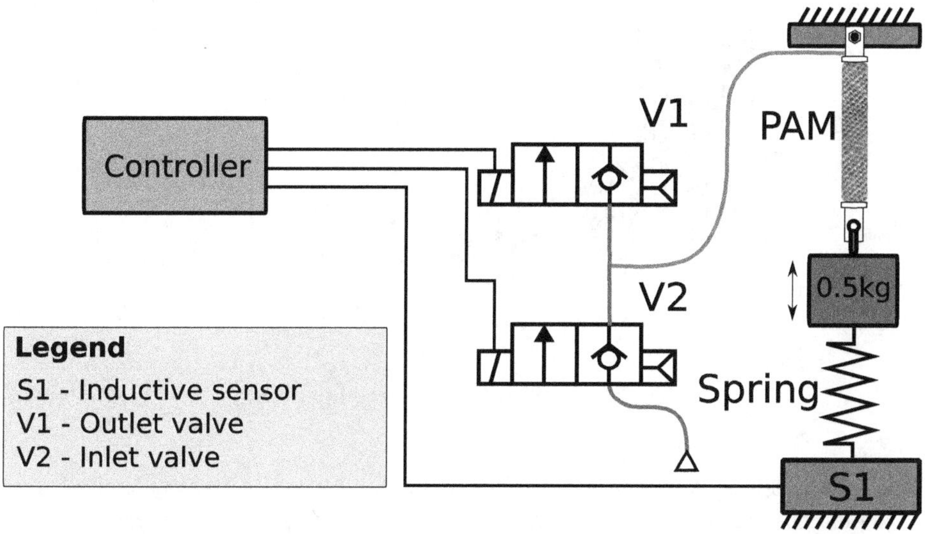

Fig. 1. PAM vertical positioning system bench schematics.

The objective is to minimize the integral of the absolute error (IAE) of a step response. SW and BO were used to learn the control parameters with a limit of 500 evaluations.

BO was used first, followed by SW and BO again (BO2). The learning process is illustrated in Fig. 2.

In this control experiment, the Shrinking Window Optimization (SWO) performed better than BO and the second run of BO (BO2), each taking around 2.7 h to finish. While BO didn't improve with further evaluations, SW was capable of improving the current solution up to the evaluation limit, producing a better answer in nearly half of the time.

Optimizing just three parameters for the PID isn't sufficient for a real world position control of a nonlinear system. Instead multiple parameters should be

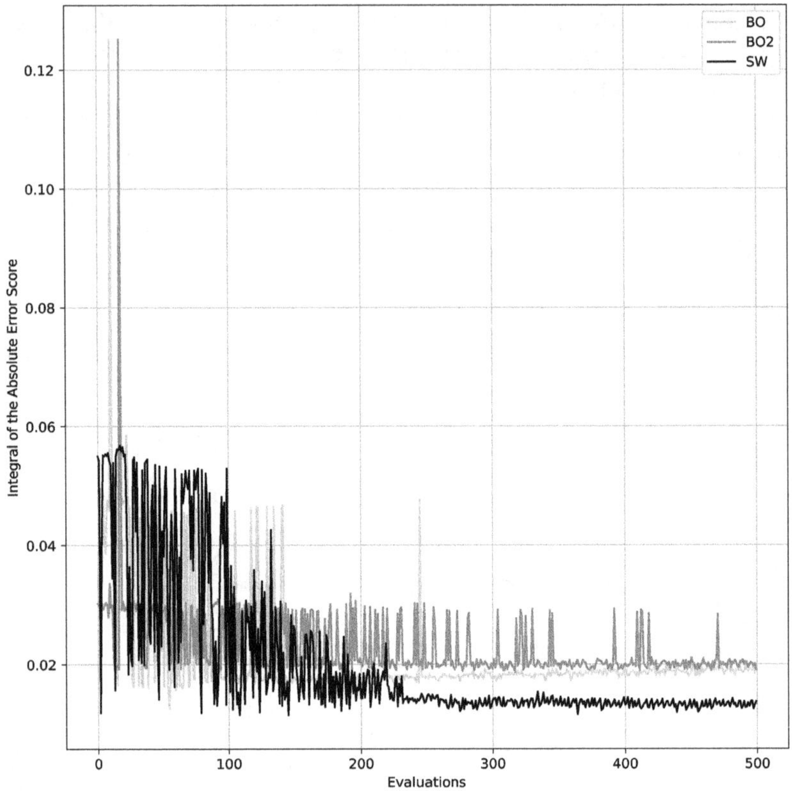

Fig. 2. Algorithms learning comparison.

used such as in a gain schedule approach. In this respect, SW was shown to be more efficient for optimizing functions with higher size of parameters (e.g. J_2). Therefore, it showed potential to be a viable solution to the control problem of PAMs.

5 Conclusions

It was shown that the SW algorithm is capable of optimizing noisy objective functions and learn the control of a nonlinear PAM positioning system. Even though it is a Monte Carlo Method, it was comparable to BO with few evaluations on synthetic functions and outperformed BO on the learning of the control parameters of the positioning system and when learning four parameters on J_2. This shows the potential to use this algorithm to solve complex control problems.

References

1. Anh, H.P.H.: Online tuning gain scheduling mimo neural pid control of the 2-axes pneumatic artificial muscle (pam) robot arm. Exp. Syst. Appl. **37**(9), 6547–6560 (2010)
2. Åström, K.J., Hägglund, T.: The future of PID control. Control Eng. Pract. **9**, 1163–1175 (2001)
3. Chan, S., Lilly, J.H., Repperger, D.W., Berlin, J.E.: Fuzzy pd+ i learning control for a pneumatic muscle. In: The 12th IEEE International Conference on Fuzzy Systems, FUZZ 2003, vol. 1, pp. 278–283. IEEE (2003)
4. Lin, L.H., Yen, J.Y., Wang, F.C.: Robust control for a pneumatic muscle actuator system. Trans. Can. Soc. Mech. Eng. **37**(3), 581–590 (2013)
5. Medrano-Cerda, G.A., Bowler, C.J., Caldwell, D.G.: Adaptive position control of antagonistic pneumatic muscle actuators. In: Proceedings 1995 IEEE/RSJ International Conference on Intelligent Robots and Systems. Human Robot Interaction and Cooperative Robots, vol. 1, pp. 378–383. IEEE (1995)
6. Portal, S.P.G.W.: Robótica é aliada na reabilitação na rede lucy montoro (2011). http://www.saopaulo.sp.gov.br/spnoticias/ultimas-noticias/robotica-e-aliada-na-reabilitacao-na-rede-lucy-montoro/
7. Repperger, D., Johnson, K., Philips, C.: Nonlinear feedback controller design of a pneumatic muscle actuator system. In: Proceedings of the 1999 American Control Conference, vol. 3, pp. 1525–1529. IEEE (1999)
8. Sakthivelu, V., Chong, S.H.: Motion control of a 1-DOF pneumatic muscle actuator positioning system. In: Proceedings 10th Asian Control Conference, Kota Kinabalu, Malaysia, pp. 1983–1987 (2015)
9. Schröder, J., Kawamura, K., Gockel, T., Dillmann, R.: Improved control of a humanoid arm driven by pneumatic actuators. In: Proceedings of International Conference Humanoid Robots, Munich, Germany, pp. 29–30 (2003)
10. Situm, Z., Herceg, S.: Design and control of a manipulator arm driven by pneumatic muscle actuators. In: 2008 16th Mediterranean Conference on Control and Automation, pp. 926–931. IEEE (2008)
11. Snoek, J., Larochelle, H., Adams, R.P.: Practical Bayesian optimization of machine learning algorithms. In: Advances in Neural Information Processing Systems, pp. 2951–2959 (2012)
12. Tang, T., Chong, S., Chan, C., Sakthivelu, V.: Point-to-point positioning control of a pneumatic muscle actuated system using improved-pid control. In: IEEE International Conference on Automatic Control and Intelligent Systems (I2CACIS), pp. 45–50. IEEE (2016)
13. Thanh, T.D.C., Ahn, K.K.: Nonlinear pid control to improve the control performance of 2 axes pneumatic artificial muscle manipulator using neural network. Mechatronics **16**(9), 577–587 (2006)
14. Wu, J., Huang, J., Wang, Y., Xing, K., Xu, Q.: Fuzzy pid control of a wearable rehabilitation robotic hand driven by pneumatic muscles. In: 2009 International Symposium on Micro-NanoMechatronics and Human Science, MHS 2009, pp. 408–413. IEEE (2009)

Aging Health Behind an Image: Quantifying Sarcopenia and Associated Risk Factors from Advanced CT Analysis and Machine Learning Technologies

Marco Recenti[1], Magnus K. Gìslason[1], Kyle J. Edmunds[1], and Paolo Gargiulo[1,2(✉)]

[1] Institute of Biomedical and Neural Engineering, Reykjavik University, Reykjavik, Iceland
paologar@landspitali.is
[2] Department of Science, Landspitali, Reykjavik, Iceland

Abstract. Sarcopenia, the progressive degeneration of aging muscle, is identified as an independent risk factor for significant morbidity, disability, and mortality in elderly individuals. In this paper Artificial Intelligence technologies, in particular Machine Learning (ML) supervised algorithms, are adopted to predict physiological parameter starting from muscle, fat and connective tissue distribution values of a mid-thigh Computer Tomography (CT) images. We developed and validated a novel method for soft tissue radiodensitometric distribution profiling, which is entitled nonlinear trimodal regression analysis (NTRA) method for soft tissue CT profiling. The work shows a comparative analysis using the NTRA method and standard soft tissue CT analysis modalities which was implemented on parameters assemblies from the 3,157 patients AGES-Reykjavik dataset. Furthermore, ML approach is used to find connections between amplitude, location, width and skewness in fat, muscle, and connective tissue and link these data to biomechanical measurements, Body Mass Index (BMI) and Cholesterol. The results highlight the specificities of each muscle quality metric to Lower Extremity functions and sarcopenic comorbidities. ML approach shows good predictive values for BMI having as most significant features connective and fat amplitude. Standardizing a quantitative methodology for myological assessment in this regard would allow for the generalizability of sarcopenia research to the indication of compensatory targets for clinical intervention.

Keywords: Machine Learning · Regression · Sarcopenia · Computed tomography · Nonlinear Trimodal Regression Analysis

1 Introduction

Progressive degeneration of aging muscle, the so-called sarcopenia, is consistently identified as an independent risk factor for mortality in elderly people [1–5]. A normative quantitative definition for both the etiology and health implications of sarcopenia remain debated in literature [6–10].

G. A. Ateshian et al. (Eds.): CMBBE 2019, LNCVB 36, pp. 188–197, 2020.
https://doi.org/10.1007/978-3-030-43195-2_15

Artificial Intelligence (AI) technologies, particularly Machine Learning (ML) algorithms, are largely used in different and numerous healthcare applications (i.e. the image processing or predictive system [11, 12]). ML has a large potential and can be applied in a wide range of situations like the investigation of medical treatments, or the identification of chronic disease states that help in the design of appropriate interventions and in the medical diagnosis or moreover the minimization of fraud and abuse by the identification of unusual patterns [13].

Age Gene/Environment Susceptibility Study (AGES) Reykjavik, Iceland, is a dataset with more than 10 thousand features, designed to examine risk factors and gene/environment interaction, in relation to disease and disability in patients between 65 to 95 years old [14, 15].

Therefore, this study demonstrates that ML approach can show predictive values for some physiological parameters like Body Mass Index and Isometric Leg Strength, using as initial features the 11 Nonlinear Trimodal Regression Analysis (NTRA) parameters extracted from an axial mid-thigh computer tomography (CT) image. Moreover, in the prediction a great importance is given to the connective tissue, often little considered at the expense of muscle tissues and fat [16–18].

2 Materials and Methods

2.1 AGES-Reykjavik Database – NTRA Parameters

The database used in this paper is denominated AGES-Reykjavik: it is composed of AGES I and AGES II (same measurements on the same patients but about 5 years after AGES I). Both have 3157 patients, so in total, AGES I + II is made up of 6314 patients.

A subset of AGES is used: 11 NTRA parameters extracted from a mid-femur CT scan plus some physiological measurements like Body Mass Index (BMI) [kg/m^2], Cholesterol (Chol) [mmol/L], Normal Gait Speed (NGait) [sec] and Isometric Leg Strength (ISO) [Newtons].

The 11 NTRA parameters developed by Edmunds et al. [19, 20], are derived from a mid-thigh CT Scans. The localized scanning region extended from the iliac crest to knee. For each patient, a single 10-mm thick transaxial mid-thigh section was used in order to generate HU distributions and calculate fat and muscle cross-sectional area extensions. For each patient HU distribution were derived from each pixel's CT number value following the expression:

$$HU = CT \times 2,26625 - 190$$

After this operation, HU values (across the range of −200 to 200 HU), were categorized into 128 bins as typical for CT assessment protocols [21–25]. Resultants histograms were smoothed to obtain underlying empirical probability density functions (PDF) for each histogram. Each PDF was then exported for the NTRA regression analysis.

The method utilized to computationally describe each HU distribution was a form of modified nonlinear regression analysis: each HU distribution is defined as a quasi-probability density function defined by three Gaussian Distribution of which one standard and two skewed:

$$\sum_{i=1}^{3} \varphi(x, N_i, \mu_i, \mu_i, \alpha_i) = \sum_{1}^{3} \frac{N_i}{\sigma_i\sqrt{2\pi}} e^{-\frac{(x-\mu_i)^2}{2\sigma_i^2}} erfc\left(\frac{\alpha_i(x-\mu_i)}{\sigma_i\sqrt{2}}\right)$$

where N is the amplitude, μ is the location, σ is the width, and α is the skewness of each of the distributions, all of which are evaluated in an iterative way at each CT bin, x. This definition is done under the hypothesis that HU distributions is trimodal: they consist of three different tissue types namely fat (i = 1) [−200 to −10 HU], connective tissue (i = 2) [−9 to 40 HU] and muscle (i = 3) [41 to 200 HU]. The central connective tissue is non-skewed, while the other two are described by, respectively, a positive and negative skewness.

This method can give as results 11 parameters that are unique to every individual's CT image: 4 related to the Fat, 4 to the Muscle, and 3 to the Connective Tissue (Fig. 1).

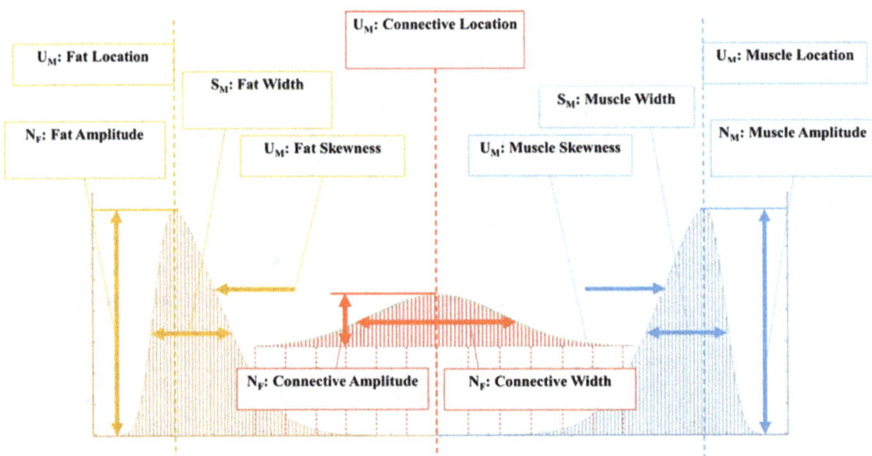

Fig. 1. 11 NTRA parameters represented on relative PDFs [19]

Some "Not a Number" (NaN) values are present in the database and must be specifically handled. NaN values cannot be read by the machine during the running of the ML algorithms, so some solutions were found to overcome these missed values in the database. Only 8 NaN are present regarding BMI, so these are substituted with the mean value of the feature itself. In the case of all the other measurements, the NaN values are too many and the mean solution would compromise the results, so, the patient is excluded from the cohort. From the definition itself of the NTRA parameters it is possible to guess that, for them, the NaN values are not present.

2.2 Machine Learning Regression Algorithms

In literature lots of supervised ML algorithms are present, whose effectivity depends on the uses and applications in healthcare [12]. In this paper, 3 Tree-Based algorithms are considered for regression analysis. They are denominated Tree-Based because the basic units on which they are built are the so-called regression trees [26]. They were selected because they do not need any normalization or standardization of the initial data and the feature importance is not difficult to extract.

The first algorithm considered is Random Forest (RF) which operate by constructing numerous decision trees during the training time and outputting the mean prediction of each of the individual trees [27, 28]. The second one is Extremely Randomized Tree, EXTRA Tree (EX-T) in which randomness goes one step further in the way splits are implemented. It consists of randomizing both attribute and cut-point while splitting a node of the tree. It can build a totally randomized tree in which the structures are independent of the output values of the learning sample [29]. The third one is Ada Boosting (ADA-B) where the principal core is to fit weak learners (such as small decision trees) on repeatedly modified versions of the training data. To give the final estimation, all the predictions from the small trees are combined through a weighted majority vote [30, 31].

The coding language used is Python (PY) with the relative ML library Scikit-Learn (SL). For each previous algorithm SL provides a specific function with default values, for the number of trees in the forest (n_estimators), and the maximum depth of the tree (max_depth) [32].

2.3 Regression

Regression algorithms need a single regressand (the continuous value that is predicted) and more regressors (the features used to predict the regressand). To estimate the performances of the regression results, the Coefficient of Determination (R^2) is taken into consideration [33].

2.4 Cross-Validation – k_Fold

In order to apply ML algorithms, the database must be divided into training and test set. The statistical procedure known as k_fold Cross-Validation [34], is used to have a better vision of all the possible results in order to select the best one and have an accurate average value of the results. The dataset is split into k consecutive folds: $k-1$ folds are used to train and 1 for the test. Training is done for each combination of folds. The mean of R^2 of all the folds can be calculated as well as the maximum value. With fewer number of folds, the train set is larger, and the test set is smaller, viceversa with more folds. An unbalanced number of folds does not allow to obtain reliable results, so k_fold division is applied using 12 folds. The results obtained with 8 folds were not relevant, compared to the $k = 12$ ones.

2.5 Methodology

Figure 2 shows the workflow of the methodology adopted in this work.

Figure 2A is the begin: linear transformation of the CT images to constituent HU absorption histograms, followed by Fig. 2B, the employment of the 11 NTRA parameters to generate distribution profile functions defined by 11 patient-specific coefficients ($p_{1\rightarrow 11}$). From Fig. 2C, where are represented the physiological parameters described on Sect. 2.1, through the discretization into subject groups for logistic regression [16] (Fig. 2D), it is possible to derive using the ML algorithms the indices (i_n) for aging health (Fig. 2E). The indices correspond to the coefficient of determination R^2 results obtained for each of the physiological parameters analysed.

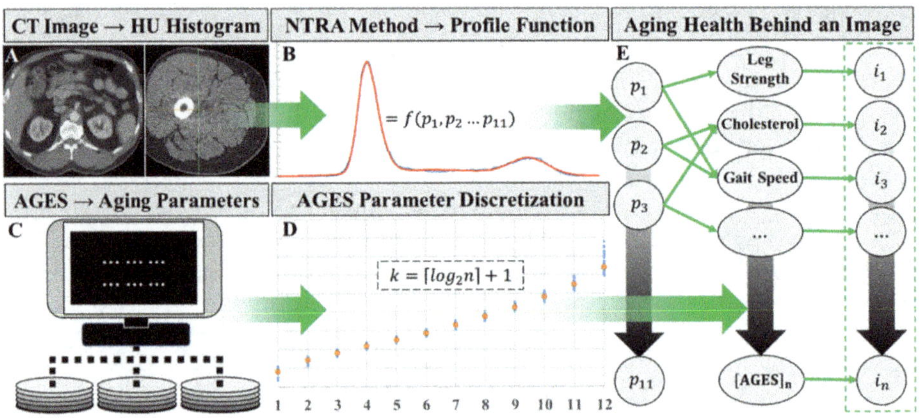

Fig. 2. Workflow, illustrating the principle behind quantifying indices for aging health using the novel NTRA method for CT image HU distribution analysis

Figure 3 shows the methodology used for the application of the ML algorithms: a focus on what was shown before in Fig. 2E. Starting from the 11 parameters used as initial features and using the k_fold cross validation with k = 12, using the 3 different tree-based regression algorithms we can have, for each aging parameter (regressand), different results in terms of R^2, which correspond to the i_n.

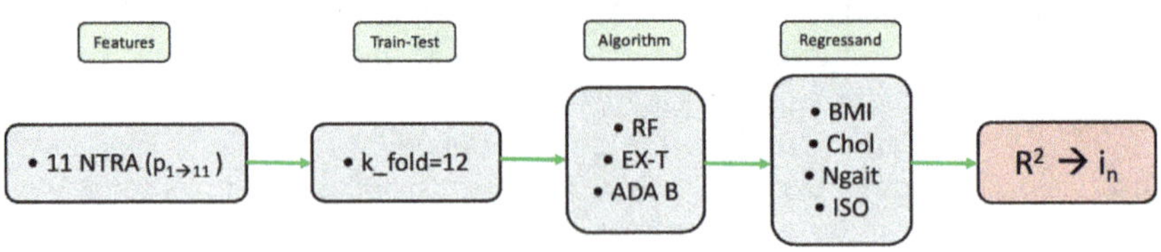

Fig. 3. Regression methodology

3 Results

In Table 1 all the prediction results, in terms of R^2 mean and maximum value, are presented: the negative ones have no significant prediction value, but they allow us to understand that there is no relation between the aging parameter itself and the starting 11 NTRA features. The good R^2 results, obtained for BMI and ISO, on the other hand, confirm that NTRA can have a relevant prediction power.

Table 1. R^2 results for each of the aging parameter

	Algorithm	R^2 mean	R^2 max
BMI	RF	0.756	0.793
	EX-T	0.756	**0.813**
	ADA-B	0.770	0.810
Chol	RF	−0.037	0.005
	EX-T	−0.050	0.021
	ADA-B	−0.010	0.063
NGait	RF	0.007	0.105
	EX-T	−0.005	0.114
	ADA-B	0.085	0.192
ISO	RF	0.483	0.599
	EX-T	0.482	**0.614**
	ADA-B	0.493	0.587

BMI can be predicted with very good accuracy: the maximum $R^2 = 0{,}813$ is obtained with EX-T while the best mean value is obtained with ADA-B ($R^2 = 0{,}77$). The most important features in the prediction shown in Fig. 4, are the amplitude of the connective tissue and of the fat (N_Conn = 0,344, N_Fat = 0,331 on a possible maximum value of 1). They, for each algorithm, always cover more than 50% of the total importance. These results, that link the amplitude of the connective tissue extracted from a CT scan to the BMI, in such an important way, suggest that even for future further application of this methodology to the prediction of other parameters, the connective tissue may have a great importance in the predictive process.

The R^2 obtained for ISO can be considered satisfactory even if they do not reach the values achieved with the BMI: it can be said that there is a link between the 11 NTRA and the strength of the leg in elderly people. The maximum $R^2 = 0{,}614$ is achieved with the EX-T algorithm, while, like BMI, the best mean value is with ADA-B ($R^2 = 0{,}493$). Muscle amplitude covers almost the 50% of the feature importance while the connective tissue, especially the location (see Fig. 1), has again a high importance in the prediction. This strengthens the results obtained with the BMI: connective tissue is again very relevant and it must be considered as one of the main prediction factors.

For what concern Chol and NGait the values of R^2 are really low, even below zero (this is possible due to the definition of R^2 itself): the maximum value is never more than 0.2 for both of them.

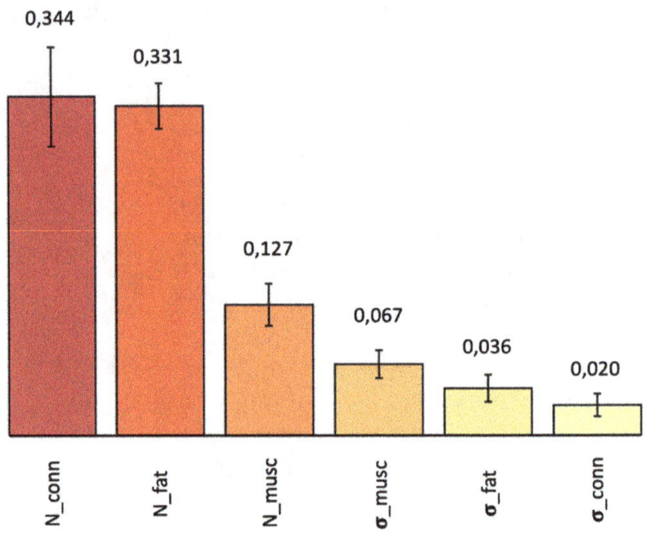

Fig. 4. Feature importance for the first 6 of 11 NTRA parameters in the BMI prediction

4 Discussion and Conclusions

This paper study was focused on the prediction of some physiological measurements and on the quantifying of indices for aging health using as initial features the 11 NTRA parameters. Usually the ML technologies applied to medical images work on the image itself in order to create masks or to do segmentations: here we try to link a CT scan to physiological parameters that are not relative to the image itself and which apparently may have not connections with a CT scan. The obtained results, especially for BMI and ISO, allow us to say that the 11 NTRA parameters can have a very significative predictive value. The three tree-based algorithms give good results, but, eventually, a future exploration of other ML algorithms can be done in order to improve, or at least confirm, the achieved results.

The results of the feature importance for the BMI and for the ISO are particularly relevant: the very high ones obtained from the connective tissue parameters deserves a special mention. Much importance is usually given to muscles and fat in the CT scan analysis, but these results confirm that in the predictive process the connective tissue has an importance that is absolutely not negligible, in some cases also primary.

Finally, this paper can be considered as a starting point for a more in-depth analysis of the AGES-Reykjavik database with the ML technologies.

Acknowledgment. We want to thank all the staff and the participants of the AGES-Reykjavik study for their important contribution: The Age, Gene/Environment Susceptibility Reykjavik Study has been funded by NIH contract N01-AG12100, the NIA Intramural Research Program, Hjartavernd (the Icelandic Heart Association), and the Althingi (the Icelandic Parliament).

Conflict of Interest Declaration. The authors declare they have no conflict of interests.

References

1. Metter, E.J., Talbot, L.A., Schrager, M., Conwit, R.: Skeletal muscle strength as a predictor of all-cause mortality in healthy men. J. Gerontol. A Biol. Sci. Med. Sci. **57**(10), B359–B365 (2002)
2. Rantanen, T.: Muscle strength and body mass index as long-term predictors of mortality in initially healthy men. J. Gerontol. **55A**(3), 168–173 (2000)
3. Newman, A.B., Kupelian, V., Visser, M., Simonsick, E.M., Goodpaster, B.H., Kritchevsky, S.B., Harris, T.B.: Strength, but not muscle mass, is associated with mortality in the health, aging and body composition study cohort. J. Gerontol. A Biol. Sci. Med. Sci. **61**(1), 72–77 (2006)
4. Goodpaster, B.H., Carlson, C.L., Visser, M., Kelley, D.E., Scherzinger, A., Harris, T.B., Newman, A.B.: Attenuation of skeletal muscle and strength in the elderly: the health ABC study. J. Appl. Physiol. **90**(6), 2157–2165 (2001). (Bethesda MD: 1985)
5. Volpi, E., Nazemi, R., Fujita, S.: Muscle tissue changes with aging. Curr. Opin. Clin. Nutr. Metab. Care **7**(4), 405–410 (2004)
6. Newman, A.B., Kupelian, V., Visser, M., Simonsick, E., Goodpaster, B., Nevitt, M., Harris, T.B.: Sarcopenia: alternative definitions and associations with lower extremity function. J. Am. Geriatr. Soc. **51**, 1602–1609 (2003). https://doi.org/10.1046/j.1532-5415.2003.51534.x
7. Maughan, R.J., Watson, J.S., Weir, J.: Strength and cross-sectional area of human skeletal muscle. J. Physiol. **338**, 37–49 (1983)
8. Brooks, S.V., Faulkner, J.A.: Skeletal muscle weakness in old age: underlying mechanisms. Med. Sci. Sports Exerc. **26**(4), 432 (1994)
9. Narici, M.V., Maffulli, N.: Sarcopenia: characteristics, mechanisms and functional significance. Br. Med. Bull. **95**, 139–159 (2010)
10. Koeun, L., Yongbin, S., Jimi, H., Yu, S.S., In-Seob, L., Kwon-Ha, Y., Kyung, W.K.: Recent issues on body composition imaging for sarcopenia evaluation. Korean J. Radiol. **20**(2), 205–217 (2019)
11. Jiang, F., Jiang, Y., Zhi, H.: Artificial intelligence in healthcare: past, present and future. Stroke Vasc. Neurol. **2**(4), 230–243 (2017)
12. Holzinger, A.: ML for health informatics. In: LNAI, vol. 9605, pp. 1–24 (2016). https://doi.org/10.1007/978-3-319-50478-0_1
13. Koh, H.C., Tan, G., et al.: Data mining applications in healthcare. J. Healthc. Inf. Manag. **19** (2), 65 (2011)
14. Harris, T.B., Launer, L.J., Eiriksdottir, G., Kjartansson, O., Jonsson, P.V., Sigurdsson, G., et al.: Age, gene/environment susceptibility–Reykjavik study: multidisciplinary applied phenomics. Am. J. Epidemiol. **165**(9), 1076–1087 (2007). https://doi.org/10.1093/aje/kwk115. PMID: 17351290

15. Johannesdottir, F., Aspelund, T., Siggeirsdottir, K., Jonsson, B., Mogensen, B., Sigurdsson, S., et al.: Mid-thigh cortical bone structural parameters, muscle mass and strength, and association with lower limb fractures in older men and women (AGES-Reykjavik study). Calcif. Tissue Int. **905**, 354–364 (2012)
16. Gargiulo, P., Helgason, T., Ramon, C., Jonsson, H., Carraro, U.: CT and MRI assessment and characterization using segmentation and 3D modelling techniques: applications to muscle, bone and brain. Eur. J. Trans. Myol. – Basic Appl. Myol. **24**(1), 55–62 (2014)
17. Engelke, K., Museyko, O., Wang, L., Laredo, J.D.: Quantitative analysis of skeletal muscle by computed tomography imaging – state of the art. J. Orthop. Transl. **15**, 91–103 (2018)
18. Mourtzakis, M., Prado, M.M.C., Lieffers, J.R., Reiman, T., McCargar, L.J., Baracos, V.E.: A practical and precise approach to quantification of body composition in cancer patients using computed tomography images acquired during routine care. Appl. Physiol. Nutr. Metab. **33**, 997–1106 (2008)
19. Edmunds, K.J., Gislason, M., Sigurðsson, S., Guðnason, V., Harris, T.B., Carraro, U., Gargiulo, P.: quantitative methods in correlating sarcopenic muscle degeneration with lower extremity function biometrics and comorbidities. PLoS ONE **13**(3), e0193241 (2018). https://doi.org/10.1371/journal.pone.0193241
20. Edmunds, K.J., Arnadottir, I., Gislason, M., Carraro, U., Gargiulo, P.: Nonlinear trimodal regression analysis of radiodensitometric distributions to quantify sarcopenic and sequelae muscle degeneration. Comput. Math. Methods Med. **2016** (2016). 8932950. PMID: 28115982. https://doi.org/10.1155/2016/8932950
21. Petursson, T., Edmunds, K.J., Gislason, M.K., Magnusson, B., Magnusdottir, G., Halldorson, G., Gargiulo, P.: Bone mineral density and fracture risk assessment to optimize prosthesis selection in total hip replacement. Comput. Math. Methods Med. (2015). ID 162481
22. Edmunds, K.J., Gislason, M.K., Arnadottir, I.D., Mercante, A., Piccione, F., Gargiulo, P.: Quantitative computed tomography and image analysis for advanced muscle assessment. Eur. J. Transl. Myol. **26**(2), 6015 (2016). https://doi.org/10.4081/ejtm.2016.6015
23. Gargiulo, P., Edmunds, K.J., Gislanson, M.K., Latour, C., Hermasson, P., Esposito, L., Bifulco, P., Cesarelli, M., Fraldi, M., Cristofolini, L., Jonsson Jr., H.: Patient-specific mobility assessment to monitor recovery after total hip arthroplasty. Proc. Inst. Mech. Eng. H **232**(10), 1048–1059 (2018)
24. Gargiulo, P., Edmunds, K.J., Arnadottir, I.D., Carraro, U., Gislason, M.K.: Muscle assessment using 3D modelling and soft tissue CT profiling. In: Rehabilitation Medicine for Elderly Patients, Practical Issues in Geriatrics (2018). https://doi.org/10.1007/978-3-319-57406-6_24
25. Gargiulo, P., Petursson, T., Magnusson, B., Bifulco, P., Cesarelli, M., Izzo, G.M., Magnusdottir, G., Haldorsson, G., Ludviksdottir, G.K., Tribel, J., Jonsson Jr., H.: Assessment of total hip arthroplasty by means of computed tomography 3D models and fracture risk evaluation. Artif. Organs **37**(6), 567–573 (2013)
26. Breiman, L., Friedman, J., Olshen, R., Stone, C.: Classification and Regression Trees. Wadsworth, Belmont (1984)
27. Ho, T.K.: Random decision forests. In: Proceedings of the 3rd International Conference on Document Analysis and Recognition, Montreal, QC, 14–16 August, 1995, pp. 278–282 (1995)
28. Ho, T.K.: The random subspace method for constructing decision forests. IEEE Trans. Pattern Anal. Mach. Intell. **20**(8), 832–844 (1998)
29. Guerts, P., Ernst, D., Wehenkel, L.: Extremely randomized trees. Mach. Learn. **63**, 3–42 (2006). https://doi.org/10.1007/s10994-006-6226-1

30. Freund, Y., Shapire, R.E.: A decision-theoretic generalization of on-line learning and an application to boosting. J. Comput. Syst. Sci. **55**, 119–139 (1997). Article no. SS971504
31. Drucker, H.: Improving regressors using boosting techniques. In: ICML 1997 (1997)
32. Pedregosa, F., Varoquaux, G., Gramfort, A., Michel, V., Thirion, B., Grisel, O., Vanderplas, J.: Scikit-learn: machine learning in Python. J. Mach. Learn. Res. **12**(Oct), 2825–2830 (2011)
33. Asuero, A.G., Sayago, A., Gonzalez, A.G.: The correlation coefficient: an overview. Crit. Rev. Anal. Chem. **36**, 41–59 (2006)
34. Stone, M.: Cross-validatory choice and assessment of statistical predictions (with discussion). J. R. Stat. Soc. Ser. B **36**, 111–147 (1974)

Toward a Patient-Specific Image Data-Driven Predictive Modeling Framework for Guiding Microwave Ablative Therapy

Michael I. Miga[1(✉)], Jarrod A. Collins[1], Jon S. Heiselman[1], and Daniel B. Brown[2]

[1] Vanderbilt University, Nashville, TN 37235, USA
Michael.Miga@vanderbilt.edu
[2] Vanderbilt University Medical Center, Nashville, TN 37232, USA

Abstract. In this work, a preliminary effort toward a novel multi-physics modeling framework is presented that combines computational approaches in soft-tissue biomechanics, and bioelectric/bioheat transport to create a patient-specific, image-data driven guidance platform to improve localization and predict thermal dose extent for microwave ablation. More specifically, a finite element modeling and optimization approach for microwave ablation delivery is driven by sparse intra-procedural geometric digitization, and pre-procedural imaging data for providing image-to-physical registration, and dielectric property estimates, respectively. In a series of mock liver phantom experiments, the framework is explored herein. Results indicate superior localization using our non-rigid registration approach, and accurate prediction of lesion formation using our image-data-driven approach to ablation forecasting. Results also provide insight on the impact of localization and material property inaccuracies with respect to therapy delivery and show systematic and considerable degradation of lesion-to-target overlap.

Keywords: Image guidance · Microwave ablation · Finite element · Registration · Modeling · Deformation

1 Introduction

1.1 Clinical Background

Hepatic tumors are a major U.S. and worldwide health care concern with the rate of primary liver cancer (HCC) continuing to rise [1]. Along with hepatocellular carcinoma, many primary neoplasms also metastasize to the liver. With respect to treatment, resection and transplant are the best options but eligibility, and scarcity still limit candidacy, respectively. For example, based on one study involving 2400 subjects, only 20% of patients were eligible for surgical resection due to risk [2]. When faced with these challenges or extensive multi-focal disease, procedures become multimodal and dynamic, e.g. physicians are exploring with staging and then among those stages combining approaches to include resection, loco-regional ablative, arterial, and conventional systemic therapies (e.g. resection combined with ablation, two-stage resection,

G. A. Ateshian et al. (Eds.): CMBBE 2019, LNCVB 36, pp. 198–207, 2020.
https://doi.org/10.1007/978-3-030-43195-2_16

ablation, radio/chemo-embolic procedures, etc.). This represents a movement toward a more chronic management viewpoint of disease with goals of facilitating surgery, bridge-to-transplant, and/or extending quality of life. Microwave ablation (MWA) is certainly one of the more promising thermal ablative technologies for the locoregional control of liver cancer [3–6]. Some studies suggest a 5-year survival rate with MWA that is comparable to surgical resection [7]. Nevertheless, with $\sim$80% of patients ineligible for resection/transplant [8], increasing MWA efficacy is certainly warranted.

1.2 Scientific Premise

Building on the promise of MWA, the underpinning premise of the work reported here is that improving MWA therapy is intrinsically dependent on the *precise localization* and *determination of dose extent* in relation to spatially-encoded disease information, i.e. anatomically-annotated, disease-related biomarkers usually provided by imaging. Without the ability to accurately localize: analysis of outcomes for determining procedural efficacy, understanding the morphology of recurrence, comparing therapeutic approaches, evaluating technique improvements, and investigating the impact of imaging biomarkers to drive therapy decisions will remain ambiguous.

With respect to *localization* specifically, in a recent n = 176 patient study looking at long-term outcomes of MWA for liver malignancies, the investigators reported a 17.6% local recurrence rate with rates increasing with tumor size, i.e. recurrence rates spanning from 1%–33% for tumors sizes ranging from <1 cm to >3 cm, respectively [9]. Clearly, outcomes are compromised with size but does the cause reside with the ability to localize delivery? or with the imaging information driving the ablation? or with soft-tissue characteristics affecting plans (e.g. material properties, deformation, etc.)? or perhaps it resides in our understanding of tumor phenotype in large lesions? The cause is likely a combination of factors. Unfortunately however, studies specifically looking at the causes of recurrence are sparse and difficult to achieve in light of a lack of precision in localization. Addressing this need is certainly a fundamental component for treatment quantification to better understand recurrence in human systems.

With respect to the *determination of dose extent*, strategies in thermography are actively under investigation within the MR (e.g. [10]) and US (e.g. [11]) communities. As a general statement, these sources of data are powerful but in the practical surgical/interventional suite are typically incomplete, cumbersome, and with varying degrees of reproducibility. Also, thermal ablation is a temporally and spatially evolving event. While thermal distributions could inform, they do not necessarily predict/protect against excessive damage to healthy liver, biliary ducts, or nearby organs. In addition, it is also important to recognize that with a dose plan based on a preoperative organ configuration, to what degree intraoperative soft tissue changes affect that delivery dose plan is unknown. From the literature, it is clear that evolving cumulative thermal dose during a hyperthermic ablative procedure is an important factor in determining tissue damage and coverage [12]. Accurate predictive dosing frameworks would allow for better control of the temporal and spatial evolution of MWA-induced thermal energy. From that arises another question, how does one 'tune' a thermal dose to a particular patient? Efforts toward this are quite sparse. It is generally accepted that variability in dielectric and thermal properties persist in healthy and diseased tissue (e.g. fatty liver, fibrosis, and malignant tissue) and are affected by temperature as well. Methods designed to address these questions in the determination of MWA dose extent are a real clinical need.

2 Methods

In this paper, a multi-physics modeling approach has been adopted to address the uncertainty in the delivery of MWA. More specifically, a biomechanical model is used to non-rigidly correct for deformations occurring intra-procedurally, and a bioelectric/bioheat transport model is used to estimate microwave thermal dose extent. When coupled together, these create a comprehensive framework to accurately forecast therapeutic delivery and extent. In the below sections, we briefly describe our approach and our experimental evaluation.

2.1 Correcting for Intra-procedural Deformations

The methods employed in this work are specifically designed to using computational biophysical models for image-to-physical non-rigid liver registration using sparse data compatible with open surgery and interventional presentations [13, 14]. The methodology is an inverse boundary condition reconstruction approach designed to match shape change as defined between preoperative and intraoperative organ states. Briefly, the methodology begins with the spatial designation of liver salient features (usually these are associated with ligament attachments and/or liver ridges) on both the subject's images and the intraoperative physical space counterpart. In the open surgery environment, the physical data can be provided by a tracked stylus being used to swab the organ surface (or with a laser scanner, stereo-camera, etc.). In the interventional environment, often the entire surface of the liver can be extracted from a computed tomography scan. In either case, the physical salient features and their preoperatively imaged counterparts are acquired which is subsequently followed by a more general acquisition of areas of the organ surface that are not salient features. This latter step captures additional intra-procedural shape characteristics. Once geometric data has been acquired, a two-step registration process begins, i.e. a rigid registration using salient features followed by non-rigid fitting process driven by all available sparse data. In the non-rigid registration process, boundary condition nodes from the biomechanical finite element model are associated with *active control surfaces* which will be allowed to drive shape change. In the current realization for open surgery, typically the posterior liver surface is designated as the *active control surface,* and in the context of interventional work, the entire liver surface is employed.

Going further, once the control surfaces are designated on the preoperative liver model, a preoperative computation phase begins where systematic perturbations to the *active control surfaces* are performed with each perturbation providing a boundary condition set to a highly resolved biomechanical finite element model; and naturally, a series of solutions is produced among the perturbations. These solutions provide an approximation to a Jacobian which is subsequently used within a least-squared sparse data/surface error minimization process. We should note that each iteration has an additional distributed loading filter to create more natural deformations. This completes the preoperative computing phase. With this determination of the Jacobian performed pre-procedurally, real time non-rigid fitting of preoperative liver images to intra-procedural sparse digitization data of the liver can ensue. While investigations are continuing in this approach, the most recently published realization can be found in [15]. It should also be noted that a conventional linear elastic model is employed to reflect the deformation behavior of the liver when subjected to intra-procedural forces. This is represented by the partial differential equation of static mechanical equilibrium, $\nabla \cdot \sigma = 0$, where σ is the mechanical stress tensor. In this description, the constitutive law that relates the mechanical stress to strain is associated with conventional linear theory (i.e. a Hookean solid) [16]. It should be noted that the non-rigid fitting phase involves both rigid and non-rigid components thus capturing some of the global rigid body motion while compensating for deformation.

2.2 Forecasting Intra-procedural Microwave Thermal Dose

The thermal dose to tissue was estimated using COMSOL Multiphysics (COMSOL Inc, Burlington, MA) modeling for simulating electromagnetic wave propagation and heat transfer. The development and absorption of electromagnetic waves radiating from the antenna within the phantom, when assuming no initial existing charge, is described by the electromagnetic wave equation $(\nabla^2 + \omega^2 \mu \varepsilon_c)\vec{E} = 0$ where ω [rad/s] is the angular frequency of the electromagnetic wave, μ [H/m] is the permeability, ε_c is the complex permittivity, and $\vec{E}$ [V/m] is the electric field strength. Heat transfer and the resulting temperature history were solved using Pennes' bioheat equation $\rho c \frac{\partial T}{\partial t} = \nabla \cdot k \nabla T + Q - Q_p + Q_m$ where ρ [kg/m^3] is mass density, c [J/kg $\cdot$ K] is specific heat capacity, k [W/m $\cdot$ K] is thermal conductivity, T [K] is temperature, Q [W/m^3] is heat generation due to absorbed electromagnetic energy, Q_p [W/m^3] is heat loss due to perfusion, and Q_m [W/m^3] is metabolic heat generation. Metabolic heat generation (Q_m) is typically neglected, perfusion (Q_p) is modeled as $m(T - T_h)$ with m as thermal perfusion transfer coefficient, and T_h as homeostatic temperature. Heat generation from power deposition by the applied electric field is calculated by $Q = \frac{1}{2}\sigma\|E\|^2$ where σ [S/m] is the electrical conductivity. The antenna is modeled as a conventional conductive core surrounded by dielectric material, surrounding catheter, with ring shaped slot cut on the outer conductor. Conductive material is not specifically realized but represented by the boundary condition, $\vec{n} \times \vec{E} = 0$. The microwave source itself is modeled as a port boundary condition which relates the field to the square root of the time average power flow in the cable and is adopted from [17]. Boundary conditions reflect a first order electromagnetic scattering condition applied to the exterior of the phantom to eliminate reflection of outgoing waves by simulating a transparent boundary. Far-field boundaries on exterior are set to homeostatic/outside-environment temperatures. Saline cooling of the antenna was simulated as a convective heat flux condition along the inner boundary of the antenna. Thermally-induced tissue damage is a function of both instantaneous temperature and thermal history. For this work, a modified Arrhenius damage integral is used to estimate the complete ablative zone (in the phantom work, protein denaturation is used as a proxy). The tissue integral takes the form of $\alpha = \int_0^t Aexp\left(-\frac{E_a}{RT(t)}\right)dt$ where α is the degree of damage at a given time, A [1/s] is the frequency factor, E_a [J/mol] is the activation energy required to damage the mock tissue, R [J/mol · K] is the universal gas constant, and $T(t)$ [K] is the temperature history of the tissue/mock-tissue. The fraction of damaged tissue was then determined by $\theta_d = 1 - e^{-\alpha}$. The current framework uses a 2-D axisymmetric model for simplicity. Additional detail can be found in [18].

2.3 Experimental Procedure

With respect to the experimental framework, a custom deformable ablation phantom in the shape of a patient liver (a heat-sensitive gel phantom consisting of liquid egg whites, and agar gel) was created and deformations similar to those experienced between diagnostic, and intra-procedural presentation were applied. With phantoms constructed, a Perseon ST (Perseon Medical, Salt Lake City, UT) microwave antenna was used to generate three separate ablations in each mock human liver and a hi-resolution MR image volume was obtained. After the application of deformation, a repeat MR was performed. This procedure was performed in n = 3 phantoms with 4 different deformation states. From this data set, mock liver surfaces were extracted and used with our image-to-physical non-rigid registration approach in the context of open surgical (sparse anterior surface), and interventional procedures (full surface). With each phantom and each deformation, a total of 9 targets could be used to quantify localization error (3 antenna tip locations, 3 antenna insertion points, and 3 ablation centroids). True ablation locations could be determined from the repeat MR imaging. It should be noted that in work not reported here, mock gross pathology has been performed to confirm physical ablation sizes are represented accurately by our MR-measured estimates. In addition, optimal dielectric properties matching the ablation predictions to measurement were previously performed using controlled localization experiments over multiple phantoms. While this approach to determining dielectric properties is not amenable to a prospective ablation, i.e. the clinically translated counterpart, this limitation is addressed in the discussion. Finally, given this experimental setup, a comparison study of lesion prediction to ground truth ablation was conducted. In addition, to understand the impact of dielectric properties, forecasted lesions were compared between the optimized properties and those estimated from volume fractions of components based on literature values. The metric used for evaluating this work was the positive predictive value (PPV) calculated by $PPV = \frac{N_{TP}}{(N_{TP} + N_{FP})}$, where N_{TP} is the volume of the model-predicted ablation zone overlapping with the observed ablation zone, and N_{FP} is the volume of the model-predicted ablation zone which does not overlap with the true ablation zone.

3 Results

With respect to localization, our biomechanically model-driven image-to-physical registration methodology to correct for deformation performed quite well. In the case of partial surface availability for registration, the average target registration error was 6.0 ± 2.3 mm and 3.7 ± 1.4 mm for rigid and non-rigid registration over all phantoms, respectively. When the full surface of the liver could be used, the average target registration error was 5.6 ± 2.3 mm and 2.5 ± 1.1 mm for rigid and non-rigid registration over all phantoms, respectively. Similarly, when comparing the predicted ablation relative to ground truth, the volumetric overlap was 67.0 ± 11.8%, and 85.6 ± 5.0% for rigid and non-rigid registration, respectively. Figure 1 *upper panel* is an example analysis from our mock phantom experiments that compares the true ablation as documented by imaging (green) with the model-predicted ablation (red) within the context of conventional rigid registration. In Fig. 1 *lower panel*, the analogous comparison is done, except in this case, our novel non-rigid registration has been employed for targeting the predicted ablation location (blue). Figure 2 is a comprehensive figure that shows the target error and PPV results over all n = 3 phantoms with 4 deformations per phantom and 9 targets per phantom. On the x-axis, the figure illuminates the localization error using sparse anterior (open surgery setting) and full surface (interventional setting) surface data and among both conventional rigid and our novel non-rigid registration methodologies. On the y-axis, the PPV for each ablation is reported which provides a sense of predicted-to-measured lesion overlap for all phantoms. In addition, located on the y-axis in red, the maximum possible PPV is provided, i.e. this is a direct model-to-physical ablation comparison with controlled localization in an idealized setup. Figure 3 shows the results from Fig. 2 with respect to optimized properties when the full surface is used for registration. As a comparator,

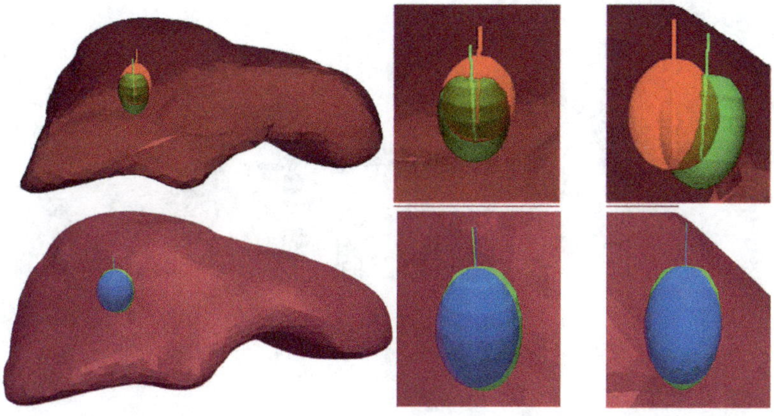

Fig. 1. Ablation model prediction example following registration with full anterior surface data. Green represents ground truth ablation. The rigidly registered ablation model is presented in the top panels (red). The registered ablation model following deformation correction is presented in the lower panels (blue). Additionally, in each panel the registered ablation antenna indicated by lines with color corresponding to the registration method.

rather than using the optimized dielectric model properties, dielectric properties based on the volume fraction of components and literature values were utilized to compare the PPV under an educated 'guess' environment.

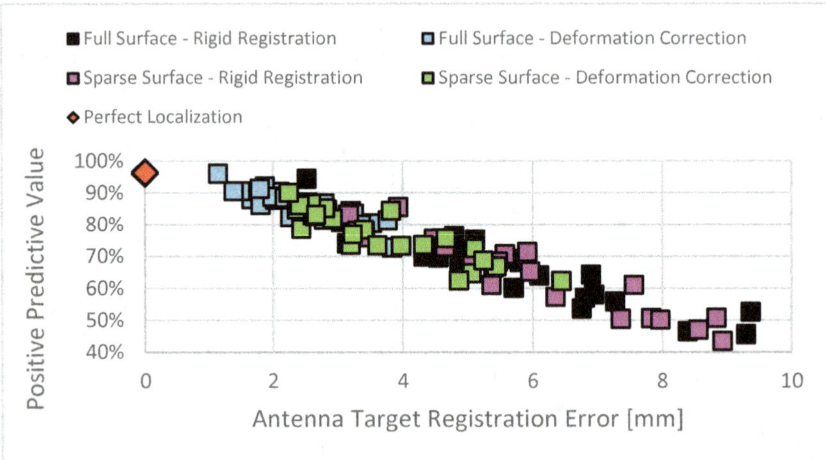

Fig. 2. Positive predictive value is presented for each registered ablation as a function of average target registration of corresponding ablation antenna. MWA model maximum PPV assuming perfect localization is in red.

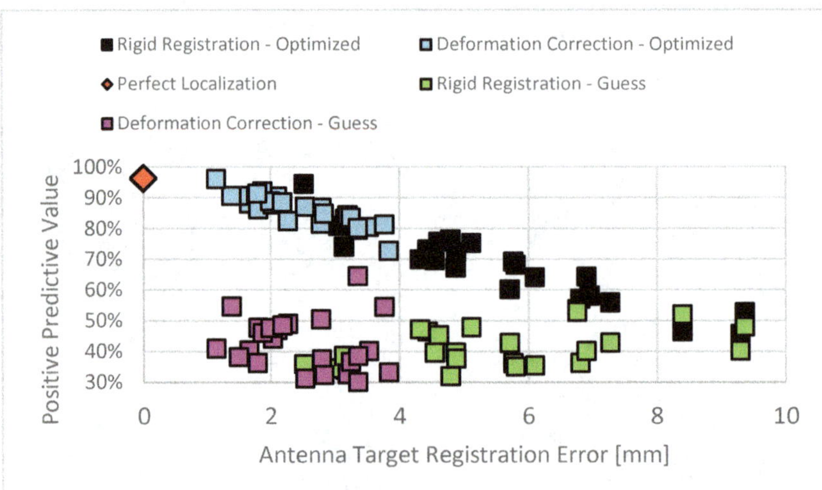

Fig. 3. Positive predictive value is presented for each registered ablation as a function of average target registration of corresponding ablation antenna. Results of the rigid, and non-rigid registration method are presented in black, and blue using image-data-driven calibrated dielectric properties. The counterpart using volume fraction components and literature values used in green, and magenta, respectively. MWA model maximum predictive power assuming perfect registration is in red.

4 Discussion

Overall the results demonstrate that our non-rigid approach to registration reduced localization error. It can also be seen that there is a difference in registration fidelity based on the level of sparsity, which is anticipated. One interesting result is that our non-rigid registration result using only partial surface data outperformed the rigid registration result using the entire surface. Another anticipated result confirmed was that as target error improved, the PPV of our MWA forecasted ablation also improved. The result in Fig. 2 is quite interesting in that it essentially represents a model of PPV degradation as a function of antenna localization error. It suggests, at least in this idealized experiment, that with each 1 mm of antenna localization error, an approximate 5–6% degradation in PPV ensues. While clearly this is an idealized experiment, it does provide some measure of uncertainty in therapeutic delivery which may be an important factor in designing advanced guidance protocols. Figure 3 is also interesting where it also demonstrates a PPV degradation but in this case as a function of the material properties in our bio-electric/heat transport model of the ablation process. It demonstrates that even in cases of precise localization, it is possible that a mismatch in model dielectric properties could result in an under-prediction of lesion forecasting by a considerable amount.

While the above is interesting, it must be put into context with respect to the limitations of the study. This is a phantom study and while our phantom properties are similar to liver, ultimately, the phantom is not structurally similar. For example, major vasculature and perfusion effects are missing. Another limitation is that this experimental setup was not achieved in a true targeting fashion. More specifically, the ablation was performed on the phantom in three locations and then imaged, and then subsequently it was deformed and re-imaged. This allows the ablation itself to become essentially a therapeutically generated target. The better experiment is to create a real physical target within the phantom itself detectable by imaging, image the phantom to find the target before the procedure, and then plan delivery, then apply deformation, and re-image to determine the organ shape in its deformed state (in addition, it provides true target location within the intraoperative presentation). Once re-imaged, using the navigation system as intended (picking a registration method), navigate to target, ablate, and then compare to ground truth target imaged prior to ablation. This would allow a much better and more therapeutically realistic comparison. Lastly, another limitation is our use of linearized biophysics for our modeling efforts. Conventional thought is that ablation is a nonlinear event and that constitutive properties will be sensitive to thermal changes among others. Considering Fig. 3, changes in properties are quite important. The work presented here is essentially a linearized fit to a nonlinear problem. It remains to be seen if a linearized forecasting approach is sufficient to provide therapeutic benefit when used within a planning system. We should note however that the concept of using image-data-driven approach to estimate dielectric properties is not remote. In work not reported here, we have performed phantom experiments similar to those above using a commercially available fat quantification sequence, mDixon Quant, to acquire images of phantoms with varying fat content and demonstrated a relationship between fat content and dielectric and thermal material property changes such that a surrogate image-based biomarker could establish appropriate values for the modeling framework in a prospective manner. In those experiments, fat content in the phantom varied between 0–10%, a range quite similar to that found in fatty liver disease, a condition on the rise in presentation and rapidly replacing viral- and alcohol-related liver disease as a major factor in HCC [19]. It is intriguing to consider imaging extending beyond anatomical information to a series of prospective image-based biomarker surrogates that could establish mechanical, electrical, thermal, and perfusion properties of liver tissue to create an accurate MWA forecasting computational environment. This does not necessarily answer the question as to the importance of nonlinear effects; however, the work here does represent first steps in being able to study this behavior.

5 Conclusions

This paper has demonstrated a complex multi-physics modeling approach to estimate MWA dose extent in liver. The approach proposes to use imaging data and biomechanical models as a means to enhance localization of MWA delivery. The approach goes further by using imaging data as a comprehensive step in model initialization of a bio-electric/heat model such that accurate MWA is forecast. While presented here in a mock tissue environment, the quantitative results are quite encouraging.

Acknowledgements. This work was supported by the National Institutes of Health with awards R01CA162477 from the National Cancer Institute and T32EB021937 from the National Institute of Biomedical Imaging and Bioengineering.

References

1. American Cancer Society: Cancer Facts and Figures 2018. American Cancer Society (2018)
2. Nikfarjam, M., Shereef, S., Kimchi, E.T., Gusani, N.J., Jiang, Y.X., Avella, D.M., Mahraj, R.P., Staveley-O'Carroll, K.F.: Survival outcomes of patients with colorectal liver metastases following hepatic resection or ablation in the era of effective chemotherapy. Ann. Surg. Oncol. **16**, 1860–1867 (2009)
3. Tanaka, K., Shimada, H., Nagano, Y., Endo, I., Sekido, H., Togo, S.: Outcome after hepatic resection versus combined resection and microwave ablation for multiple bilobar colorectal metastases to the liver. Surgery **139**, 263–273 (2006)
4. Vasnani, R., Ginsburg, M., Ahmed, O., Doshi, T., Hart, J., Te, H., Ha, T.G.V.: Radiofrequency and microwave ablation in combination with transarterial chemoembolization induce equivalent histopathologic coagulation necrosis in hepatocellular carcinoma patients bridged to liver transplantation. Hepatobiliary Surg. Nutr. **5**, 225–233 (2016)
5. Tinguely, P., Fusaglia, M., Freedman, J., Banz, V., Weber, S., Candinas, D., Nilsson, H.: Laparoscopic image-based navigation for microwave ablation of liver tumors-a multi-center study. Surg. Endosc. **31**, 4315–4324 (2017)
6. Ginsburg, M., Zivin, S.P., Wroblewski, K., Doshi, T., Vasnani, R.J., Ha, T.G.V.: Comparison of combination therapies in the management of hepatocellular carcinoma: transarterial chemoembolization with radiofrequency ablation versus microwave ablation. J. Vasc. Interv. Radiol. **26**, 330–341 (2015)

7. Andersen, I.R., Mortensen, F.V., Kirkegaard, J., Rasmussen, F., Nielsen, D.T., Kjaer, D.W.: Long-term survival following ablation of colorectal liver metastases. J. Surg. **6**, 13–18 (2018)
8. Kopetz, S., Chang, G.J., Overman, M.J., Eng, C., Sargent, D.J., Larson, D.W., Grothey, A., Vauthey, J.N., Nagorney, D.M., McWilliams, R.R.: Improved survival in metastatic colorectal cancer is associated with adoption of hepatic resection and improved chemotherapy. J. Clin. Oncol. **27**, 3677–3683 (2009)
9. Leung, U., Kuk, D., D'Angelica, M.I., Kingham, T.P., Allen, P.J., DeMatteo, R.P., Jarnagin, W.R., Fong, Y.: Long-term outcomes following microwave ablation for liver malignancies. Br. J. Surg. **102**, 85–91 (2015)
10. Ludemann, L., Wlodarczyk, W., Nadobny, J., Weihrauch, M., Gellermann, J., Wust, P.: Non-invasive magnetic resonance thermography during regional hyperthermia. Int. J. Hyperth. **26**, 273–282 (2010)
11. Ebbini, E.S., Simon, C., Liu, D.: Real-time ultrasound thermography and thermometry. IEEE Sig. Process. Mag. **35**, 166–174 (2018)
12. Yarmolenko, P.S., Moon, E.J., Landon, C., Manzoor, A., Hochman, D.W., Viglianti, B.L., Dewhirst, M.W.: Thresholds for thermal damage to normal tissues: An update. Int. J. Hyperth. **27**, 320–343 (2011)
13. Clements, L.W., Collins, J.A., Weis, J.A., Simpson, A.L., Adams, L.B., Jarnagin, W.R., Miga, M.I.: Evaluation of model-based deformation correction in image-guided liver surgery via tracked intraoperative ultrasound. J. Med. Imaging **3** (2016)
14. Clements, L.W., Collins, J.A., Weis, J.A., Simpson, A.L., Kingham, T.P., Jarnagin, W.R., Miga, M.I.: Deformation correction for image guided liver surgery: an intraoperative fidelity assessment. Surgery **162**, 537–547 (2017)
15. Heiselman, J.S., Clements, L.W., Collins, J.A., Weis, J.A., Simpson, A.L., Geevarghese, S.K., Kingham, T.P., Jarnagin, W.R., Miga, M.I.: Characterization and correction of intraoperative soft tissue deformation in image-guided laparoscopic liver surgery. J. Med. Imaging **5** (2018)
16. Boresi, A.P., Chong, K.P., Lee, J.D.: Elasticity in Engineering Mechanics. Wiley, New York (2011)
17. COMSOL Inc.: COMSOL Multiphysics Reference Manual (2013)
18. Collins, J.A., Heiselman, J.S., Clements, L.W., Brown, D.B., Miga, M.I.: Multiphysics modeling towards enhanced guidance in hepatic microwave ablation: a preliminary framework. J. Med. Imaging (2019). (in review)
19. Marengo, A., Rosso, C., Bugianesi, E.: Liver cancer: connections with obesity, fatty liver, and cirrhosis. In: Caskey, C.T. (ed.) Annual Review of Medicine, vol. 67, pp. 103–117 (2016)

Biomechanical Stress Changes on Forefoot and Hindfoot Caused by the Medializing Calcaneal Osteotomy as Adult Acquired Flatfoot Deformity Treatment

Christian Cifuentes-De la Portilla[1(✉)], Ricardo Larrainzar-Garijo[2], and Javier Bayod[1]

[1] Applied Mechanics and Bioengineering Group (AMB), Aragón Institute of Engineering Research (I3A), Universidad de Zaragoza, Zaragoza, Spain
christianjcd@gmail.com

[2] Orthopedics and Trauma Department, Surgery Department, Hospital Universitario Infanta Leonor, Madrid, Spain

Abstract. Medializing calcaneal osteotomy (MCO) is a common treatment used to correct the adult acquired flatfoot deformity (AAFD) in intermediate stages (IIa-IIb). Although structural correction achieved with this procedure is widely known, changes in the biomechanical stress that could be generated in both forefoot and hindfoot have not been sufficiently studied. The objective of this study was to evaluate the biomechanical stress generated by MCO in these foot regions, using a computational foot model. A 10 mm-MCO was generated, modifying the geometry of the calcaneus bone. The model includes the geometry of all the foot bones, the plantar fascia, cartilages, plantar ligaments, Posterior tibialis tendon, Peroneus tendons, and Achilles tendon, respecting their anatomical distribution and biomechanical properties. The simulations were carried out emulating the mid-stance phase of the gait cycle and generating some pathological scenarios related to AAFD before and after applying MCO. Results show that MCO reduces the biomechanical stress in both forefoot and hindfoot. However, as was expected, it generates some stress concentrations around the osteotomy region, which increases when it is used in patients with a diagnosis of plantar fascia weakness. Additionally, MCO generates a notable stress increase in the third and fourth metatarsals. Therefore, we conclude that MCO should be applied carefully in patients with the diagnosis of plantar fascia weakness, because it increased the risk of bone fracture around the osteotomy region. Finally, the increment of the metatarsals stress could explain the long-term pain that patients report after applying MCO as a treatment to correct AAFD.

Keywords: Biomechanics · Finite elements · Flatfoot · Osteotomy · Pes planus

G. A. Ateshian et al. (Eds.): CMBBE 2019, LNCVB 36, pp. 208–217, 2020.
https://doi.org/10.1007/978-3-030-43195-2_17

1 Introduction

Medializing calcaneal osteotomy (MCO) [1] is a common surgical procedure included within surgical treatment options to correct the progressive foot arch flattening caused by adult acquired flatfoot deformity (AAFD) in stages IIa and IIb (flexible deformity) [2, 3]. This clinical status has traditionally been related to tibialis posterior tendon (TPT) dysfunction [4, 5]. However, some recent studies have found that the weakness or failure of any of both plantar fascia (PF) or spring ligament (SL) could generate the classical signs of AAFD, which are the arch collapse and forefoot abduction [6, 7]. In the early stages, AAFD treatments are focused on reinforcing the TPT [8]. The MCO forces a supination momentum in the foot to compensate for forefoot pronation that characterizes AAFD [5, 9]. Although the foot's structural correction that could be obtained with MCO is widely known and its results are generally satisfactory, some clinical studies have shown that MCO causes some long-term side-effects such as pain in lateral metatarsals (fourth and fifth) and fractures of the calcaneus bone [10–15].

Recent studies have shown that MCO reduces the foot pronation by its own [15, 16]. However, changes in the biomechanical stress in foot bones have not been sufficiently studied, because of the difficulty to measure the bone stresses in cadavers. Patrick et al. measured the subtalar joint pressure generated by MCO using a cadaveric flatfoot model. They inserted a pressure sensor medially in the posterior facet joint, obtaining limited information about the MCO effects on hindfoot joint pressures [12]. Additionally, cadaver-based studies require high economic investment in measurement equipment, as well as meticulous control over the tested tissues to guarantee their biomechanical characteristics [17].

Finite element modeling (FEM) is a recent alternative accepted by clinicians researchers to represent the human's foot biomechanics [18–20]. Many models have been reported for studying foot biomechanics and some surgical procedures. However, these models are focused specifically on foot structure deformation and plantar pressure measurement [20, 21]. In general, these representations greatly simplify the tissue anatomy, and do not consider the biomechanical difference between cortical and trabecular bone, neither the geometry of some soft tissues as the plantar fascia, spring ligament, ligaments and tendons, which traditionally are modeled as bar elements [7].

The objective of this research was to analyze the biomechanical effects in terms of stress concentrations that MCO generates in both hindfoot and forefoot bones. The evaluation was performed simulating different pathological scenarios related to AAFD development: TPT dysfunction and a PF, SL and TPT failure.

2 Methods

This study is based on the model proposed by Cifuentes-De la Portilla et al. [7]. The model used reconstructs a healthy human unloaded foot, based on CT-images (radiographs of 0.6 mm/slide) acquired from the right foot of a 49-year-old man (weight = 75 kg, height = 1.70 m). A statement on ethical approval by a committee was not required for this work, because no intervention nor any contact was made with the volunteer whose foot was used for reconstruction and modeling.

2.1 FE Foot Model and Modifications

Tissue segmentation and the 3-D reconstruction were performed using MIMICS V. 10. To simulate the MCO, the calcaneus bone was changed, being replaced with a new calcaneus with a 10-mm and 45-degrees displacement. Internal fixation elements as screws, plates, bone graft were not included, because a complete joint fusion was considered. The complete FE model is shown in Fig. 1.

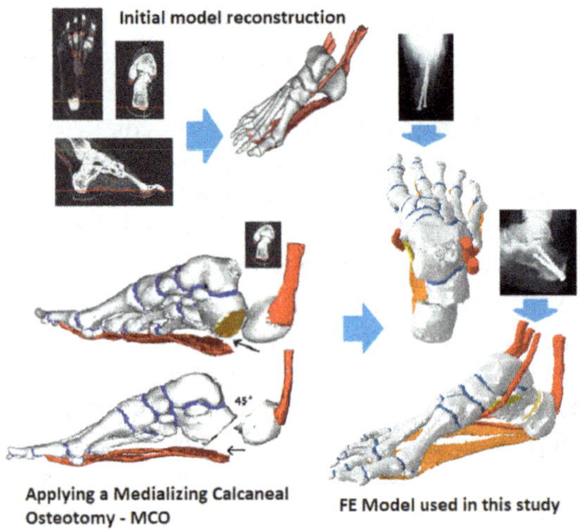

Fig. 1. Reconstruction and modifications in the model to simulate a medializing calcaneal osteotomy.

2.2 Meshing

The model's meshing was performed using the software ANSYS V.15, generating 28 cortical bone pieces, 24 trabecular bone pieces, 26 cartilage segments, 4 tendons, 3 ligaments, and the plantar fascia. A trial-error approach was employed to optimize the mesh size of each segment [19]. They suggested that in all the measured parameters, the number of inaccurate elements should be less than 5%. All simulations and post-processing were performed using the Nonlinear geometry solver of Abaqus/CAE 6.14.

The equilibrium was found with 265.547 linear tetrahedral elements (C3D4). All parameters were within good mesh quality ratios (see Table 1).

Table 1. Mesh quality metrics based on Burkhart et al. [19] recommendations.

Quality metric	Assessment criteria	Accurate elements	Inaccurate elements
Element jacobians	>0.2	99.96%	0.04%
Aspect ratio	>0.3	96.9%	3.1%
Min. angles	>30°	95.3%	4.7%
Max. angles	<120°	99.46%	0.84%

2.3 Tissue Properties

The tissue properties of the cortical bone, trabecular bone, ligaments, and plantar fascia were assigned in accordance with published data: Cortical bone (E = 17000 MPa, ν = 0.3), trabecular bone (E = 700 MPa, ν = 0.3), ligaments (E = 250 MPa, ν = 0.28) and the plantar fascia (E = 240 MPa, ν = 0.28) [6, 22, 23], being E the Young's module and ν the Poisson ratio. Tendons and cartilages were modeled as hyper-elastic materials (Ogden model), using the parameters taken from specialized articles. Its strain energy density function is shown in Eq. 1:

$$U = \frac{\mu}{\alpha^2}\left(\lambda_1^\alpha + \lambda_2^\alpha + \lambda_3^\alpha - 3\right) + \frac{1}{D} \tag{1}$$

being μ = 4.4 (initial shear modulus), α = 2 (strain hardening exponent) and D = 0.45 (compressibility parameter) [6, 24]. The tissue failures were performed applying the Isotropic Hardening theory that generates a progressive tissue weakening, resulting in a very flexible material.

2.4 Loading and Boundary Conditions

To generate a healthy normal standing load position (midstance) (our reference case) for comparisons with all the pathological scenarios simulated, the model was simulated applying the loading and boundary conditions are shown in Fig. 2, maintaining all the tissues unaltered. The load of 720 N represents the full weight of an adult of about 73 kg, leaning on one foot, which represents a traditional scenario of an AAFD diagnosis assessment. Both boundaries and loading conditions were kept without variations for all the OMC simulations [25]. The tendon traction force was included as reported by Arangio et al. [26]. Our model does not include the plantar pad or the foot muscles, because they do does not have a direct relationship with the AAFD development nor the

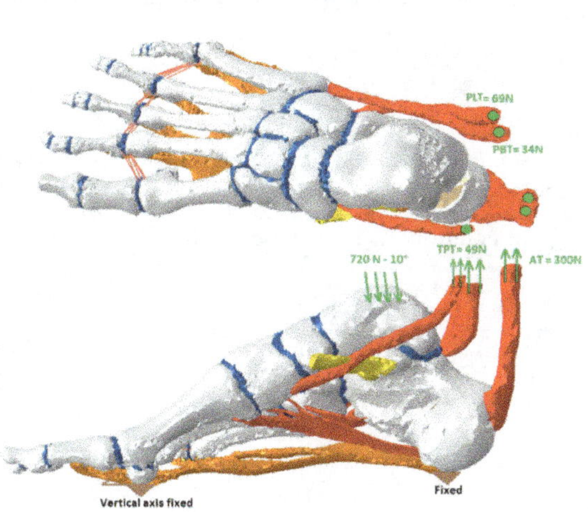

Fig. 2. Boundary and loading conditions applied in the model. Loading values correspond to the weight of a person of 70 kg.

plantar arch support [27]. Thus, to simulate the floor contact, the nodes located at the lower part of the calcaneus were fixed as well as the Z-axis displacement (vertical) of the lower nodes of the first and fifth metatarsals.

2.5 About Model Validation

This study used a model that has been previously applied to other biomechanical studies related to AAFD development [7]. It was validated following the recommendations of Tao et al. [17], measuring the displacement of some anatomical points in the sagittal view from Rx-images under two different loading conditions. Measurements performed of real patients were compared with the displacements of these points in the model. Although this study modifies the calcaneus geometry to simulate MCO, the model keeps both the biomechanical parameters and all the tissue geometry, as well as both the boundary and loading conditions of the base model.

2.6 Model Analysis and Evaluation Criteria

The biomechanical stress in the foot bones was quantified using the field output spectrum available in Abaqus/CAE. The parameter measured for evaluations was the maximum principal stress (S. Max). These values are closely related to the tensile stress that is generated in foot tissues [23].

3 Results

3.1 About Model Validation

The model generates a foot deformation close to which is expected from a healthy patient in a loading test, generating values that can be included in the inter-subject variability range under two load conditions: light loading and a normal standing load (see Table 2).

Table 2. Results of the validation process. The values correspond to the difference between the measured distance from each point to the ground, under two different loading conditions: Light loading and normal standing load.

Reference point (highest point of…)	Model's prediction (mm)	Patient's average (mm)
Talus	−0.33	−0.32 (±0.14)
Navicular	−0.27	−0.26 (±0.04)
First cuneiform	−0.26	−0.19 (±0.08)
First metatarsal's head	−0.07	0.08 (±0.003)

3.2 Stress in Forefoot Bones Generated by MCO

The first biomechanical stress analysis was performed in forefoot bones, specifically in metatarsals. The stress was quantified in Megapascals (N/mm^2). Values were normalized to 60 N/mm^2 to better visualize the results (Fig. 3). Maximum stress values are presented upper on the color scale. The color scale shows how much stress increases: Red means high stress while blue represents low stress. It can be observed that MCO increases the stress around the fourth metatarsal, and it increase even more in simulations performed with a failure of PF and/or TPT.

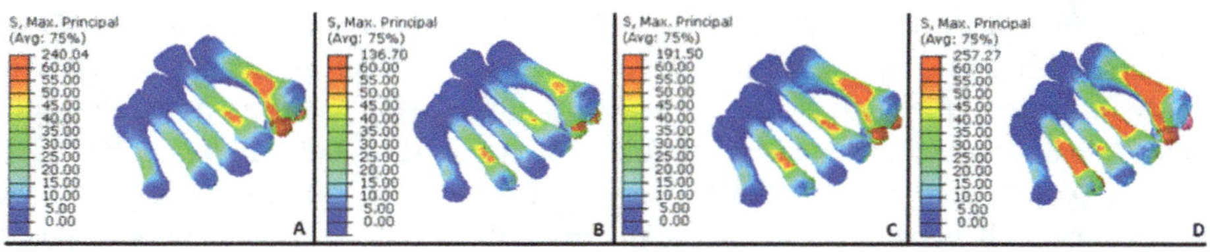

Fig. 3. Stress maximum principal in forefoot. (A) Healthy, (B) MCO, (C) MCO and TPT failure. (D) MCO and TPT, PF and SL failures.

3.3 Stress in Hindfoot Bones Generated by MCO

The stress changes in hindfoot bones were also quantified, simulating all the above-mentioned pathological cases for MCO (Fig. 4). In this case, stress values were normalized to 20 N/mm^2. It can be noted that the stress concentration around the osteotomy region increase when one of the main stabilizers of the foot arch fails (PF, TPT or SL).

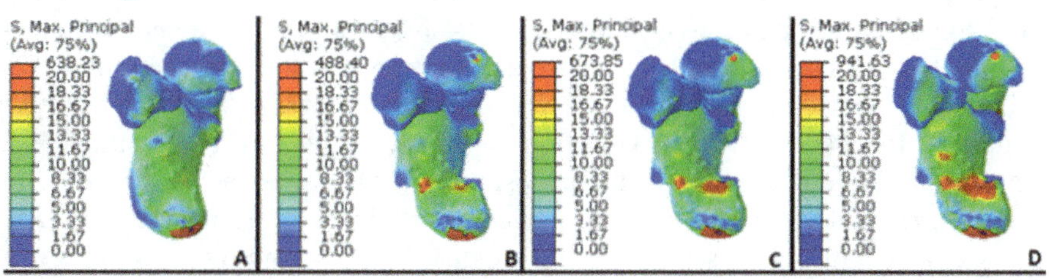

Fig. 4. Stress maximum principal in hindfoot. (A) Healthy, (B) MCO and healthy tissues, (C) MCO and TPT failure. (D) MCO and TPT, PF and SL failures.

4 Discussion

There are many treatment options nowadays used to correct the adult acquired flatfoot deformity, which depends on the arch rigidity and the stage of the condition. This pathology is characterized by a painful flattening and forefoot pronation. Recent clinical studies have shown that main stabilizers of the plantar arch are the Plantar fascia (PF), the Spring ligament (SL) and the Tibialis Posterior Tendon (TPT) [2, 5]. For stages IIa and IIb, which are related an advanced flattening but with a flexible deformity, tendon reinforcements are insufficient. Thus, the most commonly used surgical procedure is medializing calcaneal osteotomy (MCO). Some studies have been addressed to analyze the structural correction achieved with MCO, in isolation or combined with other strategies such as lateral column lengthening [10–12]. As we show, the structural correction that is achieved with MCO is widely known, but also some side-effects associated with the MCO, which includes pain in toes and fractures of the calcaneus bone fractures [10, 13, 28]. These aspects may result highly relevant for surgeons when they decide the best treatment options and evaluate the cost-benefit ratio.

The biomechanical side-effects generated by MCO in both foot bones have not been analyzed adequately, mainly because of the difficulty of measuring tissue stress in cadaveric models [15, 17]. Some experimental studies performed, use limited strategies such as plantar-print comparisons, radiological measurements and plantar pressure [29].

This background motived the use of a computational foot model to study the biomechanical changes in foot tissue stresses generated by MCO. This modern approach is a valid alternative that has been nowadays used in clinical biomechanics studies. Some computational studies have been carried out in connection with MCO modeling. Smith et al. [30] designed a computational model to evaluate the structural effect of the Evans's osteotomy, while Wang et al. [21] proposed a FE study to evaluate different variables in MCO application, such as the angle and the medializing displacement distance. Although there have been studies with significant clinical value, these proposals simplify both the anatomy of the soft tissues and the biomechanical properties of bones, which difficult its use to analyze the stress changes generated by MCO.

Our finite element model was able to reproduce the loaded foot deformity, as well as the main signs of AAFD such as foot pronation and the "too many toes" sign [3]. Additionally, this model includes the geometry of the main soft tissues related to AAFD development and differentiates the biomechanical characteristics of both trabecular and cortical bones. These characteristics offer the possibility not only to quantify the tissue stress but also to localize stress concentration points around the foot anatomy [7].

The results show that MCO reduces very well the pronation of the hindfoot typically observed in patients with AAFD (intense blue color around both astragalus and navicular bones). Additionally, results suggest that MCO compensates very well for the spring ligament failure. No significant changes were observed comparing MCO simulated with all healthy tissues versus MCO simulated with spring ligament failure. The supination momentum that MCO causes in the foot structure can explain these results. However, it is notable that when the main soft tissues that support the plantar arch (plantar fascia, spring ligament, and tibialis posterior tendon) fail, MCO by its own cannot prevent forefoot pronation. This means that, depending on the causes of the disease, MCO should be applied in combination with other strategies, such as tendon reinforcement or arthrodeses. Surgeons may analyze these clinical circumstances to avoid an under correction of the mid-foot deformity.

On the other hand, the results of the stress in the forefoot bones show that MCO reduces the bone stresses by about 43% compared to healthy cases before MCO. However, when the foot arch stabilizers fail, an increase in the stress in the second and fourth metatarsals was generated. These changes are more evident when the plantar fascia or tibialis posterior tendon fails, increasing the stress by about 36% and 40%, respectively, compared to MCO with healthy soft tissues. When these two tissues fail at the same time, the stress increases by about 88%. Again, spring ligament failure did not show significant changes in forefoot bone stresses. These changes in the stress distribution could explain the metatarsalgia and toes pain reported in patients treated with MCO to correct AAFD.

Finally, results about hindfoot bones stresses (Fig. 4), showed important stress concentrations around the osteotomy region, especially when the plantar fascia or tibialis posterior tendon fails. The stress at these cases increased by 33% and 37%, respectively. When the foot arch stabilizers fail, the stress concentration increases by about 92% (from 488 to 941 N/mm^2) compared to MCO with healthy soft tissues.

We consider that the main limitations of our model are failing to include the plantar pad or the flexor hallux longus and flexor digitorum longus tendons. However, clinical studies have shown that these tissues have a minor role in AAFD development and in the foot arch support, compared to the tissues included in the model used [26, 27]. Additionally, although the values of biomechanical stress generated cannot be assumed as true stress values for all people (because of inter-subject variability), we can analyze the relative differences generated in each case. We did not analyze bone fixation methods because we assume a complete bone healing after calcaneal displacement.

5 Conclusions

This study shows that medializing calcaneal osteotomy (MCO) is a good strategy to use when the main cause of the AAFD development is the spring ligament. MCO compensates very well for the spring ligament weakness, especially reducing forefoot pronation and hindfoot valgus. However, this surgical strategy should be applied carefully in patients with plantar fascia weakness and/or tibialis posterior tendon dysfunction, because is in these cases when the MCO noticeably increases the stress in the lateral metatarsals (probably generating long-term pain) and around the osteotomy area, which increases the risk of a calcaneal fracture.

References

1. Feibel, J.B., Donley, B.G.: Calcaneal osteotomy and flexor digitorum longus transfer for stage II posterior tibial tendon insufficiency. Oper. Tech. Orthop. **16**, 53–59 (2006)
2. Deland, J.T.: The adult acquired flatfoot and spring ligament complex: pathology and implications for treatment. Foot Ankle Clin. **6**, 129–135 (2001)
3. Deland, J.T.: Adult-acquired Flatfoot Deformity. J. Am. Acad. Orthop. Surg. **16**, 399–406 (2008)
4. Bluman, E.M., Title, C.I., Myerson, M.S.: Posterior tibial tendon rupture: a refined classification system. Foot Ankle Clin. **12**(2), 233–249 (2007)
5. Abousayed, M.M., Alley, M.C., Shakked, R., Rosenbaum, A.J.: Adult-acquired flatfoot deformity: etiology, diagnosis, and management. JBJS Rev. **5**, e7 (2017)

6. Tao, K., Ji, W.T., Wang, D.M., Wang, C.T., Wang, X.: Relative contributions of plantar fascia and ligaments on the arch static stability: a finite element study. Biomedizinische Technik/Biomed. Eng. **55**, 265–271 (2010)
7. Cifuentes-De la Portilla, C., Larrainzar-Garijo, R., Bayod, J.: Biomechanical stress analysis of the main soft tissues associated with the development of adult acquired flatfoot deformity. Clin. Biomech. **61**(c), 163–171 (2018)
8. Toullec, E.: Adult flatfoot. Orthop. Traumatol. Surg. Res. **101**, S11–S17 (2015)
9. Vora, A.M., Tien, T.R., Parks, B.G., Schon, L.C.: Correction of moderate and severe acquired flexible flatfoot with medializing calcaneal osteotomy and flexor digitorum longus transfer. JBJS **88**, 1726–1734 (2006)
10. Hadfield, M.H., Snyder, J.W., Liacouras, P.C., Owen, J.R., Wayne, J.S., Adelaar, R.S.: Effects of medializing calcaneal osteotomy on Achilles tendon lengthening and plantar foot pressures. Foot Ankle Int. **24**, 523–529 (2003)
11. Zanolli, D.H., Glisson, R.R., Nunley, J.A., Easley, M.E.: Biomechanical assessment of flexible flatfoot correction: comparison of techniques in a cadaver model. JBJS **96**, 45 (2014)
12. Patrick, N., Lewis, G.S., Roush, E.P., Kunselman, A.R., Cain, J.D.: Effects of medial displacement calcaneal osteotomy and calcaneal Z osteotomy on subtalar joint pressures: a cadaveric flatfoot model. J. Foot Ankle Surg. **55**(6), 1175–1179 (2016)
13. Jordan, T.H., Rush, S.M., Hamilton, G.A., Ford, L.A.: Radiographic outcomes of adult acquired flatfoot corrected by medial column arthrodesis with or without a medializing calcaneal osteotomy. J. Foot Ankle Surg. **50**, 176–181 (2011)
14. Iaquinto, J.M., Wayne, J.S.: Effects of surgical correction for the treatment of adult acquired flatfoot deformity: a computational investigation. J. Orthop. Res. **29**, 1047–1054 (2011)
15. Larrainzar-Garijo, R., Cifuentes-De la Portilla, C., Gutiérrez-Narvarte, B., Díez-Nicolás, E., Bayod, J.: Efecto de la osteotomía medializante de calcáneo sobre tejidos blandos de soporte del arco plantar: un estudio computacional. Revista Española de Cirugía Ortopédica y Traumatología **4**(3), 155–163 (2018)
16. Arangio, G.A., Salathé, E.P.: Medial displacement calcaneal osteotomy reduces the excess forces in the medial longitudinal arch of the flat foot. Clin. Biomech. **16**(6), 535–539 (2001)
17. Tao, K., Wang, D., Wang, C., Wang, X., Liu, A., Nester, C.J., et al.: An in vivo experimental validation of a computational model of human foot. J. Bionic Eng. **6**, 387–397 (2009)
18. Viceconti, M., Olsen, S., Nolte, L.P., Burton, K.: Extracting clinically relevant data from finite element simulations. Clin. Biomech. **20**, 451–454 (2005)
19. Burkhart, T.A., Andrews, D.M., Dunning, C.E.: Finite element modeling mesh quality, energy balance and validation methods: a review with recommendations associated with the modeling of bone tissue. J. Biomech. **46**, 1477–1488 (2013)
20. Wang, Y., Wong, D.W.C., Zhang, M.: Computational models of the foot and ankle for pathomechanics and clinical applications: a review. Ann. Biomed. Eng. **44**, 213–221 (2016)
21. Wang, Z., Kido, M., Imai, K., Ikoma, K., Hirai, S.: Towards patient-specific medializing calcaneal osteotomy for adult flatfoot: a finite element study. Comput. Methods Biomech. Biomed. Eng. **21**(4), 332–343 (2018)
22. Wright, D.G., Rennels, D.C.: A study of the elastic properties of plantar fascia. JBJS **46**, 482–492 (1964)
23. Garcia-Aznar, J.M., Bayod, J., Rosas, A., Larrainzar, R., García-Bógalo, R., Doblaré, M., et al.: Load transfer mechanism for different metatarsal geometries: a finite element study. J. Biomech. Eng. **131**, 021011 (2009)
24. Wu, L.: Nonlinear finite element analysis for musculoskeletal biomechanics of medial and lateral plantar longitudinal arch of Virtual Chinese Human after plantar ligamentous structure failures. Clin. Biomech. **22**, 221–229 (2007)

25. Bayod, J., Becerro-de-Bengoa-Vallejo, R., Losa-Iglesias, M.E., Doblaré, M.: Mechanical stress redistribution in the calcaneus after autologous bone harvesting. J. Biomech. **45**, 1219–1226 (2012)
26. Arangio, G.A., Salathe, E.P.: A biomechanical analysis of posterior tibial tendon dysfunction, medial displacement calcaneal osteotomy and flexor digitorum longus transfer in adult acquired flat foot. Clin. Biomech. **24**(4), 385–390 (2009)
27. Riddiford-Harland, D., Steele, J., Baur, L.: Medial midfoot fat pad thickness and plantar pressures: are these related in children? Int. J. Pediatr. Obes. **6**(3), 261–266 (2011)
28. Chan, J.Y., Greenfield, S.T., Soukup, D.S., Do, H.T., Deland, J.T., Ellis, S.J.: Contribution of lateral column lengthening to correction of forefoot abduction in stage IIb adult acquired flatfoot deformity reconstruction. Foot Ankle Int. **36**, 1400–1411 (2015)
29. Guha, A.R., Perera, A.M.: Calcaneal osteotomy in the treatment of adult acquired flatfoot deformity. Foot Ankle Clin. **17**, 247–258 (2012)
30. Smith, B.A., Adelaar, R.S., Wayne, J.S.: Patient specific computational models to optimize surgical correction for flatfoot deformity. J. Orthop. Res. **35**(7), 1523–1531 (2016)

The Role of Physiological Loading on Bone Fracture Healing Under Ilizarov Circular Fixator: The Effects of Load Duration and Loading Frequency

Ganesharajah Ganadhiepan(✉), Saeed Miramini, Priyan Mendis, and Lihai Zhang

The University of Melbourne, Parkville, VIC 3010, Australia
gganadhiepan@student.unimelb.edu.au,
g.ganadhiepan@gmail.com

Abstract. One of the main advantages of Ilizarov Circular Fixator (ICF) is that it enables early weight bearing, which is beneficial for patients' mobility. In addition, it is believed that early weight bearing could lead to faster and stronger fracture callus development and thus, enhance healing progression. Therefore, it is important that the early weight bearing exercises (e.g. walking) are carefully designed for the best possible conditions within the fracture site during the crucial early stage of healing. In this study, using a computational model, we explored the influence of (i) duration and (ii) frequency of dynamic loading under ICF (resulting from early weight bearing exercises), on the early fracture microenvironment. The model simulated the interactions and transport of bone cells (i.e. mesenchymal stem cells, osteoblasts, chondrocytes and fibroblasts) and growth factors (i.e. osteogenic growth factor and chondrogenic growth factor) within the early callus under dynamic loading. The changes in cell and growth factor concentrations due to different load durations (1–4 h) and load frequencies (1–4 times/day) were parametrically studied. The results suggest that continuous dynamic loading and longer durations (e.g. 4 h in a stretch) are preferable over shorter durations (e.g. 1 h) and intermittent loading for chondrogenic differentiation, which is beneficial for secondary healing. The results also suggested that, intermittent loading of longer total durations (e.g. 4 times [× 1 h]) could be more beneficial in the later stages for osteogenesis.

Keywords: Ilizarov Circular Fixator · Mesenchymal stem cells · Growth factors · Early weight bearing

1 Introduction

Bone fracture healing process involves complex cellular events that depend on a wide range of biological, chemical and mechanical variables [1–5]. The success of fracture healing depends on how these parameters influence the microenvironment at the fracture site. Plenty of studies have been carried out to understand the interactions of different parameters and their influence on fracture healing [6–11]. It is known that

G. A. Ateshian et al. (Eds.): CMBBE 2019, LNCVB 36, pp. 218–236, 2020.
https://doi.org/10.1007/978-3-030-43195-2_18

fracture healing is a cascade of events taking place naturally to restore the fractured bone to its original state [12]. In this cascade of events, chemical stimuli such as osteogenic and chondrogenic growth factors (OGF and CGF) play a key role by stimulating multipotent mesenchymal stem cells (MSC) to differentiate into bone forming cells such as osteoblasts and chondrocytes which in turn synthesise bone and cartilage tissues [7, 13]. Tissue formation is a space and time dependent process which is not just controlled by the chemical microenvironment, but also by the mechanical microenvironment [7, 9, 10]. As the fracture site experience mechanical forces, it tends to respond to the mechanical microenvironment and influence the differentiation and transport processes within the fracture site.

Experimental and clinical evidence suggest that there could be a patient specific optimal range of mechanical loading which can augment the effects of chemical stimuli and improve the bone repair process [6–8]. Therefore, a deeper understanding of the synergistic effect of chemical and mechanical stimuli could help surgeons and clinicians to treat bone fractures more effectively.

Bone has the natural capacity to heal by itself [12]. However, not all fractures can heal by itself and some external intervention is required in fracture treatments depending on the nature of the fracture. Currently, there is more focus on treatment techniques that are minimally invasive and tailorable to patient specific requirements [12, 14, 15]. This has put more focus on external fixation devices. These devices allow some degree of external control over the mechanical microenvironment of the fracture site which provides more flexibility to surgeons in treating bone fractures; because, the fixator could be adjusted throughout the healing period depending on the progress of healing.

Ilizarov circular fixator (ICF) is one such external fixator which is used to treat variety of bone defects such as fractures, non-unions, bone length discrepancies etc. [2]. Figure 1 shows a typical ICF manufactured by Smith & Nephew PLC (Memphis, Tennessee, USA) for treating tibial fractures. One of the main advantages of ICF is that, it enables early weight bearing and mobility of patients. It is also believed that early weight bearing could lead to faster and more stronger fracture callus development [1]. However, the exact mechanism by which early weight bearing enhance fracture healing under ICF remains unclear.

In one of our recent studies [16], we tested the hypothesis that dynamic loading resulting from early physiological activities (e.g. walking) could significantly enhance the cell and growth factor concentrations within the early fracture sites stabilized with ICF and thus, enhance the healing progression. We modelled the transport and interactions of MSC, osteoblasts, chondrocytes, fibroblasts, OGF and CGF and compared the results of dynamically loaded cases (mechanical stimuli + chemical stimuli) with those of the non-loaded cases (chemical-stimuli only). This allowed us to quantitatively study the impact of dynamic loading on cell differentiations and solute (i.e. cells and growth factors [GFs]) transport. The results suggested that there is positive correlation between dynamic loading and solute concentrations within the early callus, which supported the hypothesis.

It is also known that, early stage fracture microenvironment is very critical; because, the early stage of healing is when mesenchymal stem cells (MSC) commit to

Fig. 1. Ilizarov circular fixator (ICF) used for tibial fractures. This three-ring setup shows two possible ways of assembling ICF rings: (i) the top and the mid rings are connected using four threaded rods as in the conventional ICF assembly; (ii) the mid and the bottom rings are connected using six telescopic rods which is a relatively new variation to the assembly. In the latter, the rings and the telescopic rods are collectively referred to as Taylor spatial frame (TSF). In this study the conventional assembly of ICF with rods is considered. Note that pretension wires are not affixed to the bone (surrogate) at the mid and bottom ring levels in this figure. However, in real fixations pretension wires and/or half pins are used to connect the bone fragments to the fixation at each ring level.

healing pathways (i.e. osteogenic or chondrogenic) which could influence the entire healing process and the healing outcome [17]. Therefore, it is important that the early weight bearing exercises (e.g. walking) are carefully designed for the best possible healing conditions within the early stage fracture site. This requires answers to number of questions regarding the early weight bearing exercises. Most importantly, it is essential to know if the exercise duration (how long) and frequency (e.g. how many times per day) have any effect on the early fracture site.

In this study, we hypothesised that the duration of dynamic loading resulting from early weight bearing exercises (i.e. walking) under ICF and their frequencies could significantly affect the solute transport and their concentration distributions within the fracture site in a space and time dependent manner. The purpose of this study is to test this hypothesis.

2 Materials and Methods

We tested this hypothesis by adopting the methodology shown in Fig. 2, which we developed in one of our recent studies [16]. As shown in Fig. 2, a 3D computational model of a fractured tibia stabilized with a four-ring ICF was used to capture the displacements within the fracture environment under external axial loading. The model of the tibia was simplified to a cylinder with a cortical bone of 30 mm outer diameter and an inner bone marrow of 18 mm diameter. A fracture gap of 5 mm was created to represent the defect, around which, an early post inflammation stage callus was created. The geometry of the ICF was that used in the experimental study of Watson et al. [18] and all the wires were given wire pretension of 891 N (90 kg). Cortical bone, bone marrow and callus were modelled as fluid saturated poro-elastic materials and the fixator components were modelled as linear elastic stainless steel. The properties of materials used are given in Table 1.

Twelve load cases with different combinations of load durations and frequencies were considered in this study as shown in Fig. 3. Each of these load cases simulated 24-h with periods of axial loading relevant to walking (i.e. 200 N axial load at 1 Hz as described in Ganadhiepan et al. [16]) under ICF and periods of rest. Under each load case, the time dependent axial displacements within the fracture gaps were calculated using the 3D model and input as boundary conditions to a 2D-axisymmetric computational model of the fracture site, where the transport and differentiation processes were modelled.

The fracture callus in the 2D-axisymmetric model was developed as tri-phasic poro-elastic material (i.e. incorporating solid, liquid and solute phases) to describe the spatio-temporal variations in concentrations of the following six variables (solutes) during the 24-h period simulated: MSC (c^m), osteoblasts $\left(c^b\right)$, chondrocytes (c^c), fibroblasts $\left(c^{fb}\right)$, OGF $\left(g^b\right)$ and CGF (g^c). The change of concentrations of these solutes were describe using the following equations:

$$\frac{\partial \bar{c}^m}{\partial t} = -\nabla \cdot \left(-D^m \nabla \bar{c}^m + \mathbf{v}^f \bar{c}^m\right) + S_p^m - S_d^m \tag{1}$$

$$\frac{\partial \bar{c}^b}{\partial t} = -\nabla \cdot \left(-D^b \nabla \bar{c}^b + \mathbf{v}^f \bar{c}^b\right) + S_p^b - S_d^b \tag{2}$$

$$\frac{\partial \bar{c}^c}{\partial t} = -\nabla \cdot \left(-D^c \nabla \bar{c}^c + \mathbf{v}^f \bar{c}^c\right) + S_p^c - S_d^c \tag{3}$$

$$\frac{\partial \bar{c}^{fb}}{\partial t} = -\nabla \cdot \left(-D^{fb} \nabla \bar{c}^{fb} + \mathbf{v}^f \bar{c}^{fb}\right) + S_p^{fb} - S_d^{fb} \tag{4}$$

$$\frac{\partial \bar{g}^b}{\partial t} = -\nabla \cdot \left(-D^{gb} \nabla \bar{g}^b + \mathbf{v}^f \bar{g}^b\right) + S_p^{gb} - S_d^{gb} \tag{5}$$

$$\frac{\partial \bar{g}^c}{\partial t} = -\nabla \cdot \left(-D^{gc} \nabla \bar{g}^c + \mathbf{v}^f \bar{g}^c\right) + S_p^{gc} - S_d^{gc} \tag{6}$$

where, D^m, D^b, D^c, D^{fb}, D^{gb} and D^{gc} are the diffusion coefficients, S_p^m, S_p^b, S_p^c, S_p^{fb}, S_p^{gb} and S_p^{gc} are the production rates and S_d^m, S_d^b, S_d^c, S_d^{fb}, S_d^{gb} and S_d^{gc} are the degradation rates of MSC, osteoblasts, chondrocytes, fibroblasts, OGF and CGF respectively; and $\mathbf{v}^f$ is the true fluid velocity.

In this model, the diffusion coefficients of cells and growth factors were assumed to be constants (Table 2). The production rate of MSC was assumed to be constant and the production rates of osteoblasts, chondrocytes and fibroblasts were assumed to take the form $S_p^\alpha = \left(k_p^\alpha + \lambda_p^\alpha\right)\bar{c}^m$, where, α could be b, c or fb; k_p^α is the chemical stimuli-mediated production rate and λ_p^α is the mechanical stimuli-mediated production rate.

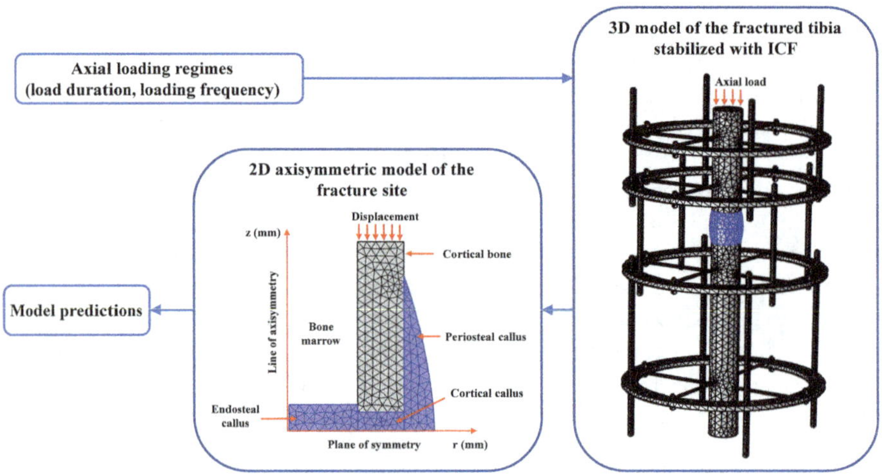

Fig. 2. Schematic diagram showing the methodology adopted in this study

The chemical stimuli mediated production rates of osteoblasts and chondrocytes (k_p^b and k_p^c) were modelled as functions of OGF and CGF concentrations respectively. The functions were based on Michaelis–Menten kinetic model, with a Michaelis constant of 10 ng/ml and maximum rates of $k_{p,max}^b$ and $k_{p,max}^c$, for osteoblast and chondrocytes, respectively [7]. However, the chemical stimuli mediated production rate of fibroblasts was assumed to be constant. The mechanical stimuli mediated productions were modelled to be controlled by the stimulation index S [19] expressed as, $S = \gamma/0.0375 + v/(3\ \mu\text{m s}^{-1})$, which is based on the mechano-regulatory theory of Prendergast et al. [10, 19]. The degradation rates of all cells were assumed to be constants. For MSC, an additional degradation rate equal to the sum of production rates of osteoblasts, chondrocytes and fibroblasts was also added, since the production of these cells is due the differentiation of MSC. For GFs, the production and degradation rates were assumed to be negligible. The cell production and degradation rates are summarised in Table 3.

2.1 Parametric Study

In this study, dynamic axial load simulating walking (i.e. 200 N cyclic load at 1 Hz as described in Ganadhiepan et al.) was applied for total duration of 1, 2, 3 or 4 h in either 1, 2 or 4 equal time sessions (i.e. loading frequency) as shown in Fig. 3. Under each load case, the temporal variation in concentrations of the six solutes considered in this study were predicted in three distinct zones of the callus, namely, periosteal, endosteal and cortical zones (Fig. 2). The 3D and the 2D axisymmetric models were meshed with solid tetrahedral and triangular elements, respectively and solved numerically using COMSOL Multiphysics® v. 5.2a (COMSOL AB, Stockholm, Sweden). Boundary conditions, initial conditions, equations of poro-elastic formulation and assumptions made while solving the model are described in detail in Ganadhiepan et al. [16]. The time dependent solver of the software was used to solve the models and the results of the parametric study were comparatively studied.

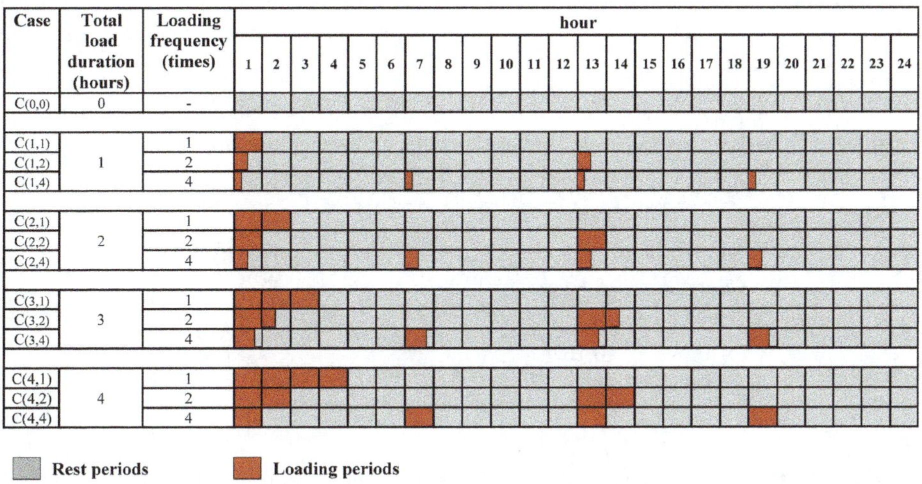

Fig. 3. Simulation cases used in this study. No loading was applied to the control case (i.e. free diffusion only) and for the loaded cases, a cyclic axial load of with a peak of 200 N was applied at 1 Hz to simulate walking cycle described in Ganadhiepan et al. [16]

3 Results and Discussion

Details on model validation and comparison of the results with experimental studies are given in detail in our previous work [16]. Therefore, these details are not discussed here. The following sections focus on the influence of load durations and loading frequencies on the concentration distributions of the six solutes considered in this study. Figures 4, 5 and 6 show the percent changes of cells and GFs relative to the control (free diffusion) case at the end of 24 h period, within the endosteal, cortical and periosteal zones of the callus, respectively. Figure 7 presents the typical temporal changes in solute concentrations due to dynamic loading, taking the 2-h (total) load duration case, as an example.

Table 1. Material properties used in this study

Material	E (MPa)	υ	ϕ	k ($m^4\ N^{-1}\ s^{-1}$)	K_s (MPa)	K_f (MPa)
Cortical bone	20000 [9]	0.3 [9]	0.04 [9]	10^{-17} [9]	2300 [9]	13920 [9]
Bone marrow	2 [9]	0.167 [9]	0.8 [9]	10^{-14} [9]	2300 [9]	2300 [9]
Granulation tissue	0.05 [20]	0.167 [9]	0.8 [9]	10^{-14} [9]	2300 [9]	2300 [9]
Stainless steel	197000 [18]	0.29 [18]	–	–	–	–

E: Elastic Modulus; υ: Poisson's ratio; $\boldsymbol{\phi}$: Porosity; ***k***: Permeability; $\boldsymbol{K_s}$: Solid compression modulus; $\boldsymbol{K_f}$: Fluid compression modulus

3.1 Changes in the Spatial Distribution of the Cells and Growth Factors

The percentage changes in the concentrations of cells and GFs compared to the free diffusion case were calculated based on the model predictions. The results suggest that, load duration and loading frequency have a significant effect on the cell and GF distributions within the early callus. The zone-specific predictions enable deeper understanding of the cell and GF transport behaviours within biomechanically different regions of the callus.

Table 2. Diffusion coefficients of solutes (i.e. cells and growth factors) used in this study

Solute	Diffusion coefficient (mm^2/day)	Reference
MSC	0.65	[11]
Fibroblasts	0.67	[13, 8]
Chondrocyte	8.64×10^{-20}	[21]
Osteoblast	8.64×10^{-20}	[21]
OGF	0.167	[7, 8]
CGF	0.167	[7, 8]

The results (Figs. 4, 5 and 6) show that there is no significant difference in the spatial distribution of MSC due to changes in load duration or loading frequency across the callus, which agrees with our previous study results where no noticeable changes (< 4%) in MSC concentrations within the callus were observed due to dynamic loading (up to 200 N at 1 Hz for 1 h). However, in vitro culture studies on electro-spun nanofibrous scaffolds have shown that dynamic loading could dramatically improve infiltration of MSC into the scaffolds, which could be attributed to the enhanced convective transport due to fluid flow. This observation is in contrast to our model predictions [22]. A possible reason for this difference could be increased differentiation of the MSCs into other types of cells (i.e. osteoblasts, chondrocytes and fibroblasts) due to enhanced mechanical stimuli and GF concentrations resulting from dynamic loading. This is corroborated by the predictions for osteoblasts, chondrocytes and fibroblasts where significant increase in concentrations were predicted. This indirectly suggests the increase of MSC and its subsequent differentiation into other types of cells due to dynamic loading, which is in agreement with the in vitro observation [22].

Table 3. Production and degradation rates of cells used in this study

Parameter	Expression	Reference
$k^{b}_{p,max}$	3.33/day	[6, 26]
$k^{c}_{p,max}$	16.67/day	[6, 26]
k^{fb}_{p}	0.0033/day	[23, 26]
λ^{b}_{p}	$(0.005 + 0.145 \times S)$/day*	[24]
λ^{c}_{p}	0.30/day	[24]
λ^{fb}_{p}	0.01/day	[11]
k^{m}_{d}	0.05/day	[11]
k^{b}_{d}	0.15/day	[11]
k^{c}_{d}	0.10/day	[11]
k^{fb}_{d}	0.05/day	[11]

*Note: S Refers to the stimulation index (Prendergast et al. [19])

This study predicts correlation between osteoblast concentration and loading regimes (load duration and loading frequency). In general, endosteal and periosteal zones exhibited significant changes in osteoblast concentration (up to 23% and 9%, respectively) due to variations in the loading regimes, whereas the changes in the cortical zone were insignificant (< 4%). In the endosteal zone, osteoblast concentrations increased from an average of 6% to 17% (compared to the free diffusion case) with the increase of load duration from 1 h to 4 h. Similarly, concentration increases up to 9% were predicted in the periosteal zone for the same increase in load duration. It is reported that low magnitudes of shear stress could induce osteogenic differentiation of MSC [25], which is probably why the highly stressed cortical zone predicted no significant changes, whereas, the moderately stressed endosteal and periosteal zones predicted significant increase of osteoblast concentration.

It was noticed that the osteoblast concentrations increased with the increase of loading frequency from 1 to 4 times (Figs. 4, 5 and 6). For example, if we consider the 4-h load duration cases, the percentage increase in osteoblast concentration within the endosteal zone (compared to the free diffusion case) increased from 12% to 23% when the loading frequency was increased from 1 (× 4-h) to 4 (× 1-h) times. This general trend of positive correlation between loading frequency and osteoblast concentration was noticed in all three regions of the callus. Sittichokechaiwut et al. [26] reported that a combination of short bouts of dynamic loading and rest periods could improve osteogenic differentiation of human MSC to levels comparable to those under pharmaceutical stimulants (e.g. Dexamethasone). Another study [27] investigating the effect of intermittent and continuous fluid stresses on human MSC in osteogenic medium demonstrated that intermittent fluid stresses are more effective in inducing osteogenic differentiation than continuous loading. Winter et al. [28], in an in vitro study, compared the effect of continuous and intermittent dynamic loading on rat osteoblast-like cells. Their study reported significantly high osteogenic activity (i.e. proliferation, differentiation and matrix synthesis) after 4, 8, 16 and 24 days under intermittent 1 h

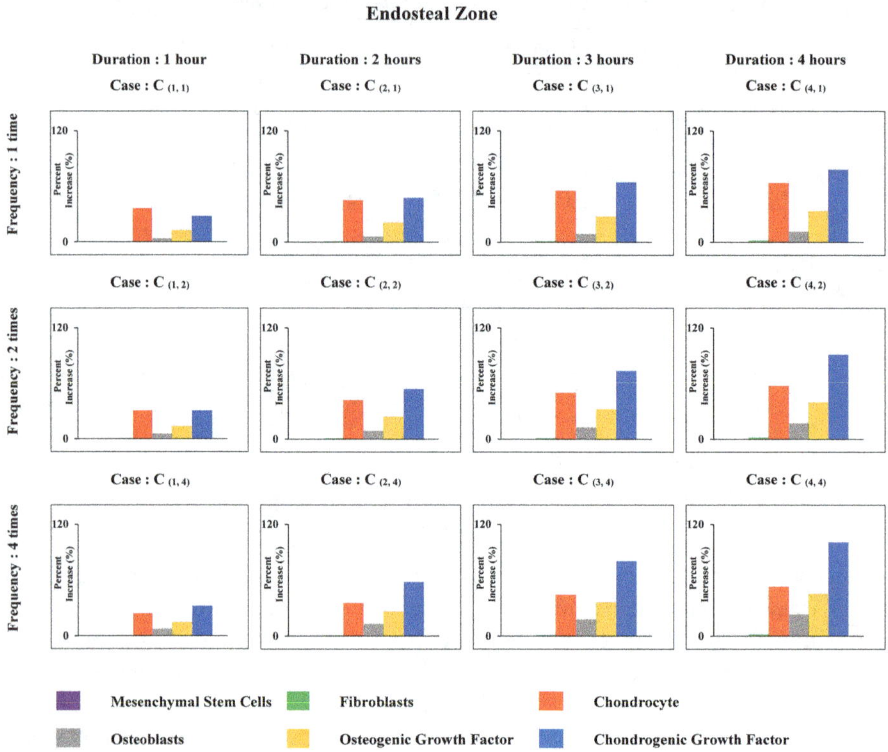

Fig. 4. Percent change of cells and growth factors within the endosteal zone of the callus under each case at the end of 24 h in comparison with the control (free diffusion) case.

dynamic loading (15-min loading followed by 15 min rest) case compared to the 1 h continuous loading and no loading cases. Numerous other studies [29, 30] have also reported positive correlations between dynamic loading frequency and osteogenic activity as predicted in this study.

Out of all four types of cells considered in this study, chondrocytes were predicted to be the most responsive to dynamic loading regimes (Figs. 4, 5 and 6). On average the content of chondrocytes increased by around 35–45% (with a maximum of 75% increase) throughout the callus due to dynamic loading compared to the free diffusion case, at the end of the 24-h period considered. The increases in the chondrocyte concentrations within the endosteal zone was significantly higher (around 45% on average) than those predicted within the cortical and periosteal zones, (around 35% on average). It was evident from the predictions that there is a strong correlation between the loading regimes and the chondrocyte concentrations within the callus. In all callus zones, a positive correlation was predicted between the load durations and chondrocyte concentrations. For instance, the average increase of chondrocyte concentration for the 4-h loading cases in the periosteal zone (compared to the free diffusion case) was predicted to be 44% which is more than twice the average increase of 21% for the 1-h loading cases. Therefore, relatively longer durations of physiological activities tend to

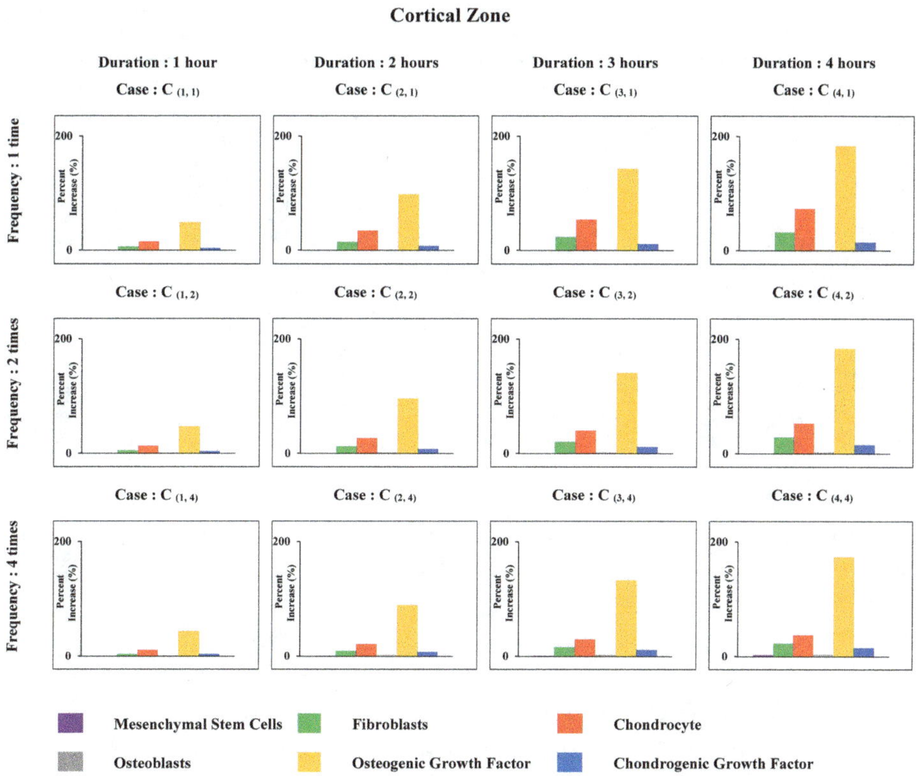

Fig. 5. Percent change of cells and growth factors within the cortical zone of the callus under each case at the end of 24 h in comparison with the control (free diffusion) case.

enhance healing by increasing the chondrocyte concentrations within the callus. There is plethora of evidence to support the idea that dynamic loading promotes chondrogenesis of MSC, as discussed in our previous study [16]. However, does loading duration affect chondrogenesis significantly as predicted in this study? is a more important question which requires answer. An in vitro study [31] examining the influence of cyclic compressive loading on human MSC reported optimal dynamic loading durations of 1–4 h (10% strain at 1 Hz) with the most chondrogenic differentiation for 4 h duration, which is in agreement with the present model predictions. Therefore, load duration appears to be strongly correlated with chondrogenesis.

Interestingly, negative correlation was observed between the loading frequency and chondrocyte concentration within the callus. For example, if we consider the 4-h load duration cases, the percentage increase in chondrocyte concentration within the cortical zone (compared to the free diffusion case) dropped from 75% to 38% when the loading frequency was increased from 1 (× 4-h) to 4 (× 1-h). This general trend of negative correlation between loading frequency and chondrocyte concentration was noticed in all three regions of the callus (Figs. 4, 5 and 6). Although relevant experimental evidence directly relating the effect of intermittent dynamic loading regimes on MSC

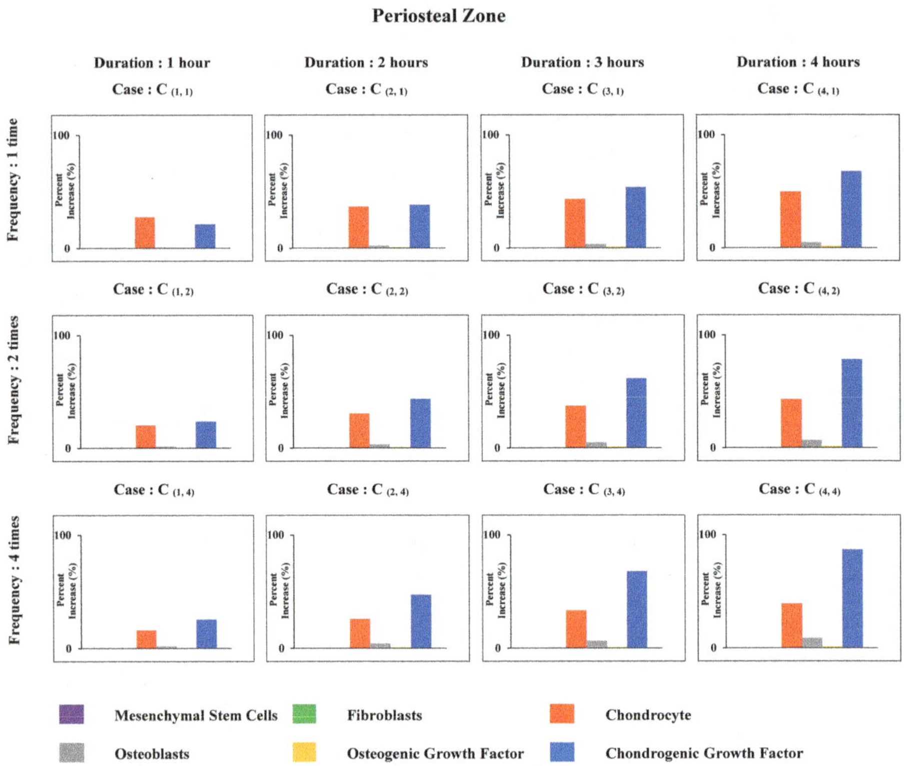

Fig. 6. Percent change of cells and growth factors within the periosteal zone of the callus under each case at the end of 24 h in comparison with the control (free diffusion) case.

differentiation into chondrocyte is not available in literature, chondrogenic activities have been reported to be inhibited by intermittent dynamic loading compared to continuous dynamic loading [32], which agrees with the present model prediction. Many studies on intervertebral discs have also shown that the right combination of dynamic load magnitude, frequency and duration could enhance the differentiation of MSC towards chondrogenic pathway [33].

Increase in fibroblast concentrations due to changes in loading regimes was predicted to be significant only within the cortical zone (Fig. 5), where increases up to 33% were observed. In the other callus zones, the changes were negligible (< 3%). This phenomenon could be easily explained considering the model. Since the chemical stimuli-mediated fibroblast differentiation was assumed to be constant in this study, the up regulation of fibroblast is directly due to the synergistic effect mechanical stimuli and advective transport resulting from dynamic loading. The highly affected cortical zone (due to loading) tends to favour fibroblastic differentiation which is augmented by the enhanced fluid flow within the zone which could give rise to the fibroblast concentrations. Though not as strong as those for chondrocytes, fibroblast concentration showed a positive correlation with the load duration and a negative correlation with the loading frequency within the cortical zone.

Osteogenic growth factor (OGF) was predicted to increase significantly in response to dynamic loading everywhere but in the periosteal callus. In the endosteal and cortical zones, OGF concentration increase compared to the free diffusion case were up to 46% and 185%, respectively. A positive correlation was predicted between load duration and OGF concentration within these zones with the endosteal and cortical zones exhibiting average increases of 41% and 182%, respectively, for the 4-h load duration cases in contrast to the 14% and 48% predicted for the 1-h load duration cases.

It was predicted that the interrelationship between loading frequency and OGF concentrations within the callus is not straightforward. For example, if we consider the 4-h load duration cases, in the endosteal zone, the percentage increase of OGF concentration (compared to the free diffusion case), rises from 34% to 46% as the loading frequency increases from 1 ($\times$ 4 h) to 4 ($\times$ 1 h) times. However, for the same change in the loading frequency, the percentage increase of OGF concentration (compared to the free diffusion case) falls from 185% to 175% within the cortical callus. Thus, both positive and negative correlations were seen between loading frequency and OGF concentrations within different regions of the callus.

On the other hand, CGF concentration was predicted to change significantly within the endosteal and periosteal zones and modestly within the cortical zone of the callus. In the endosteal, cortical and periosteal zones, CGF concentration increase compared to the free diffusion case were up to 101%, 16% and 87%, respectively. A positive correlation was predicted between load duration and CGF concentration within these zones. The endosteal, cortical and periosteal zones exhibited average increase of 91%, 16% and 78%, respectively, for the 4-h load duration cases in contrast to the 31%, 4% and 23% predicted for the 1-h load duration cases.

In contrast to the predictions of OGF, CGF concentration was predicted to have a positive correlation with loading frequency in the endosteal and periosteal regions. For example, if we consider the 4-h load duration cases, the percentage increase in CGF concentration within the endosteal and periosteal zones (compared to the free diffusion case) increased from 79% and 68% to 101% and 87%, respectively, when the loading frequency was increased from 1 ($\times$ 4-h) to 4 ($\times$ 1-h). However, no significant changes in the CGF concentrations with changes to loading frequency were predicted within the cortical zone.

The space dependent changes in OGF and CGF could be directly attributed to space dependent convective transport of solutes within the fracture callus. GFs exhibit prominent changes in concentrations compared to those of cells, since, GFs are larger molecules which have relatively smaller diffusive coefficients (D). When advective transport is of similar magnitudes to those of diffusive transport, it could result in higher relative changes in concentrations [34]. Advective transport depends on the pressure gradient. Since different zones of the callus (i.e. endosteal, cortical and periosteal) are in different mechanical microenvironments, the pressure gradients and convective transports within each region could vary. Moreover, the differences in the sources of growth factors (i.e. boundary conditions) also play an important role in influencing the spatial distribution of the GFs.

3.2 Temporal Changes in the Cell and GF Concentrations Within the Callus

Predictions for the temporal changes of cell and GF concentrations suggests that dynamic loading can accelerate the transport of cells and GFs within the callus in a space dependent manner. It was predicted that, for a given loading frequency, the general trends of temporal concentration variations of cells and GFs are similar for all load durations and the differences were only in the magnitudes. Therefore, the 2-h load duration cases (loading frequencies: 1 [× 2 h], 2 [× 1 h] and 4 [× 0.5 h]) are considered as example for the analysis (Fig. 7).

Comparing the temporal variations of MSC concentration under different loading frequencies shows that there is no significant temporal change in the MSC concentrations in any zone except in the cortical zone, where MSC concentration rises (compared to the free diffusion case) during and immediately after loading and gradually falls to zero with time as free diffusion case catches up. After each period of loading, a small bump (1, 2 and 4 bumps for frequencies = 1, 2 and 4, respectively) could be seen in the plot (Fig. 7a) where, the concentrations briefly rise and then fall to the concentration levels in the free diffusion cases. As mentioned above, one of the reasons for the rapid fall of MSC concentration upon cessation of loading could be the relatively high diffusive capacity of MSC, which makes diffusion the predominant mode of transport of it. Another reason for this prediction could be increased differentiation of MSC, directly due to mechanical stimuli and indirectly due to increased GF concentrations resulting from dynamic loading.

Fibroblasts exhibited a similar behaviour to that of MSC and showed significant temporal changes only in the cortical zone. However, it was predicted that the initial rise in fibroblast concentration was prominent with relatively higher peak compared to those of the subsequent rises due to loading. Moreover, the fall of fibroblast concentration to the base levels (free diffusion case) after initial rise was more gradual, which took relatively longer durations compared to those of MSC.

Interestingly, chondrocyte concentrations were predicted to have almost the same trend within the callus irrespective of the loading frequency. During and soon after the first loading period, the concentration of chondrocyte in the endosteal zone increased rapidly to reach a peak and then fell gradually over time to levels noticeably higher than those in the free diffusion case. A similar trend was observed in the periosteal zone as well; however, the initial peak and the subsequent fall were not as significant as those predicted in the endosteal zone. Chondrocyte concentration increase in the cortical zone appeared to be more gradual and the rate reached a somewhat steady state after the initial rise.

Osteoblasts on the other hand showed no significant temporal changes within the cortical zone. However, prominent changes were predicted in the endosteal and periosteal zones, where the concentrations rose rapidly during each loading period, reached a peak and then gradually fell to noticeable concentrations levels higher than the free diffusion case concentrations, when the loading ceased. Under all loading frequencies, the initial peaks were similar in both endosteal and periosteal zones; but, in the subsequent loading periods, endosteal zone exhibited significantly higher peaks than those predicted for the periosteal zone.

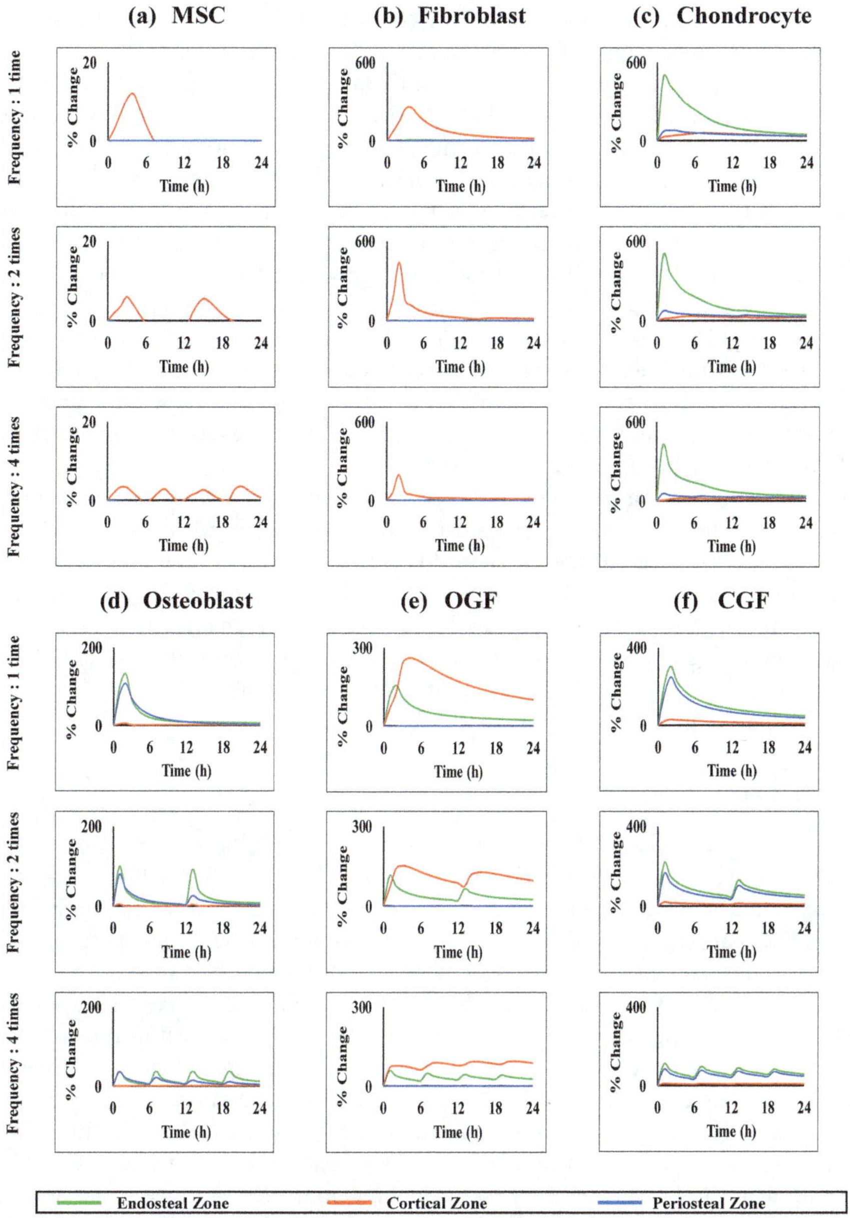

Fig. 7. Percentage change of (a) MSC, (b) fibroblasts, (c) chondrocytes, (d) osteoblasts, (e) OGF, and (f) CGF within the callus for 2-h (total) load duration cases, under different loading frequencies (i.e. 1, 2 and 4 times) during the 24 h period simulated, in comparison with the control case $C_{(0,0)}$ (free diffusion)

The model predicted notable temporal changes in the OGF concentrations only within the endosteal and cortical zones. The OGF concentrations within these two zones were predicted to be very responsive to load durations and loading frequencies. In the endosteal zone, for each loading period, OGF concentrations rose rapidly, reached a peak and then fell gradually to significantly higher concentration levels compared to those of the free diffusion case as the loading ceased. This pattern was similar in the cortical region as well, but, the peaks were relatively higher and the drops following the peaks were more gradual in the cortical zone than those in the endosteal zone.

On the other hand, predictions for CGF showed noticeable temporal changes throughout the callus; however, the changes in the endosteal and periosteal zones were prominent. Rapid rise in concentrations (compared to the free diffusion case) during loading to reach peak and gradual fall upon cessation of loading were predicted for each loading period in the endosteal and periosteal zones. The same was predicted in the cortical zone as well; however, the peaks and rates of drop were relatively very small compared to those of the other two zones.

In general load duration and loading frequency appears to influence the temporal solute concentrations within the early callus in a space dependent manner. There is noticeable change in the concentration profiles of cells and GFs after the application of dynamic loading. Each cell and GF showed some increase at least in one of the callus zones during each loading period of the loading regimes. Dynamic loading appears to provide a pumping action within the callus which can increase the cell and GF concentrations. This in turn could influence the level of differentiation of different cells within the early callus.

3.3 Summary

The results are in support of the hypothesis that the duration of early physiological activities under ICF and their frequencies could significantly affect solute transport and their concentration distributions within the fracture site in a space and time dependent manner. The pumping action due to dynamic loading does increase the advective transport of cells and GFs in a time and space dependent manner. The increase in GF concentrations are notably high as the relative solute transport increase due to advection is high for large molecules such as GFs, which have relatively smaller diffusion coefficients.

The increase in cell concentrations due to dynamic loading is due to the synergistic effects of (i) increase in mechanical stimuli-mediated differentiation (ii) increase in chemical stimuli-mediated differentiation (due to increase in GF concentrations) and (iii) increase in advective transport of cells. This synergistic effect has positive influence within the fracture site, especially inducing more MSC towards chondrogenic pathway which is beneficial in the early stages of secondary healing. As more chondrocytes could synthesise more cartilage tissue, the fracture site could be stabilized faster. It was predicted that chondrogenic differentiation of MSC is positively correlated with load duration, which suggests that physiological activities (e.g. walking) for relatively longer periods (e.g. up to 4 h) could be more beneficial for patients in the early stages. The predictions also suggested that physiological activities done in a single stretch could be more beneficial in the early stages than those done intermittently for the same cumulative durations.

It should also be noted that, longer duration loads with minimal rest, can also give rise to fibroblast concentrations within the cortical zone. However, the increase in chondrogenic differentiations of MSC in all three zones of the callus far outweighs the increase in fibroblast concentration. Moreover, in the early stages, cortical zone will any way be unstable, and it is the changes in the periosteal callus that matters the most; because, it lays the path to stabilize the fracture so that better microenvironments are achieved in later stages within the cortical zone. Increase in chondrocyte concentrations up to 75% were observed within the periosteal zone for 4-h dynamic load applied in a single stretch. This is indicative of more cartilage tissue development within the periosteal zone and faster stabilization of the fracture site.

Osteoblast differentiation, which could subsequently lead to direct bone tissue formation, was also predicted to have a positive correlation with load duration. However, the predictions suggested that intermittent loading to be more beneficial for osteogenic differentiation of MSC than continuous loading for the same duration. This leads to a complex situation in choosing between single stretch and intermittent loading. Since cartilage development in the early stages is the most important step of secondary healing, it would be ideal to opt for longer physiological loading with minimal rest periods in the early stages. The predictions also suggest that in the later stages of healing it may be better to go for intermittent loading as it could improve bone ossification. However, these conclusions need further investigations and experimental evidence.

3.4 Limitations

This study was based on number of assumptions and simplifications. The geometries used in study were simplified. Firstly, the tibial bone was assumed to be cylindrical, which enabled us to further simplify the analysis to 2D axisymmetric model. In fact, realistic bone geometries could have been used to improve the accuracy as done in one of our previous studies [35], where we used computer tomography (CT) based reconstruction of tibial geometry. However, such complicated 3D models are computationally expensive and increase the computational time drastically.

It should also be kept in mind that the model makes predictions based on our inputs. Therefore, any changes to the model parameters and boundary conditions adopted in this model could lead to totally different predictions to those presented here. In this study, the model parameters and boundary conditions were chosen based on existing literature and assumed to represent the general case. However, the parameters are often patient specific, which need be considered.

In this study, load durations only up to 4 h were considered. However, it is possible that patients may get involved in physiological activities for longer durations. Moreover, we adopted uniform loading patterns, whereas the loading durations and resting periods could be random in real situations. Finally, in vivo experimental studies investigating the influence of loading regimes on early stage differentiations are very limited if not totally unavailable. Although, there are in vitro culture studies focusing on the effects of certain aspects of dynamic loading (e.g. load duration) on MSC differentiations, they are not directly comparable with the current results which focuses on the effects of physiologically relevant loading regimes on early stage fracture healing microenvironment. Therefore, further experimental evidence is required to confirm the findings of this study. Despite these limitations, the present study provides some useful information and new insights into the relationship between physiologically relevant dynamic loading and early stage fracture site microenvironment.

4 Conclusions

This study demonstrates the potential of computational models in elucidating the effects of mechanical loading regimes on the biology of fracture healing. Moreover, models like the one used in this study could be used to test new hypotheses and plan in-vivo and in-vitro experiments more efficiently. The results of this study could serve as a good starting point for surgeons and/or clinicians in designing patient specific physiological exercises under ICF. A summary of the main outcomes of this study are as follows:

- Duration of physiological activities under ICF (e.g. walking) and their frequencies (i.e. number of times per day) in the early stages of fracture healing significantly affect MSC differentiation pathways in a time and space dependent manner, which could eventually influence the progress of healing.
- Early physiological loading relevant to walking (axial load with an amplitude of 200 N at 1 Hz) for up to 4 h could increase the chondrocyte concentrations (compared to no-load control case) by up to 65%, 75% and 50% within the endosteal, cortical and periosteal zones respectively at the end of 24-h period, which is significantly higher than the maximum increase in the osteoblast (up to 23%) and fibroblast (up to 33%) concentrations.
- The concentrations of CGF could increase up to 101% within the endosteal and periosteal zones, but to modest 16% within the cortical zone; and the concentrations of OGF could increase in the endosteal and cortical zones up to 185% by the end of 24 h because of dynamic loading relevant to walking up to 4 h.
- Early dynamic loading appears to favour chondrogenic differentiation of MSC more than osteogenic or fibrogenic differentiations, which is beneficial for secondary healing. Longer durations (up to 4 h) of continuous loading seems preferable over shorter durations and intermittent loading for chondrogenic differentiation. However, intermittent loading appears to enhance osteogenic differentiation of MSC more compared to that under continuous loading for a given total duration of loading.
- The results suggest that longer durations (up to 4 h) of continuous physiological activities is more beneficial in the early stages to enhance chondrogenesis and intermittent loading of similar durations could be more beneficial in the later stages to enhance osteogenesis.

References

1. Ai-Aql, Z., Alagl, A.S., Graves, D.T., Gerstenfeld, L.C., Einhorn, T.A.: Molecular mechanisms controlling bone formation during fracture healing and distraction osteogenesis. J. Dent. Res. **87**, 107–118 (2008)
2. Doblaré, M., Garcıa, J., Gómez, M.: Modelling bone tissue fracture and healing: a review. Eng. Fract. Mech. **71**, 1809–1840 (2004)
3. Garcia-Aznar, J., Kuiper, J., Gómez-Benito, M., Doblaré, M., Richardson, J.: Computational simulation of fracture healing: influence of interfragmentary movement on the callus growth. J. Biomech. **40**, 1467–1476 (2007)
4. Augat, P., Simon, U., Liedert, A., Claes, L.: Mechanics and mechano-biology of fracture healing in normal and osteoporotic bone. Osteoporos. Int. **16**, S36–S43 (2005)
5. Einhorn, T.A.: The cell and molecular biology of fracture healing. Clin. Orthop. Relat. Res. **355**, S7–S21 (1998)
6. Claes, L., Augat, P., Suger, G., Wilke, H.J.: Influence of size and stability of the osteotomy gap on the success of fracture healing. J. Orthop. Res. **15**, 577–584 (1997)
7. Bailon-Plaza, A., Van Der Meulen, M.C.: A mathematical framework to study the effects of growth factor influences on fracture healing. J. Theor. Biol. **212**, 191–209 (2001)
8. Bailón-Plaza, A., van der Meulen, M.C.: Beneficial effects of moderate, early loading and adverse effects of delayed or excessive loading on bone healing. J. Biomech. **36**, 1069–1077 (2003)
9. Lacroix, D., Prendergast, P.: A mechano-regulation model for tissue differentiation during fracture healing: analysis of gap size and loading. J. Biomech. **35**, 1163–1171 (2002)
10. Prendergast, P., Huiskes, R., Søballe, K.: Biophysical stimuli on cells during tissue differentiation at implant interfaces. J. Biomech. **30**, 539–548 (1997)
11. Isaksson, H., van Donkelaar, C.C., Huiskes, R., Ito, K.: A mechano-regulatory bone-healing model incorporating cell-phenotype specific activity. J. Theor. Biol. **252**, 230–246 (2008)
12. Dimitriou, R., Jones, E., McGonagle, D., Giannoudis, P.V.: Bone regeneration: current concepts and future directions. BMC Med. **9**, 66 (2011)
13. Geris, L., Gerisch, A., Vander Sloten, J., Weiner, R., Van Oosterwyck, H.: Angiogenesis in bone fracture healing: a bioregulatory model. J. Theor. Biol. **251**, 137–158 (2008)
14. Bronson, D.G., Samchukov, M.L., Birch, J.G., Browne, R.H., Ashman, R.B.: Stability of external circular fixation: a multi-variable biomechanical analysis. Clin. Biomech. **13**, 441–448 (1998)
15. Fragomen, A.T., Rozbruch, S.R.: The mechanics of external fixation. HSS J. **3**, 13–29 (2007)
16. Ganadhiepan, G., Zhang, L., Miramini, S., Mendis, P., Patel, M., Ebeling, P., Wang, Y.: The effects of dynamic loading on bone fracture healing under Ilizarov circular fixators. J. Biomech. Eng. **141**, 051005-051005-051012 (2019)
17. Epari, D.R., Taylor, W.R., Heller, M.O., Duda, G.N.: Mechanical conditions in the initial phase of bone healing. Clin. Biomech. **21**, 646–655 (2006)
18. Watson, M., Mathias, K., Maffulli, N., Hukins, D., Shepherd, D.: Finite element modelling of the Ilizarov external fixation system. Proc. Inst. Mech. Eng. [H] **221**, 863–871 (2007)
19. Huiskes, R., Van Driel, W., Prendergast, P., Søballe, K.: A biomechanical regulatory model for periprosthetic fibrous-tissue differentiation. J. Mater. Sci. - Mater. Med. **8**, 785–788 (1997)
20. McCartney, W., Mac Donald, B., Hashmi, M.: Comparative performance of a flexible fixation implant to a rigid implant in static and repetitive incremental loading. J. Mater. Process. Technol. **169**, 476–484 (2005)

21. Zhang, L., Miramini, S., Richardson, M., Mendis, P., Ebeling, P.: The role of impairment of mesenchymal stem cell function in osteoporotic bone fracture healing. Aust. Phys. Eng. Sci. Med. **40**, 603–610 (2017)
22. Nerurkar, N.L., Sen, S., Baker, B.M., Elliott, D.M., Mauck, R.L.: Dynamic culture enhances stem cell infiltration and modulates extracellular matrix production on aligned electrospun nanofibrous scaffolds. Acta Biomater. **7**, 485–491 (2011)
23. Peiffer, V., Gerisch, A., Vandepitte, D., Van Oosterwyck, H., Geris, L.: A hybrid bioregulatory model of angiogenesis during bone fracture healing. Biomech. Model. Mechanobiol. **10**, 383–395 (2011)
24. Andreykiv, A., Van Keulen, F., Prendergast, P.: Simulation of fracture healing incorporating mechanoregulation of tissue differentiation and dispersal/proliferation of cells. Biomech. Model. Mechanobiol. **7**, 443–461 (2008)
25. Lee, D.A., Knight, M.M., Campbell, J.J., Bader, D.L.: Stem cell mechanobiology. J. Cell. Biochem. **112**, 1–9 (2011)
26. Sittichokechaiwut, A., Edwards, J., Scutt, A., Reilly, G.: Short bouts of mechanical loading are as effective as dexamethasone at inducing matrix production by human bone marrow mesenchymal stem cell. Eur. Cell Mater **20**, 45–57 (2010)
27. Liu, L., Yu, B., Chen, J., Tang, Z., Zong, C., Shen, D., Zheng, Q., Tong, X., Gao, C., Wang, J.: Different effects of intermittent and continuous fluid shear stresses on osteogenic differentiation of human mesenchymal stem cells. Biomech. Model. Mechanobiol. **11**, 391–401 (2012)
28. Winter, L., Walboomers, X., Bumgardner, J., Jansen, J.: Intermittent versus continuous stretching effects on osteoblast-like cells in vitro. J. Biomed. Mater. Res. Part A: Off. J. Soc. Biomater. Japan. Soc. Biomater. Aust. Soci. Biomater. Korean Soc. Biomater. **67**, 1269–1275 (2003)
29. Turner, C.H.: Three rules for bone adaptation to mechanical stimuli. Bone **23**, 399–407 (1998)
30. Robling, A.G., Hinant, F.M., Burr, D.B., Turner, C.H.: Improved bone structure and strength after long-term mechanical loading is greatest if loading is separated into short bouts. J. Bone Miner. Res. **17**, 1545–1554 (2002)
31. Remya, N., Nair, P.D.: Mechanoresponsiveness of human umbilical cord mesenchymal stem cells in in vitro chondrogenesis—A comparative study with growth factor induction. J. Biomed. Mater. Res. Part A **104**, 2554–2566 (2016)
32. Bryant, S.J., Nicodemus, G.D., Villanueva, I.: Designing 3D photopolymer hydrogels to regulate biomechanical cues and tissue growth for cartilage tissue engineering. Pharm. Res. **25**, 2379–2386 (2008)
33. Chan, S.C., Ferguson, S.J., Gantenbein-Ritter, B.: The effects of dynamic loading on the intervertebral disc. Eur. Spine J. **20**, 1796 (2011)
34. O'Hara, B., Urban, J., Maroudas, A.: Influence of cyclic loading on the nutrition of articular cartilage. Ann. Rheum. Dis. **49**, 536–539 (1990)
35. Ganadhiepan, G., Miramini, S., Patel, M., Mendis, P., Zhang, L.: Bone fracture healing under Ilizarov fixator: influence of fixator configuration, fracture geometry, and loading. Int. J. Numer. Methods Biomed. Eng. **35**, e3199 (2019)

The Effect of Discretization on Parameter Identification. Application to Patient-Specific Simulations

Nava Schulmann(✉), Stéphane Cotin, and Igor Peterlik

University of Strasbourg, Strasbourg, France
schulmann.nava@gmail.com

Abstract. Identifying the elastic parameters of a finite element model from a dynamically acquired set of observations is a fundamental challenge in many data-driven medical applications going from soft surgical robotics to image-guided per-operative simulations. While various strategies exist to tackle the parameter-identification inverse problem (Aster et al. 2013), the effect of sub-optimal discretization, as often required in real-time applications, is largely overlooked. Indeed, the need to tune the parameter values in order to account for discretization-induced stiffening in specific models is reported in different works (e.g. Chen et al. 2015; Mira et al. 2018). However, to the best of our knowledge, no systematic study of this phenomenon exists to date, nor has any strategy to select optimal effective values been developed. Our work addresses the issue of parameter identification in coarsened meshes with special attention to the dynamical nature of the identification. We focus on the estimation of Young's moduli in simplified systems and show that the estimated stiffnesses are underestimated in a systematic manner when reducing the number of degrees of freedom. We also show that the effective stiffness of a given coarse mesh, when associated with an undersampled mesh discretization, is not constant but strongly depends on the prescribed deformations. These results show that the estimated parameters should not be considered as the true parameter value of the organ or tissue but instead are model-dependent values. We argue that Bayesian methods present a clear advantage w.r.t. classical minimization methods by their ability to efficiently adapt the local parameter values.

Keywords: Parameter identification · Finite element methods · Kalman filters · Bayesian inference · Discretization · Real-time control

1 Introduction and Related Work

In the last decade the importance of computer medical simulation in intra-operative guidance has increased considerably. In various applications, ranging from laparoscopic liver surgery to catheter navigation or interventional radiology, physically-based numerical models play the role of complementing and

G. A. Ateshian et al. (Eds.): CMBBE 2019, LNCVB 36, pp. 237–247, 2020.
https://doi.org/10.1007/978-3-030-43195-2_19

regulating the information obtained by imaging. Since preoperative imaging (or three-dimensional anatomical models reconstructed from it) and intra-operative imaging often rely on different modalities, they provide different and often complementary information on tissue structures as well as deformation changes. In that case, the fusion of intra-operative and pre-operative images into a unique coordinate frame adds significant value. This can provide an enhanced visualization, for instance of the subsurface anatomy during laparoscopy, to help determine resection planes and facilitate navigation (Heiselman et al. 2017). Generally speaking, intra-operative image augmented with pre-operative data results in easier and better guidance of an intervention, leading to reduced invasiveness and/or increased accuracy during surgical procedures (Peters and Cleary 2008]. In percutaneous ablations and tumor resections, augmented intra-operative images can facilitate the localization of small tumors. In this case, the augmented image helps follow the safety margins near the tumor while preserving a sufficient amount of healthy tissue.

However, one challenge that often hinders the ability to fuse pre- and intra-operative images comes from the tissue deformations that often take place between the two image acquisitions. Handling these deformations requires non-rigid registration methods to align the pre-operative models or images onto the intra-operative images. A very large body of work has addressed this problem, and it has been shown that biomechanical models can achieve excellent results even in the presence of large deformations (Haouchine et al. 2015) (Collins et al. 2017).

If a real-time augmentation of the data is sought, to provide augmented reality as in (Haouchine et al. 2015) or (Suwelack et al. 2014), this brings two strong constraints onto the soft tissue model. The predictive accuracy of the model should be higher than what is generally used in the context of computer-based training systems, while still being under the constraint of real-time computation. Achieving these objectives jointly is a challenge, and generally requires some trade-off. In addition, intra-operative guidance requires specific models, tailored to the patient. This is necessary to ensure simulations that are as accurate as possible. While the general constitutive law for a given organ is known, its parameters are often patient-specific. In this paper we address an important issue that stems from the association of these constraints: real-time computation, patient-specific modeling, and biomechanical modeling.

Patient-specific models can be obtained, at least partially, through pre- or intra-operative imaging. While the geometry of the organ can be extracted quite easily, and almost automatically, through pre-operative images such as computed tomography (CT) or magnetic resonance imaging (MRI), material properties are a lot more difficult to determine in a patient-specific way. In some cases, these parameters may be obtained by elastography (Ophir et al. 1991), a technique that relies on MRI or ultrasound imaging modalities. However, this solution remains limited today in a clinical context due its technical requirements. For this reason, the development of alternative identification methods, that allow for real-time parameter estimation from per-operative images, has significantly increased in the recent years (see for example (Avril and Evans 2017)).

Though different numerical approaches for resolving the mechanical behaviors of soft tissues exist, finite element (FE) methods are by far the most commonly used in the medical simulation and biomechanics communities (see (Han et al. 2011) and (Zhang et al. 2018) for instance). Different approaches have been proposed to reduce the high computational cost of FE methods in order achieve real-time performances. A first strategy consists of replacing the complex nonlinear physical laws by linear descriptions. This solution, however, is rarely appropriate for anatomical structures. Another direction is to perform model order reduction such as Proper Orthogonal Decomposition (POD) (Bialecki et al. 2005) or Proper Generalized Decomposition (PGD) (Niroomandi et al. 2013). These techniques allow for a significant computational gain, but at the expense of a pre-computation step which can be very costly. When this is not possible due to time constraints, as it can be the case for patient-specific simulations in a clinical context, alternative strategies are used. A common approach consists of maintaining the complexity of the physical model while reducing the number of degrees of freedom in the mesh. The solution obtained in this case is therefore only an approximation of the solution at convergence.

Reducing the FE mesh resolution can also have a substantial impact on the results of parameter identification. Parameter values may not converge to ground-truth values, but strongly depend on the level of discretization (Aster et al. 2013). For example, coarse meshes tend to be stiffer than fine meshes (see Fig. 1). Different authors have suggested to resolve this discretization-induced parameter dependency by tuning the parameter values in order to fit the observation (e.g. (Chen et al. 2015; Anna et al. 2018)). However, neither a systematic study of this phenomenon exists, nor a strategy to select optimal effective values has been developed. More importantly, simple tests show that stiffening depends not only on the coarsening level, but also on the type of deformation. In other words, for a given model, estimated parameters that fit the observations within a certain range of deformation, may give rise to important errors when used to simulate different manipulations, as illustrated in Fig. 1.

The goal of this work is to study this effect systematically and suggest a general strategy for an efficient dynamical parameter identification in coarse meshes. We will discuss the simulation framework and the optimization scheme in Sect. 2, experimental in-silico results are reported in Sect. 3. We conclude with a summary and discussion of future work.

2 Methodology

As stated in the introduction, in this work, we are interested in simulations of large deformations occurring during manipulations with soft tissues, e.g. during a laparoscopic surgery. Models for this type of simulations have been developed in the theory of elasticity and are parametrized by material properties.

The estimation of material properties is typically composed of three phases: (i) selection of appropriate parametric model which captures the behaviour of

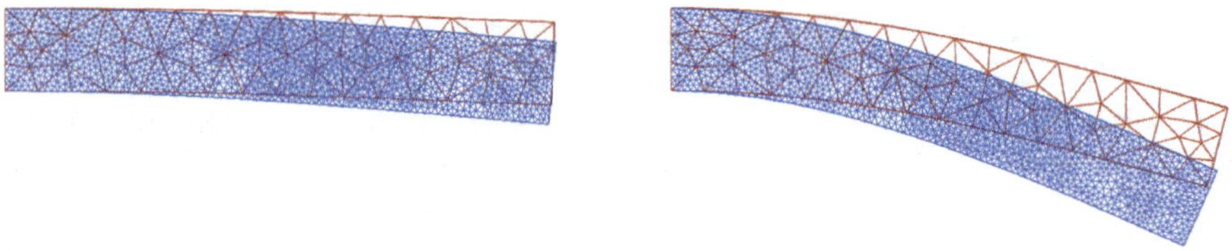

Fig. 1. Comparison between the deformations of fine and coarse meshes under small and big mechanical stresses. The fine mesh (blue) consists of 11355 nodes and the coarse mesh (red) of 239 nodes. Each sub-figure shows the deformations of the two meshes under the same stress constraints. In both cases the coarse mesh presents a stiffer behavior than the fine mesh, however, the stiffening is clearly higher when the stress is more important.

interest; (ii) experimental measurements typically performed on representative samples; (iii) parameter fitting: an optimization procedure which yields values of the selected model so that the outputs of the model correspond to the measurements.

In the context of tissue modeling for computer-aided interventions, each of the three phases listed above gives rise to specific challenges: models capturing the behaviour of tissues are highly non-linear, the accuracy of measurements is quite limited since they must be performed either on cadavers or non-invasively, parameters obtained by fitting are often difficult to generalize since biomechanical coefficients of tissues show high inter-patient variability.

Being aware of these challenges, many of which have been studied in the literature (Marchesseau et al. 2010; Rifai et al. 2011; Wittek et al. 2009), we focus solely on the problematic of parameter identification due to coarse discretization which is necessary in real-time context. To this end, we propose a methodology based on numerical experiments which were designed to be generic while maintaining the essential features related to real-time parameter identification in coarse meshes, as explained below.

2.1 Finite-Element Modeling

In order to investigate the solution of the problem introduced by the dependency of parameters on FE mesh density, we employ a model of highly-deformable homogeneous object having a simple regular geometry. The behaviour of the cantilever beam, fixed on one side and subjected to uniformly-distributed pressure, is modelled with StVenant-Kirchhoff hyperelastic constitutive law. The resulting boundary-value problem solved by the finite element method where linear tetrahedral elements (usually denoted as P1) are employed to discretize the domain of the object.

This type of elements is usually not recommended because of its high discretization error (as stress and strain are constant over the volume of an element) and locking phenomenon. However, here we must recall the real-time context of this work which puts a hard constraint on the computationalal time. From this point of view, P1 elements are optimal since they require significantly lower computational time needed to assemble the element equations when compared to higher-order and/or non-simplex elements which require advanced quadratures to perform the integration. While also in the case of P1 elements, their number used for the discretization is already limited by the real-time requirement, in the case of more complex elements, the resolution of the mesh would not allow for a reasonable approximation of the modelled object.

2.2 Identification Method

In order to account for the dynamic nature of data acquisition and for the parameter dependencies on the prescribed deformation, a certain form of adaptive filtering is needed. Classical optimization methods minimize some metrics between the model predictions and the data. The method of least squares (LS) is a standard approach in which the overall solution minimizes the sum of the error squares. Recursive methods also known as adaptive filters allow for a dynamical update of the parameter estimation. They split into Stochastic gradient filters (different variants of LS) and least squares recursive filter (RLS). In the case of time independent parameters and in the presence of large amount of data, the different identification methods converge to the same values, thus the choice between the filters is a tradeoff between efficiency and computational costs. However, when data is sparse, the convergence rates may highly affect the estimations. In order to tackle the data sparsity in the context of computer-aided interventions, we employ Bayesian methods, namely Kalman filters which present particularly good convergence rates and have proven their worth in numerous real-time applications. Bayesian methods take into account some *a priori* knowledge about the state of the system and combine it with measured data to infer an estimated state distribution. In this framework, the parameter identification is performed through a joint estimation: the system state is augmented by the parameters of the model, and the filter simultaneously updates the system state together with the parameters. This method may be interpreted as a generalized LMS where, instead of following a local gradient, a path which optimizes the global convergence is chosen.

The original Kalman filters, which were designed for linear systems, have been adapted to non-linear estimation. While *Extended Kalman filter* requires explicit derivative of the model w.r.t. the parameter which cannot be obtained easily in our case, we opt for *Unscented Kalman filter* (UKF). In a nutshell, UKF samples the probability space in each step of the estimation and thus requires forward execution of the model (i.e., FE simulation) for each sample. This approach would become computationally infeasible in the case of joint state-parameter estimation due to a very high number of samples (thus FE simulations), we rely on reduced-order version of this filter (denoted as RO-UKF) proposed by (Moireau and Chapelle 2011): in our case, only the samples related to the model parameters are propagated through the non-linear FEM.

3 Results

In this section, we report the results of our in-silico deformation experiments. We first explore the dependency between the resolution level and the estimated stiffness for one given deformation. Then, for an intermediate coarsening level, we compute the apparent stiffnesses for increasing pressures in various directions. We finally compare the shape errors between simulated objects, with and without data assimilation.

3.1 Estimated Stiffness

3.1.1 Element Size Dependence

To assess the effect of coarse meshing on parameter identification, we impose identical bending stresses to geometrically identical beams, which are modeled with different levels of discretization. The Young's modulus of each beam is estimated using a Kalman Filter. As can be seen in Fig. 2, the estimated values asymptotically approach the ground truth value of 5000 kPa when the resolution is high; however, under-sampling the mesh leads to a systematic underestimation of the stiffness. These results align well with our expectations: coarse meshes tend to present a stiffer elastic behavior than fine meshes. Accordingly, the filter 'softens' the Young's modulus of coarse meshes, in order to fit the observations.

Two important facts are highlighted by this simple experiment: first, in order to obtain realistic deformations with undersampled meshes, it is necessary to tune the Young's modulus, to compensate the discretization-induced stiffening. More generally, these results indicate that in the case of under-sampled meshes, the process of parameter identification does not give access to *true* parameter values of the elastic object. Rather, the estimated parameters are *effective* values 'chosen' by the filter to optimally match between the given, inaccurate model, and the observations.

3.1.2 Deformation Strength and Direction Dependence

The results reported in the last paragraph suggest that it is possible to offset the discretization effects by carefully selecting the model parameters. However, another series of experiments reported in Fig. 3 shows that things are not so simple. Here again we estimate the Young's modulus using a Kalman filter, but this time the applied pressure is varied for a fixed discretization. Figure 3a. shows the estimated stiffness as a function of time, in a quasi-static process, for stretching and/or bending deformations in different directions. Comparing to the ground-truth value of 5000 kPa, we see that the value of Young's modulus is not constant. The 'local' stiffness strongly depends on the direction and strength of the deformation. To highlight the spatial dependencies, the same data is presented as a heat map in Fig. 3b. The positions of the data points correspond to the positions of one node of the beam, at different stages of the deformation. The color of each point represents the local value of the estimated stiffness. These results have an important implication in our application context: when using a

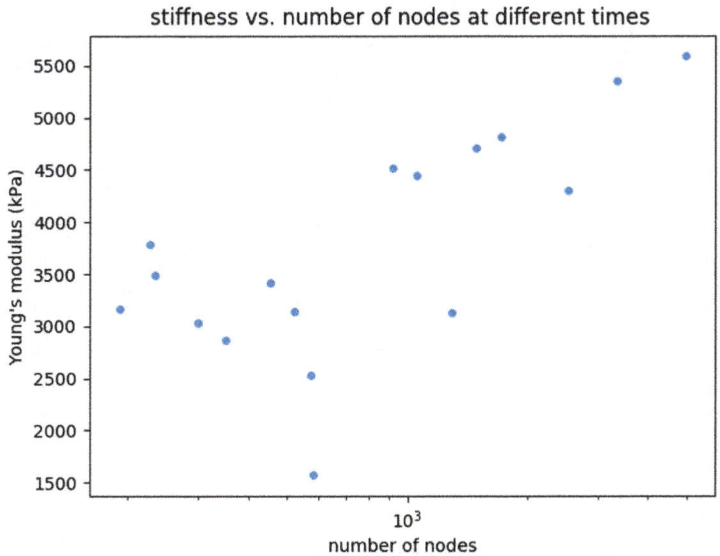

Fig. 2. Young's modulus of an homogeneous beam as a function of mesh resolution, as estimated by data assimilation. The values are underestimated for coarse meshes and approach the ground-truth value at high resolutions

coarse mesh in a real-time medical simulation, estimating the patient-specific parameters at the beginning of the intervention may not be sufficient. Parameters that fit the observations within a certain range of deformations may cause important errors when used to simulate different manipulations.

3.2 Shape Error

We have shown that a parameter identification procedure in coarse meshes gives rise to values that significantly differ from ground-truth values. Moreover, the estimated parameters not only depend on the mesh resolution but also on the nature of the deformation. This results raise a fundamental question: is there a sense in estimating parameters that are not constant but vary as a function of the system state? Does this heavy procedure contribute to the quality of the simulations? In order to answer this question, we compare in this section the simulation predictions with and without data assimilation. Using a regular grid we measure the shape error of each deformed system, the shape error being defined as the sum of the euclidian distances of the grid's nodes from the nodes of the reference system.

Figure 3 sketches the shape error for different levels of discretization. Red points represent the shape errors for the deformations obtained by direct simulations, using the ground-truth value of the Young's modulus. Green points correspond to the shapes attained by data assimilation. Apart from two system

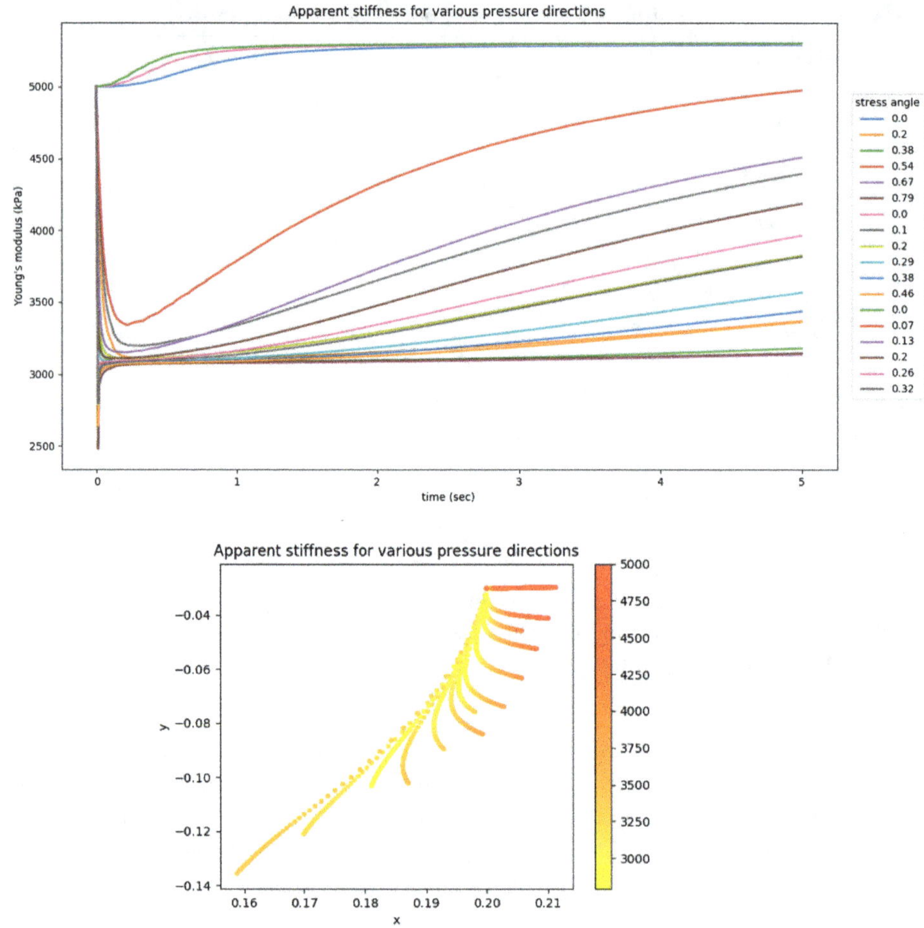

Fig. 3. Effective stiffness dependence on deformation direction and strength. (a) Effective stiffness as a function of time in a quasi-static process where pressure strength is regularly increased. Different colors correspond to different stress directions. (b) Same data represented as a heat map in the deformation space. All data points correspond to data assimilation estimations obtained for a coarse mesh of 510 nodes. Ground-truth is simulated with a fine mesh of 11394 nodes.

where the filtering process did not converge, these results clearly demonstrate the efficiency of the data assimilation process in reducing the shape errors (Fig. 4).

4 Discussion

In this paper, we studied the effect of mesh resolution on the effective stiffness of an elastic solid, in the context of real-time simulation and parameter estimation. We have shown that the estimated stiffness depends on the discretization level

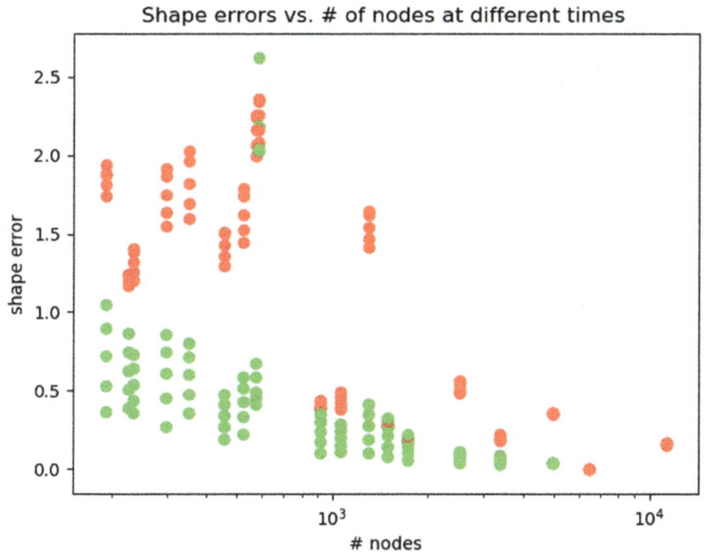

Fig. 4. Shape error as a function of mesh resolution (number of nodes) for direct (red) and data assimilation (green) simulations. The error decreases with resolution level. In coarse meshes errors are significantly reduced by data assimilation.

but also on the deformation. We also demonstrated that Kalman filters yield smaller shape errors than direct simulations parameterized with ground-truth values.

This means, among other things, that when insufficiently fine meshes are used, the estimated parameters should not be considered as the true parameter values of the organ or tissue, but instead as effective parameters that allow retrieving a shape that fits the observations. The values which are estimated can be seen as 'local', in the sense that they depend on the prescribed deformation. If the simulation is performed in real-time, and an efficient Kalman filter is used (such as a Reduced Order Unscented Kalman Filter), the filter might be able to continuously estimate a new parameter set that will allow the model to fit the data. However, if observations were to disappear then the model will start drifting from the real shape.

A clear strategy regarding parameter estimation in the interventional context arises from these findings. First, parameters should not be determined once and then used by the simulation (as it would be the case with an optimization method) but rather estimated constantly along the intervention. Secondly, a special attention should be given to the convergence rate of the identification. If in principle, any type of adaptive filter may allow for local parameter identification, in the case of interventional therapy, where little data is available at each step of the deformation, slow convergence rates may highly deteriorate the simulated prediction. Accordingly, despite their computational cost, Bayesian filters present a clear advantage w.r.t classical ones due to their highly efficient convergence.

References

Anna, M., Carton, A.-K., Muller, S., Payan, Y.: Breast biomechanical modeling for compression optimization in digital breast tomosynthesis. In: Gefen, A., Weihs, D., (eds.) Computer Methods in Biomechanics and Biomedical Engineering. Lecture Notes in Bioengineering, pp. 29–35. Springer (2018)

Aster, R.C., Borchers, B., Thurber, C.H.: Parameter Estimation and Inverse Problems, 2nd edn. Academic Press, Cambridge (2013)

Avril, S., Evans, S.: Material Parameter Identification and Inverse Problems in Soft Tissue Biomechanics, vol. 573. Springer, Heidelberg (2017)

Bialecki, R.A., Kassab, A.J., Fic, A.: Proper orthogonal decomposition and modal analysis for acceleration of transient fem thermal analysis. Int. J. Numer. Methods Eng. **62**, 774–797 (2005)

Chen, D., Levin, D.I.W., Sueda, S., Matusik, W.: Data-driven finite elements for geometry and material design. ACM Trans. Graph. **34**(4), 74:1–74:10 (2015)

Collins, J.A., Weis, J.A., Heiselman, J.S., Clements, L.W., Simpson, A.L., Jarnagin, W.R., Miga, M.I., et al.: Improving registration robustness for image-guided liver surgery in a novel human-to-phantom data framework. IEEE Trans. Med. Imaging **36**(7), 1502–1510 (2017)

Han, L., Hipwell, J.H., Tanner, C., Taylor, Z., Mertzanidou, T., Cardoso, J., Ourselin, S., Hawkes, D.J.: Development of patient-specific biomechanical models for predicting large breast deformation. Phys. Med. Biol. **57**(2), 455–472 (2011)

Haouchine, N., Cotin, S., Peterlik, I., Dequidt, J., Lopez, M.S., Kerrien, E., Berger, M.-O.: Impact of soft tissue heterogeneity on augmented reality for liver surgery. IEEE Trans. Visual. Comput. Graph. **21**(5), 584–597 (2015)

Heiselman, J.S., Clements, L.W., Collins, J.A., Weis, J.A., Simpson, A.L., Geevarghese, S.K., Kingham, T.P., Jarnagin, W.R., Miga, M.I.: Characterization and correction of intraoperative soft tissue deformation in image-guided laparoscopic liver surgery. J. Med. Imaging **5**(2), 021203 (2017)

Marchesseau, S., Heimann, T., Chatelin, S., Willinger, R., Delingette, H.: Multiplicative Jacobian energy decomposition method for fast porous visco-hyperelastic soft tissue model. In: International Conference on Medical Image Computing and Computer-Assisted Intervention, pp. 235–242. Springer (2010)

Moireau, P., Chapelle, D.: Reduced-order unscented Kalman filtering with application to parameter identification in large-dimensional systems. Control Optim. Calc. Var. **17**(2), 380–405 (2011)

Niroomandi, S., González, D., Alfaro, I., Bordeu, F., Leygue, A., Cueto, E., Chinesta, F.: Real-time simulation of biological soft tissues: a PGD approach. Int. J. Numer. Methods Biomed. Eng. **29**(5), 586–600 (2013)

Ophir, J., Céspedes, I., Ponnekanti, H., Yazdi, Y., Li, X.: Elastography: a quantitative method for imaging the elasticity of biological tissues. Ultrason. Imaging **13**(2), 111–134 (1991)

Peters, T., Cleary, K.: Image-Guided Interventions: Technology and Applications. Springer, Heidelberg (2008)

Rifai, K., Cornberg, J., Mederacke, I., Bahr, M.J., Wedemeyer, H., Malinski, P., Bantel, H., Boozari, B., Potthoff, A., Manns, M.P., et al.: Clinical feasibility of liver elastography by acoustic radiation force impulse imaging (ARFI). Dig. Liver Dis. **43**(6), 491–497 (2011)

Suwelack, S., Röhl, S., Bodenstedt, S., Reichard, D., Dillmann, R., dos Santos, T., Maier-Hein, L., Wagner, M., Wünscher, J., Kenngott, H., et al.: Physics-based shape matching for intraoperative image guidance. Med. phys. **41**(11), 111901 (2014)

Wittek, A., Hawkins, T., Miller, K.: On the unimportance of constitutive models in computing brain deformation for image-guided surgery. Biomech. Model. Mechanobiol. **8**(1), 77–84 (2009)

Zhang, J., Zhong, Y., Gu, C.: Deformable models for surgical simulation: a survey. IEEE Rev. Biomed. Eng. **11**, 143–164 (2018)

Local Stiffness Estimation of the Human Eardrum via the Virtual Fields Method

Felipe Pires[1,2](✉), Stéphane Avril[3], Julio Cordioli[4], Steve Vanlanduit[1,2], and Joris Dirckx[1]

[1] Universiteit Antwerpen, Antwerp, Belgium
felipe.pires@uantwerpen.be
[2] Vrije Universiteit Brussel, Brussels, Belgium
[3] Mines Saint-Étienne, Université Lyon, Lyon, France
[4] Federal University of Santa Catarina, Florianópolis, Brazil

Abstract. This work proposes a method from which a heterogeneous and isotropic stiffness can be estimated from human eardrums. Under normal hearing conditions, neither the bending nor the membrane strains can be neglected on such a sample. Due to this, the Virtual Fields Method was adapted by invoking specific assumptions related to the Kirchhoff Love plate model, so both strain types could be taken into account during the inverse method. Apart from this contribution, specific virtual fields were designed, so the virtual work could be computed on enclosed domains. By neglecting spatial stiffness variations within the prescribed regions, such an estimation could be performed locally. Those attempts were validated on the basis of a Finite-element model and the identified stiffness distribution was in agreement with the reference one. Due to the quality of the preliminary results, the proposed method has shown itself to be a reliable tool to identify heterogeneous material properties on shell-like structures.

Keywords: Virtual Fields Method · Finite-elements · Kirchhoff Love plate model · Eardrum

1 Introduction

The Virtual Fields Method (VFM) is widely documented in literature [1,2] and it is a method, with which material properties are estimated on the basis of full field measurements and proper virtual fields (VF). The obtained results depend not just on the quality of measurements but also on the assumptions being invoked for the studied sample. To the authors' knowledge, the works presented so far assume that a studied sample behaves in accordance with either the membrane assumption [1] or by considering that the in-plane strains are negligible when compared to the bending ones [2].

These assumptions do not describe the behavior of the human eardrum with accuracy, when it vibrates under normal hearing conditions. In such cases, it

G. A. Ateshian et al. (Eds.): CMBBE 2019, LNCVB 36, pp. 248–255, 2020.
https://doi.org/10.1007/978-3-030-43195-2_20

is suggested that both the membrane and bending strains should be taken into account. Moreover, the configuration of the eardrum's fibers suggests that the stiffness is spatially heterogeneous and, therefore, the parameter identification should be computed locally, if the VFM is to be applied.

The purpose of the this work is to present a variation of the VFM, with which the stiffness estimation is made feasible for the case of human eardrums. Here, by combining an assumption that considers both the bending and membrane strains, and by developing particular VFs, it is feasible to perform the heterogeneous material identification. The results obtained through a validation have shown that the invoked assumptions are sufficient for the constitutive parameter distribution to be assessed with accuracy and that the adapted VFM is a reliable tool to study shells with a complex geometry.

2 Theory

2.1 Statement of the Principle of Virtual Work

The strategy to solve the weak form of equilibrium related to eardrums depends on the conditions being imposed during measurements. Among other measuring techniques, the most promising to study such a sample are the Digital Stroboscopic Holography [3] and Digital Image Correlation [4], which permit a dense point cloud of displacement fields to be recorded. Such measurements can also be performed, while the eardrum is excited under harmonic and uniform pressures. If this is the case, the virtual work related to the acceleration and the external forces should be taken into account while developing the principle of virtual work (PVW). Lastly, it was mentioned that both strain types are significant in the analyzed sample and this requires the displacements and strains to vary along the through-thickness coordinate.

By taking all these complexities into account while neglecting the body forces and by defining a parameter J_e as the square of the PVW's residuals that should be minimized, it follows that

$$J_e = \sum_k \left(\underbrace{- \int_V \boldsymbol{\sigma}_k^{\zeta} : \boldsymbol{\varepsilon}_e^{\zeta} dV}_{\text{Internal forces}} + \underbrace{P_k \int_S \mathbf{u}_e^{\zeta} \cdot \mathbf{e}_3 dS}_{\text{External forces}} + \underbrace{\rho\omega_k^2 \int_V \mathbf{u}_k^{\zeta} \cdot \mathbf{u}_e^{\zeta} dV}_{\text{Acceleration}} \right)^2 \tag{1}$$

where k and e label the numbering of an experimental data and a VF, respectively; the superscript ζ is the through-thickness coordinate, which indicates that the refereed term varies with respect to (w.r.t.) ζ, ρ is the density, ω is the angular frequency, at which the middle ear is being excited, $\boldsymbol{\sigma}$ is the stress tensor, $\boldsymbol{\varepsilon}$ is the strain tensor, $\mathbf{u}$ is the displacement vector, and P is the excitation pressure.

The proposed minimization problem shown in Eq. (1) is meant to estimate local stiffness values. To achieve this goal, it was decided to create a local VF

$\mathbf{u}_e^\zeta$, whose contribution would just be meaningful in a predefined region of the eardrum. Moreover, as it is explained in Sect. 2.2, the eardrum will be represented with a Finite Element (FE) mesh, whose individual elements are flat triangles.

By taking advantage of this simplification, the number of VFs "e" will be equal to the number of FE elements. Hence, each VF is developed and meant to estimate a local stiffness value for a single element. This approach requires the minimization problem [Eq. (1)] to be solved as many times as the total number of elements in the mesh.

2.2 Assumptions

The assumptions that should properly describe the eardrum's behavior should take both the membrane and bending strains into account, since non of those strain types are dominant. Moreover, the recording of experimental data of such samples are just accessible by measuring their outer-surfaces, i.e., at $\zeta = h/2$ [3].

By taking these features into account and by considering the eardrum to be a thin structure, the Kirchhoff-Love plate model is used to describe its shape. Such an assumption allows the sample's geometry to be describe as an FE assembly made of triangular and flat elements [5], each of which obey the Kirchhoff plate theory. Therefore, by coupling the rotations with the displacements, it holds that

$$\mathbf{u}^\zeta = \mathbf{u}^0 + \zeta\boldsymbol{\theta}, \quad \mathbf{u}^0 = \begin{Bmatrix} u_1^0 \\ u_2^0 \\ u_3^0 \end{Bmatrix}, \quad \boldsymbol{\theta} = \begin{Bmatrix} \theta_1 \\ \theta_2 \\ 0 \end{Bmatrix}, \quad -\frac{h}{2} \leq \zeta \leq \frac{h}{2}, \tag{2}$$

where h is the sample's thickness, the superscript "0" refers to terms of the mid-surface, $\boldsymbol{\theta}$ is the rotation vector, and the subscripts "1", "2" and "3" are indications that the referred components are aligned with a set of local unit vectors for each element of the mesh ($\mathbf{e}_1, \mathbf{e}_2, \mathbf{e}_3$). By invoking the infinitesimal strain theory, the components of the strain tensor can be interpreted as a combination of a membrane and a bending contribution. They can be assessed with

$$\boldsymbol{\varepsilon}^\zeta = \boldsymbol{\varepsilon}^0 + \zeta\boldsymbol{\chi}, \quad \boldsymbol{\varepsilon}^0 = \frac{1}{2}\left(\nabla\mathbf{u}^0 + \mathbf{u}^0\nabla\right), \quad \boldsymbol{\chi} = \frac{1}{2}\left(\nabla\boldsymbol{\theta} + \boldsymbol{\theta}\nabla\right), \tag{3}$$

being $\boldsymbol{\chi}$, the bending strain tensor; and $\boldsymbol{\varepsilon}^0$, the membrane one. Since the Kirchhoff plate theory is invoked for the elements, all terms of the strain tensors $\boldsymbol{\varepsilon}^0$ and $\boldsymbol{\chi}$, which are related to the direction $\mathbf{e}_3$ can be neglected. Therefore, it holds that $\varepsilon_{i3}^0 = \chi_{i3} = 0$.

Lastly, by assuming a linear and isotropic elasticity, the relation between the relevant stresses and strains are expressed as follows in Voigt notation:

$$\boldsymbol{\sigma}^\zeta \overset{\text{def.}}{=} \begin{Bmatrix} \sigma_{11}^\zeta \\ \sigma_{22}^\zeta \\ \sigma_{12}^\zeta \end{Bmatrix} = C(1+\eta\mathrm{j}) \begin{bmatrix} 1 & \nu & 0 \\ \nu & 1 & 0 \\ 0 & 0 & \frac{1-\nu}{2} \end{bmatrix} \cdot \begin{Bmatrix} \varepsilon_{11}^\zeta \\ \varepsilon_{22}^\zeta \\ 2\varepsilon_{12}^\zeta \end{Bmatrix}, \tag{4}$$

being η the internal loss factor, ν the Poisson's coefficient and j is the complex unit. If a VF is available and by considering C to be the only unknown in Eqs. (4), (1) can be integrated w.r.t. ζ and the variable C can be estimated through a pseudo-inversion.

2.3 Transformation of Coordinates

It is the aim of the current work to estimate stiffness distributions on the basis of experimental data and it is often the case that those measurements are not aligned with a set of local coordinates $(\mathbf{e}_1, \mathbf{e}_2, \mathbf{e}_3)$ of the shell-like structure. Due to this, it is assumed that the "k" measurements are acquired in a predefined Cartesian coordinate system and that the only available data are the ones related to the outer-surface ($\zeta = h/2$). These displacements are labeled here as $\mathbf{U}_k^{h/2}$ and they are stored at the nodes of the FE mesh.

The rotations referenced with the same coordinate system are labeled here as $\mathbf{\Theta}_k$ and they can be computed via the normal directions of the nodes. By subtracting these normals while the surface is deformed ($\bar{\mathbf{n}}$) with the same ones from the reference geometry ($\mathbf{n}$), the rotation $\mathbf{\Theta}$ can be calculated. The components of those vectors and the ones related to the displacement at the mid-surface ($\mathbf{U}^0$) are described here as

$$\mathbf{U}^{h/2} = \begin{Bmatrix} u_x^{h/2} \\ u_y^{h/2} \\ u_z^{h/2} \end{Bmatrix}, \quad \mathbf{U}^0 = \begin{Bmatrix} u_x^0 \\ u_y^0 \\ u_z^0 \end{Bmatrix}, \quad \mathbf{\Theta} = \bar{\mathbf{n}} - \mathbf{n} = \begin{Bmatrix} \theta_x \\ \theta_y \\ \theta_z \end{Bmatrix}, \tag{5}$$

being the subscripts x, y and z indications that the vectors' components are referenced w.r.t. the Cartesian coordinate system.

A transformation of coordinates on the terms in Eq. (5) should take place, so the these variables can be aligned with the local coordinates $(\mathbf{e}_1, \mathbf{e}_2, \mathbf{e}_3)$. To that end, a transformation tensor is used, which can be implemented through the following equations:

$$\mathbf{u}^{h/2} = \mathbf{T} \cdot \mathbf{U}^{h/2}, \quad \mathbf{u}^0 = \mathbf{T} \cdot \mathbf{U}^0, \quad \boldsymbol{\theta} = \mathbf{T} \cdot \mathbf{\Theta}, \quad \mathbf{T} = \left[\mathbf{e}_1 \; \mathbf{e}_2 \; \mathbf{e}_3\right]^T. \tag{6}$$

2.4 Development of the Virtual Fields

The remaining variables in Eq. (1) that must be defined here are related to the VFs "e". It can be seen from Eq. (4) that the stresses are defined w.r.t. an uniform constitutive parameter C. However, the main task of this work is to estimate a heterogeneous stiffness over the eardrum's surface. This issue is overcome by developing a VF that activates the virtual work at a certain region of the sample's surface. The variations of the stiffness is then considered to be negligible within the prescribed domain.

A field that attends such a demand is

$$F_e(\xi_1, \xi_2) = \exp\left(-\frac{(\xi_1 - a_e)^2 + (\xi_2 - b_e)^2}{R^2}\right), \tag{7}$$

where ξ_1 and ξ_2 are newly defined parametric coordinates, the variables a_e and b_e are the barycentric coordinates of each FE element e in the parametric domain and R is a tuning parameter that modifies the radius of the field F_e. After computing Eq. (7) for every element "e", a transformation of domains takes place and all fields F_e are represented with the original coordinates (x, y, z):

$$F_e(\xi_1, \xi_2) \rightarrow F_e(x, y, z). \tag{8}$$

Lastly, it is defined that the VF is oriented towards the normal directions of the mesh's nodes $\mathbf{n}$. Hence, by combining the field F_e with $\mathbf{n}$, it follows that

$$\mathbf{U}_e^0 = \begin{Bmatrix} (f_e)_x \\ (f_e)_y \\ (f_e)_z \end{Bmatrix} = F_e(x, y, z) \cdot \mathbf{n}, \tag{9}$$

where $\mathbf{n}$ is an unit vector, which is computed from the reference configuration of the mesh, and the $(f_e)_x$, $(f_e)_y$ and $(f_e)_z$ indicate the components of the VF $\mathbf{U}_e^0$ in Cartesian coordinates.

2.5 Methodology

Sections 2.3 and 2.4 show that the input data for this application are the experimental displacements "k" and several VFs "e", i.e., $\mathbf{U}_k^{h/2}$ and $\mathbf{U}_e^0$. However, all variables shown in Eq. (1) require both displacements and related strains to be represented w.r.t. a local triad of unit vectors $(\mathbf{e}_1, \mathbf{e}_2, \mathbf{e}_3)$. The required data processing to gather all necessary variables shown in Eq. (1) are described in Table 1.

Table 1. Steps towards the assessment of the required variables in Eq. (1). Step 3 is only applicable to the experimental data "k", while it is skipped when the virtual fields "e" are processed.

Steps	Description
1	Computation of $\boldsymbol{\Theta}$ [Eq. (5)]
2	Transformation of coordinates [Eq. (6)]
3	Assessment of $\mathbf{u}^0$ [Eq. (2)]
4	Computation of the strains [Eq. (3)]

After following the steps shown in that table, the variables related to all "k" data associated with the measurements $(P_k, \omega_k, \rho, \mathbf{u}_k^\zeta, \boldsymbol{\varepsilon}_k^\zeta)$ and all "e" VFs $(\mathbf{u}_e^\zeta, \boldsymbol{\varepsilon}_e^\zeta)$ are available. With all the prerequisite variables at hand, the global least-squares minimization problem [Eq. (1)] can be solved for every VF.

3 Results

The validation of the proposed approach was carried out on a FE model of a human middle ear, and all steps described in Sect. 2.5 were applied on the eardrum of this model. It was assumed that the radial fibers of this membrane were aligned with the 1st principal curvature direction of the geometry. This allowed the local coordinate $\mathbf{e}_1$ to be estimated and its directions are represented in Fig. 1. The stiffness of the eardrum was set as isotropic and with a spatial distribution that varied along the radial fibers. Figure 1 also shows the set stiffness distribution C [MPa] on the modeled eardrum. An uniform and harmonic excitation P was applied on the eardrum's surface at several "k" frequencies. After the simulation's completion, the displacement fields related to the shell's outer-surface (i.e., $\mathbf{U}_k^{h/2}$) were exported for further processing.

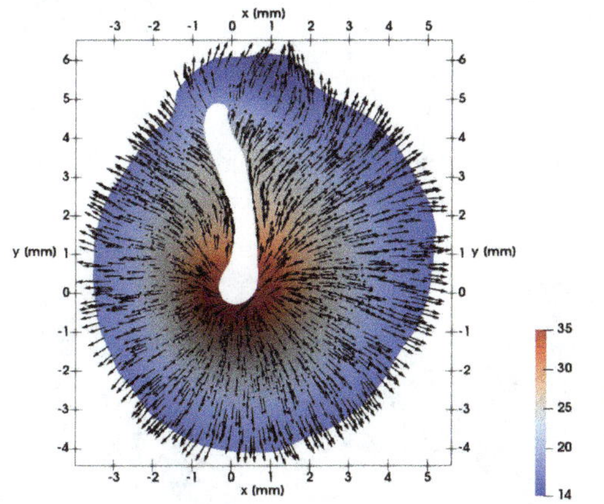

Fig. 1. Representation of an human eardrum, which is a part of a FE model of the middle ear. The vectors shown on the geometry's surface are the local coordinates $\mathbf{e}_1$ and are aligned with the 1st principal curvature of the sample. The fields on the same figure correspond to the C parameter [present in Eq. (4)]. The free-edges related to the outer-circumference of the geometry are connected to the ear canal. The inner circumference is the region where the eardrum is connected to the malleus.

The first step to create the VFs is related to the transformation of domains, i.e., to the link between the spatial coordinates (x, y, z) and the parametric coordinates (ξ_1, ξ_2) [Eq. (7)]. The new domain was computed via a conformal transformation and it is displayed in Fig. 2(a) as a 2D grid over the original surface. By having the parametric coordinates at hand, the relevant VFs could be developed, being 1 VF assigned for each element "e" of the mesh. It must be noted that the proposed VFs are not applicable for elements that are too close to either the inner or outer free-edges of the sample (see Fig. 1), since these fields are not considered to be kinematically admissible at those regions. Due to this, it was decided to tune the parameter R to be 0.4 mm: a number that is small enough and permits the stiffness identification to be performed closer to the boundaries. At the end of this process, all displacements $\mathbf{U}_e^0$ [Eq. (9)] are available and one among all virtual displacements is shown in Fig. 2(b) as an example.

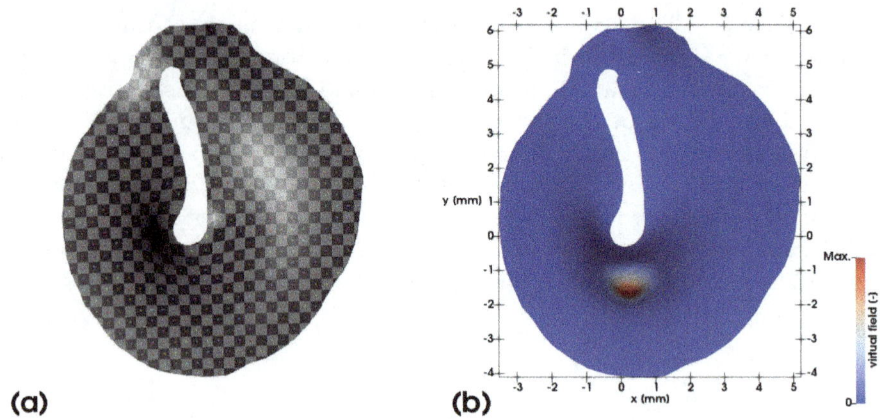

Fig. 2. (a): Representation of the parametric coordinates (ξ_1, ξ_2) as a deformed grid over the eardrum's surface. (b): Example of a single virtual field $\mathbf{U}_e^0$ that is computed via the Eqs. (7–9).

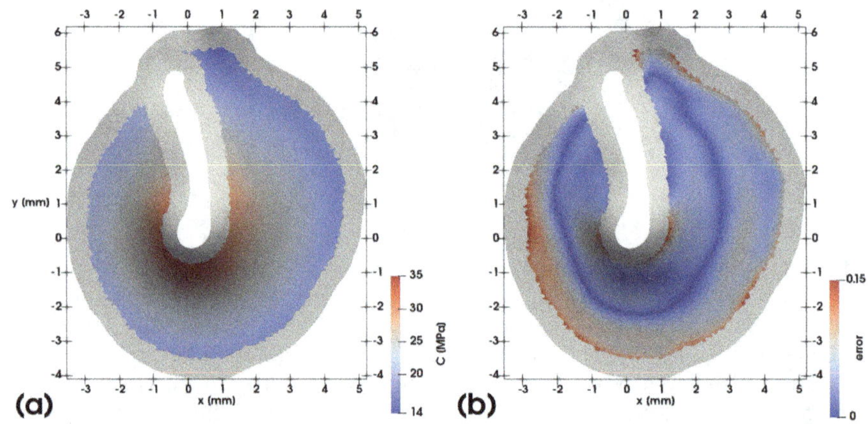

Fig. 3. (a): Estimated distribution C via the Eq. (1). (b): Error of the estimated field in (a) when compared to the reference (Fig. 1). The elements, where the C or the "error" fields are absent, are at least 0.6 mm away from the inner or outer boundaries and were, therefore, not evaluated.

Since all vectors $\mathbf{U}_k^{h/2}$ and $\mathbf{U}_e^0$ are at hand, all steps shown in Table 1 can be computed for both data types and the pseudo-inversion relative to Eq. (1) can be performed. Figure 3(a) presents the estimated C distribution that was computed via the Eq. (1), while Fig. 3(b) displays the local errors of this estimation when compared to the reference shown in Fig. 1.

4 Conclusions

By comparing Fig. 3(a) with Fig. 1, it can be seen that the estimated C is in agreement with the reference distribution. By analyzing Fig. 3(b), it can be noted that the errors are on average lower than 4.8% and that they have their lowest values in regions that are relatively distant from the outer and inner free-edges. This is due to the kinematic admissibility that is close to be satisfied on those very regions.

By considering all limitations involved from what can be measured and due to the invoked assumptions related to Kirchhoff Love plate model [5], the method was considered to be validated since the estimated C is qualitatively similar to the reference. Due to these preliminary results, the proposed method has shown itself to be reliable to estimate heterogeneous and isotropic material properties from shell-like structures.

References

1. Farzaneh, S., Trabelsi, O., Avril, S.: Inverse identification of local stiffness across ascending thoracic aortic aneurysms. Biomech. Model. Mechanobiol. **18**(1), 137–153 (2019). https://doi.org/10.1007/s10237-018-1073-0
2. Giraudeau, A., Pierron, F.: Identification of stiffness and damping properties of thin isotropic vibrating plates using the virtual fields method: theory and simulations. J. Sound Vib. **284**(3–5), 757–781 (2005). https://doi.org/10.1016/j.jsv.2004.07.009
3. De Greef, D., Soons, J., Dirckx, J.J.: Digital stroboscopic holography setup for deformation measurement at both quasi-static and acoustic frequencies. Int. J. Optomechatronics **8**(4), 275–291 (2014). https://doi.org/10.1080/15599612.2014.942928
4. Pires, F., Avril, S., Vanlanduit, S., Dirckx, J.: Structural intensity assessment on shells via a finite element approximation. J. Acoust. Soc. Am. **145**(1), 312–326 (2019). https://doi.org/10.1121/1.5087564
5. Chapelle, D., Bathe, K.J.: The Finite Element Analysis of Shells-Fundamentals. Springer, New York (2010)

Limits of Dynamic Postural Stability with a Segmented Foot Model

Carlotta Mummolo[1(✉)] and Giulia Vicentini[2]

[1] New Jersey Institute of Technology, Newark, NJ 07102, USA
cmummolo@njit.edu

[2] Sapienza University of Rome, 00185 Rome, Italy

Abstract. The human foot structure withstands complex interactions with the ground during balancing and locomotion tasks, leading to agile and stable movements. Despite this, several postural stability analysis methods are limited by the use of overly simplified foot structure and contact model. The proposed foot model has a toe link that enables multiple contact modes with the ground. A contact mode detection algorithm is created to map foot angles into discrete contact modes and associated constraints. Two surface functions provide continuous values for lower and upper center of pressure limits, as the foot traverses the non-specified sequence of discrete contact modes. The proposed foot and contact model are integrated in a planar mechanism for the analysis of whole-body postural stability in sagittal plane. Constrained optimization generates balancing trajectories and contact mode sequence for the legged system recovering from extreme velocity perturbations in the anterior-posterior direction. The entire set of initial perturbations that can be sustained in the anterior-posterior direction is numerically evaluated in the cases of flat and segmented feet, to investigate the effects of multimodal foot-ground interaction on the dynamic stability limits for human posture. By exploiting the segmented foot model and the multimodal contact sequence arising from the optimization solutions, the legged system can recover from greater velocity perturbations in the sagittal plane, as compared to the case of flat feet.

Keywords: Postural stability · Dynamic balance · Segmented foot · Foot rocking · Limits of recovery

1 Introduction

The human foot is a complex multi-segment structure that realizes various contact sequences as it interacts with the ground during walking, stepping, and push-recovery tasks. This rich interaction allows the human body to move with remarkable agility and stability [1, 2]. For example, during locomotion, the heel detaches from the ground to generate the push-off phase, critical for energy optimality, and heel-to-toe rocking is the basis of energy efficient gait [3, 4]. Similarly, various studies have investigated the important role that the foot structure plays during standing [5, 6]. During balancing tasks, the recovery action from a perturbed posture could benefit from intermittent heel and toe lifting (also called heel-to-toe rocking), as compared to the flat-foot fixed contact configuration. Only few studies have investigated this in human experiments [7] and biped

G. A. Ateshian et al. (Eds.): CMBBE 2019, LNCVB 36, pp. 256–270, 2020.
https://doi.org/10.1007/978-3-030-43195-2_21

robot design [8, 9]. While numerous work has focused on the effect of the toe link in human [10, 11] and humanoid walking [12–14], almost no work has been done to rigorously isolate and quantify the specific contribution of a complex foot model on the limits of balance stability for biped systems.

In recent work, the balancing capabilities of a legged system have been quantified by the set of maximum allowable velocity perturbations of the system's center of mass (COM) along a given direction [15]. If a balancing trajectory exists to bring the COM from a perturbed state (position and velocity) to a rest state (upright static posture), without altering the system's stance configuration, then the perturbed state is defined as balanced. Balancing trajectories that bring a legged system's COM from a series of perturbed states to a rest state can be generated to quantify the maximum allowable COM velocity perturbation along a given direction as a measure of the system's dynamic balance stability limits. To quantify the effects of a multimodal of foot-ground contact interaction on the balance ability of a legged system, balancing trajectories are generated in this work for a whole-body posture model with a segmented foot, as it recovers from extreme velocity perturbations in the anterior-posterior direction. The proposed unified approach generates balancing trajectories of a human posture model recovering from velocity perturbations in the anterior-posterior direction, while exploiting the multimodal nature of the foot-ground contact interaction.

2 Human Posture Model with Segmented Feet

A model of a human body in standing posture is described in the sagittal plane, including lower and upper body segments and two segmented stance feet.

2.1 Standing Posture Model in Sagittal Plane

The human body in sagittal plane is modeled with a six-link planar mechanism with five revolute joints (Fig. 1). Each link rigid motion is described by the attached link local frames, according to the Denavit-Hartenberg convention [16]. The z-axis of each local frame points out of the (X, Y) plane and represents the joint axis of rotation. As opposed to the traditional fixed base structure, the mechanism has a floating base structure that allows the system to freely move in the (X, Y) sagittal plane and, most importantly, the foot to make various contacts with the ground, without any specified foot point fixed to the ground. The thigh local frame, located at the hip joint, is chosen as the system's floating base and is connected to the origin O of the global reference frame $\{X, Y\}$ through three degrees-of-freedom (DOF) representing the unconstrained in-plane translations and rotation of the local frame. Therefore, the total number of DOF describing the floating-base while-body dynamics of the system is eight (three global DOF and five revolute joint DOF).

The upper body branch includes the HAT (head-arms-torso) body segment connected to the hip by a torso joint and a massless pelvis segment (Fig. 1). In the HAT segment the arms' mass distribution is considered, but they are assumed fixed with the upper body. The lower body branch is an equivalent single-chain mechanism that describes the symmetric dynamics of left and right legs (leg symmetry is a reasonable

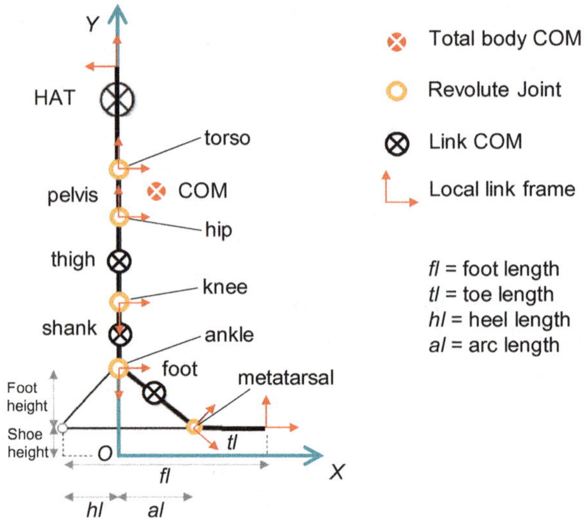

Fig. 1. Sagittal plane human posture model in double stance.

assumption for standing balancing tasks in the sagittal plane). Lower-body joints are modeled with a single joint for hip, knee, ankle, and toe motion, respectively, with rotation and actuation limits equivalent to those of two joints in parallel. Similarly, link mass for thigh, shank, and foot segment includes the values for both legs, while is assumed negligible for the toe link.

2.2 Segmented Foot Model

A two-link foot model is integrated with the whole-body mechanism for human standing posture. As for the legs, this foot structure represents the left/right symmetric feet motion. The foot has two rigid segments (a triangular element for the main foot body and a thin massless rod for the toe link) connected by the metatarsal joint and a rigid foot sole of thickness approximately equal to a shoe height ($sh = 0.03$ m) (Fig. 1).

The distributed ground reaction forces and moments are modeled through one equivalent resultant force applied at the system's time-varying center of pressure (COP), to mimic heel-to-toe ground pressure distribution. In a real segmented foot, the point of application of the COP would move from the foot to the toe segment, and vice versa, depending on the current contact configuration (herein called contact mode) with the ground and the corresponding ground reactions. Therefore, in a real system, the equivalent resultant force should be applied to a different rigid body at different times, when the COP results within either the foot or toe segment (Fig. 2). Switching the application of the resultant force between two rigid bodies would increase the numerical and computational model complexity, given that it would require additional variables to model the applied force and that, in this work, the desired sequence of contact modes is unspecified a priori.

For these reasons, the proposed foot model assumes that a single equivalent resultant force is always applied at the toe link, but at a variable COP position ${}^{5}\mathbf{P}_{COP}$ relative to the local toe link frame {5}, accurately calculated to simulate ground reactions at both the foot and toe segments. The foot model is perfectly equivalent to the real segmented system when the COP lies within the toe segment and the metatarsal is exerting negative torque (Fig. 2; right). When the real-system COP lies within the foot segment, the resultant force in the model is still applied at the toe link at the calculated ${}^{5}\mathbf{P}_{COP}$ (a point of the toe link, but now extending beyond the dimensions of the toe segment) and an equivalent effect on the foot is reproduced in the model by the reactions exchanged at the metatarsal joint (Fig. 2; left). Static equilibrium equations applied at foot and toe links show that the only approximation in the model occurs when the COP lies within the foot segment, due to an overestimation of the toe torque, which is zero in the real system. This only discrepancy has been considered by setting an arbitrarily large limit for positive metatarsal torque τ_{toe}, to avoid over constraining the COP position. This approximation will not affect the rest of whole-body dynamics; for instance, it is shown that, regardless of the COP position, the ankle joint torque τ_{ankle} is not affected by the model approximation (Fig. 2). In this proof, static equations and resultant normal force only have been considered, without loss of generality.

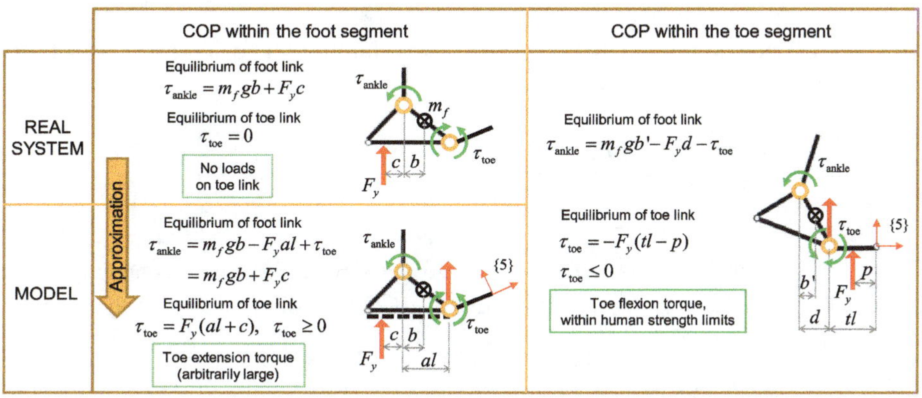

Fig. 2. Evaluation of the proposed foot model assumption, in which the resultant reaction force (with normal component F_y) is always applied at a point of the toe link and is transmitted to the foot link via the metatarsal joint. For the case of COP within the foot segment (of mass m_f), the point of application in the model belongs to an extension (dashed) of the toe link, causing the only approximation of the model, i.e., the overestimation of toe extension torque.

The model symmetry implies that left and right reaction forces are equal to half of the equivalent resultant force and that the toe and ankle strengths (i.e., maximum torques) are doubled, representing two joints in parallel (as for the knee and hip joints).

The equivalent frictional and normal ground reactions F_x and F_y and their point of application are found from the legged system's centroidal dynamics:

$$F_x = m\ddot{x}_{COM} \tag{1}$$

$$F_y = m(\ddot{y}_{COM} + g) \tag{2}$$

$$({}^O p_x - x_{COM})F_y - ({}^O p_y - y_{COM})F_x = \dot{H}_z \tag{3}$$

where m is the whole-body mass, g is the gravitational constant, x_{COM} and y_{COM} and their second time derivatives are the global coordinates and accelerations, respectively, of the system's COM, ${}^O\mathbf{P}_{COP} = [{}^O p_x \; {}^O p_y]^T$ is the point of application of the resultant force as observed by the global reference frame, and $\dot{H}_z$ is the rate of change of angular momentum generated by the limbs about the COM.

Normal force is subject to the inequality constraints $F_y \geq 0$ and frictional force is subject to the friction cone constraint, with coefficient of static friction μ, implemented as follows:

$$\begin{cases} F_x - \mu F_y \leq 0 \\ F_x + \mu F_y \geq 0 \end{cases} \tag{4}$$

The COP position calculated from the above equilibrium equations relative to the global frame $\{X, Y\}$ results as follows:

$${}^O\mathbf{P}_{COP} = \begin{bmatrix} {}^O p_x \\ {}^O p_y \end{bmatrix} = \begin{bmatrix} \dfrac{\dot{H}_z - y_{COM} m\ddot{x}_{COM}}{m(\ddot{y}_{COM} + g)} + x_{COM} \\ 0 \end{bmatrix} \tag{5}$$

where ${}^O p_y = 0$ has been used to indicate ground level. For implementation of the proposed foot model, ${}^O\mathbf{P}_{COP}$ is mapped into the COP position relative to the local toe frame $\{5\}$, as follows:

$${}^5\mathbf{P}_{COP} = {}^O R_5^T ({}^O\mathbf{P}_{COP} - {}^O\mathbf{P}_5) \tag{6}$$

which represents the variable point of application of the resultant force at the toe link in any general contact mode. Position vector ${}^O\mathbf{P}_5$ and rotation matrix ${}^O R_5$ give the relative position and orientation, respectively, of frame $\{5\}$ relative to $\{X, Y\}$. With the proposed foot model, the only remaining variables with explicit dependency on the contact mode are the time-varying COP limits:

$${}^O LB(t) \leq {}^O p_x \leq {}^O UB(t) \tag{7}$$

which can be easily expressed relative to the global frame and imposed on ${}^O p_x$ and whose complete formulation for each contact mode is presented in the next section.

3 Multimodal Foot-Ground Contact Interaction

The proposed segmented foot model can establish multiple contact modes with the ground, which allows to generate various balancing motions that exploit heel-to-toe rocking.

3.1 Foot-Ground Contact Modes Description

With two links and three possible contact points, the foot has a total of eight possible contact modes (Fig. 3). The foot and toe angles, θ_1 and θ_2 respectively, are measured counterclockwise from the positive X-axis of the global frame and are used as variables to detect at all times the foot contact mode and corresponding global COP limits, i.e., the limits of the contact region. Mode (6) is excluded from the set of possible contact modes, since no-contact (i.e., flight) phases are not of interest in the postural stability analysis considered in this work. A pure tip-toe contact (mode (3)) is also excluded, as it is considered unnatural for average metatarsal toe strength and shoe stiffness. Mode 8 is also excluded because unnatural during standing tasks; in particular, any foot configuration for which $\theta_1 \geq 0$ and $\theta_2 \geq \theta_1$ is considered infeasible.

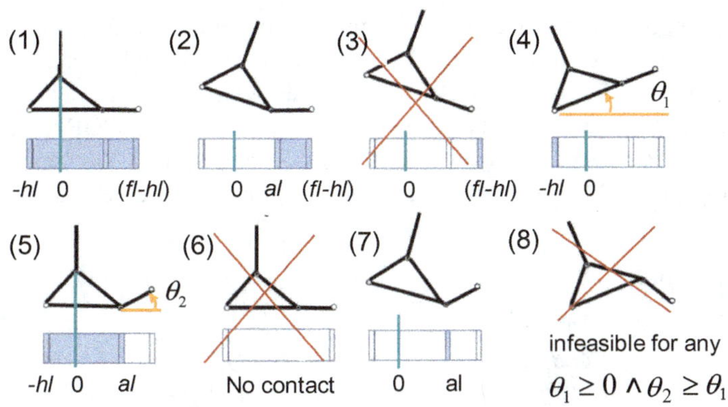

Fig. 3. Foot-ground contact modes, associated contact regions, and COP limits (shaded rectangles) expressed relative to the origin O of the global reference frame

A diagram of contact modes can be represented in the (θ_1, θ_2) plane, where solid borders indicate the discrete transitions between modes and the dashed border indicate the region of excluded (infeasible) contact modes (Fig. 4). The feasible range for the foot angle θ_1 is arbitrarily large, $[-\pi/2, \pi/2]$, while the feasible range for the toe angle θ_2 results within $\left[0, \pi/2 + \theta_{\text{toe,max}}\right]$, due to selected feasible modes and anatomical limits of metatarsal joint angle, $\theta_{\text{toe,max}}$. A small tolerance of $\theta_2 = \pm 0.01$ rad is considered for mode 2 and mode 1; in addition, a small range of $\theta_1 \subset [-0.012, 0.0628]$ (rad) allows some rotation of the skeletal foot structure during modes 1 and 5 and has been calculated from the data available on human heel pad deformation in loaded and unloaded conditions [17].

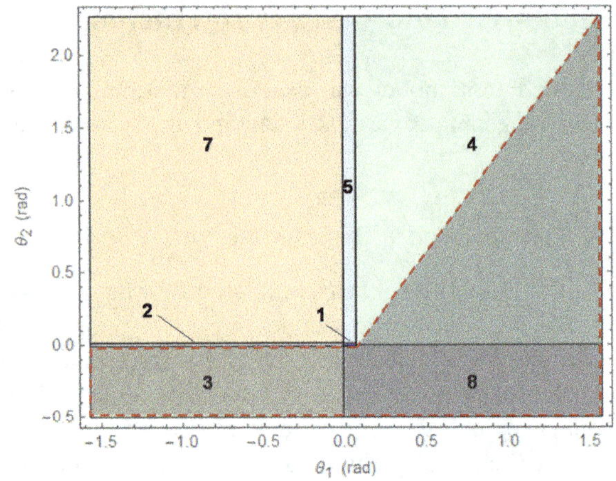

Fig. 4. Diagram of feasible contact modes with discrete transitions and angle tolerances. Infeasible contact modes are indicated by the dashed border enclosing the shaded gray area.

3.2 Continuous Model for Contact Modes Transition

Each feasible mode corresponds to different contact regions and, therefore, different feasible ranges for the COP position. The above diagram is used to establish a contact mode detection algorithm that maps at each time the foot and toe angles θ_1 and θ_2 into to corresponding contact modes and associated piece-wise constant COP lower and upper limits (Table 1). In modes 4 and 7, where the lower and upper COP limits coincide, a small relaxation ($\varepsilon = 0.01$ m) is used in the upper bound for numerical purposes.

Table 1. Scheme of contact modes and associated COP limits in the global frame

Mode	1	2	4	5	7
${}^{O}LB$	$-hl$	al	$-hl$	$-hl$	al
${}^{O}UB$	$fl - hl$	$fl - hl$	$-hl + \varepsilon$	al	$al + \varepsilon$

In order to implement the correct COP limits in a constrained optimization formulation, the transition between discrete modes is approximated by two surface functions, ${}^{O}LB = f^{LB}(\theta_1(t), \theta_2(t))$ and ${}^{O}UB = f^{UB}(\theta_1(t), \theta_2(t))$, that provide continuous values for COP lower and upper limits. This is especially critical in this work because the sequence of contact modes is unspecified a priori but is discovered by the optimization for the given balancing task. The two surface functions allow to formulate proper continuous and differentiable COP constraints not explicitly dependent on time, but dependent on the time-varying kinematic variables θ_1 and θ_2, whose time profile is found as a result of the numerical optimization algorithm. In this work, the COP limits are modeled as two-dimensional surface functions (Fig. 5) created by adding ellipsoidal volumes to the horizontal surfaces representing the desired piece-wise constant values (Table 1) for COP limits in the various regions of the contact mode diagram:

$$f^{LB}(\theta_1, \theta_2) = -hl + (\frac{al + hl}{2})(1 + \mathrm{Tanh}(\Omega^{LB}(\theta_1, \theta_2)/\delta)) \tag{8}$$

$$\begin{aligned} f^{UB}(\theta_1, \theta_2) = al + (\frac{fl - hl - al}{2})(1 + \mathrm{Tanh}(\Omega_1^{UB}(\theta_1, \theta_2)/\delta_1)) - \\ (\frac{hl + \varepsilon + al}{2})(1 + \mathrm{Tanh}(\Omega_2^{UB}(\theta_1, \theta_2)/\delta_2)) \end{aligned} \tag{9}$$

where Ω^{LB}, Ω_1^{UB}, Ω_2^{UB} are the equations of ellipses with selected parameters for principal axes rotation and origin coordinates, while the parameter $\delta_{(.)}$ is a measure of stiffness of the surfaces curvature. The parameters selection for the ellipses functions has been performed after numerical experiments, but it could be further refined using least-square based optimization for an optimal fit with the contact mode regions. Different functions can be tested in future work to reproduce similar two-dimensional step functions.

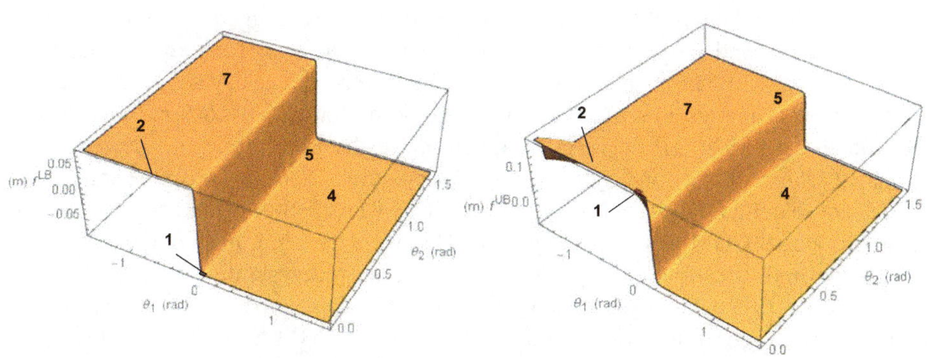

Fig. 5. Surface functions $f^{LB}(\theta_1, \theta_2)$ and $f^{UB}(\theta_1, \theta_2)$ representing the lower and upper COP limits in the proposed smooth transition model for contact modes.

4 Limits of Dynamic Postural Stability

The standing posture model with segmented feet is used to evaluate the limits of dynamic postural stability of the human body and to investigate the effects of the multimodal foot-ground contact interaction on these limits. In this work, the dynamic postural stability limits are quantified by the set of all possible initial conditions from which the system can reach an upright static equilibrium, while avoiding stepping. The resulting set of viable initial conditions is called the set of balanced states [15] and quantifies the maximum allowable velocity perturbations for the legged system during standing. The terminology dynamic stability refers to the ability quantified in this study of maintaining standing balance in presence of large perturbations, for which the ground projection of the COM exits the foot support region [18, 19].

Balance Stability Criterion. For any initial conditions within the set of balanced states, there exists at least one controlled trajectory that leads the legged system safely to a final static equilibrium, for instance, the upright standing posture [15]. If the system in a given stance has initial conditions outside of the set of balanced states, due to external or self-induced perturbations, there is no controlled trajectory with the original stance configuration that allows recovery. In this case, the system is said to be in an unbalanced state and must change its stance configuration in order to avoid a fall [15]. The set of balanced states is contact- (or stance-) dependent and can be generally calculated for any type of stance configuration (single or multi-contact support [20]). In this analysis of human postural stability, the set of balanced states is constructed for the symmetric double stance configuration of the proposed model, enhanced by the novel multimodal characteristic of the segmented foot.

Algorithm for Maximum Limits of Recovery. The set of balanced states is identified by its boundary (i.e., the set of extreme initial conditions), which represents the maximum limits of recovery for the legged system during standing. A series of constrained optimization problems is formulated to find the legged system's balanced states boundary and the corresponding optimal joint and actuator torques trajectories driving the system from its maximum limits of recovery to a rest state. Joint-space and COM-space dynamics are used to formulate joint torques, COM motion, ground reaction forces, and COP position, as functions of the joint motion to be optimized. Due to the high dimensionality of the human posture model, the extreme initial conditions are searched in the state space of the COM Cartesian position and velocity. For any given initial COM position, the initial COM velocity is minimized and maximized along a desired direction, while ensuring that the legged system (a) can reach a final rest position, (b) satisfies various system and physics constraints, and (c) avoids stepping. Balancing trajectories that drive the system's COM from the perturbed initial state (position and velocity) to a rest state (upright static posture) are generated for various initial COM positions through a series of nonlinear programming problems, solved with the sequential quadratic programming method. In this work, the maximum limits of recovery in the anterior-posterior direction are analyzed, hence the cost function for the minimization and maximization problems is the COM X velocity.

Constraints During Balancing Tasks. The joint and COM dynamics resulting from the above optimization solutions are subject to several kinematics and kinetics constraints, to ensure that the balancing tasks are physically consistent.

Kinematics Constraints. Upper and lower body joints are subject to anatomical ranges of motion. A kinematic constraint coupling torso and hip joint angles, $\theta_{\text{torso}}(t)\theta_{\text{hip}}(t) \leq 0$, is imposed at all times for more human-like results, in which torso forward lean is associated with hip flexion and torso backward lean is coupled with hip extension [21]. Next, kinematic constraints are imposed to exclude infeasible contact modes: mode 3 and

8 are excluded by imposing limits on the range of θ_2; the infeasible region of mode 4 is excluded by imposing limits on the metatarsal joint $[0, \theta_{toe,max}]$; the exclusion of mode 6 is ensured by constraining the global position of the toe joint on an arc of radius (al + hl) and centered in ($-hl$, 0), while the heel global coordinates must lie on an arc of radius (al + hl) and center (al, 0). Ground penetration is avoided by constraining the Y coordinate of each point of the foot model. Lastly, kinematic constraints at specific times (initial and final) are required to generate the balancing trajectories: (*i*) in each optimization problem a different initial COM position is assigned, within selected points of interest in the (X, Y) plane at which the velocity extrema are evaluated; (*ii*) for each assigned initial COM position, the same desired final rest state is imposed (i.e., COM final position (x_{home}, y_{home}) corresponding to upright standing posture, COM final velocity and acceleration set to zero, and final foot pose in mode 1); (*iii*) a forward lean of the torso is imposed at the initial time for the assigned initial COM positions with $x_{COM} > x_{home}$, and vice versa, to avoid unnatural poses and locally optimal solutions for the velocity extrema due to kinematic redundancy [15].

Kinetics Constraints. Upper and lower body joints are subject to anthropometric torque limits. Contact-related kinetics constraints include positive normal ground reaction force, friction cone, and the COP position within the functions $f^{LB}(\theta_1, \theta_2)$ and $f^{UB}(\theta_1, \theta_2)$ herein introduced.

5 Results and Discussion

The boundary of the set of balanced states is numerically constructed for the proposed human posture model with segmented feet, to investigate the maximum limits of recovery along the anterior-posterior direction. The human model represents a subject of total mass m = 56.7 kg and approximately 1.6 m tall. Anthropomorphic link length, mass, and foot parameters [19, 22] and joint angle [23, 24] and torque limits parameters [25, 26] are adapted from the literature.

To investigate the dynamic balance ability of the human body in various upright standing postures, the extreme initial conditions are evaluated at initial positions of the COM sampled at (x_{COM}, y_{COM}) = (x_{home} + 0.02n, y_{home} − 0.01), in meters, where (x_{home}, y_{home}) = (0.001, 1.129) m and $n = \pm 1, 2, 3, \ldots$. In other words, the initial COM position in each simulated balancing recovery is assigned to points in the (X, Y) plane sampled along the horizontal line y_{COM} = y_{home} − 1 cm, with a 2 cm spacing in the positive and negative X direction. The algorithm and balancing criterion are not limited by this sampling choice, as more points of interest can be evaluated, for example, for a broader balance stability analysis with large COM vertical displacement.

The velocity extrema at the selected points of interest are evaluated in two different conditions: stance feet constrained in mode 1 versus feet in multimodal contact interaction with unspecified contact mode sequence. The results are two different boundaries and corresponding sets of viable initial conditions in the state space of COM X position and velocity (Fig. 6) from which the system can restore the upright posture without stepping. The comparison of the sets of balanced states in mode 1 versus multimodal contact conditions allows to quantify the isolated effects of a

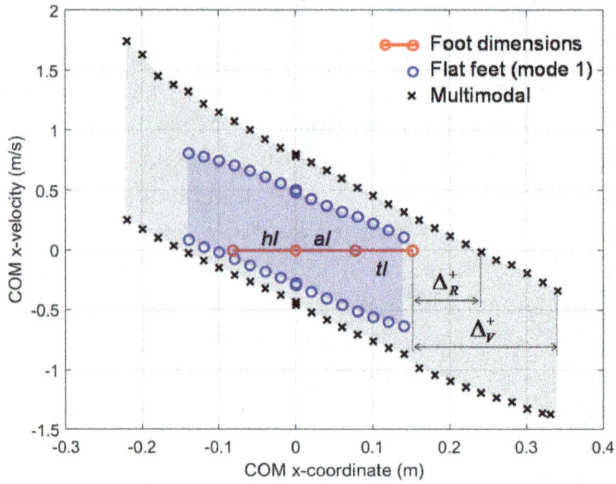

Fig. 6. Maximum limits of recovery of the human posture model from various initial COM *X*-positions and constant initial COM *Y*-position equal to (y_{home} − 1) cm. The limits are larger when the additional balancing mechanism of multimodal foot-ground interaction is enabled.

segmented foot model on the whole-body postural stability limits of the human body. The segmented foot model introduces the heel-to-toe rocking balancing strategy, in addition to the traditional ankle, hip, and upper body strategies. The heel-to-toe rocking balancing strategy is enabled by the proposed multimodal foot-ground contact model and results in a larger set of balanced states (Fig. 6).

For any initial COM position, the multimodal contact interaction allows for greater COM velocity perturbation in the anterior-posterior directions, as compared with flat-foot interaction (Fig. 6). In addition to the maximum allowable velocity perturbation, the increase in the limits of recovery can be quantified by the maximum allowable COM displacement along the *X* axis. In particular, the dynamic characteristics of postural stability can be captured by how far the COM can extend beyond the foot support region, which is rarely quantified during traditional dynamic posturography. Two margins are evaluated for this purpose: the COM reachable margin Δ_R and the COM viable margin Δ_V in both the anterior and posterior directions (indicated by a + or − superscript; Fig. 6).

The reachable margin Δ_R is the difference between the maximum position that the COM can reach with zero velocity without losing balance and the edge of the stance foot. In other words, Δ_R is the maximum displacement relative to the foot size at which the COM is still able to invert its motion (hence, zero velocity) by using only its internal actuator torques control. Its interpretation is similar to a maximum voluntary COM sway [27], but calculated relative to the foot size to capture the dynamic ability of sway control: it measures as far as the body can displace its COM outside of the footprint and come back to upright equilibrium, while standing and without any external help (e.g., stepping).

The viable margin Δ_V is the difference between the maximum position for a COM state to be viable (i.e., with some velocity within the set of balanced state) and the edge of the stance foot. In other words, Δ_V is the limiting COM displacement relative to the foot size for which there exist a set of initial velocities that allow the system to recover the upright equilibrium without stepping, enabled only by its initial conditions and internal actuator torques control. Differently from the reachable margin, the viable margin cannot be reached by the legged system's COM by means of voluntary sway in the given stance: previous steps or external pushes are responsible for driving the system's COM state within its viable margins. However, once in this region, the external help can cease and the system in the given stance can generate balance recovery trajectories by means of proper (externally imposed) initial conditions and internal actuation capacity, without stepping. The concept of viable margin quantifies another aspect of dynamic balance ability for a legged system and, in case of standing posture, can provide a map of target COM states during assisted postural stability exercises.

Table 2. Reachable and viable margins of maximum allowable COM displacement

	Posterior margin (cm)		Anterior margin (cm)	
	Δ_R^-	Δ_R^+	Δ_V^+	Δ_V^-
Mode 1	1.8	−1.1	−1.1	5.8
Multimodal	5.8	8.9	18.9	13.8

With an accuracy of ± 2 cm due to the chosen sampling rate, the reachable and viable margins are quantified (Table 2). The margins in presence of the multimodal foot-ground interaction are far greater than those obtained for the flat foot model, both in the anterior and posterior directions. The small margins in the flat foot condition are due to the stricter kinematic constraints for mode 1, as well as the absence of the additional multimodal balancing mechanism.

When more balancing mechanisms are combined, as in the multimodal case, the set of balanced states and the COM margin increase, due to the increased capability of the system to regulate angular momentum about its COM [20, 28, 29]. As demonstrative examples, balancing trajectories from initial COM position equal to $(x_{\text{home}}, y_{\text{home}} - 1\text{ cm})$ and extreme initial COM velocities are presented (Fig. 7). As expected, the maximum allowable velocity extrema in the anterior/posterior directions are greater when the heel-to-toe rocking strategy is enabled (+0.802/−0.461 m/s), as compared to the case of flat feet (i.e., mode 1; +0.484/−0.292 m/s). From the balancing animations it can be observed that in addition to the foot rocking strategy, the segmented foot model in multimodal contact interaction also leads to an increased used of knee flexion during recovery from both positive and negative velocity perturbations, while the knee joint motion is rather stiff when the foot is constrained to be flat.

A novel segmented foot and contact model have been proposed to allow the unified treatment of multiple contact modes between the feet and the ground. This model is integrated in a human standing posture mechanism to study the maximum limits of recovery from various upright standing postures. Overall, the multimodal foot-ground contact schedule naturally arises from optimization results whenever the segmented foot is allowed the rocking motion; this additional balancing strategy leads to greater

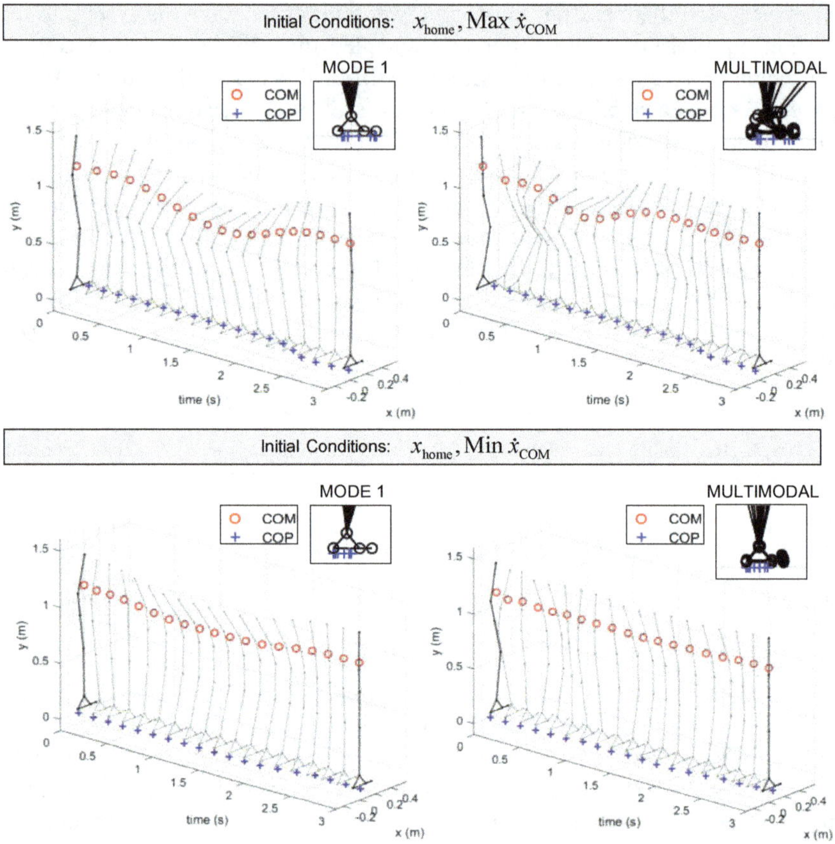

Fig. 7. Results of COP (+) and COM (o), and whole-body balancing trajectories, recovering from maximum positive and negative COM *X*-velocity perturbation from an upright standing posture.

dynamic balance capabilities in the human postural model, confirming the importance of replacing flat feet assumptions with a more human-like foot-toe complex. By introducing novel reachable and viable stability margins and the formulation of multimodal foot contact interactions with the environment, the proposed method gives an optimization-based algorithm and metrics to study the balance stability of legged systems beyond standard posturography techniques and provides important insights for the design of robot and prosthetic extremities.

Acknowledgment. The authors thank Karen Ayoub for her help in running the numerical results.

References

1. Ku, P.X., Osman, N.A.A., Yusof, A., Abas, W.A.B.W.: The effect on human balance of standing with toe-extension. PLoS ONE **7**(7), e41539 (2012)
2. Deschamps, K., Staes, F., Roosen, P., Nobels, F., Desloovere, K., Bruyninckx, H., Matricali, G.A.: Body of evidence supporting the clinical use of 3D multisegment foot models: a systematic review. Gait Posture **33**(3), 338–349 (2011)
3. Burnfield, M.: Gait analysis: normal and pathological function. J. Sports Sci. Med. **9**(2), 353 (2010)
4. Farizeh, T., Sadigh, M.J.: Effect of heel to toe walking on time optimal walking of a biped during single support phase. In: ASME 2014 International Mechanical Engineering Congress and Exposition, p. V003T03A054. American Society of Mechanical Engineers, November 2014
5. Nolan, L., Kerrigan, D.C.: Postural control: toe-standing versus heel-toe standing. Gait Posture **19**(1), 11–15 (2004)
6. Ivanenko, Y.P., Levik, Y.S., Talis, V.L., Gurfinkel, V.S.: Human equilibrium on unstable support: the importance of feet-support interaction. Neurosci. Lett. **235**(3), 109–112 (1997)
7. Murnaghan, C.D., Elston, B., Mackey, D.C., Robinovitch, S.N.: Modeling of postural stability borders during heel–toe rocking. Gait Posture **30**(2), 161–167 (2009)
8. Kouchaki, E., Wu, C.Q., Sadigh, M.J.: Effects of constraints on standing balance control of a biped with toe-joints. Int. J. Humanoid Rob. **9**(03), 1250016 (2012)
9. Vukobratović, M., Herr, H.M., Borovac, B., Raković, M., Popovic, M., Hofmann, A., Jovanović, M., Potkonjak, V.: Biological principles of control selection for a humanoid robot's dynamic balance preservation. Int. J. Humanoid Robot. **5**(04), 639–678 (2008)
10. Hughes, J., Clark, P., Klenerman, L.: The importance of the toes in walking. J. Bone Jt. Surg. Br. Vol. **72**(2), 245–251 (1990)
11. Kerrigan, D.C., Della Croce, U., Marciello, M., Riley, P.O.: A refined view of the determinants of gait: significance of heel rise. Arch. Phys. Med. Rehabil. **81**(8), 1077–1080 (2000)
12. Nishiwaki, K., Kagami, S., Kuniyoshi, Y., Inaba, M., Inoue, H.: Toe joints that enhance bipedal and fullbody motion of humanoid robots. In: Proceedings 2002 IEEE International Conference on Robotics and Automation, vol. 3, pp. 3105–3110. IEEE, May 2002
13. Yamane, K., Trutoiu, L.: Effect of foot shape on locomotion of active biped robots. In: 2009 9th IEEE-RAS International Conference on Humanoid Robots, pp. 230–236. IEEE, December 2009
14. Agarwal, S., Popovic, M.: Study of toe joints to enhance locomotion of humanoid robots. In: 2018 IEEE-RAS 18th International Conference on Humanoid Robots (Humanoids), pp. 1039–1044. IEEE, November 2018
15. Mummolo, C., Mangialardi, L., Kim, J.H.: Numerical estimation of balanced and falling states for constrained legged systems. J. Nonlinear Sci. **27**(4), 1291–1323 (2017)
16. Fu, K.S., Gonzalez, R.C., Lee, C.S.G.: Robotics – Control, Sensing, Vision, and Intelligence. McGraw Hill, Columbus (1987)
17. Wearing, S.C., Hooper, S.L., Dubois, P., Smeathers, J.E., Dietze, A.: Force-deformation properties of the human heel pad during barefoot walking. Med. Sci. Sports Exerc. **46**(8), 1588–1594 (2014)
18. Mummolo, C., Kim, J.H.: Passive and dynamic gait measures for biped mechanism: formulation and simulation analysis. Robotica **31**(4), 555–572 (2013)
19. Mummolo, C., Mangialardi, L., Kim, J.H.: Quantifying dynamic characteristics of human walking for comprehensive gait cycle. J. Biomech. Eng. **135**(9), 091006 (2013)

20. Mummolo, C., Peng, W.Z., Gonzalez, C., Kim, J.H.: Contact-dependent balance stability of biped robots. J. Mech. Robot. **10**(2), 021009 (2018)
21. Lee, R.Y., Wong, T.K.: Relationship between the movements of the lumbar spine and hip. Hum. Mov. Sci. **21**(4), 481–494 (2002)
22. Winter, D.A.: The Biomecanics and Motor Control of Human Gait: Normal. Elderly and Pathological, 2nd edn. Canadian Cataloguing in Publication Data, Waterloo (1991)
23. Wojcik, L.A., Thelen, D.G., Schultz, A.B., Ashton-Miller, J.A., Alexander, N.B.: Age and gender differences in peak lower extremity joint torques and ranges of motion used during single-step balance recovery from a forward fall. J. Biomech. **34**(1), 67–73 (2001)
24. Chao, E.Y., Laughman, R.K., Schneider, E., Stauffer, R.N.: Normative data of knee joint motion and ground reaction forces in adult level walking. J. Biomech. **16**(3), 219–233 (1983)
25. Stockdale, A.A.: Modeling three-dimensional hip and trunk peak torque as a function of joint angle and velocity (2011)
26. Anderson, D.E., Madigan, M.L., Nussbaum, M.A.: Maximum voluntary joint torque as a function of joint angle and angular velocity: model development and application to the lower limb. J. Biomech. **40**(14), 3105–3113 (2007)
27. Warnica, M.J., Weaver, T.B., Prentice, S.D., Laing, A.C.: The influence of ankle muscle activation on postural sway during quiet stance. Gait Posture **39**(4), 1115–1121 (2014)
28. Boström, K.J., Dirksen, T., Zentgraf, K., Wagner, H.: The contribution of upper body movements to dynamic balance regulation during challenged locomotion. Front. Hum. Neurosci. **12**, 8 (2018)
29. Mummolo, C., Mangialardi, L., Kim, J.H.: Identification of balanced states for multi-segmental legged robots using reduced-order model. In: 2015 IEEE-RAS 15th International Conference on Humanoid Robots (Humanoids), pp. 914–919, November 2015

Optimization of the Kinematic Chain of the Thumb for a Hand Prosthesis Based on the Kapandji Opposition Test

Antonio Pérez-González^(✉) and Immaculada Llop-Harillo

Grupo de Biomecánica y Ergonomía,
Departamento de Ingeniería Mecánica y Construcción,
Universitat Jaume I, 12071 Castellón, Spain
aperez@uji.es

Abstract. The thumb plays a key role in the performance of the hand for grasping and manipulating objects. In artificial hands the complex thumb's kinematic chain (TKC) is simplified and its five degrees of freedom are reduced to only one or two with the consequent loss of dexterity of the hand. The Kapandji opposition test (KOT) has been clinically used in pathological human hands for evaluating the thumb opposition and it has also been employed in some previous studies as reference for the design of the TKC in artificial hands, but without a clearly stated methodology. Based on this approaches, in this study we present a computational method to optimize the whole TKC (base placement, link lengths and joint orientation angles) of an artificial hand based on its performance in the KOT. The cost function defined for the optimization (MPE) is a weighted mean position error when trying to reproduce the KOT postures and can be used also as a metric to quantify thumb opposition in the hand. As a case study, the method was applied to the improvement of the TKC of an artificial hand developed by the authors and the MPE was reduced to near one third of that of the original design, increasing significantly the number of reachable positions in the KOT. The metric proposed based on the KOT can be used directly or in combination with other to improve the kinematic chain of artificial hands.

Keywords: Artificial hand · Kinematic chain · Optimization

1 Introduction

The human hand is a marvelous tool optimized in an evolutionary process since our ancestors [1, 2]. Thumb opposition is said to be one distinctive feature of the human hand. Interestingly, this dexterity can be obtained even with an important variability in the thumb anatomy among individuals [3]. The human thumb is composed of three bones (Fig. 1) [4]: the distal phalanx, the proximal phalanx and the first metacarpal bone, connected to the wrist. The interphalangeal joint (IP) is a hinge joint with one degree of freedom (DoF) whereas the metacarpo-phalangeal joint (MCP) is condylar and the carpo-metacarpal joint (CMC) is of saddle type, both with two DoFs. Therefore, the thumb's kinematic chain (TKC) can be considered as an open chain connected to the wrist with 5 DoFs, allowing a high range of positions and orientations of the

G. A. Ateshian et al. (Eds.): CMBBE 2019, LNCVB 36, pp. 271–287, 2020.
https://doi.org/10.1007/978-3-030-43195-2_22

thumb tip. It has been shown that the assumption of universal joints (two perpendicular and intersecting axes of rotation) for the CMC and MCP joints is not realistic and that a biomechanical model with five links [4], including two virtual links in these joints and considering non-orthogonal and non-intersecting axes in the joints is more realistic and represents better the anatomical evidences [3].

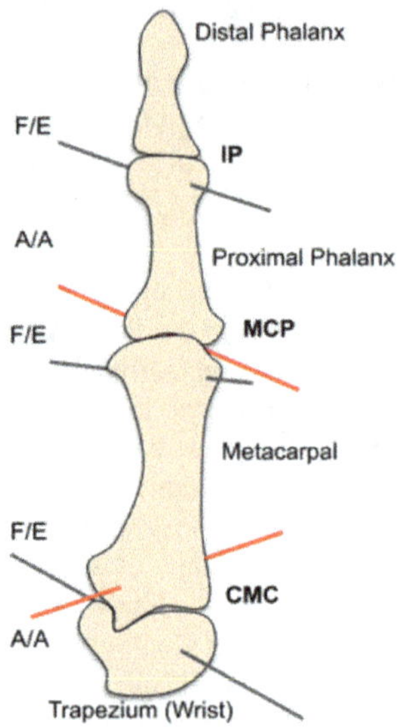

Fig. 1. Bones and joints of the human thumb [4].

The design of artificial hands, both prosthetic and robotic, is moving in last decades towards devices more anthropomorphic, to improve the functionality and the cosmetic appearance. Given the complexity of the TKC above explained, the designers of artificial hands need to introduce simplifications for this digit. These simplifications are mainly guided by the difficulty of obtaining adequate mechanical solutions for reproducing the geometry and mobility of the CMC and MCP joints, but also by the attempt to simplify the artificial hand control. Generally, the five DoFs of the human thumb are simplified in mechanical hands to achieve two basic motions: flexion/extension and circumduction. The circumduction rotation of the thumb is the movement requested to change the type of opposition of the thumb with respect to the long fingers, it allows to alternate between a lateral grasp and a power or precision grasp. In the human hand, the circumduction motion is achieved through a combination of 3 joints at the base of the thumb [5]. Belter et al. [6] reviewed the thumb design and position for different prosthetic hands. They highlighted the relevance of the relationship between the circumduction rotation axis of the thumb and the main axis of the wrist for functional grasps. In most of the prosthetic hands that Belter et al. analyzed, the thumb is actuated with a simple closing or opening (flexion/extension) and along the circumduction rotation axis, that is not always oriented parallel with the wrist rotation axis. They recommended to jointly approximate in a single DoF the thumb flexion and circumduction rotation for keeping complexity low. Ten Kate et al. [7] reviewed the kinematic specifications of 3D-printed hand prostheses and specified the range of motion for the thumb flexion and thumb circumduction of 58 devices. Three of the hands analyzed lack a thumb, 7% of the thumbs of the other hands did not perform flexion movement and 62% did not perform circumduction movement. Grebenstein et al. [8] analyzed anatomy, surgery and rehabilitation data for defining some guidelines to be used for the design of a robotic thumb for the DLR hand. They proposed a minimum of 3 DoF to allow proper manipulation.

From the literature of both robotic and prosthetic fields, we can find thumbs with different mechanical configurations, changing the number of links and DoFs. Figure 2 shows several representative examples.

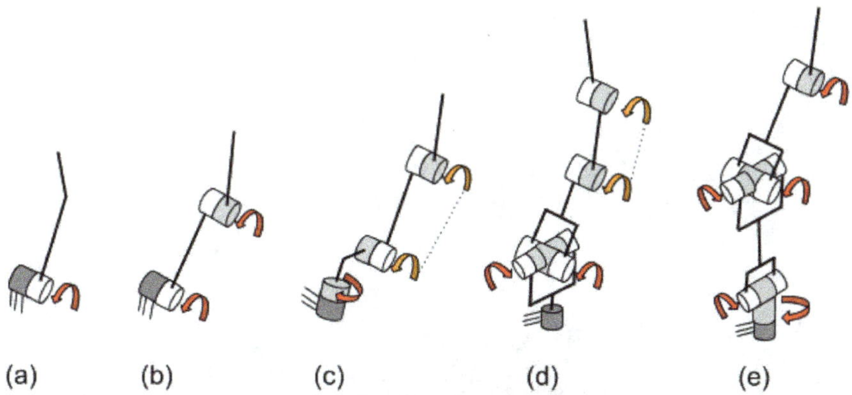

Fig. 2. Thumb's kinematic chain (TKC) for several artificial hands. a: SensorHand Speed [9], b: FRH-4 Hand [10], c: Bebionic hand [9], d: DLR/HIT Hand II [11], e: Shadow Dexterous Hand [12]. Red arrow: actuated by an independent actuator; yellow arrow: several DoFs actuated by the same actuator.

The thumb of the SensorHand Speed [9] (Fig. 2a) is a rigid bar with only 1 DoF. The thumb of the FRH-4 Hand of the mobile-assisting robot ARMAR [10] (Fig. 2b) has 2 DoFs actuated by two independent fluidic actuators that produce flexion motion of the MCP and IP joints, respectively. The thumb of the Bebionic hand [9] (Fig. 2c) has 3 DoFs, one actuator produce the flexion of the MCP and IP joints and the MCP circumduction has two selectable fixed positions, manually placed by the user of the prosthesis. The thumb of the DLR/HIT Hand II [11] (Fig. 2d) has 4 DoFs and 3 actuators, one for the CMC flexion, other for the MCP and IP flexion and other for the CMC abduction. The thumb of the Shadow Dexterous Hand [12] (Fig. 2e) has 5 independently actuated DoFs, as the human hand, but the MCP and CMC are universal joints.

It could be interesting to have objective methods to evaluate the impact of simplifications made in the thumb of artificial hands in the loss of ability to grasp in real life applications. These objective methods could help designers to obtain hand designs with improved grasping abilities. The Kapandji opposition test (KOT) [13], also called total opposition test, can be of interest for this goal. The KOT was proposed as a simple method for assessing the opposition of the thumb in pathological hands and is used in current clinical practice. It involves touching different areas of the four long fingers with the tip of the thumb (Fig. 3). The score obtained in the test ranges from 1 to 10 depending on the last reached area, being the test performed in the order indicated in Table 1.

Table 1. Scores according to the Kapandji opposition test (KOT)

Score	Finger	Area of contact
1	Index	Lateral side of the second phalanx
2	Index	Lateral side of the third phalanx
3	Index	Tip
4	Middle	Tip
5	Ring	Tip
6	Little	Tip
7	Little	DIP crease
8	Little	PIP crease
9	Little	Proximal crease
10	–	Distal volar crease of the hand

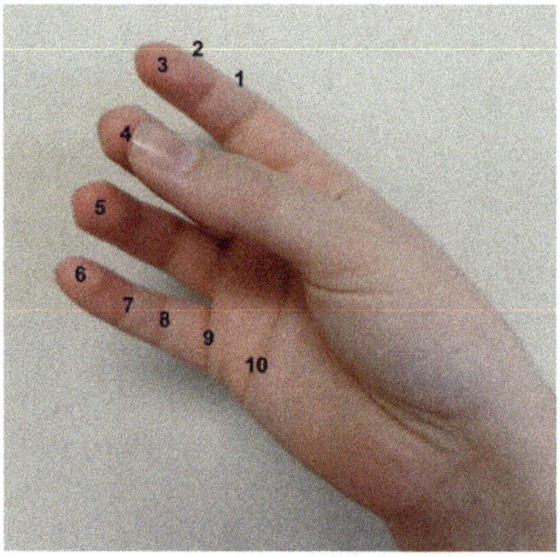

Fig. 3. Areas to touch with the thumb tip in the Kapandji opposition test (KOT)

Grebenstein et al. [8] considered that the KOT includes motion of the fingers and the thumb sufficient to evaluate the manipulation abilities. Other authors used the KOT to evaluate the functionality and anthropomorphism of artificial hands. Shin et al. [14] used the KOT to analytically analyze a new dexterous robot hand for delicate object grasping. Chalon et al. [15] used the KOT to optimize the thumb of the Awiwi Hand obtaining the maximum score at KOT. Roa et al. [16] explored the relationship between kinematic design and manipulation performance of robotic hands, to analyze it they performed the KOT on seven thumb placements of a modular robotic hand. Deimel et al. [17] also assessed the dexterity of the opposable thumb of a soft robotic hand using the KOT. Cerruti et al. [18] used the KOT to validate the thumb base placement of a kinematic model of an anthropomorphic robotic hand used for gesturing and grasping.

Some of the above mentioned studies that used the KOT made some adaptation of the test. For example, in some studies the authors did not consider the positions 1 and 2 corresponding to contact of the thumb with the lateral side of the index [16, 17] or removed some positions in the little finger [17]. Contrary, in some cases additional positions in the middle and ring fingers are included [16, 17]. In most of these studies the KOT is only used to evaluate different hand designs. In some of them the authors used the KOT to optimize the thumb base placement [15, 16, 18]. In [15] the optimization of the thumb included also as parameters the orientation of the joints, but the details about the cost function are not provided. To our knowledge, no previous study made an optimization of the TKC including simultaneously base placement, link lengths and all the joint orientation angles.

Following these approaches, in this study the objective was to define a computational method to optimize the whole TKC (base placement, link lengths and joint orientation angles) of an artificial hand based on its performance in the KOT. This method could be useful to improve the design of prosthetic and robotic hands regarding thumb opposition, leading to a better object grasping and manipulation. The application of the method involves defining an index, used in the cost function for optimization, which provides a metric for rating thumb opposition in artificial hands. To test the method we applied it to a 3D-printed prosthetic hand developed by the authors: the IMMA hand [19].

2 Materials and Methods

2.1 IMMA Hand

The IMMA hand [19] is a low-cost tendon-driven anthropomorphic prosthetic hand designed by the authors. It has five fingers with three phalanges per finger and 6 DoFs in total: independent flexion/extension in each of the four long fingers, and two independent DoFs for the thumb. The MCP and IP joints of the thumb are actuated both with the same tendon for flexion and the CMC joint is actuated by a separate tendon for circumduction. Figure 4 shows the TKC of the IMMA hand. Figure 5 shows the achievable target areas of the KOT by the right IMMA hand prototype. As is shown, its score is 4, because the positions 5 to 10 (see Table 1) cannot be reached.

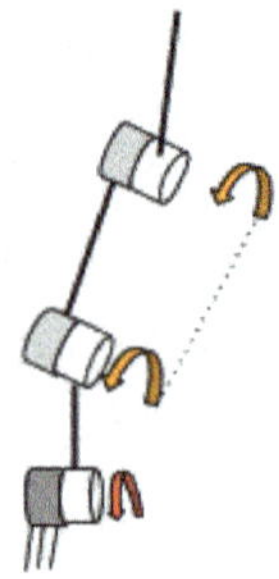

Fig. 4. Thumb's kinematic chain (TKC) of the IMMA hand.

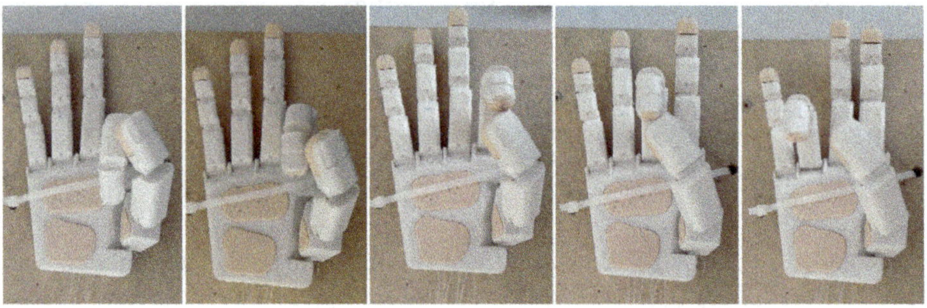

Fig. 5. IMMA hand performing the Kapandji opposition test (KOT) in positions 1 (left) to 5 (right).

2.2 Computational Model

Hand Model. The model of the hand used in this study has a maximum of four straight links for each digit and a maximum of five digits. To define the kinematic chain of the hand, a local coordinate system (LCS) for each link has to be defined. In this study the LCS of each link is located in the middle of the joint with the proximal link. The LCSs were defined under the following criteria: Z-axis coincident with the flexion/extension axis of the joint, and oriented so that the flexion motion corresponds to a positive rotation around Z, X-axis aligned in palmar-dorsal direction pointing dorsally, thus indicating the abduction/adduction axis in the joint and Y-axis defining a right-handed coordinate system with the previous ones, resulting in a distal direction, in other words, pointing in the direction towards the tip of the fingers. This convention for the orientation of the axes is similar to that proposed by the ISB [20], with the difference that the X and Y axes have opposite positive directions. With this selection the position of each LCS relative to the proximal one in the kinematic chain presents positive values in the translation along the Y axis.

Therefore, the kinematic chain of the hand is defined in the reference position with the three displacements and rotations of each LCS with respect to the immediate proximal in the chain. The wrist is taken as the fixed LCS for all the digits. Specifically, the LCS of link j (1: metacarpus, 2: proximal phalanx, 3: intermediate phalanx, 4: distal phalanx) of the digit i (1: thumb, 2: index, 3: middle, 4: ring, 5: little) is defined with the Eq. 1.

$$v_{i,j} = \begin{bmatrix} x_{i,j} & y_{i,j} & z_{i,j} & \theta_{x_{i,j}} & \theta_{y_{i,j}} & \theta_{z_{i,j}} \end{bmatrix}, \quad i = 1, 2..., 5, \quad j = 1, 2..., 5 \tag{1}$$

where the first three elements of the vector correspond to the translation vector of $LCS_{i,j}$ with respect to $LCS_{i,j-1}$ and the last three to the Euler angles with sequence of rotations XZY to orient $LCS_{i,j-1}$ as $LCS_{i,j}$. At each finger, a last LCS (j = 5) is added, positioned at its end (fingertip), with its Y axis in the proximal-distal direction and its X axis in the palmar-dorsal direction.

For each digit a maximum of six DoFs can be included in the hand model, two in CMC and MCP joints and one in the IP joints. Universal joints are considered in those with two DoFs. The hand position can be obtained straightforward by direct kinematics using the Eq. 1 and the rotation angles in the joints.

Cost Function. To optimize the TKC we defined a cost function based on the KOT postures. We defined the position error ε_i for each posture i of the KOT as the minimum possible distance between the thumb tip and the corresponding target point of the test while the hand is moved within its workspace (Eq. 2).

$$\varepsilon_i = \min(dist(p_i, p_t)) \tag{2}$$

where p_i is the target point, p_t the thumb tip point, *dist* is a function defining the distance between two points and *min* is a function obtaining the minimum possible value of *dist* when moving the hand within its workspace.

Given a hand geometry and the range of motion of their joints, the calculation of the ε_i involves an optimization. The variables for this optimization are the joint rotation angles of the hand. If these angles are coupled with a linkage, the number of variables for the optimization can be reduced, because the coupled rotation angles can be obtained from the coupling equations.

The final cost function for the optimization of the TKC was defined with Eq. 3 as a relative mean position error (MPE) for the different positions of the KOT.

$$MPE = \sum\nolimits_i \varepsilon_i w_i \tag{3}$$

where w_i is a weighting coefficient for the posture i of the KOT. Dividing the lengths of the kinematic chain by the hand length is convenient for having a non-dimensional index and making the evaluation independent of the hand size.

Optimization Algorithm. For optimizing the TKC based on the KOT, the MPE above defined (Eq. 3) has to be minimized, being the optimization variables the parameters defining the TKC: base placement, joint angles orientation and links length. Depending on the designer interest, it is also possible to limit the variables to only some of those defining the TKC. As the calculation of the minimum MPE requires the previous computation of the position errors ε_i (Eq. 2), the computational model involves two nested optimizations. Matlab was used in this study for the computation. The built-in *Genetic Algorithm* ('*ga*') was used for the optimization of the TKC whereas '*fmincon*' function for non-linear optimization was used for the nested optimization corresponding to each position of the KOT. Table 2 shows the pseudocode used in the computation model.

Table 2. Pseudocode for optimizing the TKC.

```
Define OpOpt_1=Stop_Optimization_Options_Genetic_Algorithm on MPE
Define OpOpt_2=Stop_Optimization_Options_FMINCON_Algorithm on ε_i
Define w_i for MPE
While OpOpt_1 not accomplished
   TKC updated by Genetic Algorithm
   For each KOT position i
      Initialize JA=Joint_Rotation_Angles
      While OpOpt_2 not accomplished
         JA updated by FMINCON algorithm
         Compute ε_i for TKC and JA
      End
   End
   Compute MPE from ε_i and w_i
End
Output optimum TKC
```

2.3 Case Study: Optimization of the TKC of the IMMA Hand

A simplified model of the IMMA hand [19] was created in Matlab (Fig. 6). Table 3 shows the components of the translation-rotation vectors that define the kinematic chain of the hand (vectors $v_{i,j}$, Eq. 1), where x, y, z are non-dimensional values related to the hand length (distance from the wrist to the end of the middle finger) and θ_x, θ_y, θ_z angles are the Euler rotations around the X, Y, Z axes, respectively, with rotation order XZY, expressed in radians. The joints range of motion (ROM) were defined based on the hand prototype (Fig. 5) and are shown in Table 4. For the abduction/adduction movement in the MCP joints we included a small ROM accounting for the flexibility of the joints in the prototype.

Table 3. Data for the kinematic chain of the IMMA hand according to Eq. 1 (lengths are non-dimensional values related to the hand length and Euler angles are in radians).

Links	$v_{i,j}$	Thumb	Index	Middle	Ring	Little
Metacarpal	x	0	0	0	0	0
	y	0.2169	0	0	0	0
	z	0.1577	0	0	0	0
	θx	1.5708	0	0	0	0
	θy	0	0	0	0	0
	θz	0	0	0	0	0

(*continued*)

Table 3. (*continued*)

Links	$v_{i,j}$	Thumb	Index	Middle	Ring	Little
Proximal phalanx	x	0	0	0	0	0
	y	0.1320	0.4588	0.4370	0.4192	0.3874
	z	0	0.1188	0.0016	−0.1054	−0.2004
	θx	0	0.0873	0	−0.1250	−0.2618
	θy	−0.7854	0	0	0	0
	θz	0	0	0	0	0
Intermediate phalanx	x	0	0	0	0	0
	y	0.2158	0.2456	0.2725	0.2291	0.1968
	z	0	0	0	0	0
	θx	0	0	0	0	0
	θy	0	0	0	0	0
	θz	0	0	0	0	0
Distal phalanx	x	0	0	0	0	0
	y	0.1659	0.1399	0.1623	0.1509	0.0590
	z	0	0	0	0	0
	θx	0	0	0	0	0
	θy	0	0	0	0	0
	θz	0	0	0	0	0
Fingertip	x	0	0	0	0	0
	y	0	0.1243	0.1324	0.1324	0.1135
	z	0	0	0	0	0
	θx	0	0	0	0	0
	θy	0	0	0	0	0
	θz	0	0	0	0	0

Table 4. Joints range of motion (minimum angle/maximum angle) in degrees for the IMMA hand (add/abd: adduction/abduction, ext/flex: extension/flexion).

Joint	Thumb	Index	Middle	Ring	Little
CMC add/abd	0/0	0/0	0/0	0/0	0/0
CMC ext/flex	−10/70	0/0	0/0	0/0	0/0
MCP add/abd	−1/1	−1/1	−1/1	−1/1	−1/1
MCP ext/flex	−11/55	−20/85	−17/86.3	−15/70	−20/65
PIP ext/flex	−13/55	−20/60	−17/75	−20/75	−20/65
DIP ext/flex	0/0	−15/50	−20/75	−20/70	−20/75

For the optimization of the TKC in this case study the position 10 of the KOT (see Fig. 3) was not considered because it was difficult to locate in a simplified model of the hand. The positions 1 and 2 where considered in the more proximal point of the corresponding index phalanx. In Eq. 3 positions considered were weighted equally, so we used $w_i = 1/9$ for $i = 1$ to 9 and $w_{10} = 0$. Moreover, in this case the joint rotation

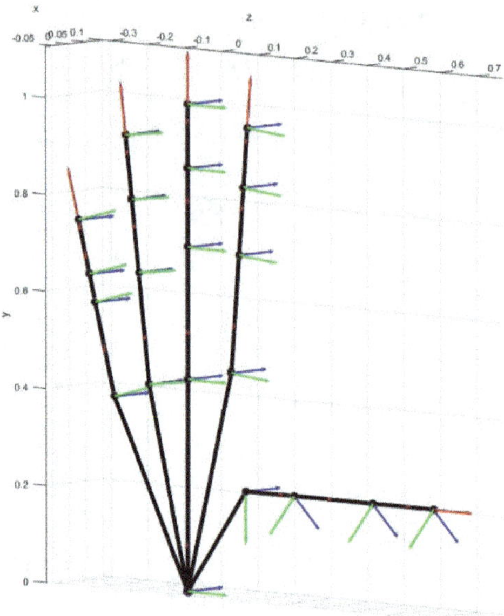

Fig. 6. Simplified model of the right IMMA hand. Colored arrows in the joints indicate local coordinate systems (LCSs): green for Z-axis, red for Y-axis, blue for X-axis.

angles of the hand were considered independent, without taking into account the coupling equations resulting from the actuation of several joints with a same tendon. The variables for the optimization were the orientation of the CMC and MCP joints, the thumb's links length and the position of the CMC joint. The feasible range of these variables, upper and lower bounds for the optimization, is shown in Table 5. The default optimization options were considered for the genetic algorithm of Matlab ('*ga*') except 'FitnessLimit' and 'FunctionTolerance' both set to 0.001 and 'PopulationSize' set to 50. For the non-linear optimization with Matlab built-in function ('*fmincon*') the default optimization options were also considered except 'MaxFunctionEvaluations' set to 10000 and 'StepTolerance' set to 0.0001.

Table 5. Lower and upper bounds (Min/Max) of the optimization variables in this case study (lengths are non-dimensional values related to the hand length and Euler angles are in radians).

$v_{i,j}$ component	Min	Max	Anatomical meaning
$x_{1,1}$	−0.3	0	CMC position
$y_{1,1}$	0.1	0.3	
$z_{1,1}$	0.1	0.3	
$\theta x_{1,1}$	$\pi/2$	π	CMC orientation
$y_{1,2}$	0.1	0.3	Metacarpal length
$\theta y_{1,2}$	$-\pi/2$	$\pi/2$	MCP orientation
$y_{1,3}$	0.1	0.3	Proximal phalanx length
$y_{1,4}$	0.1	0.3	Distal phalanx length

3 Results

Figure 7 shows the position error (ε_i, Eq. 2) for the initial IMMA hand and for the IMMA hand after optimizing the TKC following the method indicated in Table 2. According to the simplified model the position 1 of the KOT is not reachable by the original hand nor by the improved one. Without considering this position, the score in the KOT improved from 3 in the original hand to 5 in the model with optimized TKC. The position error for positions 6 to 8 improved significantly. The MPE in the optimized model was reduced to about one third with respect to the original model (0.121 to 0.035).

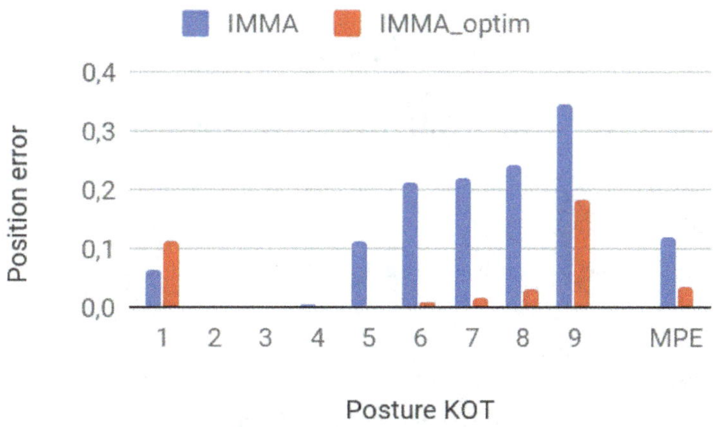

Fig. 7. Position error (ε_i, Eq. 2) for each posture of KOT and mean position error (MPE).

The kinematic chain of the improved hand can be seen in Fig. 8 and the Table 6 shows the comparison of the initial and optimized values of the parameters for the TKC. From the comparison of the thumb in Figs. 6 and 8 it can be observed that the base placement of the thumb, i.e. the CMC joint, is closer to the center of the palm in the optimized design, the orientation of the MCP joint is slightly varied and the phalanges length has changed, being the distal phalanx longer in the optimized design.

Fifty-nine generations were necessary in the genetic algorithm for the optimization of the TKC and the execution took about 10 h in an Intel Core i7 2.6 GHz processor. Figure 9 shows the evolution of the mean and best fitness values, corresponding to MPE, for the different generations of the genetic algorithm. The 50 individuals of the last generation of the genetic algorithm were all very similar among them, representing quite similar TKCs.

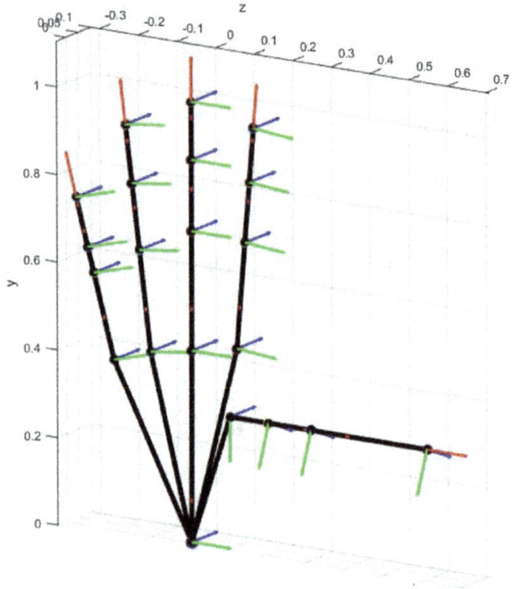

Fig. 8. Kinematic chain of the right IMMA hand with optimized TKC.

Table 6. Initial and optimized parameters of the TKC for the IMMA hand (lengths are non-dimensional values related to the hand length and Euler angles are in radians).

$v_{i,j}$ component	Initial	Optimized
$x_{1,1}$	0.000	0.000
$y_{1,1}$	0.2169	0.3000
$z_{1,1}$	0.1577	0.1002
$\theta x_{1,1}$	1.5708	1.5714
$y_{1,2}$	0.1320	0.1000
$\theta y_{1,2}$	−0.7854	−0.3855
$y_{1,3}$	0.2158	0.1094
$y_{1,4}$	0.1659	0.3000

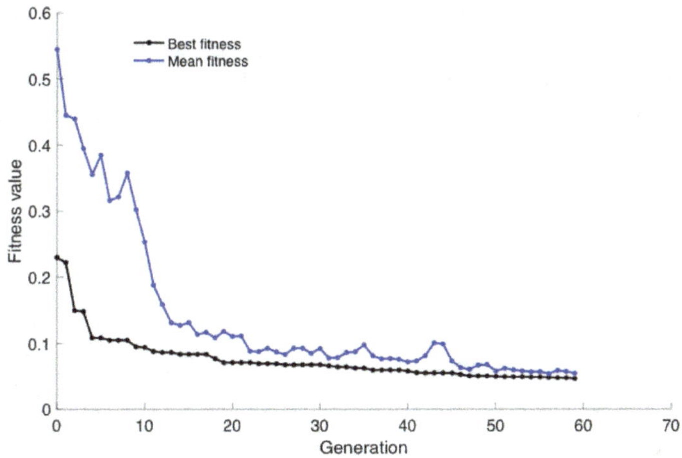

Fig. 9. Fitness value (MPE) evolution among generations of the genetic algorithm for optimizing the thumb's kinematic chain (TKC).

4 Discussion

We have presented a new method to optimize the TKC of an artificial hand based on their performance in the KOT. Despite the KOT has been taken into account in previous studies for the design of artificial hands [15, 16, 18] none of these previous studies presented a clear computational method able to do it following a formal optimization procedure and including all the parameters defining the TKC. The methodology presented here, using a double nested optimization method (Table 2) allows considering all the KOT postures for the optimization or only some of them. We have defined an index quantifying the mean weighted position error (MPE) based on the position error for each KOT posture, which can be useful as a method to compare the opposition performance of an artificial hand. The weight associated to each posture in this index can easily be tuned by the designer depending on their design specifications. Moreover, the optimization procedure allows including as variables any of the parameters defining the TKC or even other parameters of the kinematic chain of the hand.

In this study the methodology was applied, as a case study, to the optimization of the TKC of the IMMA hand, a 3D-printed cable driven hand developed by the authors. The MPE in the optimized design was reduced to less than one third of that of the original design, showing the effectiveness of the computational method. Due to the simplification of the hand model, whose segments are considered as straight lines, the computational model is only an approximation to the real prototype and some differences can be observed in the KOT score obtained with the real prototype and with the model. As Fig. 5 shows, the original prototype of the IMMA hand, can achieve the positions 1 to 4 of the KOT. Nevertheless, Fig. 7 shows a non-null position error in the model for positions 1 and 4. This could be attributed to the fact that the width and thickness of the phalanges were neglected in the simplified model and also to the fact that positions 1 and 2 where considered in the more proximal point of the corresponding index phalanx. Taking this into account, the positions 2 to 8 in the improved model can be considered as reachable in practical terms, whereas only the positions 2 to 4 are clearly reachable in the initial hand. Regarding the position 1, it could also be reachable depending on the position taken for the target point in the middle phalange of the index finger (Fig. 10).

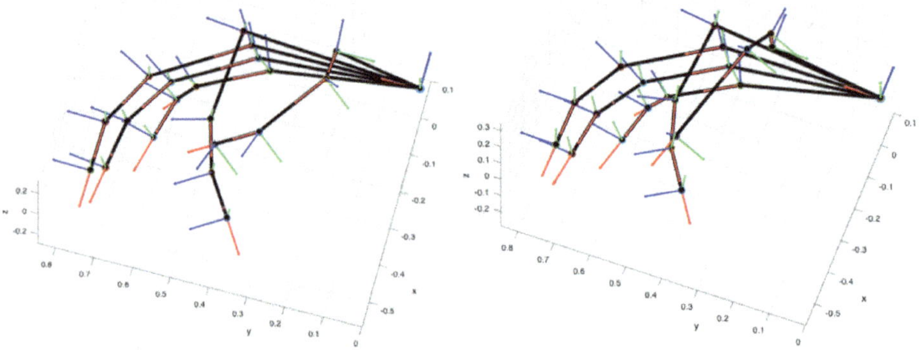

Fig. 10. Posture 1 of the Kapandji opposition test (KOT). Left: original IMMA hand. Right: optimized IMMA hand.

Some of the parameters of the optimized model (Table 6) are in the upper or lower bounds selected in the optimization (Table 5), which could indicate that widening the allowable range for the parameters could produce TKC designs able to reduce even more the MPE. This has not been analyzed in the present study but is a possible future work.

The optimized design obtained in this study has a thumb base location more distal, favoring the score in the KOT, but probably making more difficult grasping big objects. The total length of the optimized TKC is similar to that of the original design, but the proportion of the segments changed, with a longer distal phalanx and shorter metacarpal and proximal phalanges as compared to the original model. One possible reason for this is that this configuration helps to reduce the position error in the last postures of the KOT (7 to 9). We analyzed the changes in the results when the KOT positions included in the MPE are restricted to positions 1 to 6, reducing the effect of little finger opposition in the MPE. Figure 11 shows a graphical comparison of the TKC parameters of the original design and the optimized designs including positions 1 to 9 or 1 to 6 in the MPE, as well as the upper and lower bounds considered for the optimization. The results indicate that the optimized TKCs including positions 1 to 9 or 1 to 6 are very similar.

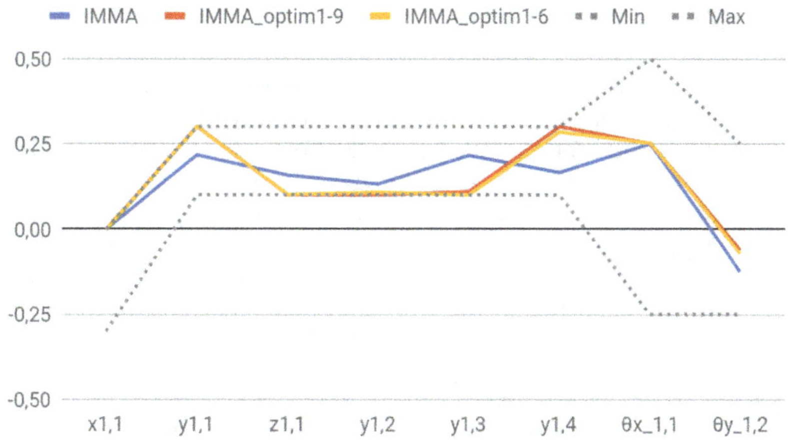

Fig. 11. Thumb's kinematic chain (TKC) parameters (angles $\theta_{x\_1,1}$ and $\theta_{y\_1,2}$ divided by 2π) for the original design of the IMMA hand and for the optimized versions obtained including positions 1–9 or 1–6 in the MPE. Upper (Max) and lower (Min) bounds for the optimization are shown with dotted lines.

Additional simulations in Matlab for the original IMMA hand and the optimized design suggest that the reduction of MPE in KOT does not guarantee a better design according to other criteria when comparing with the human hand. We compared both designs with three different anthropomorphic indexes of the kinematic chain of the whole hand [21] and the performance was similar, with differences lower to 3%, for two of them: one based on the comparison of the kinematic chain with that of the human hand; a second based on grasping postures for some primitive objects (sphere, cylinder and prism). However, the optimized design had a worse performance, 25% lower, in the index defined considering the intersection of the workspaces of the artificial hand and the human hand. This result is aligned with the observation of Roa et al. [16] about the difficulties to find direct correlations between the Kapandji test score and the size of the functional workspace. This aspect should be better investigated in the next future. The performance of the optimized design of the IMMA hand with respect to the original design should also be compared in the next future with physical prototypes, using grasping benchmarks.

The index based on the MPE defined in this study is complementary to other anthropomorphism indexes developed by the authors and existing in the literature. Probably an adequate combination of these indexes can help to improve the hand's kinematic chain and should be investigated, as well as the extension of the optimization to other parameters of the kinematic chain of the hand, not restricted to the thumb.

5 Conclusion

A straightforward methodology to analytically optimize the kinematic chain of the thumb of an artificial hand based on the performance in the KOT has been presented. The cost function defined for the optimization (MPE) is a weighted mean position error when trying to reproduce the KOT postures and can be used also as a metric to quantify thumb opposition in the hand. The application of the method to the IMMA hand thumb allowed defining a new TKC reducing the MPE to near one third of that of the original design and increasing significantly the number of reachable positions in the KOT. However, additional simulations showed that the optimized design could have a worse outcome according to other index considering the intersection between the workspace of the artificial hand and that of the human hand. Benchmarking grasping test on physical prototypes could give additional insights about the usefulness of the proposed methodology. The metric proposed based on the KOT can be used directly or in combination with other to improve the kinematic chain of artificial hands.

Acknowledgments. This work was supported by the Spanish Ministry of Economy and Competitiveness and ESF [grant number BES-2015-076005]; Spanish Ministry of Economy and Competitiveness, AEI and ERDF [grant numbers DPI2014-60635-R, DPI2017-89910-R]; and Universitat Jaume I [grant number UJI-B2017-70].

References

1. Feix, T., Kivell, T.L., Pouydebat, E., Dollar, A.M.: Estimating thumb-index finger precision grip and manipulation potential in extant and fossil primates. J. R. Soc. Interface **12**(106), 20150176 (2015)
2. Bardo, A., Vigouroux, L., Kivell, T.L., Pouydebat, E.: The impact of hand proportions on tool grip abilities in humans, great apes and fossil hominins: a biomechanical analysis using musculoskeletal simulation. J. Hum. Evol. **125**, 106–121 (2018)
3. Santos, V.J., Valero-Cuevas, F.J.: Reported anatomical variability naturally leads to multimodal distributions of Denavit-Hartenberg parameters for the human thumb. IEEE Trans. Biomed. Eng. **53**(2), 155–163 (2006)
4. Giurintano, D.J., Hollister, A.M., Buford, W.L., Thompson, D.E., Myers, L.M.: A virtual five-link model of the thumb. Med. Eng. Phys. **17**(4), 297–303 (1995)
5. Coert, J.H., van Dijke, G.A.H., Hovius, S.E.R., Snijders, C.J., Meek, M.F.: Quantifying thumb rotation during circumduction utilizing a video technique. J. Orthop. Res. **21**(6), 1151–1155 (2003)
6. Belter, J.T., Segil, J.L., Dollar, A.M., Weir, R.F.: Mechanical design and performance specifications of anthropomorphic prosthetic hands: a review. J. Rehabil. Res. Dev. **50**(5), 599–618 (2013)
7. Ten Kate, J., Smit, G., Breedveld, P.: 3D-printed upper limb prostheses: a review. Disabil. Rehabil. Assist. Technol. **12**(3), 300–314 (2017)
8. Grebenstein, M., Chalon, M., Hirzinger, G., Siegwart, R.: A method for hand kinematics designers 7 billion perfect hands. In: 1st International Conference Applied Bionics and Biomechanics (2010)

9. Ottobock: Ottobock myoelectric prosthetics (2017). https://www.ottobockus.com/prosthetics/upper-limb-prosthetics/solution-overview/myoelectric-prosthetics/. Accessed 01 March 2019
10. Gaiser, I., et al.: A new anthropomorphic robotic hand. In: Humanoids 2008 - 8th IEEE-RAS International Conference on Humanoid Robots, pp. 418–422 (2008)
11. Liu, H., et al.: Multisensory five-finger dexterous hand: the DLR/HIT Hand II. In: 2008 IEEE/RSJ International Conference on Intelligent Robots and Systems, pp. 3692–3697 (2008)
12. Shadow Robot Company: Shadow dexterous hand (2019). https://www.shadowrobot.com/products/dexterous-hand/. Accessed 01 March 2019
13. Kapandji, A.: Clinical opposition and reposition test of the thumb [Cotation clinique de l'opposition et de la contre-opposition du pouce]. Ann. Chir. la Main **5**(1), 67–73 (1986)
14. Shin, S., Choi, D., Choi, M., Moon, H., Choi, H.R., Koo, J.C.: Development of dexterous robot hand for delicate object grasping. In: 2012 9th International Conference on Ubiquitous Robots and Ambient Intelligence (URAI), pp. 462–463 (2012)
15. Chalon, M., Dietrich, A., Grebenstein, M.: The thumb of the anthropomorphic Awiwi hand: from concept to evaluation. Int. J. Humanoid Robot. **11**(3), 1450019 (2014)
16. Roa, M.A., et al.: Towards a functional evaluation of manipulation performance in dexterous robotic hand design. In: 2014 IEEE International Conference on Robotics and Automation (ICRA), pp. 6800–6807 (2014)
17. Deimel, R., Brock, O.: A novel type of compliant and underactuated robotic hand for dexterous grasping. Int. J. Robot. Res. **35**(1–3), 161–185 (2016)
18. Cerruti, G., Chablat, D., Gouaillier, D., Sakka, S.: Design method for an anthropomorphic hand able to gesture and grasp. In: 2015 IEEE International Conference on Robotics and Automation (ICRA), June 2015, pp. 3660–3667 (2015)
19. Llop-Harillo, I., Pérez-González, A.: System for the experimental evaluation of anthropomorphic hands. Application to a new 3D-printed prosthetic hand prototype. Int. Biomech. **4** (2), 50–59 (2017)
20. Wu, G., Cavanagh, P.R.: ISB recommendations for standardization in the reporting of kinematic data. J. Biomech. **28**(10), 1257–1261 (1995)
21. Llop-Harillo, I., Pérez-González, A., Andrés de la Esperanza, F.J.: Comparación de la cadena cinemática de manos antropomorfas artificiales con la de la mano humana. In: Anales de Ingeniería Mecánica. Revista de la Asociación Española de Ingeniería Mecánica. Actas del XXII Congreso Nacional de Ingeniería Mecánica, pp. 49–68 (2018)

Mathematical Model of Age-Specific Tendon Healing

Akinjide R. Akintunde[1], Daniele E. Schiavazzi[2], and Kristin S. Miller[1(✉)]

[1] Tulane University, New Orleans, LA 70112, USA
kmillell@tulane.edu
[2] University of Notre Dame, Notre Dame, IN 46556, USA

Abstract. Tendons perform unique functions in the body—transmit muscle generated force to bones for joint motion. However, decreased mechanical response is observed post-injury and during aging, which in turn limits functional capacity. While there are many strategies aimed at restoration of pre-injury mechanical properties, they fail typically because of lack of understanding of tendon healing mechanisms, particularly at the extracellular matrix level. Toward this end, mathematical models, especially those with microstructural details can be insightful. In prior study, we evaluated the ability of three constitutive models to describe uniaxial mechanical test data from murine patellar tendons excised pre- and post-injury from multiple age groups. The chosen models range from simple i.e. the Freed-Rajagopal (FR) model, to complex i.e. the Gasser-Ogden-Holzapfel (GOH) and Shearer (SHR) models. Least-squares optimization was performed to obtain model parameter values, while the models fitted the experimental data adequately, the relatively complex models exhibited low parameter identifiability evidenced by high correlation. To address the limitations observed in the prior study, we adopted a Bayesian approach using an adaptive Markov chain Monte Carlo (MCMC) to compute the posterior distribution of model parameters. Agreement of two approaches was observed only in the FR model parameters. This study highlights the trade-off between model complexity and confidence level of inferred parameters, the critical need for structural data to motivate clinical relevance of mathematical model for tendons. Addressing these needs would enhance translational research and motivate the rational design of tissue engineering strategies for better treatment outcomes.

Keywords: Tendon aging · Tendon injury · Tendon model · Bayesian inference

1 Introduction

Tendons are connective tissues that transmit muscle-generated force to bones to permit joint mobility [1, 2]. Further, tendons contribute to overall joint stability and protect muscles by absorbing external impact. However, tendon injuries are common, debilitating, and painful disorders characterized by altered tissue composition and structure [3–5]. Current treatment strategies often fail to restore tendons to pre-injury functional

G. A. Ateshian et al. (Eds.): CMBBE 2019, LNCVB 36, pp. 288–296, 2020.
https://doi.org/10.1007/978-3-030-43195-2_23

capacity [6, 7]. The underlying extracellular matrix (ECM) dynamics that drive post-injury mechanical response and influence the healing process are not fully elucidated [6, 7]. Further, aging is considered a risk factor for tendon injury and the healing potential of tendons is thought to be impaired with age [8]. The effect of aging and age-dependent healing on tendon ECM dynamics, however, is not fully elucidated [9–11]. Therefore, there is a need to elucidate the underlying ECM changes in tendons post injury, and the impact of age on tendon healing potential. Towards this end, we adopted deterministic and Bayesian approaches to parameter inference by evaluating the descriptive ability of three microstructurally motivated constitutive models fitted to uniaxial tensile data from mechanically tested murine patellar tendons.

2 Methods

2.1 Experimental Data

From IACUC-approved published studies [12–15], we obtained pre- (0 week) and post-injury (3 and 6 weeks) stress-stretch data from uniaxial extension tests performed on female murine patellar tendon at three age timepoints—120, 270 and 540 days. The studies followed an established biopsy-punch injury model to the mid-substance of the patellar tendon [16, 17].

2.2 Models

We implemented three constitutive models with varied levels of microstructural details to fit the stress-stretch data sets. Of the three models, the Shearer (SHR) model [18] is the most microstructural with parameters related to collagen organization at the fibrillar and fascicular levels. The Gasser-Ogden-Holzapfel (GOH) model [19] accounts for the dispersion of collagen fiber via its dispersion and alignment parameters. The Freed-Rajagopal (FR) model [20] has three parameters: the toe-region and linear-region moduli, and the transition stretch between the regions. Complete details on these models and our implementation can be found in Akintunde and Miller [21] and Akintunde et al. [22]. Briefly, the axial Cauchy stress (σ_{zz}) – axial stretch (λ) relationship resulting from the SHR model is:

$$\sigma_{zz}^{\mathrm{SHR}} = (1-\phi)\,\mu\left(\lambda^2 - \frac{1}{\lambda}\right) + 2\,\chi^{\mathrm{SHR}}\left(\lambda^2\cos^2\psi - \frac{\sin^2\psi}{2\lambda}\right) \tag{1}$$

where

$$\chi^{SHR} = \begin{cases} 0 & I_4 < 1 \\ \phi \dfrac{E}{6\sqrt{I_4^{SHR}} \sin^2 \theta_0} \left[2 - \left(\dfrac{3 I_4 - 1}{I_4^{SHR} \sqrt{I_4^{SHR}}} \right) \right], & 1 \le I_4 \le \lambda^{*2} \\ \phi \dfrac{E}{2\sqrt{I_4^{SHR}}} \left(2 \dfrac{1 - \cos^3 \theta_0}{3 \sin^2 \theta_0} - \dfrac{1}{\sqrt{I_4^{SHR}}} \right) & I_4 > \lambda^{*2} \end{cases} \tag{2}$$

The parameters in the model are $\phi \in (0, 1]$—collagen fiber volume fraction, μ—the shear modulus of the ground matrix which is made up of non-collagenous substances such as elastin, glycosaminoglycans, and proteoglycans, E—fibril Young's modulus, θ_0—the crimp angle of the outermost fibrils and $\lambda^* = 1/\cos \theta_0$ is the critical or transition stretch. $I_4^{SHR} = \lambda^{-1} \sin^2 \psi + \lambda^2 \cos^2 \psi$, where ψ is the fascicle alignment angle. Similarly, for the GOH model,

$$\sigma_{zz}^{GOH} = (1 - \phi)\, \mu \left(\lambda^2 - \frac{1}{\lambda} \right) + \chi^{GOH} \left[\kappa \left(\lambda^2 - \frac{1}{\lambda} \right) + (1 - 3\,\kappa) \left(\lambda^2 \cos^2 \xi - \frac{\sin^2 \xi}{2\,\lambda} \right) \right] \tag{3}$$

Where, $\chi^{GOH} = 4\,\phi\, c_1\, v\, e^{c_2 v^2}$, $v = \kappa I_1 + (1 - 3\,\kappa) I_4^{GOH} - 1$, $\kappa \in [0, 1/3]$ captures dispersion of collagen fibers about tendon's long axis and $I_4^{GOH} = \lambda^{-1} \sin^2 \xi + \lambda^2 \cos^2 \xi$. $c_1 > 0$ is modulus-like parameter, and $c_2 > 0$ is a dimensionless parameter.
For the FR model,

$$\ln \lambda = \frac{1}{\beta} \left\{ 1 - \frac{1}{\left[1 + (\beta - 1) \frac{\sigma_{zz}^{exp}}{E_e} \right]^{\frac{\beta}{\beta - 1}}} \right\} + \frac{\sigma_{zz}^{exp}}{E_c} \tag{4}$$

E_e (>0) and E_c (>E_e) are the elastic moduli of the toe and linear regions—dominated by elastin (e) and collagen fibers (c), respectively and $\beta = 1/\epsilon_e^{max} (> 1)$ is the inverse of the true strain at which the linear region starts, and the collagen fibers become straight.

2.3 Deterministic and Bayesian Approach to Inference

We employed least squares regression algorithm and Bayesian estimation to determine the optimal values of the model parameters. While details can be found in the respective articles [21, 22]. Briefly, we employed the trust region reflective algorithm to minimize normalized difference between theoretical and experimental axial Cauchy stresses (SHR & GOH) and true strain (FR). We assessed local parameter correlation via the correlation matrix and performed sensitivity analysis via the normalized stretch-dependent derivatives of the dependent model variable with respect to each model parameter.

To assess changes due to age and injury, a two-way analysis of variance (ANOVA) was performed on model parameter values.

In the Bayesian approach to estimation, uniform priors were assumed for the model parameters. Differential evolution adaptive metropolis algorithm for Markov chain Monte Carlo (MCMC) was employed to generate samples from the posterior distribution, while assessing convergence through the Gelman-Rubin metric. Maximum-a - posteriori (MAP) estimates were determined via restarted Nelder-Mead algorithm using MCMC samples with maximum posterior as initial guesses. Local identifiability was assessed via parameter correlations and learning factor $\in [0, 1]$, it is 0 when the marginal posterior variance is unchanged from its prior value and 1 when it is zero i.e. the parameter was perfected *learned* during the process. We computed local and global sensitivity indices via finite difference and Monte Carlo estimates, respectively. Collagen fibers of the patellar tendon are known to be relatively aligned along the long axis of the tendon [23]. Hence, Bayesian inference was performed including and excluding (fixed at zero) the collagen fiber (ξ) and fascicle (ψ) alignment angles. In addition, we performed Bayesian estimation using data for all the samples per group (ALL) and for the average data over all the samples per group (AVG). Uncertainty in model parameter values was assessed by computing a 10–90% confidence interval.

For both approaches, parameters were constrained to bounds motivated by theoretical and experimental values (see Table 1 of [21]).

3 Results

3.1 Deterministic Inference

The three models exhibited adequate qualitative (Fig. 1) and quantitative fits to the experimental data [R^2 (mean $\pm$ std. err), for the models are 0.992 ± 0.001 (SHR), 0.976 ± 0.001 (GOH), and 0.942 ± 0.008 (FR)].

The FR and GOH model exhibited the least and largest relative error values, respectively, particularly at lower axial stretch values. We found dependencies between the collagen fibril Young's modulus (E) and crimp angle (θ_0) of the SHR model with a correlation value ≈ -1, and between modulus-like parameter (c_1) and fiber alignment angle (ξ) of the GOH model with correlation value ≈ 1. Unsurprisingly, for SHR and GOH models, the structural parameters related to collagen organization (κ-fiber dispersion, ψ-fascicle alignment angle) were most influential to the mechanical response of the patellar tendon during aging and healing. From the ANOVA results, consistent with their significant influence, both parameters exhibited statistical significance ($p < 0.05$) as a function of age, injury and age injury interaction. We did not observe dependencies in the FR parameters. The toe- and linear-region moduli's most significant influences were localized to the toe and linear regions, respectively. In addition,

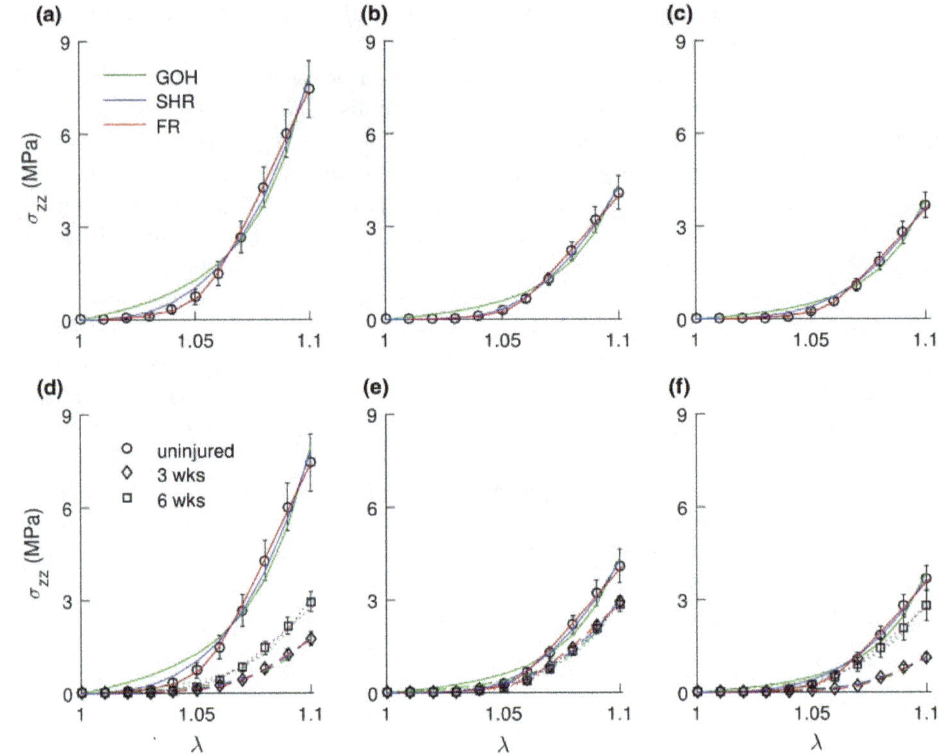

Fig. 1. The model-to-data fits for the three models. On the y-axis is the axial Cauchy stress and the axial stretch ratio is on the x-axis for (a) to (f). Top row: (a) to (c) Uninjured i.e. 0 week data for 120-, 270-, and 540-day age groups, respectively. Bottom row: (d) to (f) 0-, 3- and 6-week data and model fits for 120-, 270- and 540-day age groups, respectively. Models exhibited adequate qualitative fits to data. Decline in mechanical response observable as a function of age (a–c) and injury (d, e, f). Figure from [21].

the moduli were statistically significant ($p < 0.05$) with age, injury and their interaction. Interesting, while we observed statistically significant post-injury decrease (0 to 3 week) and subsequent increase during healing (3 to 6 week) for the moduli in the 120- and 540-day groups, the 270-day group did not exhibit a similar behavior, showing no statistically significant change post-injury and during healing.

3.2 Bayesian Inference

Learning factor values indicated that using data for all the samples (ALL) rather than average data improved parameter identifiability for all three models (Fig. 2).

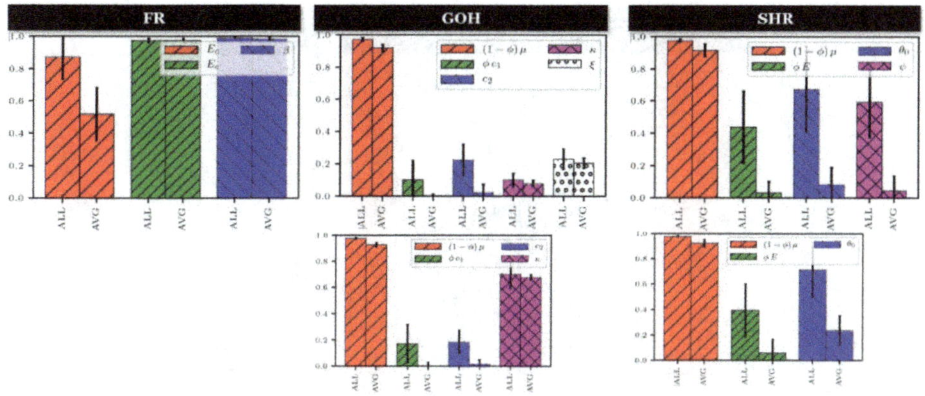

Fig. 2. The learning factor values for parameters of FR (left), GOH (middle) and SHR (right) models with (top row) collagen alignment angles included for GOH and SHR models and (bottom row) excluded. All FR parameters were identifiable (1st column). Identifiability of collagen dispersion parameter (κ) improved significantly when alignment angle is set to zero (2nd column). No improvement was observed in learning factors for SHR parameters with fascicle alignment angle set to zero (3rd column). Figure adapted from [22].

In the GOH model, fixing the alignment angle at zero improved identifiability of the collagen dispersion parameter significantly. For the SHR model, fixing the fascicle alignment angle had negligible effect on the identifiability of its other parameters (Fig. 2). Overall the FR model exhibited high identifiability evidenced by learning factor values ≈ 1. When the alignment angles are included in the inference, the structural parameter pairs of the SHR (θ_o & ψ) and GOH (κ & ξ) models exhibited strong negative correlation. When excluded, in the SHR model, the fibril Young's modulus (E) and crimp angle (θ_o) exhibited high positive correlation, similarly, the modulus-like parameter (c_1) and fiber dispersion (κ) exhibited high positive correlation in the GOH model. However, the FR model parameters did not exhibit high correlation values. Agreement between the MAP estimates and optimization derived values for parameters was observed only in the FR model. Local and global sensitivity results indicate structural parameters κ and θ_o as most influential to the patellar tendon's mechanical response during aging and healing. The FR parameters exhibited less uncertainty, i.e. shorter shaded area widths, than those of the SHR and GOH parameters (Fig. 3).

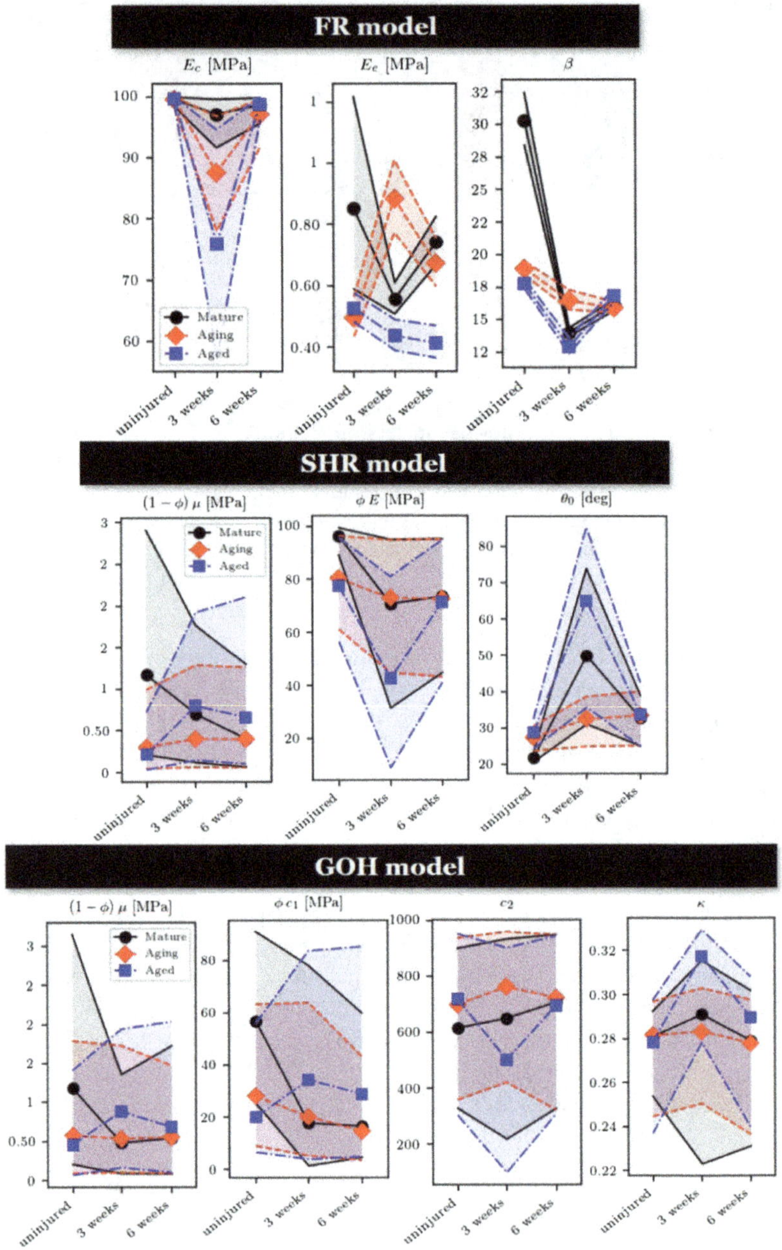

Fig. 3. Changes with injury in the median values and for all age groups. Shaded regions represent 10–90% confidence interval. Top row: FR model, middle row: SHR model and bottom row: GOH model. The FR model exhibited the least uncertainty shown by slim width of confidence interval. Figure adapted from [22].

4 Discussion and Conclusion

We evaluated three microstructurally motivated models of varied complexity in describing age and age-dependent injury-related changes in mechanical behavior and properties of murine patellar tendon. We observed a post-injury behavior in the 270-day group that is different from that of the 120- and 540-day groups. We posit that estimation methods such as Bayesian inference employed herein, provide robust frameworks to explore the entire parameter space of model parameters, their identifiability, uncertainties and their global influence on the model output rather than point estimates typically obtained via optimization. While the SHR and GOH models provide desirable microstructural information, their identifiability is relatively low when compared to the simple, phenomenological, but highly determinable FR model. The inherent dependencies due to their complexities captured by the correlation parameter values are responsible for the low parameter identifiability and high uncertainties in inference. To increase confidence in the inference made from these complex models, additional experimental data, particularly for structural parameters (considering their relatively substantial influence) would be needed to optimize and estimate the model parameter. In this regard imaging data is highly essential to motivate bounds for or fix these parameters. Future work will be devoted to multiaxial mechanical test and imaging data collection, and improved Bayesian methods for physiologically relevant inference.

Acknowledgments. The authors acknowledge support from NIH through grants R01 EB18302 (DES) and P20 GM103629 (KSM).

References

1. Butler, et al.: Biomechanics of ligaments and tendons. Exerc. Sport Sci. Rev. **6**, 125–181 (1978)
2. Józsa, L., Kannus, P.: Human Tendons: Anatomy, Physiology, and Pathology. Human Kinetics Publishers, Champaign (1997)
3. Praemer, A.: Musculoskeletal Conditions in the United States. AAOS, Rosemont (1999)
4. Sharma, P., Maffuli, N.: Biology of tendon injury: healing, modeling and remodeling. J. Musculoskelet. Neuronal Interact. **6**(2), 181–190 (2006)
5. Nourissat, G., et al.: Tendon injury: from biology to tendon repair. Nat. Rev. Rheumatol. **11**(4), 223–233 (2015)
6. Andarawis-Puri, N., et al.: Tendon basic science: development, repair, regeneration, and healing. J. Orthop. Res. **33**(6), 780–784 (2015)
7. Docheva, D., et al.: Biologics for tendon repair. Adv. Drug Deliv. Rev. **84**, 222–239 (2015)
8. Svensson, R., et al.: Effect of aging and exercise on the tendon. J. Appl. Physiol. **121**(6), 1353–1362 (2016)
9. Hubbard, R., Soutas-Little, R.: Mechanical properties of human tendon and their age dependence. J. Biomech. Eng. **106**(2), 144–150 (1984)
10. Shadwick, R.: Elastic energy storage in tendons: mechanical differences related to function and age. J. Appl. Physiol. **68**(3), 1033–1040 (1990)

11. Bailey, A.: Molecular mechanisms of ageing in connective tissues. Mech. Ageing Dev. **122**(7), 735–755 (2001)
12. Dunkman, A.A., et al.: Decorin expression is important for age-related changes in tendon structure and mechanical properties. Matrix Biol. **32**(1), 3–13 (2013)
13. Dunkman, A.A., et al.: The injury response of aged tendons in the absence of biglycan and decorin. Matrix Biol. **35**, 232–238 (2014)
14. Dunkman, A.A., et al.: The tendon injury response is influenced by decorin and biglycan. Ann. Biomed. Eng. **42**(3), 619–630 (2014)
15. Mienaltowski, M., et al.: Injury response of geriatric mouse patellar tendons. J. Orthop. Res. **34**(7), 1256–1263 (2016)
16. Lin, T., et al.: Tendon healing in interleukin-4 and interleukin-6 knockout mice. J. Biomech. **39**(1), 61–69 (2006)
17. Buckley, M., et al.: Validation of an empirical damage model for aging and in vivo injury of the murine patellar tendon. J. Biomech. Eng. **135**(4), 041005-1–041005-7 (2013)
18. Shearer, T.: A new strain energy function for modelling ligaments and tendons whose fascicles have a helical arrangement of fibrils. J. Biomech. **48**(12), 3017–3025 (2015)
19. Gasser, T., et al.: Hyperelastic modelling of arterial layers with distributed collagen fibre orientations. J. R. Soc. Interface **3**(6), 15–35 (2006)
20. Freed, A., Rajagopal, K.: A promising approach for modeling biological fibers. Acta Mech. **227**(6), 1609–1619 (2016)
21. Akintunde, A., Miller, K.: Evaluation of microstructurally motivated constitutive models to describe age-dependent tendon healing. Biomech. Model. Mechanobiol. **17**(3), 793–814 (2018)
22. Akintunde, A., et al.: Bayesian inference of constitutive model parameters from uncertain uniaxial experiments on murine tendons. J. Mech. Behav. Biomed. Mater. **96**, 285–300 (2019)
23. Shearer, T., et al.: X-ray computed tomography of the anterior cruciate ligament and patellar tendon. Muscles Ligaments Tendons J. **4**(2), 238 (2014)

Inception of Material Instabilities in Arteries

P. Mythravaruni and K. Y. Volokh(✉)

Faculty of Civil and Environmental Engineering, Technion - I.I.T, 3200003 Haifa, Israel
cvolokh@technion.ac.il,
https://cee.technion.ac.il/members/konstantin-volokh/

Abstract. We propose a theoretical approach to predict the onset of cracks in arterial wall. The arterial wall is a soft composite made up of hydrated ground matrix of proteoglycans reinforced by elastin and collagen fibers spatially dispersed in the matrix. Like any other material, the arterial tissue cannot store and dissipate strain energy above a certain threshold. This threshold value is introduced in the constitutive theory via energy limiters. The limiters naturally constrain reachable stresses and enable analysis of mathematical condition of strong ellipticity. Loss of the strong ellipticity corresponds to the juncture when superimposed waves cease to propagate due to localization of material failure into cracks perpendicular to a possible wave direction. Thus, the direction in which crack starts to appear can be analyzed in addition to its inception. We enrich the recently developed constitutive theories that account for fiber dispersion of the arterial wall by including 8 and 16 structure tensors with energy limiters. We analyze the loss of strong ellipticity in uniaxial tension in circumferential and axial directions of the arterial wall. We find that cracks appear in the direction perpendicular to tension, when the speed of the superimposed longitudinal wave vanishes. We also find that the appearance of cracks is predicted in the direction inclined (non-perpendicular) to tension, when the speed of the superimposed transverse wave vanishes.

Keywords: Failure localization · Strong ellipticity · Superimposed waves · Structure tensors

1 Introduction

In this paper, we describe a procedure to find the inception of material instabilities in arterial tissue. This requires developing constitutive models of arteries capable of failure description and examining the mathematical condition of strong ellipticity of the model for the given deformation.

G. A. Ateshian et al. (Eds.): CMBBE 2019, LNCVB 36, pp. 297–304, 2020.
https://doi.org/10.1007/978-3-030-43195-2_24

Constitutive models of the arterial wall have a great wealth of literature. Experiments [1] showed anisotropic and heterogeneous characteristics of arteries. Fung with collaborators introduced nonlinear elasticity models [2,3] to capture the large deformation response at low stresses. In these models the characteristic material directions were along the radial, circumferential and axial directions of artery. Numerous Fung-type phenomenological theories were proposed since then [4–6]. Subsequently, frame-invariant forms of these models were developed by introducing the structure tensors [7–9]. Furthermore, the anisotropy was accounted for by introducing the angular dispersion of collagen fibers in the strain energy density to develop more physically appealing structural models [10–14]. These models with analytically defined angular fiber dispersion needed angular integration on a unit sphere that was computationally intensive. The approach of generalized structure tensors (GST), which included fiber dispersion in structure tensor instead of strain energy density, was introduced to reduce the computational cost [15,16]. Inability to easily exclude the compressed fibers was the major drawback of the GST approach although it was computationally attractive. An alternative approach that allowed for an easy exclusion of compressed fibers, without forgoing the advantages of the fiber dispersion models, was developed recently in [17]. This latter approach used 16 and 8 specially chosen structure tensors to describe the fiber dispersion.

A systematic method capable of describing material failure, which was based on the introduction of the energy limiters in the strain energy functions, was proposed in [18–20]. The violation of the strong ellipticity condition can be readily examined using this approach [21].

We enhance these recently developed constitutive theories of the arterial wall including 8 and 16 structure tensors fiber dispersion models [17] with energy limiters. We analyze the loss of strong ellipticity in uniaxial tension in circumferential and axial directions of the arterial wall. We find that the imposition of the incompressibility constraint can have a significant effect on the crack direction.

2 Constitutive Theory

We assume that the arterial wall exhibits hyperelastic response. Arterial wall is made of collagen fibers dispersed in isotropic ground matrix. The strain energy function W of the intact artery wall involves two terms

$$W = g + f, \tag{1}$$

where

$$g = \frac{c}{2}(I_3^{-1/3} I_1 - 3) + K(I_3^{1/2} - 1)^2, \tag{2}$$

is the neo-Hookean strain energy for the isotropic ground matrix, where c and K are the shear and bulk modulus respectively and I_1 and I_3 are the first ad third strain invariants. In the limit of incompressibility, $I_3 = 1$ is imposed in (2).

The total energy of dispersed fibers integrated on a unit sphere is

$$f = \int \rho w d\Xi = \sum_i \gamma^{(i)} w^{(i)}, \tag{3}$$

where the cubature formula for numerical integration appears on the right hand side.

Here, $\gamma^{(i)}$ is the weight coefficient, Ξ is a solid angle; ρ is the angular density of the fiber distribution (see [17,22,23]), which is normalized as $\int \rho d\Xi = 4\pi$; w is the strain energy of single fiber per unit reference volume [7] and

$$w^{(i)} = \frac{k_1}{2k_2}(\exp[k_2 \left\langle I_4^{(i)} - 1 \right\rangle^2] - 1), \tag{4}$$

where k_1 and k_2 are material parameters; the strain invariant $I_4^{(i)} = \mathbf{C} : \mathbf{a}^{(i)} \otimes \mathbf{a}^{(i)} > 1$; and triangular brackets denote Macaulay brackets to exclude the fiber response in compression, where $\langle x \rangle = 0 \, \forall \, x < 0$ and x otherwise.

We choose a unit vector in the direction of a generic material fiber in the initial configuration as

$$\mathbf{a}^{(i)}(\Phi^{(i)}, \Theta^{(i)}) = \cos\Phi^{(i)} \sin\Theta^{(i)} \mathbf{e}_1 + \sin\Phi^{(i)} \sin\Theta^{(i)} \mathbf{e}_2 + \cos\Theta^{(i)} \mathbf{e}_3, \tag{5}$$

where $0 \leq \Phi \leq 2\pi$ and $0 \leq \Theta \leq \pi$ and integration points $(\Phi^{(i)}, \Theta^{(i)})$ are taken on the unit sphere.

Φ is the angle in the tangent plane measured from the circumferential direction $\mathbf{e}_1$ to the axial direction $\mathbf{e}_2$. Θ is the angle in the normal plane measured from the radial direction $\mathbf{e}_3$ in this plane. $\mathbf{a}^{(i)} \otimes \mathbf{a}^{(i)}$ denote a finite number of structure tensors representing fiber dispersion that account for anisotropy.

The limited bond energy of the particles in a representative volume restricts the strain energy density on the macroscopic scale. Bounded strain energy implies that a material can not sustain stress beyond a limit which leads to material failure. Thus, we introduce a limiter in the strain energy in the following form [20], to analyse the onset of failure,

$$\psi(\mathbf{F}) = \psi_{\mathrm{f}} - \psi_{\mathrm{e}}(\mathbf{F}), \tag{6}$$

where, $\psi_{\mathrm{e}}(\mathbf{F}) = \phi m^{-1} \Gamma(m^{-1}, W(\mathbf{F})^m \phi^{-m})$, $\psi_{\mathrm{f}} = \psi_{\mathrm{e}}(\mathbf{1})$, and $\Gamma(s, x) = \int_x^\infty t^{s-1} e^{-t} dt$ is the upper incomplete gamma function.

Here, ψ_{f} is the failure energy; $\psi_{\mathrm{e}}(\mathbf{F})$ is the elastic energy; $\mathbf{1}$ is identity tensor; ϕ is the energy limiter (average bond energy); and m is a material parameter. A regularized formulation as in [26,27], for example, should be used, if the failure propagation is also of interest.

3 Strong Ellipticity Condition

Incremental equations of momenta balance and constitutive equation in the Eulerian form, where the current configuration Ω is referential [19], can be written as follows

$$\rho \ddot{\tilde{\mathbf{y}}} = \mathrm{div}\tilde{\boldsymbol{\sigma}}, \quad \tilde{\boldsymbol{\sigma}} + \boldsymbol{\sigma}\tilde{\mathbf{L}}^{\mathrm{T}} = (\tilde{\boldsymbol{\sigma}} + \boldsymbol{\sigma}\tilde{\mathbf{L}}^{\mathrm{T}})^{\mathrm{T}}, \tag{7}$$

and

$$\tilde{\boldsymbol{\sigma}} = \mathbb{A} : \tilde{\mathbf{L}} + \{\Pi\tilde{\mathbf{L}}^{\mathrm{T}} - \tilde{\Pi}\mathbf{1}\}, \quad \{\tilde{\mathbf{L}} : \mathbf{1} = 0\}, \tag{8}$$

where $\rho = I_3^{-1/2}\rho_0$ is the current mass density; $\boldsymbol{\sigma} = I_3^{-1/2}\mathbf{P}\mathbf{F}^{\mathrm{T}}$ is the Cauchy stress tensor and $(\mathrm{div}\tilde{\boldsymbol{\sigma}})_i = \partial\sigma_{ij}/\partial y_j$; $\tilde{\boldsymbol{\sigma}} = I_3^{-1/2}\tilde{\mathbf{P}}\mathbf{F}^{\mathrm{T}}$ is the incremental Cauchy stress; $\tilde{\mathbf{L}} = \tilde{\mathbf{F}}\mathbf{F}^{-1}$ is the incremental velocity gradient. Note that terms in braces {...} should be considered for incompressible material models only.

The fourth order instantaneous elasticity tensor $\mathbb{A}$ has Cartesian components

$$A_{ijkl} = I_3^{-1/2} F_{js} F_{lm} \frac{\partial^2 \psi}{\partial F_{is} \partial F_{km}}. \tag{9}$$

We use the strain energy defined by (6) to calculate,

$$\frac{\partial^2 \psi}{\partial F_{is} \partial F_{km}} = \left(\frac{\partial^2 W}{\partial F_{is} \partial F_{km}} - m W^{m-1} \phi^{-m} \frac{\partial W}{\partial F_{km}} \frac{\partial W}{\partial F_{is}} \right) \exp[-W^m \phi^{-m}]. \tag{10}$$

Substitution of (10) in (9) yields

$$A_{ijkl} = I_3^{-1/2} F_{js} F_{lm} \left(\frac{\partial^2 W}{\partial F_{is} \partial F_{km}} - m W^{m-1} \phi^{-m} \frac{\partial W}{\partial F_{km}} \frac{\partial W}{\partial F_{is}} \right) \exp[-W^m \phi^{-m}]. \tag{11}$$

We choose the following form for a plane wave solution of the incremental initial-boundary-value problem

$$\tilde{\mathbf{y}} = \mathbf{r} g(\mathbf{s} \cdot \mathbf{y} - vt), \quad \{\tilde{\Pi} = \Upsilon g'(\mathbf{s} \cdot \mathbf{y} - vt),\} \tag{12}$$

where $\mathbf{r}$ and $\mathbf{s}$ are the unit vectors in the directions of wave polarization and wave propagation respectively; v is the wave speed; g' denotes the differential of g with respect to the argument of the function.

We get the incremental stress $\tilde{\boldsymbol{\sigma}}$ by substituting for $\tilde{\Pi}$ and $\tilde{\mathbf{L}} = \mathrm{grad}\tilde{\mathbf{y}} = \partial\tilde{\mathbf{y}}/\partial\mathbf{y}$ from (12) to $(8)_1$. Then, substituting this incremental stress $\tilde{\boldsymbol{\sigma}}$ and $\tilde{\mathbf{y}}$ from (12) into the linear momentum balance $(7)_1$, we get

$$\rho v^2 \mathbf{r} = \boldsymbol{\Lambda}(\mathbf{s})\mathbf{r} - \{\Upsilon \mathbf{s}\}, \tag{13}$$

where $\boldsymbol{\Lambda}(\mathbf{s})$ is the acoustic tensor with Cartesian components $\Lambda_{ik} = A_{ijkl} s_j s_l$.

Taking the scalar product of (13) with $\mathbf{r}$, we obtain for the wave speed

$$I_3^{1/2} \rho v^2 = I_3^{1/2} \mathbf{r} \cdot \boldsymbol{\Lambda} \mathbf{r} = f_1 f_2, \tag{14}$$

where

$$f_1 = f_3 - m W^{m-1} \phi^{-m} f_4^2, \quad f_2 = \exp[-W^m \phi^{-m}], \tag{15}$$

$$f_3 = s_j s_l r_i r_k F_{js} F_{lm} \frac{\partial^2 W}{\partial F_{is} \partial F_{km}}, \quad f_4 = r_k s_l F_{lm} \frac{\partial W}{\partial F_{km}}. \tag{16}$$

The mathematical condition of the strong ellipticity of the incremental initial boundary-value-problem is violated when the wave speed becomes zero and, physically, it means material fails to propagate a wave in direction $\mathbf{s}$. The latter notion can also be interpreted as the onset of a crack perpendicular to $\mathbf{s}$.

Here, we will take into account longitudinal wave (P-wave) and transverse wave (S-wave) in the plane of the arterial sheet for calculating the condition of vanishing wave speed.

We write

$$\mathbf{s} = \mathbf{r} = \cos\alpha \mathbf{e}_1 + \sin\alpha \mathbf{e}_2 \tag{17}$$

for the P-wave and

$$\mathbf{s} = \cos\alpha \mathbf{e}_1 + \sin\alpha \mathbf{e}_2, \quad \mathbf{r} = -\sin\alpha \mathbf{e}_1 + \cos\alpha \mathbf{e}_2 \tag{18}$$

for the S-wave, where α is unknown angle in the tangent plane spanned by the unit tangent vectors $\mathbf{e}_1$ in the circumferential and $\mathbf{e}_2$ in axial directions of the arterial wall, respectively.

4 Specialization of Material

We use the material models that were experimentally calibrated for the intact material behavior [17]. We enhance them with a failure description by incorporating energy limiters. The model using 16 structure tensors includes out of plane fiber dispersion while the model using 8 structure tensors does not. Parameters for these models are given in Table 1. For details on integration points and weight coefficients, reader is referred to [17].

Table 1. Material constants for models with 16 and 8 structure tensors

Structure tensors	c (kPa)	k_1(kPa)	k_2	ϕ (kPa)	m	K (kPa)
16	4	1	1.7	15.5	2.44	300
8	5.52	1	1.52	17	2.44	300

5 Results and Discussion

In this section, we present the results of the analysis of the loss of strong ellipticity for the vanishing wave speed when the arterial wall is subjected to uniaxial tension in circumferential and axial directions.

We analyze the loss of strong ellipticity for the "stiff" displacement-controlled loading. The unknown out-of-plane principal stretch λ_3 is determined in terms of the known in-plane stretches λ_1 and λ_2 using the plane stress condition for a plane sheet of arterial wall. Uniaxial tension in circumferential and axial directions are described as follows: $\lambda_2 = \lambda_1^{-0.5}$ and $\lambda_1 = \lambda_2^{-0.5}$.

The condition of the vanishing wave speed: $I_3^{1/2}\rho v^2 = f_1 f_2 = 0$; enables us to find the critical stretches that mark the loss of strong ellipticity. We obtain the results for longitudinal P-waves and transverse S-waves in slightly compressible

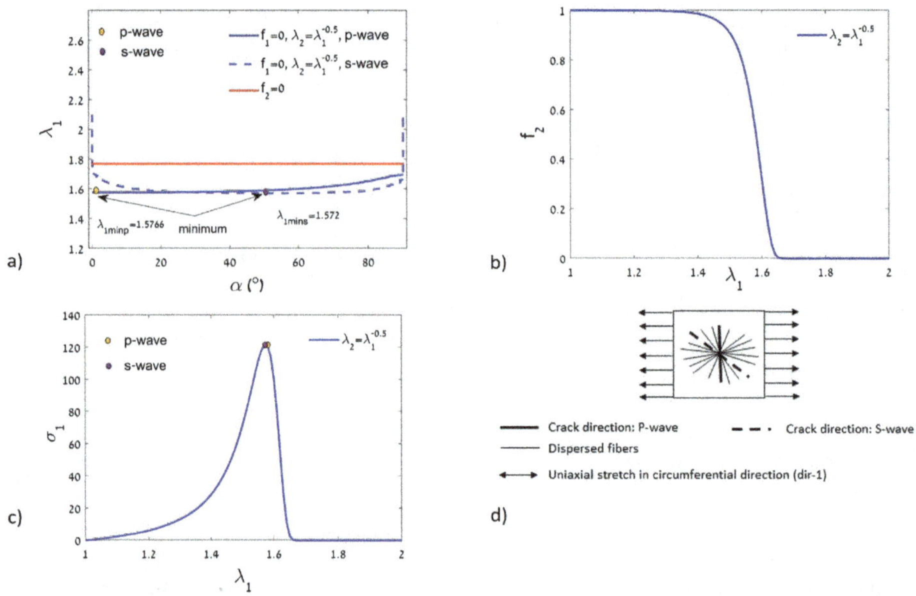

Fig. 1. Uniaxial tension in circumferential direction (the model with 16 structure tensors): (a) stretch λ_1 versus the orientation of the superimposed wave for $f_1 = 0$ and $f_2 = 0$; (b) convergence of the exponential function $f_2(\lambda_1)$ to zero; (c) Cauchy stress [kPa] versus stretch; (d) crack directions

and incompressible (for S-waves only) material models. We found that the results for the slightly compressible and incompressible materials are numerically very close.

Figure 1 shows results of the analysis of loss of strong ellipticity for the model with 16 structure tensors presented in the previous sections:

(a) Figure on top-left shows stretches as a function of direction (angle) of the propagating longitudinal (P-) or transverse (S-) waves. The minimal stretches indicate the loss of the strong ellipticity and the onset of cracks.
(b) Figure on top-right shows convergence of the exponential function f_2 to zero. Theoretically, f_2 should approach zero at infinity. However, the numerical infinity occurs fast!
(c) Bottom left shows the points on the stress-stretch curve, denoting the loss of the strong ellipticity for P- and S-waves.
(d) Bottom right presents schematic showing the loading and possible directions of the onset of cracks predicted by P- and S-waves.

The main findings can be summarized as follows:

1. The condition of the vanishing P-wave speed predicts the direction of cracks, perpendicular to tension in uniaxial tension. It should be noted that the

incompressibility constraint suppresses this prediction. This constraint eliminates the possibility of consideration of the longitudinal wave and cracks associated with it. The incompressibility constraint acts as a Trojan Horse for the study of the onset of cracks. Results akin to these have been reported in [24] for purely isotropic soft material.

2. The condition of the vanishing S-wave speed predicts the direction of cracks, inclined (non-perpendicular) to tension in uniaxial tension. Such cracks seem unreasonable at first. However, in the recent experimental work [25], peculiar form of cracks in the direction of tension in a silicone elastomer were observed. The authors of the work associated these "sideways" cracks with "microstructural anisotropy (in a nominally isotropic elastomer)".

We should note that the present approach can provide new insights in the design of experiments with cracking. Information about material anisotropy can be obtained from the character and direction of cracks - the inverse problem. However, this is outside the purview of the present work.

We emphasize that the proposed approach is suitable for the analysis of the onset of cracks only. Regularized formulations (e.g. [26,27]), necessary for monitoring the crack development were not considered in this work.

Finally, we note that more results concerning the present study can further be found in [28].

Acknowledgment. The support from the Israel Science Foundation (ISF-198/15) is gratefully acknowledged.

References

1. Roy, C.S.: The elastic properties of the arterial wall. Philos. Trans. R. Soc. B **99**, 1–31 (1880)
2. Fung, Y.C., Fronek, K., Patitucci, P.: Pseudoelasticity of arteries and the choice of its mathematical expression. Am. J. Physiol. **237**, H620–H631 (1979)
3. Choung, C.J., Fung, Y.C.: Three-dimensional stress distribution in arteries. J. Biomech. Eng. **105**, 268–274 (1983)
4. Humphrey, J.D., Strumpf, R.K., Yin, F.C.: Determination of a constitutive relation for passive myocardium. I. A new functional form. ASME J. Biomech. Eng. **112**, 333–339 (1990)
5. Wuyts, F.L., Vanhuyse, V.J., Langewouters, G.J., Decraemer, W.F., Raman, E.R., Buyle, S.: Elastic properties of human aortas in relation to age and atherosclerosis: a structural model. Phys. Med. Biol. **40**, 1577–1597 (1995)
6. Ogden, R.: Non-Linear Elastic Deformations. Dover, New York (1997)
7. Holzapfel, G.A., Gasser, T.C.: A new constitutive framework for arterial wall mechanics and a comparative study of material models. J. Elast. **61**, 1–48 (2000)
8. Zulliger, M.A., Fridez, P., Hayashi, K., Stergiopulos, N.: A strain energy function for arteries accounting for wall composition and structure. J. Biomech. **37**, 989–1000 (2004)
9. Holzapfel, G.A., Ogden, R.W.: Constitutive modeling of arteries. Proc. R. Soc. Lond. **466**, 1551–1597 (2010)

10. Lanir, Y.: Constitutive equations for fibrous connective tissues. J. Biomech. **16**, 1–12 (1983)
11. Federico, S., Gasser, T.C.: Nonlinear elasticity of biological tissues with statistical fibre orientation. J. R. Soc. Interface **7**, 955–966 (2010)
12. Saez, P., Garcia, A., Pena, E., Gasser, T.C., Martinez, M.A.: Microstructural quantification of collagen fiber orientations and its integration in constitutive modeling of the porcine carotid artery. Acta Biomater. **33**, 183–193 (2016)
13. Kassab, G.S., Sacks, M.S.: Structure-Based Mechanics of Tissues and Organs. Springer, Heidelberg (2016)
14. Gizzi, A., Pandolfi, A., Vastac, M.: Statistical characterization of the anisotropic strain energy in soft materials with distributed fiber. Mech. Mater. **92**, 119–138 (2016)
15. Freed, A.D., Einstein, D.R., Vesely, L.: Invariant formulation for dispersed transverse isotropy in aortic heart valves. Biomech. Model. Mechanobiol. **4**, 100–117 (2005)
16. Gasser, T.C., Ogden, R.W., Holzapfel, G.A.: Hyperelastic modeling of arterial layers with distributed collagen fibre orientations. J. R. Soc. Interface **3**, 15–35 (2006)
17. Volokh, K.Y.: On arterial fiber dispersion and auxetic effect. J. Biomech. **61**, 123–130 (2017)
18. Volokh, K.Y.: Hyperelasticity with softening for modeling materials failure. J. Mech. Phys. Solids **55**, 2237–2264 (2007)
19. Volokh, K.Y.: Mechanics of Soft Materials. Springer, Singapore (2016)
20. Volokh, K.Y.: On modeling failure of rubberlike materials. Mech. Res. Commun. **37**, 684–689 (2010)
21. Volokh, K.Y.: Loss of ellipticity in elasticity with energy limiters. Eur. J. Mech. A/Solid **63**, 36–42 (2017)
22. Holzapfel, G.A., Niestrawska, J.A., Ogden, R.W., Reinisch, A.J., Schriefl, A.J.: Modeling non-symmetric collagen fiber dispersion in arterial walls. J. R. Soc. Interface **12**, 20150188 (2015)
23. Schriefl, A.J., Zeindlinger, G., Pierce, D.M., Regitinig, P., Holzapfel, G.A.: Determination of the layer-specific distributed collagen fiber orientations in human thoracic and abdominal aortas and commom iliac arteries. J. R. Soc. Interface **9**, 1275–1286 (2012)
24. Mythravaruni, P., Volokh, K.Y.: On incompressibility constraint and crack direction in soft solids. J. Appl. Mech. **86**, 10 (2019)
25. Lee, S., Pharr, M.: Sideways and stable crack propagation in a silicone elastomer. Proc. Natl. Acad. Sci. U.S.A. **116**, 9251–9256 (2019)
26. Volokh, K.Y.: Fracture as a material sink. Mater. Theory **1**, 3 (2017)
27. Faye, A., Lev, Y., Volokh, K.Y.: The effect of local inertia around the crack tip in dynamic fracture of soft materials. Mech. Soft Mater. **1**, 4 (2019)
28. Mythravaruni, P., Volokh, K.Y.: On the onset of cracks in arteries. Mol. Cell. Biomech. **17**(1), 1–17 (2020)

Modelling of Abdominal Wall Under Uncertainty of Material Properties

Katarzyna Szepietowska[1(✉)], Izabela Lubowiecka[1], Benoit Magnain[2], and Eric Florentin[2]

[1] Faculty of Civil and Environmental Engineering, Gdansk University of Technology, Narutowicza 11/12, 80-233 Gdańsk, Poland
katszepi@pg.edu.pl
[2] INSA Centre Val de Loire, Univ. Orléans, Univ. Tours, LaMé, EA 7494, 18022 Bourges, France

Abstract. The paper concerns abdominal wall modelling. The accurate prediction and simulation of abdominal wall mechanics are important in the context of optimization of ventral hernia repair. The shell Finite Element model is considered, as the one which can be used in patient-specific approach due to relatively easy geometry generation. However, there are uncertainties in this issue, e.g. related to mechanical properties since the properties may vary naturally or as an effect of identification accuracy etc. The aim of the study is to include uncertainties in the modelling and investigate their influence on the model response. The parameters of Gasser-Ogden-Holzapfel hyperelastic material model including fibre orientation are treated here as random variables. The uncertainties are propagated with the use of regression based polynomial chaos expansion method. Sobol' indices are used as the measures of global sensitivity analysis and they provide information about the influence of input uncertainties on the uncertainty of the model output. Uncertainty of parameter affecting stiffness of ground substance (C_{10}) has the highest contribution to the variation of the displacement of chosen point in the center of the abdominal wall.

Keywords: Uncertainty quantification · Global sensitivity analysis · Hyperelasticity · Gasser-Ogden-Holzapfel material model · Polynomial chaos · Sobol' indices

1 Introduction

The study addresses the issues of the modelling of abdominal wall. Understanding mechanical behaviour of abdominal wall is particularly interesting in the context of ventral hernia repair. In order to improve the efficiency of hernia repair, some mechanical approaches have been employed. Various surgical meshes were investigated in the literature, e.g. [1] and models of implants were developed [2]. However, it was also acknowledged that the mechanics of abdominal wall pays a crucial role in designing implants that would be mechanically compatible with human tissue [3]. Thus the mechanical properties of abdominal wall should be known.

G. A. Ateshian et al. (Eds.): CMBBE 2019, LNCVB 36, pp. 305–316, 2020.
https://doi.org/10.1007/978-3-030-43195-2_25

The extensive review on mechanics of both abdominal wall and implant was written by Deeken and Lake in [4]. It can be noticed in their compilation of existing experimental studies on tissues of abdominal wall, that the properties of the same tissue reported in the literature varies between articles, e.g. due to different testing protocols. Existing numerical models are mainly based on *ex vivo* properties of animals [5] or human [6] samples. Some properties were also identified *in vivo*: Young's modulus [7] or parameters of isotropic hyperlastic material model on animals [8]. It can be seen that a lot of uncertainties appear in the modelling of abdominal wall, e.g. due to challenges in accurate identification of properties, natural variability of properties and so on. The aim of this study is to include those uncertainties in the modelling and study their influence on the model output.

The models mentioned above [5,6] are detailed and include various components of abdominal wall with geometry based on medical images (MRI or CT scans). Lubowiecka et al. [9] proposed simpler membrane model of abdominal wall with geometry corresponding to the external surface of abdominal wall. It was created in the perspective of patient-specific approach and *in vivo* identification of material properties by inverse methods using measurements of displacement caused by known changes of pressure during peritoneal dialysis. The behaviour of the model was compared with experiment. In [10] we propagated uncertainties related to the value of intraabdominal pressure, the parameters of linear elastic orthotropic model and the direction of orthotropy. We studied the influence of these uncertainties on the uncertainty of the output. However, soft tissues exhibit nonlinear elastic behaviour. Borzeszkowski et al. [11] applied Gasser-Ogden-Holzapfel (GOH) [12] hyperelastic material model to abdominal wall shell model and performed a parametric analysis to study the influence of various parameters. The aim of the present study is to perform uncertainty quantification and global sensitivity analysis of abdominal wall shell model including uncertainty of GOH material and structure parameters related to fibres alignment. The general purpose is to find most important and negligible variables in the context of the further identification of abdominal wall properties and optimisation under uncertainty of hernia repair parameters with the use of abdominal wall model. Polynomial chaos expansion method is used to propagate the uncertainties and to calculate Sobol' indices [13], which are global sensitivity measure of the influence of investigated uncertainties to the uncertainty of abdominal wall response. The global sensitivity of GOH was already performed in other applications [14], but since the sensitivity analysis outcomes depends on the problem and the studied quantity of interest so new results are expected.

2 Materials and Methods

2.1 Finite Element Model of Abdominal Wall

The model of abdominal wall was created in commercial Finite Element (FE) software MSC.Marc. The geometry of the model is based on measurements of human external surface of abdominal wall [15]. The model is composed of 1872 shell 4-node quadrilateral elements. Nodes on the boundary of abdominal wall (Fig. 1) have fixed translations. Thickness is assumed to be 3 cm. The model is subjected to pressure equal to 981 Pa corresponding to intraabdominal pressure caused by liquid intruded during peritoneal dialysis. The loading corresponds to experiment described in [9] when the same model (but with orthotropic linear elastic material law) was validated with experiments performed on the patient undergoing dialysis. The analysis is geometrically and physically nonlinear.

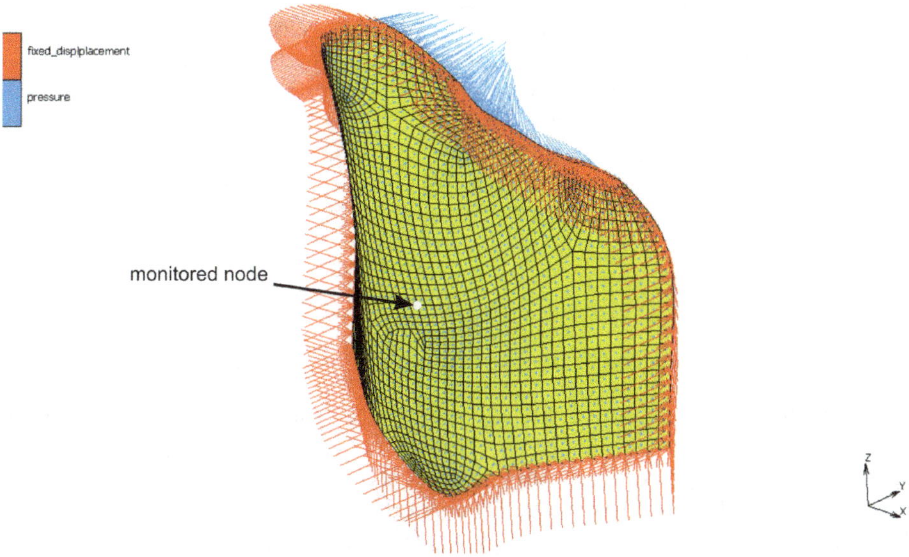

Fig. 1. FE model

2.2 Constitutive Modelling

In this work the assumed material model of abdominal wall is Gasser-Ogden-Holzapfel (GOH) model [12], which is hyperelastic anisotropic model. Although this constitutive law was developed to model arterial layers, it is also used in modelling of other soft tissues, e.g. abdominal wall tissues [16], tendons [17]. In the model a strain-energy function Ψ is assumed to be in decoupled form

$$\Psi = \Psi_{vol} + \bar{\Psi}, \tag{1}$$

where Ψ_{vol} is purely volumetric contribution Ψ_{vol} and $\bar{\Psi}$ is isochoric part. It is assumed that $\bar{\Psi}$ is superposition isotropic contribution corresponding to ground-matrix $\bar{\Psi}_g$ and contribution corresponding to embedded fibres $\bar{\Psi}_f$

$$\bar{\Psi} = \bar{\Psi}_g + \bar{\Psi}_f, \tag{2}$$

Neo-Hookean model is chosen for Ψ_g:

$$\bar{\Psi}_g = C_{10}(\bar{I}_1 - 3), \tag{3}$$

where C_{10} is the stress-like material parameter, $\bar{I}_1$ is the first invariant of the modified right Cauchy-Green deformation tensor. Only one family of fiber is assumed, then:

$$\bar{\Psi}_f = \frac{k_1}{2k_2}\left[\exp\{k_2[\kappa(\bar{I}_1 - 3) + (1 - 3\kappa)(\bar{I}_4 - 1)]^2\} - 1\right], \tag{4}$$

where k_1 is the stress-like material parameter, k_2 is the dimensionless material parameter. $\bar{I}_4$ is the invariant equal to the square of the stretch in the direction of the mean orientation α_f of the family of fibres. κ describes level of fibre dispersion and can be in the range $0 \leq \kappa \leq 1/3$, where $\kappa = 0$ corresponds to perfect alignment of fibers (transverse isotropy) and $\kappa = 1/3$ correspond to isotropy.

Parameters of the GOH model and fiber orientation are assumed to be uncertain.

2.3 Uncertainty Quantification and Sensitivity Analysis

Polynomial Chaos Expansion. Uncertainties can be included in the modelling by probabilistic approach. Since a commercial software is applied, the non-intrusive uncertainty propagation method is needed. Such methods are based on some number of deterministic calculations and do not require modification of the FE code of the model. Polynomial chaos (PC) method exists in non-intrusive variants (Non-Intrusive Spectral Projection Method [18] or regression based approach [19]) and enable performing uncertainty quantification and global sensitivity analysis method with relatively small computational costs when compared to widely-used Monte Carlo method. In the PC method the model output Y, Quantity of Interest (QoI), is expanded as follows:

$$Y \approx \sum_{\boldsymbol{\alpha} \in \mathcal{A}} a_{\boldsymbol{\alpha}} \Phi_{\boldsymbol{\alpha}}(\boldsymbol{\xi}), \tag{5}$$

where $\boldsymbol{\xi}$ is an input random vector, $a_{\boldsymbol{\alpha}}$ are coefficients, $\mathcal{A}$ is a truncation set of $\boldsymbol{\alpha}$, $\boldsymbol{\alpha}$ are M-uplets $(\alpha_1, \ldots, \alpha_M) \in \mathbb{N}^M$, and $\Phi_{\boldsymbol{\alpha}}(\boldsymbol{\xi})$ is a multivariate polynomial basis constructed by multiplying polynomials ϕ_{α_i} of order α_i

$$\Phi_{\boldsymbol{\alpha}}(\boldsymbol{\xi}) = \prod_{i}^{M} \phi^i_{\alpha_i}(\xi_i). \tag{6}$$

Polynomials have to be orthonormal with respect to a given distribution. In this case Legendre polynomials are employed because the random variables follow uniform distribution. Classic truncation has been performed, such that $\mathcal{A} = \{\boldsymbol{\alpha} \in \mathbb{N}^M : \sum_{i=1}^{M} \alpha_i \leq p\}$, where p is PC degree. Regression-based approach [19] is used to find the coefficients. The model is performed on N regression points in order to obtain the vector of exact solutions. Then the coefficients are calculated by solving least square problems. The drawback of non-intrusive methods like this one, is that the accuracy depends on the number and choice of regression points. Different strategies were compared on the hernia-related models in [20]. Based on the experience presented the points from Sobol sequence are used. $N = (M - 1)P$, where P is size of PC basis (cardinality of truncation set $\mathcal{A}$). Different techniques have been developed to control error [21]. Here, Leave-One-Out error estimate was calculated in order to evaluate PC meta-model performance because of simplicity.

Global Sensitivity Analysis. Global sensitivity analysis is a study of sensitivity of the output to variations of the input. It enables varying all variables at the same time with variations over whole domain. Sobol indices [22] are one of the global-sensitivity measures. The calculation is based on ANalysis Of VAriance (ANOVA) decomposition. Estimation of Sobol indices by MC is very expensive computationaly. It has been shown by Sudret [13] and Crestaux et al. [23] that thanks to the orthonormality of the PC basis, estimation of Sobol indices can be performed with use of the PC coefficients without additional computational cost. Sobol index $S_{i_1,\ldots,i_s}$ shows how much of the total output variance is due to the uncertainty of variables $\xi_{i_1}, \ldots, \xi_{i_s}$. To compute it by PC method a set of $\boldsymbol{\alpha}$-tuples corresponding to polynomials depending only on all input variables $\xi_{i_1}, \ldots, \xi_{i_s}$ must be found:

$$\mathcal{A}_{i_1,..,i_s} = \{\boldsymbol{\alpha} \in \mathcal{A} : \alpha_k \neq 0 \Leftrightarrow k \in \{i_1, \ldots, i_s\}\}. \tag{7}$$

Then, sobol index $S_{i_1,\ldots,i_s}$ estimated by PC is:

$$S^{PC}_{i_1,..,i_s} = \frac{1}{D} \sum_{\boldsymbol{\alpha} \in \mathcal{A}_{i_1,\ldots,i_s}} a^2_{\boldsymbol{\alpha}}, \tag{8}$$

where D is total output variance, which can be estimated by PC coefficients $D = \sum_{\boldsymbol{\alpha} \in \mathcal{A} \setminus \mathbf{0}} a^2_{\boldsymbol{\alpha}}$. Total sobol index S^{Tot}_i is the sum of all indices corresponding to a given variable i, including mixed terms. It can also be estimated using PC coefficients:

$$S^{Tot,PC}_i = \frac{1}{D^{PC}} \sum_{\boldsymbol{\alpha} \in \mathcal{A}^{Tot}_i} a^2_{\boldsymbol{\alpha}}. \tag{9}$$

where $\mathcal{A}^{Tot}_i = \{\boldsymbol{\alpha} \in \mathcal{A} : \alpha_i \neq 0\}$.

Random Variables and Quantity of Interest. Due to the uncertainty of GOH model parameters and fibre orientation, five independent uniform random variables are assumed with limits presented in Table 1. First three: C_{10}, k_1, k_2 can be classified as material parameters and the last two: κ, α_f as a structure parameters [12].

Table 1. Limits of uniform distribution $\mathcal{U}(a, b)$ of each random variable

	C_{10} [kPa]	k_1 [kPa]	k_2	κ	α_f
a	2	1	200	0	0
b	35	10	2000	0.33	π

Range of C_{10} was chosen based on reported in [24] results of shear modulus (divided by 2) of living human abdominal wall. Upper bounds of k_1 and k_2 where taken from [16]. κ are α_f may vary in their possible range of values.

The quantity of interest here is the magnitude of displacement u in chosen nodes in the central area of abdominal wall (Fig. 1). During physical experiments [9] the displacement of external surface of abdominal wall can be measured and its value may be used in the *in vivo* identification of the material parameters of living abdominal wall [7,8].

3 Results

Uncertainties have been propagated through the abdominal wall model with the use of PC method. The histogram of the QoI (the displacement of the chosen node in central area) is presented in Fig. 2. Mean equals 0.0067 m and standard deviation equals 0.0045 m.

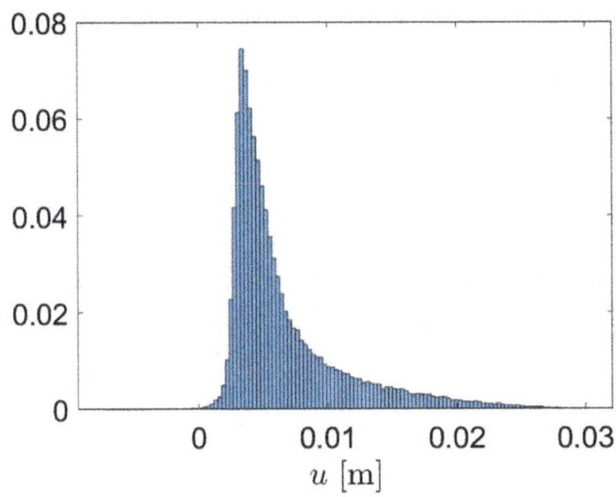

Fig. 2. Histogram of the QoI (displacement in the chosen node)

Table 2 shows the values of Sobol indices of first and second order and Table 3 shows total Sobol indices. The sensitivity indices values are also shown in Fig. 3. It can be shown that the uncertainty of C_{10} has a dominant contribution to the output variance: it has the highest first order Sobol' index $S_{C_{10}}$. Total index $S_{C_{10}}^{Tot}$ is higher mainly due to the interaction with κ and α_f (second order indices).

The first order indices of variables other than C_{10} are negligibly small. However, the total Sobol index of κ and α_f is higher due to mentioned interaction with C_{10}. The uncertainty of k_1 and k_2 has a negligible effect on the variance of investigated QoI.

Table 2. First S_i (on the table diagonal) and second order Sobol' indices S_{ij}

	C_{10}	k_1	k_2	κ	α_f
C_{10}	0.7607				
k_1	0.0038	0.0016			
k_2	0.0148	0.0002	0.0039		
κ	0.0695	0.0008	0.0013	0.0320	
α_f	0.0553	0.0006	0.0006	0.0097	0.0309

Table 3. Total Sobol' indices S_i^{Tot}

i	C_{10}	k_1	k_2	κ	α_f
S_i^{Tot}	0.9171	0.0107	0.0258	0.1249	0.1078

The dominant influence of C_{10} can also be noticed on scatter plots (Fig. 4) showing the QoI value versus each variable. The scatter plots were drawn on the exact model values, not with PC meta-model, in order to evaluate the sensitivity analysis outcomes. The visual evaluation of this graph indicates that sensitivity of the output to C_{10} is higher when C_{10} is high. It can also be interpreted as that the sensitivity to κ is slightly higher, when κ is closer to 0, so the fibers are

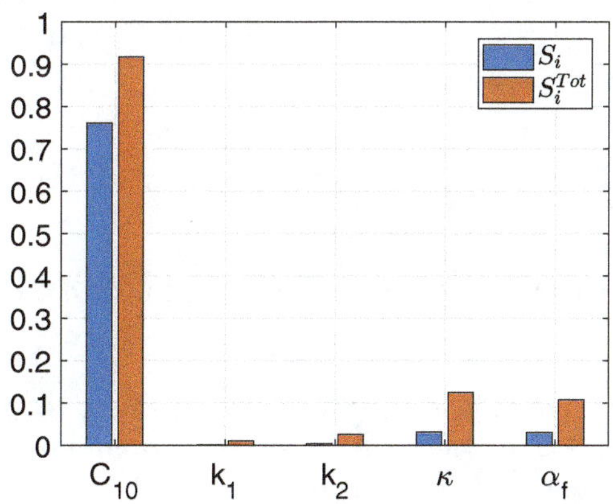

Fig. 3. Sobol indices

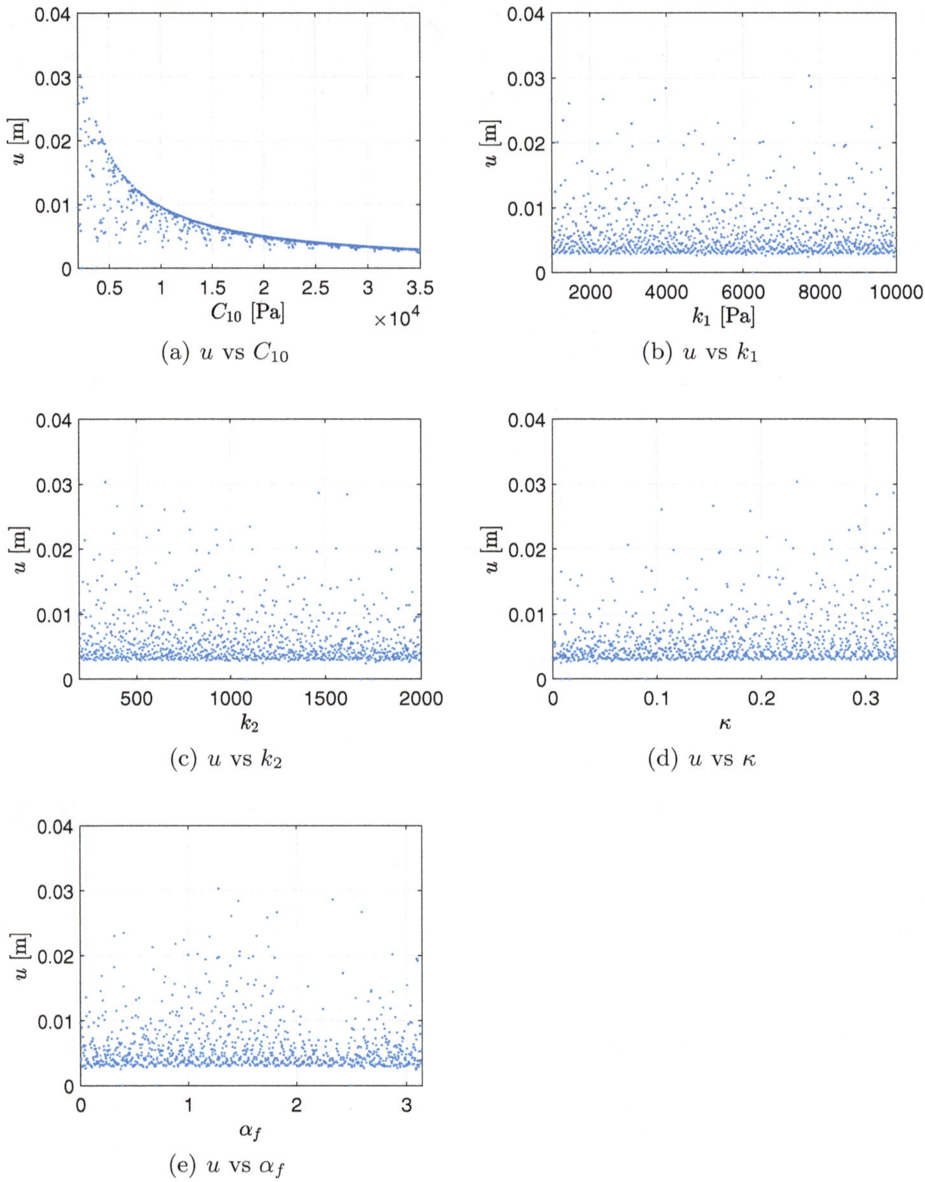

Fig. 4. Scatter plots showing sensitivity of QoI to each variable

close to perfect alignment. The sensitivity to α_f is slightly higher when the mean direction of fibers is in transverse direction of abdominal wall. The transverse direction of abdominal wall is known to be stiffer [7].

Figure 5 shows displacement of abdominal wall for two extreme values of C_{10} (limits of uniform distribution) with other the same parameters.

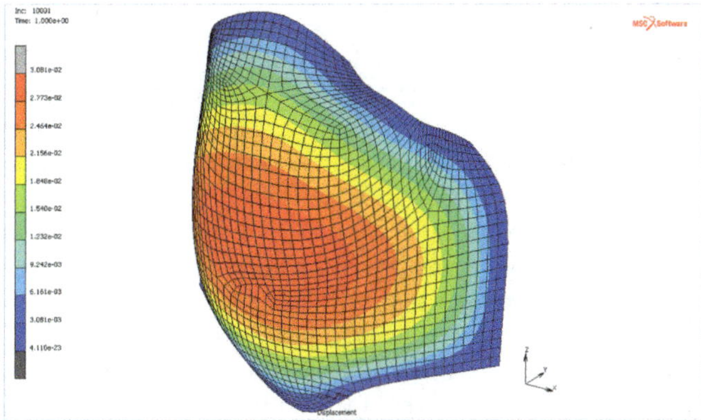

(a) C_{10}=2kPa, maximum displacement 3.081 cm

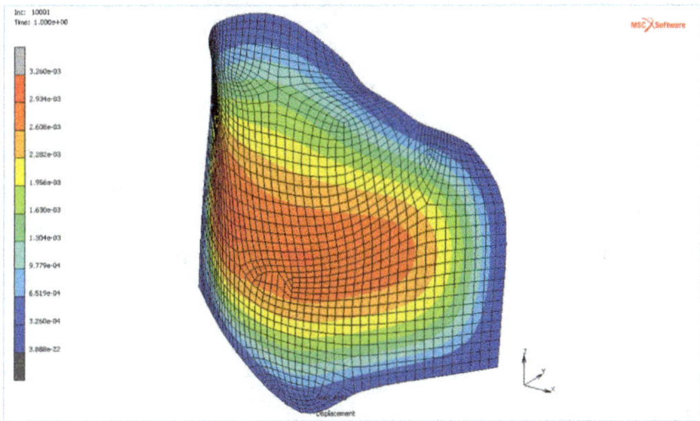

(b) C_{10}=35 kPa, maximum displacement 0.326 cm

Fig. 5. Displacement [m] of abdominal wall in case of extreme values of C_{10}

4 Conclusions

Uncertainties have been propagated in the abdominal wall and their importance has been assessed by Sobol' indices. The uncertainty of C_{10} - parameter of the isotropic part corresponding to groundmatrix has the highest dominant contribution to the variance of the output. Other variables, including one related to fibers, are much less important. This outcome can be interesting in the context of some reported postulates that despite the anisotropy of the single components of abdominal wall, the entire abdominal wall can be treated as isotropic [8]. Next most important variables are the structure ones: κ and α_f due to their interaction with C_{10} (higher order sensitivity indices). Such interaction may be considered a bit surprising and needs further research. k_1 and k_2 have negligible effect in this case. These two parameters have unclear physical interpretation [14] and low importance of their uncertainty would be beneficial in further parameter identification problems. However, it should be mentioned that the limits of their input distribution, in contrast to C_{10}, was not supported by any experimental data.

The aim of the study was o asses the importance of each structure and material parameter of the GOH model in application to modelling of human abdominal wall. Some limitations of the presented study should be mentioned. Firstly, due to the lack of large data sets or existing recommendations, distribution of random variables was assumed by our own judgement. Nevertheless, the obtained sensitivity analysis results indicate which data could be important for proper identification of probabilistic models. Secondly, as a simplification, the material and structure parameters are assumed to be constant in space and just one fiber family is considered. In reality, human abdominal wall is constructed from various components with various properties and fiber orientation. Rectus muscles covered by rectus sheath and linea alba in the midline and lateral part composed of three muscles each covered by aponeuroses, differes from each other in that sense. Spacial distribution of isotropic hyperlastic materials was presented in [8]. Abdominal wall could also be divided into regions with various properties, one corresponding to rectus muscles and one corresponding to oblique muscles as it was done in [10]. In the further research random fields could be applied to include the spatial variability.

Acknowledgments. This work was partially supported by grant UMO-2017/27/B/ST8/02518 from the National Science Centre, Poland and by subsidy for young scientists given by the Faculty of Civil and Environmental Engineering, Gdansk University of Technology. Computations were performed partially in TASK Computer Science Centre, Gdańsk, Poland.

References

1. Tomaszewska, A.: Mechanical behaviour of knit synthetic mesh used in hernia surgery. Acta Bioeng. Biomech. **18**(1), 77–86 (2016)
2. Lubowiecka, I.: Mathematical modelling of implant in an operated hernia for estimation of the repair persistence. Comput. Methods Biomech. Biomed. Eng. **18**(4), 438–445 (2015)
3. Junge, K., Klinge, U., Prescher, A., Giboni, P., Niewiera, M., Schumpelick, V.: Elasticity of the anterior abdominal wall and impact for reparation of incisional hernias using mesh implants. Hernia **5**(3), 113–118 (2001)
4. Deeken, C.R., Lake, S.P.: Mechanical properties of the abdominal wall and biomaterials utilized for hernia repair. J. Mech. Behav. Biomed. Mater. **74**, 411–427 (2017)
5. Hernández-Gascón, B., Mena, A., Pena, E., Pascual, G., Bellón, J., Calvo, B.: Understanding the passive mechanical behavior of the human abdominal wall. Ann. Biomed. Eng. **41**(2), 433–444 (2013)
6. Pachera, P., Pavan, P., Todros, S., Cavinato, C., Fontanella, C., Natali, A.: A numerical investigation of the healthy abdominal wall structures. J. Biomech. **49**(9), 1818–1823 (2016)

7. Song, C., Alijani, A., Frank, T., Hanna, G., Cuschieri, A.: Mechanical properties of the human abdominal wall measured *in vivo* during insufflation for laparoscopic surgery. Surg. Endosc. Other Interv. Tech. **20**(6), 987–990 (2006)
8. Simón-Allué, R., Calvo, B., Oberai, A., Barbone, P.: Towards the mechanical characterization of abdominal wall by inverse analysis. J. Mech. Behav. Biomed. Mater. **66**, 127–137 (2017)
9. Lubowiecka, I., Tomaszewska, A., Szepietowska, K., Szymczak, C., Lochodziejewska-Niemierko, M., Chmielewski, M.: Membrane model of human abdominal wall. simulations vs. in vivo measurements. In: Shell Structures. Theory and Applications, vol. 4, pp. 503–506 (2018)
10. Szepietowska, K., Lubowiecka, I., Magnain, B., Florentin, E.: Global sensitivity analysis of membrane model of abdominal wall with surgical mesh. In: Shell Structures: Theory and Applications, vol. 4, pp. 515–518 (2018)
11. Borzeszkowski, B., Duong, T.X., Sauer, R.A., Lubowiecka, I.: Isogeometric Shell Analysis of the Human Abdominal Wall. In: Innovations in Biomedical Engineering. Springer (2020, in press)
12. Gasser, T.C., Ogden, R.W., Holzapfel, G.A.: Hyperelastic modelling of arterial layers with distributed Collagen fibre orientations. J. R. Soc. Interface **3**(6), 15–35 (2006)
13. Sudret, B.: Global sensitivity analysis using polynomial chaos expansions. Reliab. Eng. Syst. Saf. **93**(7), 964–979 (2008). http://linkinghub.elsevier.com/retrieve/pii/S0951832007001329
14. Akintunde, A.R., Miller, K.S., Schiavazzi, D.E.: Bayesian inference of constitutive model parameters from uncertain uniaxial experiments on murine tendons. J. Mech. Behav. Biomed. Mater. **96**, 285–300 (2019)
15. Szymczak, C., Lubowiecka, I., Tomaszewska, A., Śmietański, M.: Investigation of abdomen surface deformation due to life excitation: implications for implant selection and orientation in laparoscopic ventral hernia repair. Clin. Biomech. **27**(2), 105–110 (2012)
16. Simón-Allué, R.: Towards the *in vivo* mechanical characterization of abdominal wall in animal model. Application to hernia repair. Ph.D. thesis, Universidad de Zaragoza (2016)
17. Bajuri, M., Isaksson, H., Eliasson, P., Thompson, M.S.: A hyperelastic fibre-reinforced continuum model of healing tendons with distributed collagen fibre orientations. Biomech. Model. Mechanobiol. **15**(6), 1457–1466 (2016)
18. Le Maître, O.P., Reagan, M.T., Najm, H.N., Ghanem, R.G., Knio, O.M.: A stochastic projection method for fluid flow: II. Random process. J. Comput. Phys. **181**(1), 9–44 (2002). http://www.sciencedirect.com/science/article/pii/S0021999102971044
19. Berveiller, M., Sudret, B., Lemaire, M.: Stochastic finite element: a non intrusive approach by regression. Revue européenne de mécanique numérique **15**(1-2-3), 81–92 (2006). http://remn.revuesonline.com/article.jsp?articleId=7876
20. Szepietowska, K., Magnain, B., Lubowiecka, I., Florentin, E.: Sensitivity analysis based on non-intrusive regression-based polynomial chaos expansion for surgical mesh modelling. Struct. Multidiscip. Optim. **57**(3), 1391–1409 (2018). https://doi.org/10.1007/s00158-017-1799-9

21. Chamoin, L., Florentin, E., Pavot, S., Visseq, V.: Robust goal-oriented error estimation based on the constitutive relation error for stochastic problems. Comput. Struct. **106-107**, 189–195 (2012), https://hal.archives-ouvertes.fr/hal-00776138
22. Sobol, I.M.: Global sensitivity indices for nonlinear mathematical models and their Monte Carlo estimates. Math. Comput. Simul. **55**(1–3), 271–280 (2001)
23. Crestaux, T., Le Maître, O., Martinez, J.M.: Polynomial chaos expansion for sensitivity analysis. Reliab. Eng. Syst. Saf. **94**(7), 1161–1172 (2009). Special Issue on Sensitivity Analysis
24. Tran, D., Podwojewski, F., Beillas, P., Ottenio, M., Voirin, D., Turquier, F., Mitton, D.: Abdominal wall muscle elasticity and abdomen local stiffness on healthy volunteers during various physiological activities. J. Mech. Behav. Biomed. Mater. **60**, 451–459 (2016)

Fatigue and Wear Analysis for Temporomandibular Joint Prosthesis by Finite Element Method

Edwin Rodriguez[1,2(✉)] and Angelica Ramirez-Martinez[3]

[1] Central University, Calle 21 4-40, Bogota, Colombia
edwin.rodriguezc@utadeo.edu.co
[2] Jorge Tadeo Lozano University, Cra 4 22-61, Bogota, Colombia
[3] Movement Analysis Laboratory (AM&E),
Central University, Calle 21 4-40, Bogota, Colombia

Abstract. Temporomandibular joint (TMJ) prosthesis consists of two elements: one which is solidary with the temporal bone (fossa component) and the other which replaces part of the jaw bone (condylar plate). Main problem on replacement is mechanical fail of the mandibular plate—due to cyclic muscular forces [1] and wear between the condyle and the fossa [2,3]. With the aim of supporting the virtual assessment of personalized prosthesis, this paper presents a finite element analysis (FEA) for fatigue and wear of TMJ prosthesis under muscular forces during the clenching process. The FEA model has three stages. The first one is a static model that describes the response to masticatory forces pattern including contact between the parts. Isotropic linear materials were defined: Ultra High Molecular Weight Polyethylene (UHMWPE), Titanium alloy (Ti6AlV4) and cortical bone. Second stage uses first FEA stress results as input data and evaluates high-cycle fatigue for each component. Due to mean stresses are not equal to zero, stress-life curves of materials with mean stress effects were used. The third model evaluates wear effects, the Archard linear wear model was used to calculate the lost volume during contact interaction in a quasi-static model. Wear rate was taken from experimental data [4]. The simulation showed the critical zone under fatigue and provides an understanding of the components remaining life before its construction. The wear model gives the quantity of lost volume and a preliminary affected surface that allows an evaluation of the durability of the fossa component.

Keywords: Fatigue · Wear · Temporomandibular joint prosthesis · Finite element analysis

1 Introduction

Symptoms as pain and difficulty for mandibular movement are related to TMJ alterations, which can evolve to diseases as arthritis. In United States, 12% of

G. A. Ateshian et al. (Eds.): CMBBE 2019, LNCVB 36, pp. 317–334, 2020.
https://doi.org/10.1007/978-3-030-43195-2_26

medical consultation for facial or oral pain are associated with TMJ disorders, with 10:1 predominance of female gender [4,5]. These events, added to trauma or surgical complications are cause of joint replacement.

Replacements consist in joint reconstruction with grafts from osseous and adipose tissue from same patient or joint replacement with prosthetic implants. Prosthesis have demonstrated to be more convenient than grafts principally for morbidity reduction [6], evidenced for no additional interventions for graft donations and hospitalization and surgical intervention time reduction. Prosthesis also reduce postoperative fixation needs [7] and new interventions for replacement or revisions, improving general satisfaction to procedure. Benefits for prosthesis implantation includes pain reduction up to 57%, mastication improvement of 52%, mandible opening of 58%, among others [8].

Since 1946 standard prosthetic devices have been used, it was only necessary to choose the right size for each patient [9]. Nowadays, with computed tomography images customized prosthesis development is facilitated, its geometric improvements lead to better performance. For example, detailed geometry allows to increment contact surface facilitating osseointegration process. Also, custom prosthesis have better structural integrity, because they fit anatomical and clinical patient situation [10]. On the other side, simplified geometry of standard prosthesis can generate condyle dislocation [8]. Even though as custom as standard prosthesis contribute to pain reduction and to preserve condyle translation movement, custom implant improves mandibular movement cinematic enhancing patient's wellness [8].

Main problems reported during TMJ prosthesis use are: mechanical failure of components due to muscular loads [6] or strains induced during surgical implantation [6] and excessive wear in localized contact points between articular fossa and mandibular condyle [2] allowing particle liberation, which can generate cell reactions as macrophages. This biological response produces osteoclastic bone resorption added to prosthesis micro-movements stimulates fixation elements loosening [3].

To study these problems experimental approach has been used, destructive tests *in vitro* in prosthesis components that allow to know wear rate and to extrapolate lost volume during prosthesis life time [2,9]. By other side, computational approach evaluates prosthesis structural behavior by stress and strain calculation under several load cases [10–13]. However, according to author's knowledge there are not evidence of computational analysis for fatigue and wear on TMJ prosthesis.

In this work is proposed the development of a virtual evaluation method for custom TMJ prosthesis that allows to know early its mechanical strength and endurance under load conditions and geometry defined. For that, a structural finite element model for fatigue and wear was developed.

2 Materials and Methods

2.1 Static Finite Element Model

A finite element model was built for this application using the commercial software ANSYS. Four main components are considered: Mandible, Condylar Plate, Articular Fossa and Fossa fixation. A sagittal symmetry plane was defined in order to reduce computational cost by modeling only a half of the mandible and just one TMJ. All components are considered as linear isotropic materials, as shown in Table 1.

Table 1. Mechanical properties and components for FEA static analysis

	Titanium	UHMWPE	Cortical bone	Units
Young modulus E	110	0.725	13	[GPa]
Poisson ratio ν	0.34	0.46	0.30	
Tensile strength σ_u	1.15	0.043		[GPa]
Yield strength	1.03	0.031		[GPa]
Hardness	3.4230	1.5790		[GPa]
Component	Condylar plate	Articular fossa	Mandible	
	Fossa Fixation			

Boundary conditions were applied to Fossa Fixation and Mandible components. For the first, Frictionless support in cylindrical faces of screw holes and plane face of screws heads in order to simulate this support the most realistic keeping reduced computational cost. For the second component, a symmetry plane and a zero displacement in vertical (Z) direction placed in the centroid of 1st and 2nd molars occlusal faces, but allowing movement in horizontal (XY) plane and rotations around X, Y and Z directions.

The contact pairs: Mandible-Condylar plate and Fossa Fixation-Articular Fossa were considered bonded due to osseointegration process condition between the condylar plate and mandibular bone, and a perfect gluing process between both fossa components. Condyle-Fossa contact was simulated as a frictionless pair with Augmented Lagrange algorithm, nodal normal to target contact detection and stiffness correction done each iteration, without any damping effect or offset treatment.

Muscular Forces. Muscular forces are the main source of stress in this structural analysis, where several muscles are involved in cinematic behavior of TMJ during clenching cycles. Medial Pterygoid, Masseter and Anterior Temporalis in opening phase. Digastric, Mylohyoid and Geniohyoid in closing phase are responsibles for generate the mandibular movement [1,11,14,15].

The muscular force model used in this work is based in the methodology presented by Commisso et al. [15]. In that publication the aim is to know the non implanted TMJ stress state and estimate an activation pattern for Lateral Pterygoid muscle.

Muscular force can be decomposed in two main parts: Active force that depends mainly on muscular fibers contraction and a Passive force due to stiffness of tissue.

$$F = F^A + F^P \tag{1}$$

According to Thelen [16], these two force components can be evaluated by:

- Active Force:

$$F^A(t) = a_{max} \cdot a(t) \cdot F_0^M \cdot f_L(l) \cdot f_V(v) \tag{2}$$

Where $a(t)$ is a time function that describes muscle activation level and has a range between 0 and 1 and is used to normalize maximum value a_{max}. F_0^M is maximum force in isometric contraction. $f_L(l)$ is a function that depends on muscle current length and has a value of 1 when the muscle has optimal fiber length l_0^M, is assumed that the optimal length is achieved when the interincisal space is equal to 2 mm [17] is possible to assume $f_L(l) = 1$ since the movement amplitude is small during the entire clenching cycle. $f_V(v)$ depends on contraction velocity, and because of motion slowness, $f_V(v) = 1$. In this way, the active force is simplified as:

$$F^A(t) = a_{max} \cdot a(t) \cdot F_0^M \tag{3}$$

- Passive Force:

$$F^P(t) = F_0^M \cdot \overline{F}^P(\overline{l}^M(t)) \tag{4}$$

Where $\overline{l}^M(t)$ is normalized fiber length, calculated as

$$\overline{l}^M(t) = l^M(t)/l_0^M \tag{5}$$

And $\overline{F}^P$ is normalized passive force, obtained from:

$$\overline{F}^P(\overline{l}^M(t)) = \begin{cases} 1 + \frac{k^P}{\varepsilon_0^M}\left(\overline{l}^M(t) - (1+\varepsilon_0^M)\right), & \overline{l}^M > 1 + \varepsilon_0^M \\ \frac{e^{\left(k^P(\overline{l}^M(t)-1)/\varepsilon_0^M\right)}}{e^{k^P}}, & \overline{l}^M \leq 1 + \varepsilon_0^M \end{cases} \tag{6}$$

Where $\varepsilon_0^M = 0.6$ is passive muscular strain during maximum isometric force and $k^P = 4$ is a shape factor for passive force - length relation.

Activation patterns and maximum forces are available on literature [18] and adapted by Commisso [15] for activation. In Fig. 1 are shown total muscle activation profiles $a(t) \cdot a_{max}$ during clenching cycle in ipsilateral bite. Is noticed the activation of closing muscles during this phase between 0 and 0.33 s within a range of 0 and 1.0 in left vertical scale. In the right scale, the opening muscles from 0.33 to 0.76 s between 0 and $6 \cdot 10^{-4}$. Contralateral profiles are no used due to symmetry assumed for the model. a_{max} values are extracted from Commisso's work [15] where they were modified from Hannam's publication [19] in order to adjust for an interincisal opening of 8 mm.

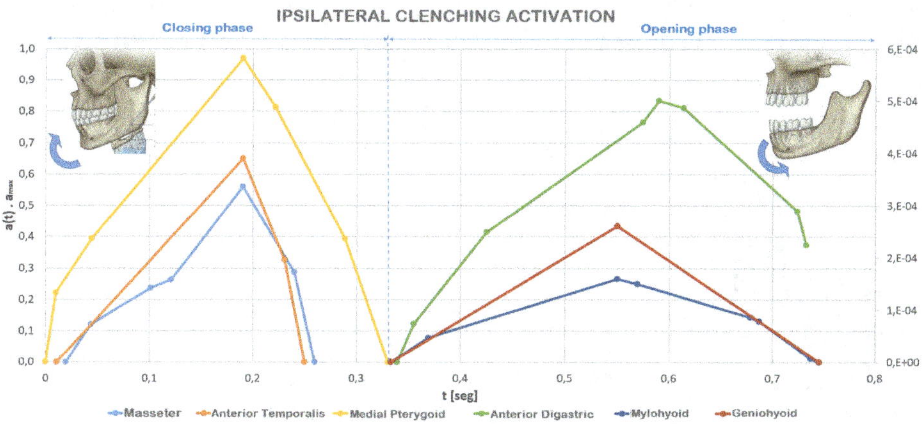

Fig. 1. Activation patterns for mastication forces calculation

Maximum force values F_0^M, muscle orientation vectors and closed mouth muscle length have been published previously by Korioth and van Eijden [20,21] for each muscle involved in mastication process. All of them were taken from Commisso [15] and are shown in Table 2.

Table 2. Muscular forces parameters. Adapted from [15]

Muscle	Length [mm]	cos-X	cos-Y	cos-Z	a_{max}	F_0^M[N]
Masseter	48	−0.207	−0.419	0.885	0.56	190.4
Anterior Temporalis	51.9	−0.149	−0.044	0.988	0.65	158
Medial Pterygoid	42.7	0.486	−0.372	0.791	0.97	174.8
Anterior Digastric	43.2	−0.576	0.815	0.053	$5 \cdot 10^{-4}$	40
Geniohyoid	40	−0.103	0.995	−0.018	$2.6 \cdot 10^{-4}$	38.8
Anterior Mylohyoid	23	0.831	0.176	−0.528	$1.6 \cdot 10^{-4}$	63.6
Posterior Mylohyoid	42.6	0.616	0.223	−0.756	$4.7 \cdot 10^{-4}$	21.2

Directions ***X***, ***Y*** and ***Z*** used in Table 2 for muscle orientation are lateral right, posterior and cranial respectively. These directions are supposed to be constants during clenching process and uniforms over force application surfaces.

With all this information it is not possible to calculate the muscular forces, is still necessary to evaluate $l^M(t)$ in a interincisal opening position between

0 and 8 mm and l_0^M of 2 mm according to the clenching movements [17,19]. In order to accomplish this task, a rigid body cinematic model was built, replicating boundary conditions from the deformable model and defining a spherical joint to connect the condylar plate and articular fossa components, very low constant ($1 \cdot 10^{-5}$ N/mm) spring elements were used to simulate the muscles and the variable opening was used to obtain the muscles elongation vs time $l^M(t)$. Temporal scale was amplified 100 times to facilitate implementation in FEA Software.

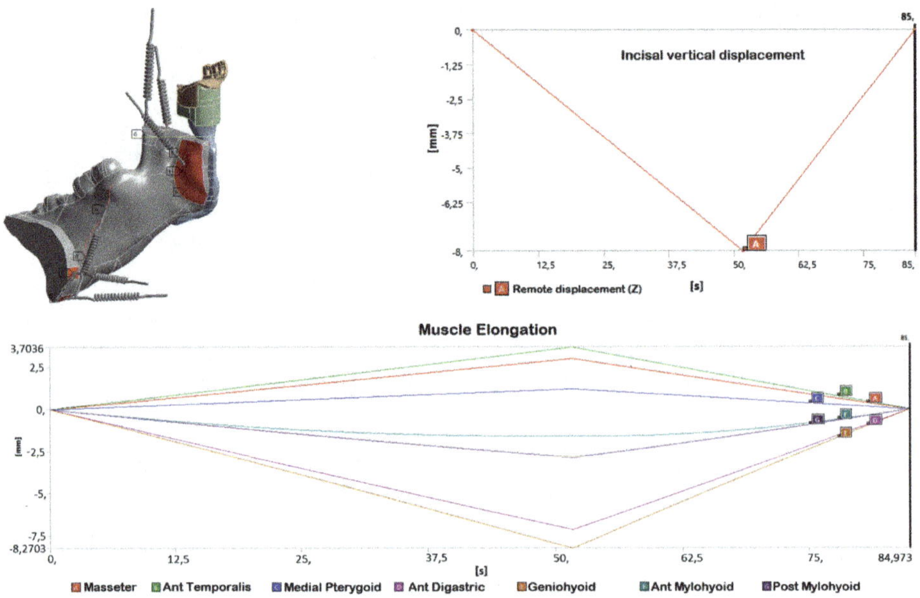

Fig. 2. Rigid body model for muscle elongation

Muscular elongation patterns are shown in Fig. 2. Opening muscles are extended while closing muscles are contracted.

In Fig. 3 calculated muscular forces are shown. This boundary conditions are used to apply time dependant forces in the mandible in the deformable model. Is noticed that opening forces are several scales greater than closing ones.

2.2 Fatigue Model

The objective of fatigue analysis is to determinate durability of components under cycling load. In this study fossa (fixation and articular) and condylar plate are individually examined under fatigue perspective to calculate remanent life, safety factor and fatigue sensitivity. TMJ prosthesis is expected to endure as much as possible in order to avoid additional surgical interventions for prosthesis replacement. During 50 years of use up to 40 million of cycles can be achieved [4,22], that is why fatigue calculations are done for this life. Stress-life (also called high-cycle) approach is the most suitable for long life fatigue requirements.

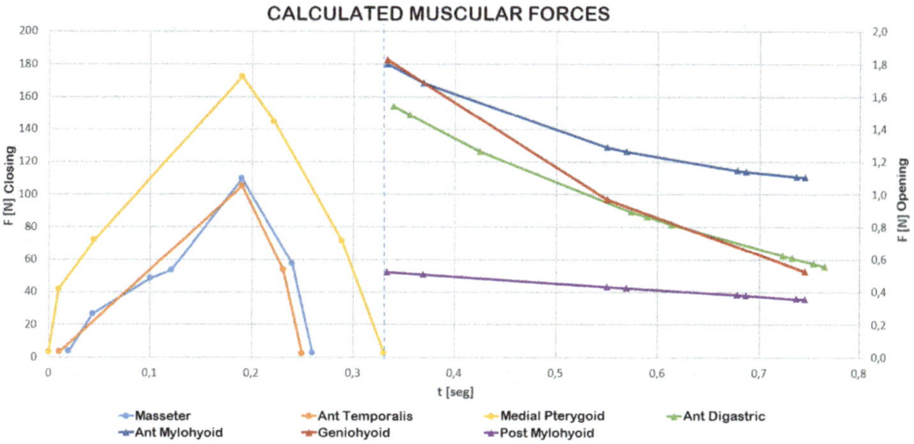

Fig. 3. Calculated muscular forces for ipsilateral mastication

Material and Fatigue Behavior. Additional to stress calculation, stress-life material behavior is the most important input data to evaluate fatigue in components. This information is test-specific and several correction factors have to be applied to fatigue calculation to conciliate differences between tests and real components [23]. The mean value between maximum and minimum stress in component during load cycles is known as mean stress, and it has an important influence in fatigue strength of materials by reducing the number of cycles achieved by a component under the same stress amplitude compared to a test specimen with a mean stress equal to zero. In this case mean stress correction is necessary in order to develop a good durability estimation.

In this work, material specific mean stress corrections have been done to estimate fatigue strength for each material. For Titanium alloy (Ti6AlV4), adjust to Walker's equation done by Dowling [24]:

$$\sigma_{aR-1} = \sigma_a \left(\frac{2}{1-R}\right)^{1-\gamma} \tag{7}$$

Where R is stress ratio $R = \sigma_{min}/\sigma_{max}$, σ_{aR-1} is alternating stress where a failure is detected in a number of cycles N under fully reversed load ($R = -1$), σ_a is the alternating stress where failure is detected at N cycles and a stress ratio R. And $\gamma = 0.064$ is an exponent for Titanium.

By this way, it is possible to generate several S-N curves for titanium based on fully reversed Stress-Life experimental data. These curves were calculated for $R = -0.5$; 0.0; 0.5; 0.99, as shown in Fig. 4.

For UHMWPE, Chandran's equation [25] is used:

$$N = \left[-\frac{1}{C_n} ln \left(\frac{\frac{2\sigma_a}{1-R} - \frac{2\sigma_{eR-1}}{1-R}}{\sigma_u - \frac{2\sigma_{eR-1}}{1-R}}\right)\right]^{\frac{1}{m_n}} \tag{8}$$

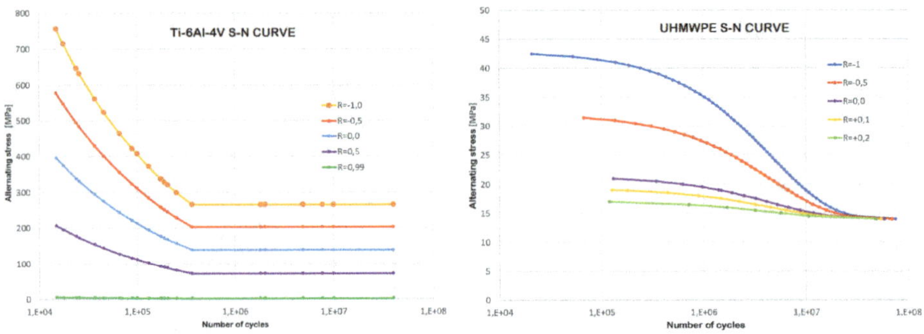

Fig. 4. S-N curves with mean stress effect for Titanium and UHMWPE

Where N is the number of cycles in failure at stress ratio R and an alternating stress σ_a, σ_{eR-1} material fatigue strength limit at $R = -1$. Approximations made for $R = -0.5$; 0.0; 0.1; 0.2 are also shown in Fig. 4.

These plots show fatigue strength under test conditions, due to differences between test and real components, is necessary to adjust the fatigue calculations by [23]:

$$K_f = C_{load} \cdot C_{size} \cdot C_{sup} \cdot C_{temp} \cdot C_{reliab} \tag{9}$$

Where all the C factors up to 1.0 define reductions because of differences in load, size, surface finish, temperature and reliability. K_f is fatigue strength factor and is an input data in fatigue simulation in ANSYS *Mechanical Workbench*.

For TMJ components, C_{sup} can be considered as 1.0 due to specimen test fabrication is made by additive manufacturing as well as real components, $C_{temp} = 1.0$ since operation temperature is below 40 °C and there are not creep phenomena, $C_{reliab} = 0.753$ for 99.9% reliability, $C_{size} = 1.0$ due to test specimen size are similar to condylar plate neck and C_{load} is defined for flexural or axial load.

$$K_f = 0.753 \cdot C_{load} \tag{10}$$

2.3 Wear

A bidimensional model was built to simulate wear between condylar plate and articular fossa components, based on Loon's experimental work [9] where a wear measuring device was developed for a TMJ prototype, its geometry is a cup-sphere pair.

In Fig. 5 are demonstrated the wear test conditions: A 8.0 mm diameter stainless steel ball is installed in an UHMWPE cup with a cavity of radius 4.0 mm. For each sphere are used two cups and a 200 N normal direction force is applied. Horizontal (Y) direction displacement applied cyclically to top bar generates a sphere rotation of 28° at 1.95 Hz during $7 \cdot 10^6$ cycles. Table 3 presents the wear results for this test.

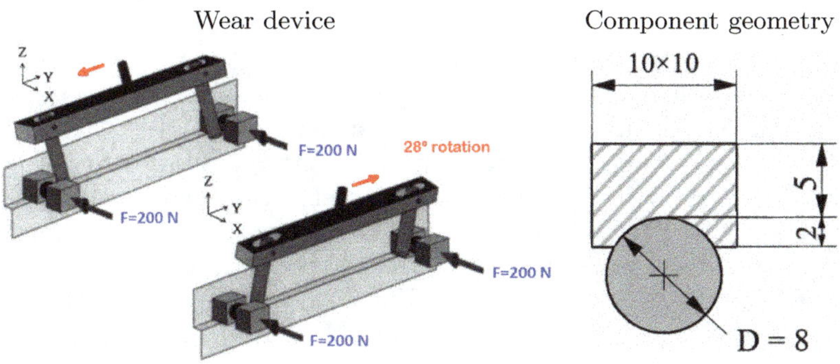

Fig. 5. Wear test for TMJ prosthesis, modificated from Loon [9]

Table 3. TMJ wear test results

Magnitude	Value	Units
Normal force	200	[N]
Sliding velocity	7.62	$[\frac{mm}{s}]$
Sliding distance	27367	[m]
Wear rate K/H	$6,8 \times 10^{-7}$	$[\frac{mm^3}{N \cdot m}]$

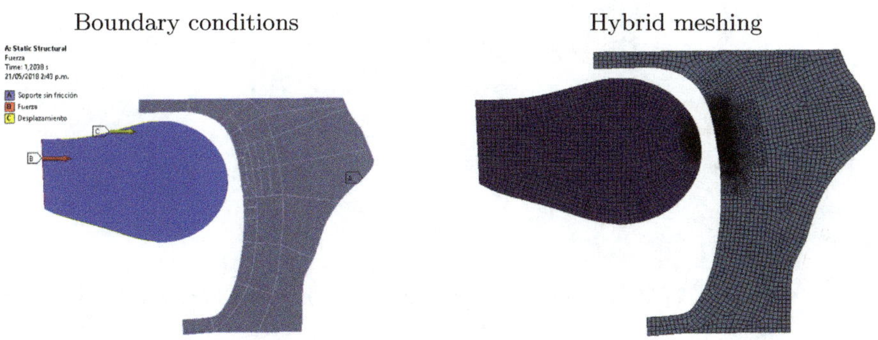

Fig. 6. Wear model configuration

A 2D geometry is extracted from the 3D model keeping contact zone between condyle and fossa, some surface splits are made in order to facilitate meshing process, as presented in Fig. 6. Wear rate was taken from Loon's test [9].

Boundary conditions are defined in two main stages as follows: The first one to close the contact pair by means of prescribed displacement and then applying a 200 N load. And the second one defined between $1.0 < t \leq 15001$ to allow components to wear under applied loads, by deleting the first stage displacement and activating the wear process.

Rezoning process was developed for this 2D model. At the end of each time step, the distance between contiguous nodes of same structural element located at contact zone are calculated due to wear. If the variation percentage of this distance related to same unworn element distance is equal or greater than a prescribed value, it is necessary to start a Rezoning process before continuing the next time step. This Rezoning process generates a new mesh in the worn zone (marked with Named Selections in ANSYS Workbench) then maps all the variables from the old mesh to the new one before continuing with the unsolved time steps. More details are presented in Fig. 7. In this case simulation was solved using $\%\_des = 45\%$ and a final time equal to 150001 s (without inertial effects).

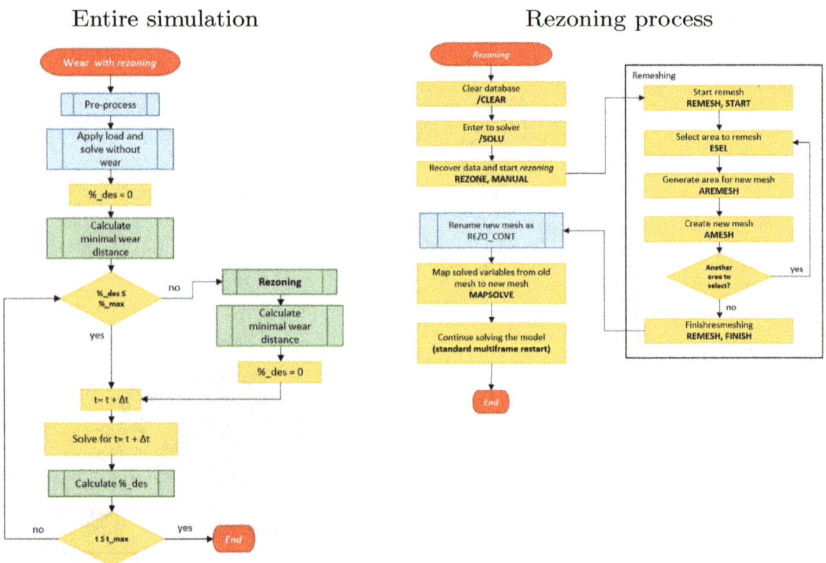

Fig. 7. Wear model and Rezoning scheme for 2D cases

Archard Wear Model. In this work Archard [26] equation is used to determine the lost volume because of wear:

$$\frac{V}{s} = K\frac{F_N}{H} \tag{11}$$

Where V is the lost volume (m^3), s is sliding distance (m), K is adimensional wear rate, F_N is normal force (N) and H is surface hardness (Pa). Dividing both sides by contact and time:

$$\dot{V} = \frac{K}{H}P^m v_{rel}^n \tag{12}$$

Where $\dot{V}$ is lost volume by time unit ($\frac{m^3}{s}$), P is contact pressure (Pa) and v_{rel} is sliding velocity ($\frac{m}{s}$), exponents m and n are used to give an nonlinear behavior to the basic equation. For this work, these were defined as 1.0.

In Eq. 12 is possible to note that when sliding velocity v_{rel} is constant, there are not time dependant effects in wear phenomenon and is not necessary to evaluate velocity as FEA result. This fact allows to reduce computational cost in simulation by calculating only contact pressure. Ordering Eq. 12 is possible to define one "total wear factor" that includes K, H and v_{rel}.

$$\dot{V} = K_{total} P \; con \; K_{total} = v_{rel} \frac{K}{H} \tag{13}$$

By this strategy is also possible to include variable velocity by modifying K_{total} in each time step according to current velocity in the pseudo time scale of finite element analysis.

3 Results

3.1 Static Structural Model

Von Mises stress were calculated for structural model that includes all components. Time dependent stresses show a maximum at time step 20 (0.18 s) for Condylar Plate and Fixation Fossa components, coincident with maximum muscular load time shown in Fig. 3. For Articular Fossa component maximum stress is shown at 0.33 s.

Fossa fixation presents a maximum Von Mises stress of 746.6 MPa at 0.18 s under yield stress value of 1100 MPa for titanium alloy located near to fixation screws in the zone of plate curvature. Titanium condylar plate has 212.7 MPa as maximum in the condyle-plate transition zone (neck). Articular fossa shows its maximum value of 52.3 MPa at 0.33 s in a small area in the contact lateral zone, above UHMWPE yield stress value of 31 MPa. At 0.18 s a local maximum is present behind the surface in the concave contact zone of 25.9 MPa.

Safety factor to static failure using Maximum Von Mises stress criteria were calculated for all components, as shown in Table 4.

Table 4. TMJ static structural analysis results

Component	Maximum stress [MPa]	Time [s]	Safety factor	Location
Fossa fixation	746.6	0.18	1.47	Screw holes
Articular fossa	52.7	0.33	0.58	Lateral
Articular fossa	25.9	0.18	1.19	Concave
Condylar plate	212.7	0.18	5.17	Neck

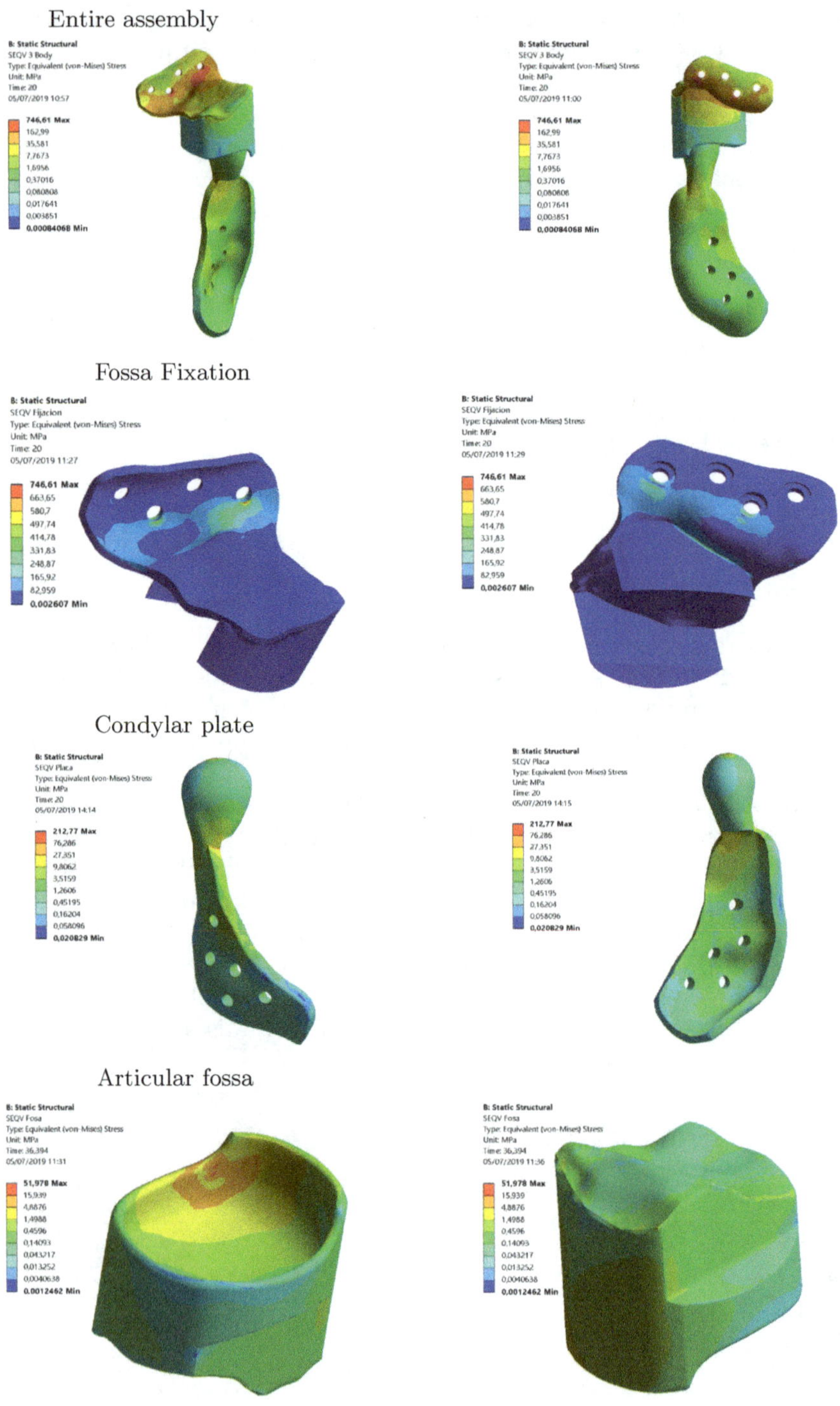

Fig. 8. Von Mises stress distribution for Static structural analysis of TMJ

3.2 Fatigue

Each TMJ component has particular fatigue conditions due to material properties and stress pattern, that is why is necessary to evaluate them individually (Fig. 8).

For fossa fixation, stress variation has no variable amplitude cycles with a maximum of 746.6 MPa at 0.18 s and a minimum of 107.63 MPa at 0.28 s. Therefore, a constant amplitude load ratio with stress results at 0.18 s is used for fatigue calculation. In this case stress ratio is:

$$R_{fix} = \frac{\sigma_{min}}{\sigma_{max}} = \frac{107.63}{746.6} = 0.14 \tag{14}$$

Figure 9 presents component calculated life results. Any color different to blue means a durability less than 40 million cycles. The lowest life is present near screw fixation holes with less than 4 cycles. Because of asymptotic behavior behind 14.6 million cycles of S-N curves for titanium, $1 \cdot 10^9$ cycles is assumed as infinite life.

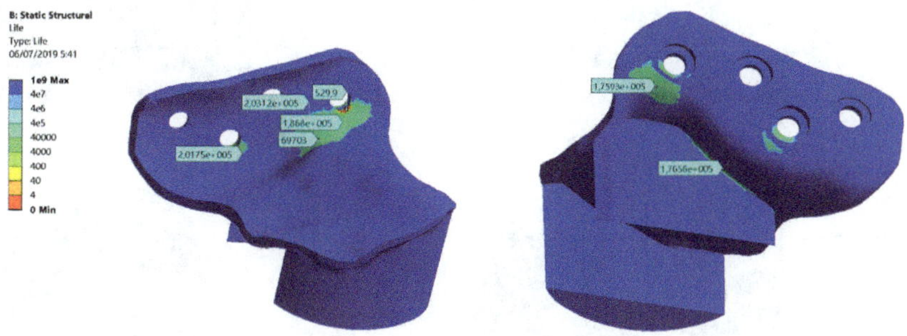

Fig. 9. Fossa fixation life

Stress ratio calculation was made for condylar plate, $R_{plate} = \frac{0}{212.7} = 0$ called zero based load. Available life is superior to 40 millions cycles in all the component. For articular fossa two zones were defined: The first one in lateral interior surface. Stress ratio is calculated for fatigue evaluation $R_{latfossa} = \frac{16.64}{52.03} = 0.32$ with $\sigma_{min} = 16.64$ MPa at 0.057 s with an available life of 40 million cycles in the volume and a singular value of zero in nodal contact zone. The second zone located in concave contact zone $R_{concfossa} = \frac{0}{25.92} = 0$ with $\sigma_{max} = 25.92$ at 0.18 s and $\sigma_{min} = 0$ MPa at 0.33 s offers a 40 million cycles life in volume and 9.55 millions in volume under contact zone. As shown in Fig. 10.

Fatigue safety factor allows to evaluate critical zones for an endurance of 40 million cycles. The lowest value found for fossa fixation is 0.19. For fossa lateral minimum value is 0.42, in fossa concave is 0.81 located at same place of minimum life. Condylar plate presents 0.65 singularly in an edge close neck zone. In Fig. 11 detailed contours for that result for all components are presented.

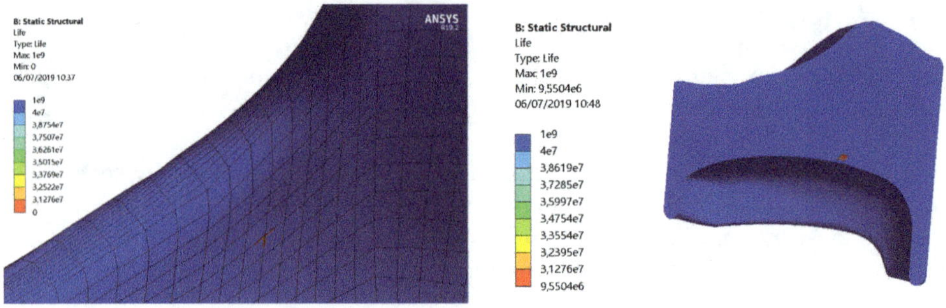

Fig. 10. Articular fossa life

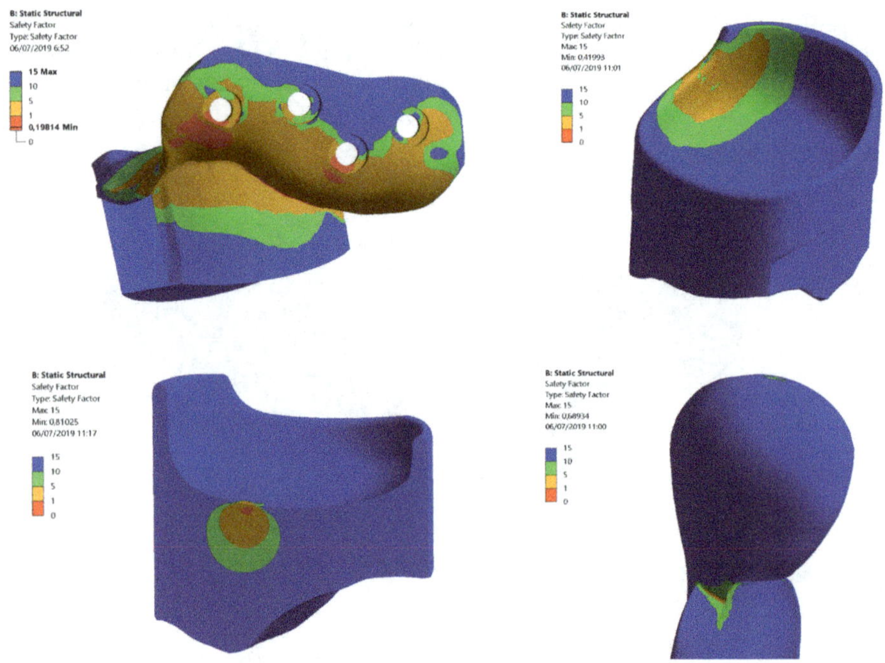

Fig. 11. TMJ components Safety factor

Fatigue sensitivity diagrams are built for all components. Stress state is affected by a factor that varies from 0.5 to 2.0 in order to predict how sensitive is each part to variations in load. In Fig. 12 is possible to note that condylar plate is able to endure 40 million cycles up to 1.125 times the stress, beyond that, values below one million cycles are achieved.

For fossa components is possible to estimate the maximum load factor to obtain a desired endurance. For example, to reach 40 million cycles is necessary to reduce stress below 70% for fossa lateral and fixation. 95% is sufficient for fossa concave contact zone.

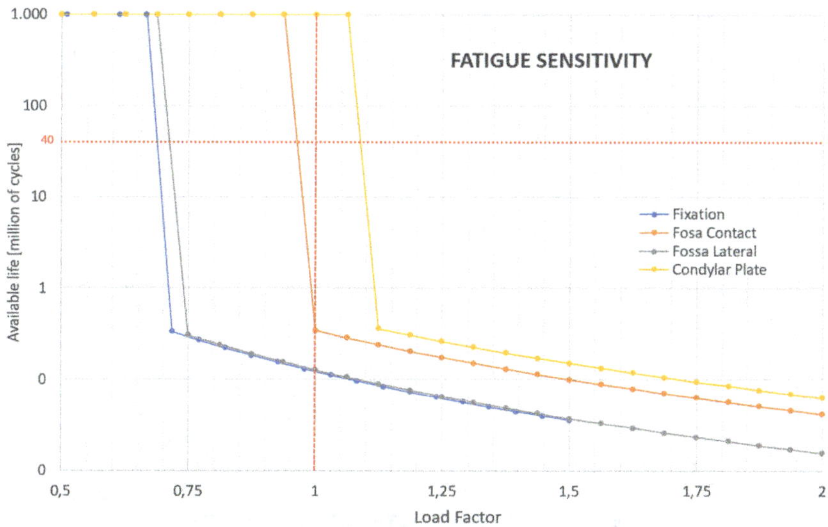

Fig. 12. Wear results for TMJ prosthesis 2D model

3.3 Wear Simulation

During 150001 s of wear simulated, at velocity of 7.62 $\frac{\text{mm}}{\text{s}}$ a linear displacement of 1143 m is achieved, that is equivalent to 0.292 million of cycles, in this range lost volume calculated by wear model in Fig. 13, in this way is possible to make an estimation of lost volume for a million of cycles:

$$Vol_{@million} = Vol \cdot \frac{1}{0.292}$$

$$= 0.1582\,\text{mm}^3 \cdot \frac{1}{0.292} = 0.5409 \frac{\text{mm}^3}{\text{million cycles}}$$

This wear value is close to $0.47 \pm 0.17\,\text{mm}^3/\text{million cycles}$ value of wear found in Loon's work [9], according to this author this is considered acceptable for TMJ prosthesis application. In Fig. 13 is presented wear distance in articular fossa using quasi static model. In this figure is shown the non linear behavior of geometric affectation of articular fossa after wear. In Fig. 14 is possible to appreciate mesh modification after 150000 s of simulation time.

4 Discussion

Static structural analysis shows that the only part with high safety factor is condylar plate. On the other hand, low safety factor for static failure (below 1.5) in fossa components were found, fossa fixation and articular fossa components in lateral and in concave locations.

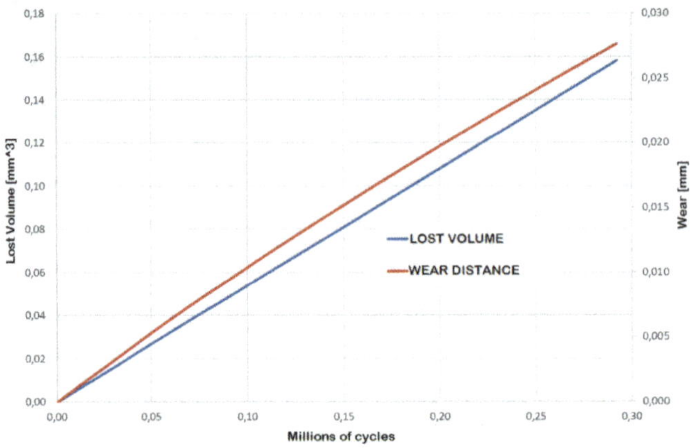

Fig. 13. Wear results for TMJ prosthesis 2D model

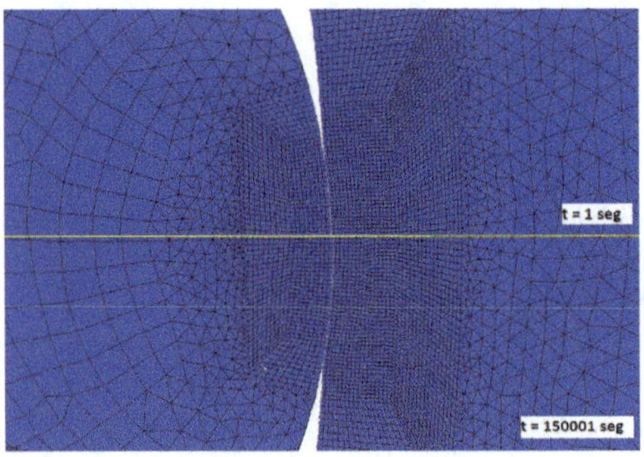

Fig. 14. Modified mesh due to wear after 150000 s

High stress of 746.6 MPa in titanium fossa fixation is calculated near screw holes in the geometric curvature done to the part to fit the Temporal bone and allow fixation by lateral screws. In FEA static model, this component is supported only by frictionless condition where displacements are restricted in each screw hole surface normal only and all the horizontal component surface allowed to move vertically with no restriction generating this high flexural condition. Maximum vertical displacement of 0.16 mm was found in calculations at time 0.18 s, this gap between bone and titanium could be reached during surgical procedure. Thus a conservative estimation is made in this case.

Low safety factor in articular fossa suggests the need to include a more detailed UHMWPE material model. Using a linear elastic model generates elevated stresses due to overestimated elastic modulus beyond yield strength. In this static analysis maximum total strain in articular fossa component reached 0.2 at 0.33 s, that is another reason to implement a great strain material model in following related projects.

Fatigue results in fixation component shows that this one is the critical part according to boundary conditions defined. Even ignoring fixation screw holes, life values as low as 40000 cycles (20 days) [4,22] appeared in curvature zone due to flexural load. If is not feasible to guarantee full contact between this part and temporal bone, is important to consider geometry modifications -thickness augmentation- in order to reduce this stress level and thus improving available life. For condylar plate fatigue results are satisfactory and can be named as an adequate design.

Articular fossa component showed available life and safety factor fatigue results according to stress state. Although ignoring peak values this results could be satisfactory, is important to remember that fatigue stress life theory is based on stress values below material yield strength. Strain life approach could be a better option to analyze this component. For that case is imperative to obtain a cyclic stress strain curve for material.

Wear result can be compared with a practical procedure published by the consumer protection federal agency of United States government: Food and Drug Administration (FDA) for a commercial TMJ prosthesis [27] in this test is shown a wear rate of $0.39\,\text{mm}^3$/million cycles after apply a load of 9 Kg-f = 88.26 N, equivalent to approximately 44% of Loon's test load 200 N [9]. Under this consideration, simulation value surpasses FDA's approval criteria by scaling results in function of applied load, lost volume would be $0.54/200 \cdot 88.26 = 0.24$ mm^3/million cycles.

References

1. Mercuri, L.G.: Temporomandibular Joint Total Joint Replacement - TMJ TJR. Springer, Cham (2016)
2. Van Loon, J.-P., Verkerke, G., De Vries, M., De Bont, L.: Design and wear testing of a temporomandibular joint prosthesis articulation. J. Dent. Res. **79**(2), 715–721 (2000)
3. Wolford, L.: Factors to consider in joint prosthesis systems. Proce. (Bayl. Univ. Med. Cent.) **19**, 232–238 (2006)
4. Sinno, H., Tahiri, Y., Gilardino, M., Bobyn, D.: Engineering alloplastic temporomandibular joint replacements. Mcgill J. Med. **13**, 63 (2011)
5. Briceño, F., Ayala, R., Delgado, K., Piñango, S.: Evaluation of temporomandibular joint total replacement with alloplastic prosthesis: observational study of 27 patients. Craniomaxillofac. Trauma Reconstr. **6**, 171–178 (2013)
6. Mercuri, L.G.: The role of custom-made prosthesis for temporomandibular joint replacement. Revista Española de Cirugía Oral y Maxilofacial **35**(1), 1–10 (2013)
7. Guarda-Nardini, L., Manfredini, D., Ferronato, G.: Temporomandibular joint total replacement prosthesis: current knowledge and considerations for the future. Int. J. Oral Maxillofac. Surg. **37**(2), 103–110 (2008)
8. Keller, E.E., Baltali, E., Liang, X., Zhao, K., Huebner, M., An, K.-N.: Temporomandibular custom hemijoint replacement prosthesis: prospective clinical and kinematic study. J. Oral Maxillofac. Surg. **70**(2), 276–288 (2012)
9. Loon, J.-P.V., Verkerke, G., de Bont, L., Liem, R.: Wear-testing of a temporomandibular joint prosthesis: UHMWPE and PTFE against a metal ball, in water and in serum. Biomaterials **20**(16), 1471–1478 (1999)

10. Abel, E.W., Hilgers, A., McLoughlin, P.M.: Finite element analysis of a condylar support prosthesis to replace the temporomandibular joint. Br. J. Oral Maxillofac. Surg. **53**(4), 352–357 (2015)
11. Huang, H.-L., Su, K.-C., Fuh, L.-J., Chen, M.Y., Wu, J., Tsai, M.-T., Hsu, J.-T.: Biomechanical analysis of a temporomandibular joint condylar prosthesis during various clenching tasks. J. Craniomaxillofac. Surg. **43**(7), 1194–1201 (2015)
12. Ramos, A., Mesnard, M.: Load transfer in Christensen® TMJ in alloplastictotal joint replacement for two different mouth apertures. J. Craniomaxillofac. Surg. **42**(7), 1442–1449 (2014)
13. Mahdian, N., Dostálová, T., Daněk, J., Nedoma, J., Kohout, J., Hubáček, M., Hliňáková, P.: 3D reconstruction of TMJ after resection of the cyst and the stress–strain analyses. Comput. Methods Programs Biomed. **110**(3), 279–289 (2013)
14. Drake, R., Mitchell, A., Vogl, A.: Gray. Anatomía para estudiantes, Elsevier Elibrary, Elsevier Health Sciences Spain (2015)
15. Commisso, M.S., Martínez-Reina, J., Ojeda, J., Mayo, J.: Finite element analysis of the human mastication cycle. J. Mech. Behav. Biomed. Mater. **41**, 23–35 (2015)
16. Thelen, D.G.: Adjustment of muscle mechanics model parameters to simulate dynamic contractions in older adults. J. Biomech. Eng. **125**, 70–77 (2003)
17. Langenbach, G., Hannam, A.: The role of passive muscle tensions in a three-dimensional dynamic model of the human jaw. Arch. Oral Biol. **44**(7), 557–573 (1999)
18. Hylander, W.L.: Functional anatomy. In: The Temporomandibular Joint: A Biologic Basis for clinical Practice. Saunders Co., Philadelphia (1992)
19. Hannam, A.G., Stavness, I., Lloyd, J.E., Fels, S.: A dynamic model of jaw and hyoid biomechanics during chewing. J. Biomech. **41**, 1069–1076 (2008)
20. Korioth, T.W., Romilly, D.P., Hannam, A.G.: Three-dimensional finite element stress analysis of the dentate human mandible. Am. J. Phys. Anthropol. **88**, 69–96 (1992)
21. Van Eijden, T.M., Korfage, J.A., Brugman, P.: Architecture of the human jaw-closing and jaw-opening muscles. Anat. Rec. **248**, 464–474 (1997)
22. van Loon, J.-P., de Bont, L.G., Boering, G.: Evaluation of temporomandibular joint prostheses: review of the literature from 1946 to 1994 and implications for future prosthesis designs. J. Oral Maxillofac. Surg. **53**(9), 984–996 (1995)
23. Norton, R.: Diseño de maquinaria. McGraw-Hill/Interamericana de España, S.A. (2012)
24. Dowling, N.E., Calhoun, C.A., Arcari, A.: Mean stress effects in stress-life fatigue and the walker equation. Fatigue Fract. Eng. Mater. Struct. **32**, 163–179 (2009)
25. Chandran, K.R.: Mechanical fatigue of polymers: a new approach to characterize the sn behavior on the basis of macroscopic crack growth mechanism. Polymer **91**, 222–238 (2016)
26. Archard, J.F., Hirst, W.: The wear of metals under unlubricated conditions. Proc. Roy. Soc. London Ser. A Math. Phys. Sci. **236**(1206), 397–410 (1956)
27. FDA: Food and Drugs Administration Premarket Approval PMA (1999). https://www.accessdata.fda.gov/scripts/cdrh/cfdocs/cfpma/pma.cfm?id=p980052. Accessed 09 May 2018

Connection Between Gait and Balance Functions in Pediatric Patients with Either Neurological or Sensory Integration Problems

Malgorzata Syczewska(✉), Ewa Szczerbik, Malgorzata Kalinowska, and Anna Swiecicka

Department of Rehabilitation, The Children's Memorial Health Institute, Al. Dzieci Polskich 20, 04-730 Warsaw, Poland
m.syczewska@ipczd.pl

Keywords: Balance · Gait · Children

1 Introduction

Locomotion and maintaining the proper body posture are the two most important functional tasks in everyday life. Their impairment handicap the patients, and often compromise the effectiveness of performance of other tasks. Therefore gait and balance are often assessed in patients who encounter problems during various functional tasks.

It is believed that gait deficits may manifest balance problems, and vice versa. In the study of Guffey et al. [1] in which in a group of healthy 2- to 4-years old children balance and gait were assessed the spatio-temporal gait parameters (together with age) explained 51% of the balance score variance. Children and adolescents with sensorineural hearing loss demonstrate worse gait performance than their healthy peers [2], and there are studies showing that loss of hearing impairs also balance [3].

Patients with many neurological illnesses demonstrate both gait and balance problems [4], similarly to children with coordination problems and sensory integration deficits [5]. Therefore a general assumption of the dependence of the gait disturbances on balance problems is accepted in clinical environment. Nevertheless there are studies which demonstrate, that such dependence is not as straightforward. In a study concerning children with unilateral cerebral palsy only partial dependence between gait disturbances and balance instability was found, as the correlations between them although statistically significant were low [6]. Based on the assumption of dependence between gait and balance the intervention rehabilitation programs often imply that improving balance will improve gait. A meta-analysis of the influence of the backward walking on balance [7] revealed some improvement of the single leg stance test, while balance and strength training in intellectually disabled patients had no impact on gait performance [8].

The aim of the present study was to evaluate gait parameters and gait indices in patients with neurological diseases (ND) and sensory integration (SI) problems, and to evaluate their dependence on the results of the balance tests.

G. A. Ateshian et al. (Eds.): CMBBE 2019, LNCVB 36, pp. 335–338, 2020.
https://doi.org/10.1007/978-3-030-43195-2_27

2 Materials and Methods

2.1 Patients

Forty seven patients were recruited to the study. There were 30 patients with neurological (ND) diseases which compromise balance abilities (Guillain-Barre Syndrome, Cerebral Palsy, polyneuropathies, etc.) and 17 patients with sensory integration problems (SI), in whom SI diagnostic tests revealed balance problems [9]. The group comprised 21 boys and 26 girls aged from 5 to 17 years old.

The study was approved by the Local Ethical Committee (Warsaw Medical Univeristy, KB/28/2014), and the parents and children over 16 years of age gave their informed consent to participate in this study. The patients with co-existing comorbidities, which can additionally affect the functional abilities were excluded from the study.

2.2 Methods

All patients underwent instrumented gait analysis and balance tests.

The gait analysis was performed using VICON MX system with 12 cameras. The Plug-In-Gait marker set and body model were used. Each subject walked with her/his preferred walking speed. Six trials were captured, processed with Nexus software and later averaged using Polygon software. From the Polygon reports the spatio-temporal parameters, and kinematics parameters were extracted. The gait speed, cadence, and step length were normalized to sex and age matched reference data.

Additionally for each patients the Gait Deviation Index, Movement Analysis Profile (MAP) and gait Profile Score (GPS) were calculated, separately for left and right legs.

GDI [10] was calculated in Nexus software for each trial of each patient, separately for left and right leg, and later averaged for the patient's session. GDI is a single number index, resulting from kinematic plots and principal component analysis. The methodology uses three dimensional angles of the pelvis and hip, at the knee and ankle joints only angles in sagittal plane are used, together with foot progression angle. GDI is transformed and scaled in such a way, that its average value for healthy subjects is 100, with a standard deviation of 10.

Gait Variable Score (GVS), another gait index, is calculated as root mean square difference between kinematic variable across gait cycle of the patient and the same reference variable, representing healthy subjects. The GVS are calculated for nine key kinematic variables: pelvic tilt, hip flexion, knee flexion, ankle dorsiflexion, pelvic obliquity, hip abduction, pelvic rotation, and foot progression, and from them a Movement Analysis Profile (MAP) can be combined. From GVS an overall index, Gait Profile Score (GPS) was calculated for each subject [11].

During the same session patients underwent two balance tests on Biodex Balance System: first one was standing on the stable platform (according to standard Biodex protocol: three repetitions, each lasting 20 s), and the second was standing on unstable platform, stability level 4 (three repetitions, each lasting 20 s). Both tests were performed with eyes open, and the overall stability indices, given by Biodex software, were used for further calculations.

The data were later analyzed using stepwise forward discriminant analysis (StatSoft), and logistic regression (MedCalc) to see the influence of disease (ND or SI), abnormal balance on stable base or unstable base on gait parameters. The cut-off value of p-level was 0.05.

3 Results

The summary of the results of statistical analysis (discriminant analysis and logistic regression) are presented in Table 1.

Table 1. The summary of the statistical analysis, predictive variables chosen by each statistical methods are presented with percent of cases correctly classified.

	Disease	Balance test on stable platform	Balance test on unstable platform
Discriminant analysis	+ Variability of GDI Pelvis MAP in sagittal plane 67,7%	+ MAP foot progression 86,9%	–
Logistic regression	–	+ MAP foot progression 86,9%	+ Pelvis MAP in transverse plane 78,3%

The median values of the predictive variables were calculated.

For ND group median value of GDI variability was 2,3, for SI group 2,8, and the median values of pelvis MAP in sagittal plane were 3,818 and 5,225 respectively.

The children with good stability index during balance test on stable platform had median value of MAP foot progression 8,029, and with abnormal balance index 7,782.

The children with good stability index during balance test on unstable platform had median value of pelvis MAP in transverse plane 13,789, and with abnormal balance index 8,817.

4 Discussion

The results indicate, that there is some connection between the ability to maintain proper balance and the gait, but this connection is rather very weak, as only two parameters were indicated as prognostic variables by statistical analysis: MAP progression angle and pelvis MAP in transverse plane. This finding, i.e. that there is a weak dependence between gait and balance is in accordance with study of Domagalska-Szopa and co-workers [6] who found that there is weak dependence between gait pattern of children with unilateral cerebral palsy, and their balance function.

The results of the statistical analysis when the patients were divided according to clinical problems suggests, that the gait of the SI patients is less repeatable than ND patients: median value of GDI variability is smaller in ND group than in SI group.

Interestingly, two used statistical approaches gave the same results when the patients were divided according to the result of the balance test on stable platform. The discriminant analysis and logistic regression are closely related to each other, but logistic regression does not require multivariate normal assumption. This example shows, that using different statistical methods can influence the final results of the analysis.

One of the major shortcomings of the present study is low number of subjects. In all three divisions one of the subgroups was smaller than the other, making the conclusions less reliable.

References

1. Guffey, K., Regier, M., Mancinelli, C., Pergami, P.: Gait parameters associated in balance in healthy 2- to 4-years old children. Gait Posture **43**, 165–169 (2017)
2. De Souza Melo, R.: Gait performance of children and adolescents with sensorineural hearing loss. Gait Posture **57**, 109–114 (2017)
3. Maes, L., de Kegel, A., van Waelvelde, H., Dhooge, I.: Association between vestibular function and motor performance in hearing-impaired children. Otol. Neurotol. **35**, 343–347 (2014)
4. Smith, M., Kurian, M.A.: Neurological gait disorders in childhood. Paediatr. Child Health **28** (10), 454–458 (2018)
5. Bril, B., Ledebt, A.: Head coordination as a means to assist sensory integration in learning to walk. Neurosci. Biobehav. Rev. **22**, 555–563 (1998)
6. Domagalska-Szopa, M., Szopa, A., Czamara, A.: Dependence of gait deviation on weight-bearing asymmetry and postural instability in children with unilateral cerebral palsy. PLoS ONE **11**(10), e0165583 (2016)
7. Wang, J., Xu, J., An, R.: Effectiveness of backward walking training on balance performance: a systematic review and meta-analysis. Gait Posture **68**, 466–475 (2019)
8. Lee, K., Lee, M., Song, C.: Balance training improves postural balance, gait and functional strength in adolescents with intellectual disabilities: single blinded randomized clinical trial. Disabil. Health J. **9**, 416–422 (2016)
9. Barton, E.E., Reichow, B., Schnitz, A., Smith, I.C., Sherlock, D.: A systematic review of sensory-based treatments for children with disabilities. Res. Dev. Disabil. **37**, 64–80 (2015)
10. Schwartz, M., Rozumalski, A.: The gait deviation index: a new comprehensive index of gait pathology. Gait Posture **28**, 351–357 (2008)
11. Baker, R., McGinley, J.L., Schwartz, M.H., Beynon, S., Rozumalski, A., Graham, H.K., Tirosh, O.: The gait profile score and movement analysis profile. Gait Posture **30**, 265–269 (2009)

Classification of Elderly Fallers and Non-fallers Using Force Plate Parameters from Gait and Balance Tasks

Ashirbad Pradhan, Sana Oladi, Usha Kuruganti, and Victoria Chester(✉)

Andrew and Marjorie McCain Human Performance Laboratory, University of New Brunswick, Fredericton, NB E3B 5A3, Canada
vchester@unb.ca

Abstract. Accidental falls are a major health concern among older adults, hence identifying biomechanical parameters from gait and balance tasks that differentiate between fallers and non-fallers are crucial. Limited studies using force platforms to assess postural control have used machine learning algorithms to classify older adult fallers. The results are diverse due to variation in task routines, biomechanical parameters and classification algorithms. Therefore, research analysing the performance of different classification algorithms is warranted. The purpose of this study was to compare the classification accuracy of different classification algorithms for identifying elderly fallers using force plate parameters measured during balance and gait tasks. Participants included 58 non-fallers (age = 72.3 ± 5.7) and 41 fallers (age = 74.0 ± 12.3) who performed balance and gait tasks on a walkway with embedded force plates (Kistler Instruments, Winterthur, Switzerland). The force plate parameters included 2D ground reaction force (GRF)-time data and centre of pressure (COP) displacement/velocity data. Using this data as input, five different classification algorithms were used to build models: Naïve Bayesian (NB), Artificial Neural Network (ANN), Linear Discriminant Analysis (LDA), Support Vector Machine (SVM) and k nearest neighbours (kNN). A maximum accuracy of 84.95% for classifying faller/non-faller categories was obtained using LDA classifier based on parameters from combined gait and balance tasks. Combining force plate parameters from gait and balance tasks resulted in higher classification accuracies of older adult fallers (>75%) for all the algorithms. The findings of this study suggest that high accuracy of classifying elderly fallers can be obtained using force plate parameters.

Keywords: Machine learning · Force plate · Gait · Balance

1 Introduction

Accidental falls are one of the leading causes of death among older adults [1]. Approximately 20% of Canadians aged 65 and older (862,000 older adults) reported a fall in 2008 [2]. Fall prevention programs, which focus on reducing future incidences of fall in older adults, employ clinical assessment tools for identifying elderly fallers.

G. A. Ateshian et al. (Eds.): CMBBE 2019, LNCVB 36, pp. 339–353, 2020.
https://doi.org/10.1007/978-3-030-43195-2_28

Although clinical tools have been used to assess balance [3, 4] and mobility [5, 6] among older adults, previous research has found them to be inaccurate in identifying an elderly faller and hence may not be effective in preventing future fall incidences [7, 8]. On the other hand, data-driven (machine learning) approaches, have found growing applications for differentiating between older adult fallers and non-fallers, which could be used for effective fall prevention programs [9, 10]. Machine learning techniques utilize certain algorithms that are capable of classifying categories of objects based on a set of feature parameters pertaining to these objects. Some studies have used these algorithms on data from force platforms as a screening tool for classification of older adult fallers and non-fallers [11, 12].

Force plates have been considered the "gold standard" for objective balance and gait assessments [13]. Force plates provide the three-dimensional components of the ground reaction force (GRF) [14] during activity. GRFs are widely used for distinguishing abnormal gait from normal gait in individuals [15], subject recognition [16] and fall risk assessment [17]. The GRF parameters from sit-to-stand tasks have been a significant indicator of previous fall incidents [17, 18]. The point of application of the ground reaction forces is termed as the centre of pressure (COP) [19]. Quantitative posturography used for assessing stability involves the measurement of COP based variables [20]. Previous research suggests that certain parameters based on COP displacement might be an important predictor of falls in older adults [11, 12].

The two most commonly reported causes of falls in older adults are impairments in balance and poor mobility [1]. Multiple studies have examined both balance and gait tasks for fall risk prediction [21]. Results obtained from these studies have been inconclusive in determining the optimal task protocol. The studies related to identifying fallers are limited and the results are diverse due to variation in task routines and classification algorithms. The performance of these algorithms varied based on the parameters used. Previous research have utilized different algorithms for classification of fallers including simple algorithms such as Linear Discriminant Analysis, k-Nearest Neighbours, Naïve Bayes, Support Vector Machines and more complex algorithms such as Artificial Neural Networks [10, 14, 22, 23]. Research analysing the performance of different classification algorithms based on these force plate parameters from balance and gait tasks is warranted to determine optimal classification methods.

Thus, the purpose of the study was twofold: (1) to compare the accuracy of different classification models for identifying older adult fallers using force plate parameters measured during gait and balance tasks, and (2) to determine whether gait tasks, balance tasks or a combination of both are optimal for identifying fallers and non-fallers.

2 Methods

2.1 Participants

Ninety-nine participants (n = 99) over the age of 65 were recruited from nursing homes, rehabilitation centers, physiotherapy clinics, orthopedic clinics, the YMCA and older adult community centers in Fredericton, New Brunswick. The participants were

Table 1. Participant demographics (mean ± std dev)

Characteristics	Faller (n = 41)	Non-faller (n = 58)
Age (years)	74.0 ± 12.3	72.3 ± 5.7
Height (cm)	164.8 ± 8.8	164.5 ± 10.0
Weight (kg)	77.2 ± 13.3	73.1 ± 13.3
Gender (#M; #F)	(13M; 28F)	(23M; 35F)

categorized into two groups based on their history of falls. Forty-one (n = 41) of the participants who had experienced one or more falls within the last 12 months were placed in the faller (F) group. All other participants were assigned to the non-faller (NF) group (n = 58). Participants who experienced falls due to inevitable events such as furniture collapse were excluded from the study. Informed consent was obtained from the participants. Demographics and data regarding the participants fall history were collected (see Table 1). The participants were asked to complete a PAR-Q questionnaire to identify any chronic conditions within the last six months. The study was approved by the University Research Ethics Board.

2.2 Instrumentation

Data was collected from six piezoelectric force plates (Kistler Instruments Ltd, Switzerland), embedded in the lab floor. The force plates were used to collect the three-dimensional forces and moments during the balance and gait tasks at a sampling frequency of 1600 Hz. The data obtained from the force plates were exported to Matlab (Matlab, Mathworks, inc. Version 2019a) for further processing.

2.3 Experimental Protocol

All data collection occurred at the Andrew and Marjorie McCain Human Performance Laboratory. Testing occurred in a single session lasting approximately 90 min. After the collection of demographic information, the participants were asked to complete gait and balance tasks. Participants were first asked to walk at a self-selected speed on a 20-ft walkway with six force plates embedded in the center of the walkway. They were asked to stand behind a start line and upon hearing the "go" command, they were instructed to walk towards the other end of the walkway, turn around and walk back to the starting line. To avoid targeting of force plates participants were provided with a visual target at a distance of 1.84 m from the walkway. The trials were repeated until three successful strikes from the left foot and the right foot were attained within the same cycle. Participants were provided with a two-minute rest period between every trial. Participants were then asked to perform two standing balance tests for 30 s:

(1) semi tandem stance (place one foot directly in front of the other) with eyes open (EOTS); (2) narrow stance (base of support less than the width of the shoulder) with eyes closed (ECNS) [24]. The tests were repeated three times each with a one-minute rest period for the EOTS and a three-minute rest period for the ECNS.

2.4 Data Analysis

Force Plate Parameters. For each of the balance and the gait tasks, the average data of the three trials was used. Ground reaction force (GRF) data was normalized to bodyweight and percent stance duration. GRF parameters for the gait task included max, min, and zero crossing data for the vertical and anterior-posterior components and time-to-peak data (Fig. 1) [25]. The medial-lateral GRF components were excluded from analysis due to their very small magnitude and high variability [26]. Center of pressure (COP) parameters, were computed relative to 4 subphases of stance as explained in Fig. 2 [27]. For each sub-phase the instantaneous displacement in the ML direction, instantaneous displacement in the AP direction, the total displacement of COP movement [27], and their variability (standard deviations) were calculated. The GRF and COP parameters are listed in Table 2.

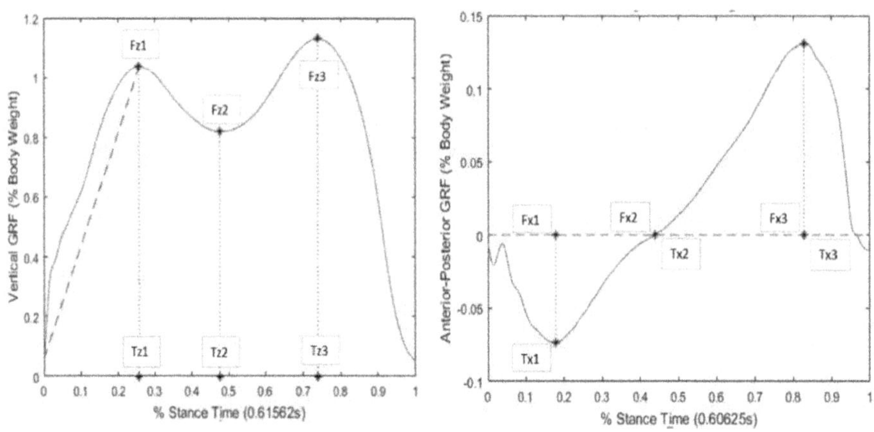

Fig. 1. (Left) Vertical GRF parameters from an individual's walking trial: Fz1 (first peak force), Fz2 (minimum force), Fz3 (second peak force), and the relative time to reach each of the forces (Tz1, Tz2, Tz3). The red line displays the loading rate, which is defined as the absolute value of the slope of the straight line connecting the origin and the peak force Fz. (Right) Anterior-posterior GRF parameters from an individual's walking trial: Fx1 (maximum breaking force), Fx2 (mid stance), Fx3 (maximum propulsion force), and the relative time to reach each of the forces (Tx1, Tx2, Tx3) [25].

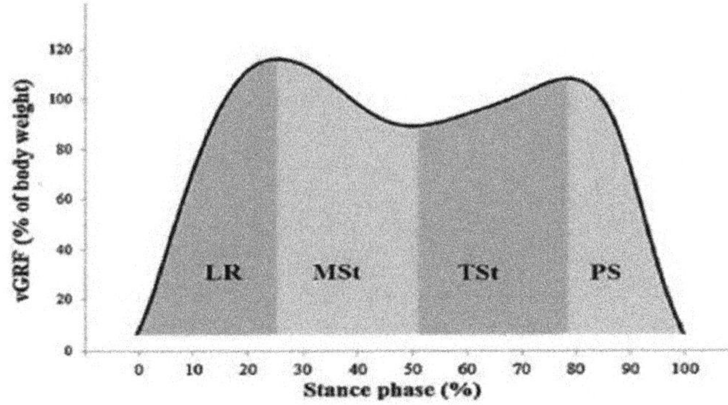

Fig. 2. The division of the stance phase into sub-phases; Loading Response (LR): the time between heel strike and first peak vGRF, Mid-Stance (MSt): the time between first peak vGRF and minimum vGRF, Terminal Stance (TSt): the time between minimum force and second peak vGRF, Pre-Swing (PS): the time between second peak and toe-off [27].

Table 2. Force plate parameters (GRF and COP) from gait tasks

Sl No.	Parameter	Description
1	Fz1 (%Bw)	First peak force in the vertical direction
2	Fz1_SD (%Bw)	Standard deviation of Fz1
3	Fz2 (%Bw)	Minimum force in the vertical direction
4	Fz2_SD (%Bw)	Standard deviation of Fz2
5	Fz3 (%Bw)	Second peak force in the vertical direction
6	Fz3_SD (%Bw)	Standard deviation of Fz3
7	Fx1 (%Bw)	Maximum breaking force in the anterior-posterior direction
8	Fx1_SD (%Bw)	Standard deviation of Fx1
9	Fx2 (%Bw)	Mid stance force in the anterior-posterior direction
10	Fx2_SD (%Bw)	Standard deviation of Fx2
11	Fx3 (%Bw)	Maximum propulsion force in the anterior-posterior direction
12	Fx3_SD (%Bw)	Standard deviation of Fx3
13	Tz1 (%stance)	Time taken to reach Fz1
14	Tz2 (%stance)	Time taken to reach Fz2
15	Tz3 (%stance)	Time taken to reach Fz3
16	Tx1 (%stance)	Time taken to reach Fx1
17	Tx2 (%stance)	Time taken to reach Fx2
18	Tx3 (%stance)	Time taken to reach Fx3
19	Loading Rate (%BW/second)	The rate at which Fz1 is reached
20	LR_COP_T_Disp	Instantaneous displacement of total COP during the loading response phase
21	LR_COP_T_Disp_SD	Standard deviation of displacement of total COP during the loading response phase
22	LR_COP_{ML}_Disp	Instantaneous displacement of mediolateral COP during the loading response phase

(*continued*)

Table 2. (*continued*)

Sl No.	Parameter	Description
23	LR_COP_{ML}_Disp_SD	Standard deviation of displacement of mediolateral COP during the loading response phase
24	LR_COP_{AP}_Disp	Instantaneous displacement of medial-lateral COP during the loading response phase
25	LR_COP_{AP}_Disp_SD	Standard deviation of displacement of medial-lateral COP during the loading response phase
26	MSt_COP_{T}_Disp	Instantaneous displacement of total COP during the mid-stance phase
27	MSt_COP_{T}_Disp_SD	Standard deviation of displacement of total COP during the mid-stance phase
28	MSt_COP_{ML}_Disp	Instantaneous displacement of medial-lateral COP during the mid-stance phase
29	MSt_COP_{ML}_Disp_SD	Standard deviation of displacement of medial-lateral COP during the mid-stance phase
30	MSt_COP_{AP}_Disp	Instantaneous displacement of anterior-posterior COP during the mid-stance phase
31	MSt_COP_{AP}_Disp_SD	Standard deviation of displacement of anterior-posterior COP during the mid-stance phase
32	TSt_COP_{T}_Disp	Instantaneous displacement of total COP during the terminal stance phase
33	TSt_COP_{T}_Disp_SD	Standard deviation of displacement of total COP during the terminal stance phase
34	TSt_COP_{ML}_Disp	Instantaneous displacement of medial-lateral COP during the terminal stance phase
35	TSt_COP_{ML}_Disp_SD	Standard deviation of displacement of medial-lateral COP during the terminal stance phase
36	TSt_COP_{AP}_Disp	Instantaneous displacement of anterior-posterior COP during the terminal stance phase
37	TSt_COP_{AP}_Disp_SD	Standard deviation of displacement of anterior-posterior COP during the terminal stance phase
38	PS_COP_{T}_Disp	Instantaneous displacement of total COP during the pre-swing phase
39	PS_COP_{T}_Disp_SD	Standard deviation of displacement of total COP during the pre-swing phase
40	PS_COP_{ML}_Disp	Instantaneous displacement of medial-lateral COP during the pre-swing phase
41	PS_COP_{ML}_Disp_SD	Standard deviation of displacement of medial-lateral COP during the pre-swing phase
42	PS_COP_{AP}_Disp	Instantaneous displacement of anterior-posterior COP during the pre-swing phase
43	PS_COP_{AP}_Disp_SD	Standard deviation of displacement of anterior-posterior COP during the pre-swing phase

1–19 are GRF parameters
20–43 are COP parameters

For both the ECNS and EOTS balance tasks, the force plate parameters from the GRF data include the standard deviations of the forces in vertical, ML force and AP direction. The COP parameters include the following: the COP total velocity, the average velocity of the COP displacement in AP and ML directions. Also, the 95% confidence ellipse area (COPEllArea) and the sample entropy of ΔΦ (SampEnt-COPΔΦ) was calculated. The parameters for the balance tasks are listed in Table 3.

Table 3. Force plate parameters (GRF and COP) from balance tasks

Sl No.	Parameters	Description
1	EOTS_FzStand_SD (%Bw)	Standard deviation of the vertical force (Fz)
2	EOTS_FyStand_SD (%Bw)	Standard deviation of the medial lateral force (Fy)
3	EOTS_FxStand_SD (%Bw)	Standard deviation of the anterior posterior force (Fx)
4	EOTS_COP_T-VelStand (cm/sec)	COP total velocity
5	EOTS_COP_{AP}-VelStand (cm/sec)	COP velocity in the anterior-posterior direction
6	EOTS_COP_{ML}-VelStand (cm/sec)	COP velocity in the medial-lateral direction
7	EOTS_COPEllArea (cm^2)	95% confidence ellipse area of the COP
8	EOTS_SampEnt-COPΔΦ	Sample entropy of ΔΦ
9	EOTS_FzStand_SD (%Bw)	Standard deviation of the vertical force (Fz)
10	ECNS_FyStand_SD (%Bw)	Standard deviation of the medial lateral force (Fy)
11	ECNS_FxStand_SD (%Bw)	Standard deviation of the anterior posterior force (Fx)
12	ECNS_COP_T-VelStand (cm/sec)	COP total velocity
13	ECNS_COP_{AP}-VelStand (cm/sec)	COP velocity in the anterior-posterior direction
14	ECNS_COP_{ML}-VelStand (cm/sec)	COP velocity in the medial-lateral direction
15	ECNS_COPEllArea (cm^2)	95% confidence ellipse area of the COP
16	ECNS_SampEnt-COPΔΦ	Sample entropy of ΔΦ

EOTS: Eyes open tandem stance ECNS: Eyes closed narrow stance
1–8 parameters are computed from the EOTS task data.
9–16 parameters are computed from the ECNS task data.
1–3 and 9–11 are GRF parameters.
4–8 and 12–16 are COP parameters.

Classification Models. Faller/non-faller classification accuracies were compared between five classification algorithms, namely: Artificial Neural Network (NN), Linear Discriminant Analysis (LDA), Support Vector Machine (SVM), Naïve Bayesian

(NB) and k-Nearest Neighbours (kNN). Classification models were generated based on each of these algorithms. The model development parameters specific to each algorithm was based on previous research related to faller/non-faller classification [10, 14, 22, 23]. NN are algorithms that are inspired by the functioning of the human brain [28]. They include a collection of nodes called neurons, which correspond to the neurons of the brain. The back-propagation trained feed-forward classification was used for the NN algorithm [29]. The number of nodes in single hidden layer was varied in between 5–20 and the NN models were generated [8]. The SVM algorithm projects the feature space onto a higher dimensional representation in such a way that the data becomes linearly separable [29]. A hyperplane is then created to maximize the distance between the hyperplane and the two classes. For the SVM models, radial basis function kernels were used to form the classification models, with varying gamma and regularization parameters [30]. The kNN algorithm compares the sample with its k-nearest neighbors and classifies it as the most represented class among those k-neighbors. The nearest neighbors were determined using the Euclidean distance. For the kNN classifier, the value of k varied between 1 and 10 and the best model was reported [14, 22]. The NB is a probabilistic classifier based on the Bayes rule which assumes that a particular feature in a class is independent of the other features in the same class [8, 22]. The Fisher linear discriminant analysis is a simple method which uses linear combinations of features to define the relationships between the classes and the features [31]. All models were developed with the PRTools library [32] using Matlab.

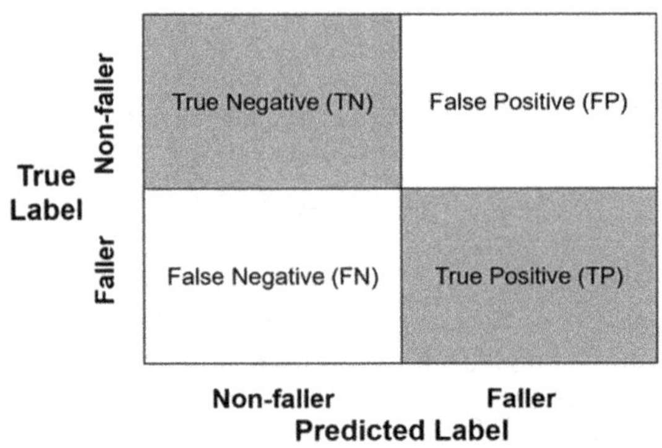

Fig. 3. The confusion matrix for faller/non-faller classification: On the left axis are the true labels for the predicted variable (non-faller/faller) and on bottom axis are the predicted labels. This matrix defines four critical categories: True Negative for which the individual is a non-faller and non-faller class is predicted, True Positive for which the individual is a faller and faller class is predicted, False Negative for which the individual is a faller and non-faller class is predicted, and False Positive for which the individual is a non-faller yet faller class is predicted.

Feature Selection and Model Evaluation. For all of the participants (N = 99), a total of 59 force plate parameters (Tables 2 and 3) represented the balance (6 GRF and 10 COP) and gait tasks (19 GRF and 24 COP). However, higher dimensionality of feature space may result in degradation of classification performance [33]. Therefore, feature selection was performed that involves selecting a subset of features for improved classification performances. Feature selection was performed using the univariate Fisher ranking score (F score), which assumes that the input features are independent and ignores correlations between them [34]. The optimal subset of features was determined by varying the number of features until maximum accuracy was reached. Classification performance was based on force plate parameters from: (1) balance task only, (2) gait task only, and (3) a combination of balance and gait tasks. In each case different subsets of the total features were selected and separately trained and tested by the selected classifiers. The subset with the greatest classification performance was reported. For the evaluation of classification models according to their performance, 75% of the participant data was randomly selected for training and the other 25% were used for testing. A stratified eight-fold cross validation was performed, and the averaged model evaluation parameters were reported. A confusion matrix was developed for the assessment of each classification model as shown in Fig. 3. The classification models were evaluated based on accuracy, specificity, sensitivity, positive predictive value (PPV), negative predictive value (NPV) (Eqs. 1–5) [35]. For instance, if a model classifies all 99 participants (N = 58; F = 41) as non-fallers, the model's accuracy is 58.5%, specificity is 100%, sensitivity is 0%, PPV is 0%, NPV is 58.5%. According to previous research, accuracy and sensitivity are the most important indexes for fall detection algorithms [36], and hence, the models were ranked in terms of their accuracy followed by sensitivity.

$$Accuracy = \frac{TP + TN}{N} \times 100(\%) \tag{1}$$

$$Specificity = \frac{TN}{TN + FP} \tag{2}$$

$$Sensitivity = \frac{TP}{TP + FN} \tag{3}$$

$$PPV = \frac{TP}{TP + FP} \tag{4}$$

$$NPV = \frac{TN}{TN + FN} \tag{5}$$

3 Results

Figure 4 illustrates the ranking of all 59 force plate parameters based on their F-score. Among the five top ranked parameters were GRF parameters: Fx3 (F-score = 0.461), Fx1 (F-score = 0.403), Fz2 (F-score = 0.351), Loading Rate (F-score = 0.378), and Tz1 (F-score = 0.311). The next 5 parameters were COP parameters ($0.127 <$ F-score < 0.269). All the parameters ranked in the top 20 (F-score > 0.06) were collected during the gait tasks.

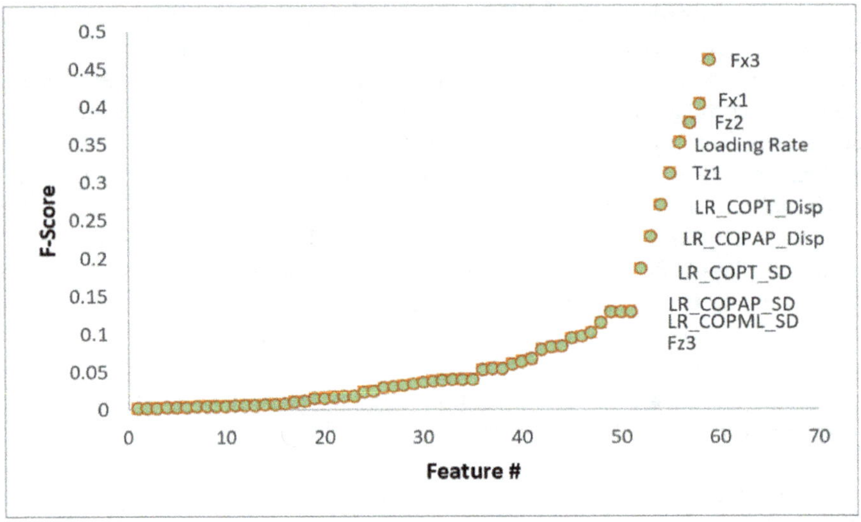

Fig. 4. The ranking of features (n = 59) based on their F-score. The F-score is used to quantify predictor importance for classification problems [30]. The top ten features are added as labels on the scatter plot.

For the balance tasks, a subset of features determined by the feature selection method was used for different classification models with NN, SVM, kNN, NB and LDA algorithms. The classification algorithms are ranked based on their accuracy (Table 4). In case of similar accuracies, sensitivity was considered for ranking. Table 4 shows the model evaluation parameters for the classifiers based on force plate parameters from the balance tasks only. The highest classification performance was shown by the kNN classifier (accuracy = 69.89%, sensitivity = 0.64). The SVM classifier was ranked second but had a very similar performance (accuracy = 69.89%, sensitivity = 0.49). The lowest performance was shown by the NB classifier (accuracy = 61.29%, sensitivity = 0.41).

Table 4. Classification models based on force plate parameters during balance tasks

Method	Accuracy (%)	Specificity	Sensitivity	NPV	PPV
kNN	69.89	0.74	0.64	0.74	0.64
SVM	69.89	0.85	0.49	0.70	0.70
LDA	65.59	0.72	0.56	0.70	0.59
NN	65.59	0.72	0.56	0.70	0.59
NB	61.29	0.76	0.41	0.64	0.55

NN: Neural Networks, SVM: Support Vector Machines, LDA: Linear Discriminant Analysis, NB: Naïve Bayes, kNN: k nearest neighbour, NPV: Negative Predictive Value: PPV: Positive Predictive Value

For the gait tasks, a subset of features determined by the feature selection method was used for different classification algorithms. Table 5 shows the model evaluation parameters for the classifiers based on a subset of force plate parameters from the gait tasks. The highest classification performance was demonstrated by the NB classifier (accuracy = 81.82%, sensitivity = 0.73). The LDA classifier had a similar performance (accuracy = 81.82%, sensitivity = 0.63) and was ranked second. The lowest performance was observed for the NN classifier (accuracy = 74.75%, sensitivity = 0.66).

Table 5. Classification models based on force plate parameters during gait tasks

Method	Accuracy (%)	Specificity	Sensitivity	NPV	PPV
NB	81.82	0.88	0.73	0.82	0.81
LDA	81.82	0.95	0.63	0.79	0.90
SVM	76.77	0.81	0.71	0.80	0.73
kNN	75.76	0.72	0.80	0.84	0.67
NN	74.75	0.81	0.66	0.77	0.71

NN: Neural Networks, SVM: Support Vector Machines, LDA: Linear Discriminant Analysis, NB: Naïve Bayes, kNN: k nearest neighbour, NPV: Negative Predictive Value: PPV: Positive Predictive Value

For the combined tasks, a subset of features determined by the feature selection method was used for different classification algorithms. Table 6 shows the model evaluation parameters for the classifiers based on a subset of force plate parameters from a combination of balance and gait tasks. The highest classification performance was shown by the LDA classifier (accuracy = 84.95%, sensitivity = 0.77). The NB classifier had a similar performance (accuracy = 83.87%, sensitivity = 0.72). The lowest performance was shown by the NN classifier (accuracy = 76.34%, sensitivity = 0.69).

Table 6. Classification models based on force plate parameters during combined tasks

Method	Accuracy (%)	Specificity	Sensitivity	NPV	PPV
LDA	84.95	0.91	0.77	0.84	0.86
NB	83.87	0.93	0.72	0.82	0.88
SVM	82.8	0.91	0.72	0.82	0.85
kNN	77.42	0.93	0.56	0.75	0.85
NN	76.34	0.81	0.69	0.79	0.73

NN: Neural Networks, SVM: Support Vector Machines, LDA: Linear Discriminant Analysis, NB: Naïve Bayes, kNN: k nearest neighbour, NPV: Negative Predictive Value: PPV: Positive Predictive Value

4 Discussion

All of the classification models based on force plate parameters are reported in Tables 4, 5 and 6. The best performing model was found to be the LDA classifier using force plate parameters from the combined tasks. The model resulted in an accuracy of 84.95%. The sensitivity was 0.77 which indicates that 77% of fallers were correctly classified, whereas 23% were classified as non-fallers. Similarly, the specificity was 0.91, indicating that 91% of non-fallers were correctly classified. The PPV and NPV was 0.86 and 0.84 respectively, implying a predicted faller has an 86% chance of correct classification and a predicted non-faller has an 84% chance of correct classification. Overall, the model can be used for accurate classification of fallers/non-fallers based on force plate parameters from both balance and gait tasks.

It was found that the force plate parameters extracted during the gait tasks ranked in the top 20 features (1–20, total = 59) based on their F-score. This is further confirmed with the results from balance and gait tasks. The force plate parameters from gait tasks had a higher classification accuracy (74.75%–81.82%) than the force plate parameters from the balance tasks (61.29%–69.89%). The kNN classifier based on data from gait tasks has the highest sensitivity, even though it is ranked fourth in terms of accuracy (accuracy = 75.76, sensitivity = 0.80). This indicates that the parameters from gait tasks have higher predictor importance than those from the balance tasks. One explanation may be that walking is the most frequently cited cause of falls [30]. However, the parameters from combined tasks contributed in the most accurate performance (76.34%–84.95%). Although the parameters during balance tasks alone resulted in lower accuracy, they improved the classification performance when combined with the parameters from gait tasks. This suggests that although the gait tasks are crucial for assessing older adult fallers, balance tasks are equally important for obtaining static posturography information for faller/non-faller identification.

For the force plate parameters from combined tasks (Table 6), the minimum accuracy obtained (76.34%) is still considered high in comparison to previous research for faller/non-faller classification [9]. Howcroft et al. [8] reported an accuracy of 84% using NN and SVM models using data from foot pressure sensors and head, pelvis and shank accelerometery. The NN and SVM models ranked higher than NB models (80%) [10]. In another study comparing classification methods using data from accelerometers, it was found that NB and kNN and NN models had a high accuracy (>95%) whereas the SVM had a relatively lower accuracy (75%) [22]. A prospective study using LDA classifiers and data from wii balance boards was able to identify multiple fallers with a high accuracy (84.9%) [23]. In a study using data from force platforms and kNN as the classification algorithm, a 100% accuracy was achieved [14]. In the present study, it was found that the classification accuracies for different methods to be similar (76.34%–84.95%). In fact, LDA, a computationally simpler method, with no overfitting issues had the highest accuracy (84.95%). This indicates that faller/non-faller classification has less dependency on the nature of the classification algorithm used and more dependency on the combination of biomechanical parameters used and on the nature of the tasks performed for assessment.

The present study included data from ninety-nine participants (F = 41 and N = 58). The large number of participants allows for a more generalized and robust model which can be used to classify new participants. Although previous studies have shown very high accuracies [37, 38], lack of robustness may cause degradation in the classification performance when the models are implemented in real world conditions. The feature selection with univariate F-score ranking shows that the top five ranked (1–5) features are GRF and the next five (6–10) are COP parameters. Previous studies have assessed both the GRF parameters [17, 18] and COP displacement [11, 12]. Although the present study used a combination of both it is important for future studies to compare the classification performance based on different biomechanical parameters.

5 Conclusion

The LDA classifier, which is a simple algorithm has a higher classification performance using force plate parameters from gait and balance tasks than more complex methods such as SVM and ANN. The findings of this study suggest that high accuracy of identifying older adult fallers can be obtained using force plate parameters. It was also shown that the parameters from a combination of gait and balance tasks resulted in a higher accuracy than gait tasks or balance tasks alone. These results could potentially have applications in fall prevention programs for older adults.

References

1. Delbaere, K., Close, J.C., Heim, J., Sachdev, P.S., Brodaty, H., Slavin, M.J., et al.: A multifactorial approach to understanding fall risk in older people. J. Am. Geriatr. Soc. **58**, 1679–1685 (2010)
2. Pearson, C., Geran, L., St-Arnaud, J.: Understanding seniors' risk of falling and their perception of risk: Statistics Canada (2014)
3. Jácome, C., Cruz, J., Oliveira, A., Marques, A.: Validity, reliability, and ability to identify fall status of the Berg Balance Scale, BESTest, Mini-BESTest, and Brief-BESTest in patients with COPD. Phys. Ther. **96**, 1807–1815 (2016)
4. Bogle Thorbahn, L.D., Newton, R.A.: Use of the Berg Balance Test to predict falls in elderly persons. Phys. Ther. **76**, 576–583 (1996)
5. Tinetti, M.E.: Performance-oriented assessment of mobility problems in elderly patients. J. Am. Geriatr. Soc. **34**, 119–126 (1986)
6. Hayes, K.W., Johnson, M.E.: Measures of adult general performance tests: The Berg Balance Scale, Dynamic Gait Index (DGI), Gait Velocity, Physical Performance Test (PPT), Timed Chair Stand Test, Timed Up and Go, and Tinetti Performance-Oriented Mobility Assessment (POMA). Arthritis Care Res.: Off. J. Am. Coll. Rheumatol. **49**, S28–S42 (2003)
7. Brauer, S.G., Burns, Y.R., Galley, P.: A prospective study of laboratory and clinical measures of postural stability to predict community-dwelling fallers. J. Gerontol. Ser. A: Biol. Sci. Med. Sci. **55**, M469–M476 (2000)
8. Howcroft, J., Kofman, J., Lemaire, E.D.: Prospective fall-risk prediction models for older adults based on wearable sensors. IEEE Trans. Neural Syst. Rehabil. Eng. **25**, 1812–1820 (2017)

9. Sun, R., Sosnoff, J.J.: Novel sensing technology in fall risk assessment in older adults: a systematic review. BMC Geriatr. **18**, 14 (2018)
10. Howcroft, J., Lemaire, E.D., Kofman, J.: Wearable-sensor-based classification models of faller status in older adults. PLoS One **11** (2016). e0153240
11. Pajala, S., Era, P., Koskenvuo, M., Kaprio, J., Törmäkangas, T., Rantanen, T.: Force platform balance measures as predictors of indoor and outdoor falls in community-dwelling women aged 63–76 years. J. Gerontol. Ser. A: Biol. Sci. Med. Sci. **63**, 171–178 (2008)
12. Piirtola, M., Era, P.: Force platform measurements as predictors of falls among older people–a review. Gerontology **52**, 1–16 (2006)
13. Hamacher, D., Singh, N., Van Dieen, J., Heller, M., Taylor, W.: Kinematic measures for assessing gait stability in elderly individuals: a systematic review. J. R. Soc. Interface **8**, 1682–1698 (2011)
14. Liang, S., Ning, Y., Li, H., Wang, L., Mei, Z., Ma, Y., et al.: Feature selection and predictors of falls with foot force sensors using kNN-based algorithms. Sensors **15**, 29393–29407 (2015)
15. Muniz, A., Nadal, J.: Application of principal component analysis in vertical ground reaction force to discriminate normal and abnormal gait. Gait Posture **29**, 31–35 (2009)
16. Moustakidis, S.P., Theocharis, J.B., Giakas, G.: Subject recognition based on ground reaction force measurements of gait signals. IEEE Trans. Syst. Man Cybern. Part B (Cybern.) **38**, 1476–1485 (2008)
17. Yamada, T., Demura, S.: Relationships between ground reaction force parameters during a sit-to-stand movement and physical activity and falling risk of the elderly and a comparison of the movement characteristics between the young and the elderly. Arch. Gerontol. Geriatr. **48**, 73–77 (2009)
18. Cheng, Y.-Y., Wei, S.-H., Chen, P.-Y., Tsai, M.-W., Cheng, I.-C., Liu, D.-H., et al.: Can sit-to-stand lower limb muscle power predict fall status? Gait Posture **40**, 403–407 (2014)
19. Winter, D.A.: ABC (Anatomy, Biomechanics and Control) of Balance During Standing and Walking. Waterloo Biomechanics, Ontario (1995)
20. Shumway-Cook, A., Woollacott, M., Kerns, K.A., Baldwin, M.: The effects of two types of cognitive tasks on postural stability in older adults with and without a history of falls. J. Gerontol. Ser. A: Biol. Sci. Med. Sci. **52**, M232–M240 (1997)
21. Ejupi, A., Lord, S.R., Delbaere, K.: New methods for fall risk prediction. Curr. Opin. Clin. Nutr. Metab. Care **17**, 407–411 (2014)
22. Caby, B., Kieffer, S., de Saint Hubert, M., Cremer, G., Macq, B.: Feature extraction and selection for objective gait analysis and fall risk assessment by accelerometry. Biomed. Eng. Online **10**, 1 (2011)
23. Howcroft, J., Lemaire, E.D., Kofman, J., McIlroy, W.E.: Elderly fall risk prediction using static posturography. PLoS One **12** (2017). e0172398
24. Brotherton, S.S., Williams, H.G., Gossard, J.L., Hussey, J.R., McClenaghan, B.A., Eleazer, P.: Are measures employed in the assessment of balance useful for detecting differences among groups that vary by age and disease state? J. Geriatr. Phys. Ther. **30**, 14–19 (2005)
25. Hasan, C.Z.C., Jailani, R., Tahir, N.M., Ilias, S.: The analysis of three-dimensional ground reaction forces during gait in children with autism spectrum disorders. Res. Dev. Disabil. **66**, 55–63 (2017)
26. Chockalingam, N., Dangerfield, P.H., Rahmatalla, A., Ahmed, E.-N., Cochrane, T.: Assessment of ground reaction force during scoliotic gait. Eur. Spine J. **13**, 750–754 (2004)
27. Bizovska, L., Svoboda, Z., Kutilek, P., Janura, M., Gaba, A., Kovacikova, Z.: Variability of centre of pressure movement during gait in young and middle-aged women. Gait Posture **40**, 399–402 (2014)

28. Zurada, J.M.: Introduction to Artificial Neural Systems, vol. 8. West publishing Company, St. Paul (1992)
29. Theodoridis, S., Pikrakis, A., Koutroumbas, K., Cavouras, D.: Introduction to Pattern Recognition: A Matlab Approach. Academic Press, Amsterdam (2010)
30. McGranaghan, R.M., Mannucci, A.J., Wilson, B., Mattmann, C.A., Chadwick, R.: New capabilities for prediction of high-latitude ionospheric scintillation: a novel approach with machine learning. Space Weather **16**, 1817–1846 (2018)
31. Martínez, A.M., Kak, A.C.: PCA versus LDA. IEEE Trans. Pattern Anal. Mach. Intell. **23**, 228–233 (2001)
32. Van Der Heijden, F., Duin, R.P., De Ridder, D., Tax, D.M.: Classification, Parameter Estimation and State Estimation: An Engineering Approach Using MATLAB. Wiley, Hoboken (2005)
33. Zhang, X., Wu, Y., Wang, L., Li, R.: Variable selection for support vector machines in moderately high dimensions. J. R. Stat. Soc.: Ser. B (Stat. Methodol.) **78**, 53–76 (2016)
34. Gu, Q.,, Li, Z., Han, J.: Generalized fisher score for feature selection. arXiv preprint arXiv: 1202.3725 (2012)
35. Lalkhen, A.G., McCluskey, A.: Clinical tests: sensitivity and specificity. Contin. Educ. Anaesth. Crit. Care Pain **8**, 221–223 (2008)
36. Huang, S., Yang, Y., Liu, W.: An enhanced fall detection approach based on cost sensitivity analysis. In: 2011 First ACIS International Symposium on Software and Network Engineering, pp. 81–85 (2011)
37. Özdemir, A., Barshan, B.: Detecting falls with wearable sensors using machine learning techniques. Sensors **14**, 10691–10708 (2014)
38. Liu, J., Lockhart, T.E.: Automatic individual calibration in fall detection–an integrative ambulatory measurement framework. Comput. Methods Biomech. Biomed. Eng. **16**, 504–510 (2013)

Accuracy of Anthropometric Measurements by a Video-Based 3D Modelling Technique

Chuang-Yuan Chiu(✉), Michael Thelwell, Simon Goodwill, and Marcus Dunn

Centre for Sports Engineering Research, Sheffield Hallam University, Sheffield, UK
{c.chiu,m.thelwell,S.r.goodwill,m.dunn}@shu.ac.uk

Abstract. The use of anthropometric measurements, to understand an individual's body shape and size, is an increasingly common approach in health assessment, product design, and biomechanical analysis. Non-contact, three-dimensional (3D) scanning, which can obtain individual human models, has been widely used as a tool for automatic anthropometric measurement. Recently, Alldieck et al. (2018) developed a video-based 3D modelling technique, enabling the generation of individualised human models for virtual reality purposes. As the technique is based on standard video images, hardware requirements are minimal, increasing the flexibility of the technique's applications. The aim of this study was to develop an automated method for acquiring anthropometric measurements from models generated using a video-based 3D modelling technique and to determine the accuracy of the developed method. Each participant's anthropometry was measured manually by accredited operators as the reference values. Sequential images for each participant were captured and used as input data to generate personal 3D models, using the video-based 3D modelling technique. Bespoke scripts were developed to obtain corresponding anthropometric data from generated 3D models. When comparing manual measurements and those extracted using the developed method, the accuracy of the developed method was shown to be a potential alternative approach of anthropometry using existing commercial solutions. However, further development, aimed at improving modelling accuracy and processing speed, is still warranted.

Keywords: Anthropometry · 3D modelling · Accuracy

1 Introduction

The use of anthropometric measurements, to understand an individual's body shape and size, is an increasingly common approach in health assessment, product design, and biomechanical analysis. For instance, Streng et al. (2018) indicated that female heart failure patients with high waist-hip ratio have a high risk of mortality. Verwulgen et al. (2018) introduce a new workflow to use 3D anthropometry to design close-fit products for users. Pandis and Bull (2017) and Smith and Bull (2018) used low-cost 3D scanning to acquired body segment parameter for biomechanical analysis.

G. A. Ateshian et al. (Eds.): CMBBE 2019, LNCVB 36, pp. 354–361, 2020.
https://doi.org/10.1007/978-3-030-43195-2_29

Manual anthropometric techniques are a traditional and widely used approach as the equipment is easy to access and calibrate (Hume et al. 2018). However, technical expertise is required to ensure the accuracy of measurement (Perini et al. 2005). Sebo et al. (2017) indicated that waist girth measurements collected by untrained general practitioners could have errors of more than 2 cm. Furthermore, this approach cannot obtain more complex anthropometric measurements, such as body volume, surface area, or body segment parameters directly.

Three-dimensional (3D) scanning systems use specific sensors, such as depth-cameras and stereo-cameras, to obtain individual human models. Simple and complex anthropometric data can be extracted virtually from the resulting scanning output (i.e. the individual 3D human models). Ma et al. (2011a, b) used this technique to obtain individual body segment inertia parameters and developed mathematical models to estimate the body segment inertia parameters from body mass and stature. Recently, advanced computer vision techniques such as 3D correspondence approaches (Groueix et al. 2018; Zuffi and Black 2015) have been developed enabling anatomical landmarks to be identified without manual palpation or placing any markers on the participants. Furthermore, some cost-effect 3D scanning systems, such as KX-16, Proscanner and Styku scanners have been developed for non-contact anthropometric measurement. Consequently, these non-contact 3D scanning techniques have been widely used as a tool to carry out anthropometric procedures.

All of these 3D scanning systems generally require specific hardware (scanning booth, depth camera and turntable). Therefore, users typically go to specialist facilities to complete 3D scanning for anthropometric assessments, which in turn reduces the accessibility of 3D scanning based anthropometry. A mobile 3D scanning solution which, could be used in flexible environments (i.e. minimal specialist equipment or facilities), would enhance the application of 3D scanning based anthropometry, particularly in health and biomechanical assessment. For instance, using a mobile 3D scanning system in their own home, patients could complete comprehensive anthropometric assessments to independently monitor health conditions such as obesity, without needing to travel to specialist facilities. Further, biomechanists could use mobile 3D scanning solutions to model the body segment inertia parameters of athletes in the field and in sports facilities, rather than requiring athletes to travel to expensive and complex laboratories. Such a technological advancement would help both users and practitioners to save time and cost, when conducting anthropometric assessments.

Recently, Alldieck et al. (2018) developed a video-based 3D modelling technique, enabling the generation of individualised human models for virtual reality purposes. As the technique is based on standard video images, hardware requirements (e.g. depth camera, turntable, etc.) are minimal, increasing the flexibility and range of the technique's applications. However, whilst good levels of accuracy for point-to-point distances have been demonstrated (Alldieck et al. 2018), the accuracy of anthropometric measurements derived using this technique must be examined before its use in anthropometric applications. Furthermore, there is no software developed to obtain body measurements from the 3D models built from this video-based 3D modelling technique. Thus, the aim of this study was to develop an automated method to acquire anthropometric measurements from the models generated using this video-based 3D modelling technique and determine the accuracy of the developed method.

2 Method

2.1 Participants

The study was approved by the ethics committee at Sheffield Hallam University. Five male and six female healthy participants were recruited (stature: 1.71 ± 0.09 m; mass: 77.2 ± 13.8 kg) in this study. All participants gave written consent before participating. They were requested to wear close-fitting clothing during the all test procedures.

2.2 Manual Anthropometric Measurement

Traditional anthropometric data, including stature, mass, waist and hip girths were measured manually by accredited operators according to the International Society for the Advancement of Kinanthropometry (ISAK) protocols (Stewart et al. 2011). Understanding the accuracy of waist and hip girths obtained from the developed method was considered as the initial stage to determine the potential for further anthropometric measurement. Thus, ISAK manual measurements of waist and hip girths were regarded as the reference values to evaluate the accuracy of the developed methods in this study.

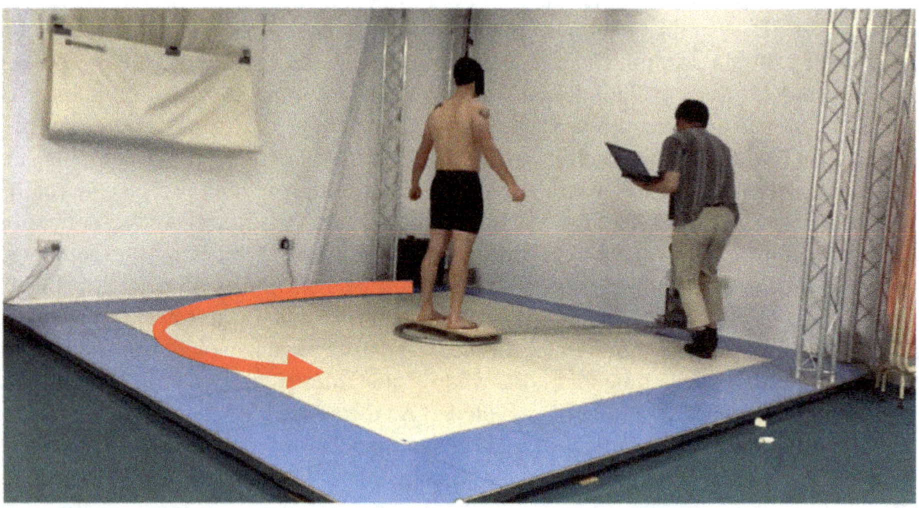

Fig. 1. A bespoke system with a moving camera was used to capture images in this study.

2.3 Anthropometric Measurement with 3D Modelling Techniques

The technique developed by Alldieck et al. (2018) enables 3D human modelling with moving participants. However, the movement artefacts resulting from participant breathing and the self-rotation data acquisition procedure might increase the amount of error of human modelling. To determine the optimal accuracy of the developed technique, a bespoke capturing system with a moving camera (Chiu et al. 2019) was used to minimize the effect of human movement and breathing during image capture. Operating the bespoke capturing system involves an operator rotating a single camera around a stationary participant whilst image data is acquired in approximately 10 s as shown in Fig. 1. Participants were requested to stand still and hold their breath at end-tidal volume during image capture.

The technique developed by Alldieck et al. (2018) can use a general camera to obtain image data and applied a convolutional neural network-based (CNN-based) program to extract silhouette and joint data for generating individual 3D models. Nevertheless, the public CNN-based models such as Deeplabv3+ (Chen et al. 2018) cannot discriminate between participants and other humans which may be present in the background within the captured images as they were not developed for the specific cases in this study. Thus, a Microsoft Kinect V2 was used as the capturing device to enable accurate silhouettes of participants to be generated for determining the accuracy of the developed method. All tasks were completed without training a new CNN-based model.

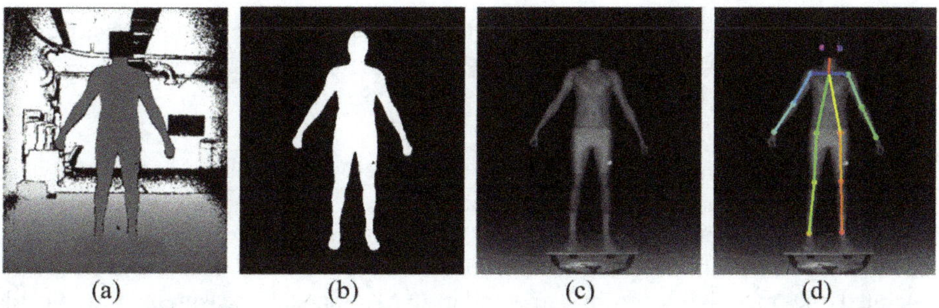

Fig. 2. Images captured from Kinect V2 and the post-processed results. (a) a depth image captured from a Microsoft Kinect V2, (b) a silhouette image extracted by applying distance threshold and random sample consensus algorithms, (c) an image captured by the infrared camera of a Microsoft Kinect V2, and (d) joint data extracted by applying OpenPose algorithms.

The depth images captured by the Microsoft Kinect V2 were processed with distance threshold and random sample consensus algorithms (Derpanis 2010) to remove background pixels (e.g. floor, wall) for extracting silhouette images as shown in Fig. 2 (a) and (b). The OpenPose algorithms[1] (Wei et al. 2016) were applied to the images captured by the infrared camera of a Microsoft Kinect V2 to extract the joint position data as shown in Fig. 2(c) and (d). The data of silhouette images and joint positions and the manual measurements of statures were then used as the input of the video-based 3D modelling technique developed by Alldieck et al. (2018) to generate individual models

[1] The algorithms was applied by referring the code provided from https://github.com/ildoonet/tf-pose-estimation.

as shown in Fig. 3(a). While applying the 3D modelling technique developed by Alldieck et al. (2018), the weights (parameter values) in the source code[2] provided by Alldieck et al. (2018) were adopted to improve the accuracy of 3D reconstruction.

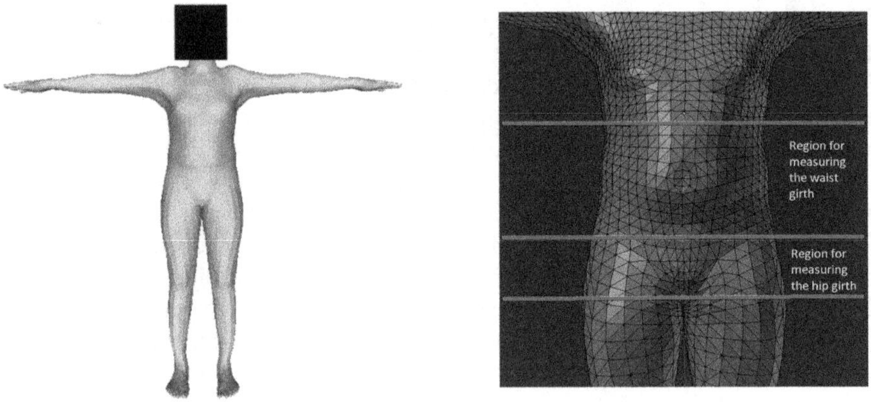

(a) (b)

Fig. 3. (a) An individual 3D model generated from the technique developed by Alldieck et al. (2018) and (b) the measuring region was detected by the bespoke scripts automatically while measuring anthropometric data from a generated 3D model.

Bespoke scripts were developed to identify the region for measuring both the waist and hip girths from the generated individual models, as shown in Fig. 3(b). The scripts obtained 2D cross-section profiles along the length of the body scan and calculated the circumference of each 2D cross-section to determine the waist and hip girth of the generated 3D models, according to similar ISAK manual measurement protocols (Stewart et al. 2011). In other words, the waist girth was measured on the level with the narrowest circumference on the torso and the hip girth was measured on the level with the most posterior prominence.

2.4 Statistical Analysis

The accuracy of anthropometric measurements obtained using the developed method was quantified according to the root mean square error (RMSE) and relative inter-method technical error of measurement (relative inter-method TEM or %TEM) compared to the reference waist and hip girths obtained using manual measurement as shown in Eqs. (1) and (2).

$$RMSE = \sqrt{\frac{\sum_{i=1}^{n}(m_i - d_i)^2}{n}} \quad (1)$$

[2] https://github.com/thmoa/videoavatars.

$$\%TEM = \frac{\sqrt{\frac{\sum_{i=1}^{n}(m_i - d_i)^2}{2 \times n}}}{\frac{\sum_{i=1}^{n}(m_i + d_i)}{2 \times n}} \times 100\% \tag{2}$$

where n is the number of participants, m_i represents the ISAK manual measurement obtained from the i-th participant, and d_i represents the measurement obtained from the i-th participant using the developed method in this study.

3 Results

Table 1 shows the results of this study. When comparing manually measured and video-based 3D modelled anthropometric data, the RMSEs for waist and hip girths were both around 5 cm. The relative inter-method TEMs were larger than 3.5%. For male participants, the accuracy of waist girths was worse than the one of hip girths. By contrast, for female participants, the accuracy of waist girths was better than the ones of hip girths.

Table 1. The accuracy of the developed method in this study.

	All participants		Male participants		Female participants	
Measurement	RMSE	%TEM	RMSE	%TEM	RMSE	%TEM
Waist girth	5.3 cm	4.49%	6.2 cm	4.84%	4.4 cm	4.01%
Hip girth	5.7 cm	3.98%	5.1 cm	3.46%	6.1 cm	4.41%

4 Discussions

The aim of this study was to develop an automated method to acquire anthropometric measurements from individual models generated using a video-based 3D modelling technique (Alldieck et al. 2018) and determine the accuracy of the developed method.

The RMSEs of waist girth measures obtained using the developed method were similar to those acquired from existing commercial solutions, such as Proscanner and Styku scanners (Bourgeois et al. 2017) as shown in Table 2. The results show that the developed method, which applies the video-based 3D modelling technique (Alldieck et al. 2018) is a potential alternative approach of performing anthropometric measurement using existing commercial solutions. However, the accuracy of hip girths obtained using the developed method was worse than those acquired using existing commercial solutions (Bourgeois et al. 2017) as shown in Table 2. The pose of the model generated from the video-based techniques (standing with feet apart) could lead to increased error of hip girth measurement, which should be performed with feet together according to ISAK protocols (Stewart et al. 2011). Thus, further development should consider correcting the pose effect to obtain more accurate hip girth measurements. A complete anthropometric validation test should be also conducted to understand which measurements can be obtained using the developed method accurately.

The relative inter-method TEM of both measurements (>3.5%) exceeded the acceptable range for some anthropometric applications (ISAK Level 1: 1.5% and Level 2: 1.0%). Furthermore, the bespoke capturing system was used in this study to minimize the movement artefacts caused by breathing and self-rotation of the participant during the scanning procedure. The accuracy of extracted measurements could be decreased when applying the developed method with self-rotating participants. Thus, further development is required to improve the accuracy of the developed method before applying this technique in applications which require accurate anthropometric measurements to be extracted from image data where the participant is moving, such as at-home health assessment and biomechanical analysis.

Table 2. Comparison accuracy (RMSE) of the developed method and the existing commercial solutions (Bourgeois et al. 2017).

Measurement	Developed method	Proscanner (Bourgeois et al. 2017)	Styku (Bourgeois et al. 2017)
Waist girth	5.3 cm	5.8 cm	6.3 cm
Hip girth	5.7 cm	4.6 cm	2.6 cm

The results of this study showed that the accuracy of male and female participants was different. Although male and female template models were used in the technique developed by Alldieck et al. (2018), the modelling algorithms to generate individual 3D human models seem the same. Further development might be required to use gender-specific penalty functions in the optimization processes while generating individual models for the application that needs accurate anthropometric data.

The use of video cameras represents a unique and flexible opportunity for estimating human morphometrics. However, the typical processing time (Azure virtual machine F1s) for generating one individual model exceeded two hours. The long processing time of the developed method might limit the potential application of this technique for general purpose or research studies. Alldieck et al. (2019) have presented a novel approach which applied machine learning to reduce the processing time from two hours to 10 s. Therefore, further development should consider applying machine learning techniques and updated development (Alldieck et al. 2019) to improve processing speed.

References

Alldieck, T., Magnor, M., Lal Bhatnagar, B., Theobalt, C., Pons-Moll, G.: Learning to reconstruct people in clothing from a single RGB camera. In: Proceedings of the IEEE Conference on Computer Vision and Pattern Recognition, pp. 1175–1186 (2019)

Alldieck, T., Magnor, M., Xu, W., Theobalt, C., Pons-Moll, G.: Video based reconstruction of 3D people models. In: Proceedings of the IEEE Conference on Computer Vision and Pattern Recognition, pp. 8387–8397 (2018)

Bourgeois, B., et al.: Clinically applicable optical imaging technology for body size and shape analysis: comparison of systems differing in design. Eur. J. Clin. Nutr. (2017). https://doi.org/10.1038/ejcn.2017.142

Chen, L.-C., Zhu, Y., Papandreou, G., Schroff, F., Adam, H.: Encoder-decoder with atrous separable convolution for semantic image segmentation. In: Proceedings of the European Conference on Computer Vision (ECCV), pp. 801–818 (2018)

Chiu, C.-Y., Thelwell, M., Senior, T., Choppin, S., Hart, J., Wheat, J.: Comparison of depth cameras for three-dimensional reconstruction in medicine. Proc. Inst. Mech. Eng. Part H: J. Eng. Med. **233**, 938–947 (2019). https://doi.org/10.1177/0954411919859922

Derpanis, K.G.: Overview of the RANSAC Algorithm. Image Rochester NY **4**, 2–3 (2010)

Groueix, T., Fisher, M., Kim, V.G., Russell, B.C., Aubry, M.: 3D-CODED: 3D correspondences by deep deformation. In: Proceedings of the European Conference on Computer Vision (ECCV), pp. 230–246 (2018)

Hume, P.A., Sheerin, K.R., de Ridder, J.H.: Non-imaging method: surface anthropometry. In: Hume, P.A., Kerr, D.A., Ackland, T.R. (eds.) Best Practice Protocols for Physique Assessment in Sport. Springer Singapore, pp. 61–70 (2018). https://doi.org/10.1007/978-981-10-5418-1_6

Ma, Y., Kwon, J., Mao, Z., Lee, K., Li, L., Chung, H.: Segment inertial parameters of Korean adults estimated from three-dimensional body laser scan data. Int. J. Ind. Ergon. **41**, 19–29 (2011a). https://doi.org/10.1016/j.ergon.2010.11.004

Ma, Y., Lee, K., Li, L., Kwon, J.: Nonlinear regression equations for segmental mass-inertial characteristics of Korean adults estimated using three-dimensional range scan data. Appl. Ergon. **42**, 297–308 (2011b). https://doi.org/10.1016/j.apergo.2010.07.005

Pandis, P., Bull, A.M.: A low-cost three-dimensional laser surface scanning approach for defining body segment parameters. Proc. Inst. Mech. Eng. Part H: J. Eng. Med. **231**, 1064–1068 (2017). https://doi.org/10.1177/0954411917727031

Perini, T.A., de Oliveira, G.L., de Ornellas, J.S., de Oliveira, F.P.: Technical error of measurement in anthropometry. Revista Brasileira de Medicina do Esporte **11**, 81–85 (2005)

Sebo, P., Herrmann, F.R., Haller, D.M.: Accuracy of anthropometric measurements by general practitioners in overweight and obese patients. BMC Obes. **4**, 23 (2017). https://doi.org/10.1186/s40608-017-0158-0

Smith, S.H.L., Bull, A.M.J.: Rapid calculation of bespoke body segment parameters using 3D infra-red scanning. Med. Eng. Phys. (2018). https://doi.org/10.1016/j.medengphy.2018.10.001

Stewart, A., Marfell-Jones, M., Olds, T., Ridderde, H.: International Standards for Anthropometric Assessment. International Society for the Advancement of Kinanthropometry, Lower Hutt (2011)

Streng, K.W., et al.: Waist-to-hip ratio and mortality in heart failure. Eur. J. Heart Fail. **20**, 1269–1277 (2018). https://doi.org/10.1002/ejhf.1244

Verwulgen, S., Lacko, D., Vleugels, J., Vaes, K., Danckaers, F., De Bruyne, G., Huysmans, T.: A new data structure and workflow for using 3D anthropometry in the design of wearable products. Int. J. Ind. Ergon. **64**, 108–117 (2018). https://doi.org/10.1016/j.ergon.2018.01.002

Wei S-E, Ramakrishna V, Kanade T, Sheikh Y Convolutional pose machines. In: Proceedings of the IEEE Conference on Computer Vision and Pattern Recognition, pp. 4724–4732 (2016)

Zuffi, S., Black, M.J.: The stitched puppet: a graphical model of 3D human shape and pose. In: Proceedings of the IEEE Conference on Computer Vision and Pattern Recognition, pp. 3537–3546 (2015)

Biomechanical Model for Dynamic Analysis of the Human Gait

Jordana S. R. Martins[1], George Sabino[2], Diego H. A. Nascimento[3], Gabriel M. C. Machado[1], and Claysson B. S. Vimieiro[1,3](✉)

[1] Graduate Program of Mechanical Engineering, Pontifícia Universidade Católica de Minas Gerais, PUC-Minas, Avenida DomJosé Gaspar, 500 - Coração Eucarístico, Belo Horizonte, MG 30535-901, Brazil
martinsjsr@gmail.com, claysson@gmail.com
[2] Graduate Program in Rehabilitation Sciences, Universidade Federal de Minas Gerais - UFMG, Av. Antônio Carlos 6627, Pampulha, MG 31270-901, Brazil
[3] Graduate Program of Mechanical Engineering, Universidade Federal de Minas Gerais - UFMG, Av. Antônio Carlos 6627, Pampulha, MG 31270-901, Brazil

Abstract. Biomechanical models are important tools in the study of human motion. The human body can be defined as an articulated system in complete static or dynamic equilibrium, where internal forces produce joint movements in the body segments. This work proposes a biomechanical model for the dynamic analysis of the lower limbs during human gait, with continuous cycle, considering kinematic and kinematic data collected from seven volunteers, with no history of pathology related to human gait, walking under a treadmill with controlled speed. The proposed model was developed in Motion View software (Altair) and is based on a kinematic chain to represent the segments of the body, connected by rotational joints with defined viscoelastic parameters. The geometry of the model is similar to the human skeleton, with dimensions based on the anthropometric data of the volunteers, and agglomerates in 8 segments, upper limbs, hips, thighs, legs and feet. The kinematic data, captured by motion analysis system Qualisys® with 8 cameras, were used to determine the joint angles, using Cardan's angular theory, and together with the ground reaction forces collected by force platforms installed under the instrumented treadmill, it is possible to estimate spatio-temporal variables and force transfer at the joints for each gait cycle. To validate the model, we compared the variables found and the previously published data.

Keywords: Biomechanical model · Human gait · Dynamic analysis · Lower limbs

1 Introduction

Biomechanics is a science derived from the natural sciences, which aims to study the movement of the biological system as well as the physical analysis of the human body. It is the science that describes, analyzes and models the movement of the bodies of living

G. A. Ateshian et al. (Eds.): CMBBE 2019, LNCVB 36, pp. 362–370, 2020.
https://doi.org/10.1007/978-3-030-43195-2_30

beings in nature. Biomechanical models have been frequently applied to study human motion dynamics since it offers a non-invasive and quick response to the assessment of internal actions and the implications on the joints of the human body [1, 2].

The gait analysis is usually an applied technology for the planning and evaluation of the treatment. In general, the characteristics of the gait, for example, pitch length, pitch width, stride length, joint angles, intersegmental forces and so on, provide important clues about the patient's condition [3, 4].

Biomechanical models for the dynamic analysis of human gait can be simplified as: an articulated system; a model of multiple bodies similar to robotics; a model of muscle mechanics; and geometric models capable of relating muscle forces to torques at the joints of multiple body models [1, 5]. The elaboration of biomechanical models that reproduce human behavior presents a great degree of difficulty, since these are non-linear models, with multiple inputs and outputs, redundant degrees of freedom, among others. Another complicating factor is the variability of the parameters related to biological systems, presenting for the same individual or for the different average groups over time [6]. The model that developed has great potential to become an excellent tool to assist in the quantitative and qualitative analysis of human gait, besides presenting data to aid dynamic simulations of prototypes of prostheses and orthoses that lead to better living conditions for the population in general [4].

This paper presents a biomechanical model of the lower limbs of the human body with ability reproduce the gait movement on the computer allowing a dynamic analysis each movement plane individually, and to the identification of possible pathologies related to gait.

Kinetic and kinematic data, collected from predefined groups, were stored on the computer where they were treated and analyzed using Matlab routines and were applying it as input data in biomechanical model for human gait analysis. The proposed model was developed in the software MotionView (Altair) and is based on a kinematic chain to represent the segments of the body, connected by rotational joints with defined viscoelastic parameters. The model will involve multi-body dynamics, inverse dynamics and geometric modeling [6].

2 Methods

2.1 Subjects

Seven healthy subjects (four males and three females, 21–31 years of age, 55–93 kg and 1.65–1.80 m) with no previous history of musculoskeletal injury were recruited. The kinematic e Kinetic data collection were approved by the ethic committee of Friedrich-Schiller- Universität Jena (0558-11/00). The clinical examination was carried out in the KIP-Labor of the Friedrich-Schiller Universität Jena, Germany according to the German standard DIN 33402 (1981) (parts I and II). All volunteers signed the consent form for the test [1].

2.2 Experimental Setup

The volunteers were instructed to a walk on a treadmill, with the controlled speed of to 0.55, 0.83 and 1.11 m/s. The kinematic and kinematic data were collected simultaneously, during the march. For the kinematic analysis, 20 reflective markers were positioned in the bony protuberances of the lower limbs, according to the International Society of Biomechanics (ISB) [7], and through the Qualisys system with eight infrared cameras the reflexive markers' 3D positions were recognized and recorded during the walking, with frequency of acquisition of 200 Hz. For kinetic data, a force platform embedded under the belt mat was used to record the ground reaction force (GRF) on the feet during the run, with a frequency of acquisition of 120 Hz [1].

For the kinematic data, we used the algorithm developed in the Matlab, to first perform, a Cubic Spline interpolation to remove empty vectors of the sample provoked by the occlusions of the markers, and, later through, Cardan Angle, was realized the estimation of the movement of a limb in relation to another [8]. The rotation sequence established for the determination of the angles by the Cardan method was defined: flexion/extension movement (α) in x, adduction/abduction (β) in y and internal/external rotation (γ) in z [6, 9].

The GRF data, decomposed into in the directions, x, y, and z, were filtered in Matlab software using a Filter Butterworth 6th order and cut-off frequency of the 120 Hz [2].

2.3 Development of the Biomechanical Model

The biomechanical model has the function of representing the behavior of the human body in the accomplishment of tasks and must present the physical characteristics of the groups under study.

To facilitate the construction and analysis of body structures, some simplifications can be made. In this model, each body segment was considered a rigid body interconnected by joints that restrict its movement.

The biomechanical model was developed in Altair's Motion Solve software and their geometric model was based skeletal system with agglutination of the bones composing each limb is proposed, considering 8 independent systems (up limbs, hip, thighs, legs, feet), as represented in Fig. 1. The dimensions of the model were based on the anthropometric data collected by physiotherapists before the beginning of the test.

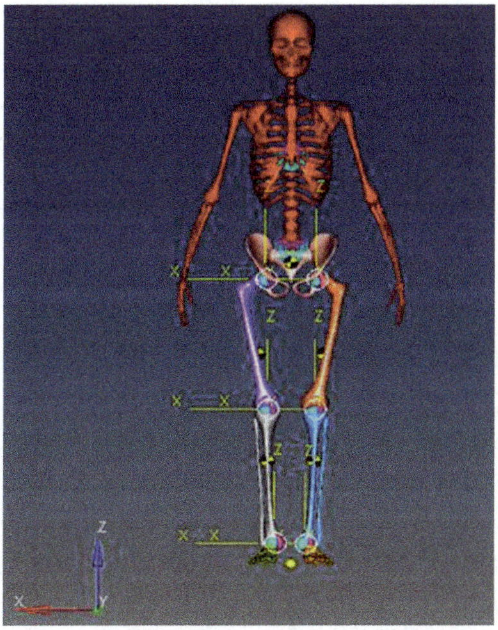

Fig. 1. Representation of the body segments of the biomechanical model

For the representation of the joints of the analyzed segments, spherical joints were used, which allow the rotation movement in all directions, restricting the movement in the desired direction with zero movement input [8]. The degrees of freedom of the joints were predefined as: joints with 3 degrees of freedom for hip and ankle, allowing rotation in the three directions; joins with 1 degree of freedom for knee, allowing flexion/extension movement [1, 2, 6].

In each joint, a damping and stiffness constant was applied, in the main directions of the movement, representing the muscular action of the limbs. These constants are presented in Table 1 [1, 6].

Table 1. Values of damping and stiffness constant.

Joint	Axis	Kt (N/m)	Ct (Ns/m)
Hip	x	1780	12,6
	y	777	6,3
	z	1296	9,1
Knee	x	2675	17,2
	y	10^{20}	0
	z	10^{20}	0
Ankle	x	1144	11,5
	y	10^{20}	0
	z	1144	11,5

The inserted angles of the joints together with the ground reaction force applied in the center of mass of foot, are fundamental to restrict movement, ensuring a more reliable representation of the movement of the lower limbs.

From this model, besides the graphical representation of the movement of the lower limbs during the human walk, capable of determining spatiotemporal variables that evaluate the gait during walking in a treadmill with controlled speed, it is also possible, to realize an analysis of the behavior of internal forces in the lower limbs.

3 Results

3.1 Kinematic and Kinematic Data

The collected kinematic data were filtered and treated in the Matlab. The Fig. 2 show the movement of the feet acquired through the Qualisys system by reading the reflective marker fixed in the intermalleolar position.

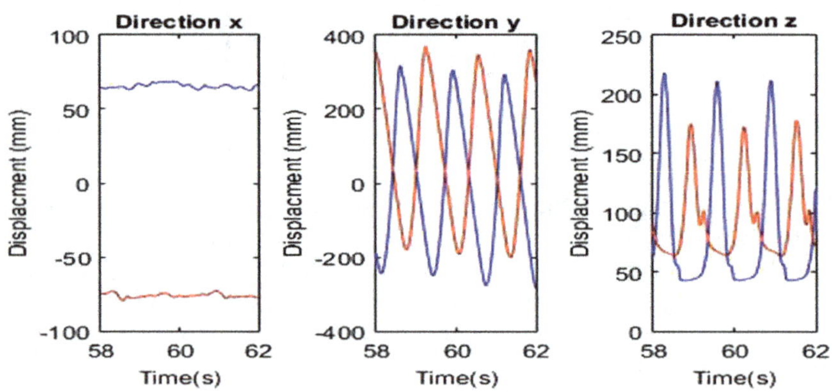

Fig. 2. Displacement of the feet, right and left, captured by Qualisys system, direction x (First figure), direction y (second image), direction z (third image)

Through the Cardan method, the angles of the lower limb joints (hip, thigh, leg and foot) were found considering the pelvis as reference. The movements found of the flexion/extension, adduction/abduction, internal/external rotation of the lower limbs, to consecutive two cycles, are shown in Fig. 3. The values were consistent with those presented in the literature [3, 10].

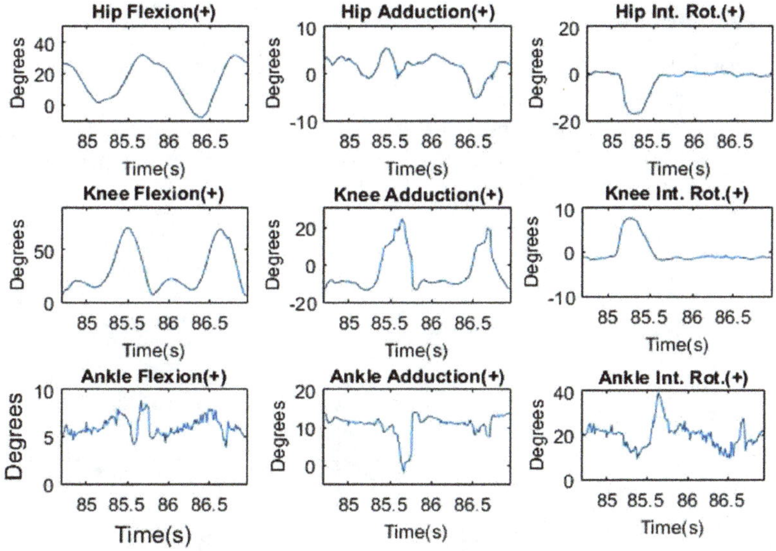

Fig. 3. Representation of joint of lower limbs angles

This kinematic data will be used as input for the joint of the model, conferred a movement for the segment connected by this joint.

The reaction force for each of the feet was separately collected in a force platform installed under the mat of the treadmill to observe the cycle of each foot individually.

The values of the ground reaction force of an subject weighing 80 kg are shown in Fig. 4.

Positive results in the z-direction, vertical force, indicate that they are in the supporting phase and in contact with the soil, already null values indicate that they are in the balance phase. By analyzing the force in the anteroposterior direction, the direction positive values indicate acceleration and negative values indicate deceleration. The force in x, mid-lateral direction, representation of supination movement, positive and prognostic values, negative values.

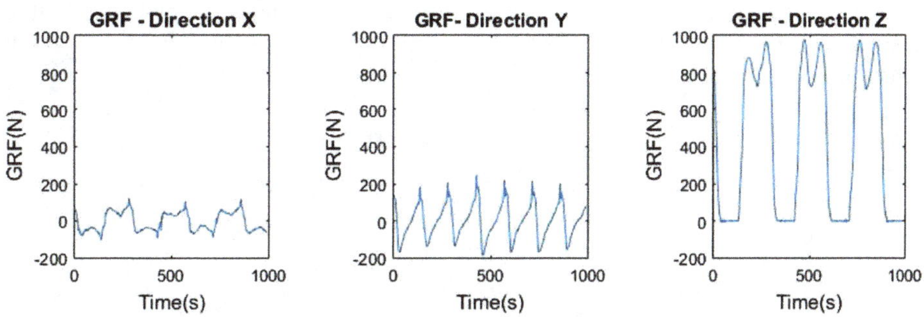

Fig. 4. Representation of the ground reaction force

3.2 Biomechanical Model

For each group, 18 curves were inserted, representing the flexion/extension movement, adduction/abduction, and internal/external rotation for each of the joints. The curves indicate the relative movement between the two members connected by the analyzed joint. It has as output kinematic data displacement, velocity and acceleration. The displacement of the feet is shown in Fig. 5.

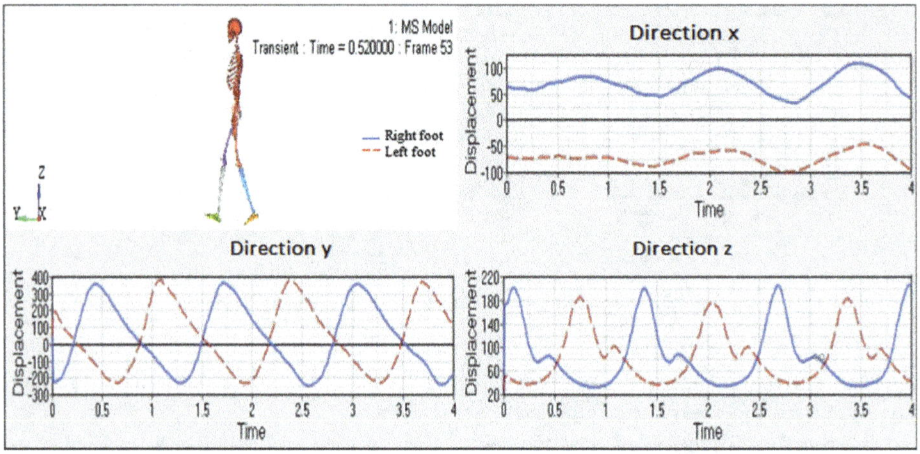

Fig. 5. Representation of displacement (mm) of the feet during the human gait

To validate the model, we measured the values of the feet displacement, as presented in Fig. 5. The values were compared to the obtained from the data using Qualisys, Fig. 2. The results show a same pattern and range of values, com maximum divergence of 7% along the curve for the directions y and z. The divergence of values in the x-curve can be justified by the pendular movement of the hip in the model, different of the movement performed during treadmill walking, with leg ajar, ensuring the capture of kinetic data from each independent foot, Therefore, these data were not passable of comparison.

The ground reaction force applied in the center of mass of that body features maximum amplitude of 978.4 N. The Fig. 6 the figure represents the forces of the lower limb joints for 1 gait cycle.

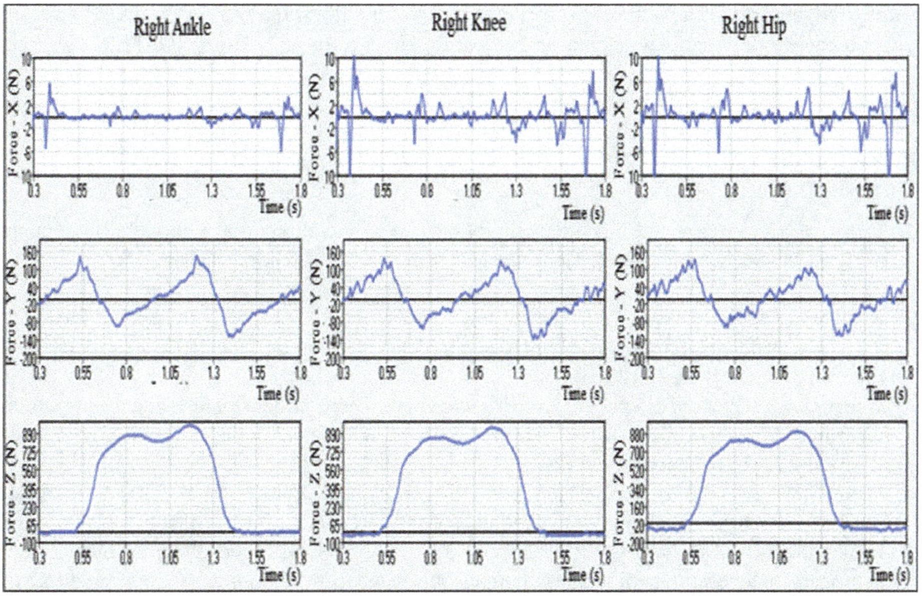

Fig. 6. Internal joint forces

Considering the magnitude of the collected ground reaction force shown in the Fig. 4, the force on the foot during the walk corresponds to 100% of this force, in the leg corresponds to 96% and in the thigh corresponds to 92%. Considering the magnitude of the ground reaction force collected in the force treadmill, shown in the figure, the force of exit of the model to the foot during the walk corresponds to 100% of this force, in the leg corresponds to 96% and in the thigh corresponds to 89%. The values were conducted with those presented in the literature. The values were consistent with those presented in the literature [4, 9].

There was a small reducing of force between the limbs caused by the damping and stiffness associated with joints and the reduction of mass of these segments.

4 Conclusion

The biomechanical model for human gait analysis was constructed from data of cinemetry and GRF, collected from healthy volunteers. The model presented similar responses to human behavior, with cyclic walking and stable parameters over time. The model obtained similar responses for the three speeds tested, with well-defined movements and consistent with the data collected by Qualysis.

The model proved to be a tool with great potential for defining variables that aid in the rehabilitation process, performance gain and injury prevention during walks or races, as well as a tool to assist in the development and testing of and prosthesis and orthosis or other studies which take into account the movement of the lower limbs.

Acknowledgements. The authors are grateful for the funding of the project by FAPEMIG, and by Coordenação de Aperfeiçoamento de Pessoal de Nível Superior - Brazil (CAPES) - Finance Code 001.

References

1. Vimieiro, C., Andrada, E., Witte, H., Pinotti, M.: A computational model for dynamic analysis of the human gait. J. Comput. Methods Biomech. Biomed. Eng. **18**(7), 799–804 (2013)
2. Rajagopal, A., Dembia, C., Demers, M., Delp, D., Hicks, J., Delp, S.: Full-body musculoskeletal model for muscle-driven simulation of human gait. IEEE Trans. Biomed. Eng. **63**(10), 2068–2079 (2016)
3. Winiarski, S., Pietraszewska, J., Pietraszewska, B.: Three-dimensional human gait pattern: reference data for young, active women walking with low, preferred, and high speeds. BioMed Research International (2019)
4. Yang, E.C., Mao, M.: Analytical model for estimating intersegmental forces exerted on human lower limbs during walking motion. Measurement **56**, 30–36 (2014)
5. Hasaneini, S.J., Macnab, C.J.B., Bertram, J.E.A., Leung, H.: The dynamic optimization approach to locomotion dynamics: human-like gaits from a minimally-constrained biped model. Adv. Rob. **27**(11), 845–859 (2003)
6. Martins, J.S.R.: Biomechanical model for Dynamic Analysis of the Human Gait. Belo Horizonte (Brazil). Department of Mechanical Engineering, Pontifícia Universidade de Minas Gerais, p. 117 (2016)
7. Wu, G., Siegler, S., Allard, P., Kirtley, C., Leardini, A., Rosenbaum, D., Whittle, M., D'Lima, D., Cristofolini, L., Witte, H., Schmid, O., Stokes, I.: ISB recommendation on definitions of joint coordinate system of various joints for the reporting of human joint motion—parts I: ankle, hip, and spine. J. Biomech. **35**(4), 543–548 (2002)
8. Begon, M., Andersen, M.S., Dumas, R.,: Multibody kinematics optimization for the estimation of upper and lower limb human joint kinematics: a systematized methodological review. J. Biomech. Eng. **140**(3) (2018)
9. Baker, R.: Globographic visualization of three dimensional joint angles. J. Biomech. **44**, 1885–1891 (2011)
10. Falisse, A., Rossom, S.V., Gijsbers, J., Steenbrink, F., Basten, B.J.H., Jonkers, I., Bogert, A.J., Groote, F.: OpenSim versus human body model: a comparison study for the lower limbs during gait. J. Appl. Biomech. **34**, 496–502 (2018)

The Effect of Non-linear Spring-Loaded Knee Orthosis on Lower Extremity Biomechanics

Christine D. Walck[1,2(✉)], Yeram Lim[1], Tyler J. Farnese[1], Victor Huayamave[1], Daryl C. Osbahr[3], and Todd N. Furman[3]

[1] Embry-Riddle Aeronautical University, Daytona Beach, FL 32114, USA
Christinedwalck@gmail.com
[2] U.S. Naval Research Laboratory, Washington, DC 20375, USA
[3] Department of Sports Medicine, Orlando Health, Orlando, FL 32806, USA

Abstract. Conventional knee orthosis are prescribed for people suffering from symptomatic knee osteoarthritis and from anterior cruciate ligament injuries. This case study aims to quantify the biomechanical response of the tibiofemoral joint complex to a non-linear spring-loaded (NLSL) knee joint orthosis (KJO). Joint angles and forces obtained from two dynamic trials were applied to a customized OpenSim computational musculoskeletal model, and a static equilibrium problem was solved at each instant during the squat cycle, with and without a NLSL KJO, to find individual muscle forces of the lower extremity. The NLSL KJO was seen to increase the gluteus maximus muscle force while decreasing the rectus femoris and soleus muscle forces. Due to the increased activation occurring in the gluteus maximus during the brace-on descent, the knee joint axis moved in a less anterior direction then what was seen in the brace-off descent. As a result, the pelvis translated in a more posterior direction due to the tension supplied by the gluteus maximus and the ease of the soleus. Furthermore, knee flexion was decreased in the upright position during the brace-on conditions. Results suggest that the NLSL KJO could be used as a performance tool to encourage a more balanced synergy that employs the posterior chain musculature versus a quadriceps dominant strategy while preventing hyperextension tendencies.

Keywords: Knee joint orthosis · Musculoskeletal modeling · Performance knee brace · Functional knee brace · Rehabilitation knee brace

1 Introduction

Knee orthotics can be categorized into four separate groups: prophylactic, functional, rehabilitation, and patellofemoral (PF) [1]. By design, prophylactic braces are intended to help prevent or reduce the likelihood of knee injuries by potentially decreasing peak tension magnitude and impulse responses of knee ligaments without reducing knee mobility [2]; functional braces are said to provide stability by reducing rotation in the knee and provide support to an already injured knee [3]; rehabilitation braces, sometimes referred to as a postoperative brace is prescribed in an attempt to decrease the recovery period by restraining the motion of the knee in some way [4]. This type of brace is also used to protect a reconstructed/repaired ligament and to allow for early

G. A. Ateshian et al. (Eds.): CMBBE 2019, LNCVB 36, pp. 371–383, 2020.
https://doi.org/10.1007/978-3-030-43195-2_31

controlled motion [5]; and PF braces, sometimes termed unloader knee brace seek to relieve anterior knee pain associated with osteoarthritis (OA) by reducing total ROM, and internal peak extension moment of the sagittal plane, as well as total ROM and adduction angle in the frontal plane [6].

The prophylactic and functional groups are referred to as preventative braces while the rehabilitation and PF groups are denoted as post-injury braces. This study uses a knee joint orthosis that utilizes a non-linear spring-loaded element and could be placed under the functional and the rehabilitation brace groups, thus acting as both a preventative tool and a rehabilitation aid [7]. This type of dual purpose is relevant and of particular interest because it's becoming the method of treatment for rehabilitation and injury prevention associated with symptomatic knee OA and anterior cruciate ligament (ACL) injuries [8, 9].

However, the biomechanical effects of these knee orthotics, including recent designs which propose the use of springs to store energy during knee flexion and provide assistance to the lower extremity during knee extension, have not yet been quantified. Instead, the current knee orthosis research is based off empirical and statistical observations of the knee complex in extension, anterior-posterior translation, and varus-valgus rotation [9]. Such qualitative methods can potentially leave the patient vulnerable to further injury. Therefore, the purpose of this case study is to quantify the biomechanical responses of the knee complex to a newly proposed NLSL KJO.

The biomechanical responses of the patient's lower extremity, with and without the NLSL KJO, were investigated during a deep squat motion. It was the assumption that the non-linear spring technology in the KJO would release stored energy, requiring less energy exertion from the wearer during knee extension causing a decrease in the quadriceps muscle activation.

2 Methods

To investigate the biomechanical responses of the knee complex to a NLSL KJO, a measured kinematic three-dimensional (3-D) motion capture analysis subjected to kinetic parameters during two separate trials (brace-off and brace-on) was used to drive a subject-specific musculoskeletal (MSK) model developed within OpenSim. The results from this type of analysis gave promising insight into the NLSL KJO's performance when compared to its designed parameters, and more specifically its effect on an individual's joint ROM and muscle force response.

2.1 Trial Movement

A squat exercise ranging from 0° knee flexion to no less than 115° knee flexion was performed during two trials: (1) brace-off: when the subject performs the movement without the NLSL KJO, and (2) brace-on: when the subject performs the movement while wearing the NLSL KJO. This movement was performed at a comfortable cadence set by the subject for five repetitions for each trial. A five-minute minimum rest period between trials was mandated to ensure fatigue was not a factor [10].

2.2 Subject

A healthy male (71.6 kg; 1.8 m tall) who was familiarized with squat techniques, signed a written consent form and volunteered to participate in the present case study. The subject, free of all known MSK disorders, performed each trial barefooted to limit the effects of shoes on the ground reaction forces (GRFs).

2.3 Non-linear Spring-Loaded Knee Joint Orthosis

The NLSL KJO used in this case study was custom fitted to the subject and is considered by the design company to be capable of providing the objectives of both a functional and post-surgery knee brace (Fig. 1). It features varying radius spring (VRS) technology which stores energy created by a bending moment during flexion and then releases said energy during extension. The release of such energy is believed to offload the quadriceps muscular activation, requiring less exerted energy from the wearer during extension [7].

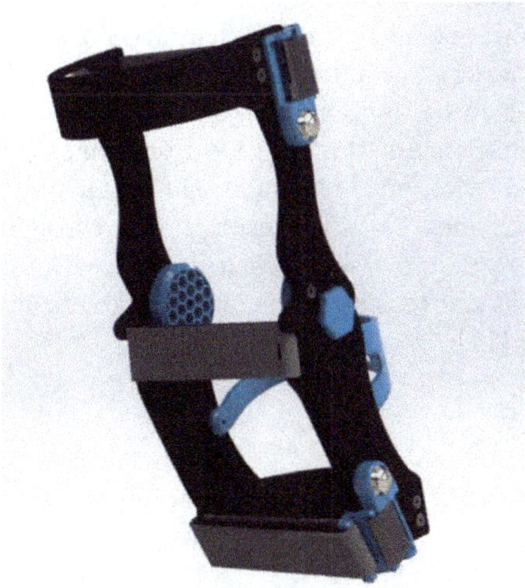

Fig. 1. A rendering of the non-linear spring-loaded knee joint orthosis used in this case study.

2.4 Measured Kinematic and Kinetic Data

Kinematic data was captured using the VICON Motion Capture System while kinetic data was collected using an AMTI OR6-6-OP force plate located under the subject's right foot. An additional AMTI filler plate was utilized to level the ground located under the left foot. Symmetry between the lower extremities was assumed.

Muscle activation was measured using Delsys Electromyography (EMG) surface sensors placed on the subject's right gluteus maximus (GMX), rectus femoris (RF), and soleus (SOL). Delsys's trigger module integrated with VICON NEXUS's proprietary software for motion analysis synced all data in the time domain.

2.5 Three-Dimensional Musculoskeletal Modeling: OpenSim

OpenSim GAIT 2392 3-D MSK model (NCSSR, Stanford, Palo Alto, CA) was used for this study's dynamic simulations. This specific model contains 12 body segments and 76 muscles and is designed for simulations of movements which are leg dominated. Such model best represents the joints and muscle groups that are of most interest to this study without adding complexity of unrelated parts such as the upper extremities [11].

However, to further accommodate the current study, the GAIT 2392 model was simplified by removing the torso and the left femur, tibia, fibula, talus, foot, and toes resulting in a model containing just six body segments and 43 muscles. These simplifications avoided misrepresentation of the left external forces.

2.6 Inverse Analysis: Pipeline SimTrack

OpenSim offers a modeling and simulation framework through a standardized pipeline called SimTrack which allows users to run dynamic simulations to estimate parameters such as joint moments and muscle force.

Starting with a generic model (i.e. simplified GAIT 2392) and experimental static kinematics as inputs, the pipeline utilized four main tools: (1) scale tool: scales the generic model according to the participant's anthropometrics and experimental static kinematic trial; (2) inverse kinematics (IK) tool: calculates joint kinematics using an experimental dynamical trial and weighted least squares algorithm; (3) inverse dynamics (ID) tool: estimates general muscle forces through an ID approach; and (4) static optimization (SO) tool: an extension of the ID tool that further resolves general forces into individual muscle forces utilizing the residual reduction algorithm (RRA) module (Fig. 2) [11].

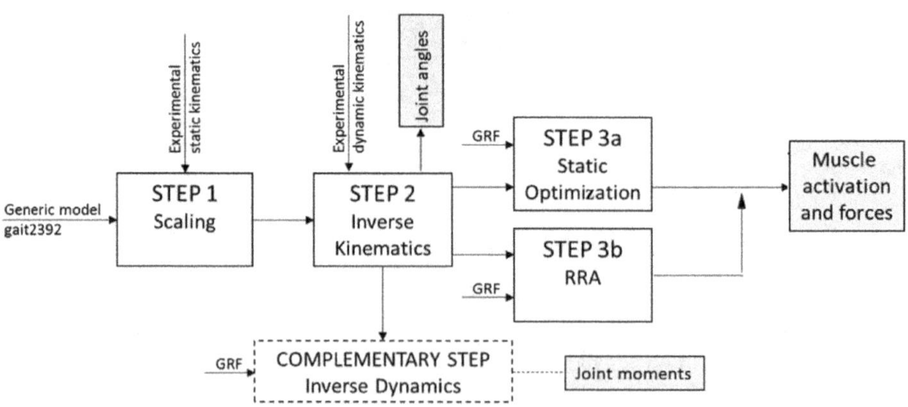

Fig. 2. SimTrack (inverse analyses) pipeline adopted from OpenSim [11].

Scaling

The default, unscaled version of the GAIT 2392 model represents a subject that is 1.8 m in height; weighing 75.16 kg. To match the subject in the current study, the model was scaled to 1.803 m and 71.6 kg with a low pass filter at 6 Hz applied to the static kinematic file. The body segments were defined using marker pairs as described in Tables 1 and 2.

Table 1. Body segments defined using marker pairs and scale factors for brace-off trial.

Body name/Measurement used	Marker pairs		Applied scale factor
Pelvis/Pelvis	R.ASIS	L.ASIS	1.066
Femur_R/Femur_R	R.ASIS	R.Upper.Knee.Lat	1.048
Tibia_R/Tibia_R	R.Upper.Knee.Lat	R.Ankle.Lat	1.050
Talus_R/Unassigned	N/A	N/A	1.0
Calcn_R/Calcn_R	R.Heel	R.Toe.Tip	0.936
Toes_R/Unassigned	N/A	N/A	1.0

Table 2. Body segments defined using marker pairs and scale factors for brace-on trial.

Body name/Measurement used	Marker pairs		Applied scale factor
Pelvis/Pelvis	R.ASIS	L.ASIS	1.068
Femur_R/Femur_R	R.ASIS	R.Upper.Knee.Lat	1.092
Tibia_R/Tibia_R	R.Upper.Knee.Lat	R.Ankle.Lat	1.008
Talus_R/Unassigned	N/A	N/A	1.0
Calcn_R/Calcn_R	R.Heel	R.Toe.Tip	0.933
Toes_R/Unassigned	N/A	N/A	1.0

Inverse Kinematics (IK)

Experimentally-measured kinematic data from the brace-on and brace-off dynamic trials was used as an input into the IK tool. As an output, a motion file containing the generalized coordinate trajectories (joint angles and/or translations) was provided and later used as an input file to the ID tool. This motion file also provided a method for tuning the model as well a visual feedback (Fig. 3).

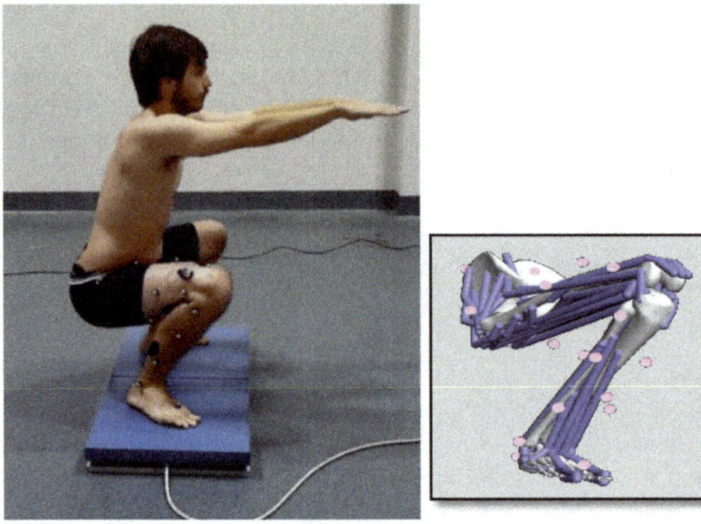

Fig. 3. From experimental motion capture data (left), OpenSim computed IK simulation (right).

Static Optimization
Due to experimental errors and modeling assumptions, kinematics is not always dynamically consistent with GRFs, indicating the existence of residual forces. In the current study, the residuals were larger then desired. This was mainly due to two factors:

(1) The GAIT 2329 model was developed for gait analysis which made it difficult to analyze movements that require higher degrees of flexion, specifically angles greater than 120° knee flexion and ±90° pelvic tilt.
(2) The missing left GRFs caused the model to assume the right lower extremity was supporting the left pelvis and lumbar.

In efforts to reduce the error, the RRA module was used to apply residual actuators at the pelvis to achieve the required flexion angles and left pelvis and lumbar muscle actuators without adding the complexity of additional body segments. The module was run twice before a suitable adjusted model was produced and used to find individual muscle forces.

3 Results

Joint angles, moments, and forces obtained from dynamic trials were applied to the simplified GAIT 2392 model, and a static equilibrium problem was solved at each instant during the squat cycle to find individual muscle forces. Position errors for each of the model's generalized coordinates and the average residual values were used to tune each model. Then a direct comparison between the brace-off (WOB) and brace-on (WB) trials was made to provide insight into how the NLSL KJO affects knee biomechanics and muscle forces. The brace-off trial results were also compared to relevant literature for model verification and validation.

A summary of average knee joint positions and moments can be found in Table 3 while a summary of muscle forces can be found in Table 4. Sign convention in the sagittal plane for knee extension (E) and knee flexion (F) are positive (+) and negative (−) values respectively.

3.1 Knee Joint Angles

The knee joint consists of the tibiofemoral joint, which carries out sagittal plane angular movement from 1.7° to 0° of extension and from 0° to approximately 160° of flexion [12–14]. The subject in the current study performed each brace-off squat at an average of 4.5° ± 1.10° in the upright position indicating a slight degree of knee flexion and an average of 133° ± 1.61° at the bottom of the squat.

During the brace-on trial, knee flexion in the upright position was seen to decrease to 0.5° ± 1.25° suggesting that the NLSL KJO encourages a more neutral stance. At the bottom of the squat knee flexion was also seen to decrease (average 125° ± 4.00° WB) however, this decrease of 8° could be due to the physical size and placement of the brace on the leg and not a consequence of the non-linear spring influence. No change in either trail was observed in knee flexion ranges seen in day to day activities [15] (Fig. 4).

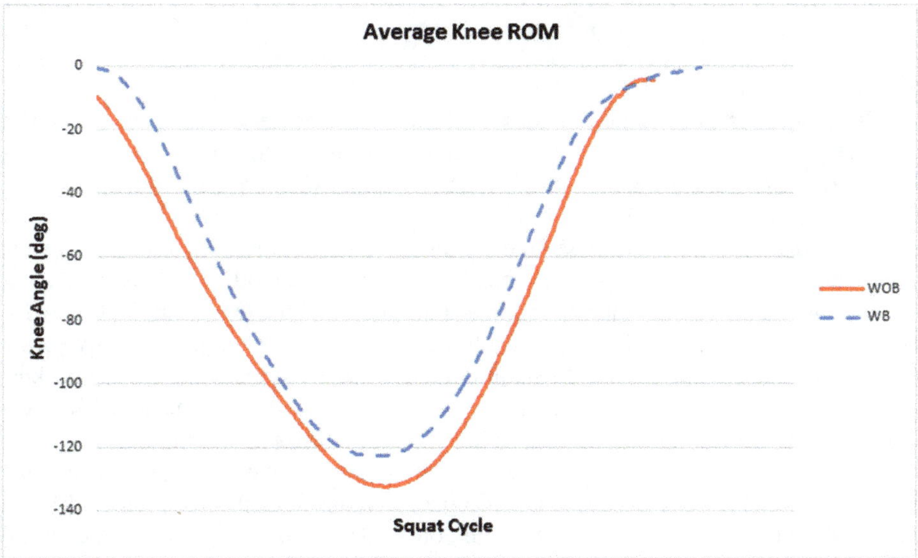

Fig. 4. Difference between the brace-off (red) and brace-on (dashed-blue) knee joint range of motion in the sagittal plane. Results show a 4° reduction in knee flexion at the top of the squat (Average 4.5° ± 1.1° WOB, 0.5° ± 1.25° WB) and a 8° reduction in knee flexion at the bottom of the squat (Average 133° ± 1.61° WOB, 125.1° ± 4.0° WB).

Table 3. A summary of average joints positions (deg) and joint moments (Nm).

	Position (deg) at full knee extension (start position)		Position (deg) at 60° knee flexion (descent)		Position (deg) at full knee flexion (bottom position)		Position (deg) at 60° knee flexion (ascent)	
Joint	Average (WOB)	Average (WB)	Average (WOB)	Average (WB)	Average (WOB)	Average (WOB)	Average (WOB)	Average (WB)
Knee	4.45 ± 1.10 (F)	0.50 ± 1.25 (F)	52.58 ± 3.41 (F)	58.18 ± 3.78 (F)	132.94 ± 1.61 (F)	125.05 ± 4.0 (F)	56.21 ± 3.62 (F)	56.48 ± 15.75 (F)
	Joint moment (Nm) at full knee extension (start position)		Joint moment (Nm) at 60° knee flexion (descent)		Joint moment (Nm) at full knee flexion (bottom position)		Joint moment (Nm) at 60° knee flexion (ascent)	
Joint	Average (WOB)	Average (WB)	Average (WOB)	Average (WB)	Average (WOB)	Average (WOB)	Average (WOB)	Average (WB)
Knee	10.78 ± 5.29	11.4 ± 4.13	46.02 ± 6.76	54.4 ± 5.32	77.04 ± 2.72	82.86 ± 2.23	45.49 ± 3.43	45.76 ± 12.40

3.2 Residual Errors

To evaluate the RRA results discussed above, OpenSim recommends a set of thresholds error values (Table 4). However, the size of residuals depends on the type of motion being studied. For example, residuals for high-speed activities like sprinting will typically be larger than walking. Residuals will also be larger if there are external forces unaccounted for, such as, in this study, the subject's left GRFs [11].

Table 4. Threshold values used to evaluate RRA results for full-body simulations of gait.

Thresholds	Good	Okay	Bad
MAX residual force (N)	0–10	10–25	>25
MAX residual moment (Nm)	0–50	50–75	>75

The model errors presented in this study are on average within these guidelines however some values are larger than the maximum threshold especially during deep knee flexion. Residual force and moment errors are shown in Figs. 5 and 6 below for brace-off and brace-on models.

Due to the large error values seen in the direction of motion (i.e. y-direction) results found during knee flexion angles less than 25° or greater than 60° could be misleading. These errors agree with the ones found in a study that used the OpenSim GAIT 2392 to calculate the tibiofemoral joint contact forces (JCFs) in six subjects for five squat repetitions of squats. While the MSK simulations underestimated the measured knee JCFs at low flexion angles, an average error of less than 20% was achieved between approximately 25°–60° knee flexion. With an average error that behaved almost linearly with knee flexion angle, an overestimation of approximately 60% was observed at deep flexion. Such data indicates that loading estimations from this particular MSK gait model at both high and low knee joint flexion angles should be interpreted carefully [16]. For these reasons, the following subsections focus on the internal forces seen at 60° knee flexion during both the descent and ascent phase.

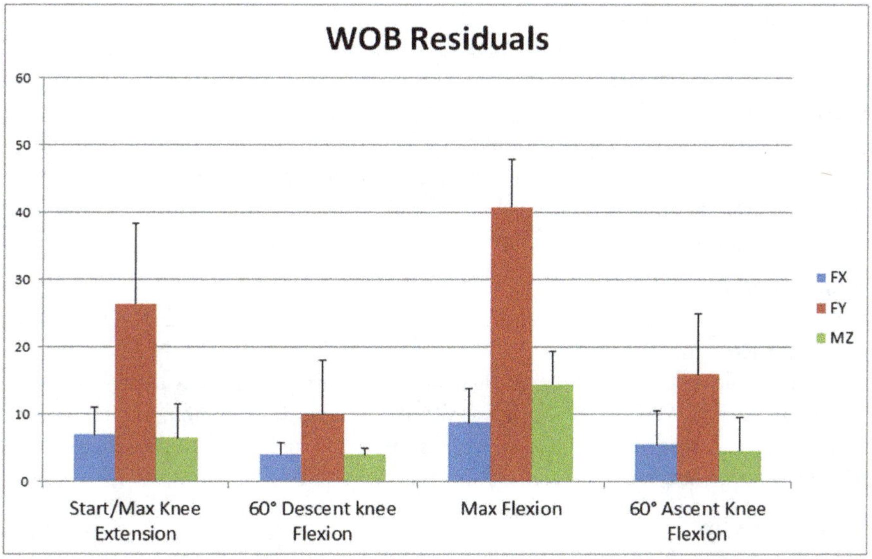

Fig. 5. Brace-off model residual forces in the x-direction (blue) and y-direction (red) and the torque about the z-axes (green).

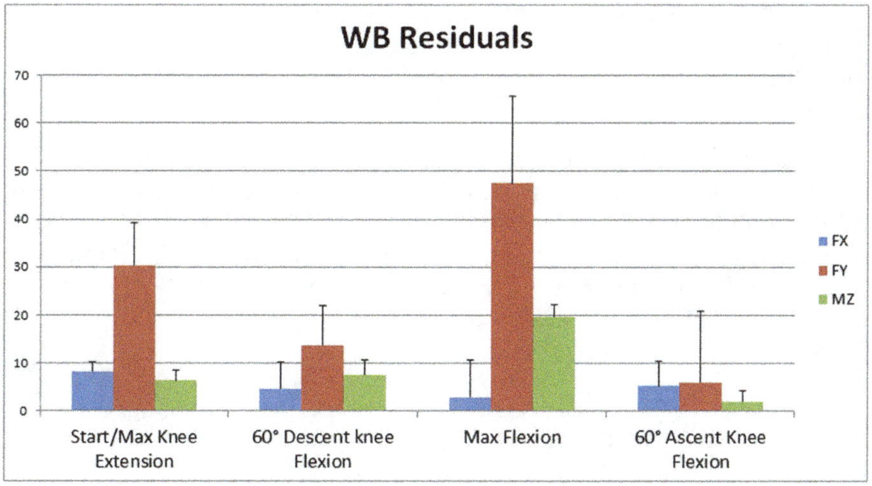

Fig. 6. Brace-on model residual forces in the x-direction (blue) and y-direction (red) and the torque about the z-axes (green).

3.3 Muscle Force

The squat exercise is a closed-chain exercise that uses both primary and synergist muscles. The primary muscles acting about the knee are the quadriceps femoris, which carry out concentric knee extension, as well as eccentrically resisting knee flexion.

These muscles also act as a flexor to the hip [17]. The synergist muscles include the hamstrings and the SOL which behave paradoxically and co-contract with the quadriceps. The synergistic action has important implications for enhancing the integrity of the knee joint in squat performance. Specifically, the hamstrings exert a counter-regulatory pull on the tibia, helping to neutralize the anterior tibiofemoral shear imparted by the quadriceps [12]. In addition to this, the hamstrings act as a hip extensor [17]. To further analyze this action, the RF, SMX, and the SOL were investigated and are discussed below.

Rectus Femoris

At the 60° knee flexion position during the descent phase, the NLSL KJO was seen to decrease the RF muscle force by 13 N (average 45.7 WOB, 32.6 N WB). However, during the ascend phase at 60° knee flexion almost no change was observed (average 31.3 WOB, 32.7 N WB). Interestingly, for the brace-off trail, the magnitude in RF force increased during the last three repetitions while the magnitude remained fairly consistent throughout the entire set of the brace-on trial.

Gluteus Maximus

The GMX is a powerful hip extensor, acting eccentrically to control squat descent and concentrically to overcome external resistance on the ascent. Given its attachment at the iliotibial band, the GMX is also thought to play a role in stabilizing knee and pelvis during squatting. GMX has been shown to produce a peak hip extensor force at approximately 90° and below [12]. This agrees with the data collected in this study.

During the brace-on trial, the GMX was seen to increase throughout the movement. Specifically, at 60° knee flexion the NLSL KJO increases the GMX (descent average 5.9 N WOB, 7.2 N WB and ascent average 4.5 N WOB, 5.8 N WB).

Soleus

The SOL is a pure plantar flexor with proximal attachments at the tibia and fibula and distal attachments at the calcaneus [12]. It is an underestimated muscle in the squat exercise yet produces a significant about of force.

At 60° knee flexion during the descent and ascent phase the NLSL KJO was seen to decrease the SOL force by 67.4 N (average 123.5 N WOB, 56 N WB) and by 3.3 N (average 47 N WOB, 43.7 N WB) respectively (Table 5).

Table 5. A summary of average individual muscle forces (N) at 60° knee flexion during descent and ascent for brace-off (WOB) and brace-on (WB) conditions.

	Muscle Force (N) at 60° knee flexion (descent)		Muscle force (N) at 60° knee flexion (ascent)	
Muscle	Average (WOB)	Average (WB)	Average (WOB)	Average (WB)
GMX	5.85 ± 0.49	7.15 ± 1.22	4.5 ± 0.61	5.82 ± 0.38
RF	45.67 ± 6.26	32.61 ± 4.36	31.27 ± 5.75	32.75 ± 6.38
SOL	123.51 ± 38.5	56.15 ± 8.71	46.98 ± 6.92	43.68 ± 9.06

4 Electromyographic Response

EMG responses of the GMX, RF, and SOL using a root mean square analysis were recorded (Figs. 7, 8 and 9). These recordings were used as a validation technique when comparing trends to the models activation and force production calculated from OpenSim's SO which was filtered at a low pass 6 Hz.

Due to sensor placement and the NLSL KJO placement, motion artifacts were noticed in the RF data. Specifically, the NLSL KJO was placed just below the belly of the RF. After the KJO was tightened, it created a distributed load on the leg causing interference with the sensor and inaccurate RF muscular activation data (Fig. 8).

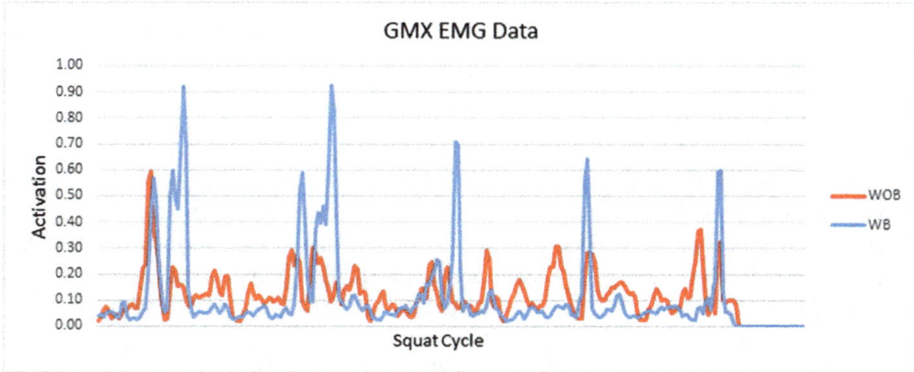

Fig. 7. GMX normalized EMG activation data for brace-off (red) and brace-on (blue) trials.

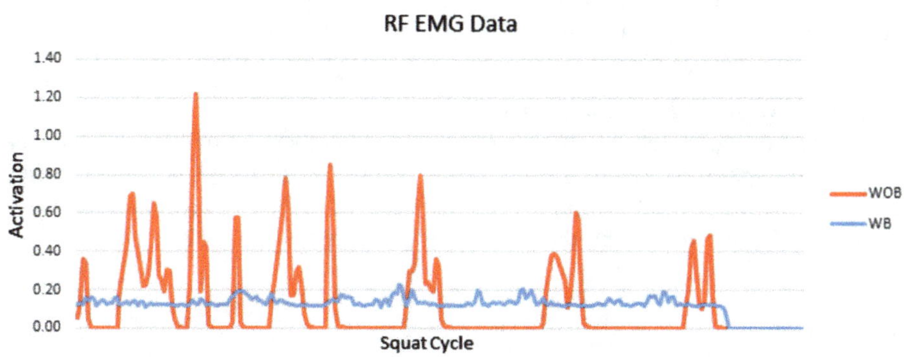

Fig. 8. RF normalized EMG activation data for brace-off (red) and brace-on (blue) trials.

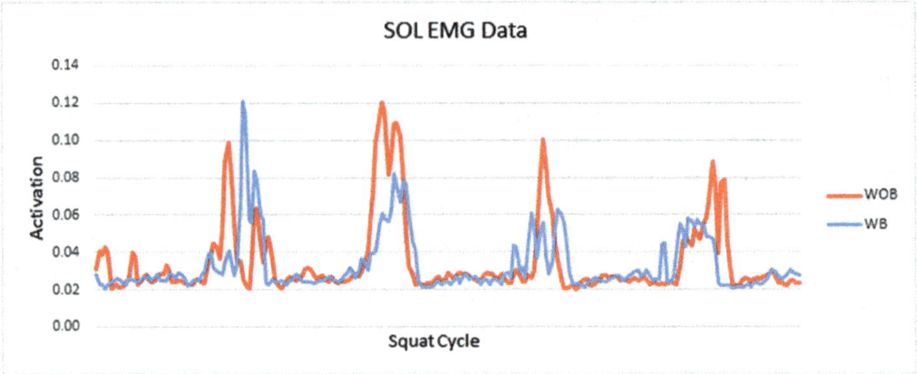

Fig. 9. SOL normalized EMG activation data for brace-off (red) and brace-on (blue) trials.

5 Discussion

Preliminary results show the NLSL KJO increases the GMX muscle force while decreasing the RF and SOL muscle forces. Due to the increased activation occurring in the GMX during descent, the knee joint axis moved in a less anterior direction then what was seen in the brace-off descent. As a result, the pelvis translated in a more posterior direction due to the tension supplied by the GMX and the ease of the SOL. These results suggest that the NLSL KJO adjusts the patient's quadriceps dominant synergy to a more balanced movement and could be effective in aiding muscular offloading about the knee joint.

The NLSL KJO also affected the posture in the upright position. Decreases were seen in the brace-on condition at full knee extension suggesting the brace is functioning as a performance training tool that aids in maintaining a neutral position. In other words, the NLSL KJO may have a performance influence through hyperextension prevention and balance synergy encouragement.

Compliance with Ethical Standards
The procedures described in this work involving human participation have been reviewed and approved by the Institutional Review Committee of the Embry-Riddle Aeronautical University. The participant gave his consent to participate and being recorded (audio and image) by the Embry-Riddle Aeronautical University.

References

1. France, E.P., Paulos, L.E.: Knee bracing. J. Am. Acad. Orthop. Surg. **2**(5), 281–287 (1994)
2. Sinclair, J.K., Vincent, H., Richards, J.D.: Effects of prophylactic knee bracing on knee joint kinetics and kinematics during netball specific movements, pp. 93–98 (2017)
3. Smith, S.D., et al.: Functional bracing of ACL injuries: current state and future directions. Knee Surg. Sports Traumatol. Arthrosc. **22**(5), 1131–1141 (2014)

4. Möller, E., et al.: Bracing versus nonbracing in rehabilitation after anterior cruciate ligament reconstruction: a randomized prospective study with 2-year follow-up. Knee Surg. Sports Traumatol. Arthrosc. **9**(2), 102–108 (2001)
5. Van Thiel, G.S.B., Back, Jr.: Musculoskeletal key fastest musculoskeletal insight engine, knee bracing for athletic injuries (2016). https://musculoskeletalkey.com/knee-bracing-for-athletic-injuries/
6. Doslikova, K., et al.: The influence of a patellofemoral knee brace on knee joint kinetics and kinematics in patients with knee osteoarthritis during stair negotiation. Osteoarthritis Cartilage **22**, S89 (2014)
7. Geldart, M., Hobson, W., Farnese, T.: Nonlinear spring effect on muscular activation. In: 2018: 8th World Congress of Biomechanics, Dublin, Ireland (2018)
8. Podraza, J.T., White, S.C.: Effect of knee flexion angle on ground reaction forces, knee moments and muscle co-contraction during an impact-like deceleration landing: Implications for the non-contact mechanism of ACL injury. Knee **17**(4), 291–295 (2010)
9. Cawley, P.W., France, E.P., Paulos, L.E.: Comparison of rehabilitative knee braces: a biomechanical investigation. Am. J. Sports Med. **17**(2), 141–146 (1989)
10. Rahimi, R.: Effect of different rest intervals on the exercise volume completed during squat bouts. J. Sports Sci. Med. **4**(4), 361–366 (2005)
11. Research, N.C.f.S.i.R. OpenSim User's Guide (2018). Accessed 3 Mar 2018. 4.0. https://simtk-confluence.stanford.edu:8443/display/OpenSim33/User%27s+Guide
12. Schoenfeld, B.J.: Squatting kinematics and kinetics and their application to exercise performance. J. Strength Cond. Res. **24**(12), 3497–3506 (2010)
13. Hwang, S., Kim, Y., Kim, Y.: Lower extremity joint kinetics and lumbar curvature during squat and stoop lifting. BMC Musculoskelet. Disord. **10**, 15 (2009)
14. Roaas, A., Andersson, G.B.J.: Normal range of motion of the hip, knee and ankle joints in male subjects, 30–40 years of age. Acta Orthop. Scand. **53**(2), 205–208 (1982)
15. Mockford, B.J., et al.: Does a standard outpatient physiotherapy regime improve the range of knee motion after primary total knee arthroplasty? J. Arthroplasty **23**(8), 1110–1114 (2008)
16. Schellenberg, F., et al.: Evaluation of the accuracy of musculoskeletal simulation during squats by means of instrumented knee prostheses. Med. Eng. Phys. **61**, 95–99 (2018)
17. Dahlkvist, N.J., Mayo, P., Seedhom, B.B.: Forces during squatting and rising from a deep squat. Eng. Med. **11**(2), 69–76 (1982)

Evaluation of a 1-DOF Hand Exoskeleton for Neuromuscular Rehabilitation

Xianlian Zhou(✉), Ashley Mont, and Sergei Adamovich

New Jersey Institute of Technology, Newark, NJ 07102, USA
alexzhou@njit.edu

Abstract. A low-cost 1-DOF hand exoskeleton for neuromuscular rehabilitation has been designed and assembled. It consists of a base equipped with a servo motor, an index finger part, and a thumb part, connected through three gears. The index part has a tri-axial load cell and an attached ring to measure the finger force. An admittance control paradigm was designed to provide intuitive control and positive force amplification to assist the user's finger movement. To evaluate the effects of different control parameters on neuromuscular response of the fingers, we created an integrated exoskeleton-hand musculoskeletal model to virtually simulate and optimize the control loop. The exoskeleton is controlled by a proportional derivative controller that computes the motor torque to follow a desired joint angle of the index part, which is obtained from inverse kinematics of a virtual end-effector mass driven by the finger force. We conducted parametric simulations of the exoskeleton in action, driven by the user's closing and opening finger motion, with different proportional gains, end-effector masses, and other coefficients. We compared the interaction forces between the index finger and the ring in both passive and active modes. The best performing assistive controller can reduce the force from around 1.45 N (in passive mode) to only around 0.52 N, more than 64% of reduction. As a result, the muscle activations of the flexors and extensors were reduced significantly. We also noted the admittance control paradigm is versatile and can also provide resistance (e.g. for strength training) by simply increasing the virtual end-effector mass.

Keywords: Hand exoskeleton · Neuromuscular rehabilitation · Musculoskeletal model

1 Introduction

Stroke, one of the leading causes of adult disability, affects approximately 800,000 individuals each year in the United States [1]. Nearly 80% of stroke survivors suffer from hemiparesis of the upper arm and thus impaired hand function, which is integral to most activities of daily living. It is well established that highly repetitive training can aid in the recovery of motor function of the hand however this can be labor intensive for the providing physical therapist in addition to the cost. In the past decade, more

A draft version of this paper was uploaded to the arXiv website at https://arxiv.org/abs/1907.07311.

G. A. Ateshian et al. (Eds.): CMBBE 2019, LNCVB 36, pp. 384–397, 2020.
https://doi.org/10.1007/978-3-030-43195-2_32

robotic hand rehabilitation devices have been introduced to help patients recover hand function through assistance during repetitive training of the hand [2–5].

In a comprehensive review by Heo et al. [2], hand exoskeleton technologies for rehabilitation and assistive engineering, from basic hand biomechanics, neurophysiology, sensors and actuators, physical human-robot interactions and ergonomics, are summarized. Different types of actuators and control schemes have been used for hand exoskeletons. In some control schemes, the robotic device will move the user passively through a preprogrammed trajectory for continuous passive movement (CPM) therapy. These devices can be beneficial for severely impaired individuals who may not have the ability to generate the forces required for specific finger or hand movement or for individuals who have abnormal muscle synergies preventing continuous movement. A few devices such as the Kinetic Maestra and Vector 1 are commercially available devices that are used for CPM [6, 7]. These devices allow for passive movement through the range of motion for individual fingers. However, as there is no active participation by the user, this device on its own may not promote neurorehabilitation. These devices can be combined with other simulations or control schemes that require active participation by the user. One commercially available hand exoskeleton that has been used extensively by our lab to provide haptics to virtual simulations is the CyberGrasp [8]. The CyberGrasp is a cable driven exoskeleton that weighs 450 grams and can provide up to 12 N of force on each finger and can be used to provide assistance for extension of the user's fingers. In one study, this was used in combination with a virtual reality simulation to train finger individuation as the user played a virtual piano [3]. The CyberGrasp was used to resist finger flexion of the inactive fingers, promoting movement of the active independent finger. Similarly, the eXtension Glove (X-Glove) was developed to be used for cyclical stretching in addition to active movement training [9–11]. This cable driven design is actuated using linear servos allowing for individual finger movement in both extension and flexion. In addition to this, each cable is integrated with a tension sensor which allows the force of each digit to be monitored. This device has two modes that can be used for rehabilitation, the first mode cyclically extends and flexes the fingers. The second mode is an active training mode in which the glove provides constant extension assistance so that the user can complete flexion tasks as long as they overcome the force required to keep the finger extended. In a further attempt to integrate user control with the exoskeleton, an external input from the user such as force or electromyography (EMG) has been incorporated into some designs such as the Helping Hand [12]. This soft robotic device allows for active assistance for each finger individually, in addition to the ability to follow control states triggered by EMG.

In this paper, we introduce a low cost 1-DOF hand exoskeleton for neuromuscular rehabilitation of individual fingers. This exoskeleton consists of a base equipped with a servo motor, an index finger part and a thumb part connected with gears. The exoskeleton's control system was designed to generate suitable actuation torques based on the interaction force between the user's finger and the exoskeleton's index part. The goal of this study is to model the exoskeleton interacting with a neuromuscular hand model in order to evaluate the effectiveness of an intuitive admittance control algorithm on providing different levels of assistance or resistance during hand rehabilitation.

2 Methods

2.1 The 1-DOF Exoskeleton and Hand Model

This exoskeleton consists of a base stationed with a servo motor (Dynamixel XM430), an index finger part and a thumb part, which are connected through 3 gears of equal sizes as shown in Fig. 1. The motor drives the top gear which in turn rotates the gear attached to the index part and then the gear attached the thumb part. The index and thumb parts both have rings for the fingers, and an OptoForce tri-axial load cell or force sensor (OnRobot, Denmark) is attached to the index ring. All parts are 3D printed with a carbon fiber reinforced nylon material called Onyx (Markforged, USA). The total weight of this exoskeleton is 0.158 kg and the mass and inertia properties of its components, which were either measured or computed based on material and part geometry, are listed in Table 1.

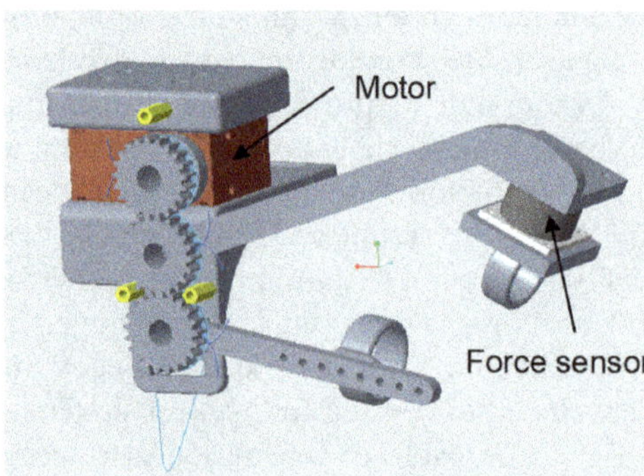

Fig. 1. The design of the 1-DOF hand exoskeleton.

Table 1. Mass and moment of inertia (MOI) properties of exoskeleton components. x: fore-aft; y: vertical; z: lateral. (MOI unit: $kg \cdot cm^2$)

Exo-Part	Mass (kg)	I_{xx}	I_{yy}	I_{zz}
Base	1.22e-1	6.62e-1	8.39e-1	6.94e-1
Index	2.95e-2	6.11e-2	5.56e-1	5.58e-1
Thumb	6.40e-3	9.20e-3	6.51e-2	6.43e-2

The exoskeleton was modeled as an articulated rigid body system with just one true rotational DOF at the motor (top) gear and its base fixed on the human hand. To model the interaction between the human hand and the exoskeleton, we used a generic hand musculoskeletal model adapted from the one developed by Lee et al. [13]. The adapted hand model, shown in Fig. 2, has six finger muscles including: extensor digitorum communis (EDC), extensor indicis (EI), flexor digitorum superficialis (FDS), flexor digitorum profundus (FDP), bipennate first dorsal interosseous (FDI) on the radial side and the ulnar side. In the original model by Lee et al., only the radial side of the FDI, which connects the radial side of the second metacarpal to the radial side of the base of the second proximal phalanx (index finger), was modeled. Here we added the opposite (ulnar) side of the FDI that connects the proximal half of the ulnar side of the first metacarpal (thumb) to the index finger as we believe this branch could be important for finger closing and opening motion. Among these muscles, EDC and EI are extensors and FDS, FDP and FDI are flexors. To assemble with the hand musculoskeletal model (Fig. 2), the exoskeleton base part was fixed on the back of the hand through a fixed joint while the ring on the index part was put on the index finger, and similarly for the ring on the thumb part.

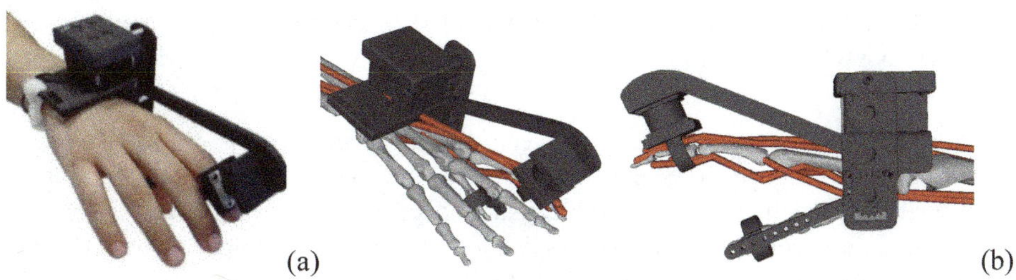

Fig. 2. (a) The hand exoskeleton device on a human hand. (b) The assembled model of the hand musculoskeletal system and the exoskeleton.

2.2 Admittance Control Paradigm

To help users with muscle weakness during training, we designed an admittance control paradigm, shown in Fig. 3, to provide intuitive control and positive force amplification to assist the user's finger movement. This admittance control transfers the force from the load cell into the motion of a virtual end-effector mass in the task space. The admittance control framework integrates the motion of this end-effector mass initially placed at the center of the index ring and moved by the force applied by the index finger. The desired angle of the motor (θ_d) is obtained from an inverse kinematics (IK) computation based on the desired end-effector position, and the motor is then commanded to achieve this position. In hardware implementation, the end-effector force is measured by a force/torque sensor and the desired motor angle θ_d can be achieved by either torque based control or position based control of the Dynamixel motor. In this work, we evaluated this admittance control loop with computer simulations. The interaction forces between the finger and the ring were modeled and will be described shortly. And a proportional derivative (PD) controller was used to prescribe a desired joint torque (τ) to the motor:

$$\tau = k_p(\theta_d - \theta) - k_d\dot{\theta} \quad (1)$$

where k_p and k_d are tunable parameters, θ and $\dot{\theta}$ are the current angle and angular velocity of the motor, respectively.

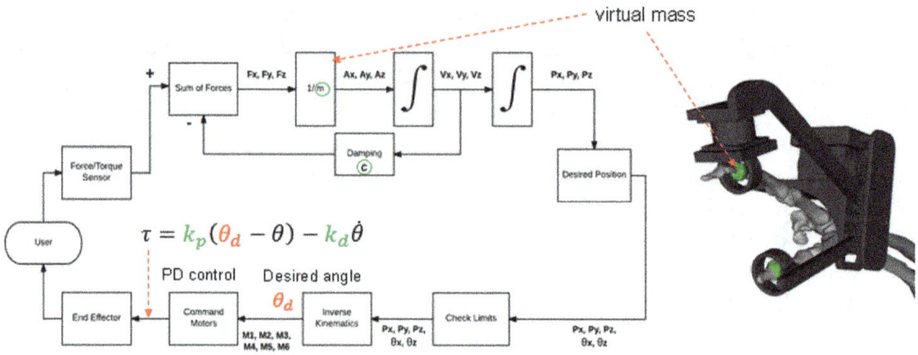

Fig. 3. The designed admittance control framework. The parameters in green or circled in green are tunable control parameters.

To model the interaction forces between the finger and ring, we introduced a tri-directional spring-damper force element that mimics the contact between them. The tri-directional force element was introduced at the center of the ring and can predict directional differences in force responses due to the relative movement of the finger and the ring. The force element computes three (XYZ) directional distances between a point on the exoskeleton and its counterpart on the finger and generates (either positive or negative) forces along these directions during their relative movement. At the initial assembly, these two points are coincident to each other and generate zero force $(x_0 = y_0 = z_0 = 0)$. The forces generated by the force element were modeled by linear damped springs:

$$\begin{cases} f_x = k_x(x - x_0) + c_x\dot{x} \\ f_y = k_y(y - y_0) + c_y\dot{y} \\ f_z = k_z(z - z_0) + c_z\dot{z} \end{cases} \tag{2}$$

The stiffness and damping constants of the directional force element are listed in Table 2. The stiffness in the lateral direction (YZ) is assumed to be 20 times of that in the X direction to mimic the behavior of harder resistance in the lateral directions and softer resistance in the sliding directions.

Table 2. Stiffness and damping of the direction spring between finger and ring.

Directional springs	Stiffness (N/m)			Damping (Ns/m)		
	k_x	k_y	k_z	c_x	c_y	c_z
Finger-Ring	500	10000	10000	80	200	200

2.3 Simulation Methods

With the developed model and simulated controlled framework, we can perform computer simulations to study the co-operation between fingers and the exoskeleton. In our simulations, we assumed the index finger moves to actively track a given closing and opening motion. The thumb is passive and moved by the thumb ring and the ulnar side of the FDI muscle. The motion of the exoskeleton is driven by the motor torque and the forces between the fingers and the rings. In the passive mode, the active motor torque was set to zero and a small velocity dependent damping was applied.

All simulations in this study were performed with our in-house extended version of the musculoskeletal simulation code, CoBi-Dyn, initially developed at CFD Research Corporation (Huntsville, AL). A hybrid inverse dynamics (ID) and forward dynamics (FD) simulation framework similar to the one presented in [14] was employed. The human finger joints were classified as ID joints such that their motions can be prescribed to track an input motion. The exoskeleton joints were classified as FD joints such that their motions were driven by the actuation forces and finger-ring interaction forces. At each time step, the hybrid dynamics framework predicted joint torques for all finger joints and accelerations for all exoskeleton joints. The predicted finger joint torques were the target or desired torques that ideally shall be generated from muscles spanning these joints. To compute muscle forces, one of the goals was to find an appropriate muscle force combination that contributed to generate the desired joint torques as closely as possible. Due to the redundancy of the muscles, there could be many such combinations and thus muscle forces were determined by solving an optimization problem. The final objective of this optimization problem was to minimize an objective function, defined as

$$\sum_{i=1}^{n} \left(\frac{f_i}{f_i^{max}} \right)^p + w\boldsymbol{C}^T\boldsymbol{C} \tag{3}$$

where f_i was the force of the ith muscle, f_i^{max} was the maximum attainable muscle force at its current state, $\boldsymbol{C}$ was the difference vector between the desired joint moments and the moments generated by spanning muscles ($\boldsymbol{C}$ is often called the residual torque), w was a weighting or penalty factor, and $\frac{f_i}{f_i^{max}}$ can be considered as the muscle activation or effort for simplicity. For all of our simulations, $p = 2$ and $w = 100$ were utilized.

2.4 Experimental Data Collection

For calibration of model parameters and validation of simulations, we collected experimental data of finger motion with the exoskeleton under passive mode (no torque control). MATLAB 2018a was used to control the exoskeleton as well as obtain data from the motor and tri-axial load cell. The Matlab script reads the force between the index finger and the exoskeleton measured by the OptoForce three-degree-of-freedom transducer (OnRobot, Denmark). The force, as well as the rotation angle of the motor, were read in a loop at 155 Hz. The EMG data were collected using the Trigno wireless system with Quattro electrodes (Delsys, USA) at 2000 Hz, and synchronized in time with the motor and tri-axial load cell collection in MATLAB using an external trigger.

EMG electrodes were placed on the FDI, EI, EDC, and FDS muscles. During the experiments, subjects were seated comfortably in a chair with armrests so that he or she could rest his or her hand in between sessions. Subjects were asked to wear the exoskeleton on their right hand, with their index finger and thumb placed comfortably in the respective rings and the gear axes approximately aligned with the gravity direction. To calibrate the motor positions, the subjects were asked to fully extend their fingers to an open position and then to flex their fingers to the position where their thumb met their index finger. These values were recorded and set as limits so that the motor would not exceed these limits as a safety precaution in addition to the mechanical stops. The force sensor was then calibrated by collecting 1000 samples and taking the average of these samples as a bias reading. This value was subtracted from the force readings during the data collection to reduce the inherent bias of the ring attachment. For each session, the subject was asked to complete 15 to 20 extension-flexion cycles in synchronization with a metronome set to four of the following speeds: 40 bpm, 50 bpm, 70 bpm, 100 bpm or 150 bpm. Under the no torque condition, power was provided to the motor but no torque was applied, allowing this to act as a passive exoskeleton where the user had full control over movement. For each session, EMG, force, and motor position were collected. The motor position is determined by the smart servo rotary motor, and through the provided conversion factors the motor joint angle can be calculated in radians. For the tri-axial load cell, we mainly looked at the perpendicular force (Z force), which is aligned with the direction of movement for the index finger.

3 Results and Discussion

3.1 Experimental Data of Passive Exoskeleton

Using MATLAB 2018a, a custom script was used to process and analyze all session data from one subject. Each session consisted of approximately 15 to 20 open and close cycles and in order to further analyze the data, this had to be divided into individual cycles using peak detection of the motor position. Motor position was resampled to 350 Hz and filtered with a 4th order lowpass Butterworth filter with a cutoff of 8 Hz. The start of the cycle was considered to be the instant when the motor position was at its extreme and the index finger was fully extended. The entire cycle included the flexion to extension movements and the velocity was derived from the position values. The force data were also resampled to 350 Hz and filtered with a 4th order lowpass Butterworth filter with a cutoff of 10 Hz. The position, velocity, and force data were then split into individual time synchronized cycles, and the average values were obtained. The EMG data were filtered with a 4th order highpass Butterworth filter with a cutoff of 20 Hz and a 4th order lowpass Butterworth filter with a cutoff of 500 Hz. The data were then resampled to 350 Hz, rectified, and split into individual cycles that align in time with the position, velocity, and force data. Further, for each muscle, the root mean square envelope was calculated using a sliding window of 30 samples, with an overlap of 29 samples. As a maximum voluntary contraction was not obtained, the root mean square (RMS) envelope was normalized to the maximum mean value of the FDI muscle.

During a no torque session at 40 bpm, on average, the subject was able to move through a rotational motor angle of approximately 16° (Fig. 4) within approximately 1.5 s. At the peak of the flexion phase, the user is applying approximately 1.2 N, and at the peak of the extension phase the user is applying 0.5 N (Fig. 5). This greater force applied during flexion can be mainly attributed to a faster velocity during flexion than that during extension. Although the subject was listening to the metronome, movement speed varied between flexion and extension due to inexact muscle control and no enforcement of velocity constraint. The muscle activity of the FDI, EI, FDS, and EDC muscles can be seen in Fig. 6. As expected, we see an onset of activation of the index flexor muscle, FDI, during finger flexion, and relaxation during the extension phase. Similarly, for the EDC muscle we see an onset during the extension phase of movement, and relaxation during flexion.

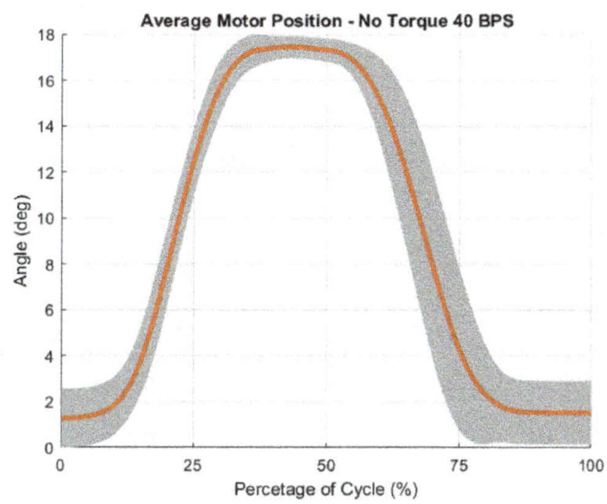

Fig. 4. Average motor position of the exoskeleton during the closing and opening cycles. The shaded area represents the standard deviation.

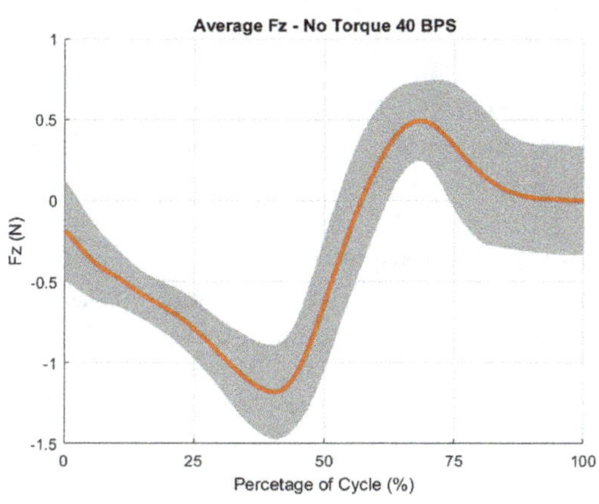

Fig. 5. Average measured force from the tri-axial force sensor during the closing and opening cycles with the shaded grey area representing the standard deviation.

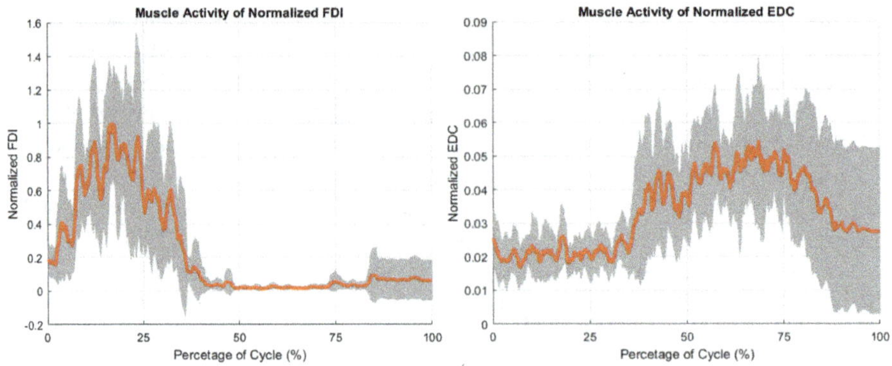

Fig. 6. Average RMS muscle EMG (normalized by the maximum RMS value of FDI) during the closing and opening cycles, with shaded grey area indicating the standard deviation.

3.2 Simulation Results

We conducted parametric simulations of the exoskeleton in action, driven by the finger's closing and opening movement and torque control of the motor, with different combinations of the proportional gain and damping of the PD controller, the end-effector mass, and the end-effector damping coefficient. The virtual end-effector mass (m) values were selected from 0.01 to 10 kg. For simplicity, the unit may be ignored in the context below. For values smaller than 0.01 (e.g. 0.001), the control tends to be highly oscillatory and has deteriorated performance. The damping coefficient (c) is chosen to be 0.01 for all cases presented here. According to our numerical tests, varying the end-effector damping coefficient seems to have non-significant effects on the results. And the PD controller's damping coefficient k_d seems to have a negative effect on the controller's performance and thus only zero damping cases were presented here.

During the simulations, the index finger metacarpophalangeal (MCP) joint tracks a closing and opening motion. The other two joints of the index finger, the proximal interphalangeal (PIP) and distal interphalangeal (DIP) joints, were assumed to be stationary (zero angle). All three joints are ID joints, which means their motions are given and torques are computed from ID. Figure 7 shows the angle and angular velocity of the input motion for the MCP joint, which symmetrically increases from 0 to 25° during closing and decreases from 25 to 0 degree during opening, with a duration of 1.5 s. The angle of the motor, which was computed from the FD, is also shown in Fig. 7. As it can be seen, it is slightly less than 14°, which is close to the experimental measurement. The difference in angles between the finger and motor was caused by their geometry differences and joint locations.

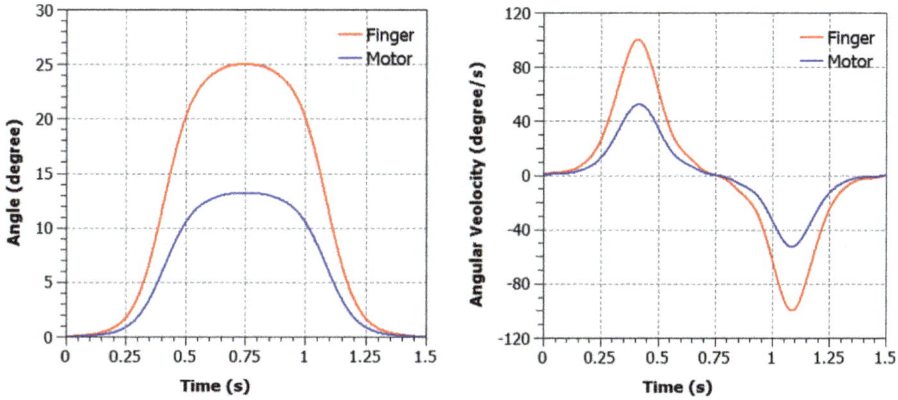

Fig. 7. Angles and angular velocities of the index finger and the motor.

During the tracking of the input motion, the finger joint torques were computed and subsequently muscle force and activation were predicted from the optimization routine. The interaction force between the finger and the ring is the main factor that affects the torques and thus muscle force prediction. In Fig. 8, interaction forces between the index finger and the ring for different control modes and parameters are presented. For the passive mode, the interaction force is slightly less than 1.45 N, within the variance of our experimental measurement. For end-effector mass $m = 0.1$, different PD proportional gains $(k_p = 0.1, 1, 2)$ were tested. Compared to the passive case, increasing k_p tends to increase the control performance as it reduces the maximum interaction force. However, once it exceeds a certain value, e.g. $k_p = 2$, the controller will start to become unstable and oscillate. Similarly, for end-effector mass $m = 0.01$, a reasonable k_p shall be smaller than 1, beyond which unstable oscillation starts. Comparing these controller parameters, the combination of $m = 0.01, k_p = 1$ produced the smallest maximum interaction force at around 0.52 N.

We further increased the end-effector mass values to 1 and 10 and compared the performance of the controllers. In Fig. 9, the interaction forces for the four cases $(m = 0.01, 0.1, 1, 10)$ are presented. Clearly, the mass parameter that provides the best assistance performance is the smallest one $(m = 0.01)$, while increasing the end-effector mass decreases the assistive performance and even produces a resistive force when $m = 10$. With further increase of the mass parameter, the resistance will keep increasing until it reaches the maximum capacity of the motor. This indicates that by simply adjusting the mass parameters, the user can tune the assistance or resistance level as desired.

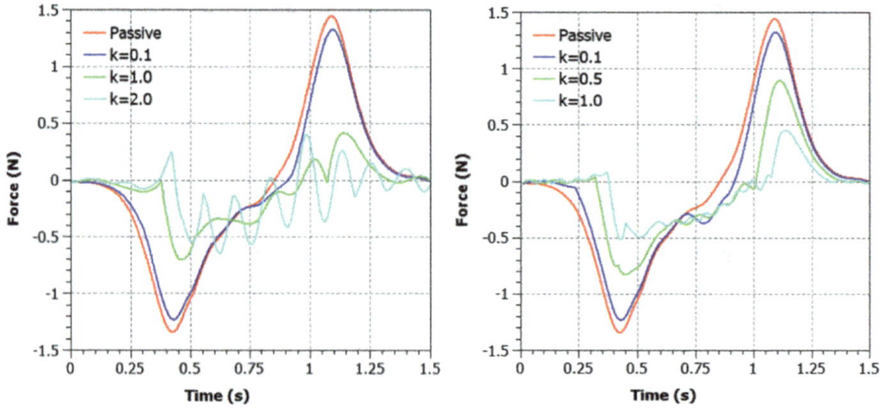

Fig. 8. Predicted finger-ring interaction forces for different PD proportional gain (k_p or k in the figure); Left: $m = 0.1$; Right: $m = 0.01$.

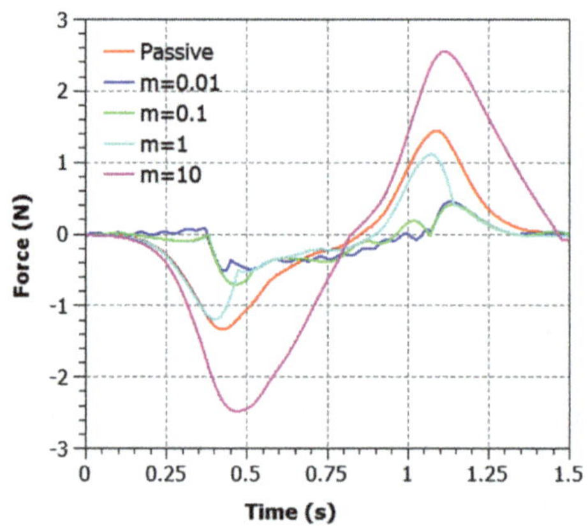

Fig. 9. Predicted finger-ring interaction forces for different mass values. $k_p = 1$ for all masses.

In Fig. 10, we compared the active motor torques for four sets of control parameters, three of which have the same $m = 0.01$ and different gains $\left(k_p = 0.1, 0.5, 1\right)$ and the other with a large $m = 10$ and $k_p = 1$. As k_p increases for $m = 0.01$, the active motor torque increases and provides better assistance to the finger's closing and opening motion, as is evident from the reduction of interaction forces shown in Fig. 8. For the large mass $(m = 10)$, the active torque is similar in magnitude as the best assistive torque $(m = 0.01, k = 1)$ but with a negative sign (reversed direction), indicating resistance instead of assistance. Considering an extreme case of near infinite mass value of the virtual end-effector, its movement is very slow and stays near the initial position and consequently, the IK based admittance controller will try to pull the exoskeleton back from its movement direction to the initial position (i.e. resisting any movement).

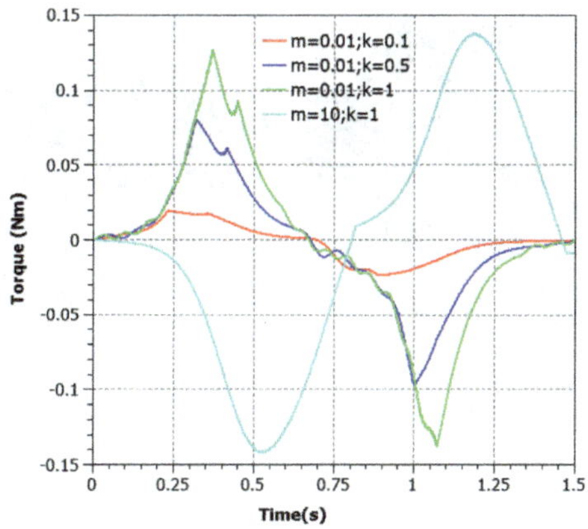

Fig. 10. Predicted motor torques for different control parameters. Assistive torques were provided for $m = 0.01$ and a resistive torque was provided for $m = 10$ as is evident from their sign difference.

We also looked at the average muscle activations of the flexors and extensors for these four controllers, as shown in Fig. 11. The flexors are active mostly during the finger closing phase while the extensors are active during the opening phase. In Fig. 12, snapshots of the hand-exoskeleton in motion are shown, with muscle activation rendered in color. For the three assistive controllers with $m = 0.01$, muscle activations of both muscle groups were reduced due to the decreased finger-ring interaction force. For the resistive controller ($m = 10$), the muscle activations for both groups have increased significantly when compared to the passive case.

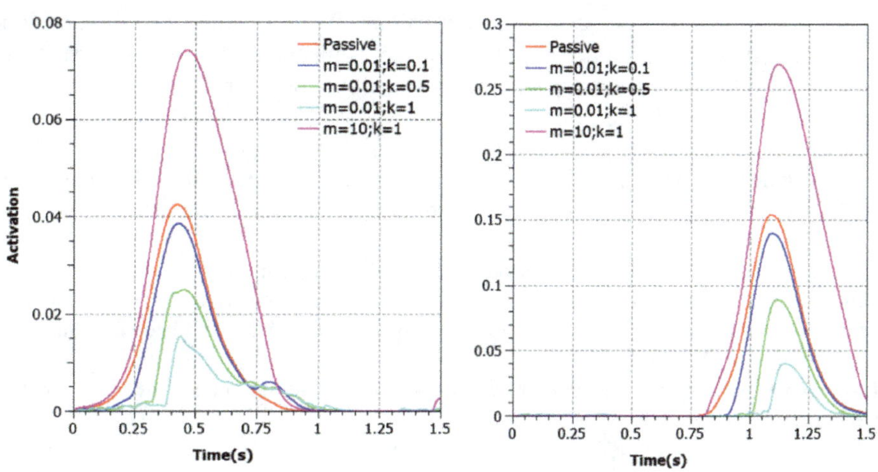

Fig. 11. Predicted average muscle activations for different control modes and parameters. Left: flexors; right: extensors.

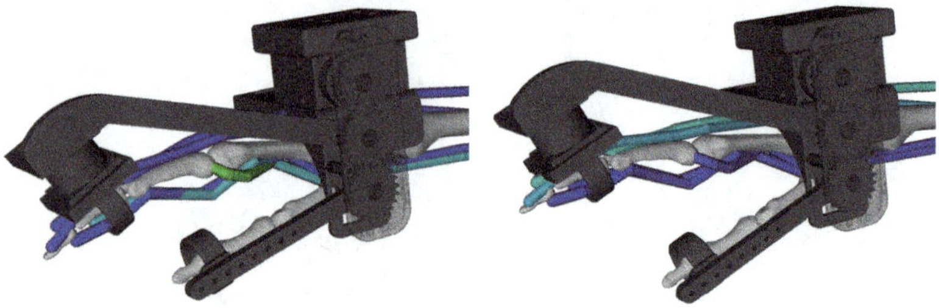

Fig. 12. Snapshots of hand-exoskeleton motion and muscle activation during finger closing (left) and opening (right) movement. Light blue indicates activation.

4 Conclusion

Through the design and modeling of a 1-DOF hand exoskeleton and its interaction with a hand musculoskeletal model, we were able to evaluate the effectiveness of an admittance control method. The results demonstrated that the assistance provided by the motor reduces muscle activation significantly as a result of reduced interaction forces. Under the current admittance control paradigm, increasing the PD controller's proportional gain k_p often results in better assistive performance until it produces overshoot oscillation. Decreasing the virtual mass seems to achieve better assistance performance as well until the occurrence of unstable oscillation. Resistance can also be achieved by simply increasing the value of the virtual end-effector mass beyond a certain value. In conclusion, modeling can help to predict the feasibility of the admittance control framework, guide the tuning of control parameters, and evaluate the exoskeleton's effectiveness for hand rehabilitation. We are currently implementing the present admittance controller in hardware and conducting tests to calibrate model parameters and validate the simulation predictions. We hope that, with the developed models and additional parametric simulations, it will enable us to fine tune control parameters, explore the design space, and devise novel or optimal control schemes for implementation in hardware.

Acknowledgment. This work was supported by the NIDILRR funded Rehabilitation Engineering Research Center grant# 90RE5021 and by the NIH grant R01HD58301.

Ethical Approval:. All procedures performed in studies involving human participants were in accordance with the ethical standards of the Institutional Review Board of the New Jersey Institute of Technology and with the 1964 Helsinki declaration and its later amednments or comparable ethical standards.

Informed Consent: Informed consent was obtained from all individual participants included in the study.

References

1. Roger, V.L., et al.: Heart disease and stroke statistics—2011 update: a report from the American Heart Association. Circulation **123**(4), e18–e209 (2011)
2. Heo, P., Gu, G.M., Lee, S.-J., Rhee, K., Kim, J.: Current hand exoskeleton technologies for rehabilitation and assistive engineering. Int. J. Precis. Eng. Manuf. **13**(5), 807–824 (2012)
3. Adamovich, S.V., Fluet, G.G., Mathai, A., Qiu, Q., Lewis, J., Merians, A.S.: Design of a complex virtual reality simulation to train finger motion for persons with hemiparesis: a proof of concept study. J. Neuroeng. Rehabil. **6**(1), 28 (2009)
4. Abbruzzese, K., Foulds, R.: Assessment of a 7-DOF hand exoskeleton for neurorehabilitation. In: 2nd International Symposium on Wearable Robotics, Segovia, Spain. Springer, Heidelberg (2016)
5. Thielbar, K.O., et al.: Training finger individuation with a mechatronic-virtual reality system leads to improved fine motor control post-stroke. J. Neuroeng. Rehabil. **11**(1), 171 (2014)
6. Kinetec Maestra (2016). Accessed 12 July 2019. https://kinetecuk.com/brands/kinetec/cpm/continuous-passive-motion-hand-and-wrist/kinetec-maestra/kinetec-maestra
7. Vector 1 Hand CPM (2016). Accessed 12 July 2019. https://lantzmedical.com/vector1-hand-cpm/
8. CyberGrasp (2017). Accessed 12 July 2019. http://www.cyberglovesystems.com/cybergrasp
9. Fischer, H.C., et al.: Use of a portable assistive glove to facilitate rehabilitation in stroke survivors with severe hand impairment. IEEE Trans. Neural Syst. Rehabil. Eng. **24**(3), 344–351 (2016)
10. Triandafilou, K.M., Kamper, D.G.: Carryover effects of cyclical stretching of the digits on hand function in stroke survivors. Arch. Phys. Med. Rehabil. **95**(8), 1571–1576 (2014)
11. Triandafilou, K.M., Ochoa, J., Kang, X., Fischer, H.C., Stoykov, M.E., Kamper, D.G.: Transient impact of prolonged versus repetitive stretch on hand motor control in chronic stroke. Top Stroke Rehabil. **18**(4), 316–324 (2011)
12. Zhao, H., Jalving, J., Huang, R., Knepper, R., Ruina, A., Shepherd, R.: A helping hand: soft orthosis with integrated optical strain sensors and EMG control. IEEE Rob. Autom. Mag. **23**(3), 55–64 (2016)
13. Lee, J.H., Asakawa, D.S., Dennerlein, J.T., Jindrich, D.L.: Finger muscle attachments for an opensim upper-extremity model. PLoS ONE **10**(4), e0121712 (2015)
14. Zhou, X., Whitley, P., Przekwas, A.: A musculoskeletal fatigue model for prediction of aviator neck maneuvering loadings. Int. J. Hum. Factors Model. Simul. **4**(3–4), 191–219 (2014)

Recursive Filtering of Kinetic and Kinematic Data for Center of Mass and Angular Momentum Derivative Estimation

François Bailly[1(✉)], Justin Carpentier[2,3], Bruno Watier[4,5], and Philippe Souères[4]

[1] Laboratoire de Simulation et Modélisation du Mouvement, Faculté de Médecine, Université de Montréal, Laval, QC, Canada
francois.bailly@umontreal.ca
[2] Département d'informatique de l'ENS, École Normale Supérieure, CNRS, PSL Research University, 75005 Paris, France
[3] Inria, Rocquencourt, France
[4] LAAS-CNRS, 7 Avenue du Colonel Roche, 31400 Toulouse, France
[5] Université de Toulouse III, UPS, LAAS, 31400 Toulouse, France

Abstract. Estimating the center of mass position and the angular momentum derivative of the human body is an important topic in biomechanics, since both quantities are essential to the dynamic description of the motion. In this work, we introduce a novel recursive algorithm to accurately estimate them, by fusing kinetic and kinematic measurements, based on a spectral description of the noise carried by each signal. This method exploits the mathematical relationships that links the center of mass position and the angular momentum derivative to recursively improve their estimation. The effectiveness of the approach is demonstrated on a simulated humanoid avatar, where access to ground truth data is granted. The results show that our method reduces the estimation error on the center of mass position with regard to kinematic estimation alone, in addition to providing a good estimate of the angular momentum variation. The proposed framework is finally applied to a recorded human walking motion in order to illustrate its applicability to real motion analysis data.

1 Introduction

Estimating the center of mass (CoM) position and of the angular momentum derivative is an important challenge which has various applications in human motion analysis for biomechanics studies or medical disorders diagnosis in humans. The main difficulty of this estimation problem lies in the fact that the two quantities are not directly measurable, and that they depend on almost fixed parameters (e.g. mass distribution of the body, dimensions of the limbs, etc.) as well as varying quantities (e.g. joints configurations, external forces, etc.).

G. A. Ateshian et al. (Eds.): CMBBE 2019, LNCVB 36, pp. 398–410, 2020.
https://doi.org/10.1007/978-3-030-43195-2_33

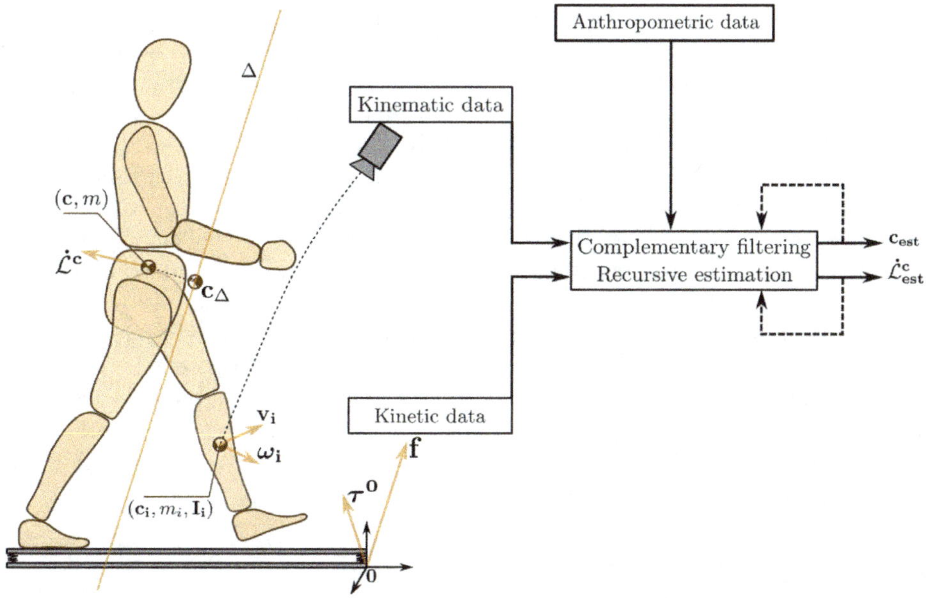

Fig. 1. Illustration of the measurement apparatus. The several physical quantities involved in the estimation framework are displayed, as well as a simplified sketch of the estimation framework. *[©2020 IEEE. Adapted, with permission, from [1]].*

1.1 State of the Art

In order to estimate the CoM position, several methods exploiting different sources of information are commonly applied [17]. They can be grouped into two broad categories, namely *kinematic methods* which only exploit kinematic measurements and *kinetic methods* which estimate the CoM position from kinetic signals. In the following, estimation methods from both biomechanics and humanoid robotics are reviewed as they provide a thorough overview of the existing techniques and because the goals and means of the two disciplines are very similar on this topic.

For kinematic methods, the required segments' masses and configurations can be obtained from anthropometric tables and motion capture (MoCap) systems [10, 13]. Then, computing the CoM position boils down to a weighted average calculation accounting for the mass distribution and the position of each segment's CoM. It is worth noticing that kinematic methods are subject to noises of various types, mostly due to uncertainties about the mass distribution of the body (statistical anthropometric tables in humans) and soft tissue artifacts [4, 22].

As for kinetic methods, the estimation of the CoM position is achieved by measuring either the Center of Pressure (CoP) or the contact forces from force sensors signals. For instance, if the body in contact is modeled as an inverted pendulum, the CoM position can be estimated from the mechanical relationship between the CoM and the CoP [16, 21]. This approach then assumes that the CoM trajectory can be assimilated to the oscillations of an inverted pendulum, and the acceptable range of motion of such a method is mostly limited to balance assessments. In [5], it was proposed to filter the CoP trajectory with a low pass filter to estimate the horizontal components of the CoM. Once again, the accuracy of this approach is limited to slow movements. Finally, by assuming a constant-height CoM, performing a double integration of external forces together with an estimation of the angular momentum derivative expressed at the CoM, [23] and [18] have provided another way to estimate the horizontal components of the CoM.

Concerning the estimation of the angular momentum derivative, to the best of our knowledge, only kinematic estimation has been performed [3, 9], affected by the same limitations regarding mass distribution errors, with additional uncertainties coming from the numerical derivation of segments poses. Kinetic data can also be used as suggested in [14], by exploiting the relation between the angular momentum and the external torques applied to the system. In most cases, however, the derivative of angular momentum is neglected for the sake of simplicity, as in a previous work [6], where the authors relied on a complementary filter in the spectral domain, to merge kinematic and kinetic data in order to estimate the CoM position, by assuming that angular momentum variations were negligible.

1.2 Contribution

This paper is an extension of a recently accepted study on humanoid robots [1], in which we proposed a generic framework to simultaneously and accurately estimate the CoM position together with the angular momentum derivative by exploiting the physical properties that link both quantities (see Fig. 1 for a simplified sketch of the approach). In this method, a recursive strategy was exploited in order to improve the estimate of the CoM trajectory from the angular momentum trajectory and vice-versa. In the present work, the theoretical and simulation parts that are presented are essentially taken from [1]. The consistency of the approach is validated on a simulated humanoid avatar, by assessing the accuracy of our estimated quantities against ground truth data. Finally, a novel case of application of the proposed framework to human locomotion data is presented in order to illustrate the efficiency of our approach on experimental data.

2 Estimation Algorithm

This section recalls how the position of the CoM (denoted by $\mathbf{c}$) and the angular momentum variation expressed at the CoM (denoted by $\dot{\mathcal{L}}^{\mathbf{c}}$) are linked together through the wrench of external forces. Then, we identify the different sources of information which contribute to the estimation of $\dot{\mathcal{L}}^{\mathbf{c}}$ and $\mathbf{c}$, and we explain how to efficiently fuse them by exploiting their level of accuracy in the spectral domain.

2.1 Estimated Variables Coupling

Let m be the mass of the body, and $\mathbf{c}$, $\dot{\mathbf{c}}$ and $\ddot{\mathbf{c}}$ its CoM position, velocity and acceleration. Let $\mathbf{f}$ and $\tau^{\mathbf{0}}$ be the resulting forces and moments exerted on the body by the

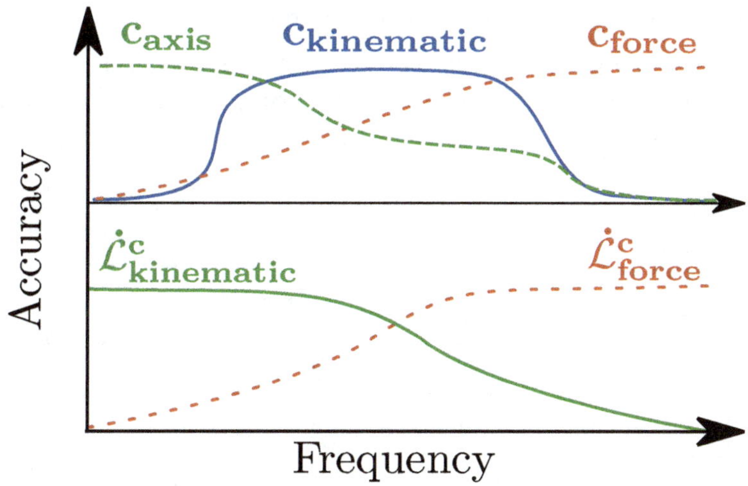

Fig. 2. Schematic representation of the spectral accuracy of the different input signals to the estimation of $\mathbf{c}$ and $\dot{\mathcal{L}}^{\mathbf{c}}$. [*©2020 IEEE. Reprinted, with permission, from [1]*].

environment, expressed at the origin **0**. They constitute the socalled external contact wrench that can be expressed at any other point of the space using the Varignon formula. It follows that the external contact wrench at the CoM is given by:

$$\tau^{\mathbf{c}} = \tau^{0} + \mathbf{f} \times \mathbf{c} \tag{1}$$

According to Euler's equations of motion, the angular momentum variation at the CoM is equal to the moment of the external action expressed at the CoM:

$$\dot{\mathcal{L}}^{\mathbf{c}} = \tau^{0} + \mathbf{f} \times \mathbf{c} \tag{2}$$

Equation (2) provides the coupling between the two quantities we want to estimate, namely the CoM position $\mathbf{c}$ and the angular momentum derivative $\dot{\mathcal{L}}^{\mathbf{c}}$.

2.2 Measurements

Using a MoCap system associated with 6D force measurement units, one can measure kinematic and kinetic data related to the physical quantity we aim to estimate:

- the resulting wrench of contact ($\mathbf{f}$ and τ^{0}) expressed at the origin, via force sensors (6-axis force plates);
- the position of the CoM, denoted by $\mathbf{c}_{\textbf{kinematic}}$, deduced from the MoCap analysis and anthropometric tables;
- the angular momentum at the CoM, denoted by $\mathcal{L}^{\mathbf{c}}_{\textbf{kinematic}}$, which can be obtained from the angular velocities and the inertias of the segments

2.3 Multi-source Estimation of the CoM Position

This section explains how to compute and process the three different input signals that carry information about the position of the CoM. The spectral accuracy of each signal is depicted in Fig. 2.

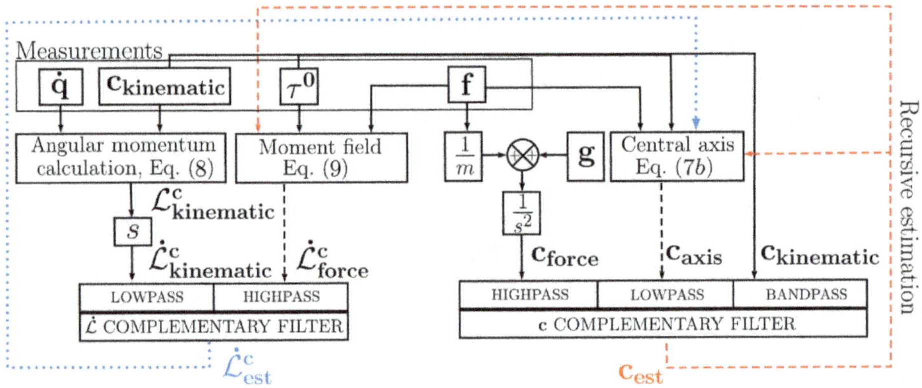

Fig. 3. Flow chart for the recursive complementary estimation framework. s is the Laplace variable. Dotted lines represent the update step of the algorithm. Complementary filtered components are summed up in order to output $\dot{\mathcal{L}}^c_{\text{est}}$ and $\mathbf{c}_{\text{est}}$, estimates of $\dot{\mathcal{L}}^c$ and $\mathbf{c}$ respectively. *[©2020 IEEE. Adapted, with permission, from [1]].*

From Kinetic Data
The second law of Newton states that:

$$\ddot{\mathbf{c}} = \frac{\mathbf{f}}{m} + \mathbf{g}. \tag{3}$$

By double integration of the right-hand side of this equation one can estimate $\mathbf{c}_{\text{force}}$ thanks to the information provided by the force sensors. However without the knowledge of integration constants, there is a quadratic drift visible in low and medium frequencies, which disturbs the estimation.

From Kinematic Data
Kinematic computations using reflective markers 3-D positions combined with anthropometric tables are usually used to estimate the CoM position. This source of data ($\mathbf{c}_{\text{kinematic}}$) suffers mainly from low-frequency biases due to modeling errors of the mass distribution. It can also be altered by the high frequency sensor noise due to the MoCap technology [24]. Then, the error between this signal and the real position of the CoM can be considered to lie in low and high frequency domains.

From the Central Axis of Contact Wrench

Considering the contact moment field, there exists one unique axis called the central axis of contact wrench and denoted by Δ [2, 11], where the moment of contact forces is collinear to $\mathbf{f}$. The projection $\mathbf{c}_\Delta$ of the CoM $\mathbf{c}$ onto Δ is given by:

$$\mathbf{c}_\Delta = \frac{\mathbf{f} \times \tau^0 + (\mathbf{c} \cdot \mathbf{f})\mathbf{f}}{\|\mathbf{f}\|^2}. \tag{4}$$

This quantity can be computed thanks to the force and moment signals provided by the force sensors and the current estimate of the CoM, denoted by $\mathbf{c}_{\mathbf{est}}$ as follows:

$$\mathbf{c}_\Delta = \frac{\mathbf{f} \times \tau^0}{\|\mathbf{f}\|^2} + (\mathbf{c}_{\mathbf{est}} \cdot \mathbf{n})\mathbf{n}, \tag{5}$$

where $\mathbf{n}$ is the normalized direction of $\mathbf{f}$. Knowing an estimate of $\dot{\mathcal{L}}^{\mathbf{c}}$, denoted by $\dot{\mathcal{L}}^{\mathbf{c}}_{\mathbf{est}}$, and $\mathbf{f}$, the difference between $\mathbf{c}$ and $\mathbf{c}_\Delta$ can be computed as:

$$\mathbf{c} - \mathbf{c}_\Delta = \frac{\dot{\mathcal{L}}^{\mathbf{c}}_{\mathbf{est}} \times \mathbf{f}}{\|\mathbf{f}\|^2}. \tag{6}$$

This leads to a third estimate of the CoM position that will be denoted by $\mathbf{c}_{\mathbf{axis}}$ and expressed as:

$$\mathbf{c}_{\text{axis}} = \mathbf{c}_\Delta + \frac{\dot{\mathcal{L}}^{\mathbf{c}}_{\mathbf{est}} \times \mathbf{f}}{\|\mathbf{f}\|^2}, \tag{7a}$$

$$\mathbf{c}_{\text{axis}} = \frac{\mathbf{f} \times \tau^0}{\|\mathbf{f}\|^2} + (\mathbf{c}_{\mathbf{est}} \cdot \mathbf{n})\mathbf{n} + \frac{\dot{\mathcal{L}}^{\mathbf{c}}_{\mathbf{est}} \times \mathbf{f}}{\|\mathbf{f}\|^2}. \tag{7b}$$

Complementary Filter of the CoM

Similarly to [6], we use a complementary filtering approach [15] to fuse these three signals ($\mathbf{c}_{\mathbf{force}}$, $\mathbf{c}_{\mathbf{kinematic}}$ and $\mathbf{c}_{\mathbf{axis}}$) in order to retrieve an accurate estimate of the CoM position. The main idea of this filter is to exploit the accuracy of these signals in the spectral domain. Indeed, each of these signals conveys a certain level noise, but these noises do not have the same spectral distribution. This property is exploited to design the complementary filter on the CoM estimates.

The theory of complementary filtering provides a way to make these three different sources contribute to the estimation of the CoM position by taking profit from the information about the spectral distribution of noise they suffer from. In the Laplace domain, 3 complementary filters are designed (high-pass, band-pass and low-pass, see Fig. 3), so that their transfer functions sum to 1. The sum of these filters maintains the energy of the signal with zero phase modification [15].

2.4 Multi-source Estimation of the Derivative of Angular Momentum

In this section, we present how to compute and process the two input signals for estimating $\dot{\mathcal{L}}^{\mathbf{c}}$. More precisely, we show how an estimate of the angular momentum quantity can be built from kinematic and kinetic signals (see Fig. 3).

From Kinematic Data
Knowing the angular and linear velocities of the segments and their inertia matrices, one can compute the angular momentum of the system by exploiting the following relation:

$$\mathcal{L}^{\mathbf{c}}_{\mathbf{kinematic}} = \sum_i \mathbf{I_i}\omega_i + \overrightarrow{\mathbf{cc_i}} \times m_i \mathbf{v_i} \tag{8}$$

where $\mathbf{I_i}$, ω_i, $\mathbf{c_i}$, m_i and $\mathbf{v_i}$ are the moment of inertia, angular velocity, CoM position, mass and relative velocity of the ith segment respectively. Taking the derivative of this quantity provides an estimation of $\dot{\mathcal{L}}^{\mathbf{c}}$ from this first source of information at the price of amplifying high-frequency noise. The modeling error of mass and inertias distribution produces a bias on the estimation of $\mathcal{L}^{\mathbf{c}}$. Theoretically, this bias depends on the posture of the body but in practice, as this posture remains fairly constant during locomotion, the bias can be considered constant. Hence, by taking the derivative of $\mathcal{L}^{\mathbf{c}}_{\mathbf{kinematic}}$, this measurement can be trusted in the low frequency domain.

From the Moment Field Properties
The second source of information for the estimation of $\dot{\mathcal{L}}^{\mathbf{c}}$, lies in Eq. (2):

$$\dot{\mathcal{L}}^{\mathbf{c}}_{\mathbf{force}} = \tau^0 + \mathbf{f} \times \mathbf{c} \tag{9}$$

and is provided by force sensors measurements with an estimation of $\mathbf{c}$.

Complementary Filter for the Derivative of Angular Momentum
In the same way as before, complementary filtering is used to add both contributions to the estimation of $\dot{\mathcal{L}}^{\mathbf{c}}$. In the Laplace domain, two complementary filters are designed (high-pass and low-pass, see Fig. 3) such that their transfer functions sum to 1.

2.5 Recursive Estimation of the CoM Position and of Derivative of Angular Momentum

Our recursive approach is motivated by the fact that the estimation of $\dot{\mathcal{L}}^{\mathbf{c}}_{\mathbf{force}}$ depends on an estimation of $\mathbf{c}$ (see Eq. (9)) and that the estimation of $\mathbf{c_{axis}}$ depends on an estimation of $\dot{\mathcal{L}}^{\mathbf{c}}$ (see Eq. (7b)). Hence, the estimation of $\mathbf{c}$ can be refined by refining the estimation of $\dot{\mathcal{L}}^{\mathbf{c}}$, and vice versa. In Fig. 3, a detailed diagram of the proposed estimation framework is displayed.

3 Experimental Validation of the Estimation Framework

In this section, the efficiency of the estimation framework is demonstrated on simulated data. Then, experimental data are processed to illustrate the performance of the proposed method on a real application scenario. In both cases, the estimation is performed recursively as sketched in Fig. 3.

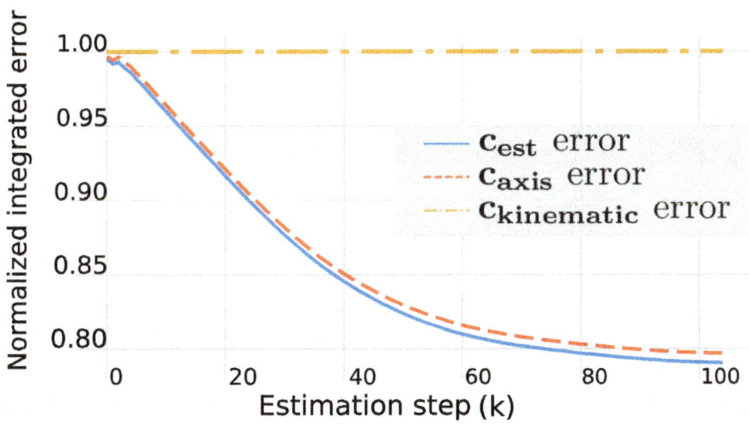

Fig. 4. *Simulation*. Evolution of the error 2-norm integrated over the whole trajectory for the different estimates of **c**. The integrated errors are normalized with regard to the error of c kinematic, which is the initialization of the algorithm and is not updated throughout the iterations (and therefore constantly 1). *[©2020 IEEE. Reprinted, with permission, from [1]].*

3.1 Generation of Ground-Truth and Noisy Measures in Simulation

In this section, a dynamically consistent walking motion was simulated for an avatar. Then, by adding noise to the several quantities involved in the estimation framework we were able to compare our estimation to ground truth data.

The simulated avatar consisted of a simplified humanoid model endowed with 36 degrees of freedom. The simulated motion was a walking task on horizontal ground at 0.5 m/s using the dynamically consistent approach introduced in [7]. Ground truth kinetic data were computed at 200 Hz using the modeling software Pinocchio which includes rigid body dynamics algorithms for poly-articulated systems [8]. Band-limited centered white noise was added to each signal, in accordance with our analysis of the noise spectral density. The low, medium and high frequency domains were delimited from 0 to 4 Hz, 16 to 24 Hz and 92 to 100 Hz respectively, in order to be able to discriminate the effect of each noise and to limit the slopes and the order of the filters. The final user would only have to adjust the cutting frequencies inside the complementary filters if needed, which depend on sampling frequency as well as on the nature of the recorded motion. The standard deviations of the noise were 1 N and 1 Nm in the low and medium frequency domains for the force and moment measurements respectively, 0.1 m and 0.1 Nm in the high frequency domain for $\mathbf{c_{kinematic}}$ and

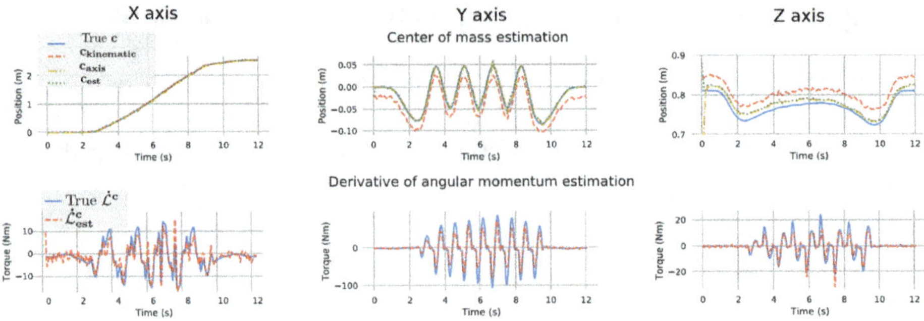

Fig. 5. Estimation of **c** and $\dot{\mathcal{L}}^{\mathbf{c}}$ for a horizontal walking motion of a simulated humanoid avatar. Top row: true **c**, $\mathbf{c}_{\mathbf{kinematic}}$ and $\mathbf{c}_{\mathbf{est}}$. Bottom row: true $\dot{\mathcal{L}}^{\mathbf{c}}$ and $\dot{\mathcal{L}}^{\mathbf{c}}_{\mathbf{est}}$. *[©2020 IEEE. Reprinted, with permission, from [1]].*

$\dot{\mathcal{L}}^{\mathbf{c}}_{\mathbf{kinematic}}$ respectively, in order to replicate the errors due to the MoCap technology. Inertial data (mass, CoM position and inertia matrix) of each segment was randomly biased in order to simulate the error due to mass distribution uncertainties (multiplicative random value following a 1-centered normal distribution with standard deviation of 1). The cut-off frequencies were 1 Hz and 25 Hz for the low-pass and high-pass second order filters of $\mathbf{c}_{\mathbf{est}}$ respectively. The band-pass filter was deduced by complementarity. The cut-off frequency for the low pass filter of $\dot{\mathcal{L}}^{\mathbf{c}}_{\mathbf{est}}$ was 2 Hz, the high-pass filter was deduced by complementarity too.

The estimation procedure was initialized with the value of $\mathbf{c}_{\mathbf{kinematic}}$ before filtering, because it is the best guess one can make before adding further data. The recursion was stopped when the last estimates of $\dot{\mathcal{L}}^{\mathbf{c}}$ and **c** were close (10^{-3}) to the current ones. Figure 4 depicts the error of $\mathbf{c}_{\mathbf{kinematic}}$, $\mathbf{c}_{\mathbf{axis}}$ and $\mathbf{c}_{\mathbf{est}}$ integrated over the whole trajectory and normalized with regard to the error of $\mathbf{c}_{\mathbf{kinematic}}$. This result shows how the complementary filter improves the initial guess $\mathbf{c}_{\mathbf{kinematic}}$ by reducing the norm of the error with regard to the ground-truth values (which are exactly known in simulation).

In more details, Fig. 5 shows the 3 components of **c** and of $\dot{\mathcal{L}}^{\mathbf{c}}$ for the different estimates, in addition to the real values of the simulation, without noise. On the top row of Fig. 5, $\mathbf{c}_{\mathbf{kinematic}}$ is displayed, to exhibit the improvement provided by our method. A low frequency bias between $\mathbf{c}_{\mathbf{kinematic}}$ and the ground truth value is noticeable, in particular on the Y and Z components because of the scale of the plots. This bias comes from the error on the inertial parameters that we purposely introduced in the simulated data. Figure 5 also shows that the final estimation $\mathbf{c}_{\mathbf{est}}$ is better than the initial guess, as corroborated by the error analysis in Fig. 4. For instance, on the Z axis, the average absolute error of the kinematic estimate is of 3.6 cm, whereas it is reduced to 1.1 cm for the final estimate. On the Y axis the average absolute error is reduced from 2.1 cm to 2.9 mm by the presented framework. On the X axis, which is the main direction of locomotion, no significant improvement is measured: the average absolute error remains close to 6 mm. The contribution of $\mathbf{c}_{\mathbf{force}}$ in the high frequency domain was quasi-zero, because the simulated motion was smooth, thus the trajectory was limited to low and medium frequency components. On the bottom row of Fig. 5, the estimation of $\dot{\mathcal{L}}^{\mathbf{c}}$ is displayed, and the result follows the dynamics of the ground truth trajectory. Some overshooting can be noted, however, on the X and Z axes notably.

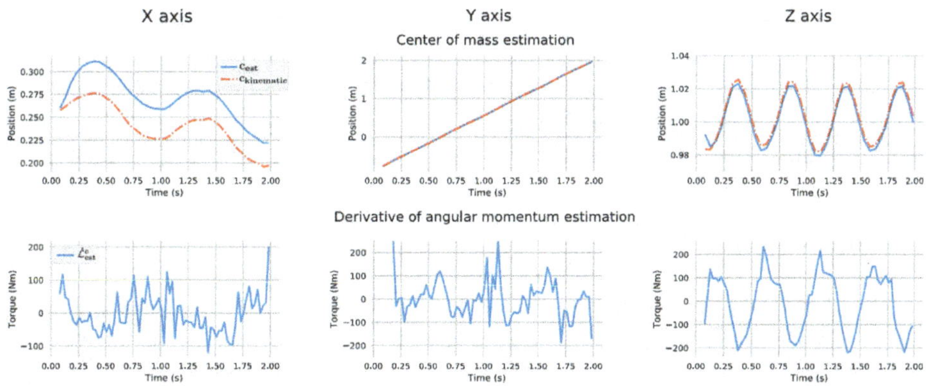

Fig. 6. Estimation of $\mathbf{c}$ and $\dot{\mathcal{L}}^{\mathbf{c}}$ on human recorded data. Top row: $\mathbf{c}_{\mathbf{kinematic}}$ and $\mathbf{c}_{\mathbf{est}}$. Bottom row: $\dot{\mathcal{L}}^{\mathbf{c}}_{\mathbf{est}}$.

3.2 Application to Human Data

In this part, we applied the estimation framework to real kinematic and kinetic data obtained during walking experiments, in order to retrieve the estimates of $\mathbf{c}$ and $\dot{\mathcal{L}}^{\mathbf{c}}$.

One male participant (age: 23 y, mass: 64 kg, height: 1.81 m) volunteered for this study and signed an informed written consent. Experiments were conducted in accordance with the standards of the Declaration of Helsinki (rev. 2013) and approved by a local ethic evaluation committee. The participant was asked to freely walk across the two force plates at a comfortable walking speed. Whole body 3D kinematic data were collected by 13 infrared cameras sampled at 200 Hz, that recorded the motion of 43 reflective markers placed on the subject, following the [25, 26] and [12] recommendations. A whole-body 3D model including 42 degrees of freedom for 15 segments was used to reconstruct the movements of the walker. Two force plates embedded into the floor were used to record ground reaction forces at 1 kHz. Inertial parameters were computed thanks to anthropomorphic tables [12], resulting in the computation of $\mathbf{c}_{\mathbf{kinematic}}$ and $\mathcal{L}^{\mathbf{c}}_{\mathbf{kinematic}}$, see Fig. 3.

Figure 6 displays a variety of curves relative to the estimation of $\mathbf{c}$ and $\dot{\mathcal{L}}^{\mathbf{c}}$ for this experimental walking on horizontal ground. First, one can notice that the CoM position profiles were slightly different from the simulated ones, presented in Sect. 3.1, especially for the vertical component. This is due to the reduced number of degrees of freedom of the simulation avatar and to the motion in simulation which was designed to keep the humanoid close to a half sitting position while walking, in order to avoid joint limits that lead to kinematic singularities. In the case of human data experiment, there was no ground truth values available to assess the quality of the estimation. On the top row of Fig. 6, the average absolute differences between the final estimates and $\mathbf{c}_{\mathbf{kinematic}}$ are of 3.1 cm, 3.5 mm and 5 mm on the X, Y and Z axes, respectively. These error values are consistent with the improvement towards the ground-truth values which were computed in simulation. This result suggests two remarks: the levels of noise introduced in simulation were realistic and the differences observed on real data after recursion correspond to an improvement in the estimate. Indeed, the low frequency contribution of $\mathbf{c}_{\mathbf{axis}}$ seems to compensate for the low frequency drift of $\mathbf{c}_{\mathbf{kinematic}}$ band-pass filtered and, as noticed in simulation, it also provides a low-frequency correction of the initial guess $\mathbf{c}_{\mathbf{kinematic}}$. On the bottom row of Fig. 6, the estimation of $\dot{\mathcal{L}}^{\mathbf{c}}$ is rather noisy, but the values of the estimates are consistent with the dynamics of the motion.

4 Conclusion

In this paper, we have recalled a previously introduced recursive filter to simultaneously estimate the CoM position and the derivative of angular momentum expressed at the CoM from kinematic and kinetic data. This approach exploits the mathematical relationship that exists between the CoM position and the angular momentum derivative at the CoM. Unlike previous works, it does not rely on any simplifying assumption, such as the negligible angular momentum variations hypothesis [6], or the coplanarity of contact surfaces [18, 21].

On a simulated walking humanoid, the results showed that the proposed framework enabled us to reduce the norm of the error between the estimation of the CoM position and its real value, when compared to kinematic estimation alone. This improvement is a direct consequence of the exploitation of the spectral properties of each measurement and of the proposed elaborate fusion algorithm. More precisely, this improvement was mainly due to a low frequency bias correction brought by a better estimation of the angular momentum variation, and exploited in the geometrical link between the CoM position and the central axis of the contact wrench. Results in simulation showed that the estimation of the derivative of the angular momentum was not perfect, and this is one explanation to the small bias that remained on the estimation of the CoM position after convergence of the algorithm, in addition to its non-observability on the Z axis. The overall estimation could be enhanced by improving the estimation of the angular momentum variation which is left as a future work.

The application of the method to human walking data also provided a difference between the raw estimation coming from the MoCap alone and the complementary estimation at a low frequency level. Although one cannot compare the quality of these two estimates on real data, results in simulation suggest that the low frequency correction was an improvement in favor of the complementary estimation, which could provide an efficient means for improving CoM and angular momentum estimation, usually biased by anthropometric tables errors. Thus, the presented method is of great interest in human and animal motion analysis [20] and sports biomechanics [19], as it permits to improve traditional CoM and derivative of angular momentum estimates, as illustrated by the experimental work presented in this study.

References

1. Bailly, F., Carpentier, J., Benallegue, M., Watier, B., Souères, P.: Estimating the center of mass and the angular momentum derivative for legged locomotion—a recursive approach. IEEE Robot. Autom. Lett. **4**(4), 4155–4162 (2019)
2. Bailly, F., Carpentier, J., Pinet, B., Souères, P., Watier, B.: A mechanical descriptor of human locomotion and its application to multi-contact walking in humanoids. In: 2018 7th IEEE International Conference on Biomedical Robotics and Biomechatronics (Biorob), pp. 350–356. IEEE (2018)
3. Bruijn, S.M., Meijer, O.G., Van Dieen, J.H., Kingma, I., Lamoth, C.J.: Coordination of leg swing, thorax rotations, and pelvis rotations during gait: the organisation of total body angular momentum. Gait Posture **27**(3), 455–462 (2008)
4. Cappozzo, A., Catani, F., Leardini, A., Benedetti, M., Della Croce, U.: Positionand orientation in space of bones during movement: experimental artefacts. Clin. Biomech. **11** (2), 90–100 (1996)
5. Caron, O., Faure, B., Brenière, Y.: Estimating the centre of gravity of the body on the basis of the centre of pressure in standing posture. J. Biomech. **30**(11), 1169–1171 (1997)
6. Carpentier, J., Benallegue, M., Mansard, N., Laumond, J.P.: Center-of-massestimation for a polyarticulated system in contact—a spectral approach. IEEE Trans. Rob. **32**(4), 810–822 (2016)
7. Carpentier, J., Tonneau, S., Naveau, M., Stasse, O., Mansard, N.: A versatile and efficient pattern generator for generalized legged locomotion. In: International Conference on Robotics and Automation (ICRA), pp. 3555–3561. IEEE (2016)
8. Carpentier, J., Valenza, F., Mansard, N., et al.: Pinocchio: fast forward and inverse dynamics for poly-articulated systems (2015–2017). https://stack-of-tasks.github.io/pinocchio
9. Dapena, J.: A method to determine the angular momentum of a human body about three orthogonal axes passing through its center of gravity. J. Biomech. **11**(5), 251–256 (1978)
10. De Leva, P.: Adjustments to Zatsiorsky-Seluyanov's segment inertia parameters. J. Biomech. **29**(9), 1223–1230 (1996)
11. Dimentberg, F.M.: The screw calculus and its applications in mechanics. Technical report, DTIC Document (1968)
12. Dumas, R., Cheze, L., Verriest, J.P.: Adjustments to Mcconville et al. and Younget et al. body segment inertial parameters. J. Biomech. **40**(3), 543–553 (2007)
13. Farley, C.T., Ferris, D.P.: Biomechanics of walking and running: center of mass movements to muscle action. Exerc. Sport Sci. Rev. **26**(1), 253–286 (1998)
14. Herr, H., Popovic, M.: Angular momentum in human walking. J. Exp. Biol. **211**(4), 467–481 (2008)
15. Higgins, W.T.: A comparison of complementary and Kalman filtering. IEEE Trans. Aerosp. Electron. Syst. **3**, 321–325 (1975)
16. Kajita, S., Kanehiro, F., Kaneko, K., Yokoi, K., Hirukawa, H.: The 3D linear inverted pendulum mode: a simple modeling for a biped walking pattern generation. In: Proceedings 2001 IEEE/RSJ International Conference on Intelligent Robots and Systems, vol. 1, pp. 239–246. IEEE (2001)
17. Lafond, D., Duarte, M., Prince, F.: Comparison of three methods to estimate the center of mass during balance assessment. J. Biomech. **37**(9), 1421–1426 (2004)
18. Levin, O., Mizrahi, J.: An iterative model for estimation of the trajectory of center of gravity from bilateral reactive force measurements in standing sway. Gait Posture **4**(2), 89–99 (1996)

19. Maldonado, G., Bailly, F., Souères, P., Watier, B.: Angular momentum regulation strategies for highly dynamic landing in Parkour. Comput. Methods Biomech. Biomed. Eng. **20**(Suppl. 1), 123–124 (2017)
20. Minetti, A.E., Cisotti, C., Mian, O.S.: The mathematical description of the body centre of mass 3D path in human and animal locomotion. J. Biomech. **44**(8), 1471–1477 (2011)
21. Morasso, P.G., Spada, G., Capra, R.: Computing the COM from the COP in postural sway movements. Hum. Mov. Sci. **18**(6), 759–767 (1999)
22. de Rosario, H., Page, A., Besa, A., Mata, V., Conejero, E.: Kinematic description of soft tissue artifacts: quantifying rigid versus deformation components and their relation with bone motion. Med. Biol. Eng. Compu. **50**(11), 1173–1181 (2012)
23. Shimba, T.: An estimation of center of gravity from force platform data. J. Biomech. **17**(1), 5359–5760 (1984)
24. Winter, D.A.: Biomechanics and Motor Control of Human Movement. Wiley, Hoboken (2009)
25. Wu, G., van der Helm, F.C., (DirkJan) Veeger, H., Makhsous, M., Van Roy, P., Anglin, C., Nagels, J., Karduna, A.R., McQuade, K., Wang, X., Werner, F.W., Buchholz, B.: ISB recommendation on definitions of joint coordinate systems of various joints for the reporting of human joint motion—Part II: shoulder, elbow, wrist and hand. J. Biomech. **38**(5), 981–992 (2005)
26. Wu, G., Siegler, S., Allard, P., Kirtley, C., Leardini, A., Rosenbaum, D., Whittle, M., D'Lima, D.D., Cristofolini, L., Witte, H., Schmid, O., Stokes, I.: ISB recommendation on definitions of joint coordinate system of various joints for the reporting of human joint motion—Part I: ankle, hip, and spine. J. Biomech. **35**(4), 543–548 (2002)

An Index Finger Musculoskeletal Dynamic Model

Jumana Ma'touq(✉)

Institute of Automatic Control, Gottfried Wilhelm Leibniz Universität Hannover, Hanover, Germany
matouq@irt.uni-hannover.de, jumana.matouq@gmail.com

Abstract. Neuromusculoskeletal models provide a mathematical tool for understanding and simulating human neuromechanics and motor control. The aim of this work is to extend our previously developed models and propose an index finger musculoskeletal dynamic model that serves as a tool in studying and replicating human behaviour. In particular, the focus of this work is to develop a skeletal dynamic model, a musculotendon dynamic model (Hill-type muscle model), and an activation estimation model. The Hill-type muscle model estimates musculotendon forces for given musculotendon lengths, length change rates, and muscle activations. The parameters of the Hill-type model were estimated so that the resulting normalised muscle length is within the operating muscle length and the resulting forces/torques are comparable to experimental data from the literature. In the activation estimation model, muscle activations are optimised by minimising the difference between the resulting torque from the musculotendon dynamic model and the skeletal dynamic model. In the skeletal dynamic model, the torques due to the passive joint properties and gravitational and inertial forces are modelled. Using the estimated Hill-type muscle model parameters, the resulting normalised length for all index muscles ranged between 0.97 and 1.03 in resting posture and between 0.5 and 1.5 in flexion/extension task. The resulting muscle activations ranged between 0 and 1 and related to the activation/deactivation of muscles during the motion task. Finally, the overall consistency between the proposed models is demonstrated and underlines the quality of the developed models.

Keywords: Index finger · Musculotendon dynamic · Muscle activation · Hill-type muscle model

1 Introduction

Understanding human hand neuromechanics is the first step in implementing human hand capabilities and behaviour in human-like smart prostheses and articulated robotic hands. In this paper, a musculoskeletal dynamic model of the index finger, which serves as a tool in studying and replicating human behaviour,

G. A. Ateshian et al. (Eds.): CMBBE 2019, LNCVB 36, pp. 411–436, 2020.
https://doi.org/10.1007/978-3-030-43195-2_34

is proposed. This model consists of five sub-models, i.e. skeletal kinematics, skeletal dynamics, musculotendon kinematics, musculotendon dynamics, and muscle activation estimation (Fig. 1). The skeletal kinematic model and the musculotendon kinematic model are based on our previous work [1,2] while the other three models (skeletal dynamics, musculotendon dynamics, and muscle activation estimation) are developed in this paper.

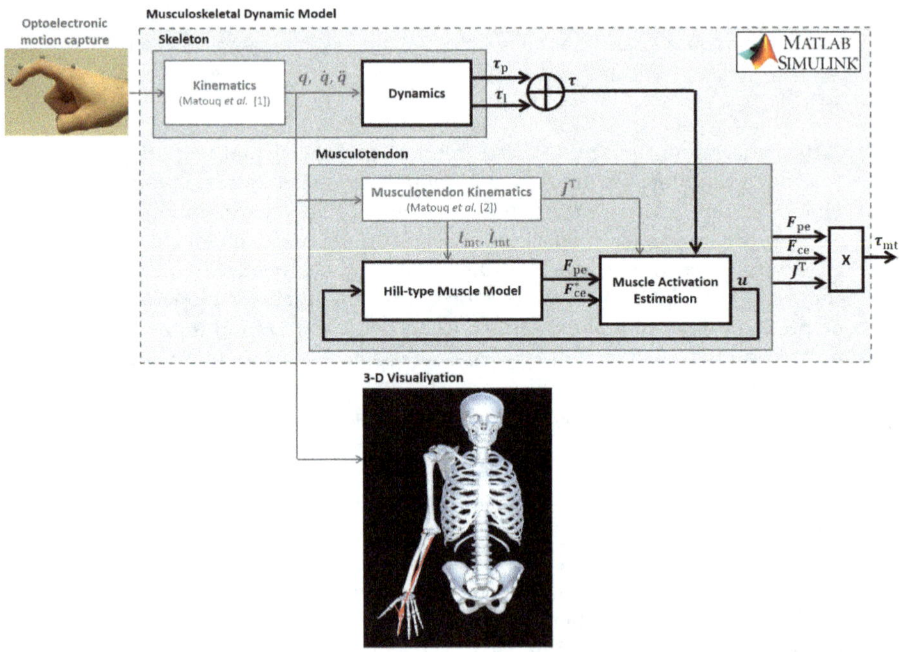

Fig. 1. The overall musculoskeletal dynamic model of the index finger. The bold blocks represent the models of the current paper. The mathematical model is implemented in MATLAB/Simulink (The MathWorks Inc., USA) and a 3-D visualisation is implemented in Unity (Unity Technologies, USA). The symbols $\boldsymbol{q}$, $\dot{\boldsymbol{q}}$, $\ddot{\boldsymbol{q}}$ denote the angular position, velocity, and acceleration, respectively; $\boldsymbol{\tau}_{\mathrm{l}}$ is the link torque; $\boldsymbol{\tau}_{\mathrm{p}}$ is the passive torque; $\boldsymbol{\tau}$ is the total torque computed from the skeletal dynamic model; $\boldsymbol{J}^{\mathsf{T}}$ is the matrix of musculotendon excursion moment arms; $\boldsymbol{l}_{\mathrm{mt}}$ and $\dot{\boldsymbol{l}}_{\mathrm{mt}}$ are the musculotendon unit length and shorting/lengthening velocity, respectively; $\boldsymbol{F}_{\mathrm{pe}}$ and $\boldsymbol{F}_{\mathrm{ce}}$ are the passive element force and contractile element force, respectively; $\boldsymbol{u}$ is the estimated muscle activation; and $\boldsymbol{\tau}_{\mathrm{mt}}$ is the estimated musculotendon unit torque.

1.1 Skeletal Dynamics

Skeletal dynamics describes the relationship between skeletal motion and muscle forces or joint torques occurring during the movements. The description of multi-joint dynamic movement of the hand is rarely studied in the literature and most of the work is focused on the dynamic movement of the arm [3]. The dynamic analysis of the human hand literature either focuses on the torque due to the inertial and gravitational forces (link torque $\boldsymbol{\tau}_{\mathrm{l}}$) only [4,5], or on the torque due to the visco-elastic properties of the joints (passive torque $\boldsymbol{\tau}_{\mathrm{p}}$) only [6–8]. However, Deshpande et al. [9] pointed out that considering $\boldsymbol{\tau}_{\mathrm{l}}$ only would be correct for motions where it is dominant, e.g. shoulder-elbow movements [9–12], and not in human hand movements where $\boldsymbol{\tau}_{\mathrm{p}}$ is more dominant. Additionally, the role of the passive joint properties in hand control is critical for hand surgery, hand therapy, and mathematical modelling of the hand [7].

The analysis of forces and moments that occur during movement has been performed either using external devices [13], mathematical models [14,15], or simulation models in MATLAB® Simulink® (The MathWorks, Inc., USA) Toolbox (The MathWorks, Inc., USA) and MATLAB® SimMechanics™ Toolbox (The MathWorks, Inc., USA) [5,16–20]. SimMechanics™ is a multibody simulation environment for 3D mechanical systems in which dynamic solutions are carried out in accordance with the laws of Newton mechanics [18]. Despite the fact that SimMechanics™ allows kinematic and kinetic analysis of movements without complex equations or external devices, which can limit the movement and hinder accurate measurement [5,18], it provides $\boldsymbol{\tau}_{\mathrm{l}}$ only and lacks $\boldsymbol{\tau}_{\mathrm{p}}$. The torque $\boldsymbol{\tau}_{\mathrm{l}}$ is defined as

$$\boldsymbol{\tau}_{\mathrm{l}} = \boldsymbol{M}(\boldsymbol{q})\ddot{\boldsymbol{q}} + \boldsymbol{c}(\boldsymbol{q}, \dot{\boldsymbol{q}}) + \boldsymbol{g}(\boldsymbol{q}), \tag{1}$$

where $\boldsymbol{q}$, $\dot{\boldsymbol{q}}$, $\ddot{\boldsymbol{q}}$ are the angular position, velocity, and acceleration, respectively and are obtained from the kinematic model in Sect. 2.1, $\boldsymbol{M}(\boldsymbol{q})$ is the mass matrix, $\boldsymbol{c}(\boldsymbol{q}, \dot{\boldsymbol{q}})$ is the centrifugal and Coriolis vector, and $\boldsymbol{g}(\boldsymbol{q})$ is the gravitational torque.

Incorporating human-like passive compliance could improve grasping and object manipulation abilities of robotic hands [7] as the passive properties of the finger and wrist musculature influence the hand posture and movement patterns [3,13,21]. The passive torque $\boldsymbol{\tau}_{\mathrm{p}}$ is defined as

$$\boldsymbol{\tau}_{\mathrm{p}} = \boldsymbol{\tau}_{\mathrm{d}} + \boldsymbol{\tau}_{\mathrm{s}}, \tag{2}$$

[3,9] where $\boldsymbol{\tau}_{\mathrm{d}}$ is the damping passive joint torque, and $\boldsymbol{\tau}_{\mathrm{s}}$ is the stiffness passive joint torque, which represents more than 90% of the total passive torque based on two-link planar model dynamics during repetitive hand movements [3].

In modelling the human hand, each finger is considered as a flexible joint robotic arm and the skeleton equation of motion can be rewritten as

$$\boldsymbol{\tau}_{\mathrm{mt}} + \boldsymbol{\tau}_{\mathrm{ext}} = \boldsymbol{\tau}_{\mathrm{l}} + \boldsymbol{\tau}_{\mathrm{p}}, \tag{3}$$

and with assuming that the joint torque due to environmental interaction force $\boldsymbol{\tau}_{\mathrm{ext}}$ to be zero, i.e. $\boldsymbol{\tau}_{\mathrm{ext}} = \boldsymbol{0}$, and by incorporating the passive visco-elastic torque (Eq. 2), Eq. 3 can be rewritten as

$$\boldsymbol{\tau}_{\mathrm{mt}} = \boldsymbol{M}(\boldsymbol{q})\ddot{\boldsymbol{q}} + \boldsymbol{\tau}_{\mathrm{d}} + \boldsymbol{\tau}_{\mathrm{s}} + \boldsymbol{c}(\boldsymbol{q}, \dot{\boldsymbol{q}}) + \boldsymbol{g}(\boldsymbol{q}), \tag{4}$$

[3,9], where $\boldsymbol{\tau}_{\mathrm{mt}}$ is the joint torque due to the total musculotendon force.

1.2 Musculotendon Dynamics

Muscle contraction dynamics is concerned with the transformation of muscle activation into muscle force [22]. Mathematical models are used to simulate muscle contraction dynamics and describe the resulting musculotendon force for a specific motion. The mathematical description of muscle dynamics depends on the type of problem to solve [23]. Despite that the Huxley model [24] describes precisely the chemical and mechanical processes that take place during muscle contraction, it is not recommended in studies that include several muscle actuators as the model complexity increases considerably [23]. Alternatively, the Hill-type muscle model [25] is used to calculate musculotendon forces in such studies because of its simplicity compared to the Huxley model [7,19,20,23,26–37].

The classical structure for the Hill-type muscle model, which is thoroughly described in Zajac [22] and Winters [38], consists of a contractile element (CE), a passive series element (SE) representing the tendon, and a passive parallel element (PE) representing the passive muscle stiffness (Fig. 2). While the CE is the active force generation element in the muscle and generates force $\boldsymbol{F}_{\mathrm{ce}}$ that has both force-length and force–velocity properties, the PE describes the passive elastic properties of the muscle fiber that has force–length property and generates force $\boldsymbol{F}_{\mathrm{pe}}$. The SE can be neglected for simplicity similar to Zajac [22] and Silva [32].

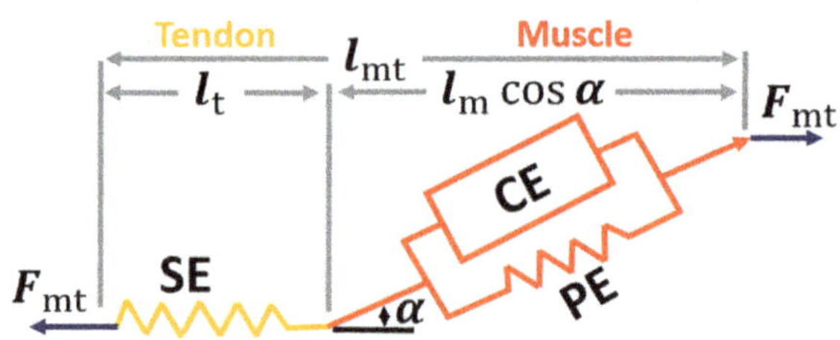

Fig. 2. Representation of the classical structure of Hill-type muscle model, with a contractile element (CE), a series element (SE, larger spring), and a parallel element (PE).

1.3 Muscle Activation

In order to calculate $\boldsymbol{F}_{\mathrm{ce}}$, muscle activations $\boldsymbol{u}$ are needed alongside the musculotendon lengths $\boldsymbol{l}_{\mathrm{m}}$ and their change rates $\dot{\boldsymbol{l}}_{\mathrm{m}}$ (Eq. 11). Both $\boldsymbol{l}_{\mathrm{m}}$ and $\dot{\boldsymbol{l}}_{\mathrm{m}}$ are obtainable using the musculotendon kinematic model [2]. On the other hand, the muscle activations $\boldsymbol{u}$ can not be measured non-invasively for all hand muscles, e.g. for deep or intrinsic muscles by means of surface electromyography (sEMG). Alternatively, the straight forward option to calculate $\boldsymbol{F}_{\mathrm{ce}}$ is to solve

$$\boldsymbol{\tau}_{\mathrm{mt}} = \boldsymbol{J}^{\mathsf{T}}(\boldsymbol{F}_{\mathrm{ce}} + \boldsymbol{F}_{\mathrm{pe}}), \tag{5}$$

where $\boldsymbol{J}^\top$ is the matrix of musculotendon excursion moment arms. However, Eq. 5 presents two main challenges: (1) the matrix $\boldsymbol{J}^\top$ is not invertible and thus the direct solution of multiplying by its inverse is not feasible, and (2) the system is underdetermined, i.e. the number of muscles is greater than number of joints, which results in an infinite number of solutions for muscle forces and joint reactions in each position [39]. Alternatively, optimisation approaches have been applied to solve this problem and estimate muscle forces [39–46].

1.4 Contribution

Modelling and simulating the dynamic behaviour of a musculoskeletal system require a musculoskeletal force model and the transformation from musculotendon forces to joint torques (the matrix of musculotendon excursion moment arms $\boldsymbol{J}^\top(\boldsymbol{q})$). In a previous work a musculotendon kinematic model that provides $\boldsymbol{J}^\top(\boldsymbol{q})$ for a given joint configuration $\boldsymbol{q}$ has been proposed [2]. The aim of this work is therefore to extend the previously developed models [1,2] and develop an index finger musculoskeletal dynamic model that serves as a tool in studying and replicating human behaviour (Fig. 1). In particular, the focus of this work is to:

1. Develop a **skeletal dynamic model** that estimates joint torques for a given motion configuration. This model includes both torques, i.e. the one resulting from gravitational and inertial forces as well as the one resulting from passive joint properties. The output of this model is fed into the activation estimation model to optimise the muscle activation.
2. Propose a **musculotendon dynamic model**, i.e. a Hill-type muscle model alongside its parameters. This model computes the musculotendon forces of all extrinsic and intrinsic index finger muscles for given musculotendon lengths and their change rates, which are obtained from the musculotendon kinematic model, alongside muscle activations.
3. Develop a **muscle activation model** that is able to estimate muscle activation for given musculotendon forces, joint torques, and musculotendon excursion moment arms matrix.

2 Methods

The complete proposed index finger musculotendon model consists of five models; skeletal kinematics, skeletal dynamics, musculotendon kinematics, musculotendon dynamics, and muscle activation estimation (Fig. 1). These models are explained in the following.

2.1 Skeletal Kinematic Model

Inverse kinematic model maps a set of surface Cartesian positions $\{\boldsymbol{X}_i\}$, which is obtained using optoelectronic motion capture system, to skeletal joint configuration vector $\boldsymbol{q}$. The index finger kinematic model of our previous work [1] is used in this paper. The index finger is modelled as four degrees of freedom (DoFs) of 2 DoFs for the metacarpophalangeal (MCP) joint flexion/extension (F/E) and abduction/adduction (Ab/Ad) and 2 DoFs for flexion/extension of the proximal interphalangeal (PIP) and the distal interphalangeal (DIP) joints.

2.2 Skeletal Dynamic Model

The dynamics of the index finger is modelled using Eq. 4. The left-hand side of Eq. 4 is modelled in Sect. 2.4 while the components of the right-hand side are modelled in the following.

2.2.1 Link Torque

In the proposed model, the link torque $\boldsymbol{\tau}_{\mathrm{l}}$ (Eq. 1) is calculated for joint kinematic data as an input using the MATLAB® SimMechanics™, which has been widely used to perform the dynamic analysis of the human motion [5,16–20]. In order to implement the index finger model in SimMechanics™, a kinematic model and anthropometric properties should be defined. The kinematic structure of the index finger is described as an open chain link-segment model that consists of 4 segments and 4 revolute joints. The index finger phalanges (bones) are assumed to be rigid cylinders with uniform mass distribution moving in a fixed plane. The phalanx centre of mass is assumed to be in the midpoint of the phalanx. Consequently, the anthropometric properties (phalanx body mass m_i and inertia tensor I_i) for each phalanx are calculated as

$$m_i = \rho V_i, \tag{6a}$$

$$I_i = \begin{pmatrix} \frac{1}{2} m_i r_i^2 & 0 & 0 \\ 0 & \frac{1}{12} m_i (3r_i^2 + l_i^2) & 0 \\ 0 & 0 & \frac{1}{12} m_i (3r_i^2 + l_i^2) \end{pmatrix}, \tag{6b}$$

where i indicates finger phalanx, ρ is the human body density 1.1 g/cm^3 [21,47–49], V_i is the volume of the phalanx body and considered as the volume of a cylinder, r_i and l_i are the phalanx body radius and length, respectively. Thus the anthropometric parameters (r_i and l_i) should be defined for each phalanx to calculate the anthropometric properties in Eq. 6. With defining these parameters, SimMechanics™ computes the joint link torque for a given motion input (Table 1).

2.2.2 Passive Torque

The passive properties of the finger joints are critical in all human movements as they allow us to interact safely, reliably, and efficiently with the world around us [50]. Despite the importance of the joint passive properties in hand movements during grasping and manipulation [3,9,50], very few robotic hands seek to copy human-like passive properties [50–52]. The passive torque $\boldsymbol{\tau}_{\mathrm{p}}$ is modelled as a summation of the damping torque $\boldsymbol{\tau}_{\mathrm{d}}$ and the stiffness torque $\boldsymbol{\tau}_{\mathrm{s}}$ (Eq. 2). These two torques are explained in the following.

Table 1. The anthropometric properties of the index finger.

Phalanx i	r_i [mm]	l_i [mm]	m_i [g]	I_i [g.cm^2]		
				$I_{i,x}$	$I_{i,y}$	$I_{i,z}$
Metacarpal (MC)	9.9	64.6	20.9	70.7716	70.7716	10.0128
Proximal (PP)	9.9	44.1	14.2	20.6548	20.6548	6.9121
Medial (MP)	9.1	25.7	7.0	5.3136	5.3136	2.8749
Distal (DP)	8.2	20.5	4.6	2.3683	2.3683	1.5402

A. Damping Torque. The damping torque $\boldsymbol{\tau}_\mathrm{d}$ is modelled to be dependent on the joint velocity $\dot{\boldsymbol{q}}$ only as

$$\boldsymbol{\tau}_\mathrm{d} = \boldsymbol{K}\dot{\boldsymbol{q}}, \tag{7}$$

[3,6,9], where $\boldsymbol{K}$ is the joint damping coefficient vector, which is considered to be constant and can be chosen based on previous studies [3,6,9,48,53–57]. The damping coefficients of the index finger joints that were used are shown in Table 2.

B. Stiffness Torque. The state-of-the-art stiffness torque models were developed based on three modelling functions, i.e. linear [55,57], polynomial [48], and double exponential [3,6–9,50]. Among these models, the double exponential model is preferred in the literature and is used in this work as it: (1) provides a good fits to many types of passive moment-angle or force-length data [13,58,59], and (2) results in a natural finger movement in simulation [60]. In the proposed model, the stiffness torque $\boldsymbol{\tau}_\mathrm{s}$ is modelled for the F/E of the three joints of the index finger as in the following.

The passive stiffness torque at the index MCP F/E joint increases exponentially as the joint angle extends or flexes [50]. Based on human subject experiment [8], the passive stiffness torque at the index MCP F/E joint showed a double exponential response. Thus, it is modelled as a double exponential function in the form of

$$\tau_{\mathrm{s},j}(q_j) = A_j(e^{-B_j(q_j-E_j)} - 1) - C_j(e^{D_j(q_j-F_j)} - 1), \tag{8}$$

[6–9,13,59,61], where j indicates the finger joint (the index MCP F/E in this case) and A_j, B_j, C_j, D_j, E_j, and F_j are the parameters of the double exponential joint stiffness torque (Table 2). A similar double exponential nature was also noticed experimentally at the PIP joint [6,62]. Therefore, the double exponential model can be used for all finger joint F/E DoF for consistency [6].

2.3 Musculotendon Kinematic Model

The musculotendon kinematic model computes musculotendon lengths $\boldsymbol{l}_\mathrm{mt}(\boldsymbol{q})$, length change rates $\dot{\boldsymbol{l}}_\mathrm{mt}(\boldsymbol{q}, \dot{\boldsymbol{q}})$, and musculotendon excursion moment arms $\boldsymbol{J}^\mathsf{T}(\boldsymbol{q})$

Table 2. Index finger joints damping and stiffness coefficients. The damping coefficient in Ab/Ad DoF is assumed to be identical to that in F/E DoF [60].

Joint j	Damping coefficient [57]	Stiffness coefficients [6]					
	K_j [N.mm.s/°]	h_j [N.mm/°]			$q_{j,0}$ [°]		
		A_j	B_j	C_j	D_j	E_j	F_j
MCP Ab/Ad	1.27	n.a	n.a	n.a	n.a	n.a	n.a
MCP F/E	1.27	1.01	0.05	3.39	0.05	70.96	13.68
PIP F/E	0.419	0.70	0.05	2.35	0.05	36.3	23.31
DIP F/E	0.237	0.31	0.05	1.04	0.05	−10.99	06.06

for a given joint configuration. The index finger musculotendon model of our previous work [2] is used in this paper. The musculotendon model includes all extrinsic and intrinsic index finger muscles, i.e. flexor digitorum profundus (FDP), flexor digitorum superficialis (FDS), extensor digitorum (communis) (EDC), extensor indicis (proprius) (EIP), palmar interossei (PI), dorsal interossei (DI), and lumbricals (LU). As the the DI muscle has two origins, i.e. the first and second metacarpals, it was modelled with two subregions, i.e. DI1-I and DI1-T, to represent each origin-insertion path. The output of the musculotendon kinematic model is fed into musculotendon dynamic model and activation estimation model to calculate musculotendon forces and muscle activations (Fig. 1).

2.4 Musculotendon Dynamic Model

The dynamic behaviour of the musculotendon unit is described using a Hill-type muscle model [25]. In this section the mathematics of the Hill-type muscle model as well as parameter estimation are discussed.

2.4.1 Mathematical Model

Two assumptions are made to simplify the used Hill-type muscle model in this paper. These assumption are:

1. The pennation angle $\boldsymbol{\alpha}_{\mathrm{m}}$ is neglected for model simplicity. The pennation angle is one of the muscle architecture parameters that affects its force–generating characteristics and is derived from dissection studies (cadaver studies). It has little effect if it is less than 20° [22,63,64], which is the case in the index finger muscles based on cadaver data [65,66].
2. The CE and PE are modelled with respect to all their properties while the SE is assumed to be rigid [19,67]. Thus, with assuming $\boldsymbol{\alpha}_{\mathrm{m}} = \mathbf{0}$, the muscle length $\boldsymbol{l}_{\mathrm{m}}$ and its shorting/lengthening velocity $\dot{\boldsymbol{l}}_{\mathrm{m}}$ are defined as

$$\boldsymbol{l}_{\mathrm{m}} = \boldsymbol{l}_{\mathrm{mt}} - \boldsymbol{l}_{\mathrm{t,o}}, \tag{9a}$$

$$\dot{\boldsymbol{l}}_{\mathrm{m}} = \dot{\boldsymbol{l}}_{\mathrm{mt}}, \tag{9b}$$

[20], where l_{mt} and $\dot{l}_{mt}$ are the musculotendon unit length and shorting/lengthening velocity, respectively, and $l_{t,o}$ is the tendon slack length, which is equal to the SE length and estimated in Sect. 2.4.2.

The force of a musculotendon unit $\boldsymbol{F}_{mt}$ is calculated as the summation of the forces generated by CE and PE, which are indicated by $\boldsymbol{F}_{ce}$ and $\boldsymbol{F}_{pe}$, respectively. The force $\boldsymbol{F}_{ce}$ depends on the muscle length $\boldsymbol{l}_m$, muscle activation $\boldsymbol{u}$, and the muscle shorting/lengthening velocity $\dot{\boldsymbol{l}}_m$ while the force $\boldsymbol{F}_{pe}$ depends on the muscle length $\boldsymbol{l}_m$ only. Thus, the total musculotendon force produced by the eight muscles of the index finger $\boldsymbol{F}_{mt} \in \mathbb{R}^8$ is given by

$$\boldsymbol{F}_{mt}(\boldsymbol{u}, \boldsymbol{l}_{mt}, \dot{\boldsymbol{l}}_{mt}) = \boldsymbol{F}_{ce}(\boldsymbol{u}, \boldsymbol{l}_m, \dot{\boldsymbol{l}}_m) + \boldsymbol{F}_{pe}(\boldsymbol{l}_m), \tag{10}$$

where $\boldsymbol{F}_{mt}$, $\boldsymbol{F}_{pe}$, $\boldsymbol{F}_{ce}$, $\boldsymbol{u}$, $\boldsymbol{l}_m$, and $\dot{\boldsymbol{l}}_{mt} \in \mathbb{R}^8$. For a *single* musculotendon unit the force is calculated as

$$\begin{aligned} F_{mt} &= F_{ce} + F_{pe}, \\ F_{mt} &= F_{max}\, u\, f_l^{ce}(l_m)\, f_v^{ce}(u, \dot{l}_m) + F_{max}\, f_l^{pe}(l_m), \\ F_{mt}(u, l_{mt}, \dot{l}_{mt}) &= F_{max}\left(u\, f_l^{ce}(l_m)\, f_v^{ce}(u, \dot{l}_m) + f_l^{pe}(l_m)\right), \end{aligned} \tag{11}$$

[68], where $f_l^{ce}(l_m)$ and $f_v^{ce}(u, \dot{l}_m)$ are the force-length and force-velocity relationships of the CE, respectively, $f_l^{pe}(l_m)$ is the force-length relationship of the PE, and F_{max} is the maximum voluntary isometric contraction force, which is proportional to the physiological cross-sectional area PCSA and calculated as

$$F_{max} = \mathrm{S} \cdot \mathrm{PCSA}, \tag{12}$$

[22,64,65,69–72], where the constant S represents the maximum muscle stress and ranges from 22 to 137 $\mathrm{N.cm}^{-2}$ [64,69–71,73–75]. The estimation of PCSA is discussed in Sect. 2.4.2.

The force–length relationship of the CE f_l^{ce} in Eq. 11 is modelled as a piecewise linear function based on the theoretical form of the normalised force–length relationship [64,76] with the ascending and descending parts as a sigmoid function [20]. Thus, f_l^{ce} is obtained from

$$\tilde{l}_m = \frac{l_m}{l_{m,o}}, \tag{13a}$$

$$f_l^{ce}(l_m) = \begin{cases} \dfrac{0.7}{1+\exp\left(-23.3(\tilde{l}_m - 0.545)\right)} & \tilde{l}_m < 0.545 \\ 4.436\,\tilde{l}_m - 2.0674 & \text{if } 0.545 \leqslant \tilde{l}_m < 0.625 \\ 0.8697\,\tilde{l}_m + 0.1574 & \text{if } 0.625 \leqslant \tilde{l}_m < 0.969 \\ 1 & \text{if } 0.969 \leqslant \tilde{l}_m < 1.032 \\ -1.9056\,\tilde{l}_m + 2.965 & \text{if } 1.032 \leqslant \tilde{l}_m < 1.4 \\ 0.6 - \dfrac{0.6}{1+\exp\left(-12(\tilde{l}_m - 1.40)\right)} & \text{otherwise,} \end{cases} \tag{13b}$$

[20], where $\tilde{l}_{\mathrm{m}}$ is the normalised muscle length, $l_{\mathrm{m,o}}$ is the optimal muscle length, which is defined as the optimal resting length for producing the maximal tension (the actin-myosin overlap reaches the maximum), and $l_{\mathrm{t,o}}$ is the tendon slack length. Both $l_{\mathrm{m,o}}$ and $l_{\mathrm{t,o}}$ are estimated in Sect. 2.4.2.

The force–velocity relationship of the CE $f_{\mathrm{v}}^{\mathrm{ce}}$ is calculated as

$$f_{\mathrm{v}}^{\mathrm{ce}}(u,\dot{l}_{\mathrm{m}}) = \frac{0.1433}{0.1074 + exp\left(-1.3\,sinh\left(2.8\frac{\dot{l}_{\mathrm{m}}}{\dot{l}_{\max}(u)} + 1.64\right)\right)}, \tag{14}$$

[77], where $\dot{l}_{\max}$ is the maximum contraction velocity above which the muscle can not produce force [22] and calculated as

$$\dot{l}_{\max}(u) = \frac{5}{2}(u+1)l_{\mathrm{m,o}}, \tag{15}$$

[20].

The force–length relationship of the PE $f_{\mathrm{l}}^{\mathrm{pe}}$ is calculated as

$$f_{\mathrm{l}}^{\mathrm{pe}}(l_{\mathrm{m}}) = \begin{cases} \frac{\exp\left(\frac{k_{\mathrm{pe}}}{\varepsilon_{\mathrm{m}}}(\tilde{l}_{\mathrm{m}}-1)\right)-1}{\exp(k_{\mathrm{pe}})-1} & \text{if } \tilde{l}_{\mathrm{m}} \geqslant 1 \\ 0 & \text{if } \tilde{l}_{\mathrm{m}} < 1 \end{cases}, \tag{16}$$

[19,20,78], where $\tilde{l}_{\mathrm{m}}$ is the normalised muscle length (Eq. 15), k_{pe} is the shape parameter and ε_{m} is the passive muscle strain. The values of k_{pe} and ε_{m} were chosen such that the resulting F_{pe} and τ_{pe} are comparable to experimental values from the literature [8].

The proposed Hill-type model (Eqs. 10–16) requires three main parameters, i.e. $\boldsymbol{F}_{\max}$, $\boldsymbol{l}_{\mathrm{m,o}}$, and $\boldsymbol{l}_{\mathrm{t,o}}$. These parameters are discussed in the next section.

2.4.2 Model Parameter Estimation

The parameters of the Hill-type muscle model are related to muscle architecture. These parameters are either derived mainly from cadaver studies [65,66,79–83], or estimated mathematically [64]. The values of the index finger muscles parameters appearing in the literature vary widely for even the same muscle in humans (Table 4). In the proposed model, the parameters of [65,66] were modified to suit the mathematical description of the proposed model as follows.

A. Physiological Cross-Sectional Area. The force $F_{\max}$ is proportional to the physiological cross-sectional area (PCSA) (Eq. 12). The PCSA is defined as the ratio between muscle volume and the optimal muscle length $l_{\mathrm{m,o}}$. Thus, for *one* muscle the PCSA is given by

$$\begin{aligned} \mathrm{PCSA} &= \frac{Vol_{\mathrm{m}}}{l_{\mathrm{m,o}}}, \\ Vol_{\mathrm{m}} &= \frac{m_{\mathrm{m}}}{\rho}, \end{aligned} \tag{17}$$

[64], where Vol_{m} and m_{m} are muscle volume and mass, respectively, ρ is muscle density and assumed to be equal to 1.0567 g/mm^3 [83, 84].

B. Optimal Length and Tendon Slack Length. Tendon slack length $\boldsymbol{l}_{\mathrm{t,o}}$ is not directly measurable and can be selected to match operating lengths [85] when available, or to match muscle active and passive moment torque measurements [71]. Thus, $\boldsymbol{l}_{\mathrm{m,o}}$ was adopted from the literature while $\boldsymbol{l}_{\mathrm{t,o}}$ was calculated by solving

$$\frac{\boldsymbol{l}_{\mathrm{m}}}{\boldsymbol{l}_{\mathrm{m,o}}} = \frac{\boldsymbol{l}_{\mathrm{mt}} - \boldsymbol{l}_{\mathrm{t,o}}}{\boldsymbol{l}_{\mathrm{m,o}}} = \mathbf{1}, \tag{18}$$

where the musculotendon lengths $\boldsymbol{l}_{\mathrm{mt}}$ are calculated using the musculotendon kinematic model (Sect. 2.3) for a simulated resting posture as an input. After that, the resulting $\boldsymbol{l}_{\mathrm{m,o}}$ were checked and tuned, if necessary, while taking into consideration:

1. The normalised muscle length is in the overall muscle operating range, i.e. $\mathbf{0.5} \leq \boldsymbol{l}_{\mathrm{m}}/\boldsymbol{l}_{\mathrm{m,o}} \leq \mathbf{1.7}$, in the joints full range of motion.
2. Comparable $\boldsymbol{F}_{\mathrm{mt}}$ and $\boldsymbol{\tau}_{\mathrm{mt}}$ to experimental values from the literature when available.

2.4.3 Model Validation

The proposed musculotendon model is validated by:

1. Validating the force–length relationships ($\boldsymbol{f}_{\mathrm{l}}^{\mathrm{ce}}$ **and** $\boldsymbol{f}_{\mathrm{l}}^{\mathrm{pe}}$), and force–velocity relationship ($\boldsymbol{f}_{\mathrm{v}}^{\mathrm{ce}}$). In this point the proposed mathematical model is validated for normalised force and normalised muscle length, i.e. without the influence of model parameters.
2. Validating the estimated Hill-type muscle model parameters. The used PCSA is validated by comparing $\boldsymbol{F}_{\mathrm{pe}}$ and the passive musculotendon torque $\boldsymbol{\tau}_{\mathrm{mt,pe}}$. On the other hand, the validation of the selected $\boldsymbol{l}_{\mathrm{m,o}}$ and $\boldsymbol{l}_{\mathrm{t,o}}$ is done by checking if the resulting $\boldsymbol{l}_{\mathrm{m}}/\boldsymbol{l}_{\mathrm{m,o}}$ is within the muscle physiological operating range, i.e. between 0.5 and 1.7, for joint full RoM. For index finger joints, the flexion/extension RoM is $[-10\ 90]^\circ$, $[0\ 110]^\circ$, $[0\ 80]^\circ$ for MCP joint, PIP joint, and DIP joint, respectively. These RoM are based on Kapandji [86], Cobos et al. [87], and Deshpande et al. [88].

3. Comparing the modelled $\boldsymbol{F}_{\mathrm{pe}}$ and $\boldsymbol{\tau}_{\mathrm{mt,pe}}$ to experimental values from the literature when available. This also implies the validation of the previous two points; the mathematical model and model parameters. The optimal approach would be comparing the musculotendon torque $\boldsymbol{\tau}_{\mathrm{mt}}$ resulting from the modelled musculotendon force $\boldsymbol{F}_{\mathrm{mt}}$ with the measured one. However, the activation $\boldsymbol{u}$ is needed to calculate $\boldsymbol{F}_{\mathrm{ce}}$ which required to calculate $\boldsymbol{F}_{\mathrm{mt}}$ (Eq. 11). As it is not possible to measure $\boldsymbol{u}$ for intrinsic muscles non-invasively, the model is validated by comparing the modelled $\boldsymbol{F}_{\mathrm{pe}}$ and $\boldsymbol{\tau}_{\mathrm{mt,pe}}$ to experimental values from the literature when available instead of $\boldsymbol{F}_{\mathrm{mt}}$ and $\boldsymbol{\tau}_{\mathrm{mt}}$.

2.5 Muscle Activation Estimation

Static optimisation, which is defined as the the process of minimising or maximising the cost/benefit of an objective function for one instant in time only, is used to estimate muscle activation in this paper. Despite that the human and its muscles are dynamic motors of the musculoskeletal system, the static optimisation is often relevant and gives satisfactory results, depending on desired evaluation and purpose of study [43,89]. Thus, dynamic forces are then seen as static in every time instant. Static optimisation is computationally efficient, allows full 3-D motion, and generally incorporates many muscles, e.g. 30 or more muscles per leg in gait studies [43].

In the proposed model, the optimisation problem is formulated as: "*find a musculotendon force* $\boldsymbol{F}_{\mathrm{mt}}$*, while minimising the sum of squared muscle activation* $\boldsymbol{u}^2$*, to satisfy the joint torque* $\boldsymbol{\tau}$ *at each joint*". Mathematically this optimisation problem is formulated as

$$\begin{aligned} &\text{minimise} && \sum_{m=1}^{8} u_m^2, \\ &\text{subject to} && \boldsymbol{\tau} - \boldsymbol{\tau}_{\mathrm{mt}} \leq 10^{-5}, \end{aligned} \tag{19}$$

[40,42,43], where m indicates the number of the musculotendon units passing through the joint, $\boldsymbol{\tau}$ is the joint torques calculated from the skeletal dynamic model (Sect. 2.2), and $\boldsymbol{\tau}_{\mathrm{mt}}$ is the estimated musculotendon unit torque (Eq. 5). This optimisation is performed using the fmincon function from MATLAB® Optimization Toolbox™ (The MathWorks, Inc., USA). Once the optimisation is performed, the estimated $\boldsymbol{u}$ is fed into Eq. 15 to calculate $\boldsymbol{F}_{\mathrm{ce}}$ (Fig. 1).

3 Results and Discussion

For the current discussion a desired task of a repetitive F/E movement of the three joints of the index finger is simulated using sinusoidal functions, which are defined as

$$\begin{aligned} q_{\mathrm{MCP}} &= 55^\circ \sin(2t + 5.5) + 45^\circ, \\ q_{\mathrm{PIP}} &= 55^\circ \sin(2t + 5.5) + 55^\circ, \\ q_{\mathrm{DIP}} &= 40^\circ \sin(2t + 5.5) + 40^\circ. \end{aligned} \tag{20}$$

This simulated task consists of three main phases; flexion, extension, and neutral (Fig. 3). The flexion phase represents the phase where the flexion angle is increasing (e.g. between 2.7 and 3.8 s) while the extension phase is the phase where the flexion angle is decreasing (e.g. between 0.8 and 2.0 s). Finally, the neutral phase represents the phase where the flexion angle is around zero (e.g. between 2.2 and 2.7 s).

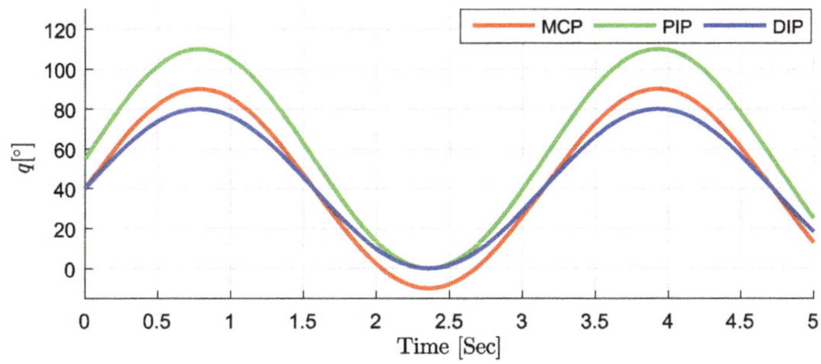

Fig. 3. The simulated flexion/extension task at the three joints of the index finger, which is obtained using sinusoidal functions as in Eq. 20.

3.1 Skeletal Dynamic Model

The highest peak-to-peak link torque $\boldsymbol{\tau}_{\mathrm{l}}$ is observed at the MCP joint followed by the PIP joint and DIP joint with 0.013 N.m, 0.005 N.m, and 0.0005 N.m, receptively (Fig. 4). This is an expected result as the MCP joint is influenced by the whole finger weight while the PIP joint is affected by the weight of the middle and the distal phalanges and the DIP joint is affected by the distal phalanx weight (Table 1).

The resulting damping torque $\boldsymbol{\tau}_{\mathrm{d}}$ at the three joints of the index finger for the same simulated task in Eq. 20 is shown in Fig. 4. The maximum damping torque is 0.14 N.m, 0.05 N.m, and 0.02 N.m at the MCP joint, the PIP joint, and the DIP joint, respectively. The torque $\boldsymbol{\tau}_{\mathrm{d}}$ is linearly dependent on $\dot{\boldsymbol{q}}$, which is calculated as the derivative of the joint configuration $\boldsymbol{q}$. Consequently, $\boldsymbol{\tau}_{\mathrm{d}}$ is positive when $\dot{\boldsymbol{q}}$ is positive due to the increase of $\boldsymbol{q}$ (joint flexion) and negative when $\dot{\boldsymbol{q}}$ is negative due to the decrease of $\boldsymbol{q}$ (joint extension).

The modelled stiffness torque $\boldsymbol{\tau}_{\mathrm{s}}$ of the three index finger joints as a function of F/E DoF for the same simulated task in Eq. 20 is shown in Fig. 5. The torque at the MCP joint ranged between −0.15 N.m and 0.05 N.m. The torque $\boldsymbol{\tau}_{\mathrm{s}}$ shows exponential decrease/increase with joint increase/decrease. The torque of PIP joint and DIP joint shows almost a similar response between 0° and 80°. This might be due to the coupling between these two joints where a strong correlation between DIP and PIP joints exist [90].

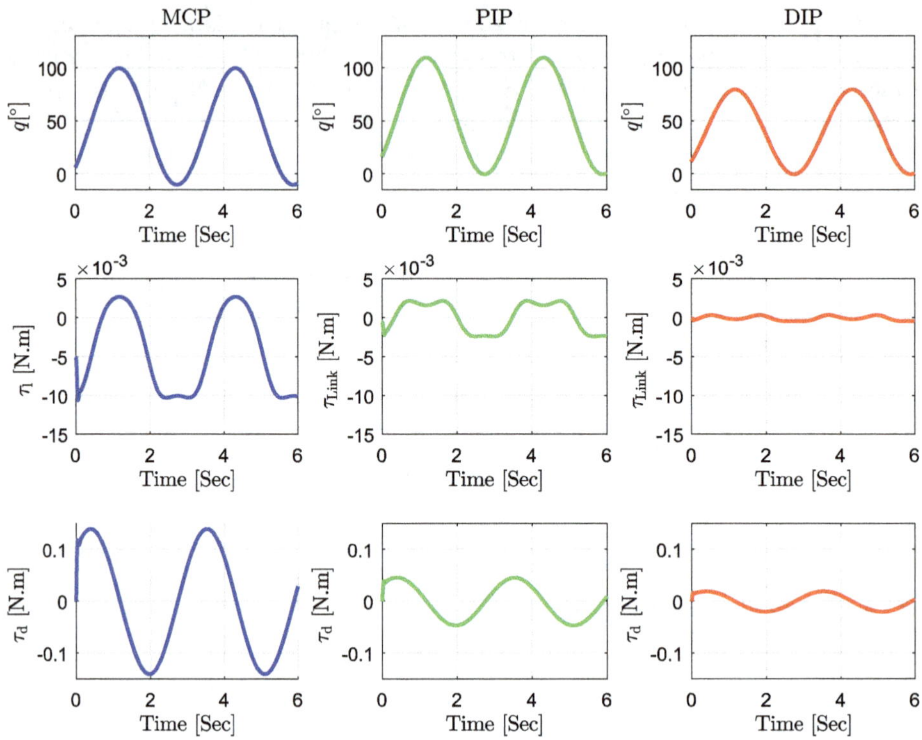

Fig. 4. The resulting torques at the three index finger joints for a simulated F/E task. Top: index finger joint configuration for a simulated F/E task given by Eq. 20. Middle: the resulting link torque $\boldsymbol{\tau}_l$. Bottom: the resulting damping torque $\boldsymbol{\tau}_d$.

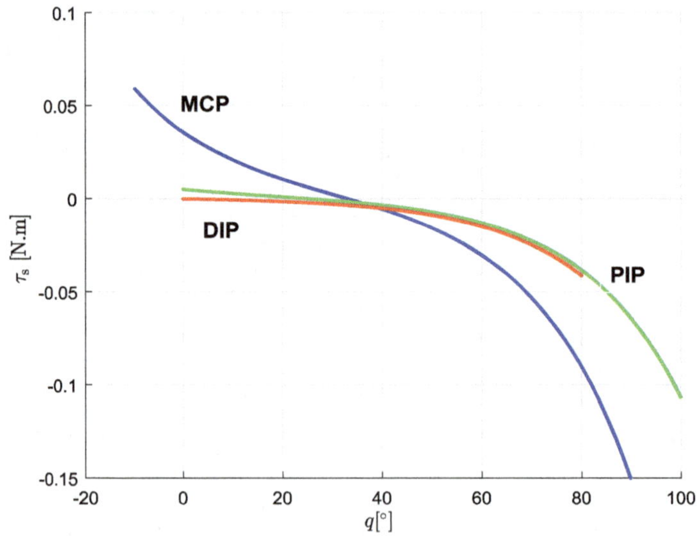

Fig. 5. The modelled joint stiffness torque $\boldsymbol{\tau}_s$ at the three joints of the index finger.

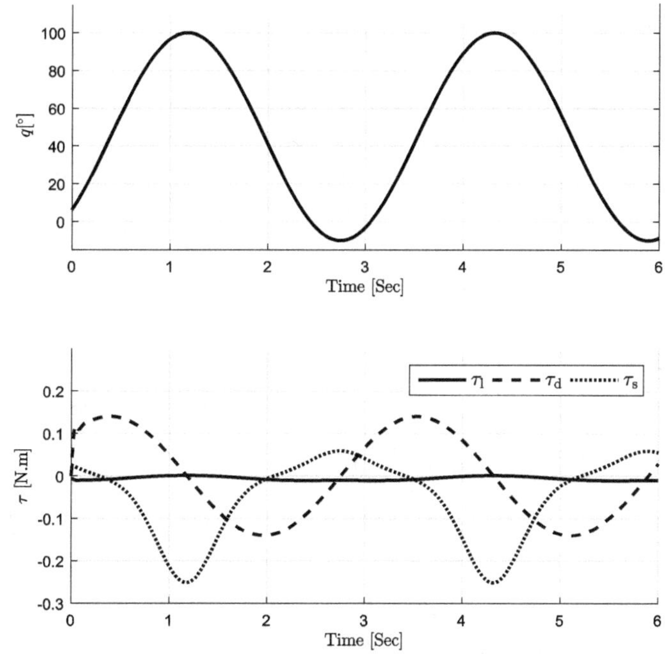

Fig. 6. Torque components at the index MCP joint for a given joint configuration. Top: the desired simulated MCP joint F/E task (Eq. 20). Bottom: the resulting link torque τ_l, damping torque τ_d, and stiffness torque τ_s.

Finally, the modelled three torques, i.e. τ_l, τ_d, and τ_s, at the index finger MCP joint for the same simulated MCP F/E task in Eq. 20 are shown in Fig. 6. The peak-to-peak torques are 0.011 mN.m, 0.28 N.m, and 0.3 N.m for τ_l, τ_d, and τ_s, respectively. The torque τ_l is significantly smaller than the other two torques (τ_d and τ_s). This is in agreement with Deshpande et al. [9] who pointed out that the passive visco-elastic component (τ_d and τ_s) is dominant over the dynamic component (τ_l). This emphasises the importance of integrating the passive torque model in skeletal dynamic modelling.

3.2 Musculotendon Dynamic Model

3.2.1 Mathematical Hill-Type Muscle Model

The total musculotendon force produced by the contractile and passive elements (Eqs. 11–16) is graphically represented in Fig. 7. In the CE force-length relationship, the maximum force is developed when the fully-activated muscle ($u = 1$) is held at its optimal length $l_m = l_{m,o}$. While the musculotendon active force is developed between $0.5 l_{m,o}$ and $1.5 l_{m,o}$, the passive force is developed when the muscle is stretched beyond its optimal length $l_m > l_{m,o}$. These curves are similar to the theoretical force–length and force–velocity curves which underlines the correctness of the proposed mathematical model for normalised force and normalised muscle length without the effect of model parameters.

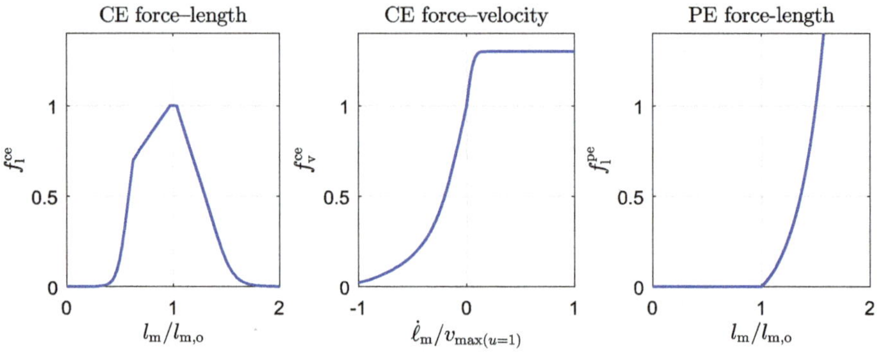

Fig. 7. The Hill-type muscle model characteristics; CE and PE force–length and force–velocity relationships.

3.2.2 Model Parameter Estimation

The estimated Hill-type muscle model parameters are shown in Table 3. Based on these parameters, the resulting normalised muscle length $\boldsymbol{l}_{\mathrm{m}}/\boldsymbol{l}_{\mathrm{m,o}}$ for extrinsic and intrinsic index muscles is ranged between 0.97 and 1.03 in resting posture and between 0.5 and 1.5 in the MCP2 joint F/E task (Fig. 8).

During the flexion phase, the extrinsic extensor muscles (EDC2 and EIP) extend to reach a normalised length of 1.32 while the extrinsic flexor muscles (FDS2 and FDP2) contract to reach a normalised length of 0.50–0.62 (Fig. 8). In the extension phase, the extrinsic extensor muscles contract while the extrinsic flexor muscles extend to return to the normalised length of 1.0 as in the neutral phase. The extrinsic muscles show a higher length change of 32–50% compared to the intrinsic muscles with 3–27%. This is in agreement with Kuo and Deshpande [8] who pointed out that the extrinsic muscles have longer length stretching compared to the intrinsic muscles.

3.2.3 Musculotendon Force and Torque

The resulting $\boldsymbol{F}_{\mathrm{pe}}$ and $\boldsymbol{\tau}_{\mathrm{mt,pe}}$ at the index finger MCP joint are shown in Fig. 9 and Fig. 10 for extrinsic muscles and intrinsic muscles. Extrinsic muscles (EDC2, EIP, FDS2, and FDP2) produce a higher $\boldsymbol{F}_{\mathrm{pe}}$ (between 1 N and 4 N) compared to intrinsic muscles (DI1-I, PI, LU1, and DI1-T) with $\boldsymbol{F}_{\mathrm{pe}} < 1$ N. These results are in agreement with the results of Kuo and Deshpande [8] who related the high force production of extrinsic muscles to their longer length stretching compared to the intrinsic muscles. In the extrinsic muscles, when the finger is flexed from

Table 3. The estimated Hill-type muscle model parameters

Muscle	PCSA [mm^2]	$l_{m,o}$ [mm]	$l_{t,o}$ [mm]
FDS2	190	84	318
FDP2	200	75	335
EDC2	55	70	381
EIP	56	59	215
DI1-I	150	32	35
DI1-T	50	32	35
PI	75	55.1	28
LU1	11	55	37

neutral position, i.e. from 0° to 100°, the extensor muscles (EDC2 and EIP) are extended which results in producing up to 4 N passive force. When the finger extended beyond the neural position, i.e. from 0° to −20°, the flexor muscles (FDS2 and FDP2) are extended more than the optimal length and thus produce up to 2.5 N passive force. The resulting passive forces in extrinsic muscles increase exponentially with stretching length and thus the resulting moment has an exponential dependency on joint configuration [8].

As the force $\boldsymbol{F}_{pe}$ is zero for flexors during flexion (from 0° to 100°) and extensors during extension (from 0° to −20°), the resulting $\boldsymbol{\tau}_{mt,pe}$ is also zero for flexors during flexion and extensors during extension. A comparison of the $\boldsymbol{\tau}_{mt,pe}$ resulting from the proposed model and from Kuo and Deshpande [8] model is shown in Fig. 9. In this figure, a little difference is noticed between the two compared torques which might be due to the differences between subjects in this model and literature model. These differences affect $\boldsymbol{J}^{\top}$ and thus the resulting $\boldsymbol{\tau}_{mt,pe}$.

3.2.4 Muscle Activation Estimation

The resulting muscle activations, musculotendon forces, and musculotendon torques from the proposed muscle activation model are shown in Fig. 10 for a simulated MCP2 F/E task. At the beginning of the simulated task, the index finger flexes which results in an increase in the flexion angle to reach 100°. During this phase, the flexor muscles are activated with muscle activation is nearly up to 0.1. After that, the index finger starts to extend and accordingly the flexion angle decreases. At this extension phase, the extensor muscles are activated with muscle activation is nearly up to 0.6. The activation during extension phase is larger than during flexion phase. This might be explained due to moving the

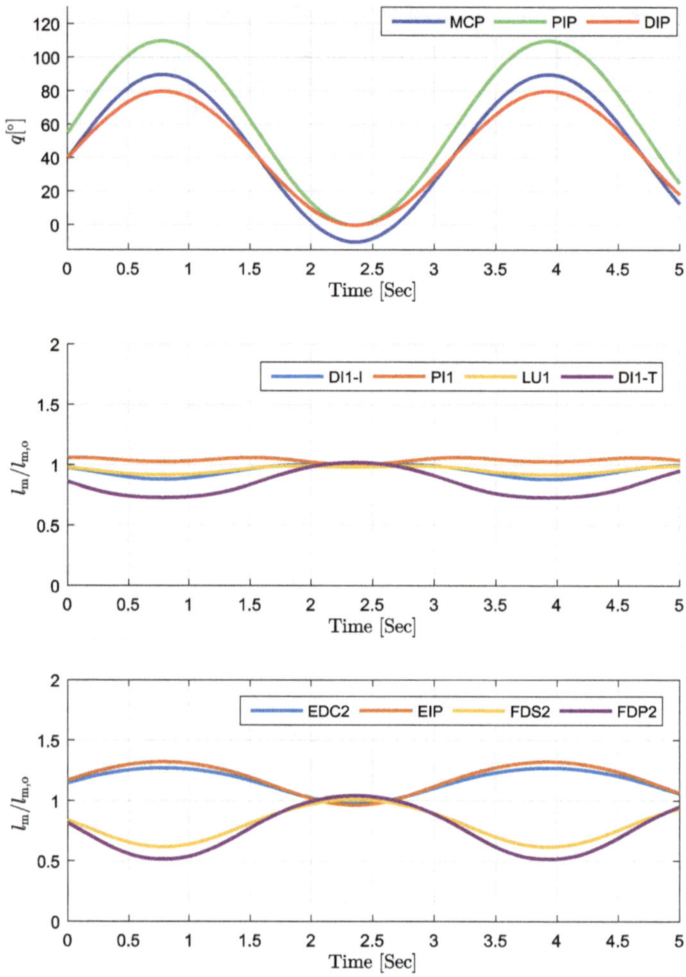

Fig. 8. Normalised muscle length $l_{\mathrm{m}}/l_{\mathrm{m,o}}$ for a simulated index finger joint flexion/extension task. Top: the simulated flexion/extension motion, which is obtained using sinusoidal functions as in Eq. 20. Middle and bottom: the normalised muscle length during the simulated task for the extrinsic and intrinsic index finger muscles, respectively.

finger against the gravity during extension where greater force is needed and thus greater activation. The estimated $\boldsymbol{u}$ is fed to the proposed Hill-type muscle model (Eqs. 11–16) to calculate musculotendon forces. Finally, the torque resulting from the proposed Hill-type muscle model τ_{mt} ranges between -0.26 N.m and 0.14 N.m and similar to the torque from the skeletal dynamic model τ.

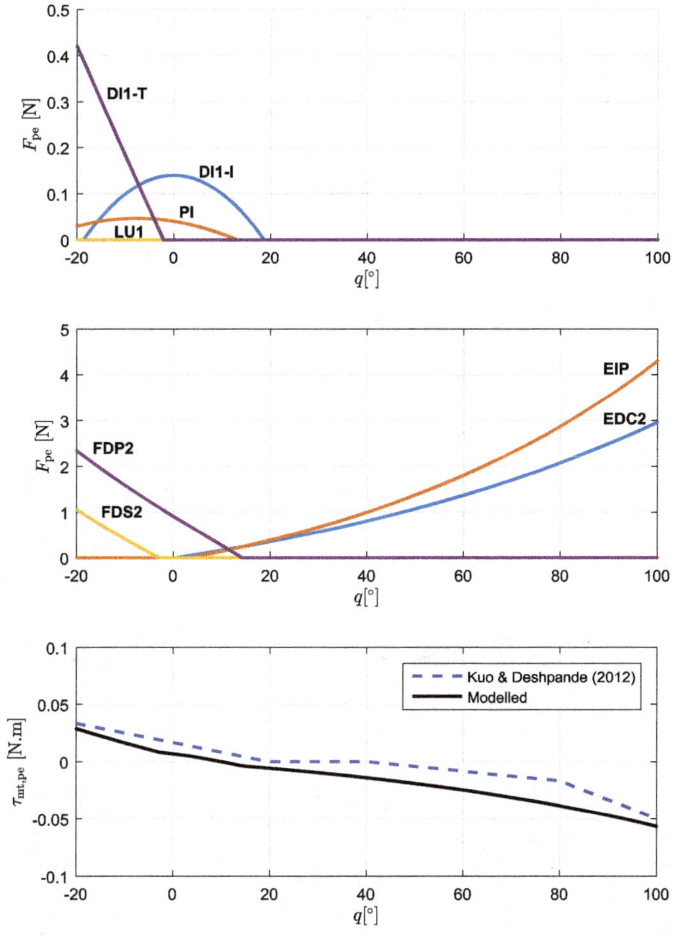

Fig. 9. The resulting $\boldsymbol{F}_{\mathrm{pe}}$ and $\boldsymbol{\tau}_{\mathrm{mt,pe}}$ at the index finger MCP joint. The passive musculotendon torque resulting from the proposed model is compared to the modelled torque from the literature [8].

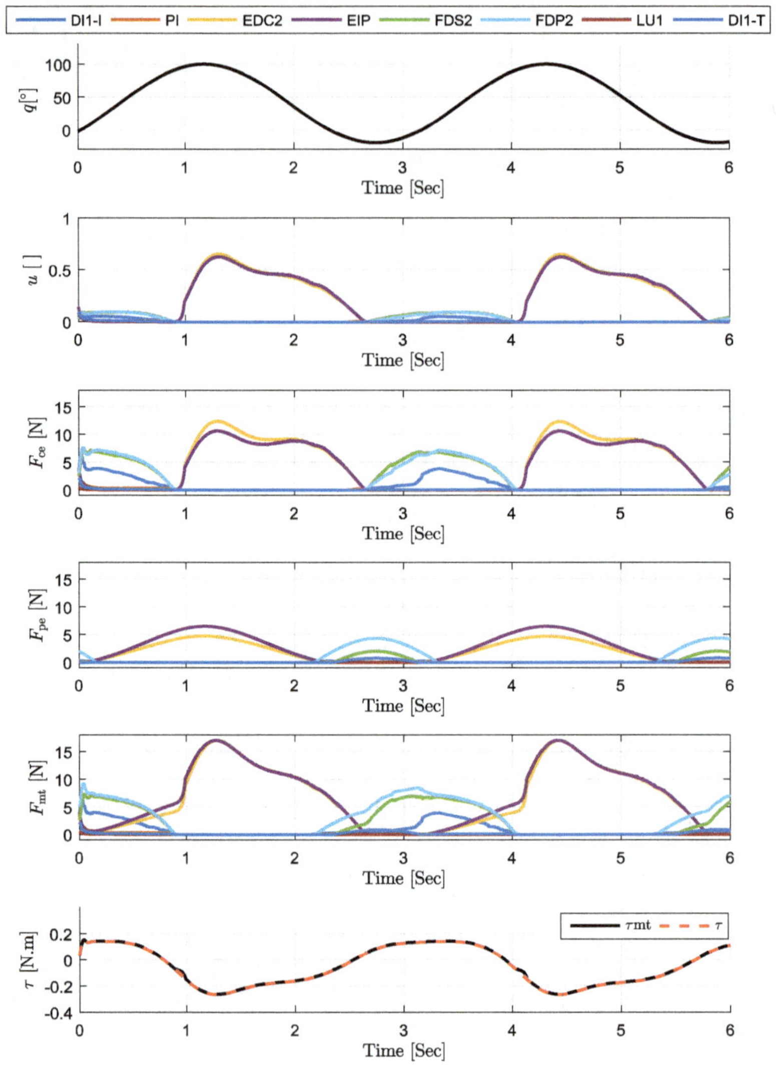

Fig. 10. The resulting muscle activations, forces, and torques for a simulated MCP2 F/E task. This simulated task is obtained using sinusoidal functions (Eq. 20).

4 Conclusion

In this paper a musculoskeletal dynamic model of the index finger that serves as a tool in studying and replicating human behaviour is proposed. This model computes joint torques, musculotendon forces, and muscle activations for a given motion. It has the limitations of: (1) modelling the stiffness passive joint torque for the flexion/extension DoF of the three joints of the index finger only, and (2) estimating the muscle activation for MCP joint flexion/extension DoF only.

Therefore, future work will address (1) extend the skeletal dynamic model to include stiffness torque of the MCP joint abduction/adduction DoF, (2) extend the muscle activation model to include the 4 DoFs of the index finger, (3) experimental validation of the proposed model using synchronised EMG and optoelectronic motion capture data, and finally (3) extend the index finger model to include the complete human hand musculoskeletal dynamics and human motor control.

Acknowledgement. I would like to express my appreciation to Prof. Christof Hurschler for his expert advice and extraordinary support during this research. I am grateful to Dr. Torsten Lilge who provided insight and expertise that greatly assisted the research. Also, I would like to thank Prof. Matthias Müller and Prof. Bernardo Wagner for their support.

Appendix

A State-of-the-Art of Muscle Modelling Parameters

Table 4. The state-of-the-art of muscle modelling parameters.

Study	Muscle							Method/Reference
	FDS2	FDP2	EDC2	EIP	DI1	PI1	LU1	
PCSA [mm^2]								
Amis et al. [79]	620	343	94					Cadaver study
Lieber et al. [65]	171	177	52	56				Cadaver study
Jacobson et al. [66]					150	75	11	Cadaver study
Holzbaur et al. [71]	140	150	40	50				Lieber et al. [65]
Ward et al. [81]	217	174						Cadaver study
Wu et al. [72]	479	479	139	112	353	280	28	Brand and Hollister [91]
Vignais and Marin [57]	171	177	52	56	150	75	11	Lieber et al. [65] and Jacobson et al. [66]
Lee [70]	140	150	40		150	80	10	Holzbaur et al. [71] and Brand and Hollister [91]
Mirakhorlo et al. [83]	183	138.5	48.9	61	62.1	51.7	7.2	Cadaver study
$l_{m,o}$ [mm]								
Amis et al. [79]	32	57	55					Cadaver study
Lieber et al. [65]	68	61	57	48				Cadaver study
Jacobson et al. [66]					38.9	30.7	68	Cadaver study
Holzbaur et al. [71]	84	75	70	59				Lieber et al. [65]
Ward et al. [81]	96.4	94.7						Cadaver study
Wu et al. [72]	70	66	60	60	14	15	66	Brand and Hollister [91]
Kuo and Deshpande [7]	83.5	74.9	70	58.9	61.9	55.1	64.9	Derived from Lieber et al. [65] and Jacobson et al. [66]

(continued)

Table 4. (*continued*)

Study	Muscle							Method/Reference
	FDS2	FDP2	EDC2	EIP	DI1	PI1	LU1	
Kuo and Deshpande [8]	72.7	67.2	61.9	52.1	38.9	30.7	68	Lieber et al. [65], Jacobson et al. [66], Holzbaur et al. [71], and Wu et al. [72]
Lee [70]	84	75	70		32	25	55	Holzbaur et al. [71] and Brand and Hollister [91]
Mirakhorlo et al. [83]	52.1 34.9	90.9	42.6	45	39.6 19.6	25.6	78.4	Cadaver study
$l_{t,o}$ [mm]								
Holzbaur et al. [71]	275	294	322	186				Operating length Murray et al. [85] and moment measurements
Ward et al. [81]	229	292.6						Cadaver study
Kuo and Deshpande [7]	338	322	385	248.04	31.7	25	55.4	Derived from Lieber et al. [65] and Jacobson et al. [66]
Kuo and Deshpande [8]	247.5	265.2	289.9	167.7	31.7	25	55.4	Lieber et al. [65], Jacobson et al. [66], Holzbaur et al. [71], and Wu et al. [72]
Lee [70]	275	294	322		296	249	228	Garner and Pandy [64] and Holzbaur et al. [71]
Mirakhorlo et al. [83]	215	260	153	206	10.5 21.5	25		Cadaver study

References

1. Ma'touq, J., Hu, T., Haddadin, S.: Sub-millimetre accurate human hand kinematics: from surface to skeleton. Comput. Methods Biomech. Biomed. Eng. **21**(2), 113–128 (2018)
2. Ma'touq, J., Hu, T., Haddadin, S.: A validated combined musculotendon path and muscle-joint kinematics model for the human hand. Comput. Methods Biomech. Biomed. Eng. **22**(7), 727–739 (2019)
3. Gialias, N., Matsuoka, Y.: A musculotendon contribution for multijoint hand control. In: 2006 International Conference of the IEEE Engineering in Medicine and Biology Society, pp. 4482–4485 (2006)
4. Matsuoka, Y., Afshar, P.: Neuromuscular strategies for dynamic finger movements: a robotic approach. In: Conference Proceedings : Annual International Conference of the IEEE Engineering in Medicine and Biology Society. 26th Conference IEEE Engineering in Medicine and Biology Society, pp. 4649–4652 (2004)
5. Serbest, K., et al.: Development of a human hand model for estimating joint torque using MATLAB tools. In: 2016 6th IEEE International Conference on Biomedical Robotics and Biomechatronics (BioRob), pp. 793–797 (2016)
6. Agarwal, P., Hechanova, A., Deshpande, A.A.: Kinematics and dynamics of a biologically inspired index finger exoskeleton. In: 13th ASME Dynamic Systems and Control Conference, Palo Alto, CA, USA (2013)
7. Kuo, P.-H., Deshpande, A.D.: Contribution of passive properties of muscle-tendon units to the metacarpophalangeal joint torque of the index finger. In: 2010 3rd IEEE RAS & EMBS International Conference on Biomedical Robotics and Biomechatronics, pp. 288–294 (2010)

8. Kuo, P.-H., Deshpande, A.D.: Muscle-tendon units provide limited contributions to the passive stiffness of the index finger metacarpophalangeal joint. J. Biomech. **45**(15), 2531–2538 (2012)
9. Deshpande, A.D., Gialias, N., Matsuoka, Y.: Contributions of intrinsic viscoelastic torques during planar index finger and wrist movements. IEEE Trans. Biomed. Eng. **59**(2), 586–594 (2012)
10. Hollerbach, J.M., Flash, T.: Dynamic interactions between limb segments during planar arm movement. Biol. Cybern. **44**(1), 67–77 (1982)
11. Gribble, P.L., Ostry, D.J.: Compensation for interaction torques during single- and multijoint limb movement. J. Neurophysiol. **82**(5), 2310–2326 (1999)
12. Dounskaia, N., Wisleder, D., Johnson, T.: Influence of biomechanical factors on substructure of pointing movements. Exp. Brain Res. **164**(4), 505–516 (2005)
13. Esteki, A., Mansour, J.M.: An experimentally based nonlinear viscoelastic model of joint passive moment. J. Biomech. **29**(4), 443–450 (1996)
14. Fok, K.S., Chou, S.M.: Development of a finger biomechanical model and its considerations. J. Biomech. **43**(4), 701–713 (2010)
15. Yun, M., Jun Eoh, H., Cho, J.: A two-dimensional dynamic finger modeling for the analysis of repetitive finger flexion and extension. Int. J. Ind. Ergon. **29**, 231–248 (2002)
16. Daumas, B., Xu, W., Bronlund, J.: Jaw mechanism modeling and simulation. Mech. Mach. Theory **40**(7), 821–833 (2005)
17. Jamshidi, N., et al.: Modelling of human walking to optimise the function of ankle-foot orthosis in Guillan-Barré patients with drop foot. Singapore Med. J. **50**(4), 412–417 (2009)
18. Serbest, K., Cilli, M., Eldogan, O.: Biomechanical effects of daily physical activities on the lower limb. Acta Orthop Traumatol Turc **49**(1), 85–90 (2015)
19. Hu, T.: Analysis and modeling of human arm withdrawal reflexes using surface electromyography. MA thesis. Leibniz Universität Hannover (2015)
20. Hu, T.: A physiological arm model based on the analysis of human withdrawal reflexes using surface electromyography. MA thesis. Leibniz Universität Hannover (2016)
21. Esteki, A., Mansour, J.M.: A dynamic model of the hand with application in functional neuromuscular stimulation. Ann. Biomed. Eng. **25**(3), 440–451 (1997)
22. Zajac, F.: Muscle and tendon: properties, models, scaling, and application to biomechanics and motor control. Crit. Rev. Biomed. Eng. **17**, 359–411 (1989)
23. Romero, F., Alonso, F.J.: A comparison among different Hill-type contraction dynamics formulations for muscle force estimation. Mech. Sci. **7**(1), 19–29 (2016)
24. Huxley, A.F.: Muscle structure and theories of contraction. Progress Biophys. Biophys. Chem. **7**, 257–318 (1958)
25. Hill, A.V.: The heat of shortening and the dynamic constants of muscle. Proc. R. Soc. Lond. Ser. B Biol. Sci. **126**(843), 136–195 (1938)
26. Audo, M.L., Davy, D.T.: The influence of muscle model complexity in musculoskeletal motion modeling. J. Biomech. Eng. **107**(2), 147–157 (1985)
27. Van Soest, A.J., Bobbert, M.F.: The contribution of muscle properties in the control of explosive movements. Biol. Cybern. **69**, 195–204 (1993)
28. de Jager, M.: Mathematical head-neck models for acceleration impacts. English. Ph.D. thesis. Eindhoven University of Technology (1996)
29. Martins, J.A.C., et al.: A numerical model of passive and active behavior of skeletal muscles. Comput. Methods Appl. Mech. Eng. **151**(3), 419–433 (1998)

30. Rosen, J., Fuchs, M.B., Arcan, M.: Performances of Hill-type and neural network muscle models – toward a myosignal-based exoskeleton. Comput. Biomed. Res. **32**(5), 415–439 (1999)
31. Kaplan, M.: Efficient optimal control of large-scale biomechanical systems. Ph.D. thesis. Stanford University (2000)
32. Silva, M.: Human motion analysis using multibody dynamics and optimisation tools. Ph.D. thesis. Instituto Superior Técnico, Universidade Técnica De Lisboa, Lisbon, Portugal (2003)
33. Vilimek, M.: Musculotendon forces derived by different muscle models. Acta Bioeng. Biomech. **9**(2), 41–47 (2007)
34. Haeufle, D., et al.: Hill-type muscle model with serial damping and eccentric force-velocity relation. J. Biomech. **47**(6), 1531–1536 (2014)
35. Jovanovic, K., Vranic, J., Miljkovic, N.: Hill's and Huxley's muscle models - tools for simulations in biomechanics. Serbian J. Electr. Eng. **12**, 53–67 (2015)
36. Kuthe, C., et al.: Mathematical model for skeletal muscle to simulate the concentric and eccentric contraction. Mol. Cell. Biomech. MCB **12**, 1–16 (2015)
37. Hamouda, A., Kenney, L., Howard, D.: Dealing with time-varying recruitment and length in Hill-type muscle models. J. Biomech. **49**(14), 3375–3380 (2016)
38. Winters, J.M.: Hill-based muscle models: a systems engineering perspective. In: Winters, J.M., Woo, S.L. (eds.) Multiple Muscle Systems: Biomechanics and Movement Organization, pp. 69–93. Springer, New York (1990). Chap. 5
39. Collins, J.J.: The redundant nature of locomotor optimization laws. J. Biomech. **28**, 251–267 (1995)
40. Crowninshield, R.D., Brand, R.A.: A physiologically based criterion of muscle force prediction in locomotion. J. Biomech. **14**(11), 793–801 (1981)
41. An, K.N., et al.: Determination of muscle and joint forces: a new technique to solve the indeterminate problem. J. Biomed. Eng. **106**(4), 364–367 (1984)
42. Kaufman, K.R., et al.: Physiological prediction of muscle forces-II. Application to isokinetic exercise. Neuroscience **40**, 793–804 (1991)
43. Anderson, F., Pandy, M.: Static and dynamic optimization solutions for gait are practically equivalent. J. Biomech. **34**, 153–161 (2001)
44. Anderson, F., Pandy, M.: Individual muscle contributions to support in normal walking. Gait Posture **17**, 159–169 (2002)
45. Thelen, D.G., Anderson, F.C., Delp, S.L.: Generating dynamic simulations of movement using computedmuscle control. J. Biomech. **36**, 321–328 (2003)
46. Thelen, D.G., Anderson, F.C.: Using computed muscle control to generate forward dynamic simulations of human walking from experimental data. J. Biomech. **39**(6), 1107–1115 (2006)
47. Sancho-Bru, J.L., et al.: A 3-D dynamic model of human finger for studying free movements. J. Biomech. **34**(11), 1491–1500 (2001)
48. Kamper, D.G., Hornby, T.G., Rymer, W.Z.: Extrinsic flexor muscles generate concurrent flexion of all three finger joints. J. Biomech. **35**(12), 1581–1589 (2002)
49. Qiu, D., Kamper, D.G.: Orthopaedic applications of a validated force-based biomechanical model of the index finger. In: 2014 36th Annual International Conference of the IEEE Engineering in Medicine and Biology Society, pp. 4013–4016 (2014)
50. Deshpande, A.D., et al.: Control strategies for the index finger of a tendon-driven hand. Int. J. Robot. Res. **32**, 115–128 (2013)
51. Pylatiuk, C., et al.: Progress in the development of a multifunctional hand prosthesis. In: Annual International Conference of the IEEE Engineering in Medicine and Biology Society, vol. 6, pp. 4260–4263. IEEE Engineering in Medicine and Biology Society (2004)

52. Lotti, F., et al.: Development of UB hand 3: early results. In: Proceedings of the 2005 IEEE International Conference on Robotics and Automation, pp. 4488–4493 (2005)
53. Hajian, A.Z.: A characterization of the mechanical impedance of human hands. Ph.D. thesis. Harvard University (1997)
54. Hajian, A.Z., Howe, R.D.: Identification of the mechanical impedance at the human finger tip. J. Biomech. Eng. **119**, 109–114 (1997)
55. Jindrich, D.L., Balakrishnan, A.D., Dennerlein, J.T.: Finger joint impedance during tapping on a computer keyswitch. J. Biomech. **37**(10), 1589–1596 (2004)
56. Xu, Z., et al.: Design and analysis of an artificial finger joint for anthropomorphic robotic hands. In: 2011 IEEE International Conference on Robotics and Automation, pp. 5096–5102 (2011)
57. Vignais, N., Marin, F.: Analysis of the musculoskeletal system of the hand and forearm during a cylinder grasping task. Int. J. Ind. Ergon. **44**(4), 535–543 (2014)
58. Keir, P.J., Wells, R.P., Ranney, D.A.: Passive properties of the forearm musculature with reference to hand and finger postures. Clin. Biomech. **11**(7), 401–409 (1996)
59. Knutson, J.S., et al.: Intrinsic and extrinsic contributions to the passive moment at the metacarpophalangeal joint. J. Biomech. **33**, 1675–1681 (2000)
60. Tkany, C.: Modeling and control of the human hand identification testbed. Student project, Institute of Automatic Control, Leibniz Universität Hannover (2018)
61. Silder, A., et al.: Identification of passive elastic joint moment-angle relationships in the lower extremity. J. Biomech. **40**(12), 2628–2635 (2007)
62. Dionysian, E., et al.: Proximal interphalangeal joint stiffness: measurement and analysis. J. Hand Surg. **30**(3), 573–579 (2005)
63. Lemay, M.A., Crago, P.E.: A dynamic model for simulating movements of the elbow, forearm, and wrist. J. Biomech. **29**(10), 1319–1330 (1996)
64. Garner, B.A., Pandy, M.G.: Estimation of musculotendon properties in the human upper limb. Ann. Biomed. Eng. **31**(2), 207–220 (2003)
65. Lieber, R.L., et al.: Architecture of selected muscles of the arm and forearm: anatomy and implications for tendon transfer. J. Hand Surg. **17**(5), 787–798 (1992)
66. Jacobson, M.D., et al.: Architectural design of the human intrinsic hand muscles. J. Hand Surg. **17**(5), 804–809 (1992)
67. Millard, M., et al.: Flexing computational muscle: model and simulation of musculotendon dynamics. J. Biomech. Eng. **135**, 1–11 (2013)
68. Lloyd, D.G., Besier, T.F.: An EMG-driven musculoskeletal model to estimate muscle forces and knee joint moments in vivo. J. Biomech. **36**(6), 765–776 (2003)
69. Buchanan, T.S., et al.: Neuromusculoskeletal modeling: estimation of muscle forces and joint moments and movements from measurements of neural command. J. Appl. Biomech. **20**(4), 367–395 (2004)
70. Lee, J.H.: A musculoskeletal model of the human hand to improve human device interaction. Ph.D. thesis. Arizona State University (2014)
71. Holzbaur, K.R.S., Murray, W.M., Delp, S.L.: A model of the upper extremity for simulating musculoskeletal surgery and analyzing neuromuscular control. Ann. Biomed. **33**(6), 829–840 (2005)
72. Wu, J.Z., et al.: Analysis of musculoskeletal loading in an index finger during tapping. J. Biomech. **41**(3), 668–676 (2008)
73. Spector, S.A., et al.: Muscle architecture and force-velocity characteristics of cat soleus and medial gastrocnemius: implications for motor control. J. Neurophysiol. **44**(5), 951–960 (1980)
74. Wickiewicz, T.L., et al.: Muscle architecture and force-velocity relationships in humans. J. Appl. Physiol. **57**(2), 435–443 (1984)

75. Weijs, W.A., Hillen, B.: Cross-sectional areas and estimated intrinsic strength of the human jaw muscles. Acta Morphol. Neerl.-Scand. **23**, 267–274 (1985)
76. Rassier, D.E., Macintosh, B.R., Herzog, W.: Length dependence of active force production in skeletal muscle. J. Appl. Physiol. **86**, 1445–1457 (1999)
77. Cavallaro, E., et al.: Hill-based model as a myoprocessor for a neural controlled powered exoskeleton arm - parameters optimization. In: Proceedings of the 2005 IEEE International Conference on Robotics and Automation, pp. 4514–4519 (2005)
78. Thelen, D.: Adjustment of muscle mechanics model parameters to simulate dynamic contractions in older adults. J. Biomech. Eng. **125**, 70–77 (2003)
79. Amis, A.A., Dowson, D., Wright, V.: Muscle strengths and musculo-skeletal geometry of the upper limb. Eng. Med. **8**, 41–48 (1979)
80. Brand, P.W., Beach, R.B., Thompson, D.E.: Relative tension and potential excursion of muscles in the forearm and hand. J. Hand Surg. **6**, 209–219 (1981)
81. Ward, S.R., et al.: High stiffness of human digital flexor tendons is suited for precise finger positional control. J. Neurophysiol. **96**(5), 2815–2818 (2006)
82. Sharma, M., et al.: A study of muscle morphology of anterior group of forearm muscles. J. Exerc. Sci. Physiother. **5**, 34–37 (2009)
83. Mirakhorlo, M., et al.: Anatomical parameters for musculoskeletal modeling of the hand and wrist. Int. Biomech. **3**(1), 40–49 (2016)
84. Borst, J., et al.: Muscle parameters for musculoskeletal modelling of the human neck. Clin. Biomech. **26**(4), 343–351 (2011)
85. Murray, W.M., Buchanan, T.S., Delp, S.L.: The isometric functional capacity of muscles that cross the elbow. J. Biomech. **33**, 943–952 (2000)
86. Kapandji, A.I.: The Physiology of the Joints: Annotated Diagrams of the Mechanics of the Human Joints, vol. 2, 6th edn. Churchill Livingstone, London (2007)
87. Cobos, S., et al.: Human hand descriptions and gesture recognition for object manipulation. Comput. Methods Biomech. Biomed. Eng. **13**(3), 305–317 (2010)
88. Deshpande, A.D., et al.: Acquiring variable moment arms for index finger using robotic testbed. IEEE Trans. Biomed. Eng. **57**(8), 2034–2044 (2010)
89. Heintz, S.: Muscular forces from static optimization. Technical report, Royal Institute of Technology, KTH Mechanics, Stockholm, Sweden (2006)
90. Troncossi, M., Mozaffari-Foumashi, M., Parenti-Castelli, V.: An original classification of rehabilitation hand exoskeletons. J. Robot. Mech. Eng. Res. **1**, 17–29 (2016)
91. Brand, P.W., Hollister, A.M.: Clinical Mechanics of the Hand, 3rd edn. Mosby Inc., St. Louis (1999). English

4D Point Cloud Registration for Tumor Vascular Networks Monitoring from Ultrasensitive Doppler Images

E. Cohen[1,2](✉), T. Deffieux[1], C. Demené[1], L. D. Cohen[2], and M. Tanter[1]

[1] Institut Langevin, ESPCI ParisTech, PSL Research University, CNRS, UMR 7587, INSERM U979, 75005 Paris, France
cohen.emm@gmail.com,
{thomas.deffieux,charlie.demene,mickael.tanter}@espci.fr

[2] University Paris Dauphine, PSL Research University, CNRS, UMR 7534, CEREMADE, 75016 Paris, France
cohen@ceremade.dauphine.fr

Abstract. Numerous diseases find their origin and their diagnosis in the physiological behavior of vascular networks [10]. In particular, to understand the architecture and growth of a tumor the study of blood flows is crucial. Recently, Ultrasensitive Doppler has enabled 4D ultrasound imaging of tumor micro-vasculature in mice [5]. In this study, we propose new computational tools to monitor the growth of a tumor vascular network by registering in time and space this new highly sensitive temporal data. We first quantify the acquired data using the minimal-path based framework we introduced in [4]; the vascular network paths around the tumor are segmented from images obtained for four days of observation; local geometrical parameters such as diameters are also estimated. Then, using a point cloud representation of the segmented vascular networks, we develop point cloud registration algorithms that automatically align similar vascular structures, thus allowing a better visualization of the growth and the evolution of the tumor vascular network. A rigid registration model is first considered by manually selecting similar features from two temporal different observations of the tumor. More accurate results are then obtained by automatically extracting invariant vascular patterns. Finally, combining rigid transformations to non-linear deformation models produce a very accurate time matching between invariant vascular structures.

1 Introduction

1.1 Context

Medical ultrasound imaging counts among the most used clinical imaging techniques, very appreciated for its portability, real time working and low-cost. It provides an anatomical imaging of high quality, usually associated to a Doppler

G. A. Ateshian et al. (Eds.): CMBBE 2019, LNCVB 36, pp. 437–456, 2020.
https://doi.org/10.1007/978-3-030-43195-2_35

examination for blood flows observation and quantification. In the last few years, ultrafast ultrasound imaging [23] revolutionized the capacity to observe very fast physiological variations, such as shear waves propagation in the body. In fact, such waves propagate at typical speeds of a few tens of $m \cdot s^{-1}$, involving very high frame rates ~1 kHz (1000 images/sec) to be detected. Instead of using focused ultrasounds and forming the final image column by column, ultrafast imaging sends an ultrasonic plane wave that propagates and reflects into the whole image plan. Thus, echoes of all the structures back-propagate in the same time, and the final image can be reconstructed in only one ultrasonic emission. Ultrasensitive Doppler is one of these new ultrasensitive techniques which allows high sensitive acquisition of small vascular features without contrast agent.

Beyond anatomical vascular imaging, ultrasensitive Doppler allows for the first time to perform functional imaging with ultrasound (functional ultrasound or fUS). Mace et al. [15] demonstrated a first proof of concept by observing significant variations of blood flux in the rat brain, following a stimulation of the rat whiskers or an epileptiform seizures.

Being tens of times more sensitive than conventional Doppler [16], ultrasensitive Doppler allows very fine imaging of blood flows. Using a 15 MHz linear ultrasound probe, typical 2D spatial resolution obtained in the plane of imaging is $100\,\mu m^2$. Yet, resolution in the out-of-plane direction remains higher, typically equal to $500\,\mu m$ for rodent brain imaging. Thus, Demene et al. [6] proposed an ultrafast Doppler tomography based system (UFD-T) able to produce 3D microvascular maps of rodent brain, with high spatio-temporal resolution of $100\,\mu m^3 \times 10\,ms$. The acquisition is achieved in vivo via a simple tomographic mechanical set-up. UFD-T scanning times, ~20 min for the whole rodent brain surface, are slightly below MRI and CT performances. For real time applications, other approaches using probes with multidimensional arrays of transducers could be more advantageous (e.g. [18]). However, due to high temporal resolution of ultrafast imaging, 3D reconstructed volumes by UFD-T contain a large quantity of temporal information, revealing how promising is the technique in the context of 4D imaging. Nowadays, with the new design of matrix-array ultrasonic probes, real 3D ultrafast imaging can be achieved *in vivo* within a single

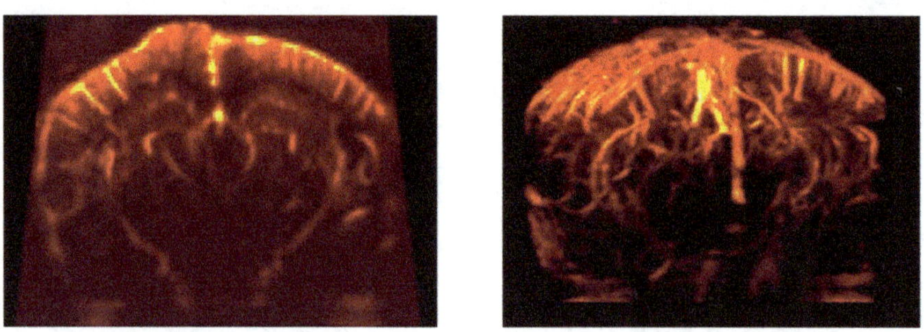

Fig. 1. Ultrasensitive Doppler imaging of the rat brain. Left: 2D ultrasensitive Doppler coronal plane. Right: 3D tomographic reconstruction UFD-T [6].

ultrafast acquisition. It was demonstrated for 3D shear-wave imaging, 3D ultrafast Doppler imaging [18] or functional imaging in rodents [19] (Fig. 1).

1.2 Motivations

Ultrafast ultrasound has introduced a new way of high resolution vascular imaging that benefits of all the convenient advantages of ultrasonic devices (real time, portability, low cost). Numerous of applications are currently in development with the production of new available data. This imaging data could be then analyzed by the implementation of new algorithms in the field of image analysis and computer vision.

In the past, Folkman [10] has already explained that tumor growth is deeply connected to angiogenesis, the creation of new blood vessels responsible for the provision in nutriments and oxygen to the tumor. Many cancers are treated with anti-angiogenetic strategies acting on tumor vasculature [14]. Monitoring of tumors and their surrounding vascular networks is thus essential to understand tumor growth and architecture.

Vascular networks can be very complex structures with many different type of vessels. In particular, tumors are characterized by highly tortuous vessels that can appear spontaneously during the tumor growth. The variability in time of those networks makes also their analysis very challenging.

UFD-T enables 3D imaging of tumors microvasculature. Demene et al. [5] have imaged the temporal evolution of a tumor implanted in the back of mice. Such impressive data need rigorous numerical analysis to be quantified. Thus, we used our minimal path based framework [4] to segment vascular networks surrounding the tumor and estimate local geometrical parameters (diameter, curvature, etc.). Such extraction of the vasculature allowed us to perform 3D

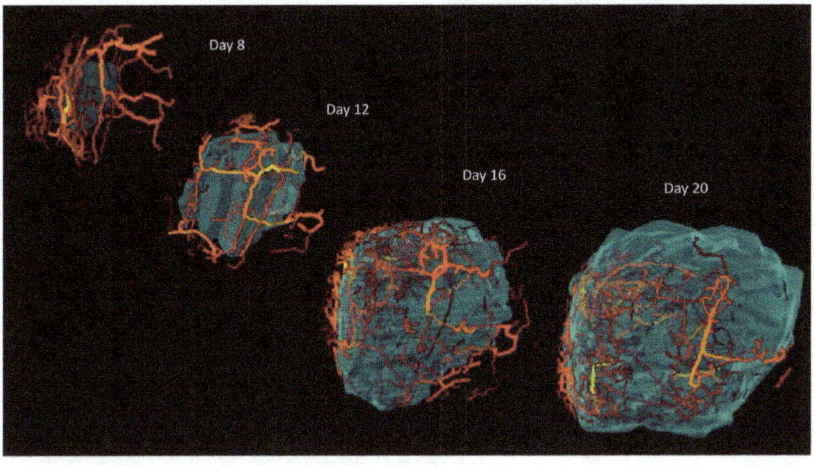

Fig. 2. Tumor growth reconstruction and visualization over four days of observations after implantation in a mouse back using our segmentation and registration pipeline.

registration of the temporal data, leading to more visual understanding of the tumor evolution and a better description of local variations. We implemented a 3D reconstruction and rendering algorithm showing the registered tumor data and its evolution (see Fig. 2). We also investigated non-linear models to match similar tumor vascular features and try to detect temporal repeatability despite the disordered evolution of a tumor vascular network.

2 Material and Mathematical Background

2.1 Material

The data used for this study was acquired by UFD-T on four mice, in the flank of which were implanted subcutaneously 10 mm^3 tumor fragments, coming from mouse Lewis lung carcinoma cells. All experiments were approved by the French Ministry of Agriculture (protocol authorization: Ce5/2012/082). More details can be found in [5]. The data has been jointly acquired on the four mice on days 8, 12, 16 and 20 after the implantation of the tumor.

We would like to understand how the tumor evolves in time from days 8 to 20. The growth of the tumor causes a global proliferation of the vascular network that is well observable, but a local comparison of the networks is hard without the use of other methods.

2.2 Rigid Point Cloud Registration

Point cloud registration is a well-known issue in the field of robotics, computer vision and graphics, and the medical field e.g. for image-guided interventions [17]. The underlying mathematical problem is to find the best geometric transformation describing the relationship between two sets of points or point clouds. The first question to answer before solving the problem is how far the two point clouds are from each other; in other words, how to model the geometric transformation between them? A first approach is to use a rigid, or linear, displacement model involving in general three kinds of operations: translation, rotation and scale. Let us note that there may be no exact solution to this problem, in particular when data are complicated involving non-linear deformation. Yet, it is still interesting to obtain a preliminary approximation that globally match the points, and further envisage a more refined model.

2.2.1 The Procrustes Problem

Let $\mathbf{X} = (x_i)$ and $\mathbf{Y} = (y_i)$ be two 3D point clouds of respectively N_x and N_y points. We will consider that $\mathbf{X}$ is the moving data set and $\mathbf{Y}$ the static model, and will search for the best rigid transformation T such as $\mathbf{Y} \sim T(\mathbf{X})$. A simple case, called Procrustes problem, is when the correspondences between points is known, i.e. when $N_x = N_y = n$ and each point x_i corresponds to the same indexed point y_i. Schonemann et al. [22] and Horn [13] have both proved least-square solutions of this problem. In what follows, we sum up the main theoretical steps. The idea is to find the transformation minimizing the sum of square errors between points. The least-square functional to be minimized is

$$F(t, R, s) = \sum_{i=1}^{n} \|y_i - sRx_i - t\|^2, \tag{1}$$

where the rigid model parameters that must be found are t, R and s, respectively denoting a translation vector, a rotation matrix and a scale scalar parameter. Let us define the centroids of each point cloud by

$$\bar{x} = \frac{1}{n}\sum_{i=1}^{n} x_i, \qquad \bar{y} = \frac{1}{n}\sum_{i=1}^{N_x} y_i, \tag{2}$$

and denote new demeaned coordinates

$$x'_i = x_i - \bar{x}, \qquad y'_i = y_i - \bar{y}. \tag{3}$$

The minimization function can be modified as follows

$$F(t, R, s) = \sum_{i=1}^{n} \|y'_i - sRx'_i - t'\|^2, \tag{4}$$

with $t' = t - \bar{y} - sR\bar{x}$. Expanding the last equation gives

$$\sum_{i=1}^{n} \|y'_i - sRx'_i\|^2 - 2\left\langle t', \sum_{i=1}^{n} [y'_i - sRx'_i] \right\rangle + n\|t'\|^2 \tag{5}$$

By noting that $\sum_{i=1}^{n} x'_i = \sum_{i=1}^{n} y'_i = 0$, it comes that the middle term of this expression is zero. This simplification of F allows to easily conclude that t' should be equal to zero to minimize F. Indeed, the first term does not depend on t' and the third one is a square so is positive. Thus, we have already found the translation

$$t = \bar{y} + sR\bar{x} \tag{6}$$

To determine the scale, we keep on expanding F

$$\sum_{i=1}^{n} \|y'_i - sRx'_i\|^2 = \sum_{i=1}^{n} \|y'_i\|^2 - 2s\sum_{i=1}^{n} \langle y'_i, Rx'_i\rangle + s^2\sum_{i=1}^{n} \|Rx'_i\|^2 \tag{7}$$

Since the rotation R is an orthogonal operator, it preserves the euclidean norm, allowing to write $\|Rx'_i\| = \|x'_i\|$. We then rewrite the last expression as follows

$$S_y - 2sD + s^2 S_x, \tag{8}$$

where $S_y = \sum_{i=1}^{n} \|y'_i\|^2$, $S_x = \sum_{i=1}^{n} \|x'_i\|^2$, and $D = \sum_{i=1}^{n} \langle y'_i, Rx'_i\rangle$. Finally, by factorizing this quadratic equation in s as

$$\left(s\sqrt{S_x} - \frac{D}{\sqrt{S_x}}\right)^2 + \frac{S_y S_x - D^2}{S_x}, \tag{9}$$

and noting that $S_y S_x - D^2 \geq 0$ (Cauchy–Schwarz inequality), the first term should be equal to zero to minimize F and we get the optimal value that s should take

$$s = \frac{D}{S_x} = \frac{\sum_{i=1}^{n} \langle y_i', Rx_i' \rangle}{\sum_{i=1}^{n} \|x_i'\|^2}. \tag{10}$$

The determination of the rotation is more complicated and a prove can be found in [21] using the singular value decomposition (SVD) method, or in [13] using the quaternion representation of rotations. We give here the SVD-based result. Defining the cross-covariance 3 matrix

$$B = \sum_{i=1}^{n} y_i' x_i'^T, \tag{11}$$

and realizing its SVD decomposition $B = U \Sigma V^T$, where U, V are two unitary matrices and Σ a diagonal matrix, the rotation is given by

$$R = U \mathrm{diag}([1, 1, \det(UV)]) V^T. \tag{12}$$

The operator diag transforms an input vector into a diagonal matrix whose diagonal is composed of this vector components. An other expression of s using Σ can also be found in [22]

$$s = \frac{\mathrm{trace}(\Sigma)}{\sum_{i=1}^{n} \|x_i'\|^2}. \tag{13}$$

Considering that s is a scalar, the above problem is called the isotropic Procrustes problem. Several authors have proposed to solve the anisotropic Procrustes problem which can be express as the minimization of the following least-square functional

$$F(t, R, s) = \sum_{i=1}^{n} \|y_i - RSx_i - t\|^2, \tag{14}$$

where S is a diagonal matrix modeling the scale anisotropy. The last expression refers to the *pre-scaling* problem where scaling S precedes the rotation R, while the *post-scaling* problem minimizes

$$F(t, R, s) = \sum_{i=1}^{n} \|y_i - SRx_i - t\|^2. \tag{15}$$

To solve the pre-scaling problem, Gower et al. [11] used an iterative procedure called block relaxation (BR). Each iteration has two steps. First, for fixed R the computation of the three diagonal coefficients of S

$$\lambda_i = \frac{(B^T R)_{ii}}{\left(\sum_{i=1}^{n} x_i' x_i'^T\right)_{ii}}, \tag{16}$$

then for fixed S, the computation of R

$$R = UV^T, \tag{17}$$

where $U\Sigma V$ is now the singular value decomposition of BS. The algorithm iterates until the change in least-square error falls below a certain threshold. The authors used a similar BR algorithm for the post-scaling problem. Dosse et al. [7] pointed out that the BR procedure can lead to local minima corresponding to incorrect solutions. They proposed an other solution less sensitive to this kind of problem. In this study, we will always refer to the pre-scaling anisotropic problem.

2.2.2 Iterative Closest Point Algorithm

Let us now assume that the correspondence between points is unknown. The problem becomes much more complicated. Indeed, obtaining an exact solution by solving a least-square problem is not possible anymore because we do not know the pairs of points whose square errors should be minimized. Thus, Besl et al. [1] introduced a new iterative method, called iterative closest point (ICP), to register 3D point clouds which attempts to match the points in $\mathbf{X}$ and $\mathbf{Y}$ by using a closest point approach.

Let us describe an iteration k of the algorithm. We denote by $\mathbf{X}^k$ the transformed point cloud issued from $\mathbf{X}$ at the k-th iteration. Firstly, the algorithm determines the closest points $\tilde{y}_i$ in $\mathbf{Y}$ to each $x_i \in \mathbf{X}^k$, by minimizing the euclidean distance

$$\tilde{y}_i = \arg\min_{y \in \mathbf{Y}} \|y - x_i\| \tag{18}$$

Note that, as soon as a point in $\mathbf{Y}$ is designated as a closest point, it cannot be selected again. As a consequence, it should be imposed that $N_y \geq N_x$, for instance by always choosing the moving point cloud $\mathbf{X}$ to be the one with the fewest points. Then, the algorithm solves a Procrustes problem to register $\mathbf{X}^k$ on the set of closest points $\tilde{\mathbf{Y}}^k$. The Procrustes problem can be isotropic or anisotropic as proposed by [3] for the registration of medical ultrasound images. The rigid transformation found is finally applied to $\mathbf{X}^k$ to obtain the next iteration $\mathbf{X}^{k+1}$. The algorithm stops the iterations when the change in least-square error falls below a threshold parameter.

2.3 Deformable Models

When dealing with real data, especially with tumor data where vascular networks may greatly vary in time, geometrical transformations that map corresponding points can be very complicated. Therefore, deformable models are usually preferred to get a more precise estimation of the transformation. A deformable model aims to find a local transformation. Minimizing an energy functional, a set of parameters is estimated locally for each pair of corresponding points, allowing to map the points with many degrees of freedom and making the transformation more realistic.

A very famous deformable model is the *Thin-Plate-Spline* (TPS) transformation introduced by [2]. Given two sets of n corresponding points $\mathbf{X} = (X_i)$ and $\mathbf{X}' = (X_i')$, the method tries to find the best mapping function $f : \mathbf{X} \to \mathbf{X}'$ minimizing the following energy

$$\int\int_{\mathbb{R}^2} \left(\left(\frac{\partial^2 f_i}{\partial x^2}\right)^2 + 2\left(\frac{\partial^2 f_i}{\partial x \partial y}\right)^2 + \left(\frac{\partial^2 f_i}{\partial y^2}\right)^2 \right) dxdy, \tag{19}$$

in the case of 2D points, where $i \in \{1, ..., 2\}$, or

$$\int\int_{\mathbb{R}^2} \left(\left(\frac{\partial^2 f_i}{\partial x^2}\right)^2 + \left(\frac{\partial^2 f_i}{\partial y^2}\right)^2 + \left(\frac{\partial^2 f_i}{\partial z^2}\right)^2 + \cdots \right.$$
$$\left. 2\left(\frac{\partial^2 f_i}{\partial x \partial y}\right)^2 + 2\left(\frac{\partial^2 f_i}{\partial y \partial z}\right)^2 + 2\left(\frac{\partial^2 f_i}{\partial x \partial z}\right)^2 \right) dxdydz, \tag{20}$$

in 3D, where $i \in \{1, ..., 3\}$ and f_i denotes the Cartesian coordinates of f. It can be shown that minimizers exist and take the following respective 2D and 3D forms

$$\begin{aligned} f_1(x_i, y_i) &= a_0 + a_1 x_i + a_2 y_i + \sum_{j=1}^{n} F_j r_{ij}^2 \ln(r_{ij}^2) \\ f_2(x_i, y_i) &= b_0 + b_1 x_i + b_2 y_i + \sum_{j=1}^{n} G_j r_{ij}^2 \ln(r_{ij}^2), \end{aligned} \tag{21}$$

and

$$\begin{aligned} f_1(x_i, y_i, z_i) &= a_0 + a_1 x_i + a_2 y_i + a_3 z_i + \sum_{j=1}^{n} F_j r_{ij} \\ f_2(x_i, y_i, z_i) &= b_0 + b_1 x_i + b_2 y_i + b_3 z_i + \sum_{j=1}^{n} G_j r_{ij} \\ f_3(x_i, y_i, z_i) &= c_0 + c_1 x_i + c_2 y_i + c_3 z_i + \sum_{j=1}^{n} H_j r_{ij}, \end{aligned} \tag{22}$$

for all points $i \in \{1, ..., n\}$, and where $r_{ij} = \parallel X_i - X_j \parallel$. Inspired by [9], let us give some technical details in the case of 2D TPS about how the unknown parameters a_0, a_1, a_2, b_0, b_1, b_2, $F_1, \cdots, F_n$ and $G_1, \cdots, G_n$ can be computed. If those parameters are determined, we found a solution to the corresponding minimization problem. If we introduce the following matrices and vectors

$$P = \begin{pmatrix} 1 & x_1 & y_1 \\ \vdots & \vdots & \vdots \\ 1 & x_n & y_n \end{pmatrix}, \tag{23}$$

$$K = \begin{pmatrix} 0 & r_{12}^2 \ln(r_{12}^2) & \cdots & r_{1n}^2 \ln(r_{1n}^2) \\ r_{21}^2 \ln(r_{21}^2) & 0 & \cdots & r_{2n}^2 \ln(r_{2n}^2) \\ \vdots & \vdots & \vdots & \vdots \\ r_{n1}^2 \ln(r_{n1}^2) & r_{n2}^2 \ln(r_{n2}^2) & \cdots & 0 \end{pmatrix}, \tag{24}$$

$$L = \left(\begin{array}{c|c} P & K \\ \hline 0 & P^T \end{array}\right) \tag{25}$$

$$w_1 = \begin{pmatrix} a_0 \\ a_1 \\ a_2 \\ F_1 \\ \vdots \\ F_n \end{pmatrix}, \quad w_2 = \begin{pmatrix} b_0 \\ b_1 \\ b_2 \\ G_1 \\ \vdots \\ G_n \end{pmatrix}, \tag{26}$$

$$x' = \begin{pmatrix} x'_1 \\ \vdots \\ x'_n \\ 0 \\ 0 \\ 0 \end{pmatrix}, \quad y' = \begin{pmatrix} y'_1 \\ \vdots \\ y'_n \\ 0 \\ 0 \\ 0 \end{pmatrix}, \tag{27}$$

the system of Eq. 21 can be expressed in terms of matrices as follows

$$\begin{aligned} x' &= Lw_1 \\ y' &= Lw_2. \end{aligned} \tag{28}$$

Notice that the last three lines of the system correspond to 6 additional equations given by

$$\begin{aligned} \sum_1^n F_i = 0, \quad \sum_1^n G_i = 0 \\ \sum_1^n x_i F_i = 0, \quad \sum_1^n x_i G_i = 0 \\ \sum_1^n y_i F_i = 0, \quad \sum_1^n y_i G_i = 0. \end{aligned} \tag{29}$$

They can be interpreted as some boundary conditions that force f to be an affine transformation in the case x and y tend to infinity. The affine part of the transformation is parametrized by the a_i, b_i coefficients. Thus, by inverting the matrix L, we can derive all the unknown parameters and finally solve the problem

$$\begin{aligned} w_1 &= L^{-1} x' \\ w_2 &= L^{-1} y'. \end{aligned} \tag{30}$$

3 Methods and Results

Once vascular networks have been segmented, they can be used to accurately register 3D images. It is clear that a sparse representation of vascular networks offers many more accurate possibilities of quantification and registration. Numerous works can be found in the literature dealing with vascular structures registration, e.g. [8,12,20].

The model used to represent vascular networks is essential for the registration to be successful. We proposed a representation in graph structure as described in [4]. This graph representation is very useful to understand how the network is organized, and to decompose in a logically and organized order the segmented structures. However, this model is somewhat limited by the errors introduced by the algorithm of segmentation. Due to noise, complexity of data, or lack of precision, robustness in the method, there should be aberrant bifurcations or edges in the resulting segmentation. Therefore, a more general representation in point clouds should be associated to the graph-based model. For registration of point clouds, our work is based on a modified version of ICP and inspired by [20].

The methods are presented and applied on 3D data acquired on rodent brains and tumors. They have been developed to solve several issues: the detection of vascular invariant structures in the brain for the production of cerebral atlas; the detection of invariant structures in the vasculature of tumors for a better understanding of their architecture; the visualization of tumors growth.

3.1 Rigid Initialization

Our goal is to realize the temporal registration of the UFD-T 3D tumor images. Here, we will start finding only rigid transformations between the data, being an initial guess into our global registration process. For each day of observation, we apply the segmentation framework presented in [4], and obtain a complete description of the vascular network including centrelines, bifurcations, and local diameters (Fig. 3). We also get the corresponding local blood flow intensities for each point of the segmented network, by interpolating the values on the original image.

3.1.1 Manual Selection of Feature Points

Now visualizing the segmented networks, it is possible to recognize some vascular features which seem to repeat themselves in the different days of observation of the same tumor. They generally correspond to vessels with high diameters and high blood flows. Our assumption is that it should always exist such time-invariant vessels that feed the tumor and from which the tumor builds a new multitude of small and tortuous vessels. A first attempt to register two vascular networks is to select manually on each network few feature points characterizing the shape of the recognizable vessels, and try to find the rigid transformation between each point set. In fact, we select pairs of feature points corresponding

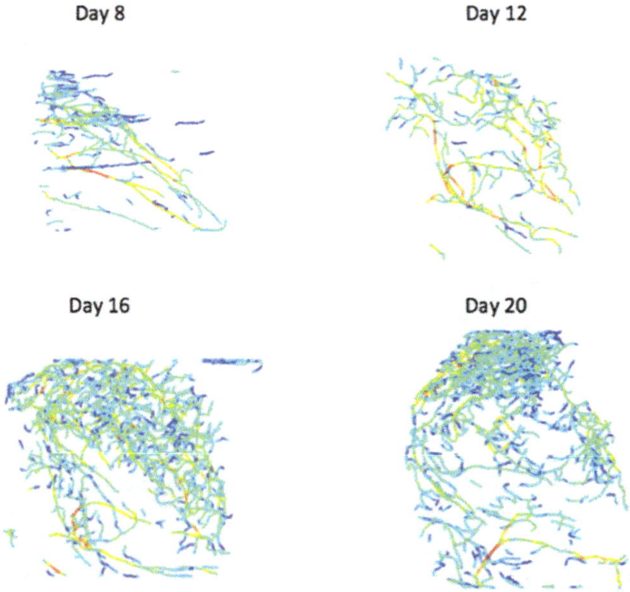

Fig. 3. Segmentation of the tumor data for each day of observation. The color map corresponds to diameters, with large and small vessels respectively in red and blue.

to the same approximate positions on each vascular network. This allows us to find the translation, rotation and scale parameters by resolving a Procrustes problem with 4 pairs of points. Figure 4 shows the segmented networks of two days of observation of the same tumor, respectively in green (day 8) and green (day 12). Four points are manually selected on each data set around the shape of interest (in bold on the figure). Once the rigid transformation is computed, the transformation is then applied to all the moving network. We observe that the shapes in bold do not exactly overlap but globally the networks have been brought closer to each other.

If we apply the same process to each day of observation of the same tumor, trying now to register day 16 on 12 and day 20 on 16, we obtain a global visualization of the tumor growth shown by Fig. 5. The dark points are the feature points manually selected on each vascular network.

3.1.2 Automatic Detection of Features

The previous registration based on manual selection of feature points is clearly not enough accurate, and one must take benefit of more feature points or shapes. Thus, we turn to a point cloud representation of the segmented vascular networks. The major question to answer is how to reject non-relevant points and keep only repeatable vascular patterns that will make the registration successful. Besides, the correspondences between points are now unknown, so we should use

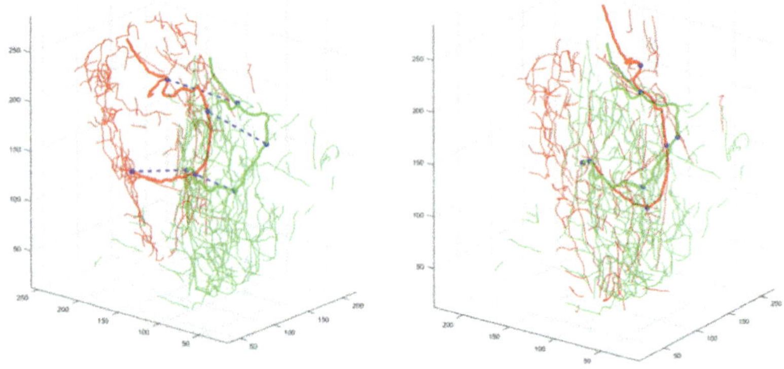

Fig. 4. Registration of tumor temporal data by manual selection of features. Two days of observations of the same tumor, days 8 and 12, are respectively in red and green. Left: four pairs of matching points are manually selected and linked with dot lines on the figure. Right: rigid registration of the red (moving) network onto the green (static) network, computing the transformation using the matching pairs of points in (a).

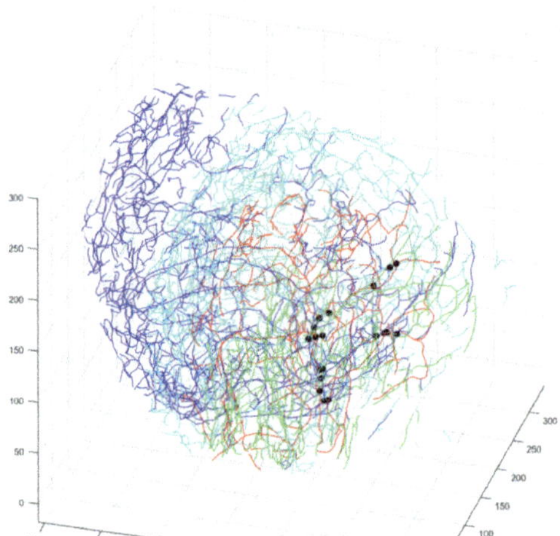

Fig. 5. Tumor growth visualization using rigid registration with manual selection of features. The features points are in dark. Observation days 8, 12, 16 and 20 of the tumor are respectively in green, red, cyan and blue.

the ICP algorithm to iteratively match the selected points and converge towards an acceptable solution.

Using ICP directly on the whole point clouds, the presence of outliers and vascular structures of high temporal variability may cause similar or even worse results than those of the previous manual selection model. Even if such result has

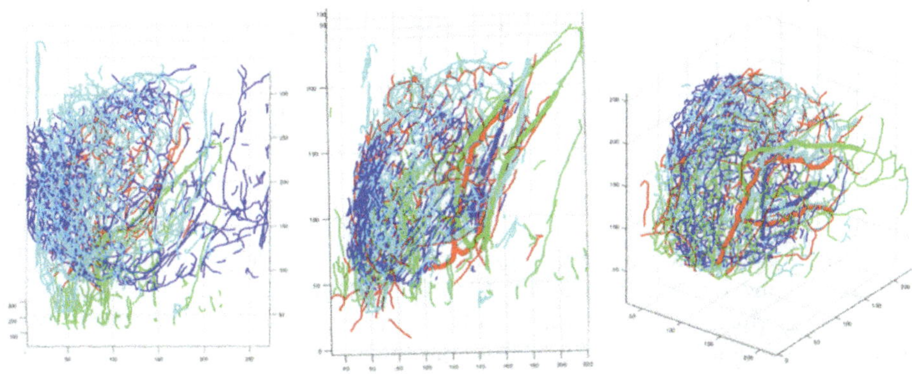

Fig. 6. Non-rigid ICP registration with scale parameter using automatic pre-selection of features. From top to bottom: three different viewing angles of the registered tumor data. Observation days 8, 12, 16 and 20 of the tumor are respectively in green, red, cyan and blue. Some invariant vascular patterns were manually annotated in bold to simply evaluate the quality of the registration.

the advantage to be an automatic procedure without manually selecting points, we must improve it in accuracy. Therefore, we pre-select the points used for ICP by choosing only those with high diameters and blood flow intensities. This is based on the assumption that tumors grow from an invariant set of large vessels. We order points according to diameter and blood flow intensity and only keep the union of points with the highest values. It is remarkable that considering vessel segments instead of points leads to better ICP performance, probably due to the continuity property of a segment allowing the algorithm to better identify vascular structures.

Thus, we apply the ICP algorithm on the pre-selected vessel segments with a non-rigid transformation adding a scale parameter to the model. The result, presented in Fig. 6, gives the great impression that the data could be finely aligned and proves that only four observations allow to describe the tumor growth dynamic. Some vascular structures, whose the repeatability is easily recognizable on the different days of observation of the tumor, are manually highlighted in bold. After registration, we observe that those structures are mainly situated in the same area, with around them the proliferation of a big network of small and more unstable vessels.

3.1.3 Automatic Detection of Local Invariant Patterns

The previous method remains particularly unstable with the possible detection of false local minima in the registration point cloud algorithm. The registration should be much more improved. In order to accurately align the repeatable structures, we propose a more local approach. In fact, the pre-selected segment features previously used from each tumor data can be partitioned into sets of connected segments, enabling to locally register smaller vascular shapes one on the other instead of performing the global registration of several shapes. To detect if two segments are connected, we take advantage of the graph representation used in [4] for the segmentation of vascular networks. When two segments E_1 and E_2 have a node in common, we connect and include them into a subgraph

$$G = \bigcup_i E_i \tag{31}$$

We continue to fill each subgraph until no more pairs of common nodes can be detected. Figure 7 shows on the left the pre-selected vascular segments from a one-day-observation tumor data. We observe the presence of several groups of connected segments, especially a large one looking like half of an ellipse. This large vascular structure corresponds to a large vessel repeated on each day of observation of the tumor. The other structures seem to be more variable and may not be so much useful for the alignment of the different tumor data. On the right of the same figure, the connected components are separated by individual colors and will be now analyzed separately. We propose to identify among those subsets of vessels the most invariant structures when we observe the tumor along its days of growth. For this purpose, we must find some features characterizing the components that can be best matched as for instance the observable semi-elliptic pattern. As a first assumption, we claim that the longer and the largest a pattern G is, the more invariant it will be. Therefore, we use as a first feature the total curve length of the connected components and as a second one, a simple estimator of the space extension of the pattern

$$\text{Max Distance}(G) = \max_{(E_i, E_j) \in G} \|m(E_i) - m(E_j)\|, \tag{32}$$

where $m(E_i)$ is the geodesic middle of the segment E_i. By plotting on a graph, Fig. 8, those two features for each of the four days of observation of the same

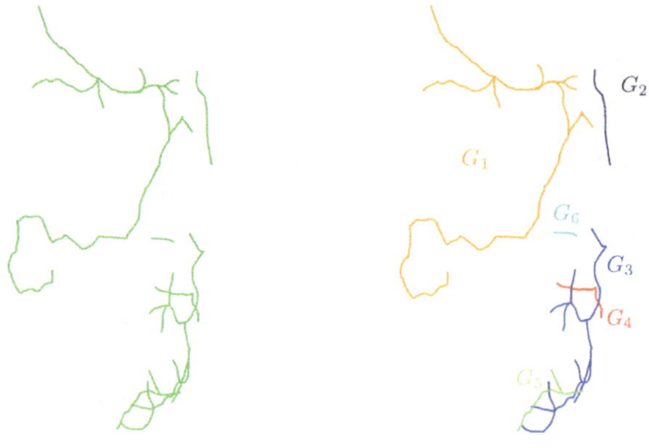

Fig. 7. Decomposition of selected features into connected sub-graphs of paths that can be processed separately. Left: selected features. Right: each connected sub-graphs of paths is plotted with a particular color.

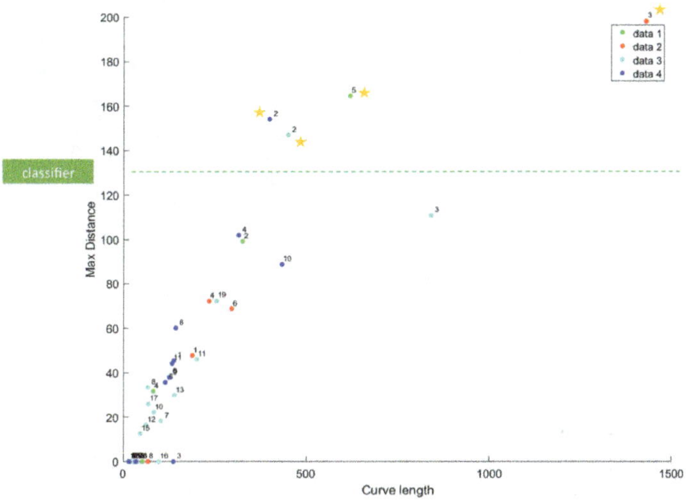

Fig. 8. Classification of sub-graphs according to their max-distance values. Each temporal data has its own color. Each point in the graph corresponds to a connected sub-graph. The points annotated with a star are the invariant detected structures, presenting highest max-distance.

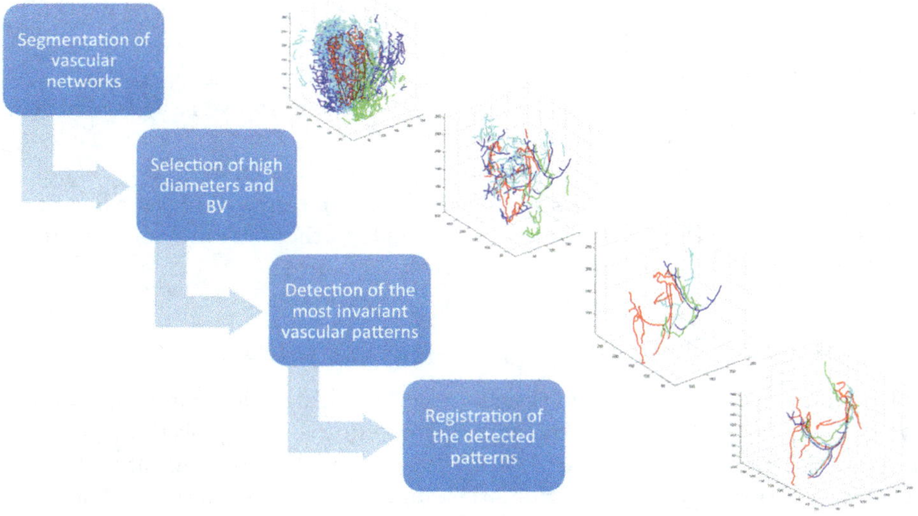

Fig. 9. The proposed registration pipeline. Observation days 8, 12, 16 and 20 of the tumor are respectively in green, red, cyan and blue

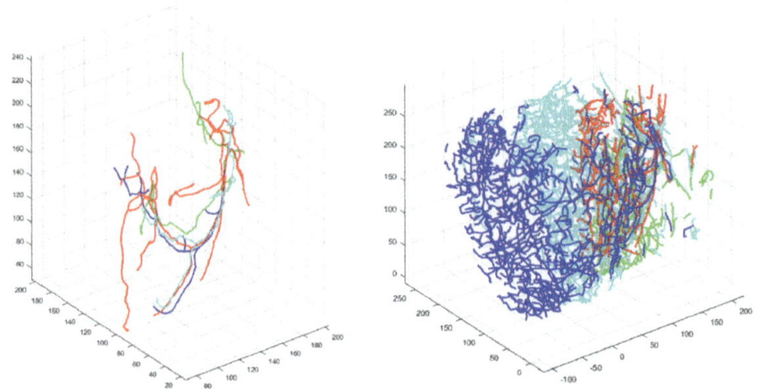

Fig. 10. Best invariant structures rigid registration. Left: similar connected components are automatically extracted from each days of observation of the tumor, and finally registered. Right: the transformation found is applied to all the point clouds in order to nicely visualize the tumor growth.

tumor (one color corresponds to one day of observation, data 1 to 4), we observe that one specific component designated by a gold star can be linearly separated from the rest of the patterns. This specific component is exactly the half elliptic form mentioned above. This method allows us to detect the most proper component in each day of observation and use it to register all the data.

The process of pre-selection of high diameters and blood volume and the detection of the best characterizing component is shown on Fig. 9. At the end, we obtain four components looking very similar. We then apply the ICP algorithm with a rigid transformation to register those four vascular patterns. The rigid transformation found by the algorithm is then applied to all the data points, leading finally to a global rigid registration of the four days of observation of the tumor (Fig. 10). The result is much more accurate than the previous attempts, and the growth of the tumor can be properly visualized.

3.2 Non-linear Registration

Once initial rigid transformations have been found, the different temporal data can be visualized in the same coordinates system and the growth of the tumor is more clearly depicted (Fig. 10). It is now reasonable to think about a non-rigid registration step, in order to match similar vascular structures with higher accuracy. This will help us to understand how the tumor is changing locally during its growth, improving the detection of invariant or unstable vascular areas.

For this purpose, we looked for a TPS deformable transformation that would best model tumor evolution. The TPS operator outputs a non-linear mapping between the matching points of two different point clouds. Since the correspondence between matching points is *a priori* unknown, we must incorporate the

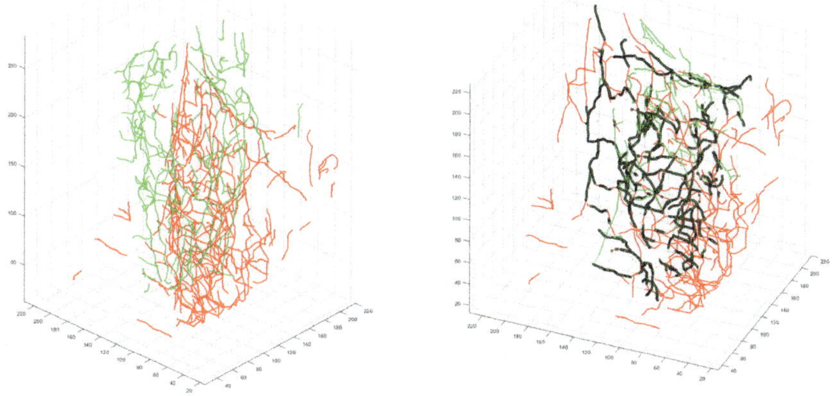

Fig. 11. Non-linear ICP registration with TPS. Left: unregistered tumor data. Right: registered tumor data using TPS. The green point cloud is the moving one. Black points are the transformed points whose distance from the red network is lower than one pixel. Those points most probably belong to vessel invariant structures.

TPS model into an ICP procedure. Thus, we apply a modified version of the ICP algorithm in which the Procrustes problem solving step is replaced by the computation of the TPS transformation.

The result is very satisfying as many vessels are accurately matched. Figure 11 shows a tumor vascular network at two different days of observation. On the left, the networks have only been aligned according to the initial rigid transformation computed in the last section, but they are not completely registered yet. On the right, the non-linear component of the transformation, including scaling and other local deformation, is found using the TPS modified ICP version. In black are represented the TPS transformed points of the green vascular network (the green is the moving point cloud while the red remains static) whose distance from the red network is lower than one pixel. For those points, the matching seems to be very accurate.

To be sure that vascular patterns are consistently matched, we plotted Fig. 12 a color map showing the similarity between the two networks found by TPS. This map can be easily obtained by plotting on the left graph the inverse coordinates (i.e. their original coordinates before TPS transformation) of the best matching points colored in black on Fig. 11, and on the right the corresponding matching points of the red network. We can observe that many vascular patterns owning the same color in both maps own also very similar shapes. This confirms at least qualitatively that the TPS model performs well.

3.3 3D Reconstruction of Tumor Growth

Once tumor data has been acquired at several different times, one difficult task is to adopt the proper display to understand the temporal dynamic of the data.

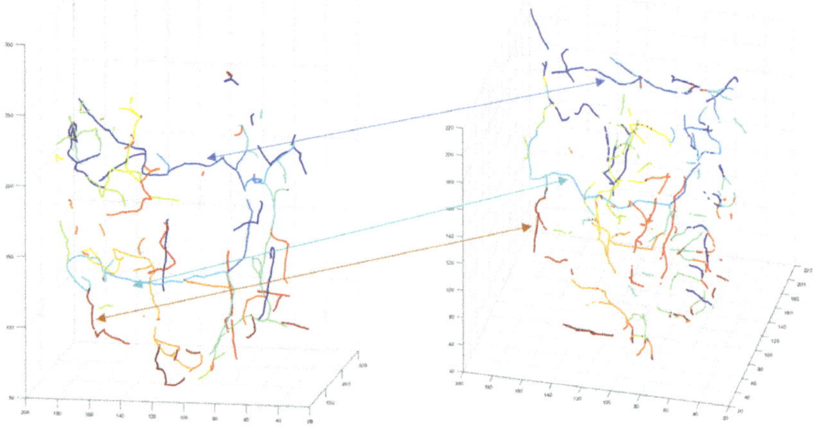

Fig. 12. Matching of vascular structures from temporal tumor data. Left: observation day 8. Right: observation day 12. Matching structures are drawn with the same color on both graphs.

The acquired images are *a priori* not properly registered, therefore it is hardly possible to identify and follow by eye the expansion and development of the tumor. Our rigid registration algorithm allows us to align the data in the same coordinate system and then to create a film displaying clearly the evolution of the tumor. Note that given the local diameter and the blood flow intensity at each point of the segmented vascular network, we can reconstruct the data in 3D using a color map corresponding to the blood flow. Figure 2 shows the 3D reconstruction of a tumor observed at 4 different days. Those images are screen shots of a film showing the data aligned and growing from a fixed point, allowing to clearly observe the expansion of the tumor, the proliferation of small unstable vessels, and the invariance of several big vascular patterns. The part of the tumor represented in blue is actually the real tumor tissue that was segmented from the same ultrasensitive Doppler acquisition using B-mode as explained in [5].

4 Conclusion

In this study, we have developed advanced numerical methods for the analysis of new medical ultrasensitive Doppler data. We used our segmentation framework [4] as a basis for the development of tumor data monitoring in 3D. Extracted tumor networks are represented as point clouds that can be automatically registered using rigid and non-rigid transformations.

For the detection of vascular brain invariant structures and the possible creation of a brain vascular atlas, our framework of segmentation represents an important tool to extract the geometry parameters and get a fine description of vascular networks. Using unsupervised machine learning techniques could be a very interesting way to cluster groups of segmented vessels and automatically detect invariant patterns. Further, by acquiring a significant amount of data, one could build a statistical atlas by discriminating the invariant from the variable parts of the networks. The non-linear registration methods used in this study are a good start to extract the invariant part of a network by selecting the best matching structures.

Finally, supervised machine and deep learning methods would be of great interest if we could build large labeled datasets from microvascular ultrasensitive Doppler data. Techniques of data augmentation could improve performances if the number of data is not sufficient and one could train a model to automatically detect repeatable vascular shapes on new data.

References

1. Besl, P.J., McKay, N.D.: Method for registration of 3-D shapes. In: Sensor Fusion IV: Control Paradigms and Data Structures, vol. 1611, pp. 586–607. International Society for Optics and Photonics (1992)
2. Bookstein, F.L.: Principal warps: thin-plate splines and the decomposition of deformations. IEEE Trans. Pattern Anal. Mach. Intell. **11**(6), 567–585 (1989)
3. Chen, E.C., McLeod, A.J., Baxter, J.S., Peters, T.M.: Registration of 3D shapes under anisotropic scaling. Int. J. Comput. Assist. Radiol. Surg. **10**(6), 867–878 (2015)
4. Cohen, E., Cohen, L.D., Deffieux, T., Tanter, M.: An isotropic minimal path based framework for segmentation and quantification of vascular networks. In: International Workshop on Energy Minimization Methods in Computer Vision and Pattern Recognition, pp. 499–513. Springer, Heidelberg (2017)
5. Demene, C., Payen, T., Dizeux, A., Barrois, G., Gennisson, J.L., Bridal, L., Tanter, M.: Comparison of tumor microvasculature assessment via ultrafast doppler tomography and dynamic contrast enhanced ultrasound. In: 2014 IEEE International Ultrasonics Symposium (IUS), pp. 421–424. IEEE (2014)
6. Demené, C., Tiran, E., Sieu, L.A., Bergel, A., Gennisson, J.L., Pernot, M., Deffieux, T., Cohen, I., Tanter, M.: 4D microvascular imaging based on ultrafast doppler tomography. NeuroImage **127**, 472–483 (2016)
7. Dosse, M.B., Ten Berge, J.: Anisotropic orthogonal procrustes analysis. J. Classif. **27**(1), 111–128 (2010)
8. Dufour, A.: Segmentation et modélisation des structures vasculaires cérébrales en imagerie médicale 3D. Ph.D. thesis, Université de Strasbourg (2013)
9. Fitzpatrick, J.M., Hill, D.L., Maurer, C.R., et al.: Image registration. Handb. Med. Imaging **2**, 447–513 (2000)
10. Folkman, J.: Tumor angiogenesis: therapeutic implications. N. Engl. J. Med. **285**(21), 1182–1186 (1971)
11. Gower, J.C., Dijksterhuis, G.B.: Procrustes Problems, vol. 30. Oxford University Press on Demand (2004)
12. Groher, M.: 2D-3D registration of vascular images. Ph.D. thesis, Technische Universität München (2008)
13. Horn, B.K.: Closed-form solution of absolute orientation using unit quaternions. JOSA A **4**(4), 629–642 (1987)
14. Jain, R.K.: Normalization of tumor vasculature: an emerging concept in antiangiogenic therapy. Science **307**(5706), 58–62 (2005)

15. Macé, E., Montaldo, G., Cohen, I., Baulac, M., Fink, M., Tanter, M.: Functional ultrasound imaging of the brain. Nat. Methods **8**(8), 662 (2011)
16. Mace, E., Montaldo, G., Osmanski, B.F., Cohen, I., Fink, M., Tanter, M.: Functional ultrasound imaging of the brain: theory and basic principles. IEEE Trans. Ultrason. Ferroelectr. Freq. Control **60**(3), 492–506 (2013)
17. Peters, T., Cleary, K.: Image-Guided Interventions: Technology and Applications. Springer, Heidelberg (2008)
18. Provost, J., Papadacci, C., Arango, J.E., Imbault, M., Fink, M., Gennisson, J.L., Tanter, M., Pernot, M.: 3D ultrafast ultrasound imaging in vivo. Phys. Med. Biol. **59**(19), L1 (2014)
19. Rabut, C., Finel, V., Correia, M., Pernot, M., Deffieux, T., Tanter, M.: Full 4D functional ultrasound imaging in rodents using a matrix array. In: 2017 IEEE International Ultrasonics Symposium (IUS), p. 1. IEEE (2017)
20. Reinertsen, I., Descoteaux, M., Siddiqi, K., Collins, D.L.: Validation of vessel-based registration for correction of brain shift. Med. Image Anal. **11**(4), 374–388 (2007)
21. Schönemann, P.H.: A generalized solution of the orthogonal procrustes problem. Psychometrika **31**(1), 1–10 (1966)
22. Schönemann, P.H., Carroll, R.M.: Fitting one matrix to another under choice of a central dilation and a rigid motion. Psychometrika **35**(2), 245–255 (1970)
23. Tanter, M., Fink, M.: Ultrafast imaging in biomedical ultrasound. IEEE Trans. Ultrason. Ferroelectr. Freq. Control **61**(1), 102–119 (2014)

Knee Medial and Lateral Contact Forces Computed Along Subject-Specific Contact Point Trajectories of Healthy Volunteers and Osteoarthritic Patients

Raphael Dumas[1(✉)], Ali Zeighami[2,3], and Rachid Aissaoui[2,3]

[1] Univ Lyon, Université Claude Bernard Lyon 1, IFSTTAR, LBMC UMR_T9406, Lyon, France
raphael.dumas@ifsttar.fr
[2] Laboratoire de Recherche en Imagerie et Orthopédie, Centre de Recherche du CHUM, Montréal, Canada
[3] Département de génie des systèmes, École de Technologie Supérieure, Montréal, Canada

Abstract. Tibiofemoral medial and lateral contact forces, analysed on healthy volunteers, seem to depend on the trajectories of the contact points. Contact point trajectories in OA patients are shifted in the medial direction and contact forces are generally reported to be slightly reduced.

The present study compares medial and lateral contact forces between OA patients and healthy volunteers. The forces are estimated during gait using a musculoskeletal model with subject-specific contact point trajectories obtained from biplane X-ray images at different flexion angles. Large inter-subject variability was found in the contact point trajectories and the contact forces. Significant but weak correlations were found between the positions of the contact points in the medial-lateral direction and the peaks of medial contact forces for both healthy volunteers and OA patients: the more medial the contact points the lower the forces.

In the literature, when computing the contact forces considering a frontal equilibrium only, the correlation is obviously strong. Relationship between the positions of the contact points and the contact forces is more controversial in studies using deformable knee models. The interactions between altered contact points and contact forces should be further investigated with subject-specific musculoskeletal models.

Keywords: Gait analysis musculoskeletal modelling · Tibiofemoral loading

1 Introduction

Tibiofemoral medial and lateral contact forces estimated from musculoskeletal modelling depend on the trajectories of the contact points [1–3].

The partition of the total contact forces is generally based on the adduction-adduction moment, the resultant of the musculo-tendon forces spanning the joint, and the position of the contact points in the frontal plane of the tibia. For a given adduction-adduction

G. A. Ateshian et al. (Eds.): CMBBE 2019, LNCVB 36, pp. 457–463, 2020.
https://doi.org/10.1007/978-3-030-43195-2_36

moment (i.e. obtained by inverse dynamics), and a given resultant of the musculo-tendon forces (i.e. obtained by static optimisation or any alternative method), the medial and lateral contact forces are then fully determined by the position of the contact points. Moving the contact points in the medial-lateral direction results in altering the contact forces by a gradient up to 0.06 BW/mm and may also result in unloading one compartment [1, 2]. When considering this frontal plane equilibrium, the tibiofemoral alignment can be also taken into account. Both contact point positions and alignment can be informed through radiographic analysis [1]. Considering a tri-dimensional (3D) equilibrium of the tibiofemoral joint (i.e. with flexion-extension, adduction-abduction, and internal-external rotation moments) and computing the musculo-tendon forces and contact forces simultaneously somewhat mitigate the dependence of the medial and lateral contact forces on the position of the contact points [3]. This was reported for subject-specific contact point trajectories of healthy volunteers which remain aligned with the medial and lateral plateau centrelines. In the pathologic case such as osteoarthritis (OA), the contact point trajectories appear to be shifted in the medial direction [4] and to demonstrate a lower excursion in the lateral compartment [5].

In the same time, the medial and lateral contact forces have been reported to be generally similar or slightly reduced in OA patients compared to healthy volunteers [6–9]. Only Meireles et al. reported slightly higher forces in OA patients [10]. In these studies, the contact point trajectories were not subject-specific, either set at two fixed positions in the frontal plane [6, 7] or computed by a generic deformable model of the knee [8–10]. Interestingly, Meireles et al. found that the medial contact point estimated by the deformable knee model at the timing of the first peak had a more posterior-lateral location in severe OA while Van Rossom et al. reported that the contact locations were not significantly different between OA patients and healthy volunteers.

The objective of the present study is to compare medial and lateral contact forces between OA patients and healthy volunteers using a lower limb musculoskeletal model which include subject-specific contact point trajectories to better understand how they differ and how they can be differently affected by the contact point positions.

2 Materials and Methods

This study includes the data of 10 healthy volunteers (6 men, 4 women, 55 yrs., 1.68 m, 71 kg) and 12 severe OA patients (2 men, 10 women, 59 yrs., 1.61 m, 85.53 kg, K-L grade 4). All subjects signed an informed consent form and the experimental protocol was approved by the ethics committees of the *Centre de Recherche, Centre Hospitalier de l'Université de Montréal* (CRCHUM) and *Ecole de Technologie Supérieure de Montréal* (ETS). All the research and methods in this study were performed in accordance with the CRCHUM and ETS ethics committee guidelines as well as with the Helsinki Declaration of 1975, as revised in 2000.

The 3D models of their tibia and femur were reconstructed and registered using EOS^{TM} low-dose biplane X-ray images at 0°, 15°, 30°, 45°, and 70° of knee flexion.

A lower limb musculoskeletal model with five joint degrees of freedom and 43 muscle lines of action [11] was scaled to the subject anthropometry using the segment lengths. The subject-specific contact point trajectories were approximated using a weighted centre of the bone-to-bone distances [4] and were expressed as a function of the knee flexion angle. These subject-specific contact point trajectories were introduced as kinematic constraints into the musculoskeletal model [3]. The corresponding Lagrange multipliers were the medial and lateral contact forces computed simultaneously with the musculo-tendon forces.

All forces were estimate during 45 s of gait at comfortable speed using the data recorded by an instrumented split-belt treadmill (AMTI) and reflective markers placed on the body segments and on the kneeKG^{TM} system tracked by a 12-camera Vicon system. The contact forces were normalized to body weight (BW) and to 100% of gait stance.

Linear regressions were computed between the peaks of medial and lateral contact forces and the positions of the contact points in the medial-lateral direction at the corresponding timings.

3 Results

3.1 Contact Point Trajectories and Contact Forces

Figure 1 displays the individual contact point trajectories during weight-bearing quasi-static squat (plotted over the generic bone of the tibia used for the 3D reconstruction [4]) and the individual medial and lateral contact forces during gait for all healthy volunteers and OA patients.

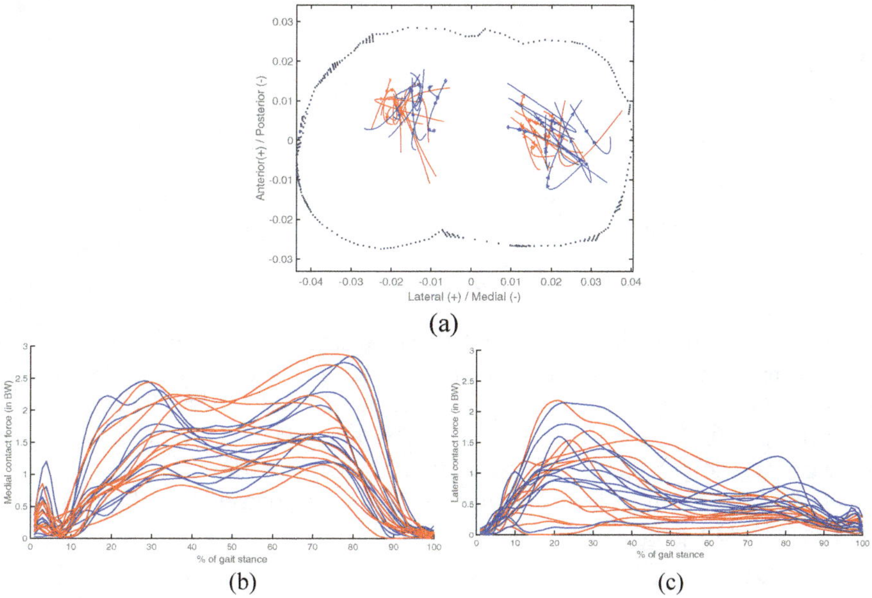

Fig. 1. (a) Contact point trajectories (in m), (b) medial contact force (in BW), (c) lateral contact force (in BW) for healthy volunteers (in blue) and OA patients (in red).

The contact point trajectories demonstrated a large inter-subject variability. In the medial compartment, on the medial-lateral axis, the trajectories fell between −6.4 and −26.8 mm with respect to the tibia centerline for the healthy volunteers and between −5.3 and −24.8 mm for the OA patients. In the lateral compartment, the trajectories fell between 9.1 and 34.8 mm with respect to the tibia centerline for the healthy volunteers and between 10.2 and 37.6 mm for the OA patients.

The contact forces also demonstrated some inter-subject variability. The medial contact force at first peak was 1.6 ± 0.5 BW (mean and standard deviation) for the healthy volunteers and 1.6 ± 0.5 BW for the OA patients. At second peak, the medial contact force was 1.9 ± 0.6 BW and 1.7 ± 0.6 BW, respectively. The lateral contact force was 1.0 ± 0.5 BW for the healthy volunteers and 1.0 ± 0.6 BW for the OA patients at first peak, 0.6 ± 0.3 and 0.5 ± 0.2 BW, respectively, at second peak.

3.2 Linear Regressions

Figure 2 displays the linear regressions obtained between the first and second peaks of medial and lateral contact forces and the positions of the contact points in the medial-lateral direction at the corresponding timings. The slopes obtained in the literature by sensitivity analysis on the positions of the contact points (between ±16 mm and ±24 mm [1] and between ±15 mm and ±35 mm [2]) are also summarized.

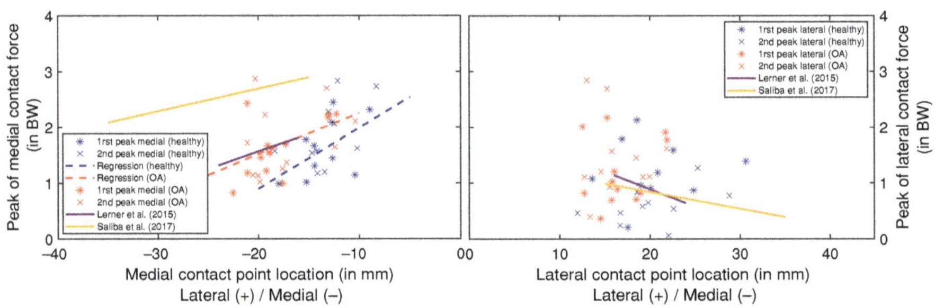

Fig. 2. Regression between first and second peaks of medial and lateral contact forces (in BW) and positions of the contact points in the medial-lateral direction (in mm) at the corresponding timings.

Linear regressions were found for the relationship between the positions of the contact points in the medial compartment and the first and second peaks of medial contact forces for both healthy volunteers ($R^2 = 0.24$, $p = 0.001$) and OA patients ($R^2 = 0.13$, $p = 0.01$). The slopes were 0.13 BW/mm and 0.07 BW/mm, respectively. No linear regression was found for the relationship between the positions of the contact points in the lateral compartment and the peaks of lateral contact forces for both healthy volunteers and OA patients.

4 Discussion

The present study compares medial and lateral contact forces between healthy volunteers and OA patients when introducing subject-specific contact point trajectories in the musculoskeletal model. The mean estimated contact forces were comparable to the forces reported in the literature for similar case-control studies [6–10] and demonstrated slightly lower forces for the OA patient (Table 1).

Table 1. Medial and lateral contact forces reported in the case-control studies of the literature

Authors		n	Medial		Lateral	
			1rst peak	2nd peak	1rst peak	2nd peak
Kumar et al. [7]	Healthy	16	2.4	1.8	1.3	0.5
	OA	12	2.6	2.1	0.9	0.1
Sritharan et al. [8]	Healthy	19	2.0	3.0	0.3	0.5
	OA	39	2.1	3.2	0.3	0.6
Dell'Isola et al. [6]	Healthy	18	2.1		1.4	
	OA	17	2.0		1.1	
Meireles et al. [10]	Healthy	17	2.2	1.8	1.2	1.2
	Early OA	14	2.6	1.9	1.4	1.5
	OA	16	2.8	2.1	1.3	1.6
Van Rossom et al. [9]	Healthy	19	1.8	1.9	1.3	1.0
	OA (Medial)	8	1.9	1.8	1.1	0.8
	OA (Lateral)	7	1.5	1.6	1.2	1.0
Present study	Healthy	10	1.6	1.9	1.0	0.6
	OA	12	1.6	1.7	1.0	0.5

For both healthy volunteers and OA patients, a large inter-subject variability was found in the contact point trajectories and the contact forces but without a strong covariation. This may be due to the fact that the contact point trajectories were measured during weight-bearing quasi-static squat while the contact forces are estimated during gait. Yet, a recent systematic review showed that the contact point trajectories obtained by biplane X-ray images during quasi-static squat matched the contact point trajectories obtained by fluoroscopy, MRI or CT during other activities (gait, step-up, kneeling, lunge) [5]. It is possible, in both healthy volunteers and OA patients, to measure the contact point trajectories during gait using bi-plane fluoroscopy [12], however, to the authors' knowledge, this information has never been introduced in subject-specific musculoskeletal model so far.

When computing the contact forces considering a frontal equilibrium only, the contact forces are directly determined by the contact point positions (proportional to the inverse of the contact point positions). Conversely, in the present study, weak correlations were found in the medial compartment and no correlation at all in the lateral compartment. Still, the slopes obtained by sensitivity analysis [1, 2] and the one obtained in the present study are comparable, although higher for the healthy volunteers. The fact Meireles et al. (2017) found, in the medial compartment for OA patients, higher contact forces and, in the same time, more lateral contact points is also consistent in terms of variation.

The weak correlation and lower slope for OA demonstrate that the contact forces are potentially governed by other parameters such as joint moments and knee alignment as reported in the literature [13–15]. Because the contact point trajectories are introduced as kinematic constraints into the lower limb musculoskeletal model, not only the position of the contact points but also the joint kinematics (i.e. including the adduction-adduction angle) is subject-specific. The relationship between knee alignment and the contact forces is therefore not expected to be strong for the OA patients of the present study but this should be tested in further analyses. Another sensitivity analysis in the literature [16] also reported a relationship between the position of the contact points in the anterior-posterior direction and the contact forces and, again, this should be tested in further analyses.

This study has some limitations. The study sample is small. It ranges between 12 and 39 (Table 1) in the previous studies comparing medial and lateral contact forces between healthy volunteers and OA patients [6–9]. Only the contact point trajectories were subject-specific. The musculoskeletal geometry was only scaled. However, it is important to keep in mind that the contact point trajectories were introduced as kinematic constraints leading also to subject-specific tibiofemoral kinematics. These contact point trajectories (and the kinematics) were measured during weight-bearing quasi-static squat and not during walking but it can be expected that the same inter-subject variability (especially for the OA patients) could be observed during gait.

5 Conclusion

In the present study, like in most case-control studies of the literature, medial and lateral contact forces during gait appears to be slightly lower in OA patients compared to healthy volunteers. Based on the subject-specific contact point trajectories, weak correlations were found between the positions of the contact points and the peaks of contact forces for both healthy volunteers and OA patients: the more medial the contact points the lower the forces in the medial compartment.

In the literature, when computing the contact forces considering a frontal equilibrium only, the correlation is obviously strong. Relationship between the positions of the contact points and the contact forces is more controversial in studies using deformable knee models. The interactions between altered contact points and contact forces should be further investigated with subject-specific musculoskeletal models.

Acknowledgments. Fonds de Recherche du Quebec en Sante (FRQ-S), Fonds de Recherche du Quebec en Nature et Technologie (FRQNT), Natural Science and Research Council of Canada (NSERC), LABEX PRIMES (ANR-11-LABX-0063) of Université de Lyon.

References

1. Lerner, Z.F., DeMers, M.S., Delp, S.L., Browning, R.C.: How tibiofemoral alignment and contact locations affect predictions of medial and lateral tibiofemoral contact forces. J. Biomech. **48**, 644–650 (2015)
2. Saliba, C.M., Brandon, S.C.E., Deluzio, K.J.: Sensitivity of medial and lateral knee contact force predictions to frontal plane alignment and contact locations. J. Biomech. **57**, 125–130 (2017)
3. Zeighami, A., Aissaoui, R., Dumas, R.: Knee medial and lateral contact forces in a musculoskeletal model with subject-specific contact point trajectories. J. Biomech. **69**, 138–145 (2018)
4. Zeighami, A., Dumas, R., Kanhonou, M., Hagemeister, N., Lavoie, F., de Guise, J.A., Aissaoui, R.: Tibio-femoral joint contact in healthy and osteoarthritic knees during quasi-static squat: a bi-planar X-ray analysis. J. Biomech. **53**, 178–184 (2017)
5. Scarvell, J.M., Galvin, C.R., Perriman, D.M., Lynch, J.T., van Deursen, R.W.M.: Kinematics of knees with osteoarthritis show reduced lateral femoral roll-back and maintain an adducted position. A systematic review of research using medical imaging. J. Biomech. **75**, 108–122 (2018)
6. Dell'Isola, A., Smith, S.L., Andersen, M.S., Steultjens, M.: Knee internal contact force in a varus malaligned phenotype in knee osteoarthritis (KOA). Osteoarthr. Cartil. **25**, 2007–2013 (2017)
7. Kumar, D., Manal, K.T., Rudolph, K.S.: Knee joint loading during gait in healthy controls and individuals with knee osteoarthritis. Osteoarthr. Cartil. **21**, 298–305 (2013)
8. Sritharan, P., Lin, Y.-C., Richardson, S.E., Crossley, K.M., Birmingham, T.B., Pandy, M.G.: Musculoskeletal loading in the symptomatic and asymptomatic knees of middle-aged osteoarthritis patients. J. Orthopaedic Res. **35**, 321–330 (2017)
9. Van Rossom, S., Khatib, N., Holt, C., Van Assche, D., Jonkers, I.: Subjects with medial and lateral tibiofemoral articular cartilage defects do not alter compartmental loading during walking. Clin. Biomech. **60**, 149–156 (2018)
10. Meireles, S., Wesseling, M., Smith, C.R., Thelen, D.G., Verschueren, S., Jonkers, I.: Medial knee loading is altered in subjects with early osteoarthritis during gait but not during step-up-and-over task. PLoS ONE **12**, e0187583 (2017)
11. Moissenet, F., Chèze, L., Dumas, R.: A 3D lower limb musculoskeletal model for simultaneous estimation of musculo-tendon, joint contact, ligament and bone forces during gait. J. Biomech. **47**, 50–58 (2014)
12. Farrokhi, S., Voycheck, C.A., Gustafson, J.A., Fitzgerald, G.K., Tashman, S.: Knee joint contact mechanics during downhill gait and its relationship with varus/valgus motion and muscle strength in patients with knee osteoarthritis. Knee **23**, 49–56 (2016)
13. Manal, K., Gardinier, E., Buchanan, T.S., Snyder-Mackler, L.: A more informed evaluation of medial compartment loading: the combined use of the knee adduction and flexor moments. Osteoarthr. Cartil. **23**, 1107–1111 (2015)
14. Richards, R.E., Andersen, M.S., Harlaar, J., van den Noort, J.C.: Relationship between knee joint contact forces and external knee joint moments in patients with medial knee osteoarthritis: effects of gait modifications. Osteoarthr. Cartil. **26**, 1203–1214 (2018)
15. Trepczynski, A., Kutzner, I., Bergmann, G., Taylor, W.R., Heller, M.O.: Modulation of the relationship between external knee adduction moments and medial joint contact forces across subjects and activities. Arthritis Rheumatol. **66**, 1218–1227 (2014)
16. Nissan, M.: Review of some basic assumptions in knee biomechanics. J. Biomech. **13**, 375–381 (1980)

Consideration of Structural Behaviour of Bones in a Musculoskeletal Simulation Model

Robert Eberle[1(✉)] and Dieter Heinrich[2]

[1] Unit of Engineering Mathematics, University of Innsbruck, 6020 Innsbruck, Austria
robert.eberle@uibk.ac.at

[2] Department of Sport Science, University of Innsbruck, 6020 Innsbruck, Austria

Abstract. Bone and ligament injuries are one of the most common serious injuries in sports. Musculoskeletal simulation models are popular to investigate injury situations on a computer. The existing musculoskeletal simulation models typically consist of a rigid multi-body model (representing the skeleton) and muscles, which actuate the movement of the skeleton. Rigid segments prevent certain mechanical behaviour of bones from being taken into account. Further, bone injuries or effects on the bone structure during the simulated movements cannot be analysed.

In this study a computational approach has been developed that allows for investigation of human movements with a simulation model considering the structural behaviour of bones. The rigid bones in the multi-body model have been replaced by Euler-Bernoulli beams. The Euler-Bernoulli beams were included in the simulation model using the floating frame of reference formulation.

The developed approach has been applied to simulate a fall from a height of 1.7 m, in which a person lands with the ulna on the ground.

Keywords: Musculoskeletal model · Floating frame of reference · Bone fracture · Ulna

1 Introduction

Bone and ligament injuries are one of the most common serious injuries in sports. To prevent such injuries, it is necessary to understand the underlying injury mechanism and related risk factors. In the last decades musculoskeletal simulation models have become more and more popular to investigate injury situations on a computer [1]. A musculoskeletal simulation model consists of a multi-body model (representing the skeleton) and muscles, which actuate the movement of the skeleton. Most of the existing musculoskeletal simulation models consist of a rigid multi-body model. If bones are incorporated as rigid segments and not deformable segments, some mechanical behaviour of bones are not considered in the simulations. Further, bone injuries or effects on the bone structure during the simulated movements cannot be analysed.

G. A. Ateshian et al. (Eds.): CMBBE 2019, LNCVB 36, pp. 464–469, 2020.
https://doi.org/10.1007/978-3-030-43195-2_37

In this work, a computational approach has been developed that takes into account the structural behaviour of bones in a multi-body simulation model to study human movements.

The approach was applied to examine a fracture of the ulna during a fall.

2 Methods

2.1 Simulation Model

A simulation model with two segments (one for the humerus and one for the ulna) was developed in Matlab that takes into account the structural behaviour of bones. Especially, the bones were modelled as Euler-Bernoulli beams, which were incorporated into the simulation model using the floating frame of reference formulation [2]. The simulation model resulted in a system of differential algebraic equations (DAE):

$$\begin{aligned} M\ddot{q} + D\dot{q} + Kq + C_q^T\lambda &= Q_v + Q_e \\ C(q,t) &= 0. \end{aligned} \tag{1}$$

Here, M denotes the mass matrix, q the generalized coordinates (which include elastic coordinates), D the damping matrix, K the stiffness matrix, C_q^T the transposed Jacobian matrix w.r.t. q of the constraints $C(q,t)$, λ the Lagrange multipliers, Q_v the quadratic velocity vector and Q_e the vector of external generalized force.

The movement of the humerus and ulna was actuated by joint moments acting at the shoulder and elbow. In the simulation model, this moments were considered in the vector of external generalized force Q_e.

The humerus and the ulna were modelled as Euler-Bernoulli beams. The rest of the body was assumed to be a mass point considered in the vector of external generalized force Q_e.

2.2 Bone Model

The humerus and the ulna were modelled as Euler Bernoulli beams. The dynamics of an Euler Bernoulli beam are given by the partial differential equation (equation of a vibrating beam, see [3])

$$\frac{\partial^2}{\partial x_b^2}(EI(x_b)\frac{\partial^2 w}{\partial x_b^2}) + \rho A(x_b)\frac{\partial^2 w}{\partial t^2} = f(x_b, t, w). \tag{2}$$

Here, x_b denotes the longitudinal coordinate of the bone, t the simulation time, $EI(x_b)$ the flexural stiffness, $w(x_b,t)$ the deflection of the bone perpendicular to the longitudinal axis x_b, ρ the mass density, and $A(x_b)$ the cross sectional area of the bone (see Fig. 1). The term f denotes the applied load on the bone. The boundary conditions of the bone depended on the kind of joint and were hinged on one end

$$w(x_b,t) = 0, \qquad \frac{\partial^2 w(x_b,t)}{\partial x_b^2} = 0 \tag{3}$$

and free on the other end

$$\frac{\partial^2 w(x_b,t)}{\partial x_b^2} = 0, \quad \frac{\partial^3 w(x_b,t)}{\partial x_b^3} = 0. \tag{4}$$

The initial conditions were given by the initial deflection of the bone and the initial velocity, which were both set to zero

$$w(x_b,0) = 0, \quad \frac{\partial w(x_b,0)}{\partial t} = 0. \tag{5}$$

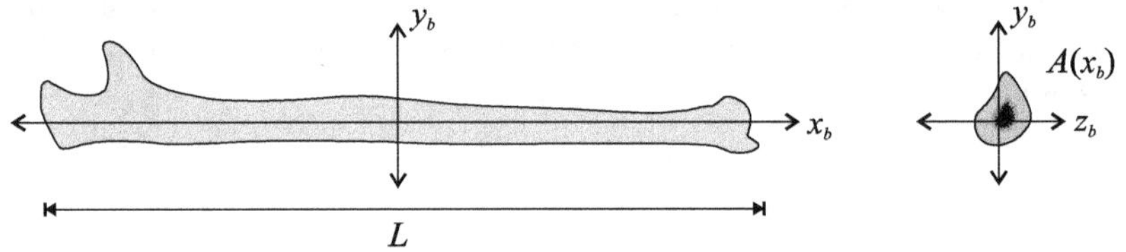

Fig. 1. Coordinate systems of the ulna: longitudinal (left) and cross cut (right).

The floating frame of reference formulation requires shape functions to represent the deformation of an elastic segment. The shape functions of the bones were represented by the eigenmodes of the Euler-Bernoulli beam (2) fulfilling the boundary conditions (3) and (4).

The geometrical data and the stiffness parameters were taken from the literature [4].

2.3 Hard Landing on the Forearm

The developed model was used to simulate a 22-A1.2 fracture. In particular, a drop (free fall) from a height of 1.7 m was simulated in which a person landed on the ground with the ulna (see Fig. 2). It was assumed that the ulna fractures after exceeding an assumed ultimate bending stress of 100 MPa.

The movement of the persons arm during the fall was actuated by predetermined torque moments acting at the shoulder and elbow. The ground was modelled using a Kelvin-Voigt constitutive equation consisting of a damper and a spring connected in parallel [5]. The impact at the ground was considered in the vector of external generalized force Q_e.

2.4 Numerical Solution

The system of DAEs (1) was solved numerically by the code `Radau5` [6]. This code solves DAEs and is based on an implicit Runge-Kutta method of order 5 including a step size control and continuous output.

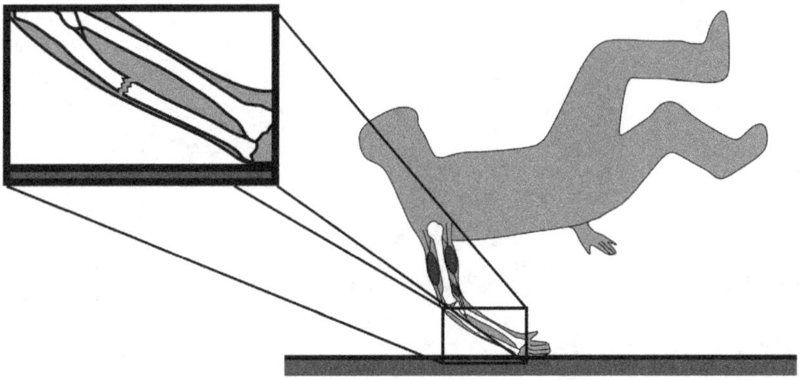

Fig. 2. A drop in which a person lands on the ground with the forearm suffering a fracture of the ulna.

3 Results

Figure 3 shows the stick-figures (grey lines) of the humerus and ulna in the simulated drop at times t = 0, 0.2, 0.21 and 0.22 s. In the simulation the ulna broke at time $t = 0.22$ s. In Fig. 3 the black star (*) depicts the position, where the ulna broke with a bending stress of 111 MPa (exceeding an assumed ultimate bending stress of 100 MPa).

4 Discussion

In this first attempt the radius was not considered in the simulation. However, it was assumed that the person lands on the ulna first, so that the radius was not loaded.

In musculoskeletal simulation, the movements are actuated by muscles, which can be applied as torque moments in the joints [7]. In this study, the movement was controlled by predetermined moments and not by moments generated by muscles. However, with this approach, muscles can be easily implemented by modelling the force generation of the muscle with a three-element Hill model [8]. The predetermined moments were chosen so that the drop corresponds to a typical case. In the future, the movement will be controlled by control strategies (see [9]).

The bones were modelled as Euler-Bernoulli beams. First cadaver studies at the Department of Trauma Surgery, Medical University of Innsbruck have shown that the ulna can be modelled very well as an Euler-Bernoulli beam. The ultimate bending stress of the ulna was assumed as 100 MPa. This value results from a three-point bending test of a single ulna, which was loaded until fracture. In a next step, the unknown parameters of the bones, such as geometry and stiffness, will also be determined based on measurements. The Euler-Bernoulli beams were incorporated into the simulation model using the floating frame

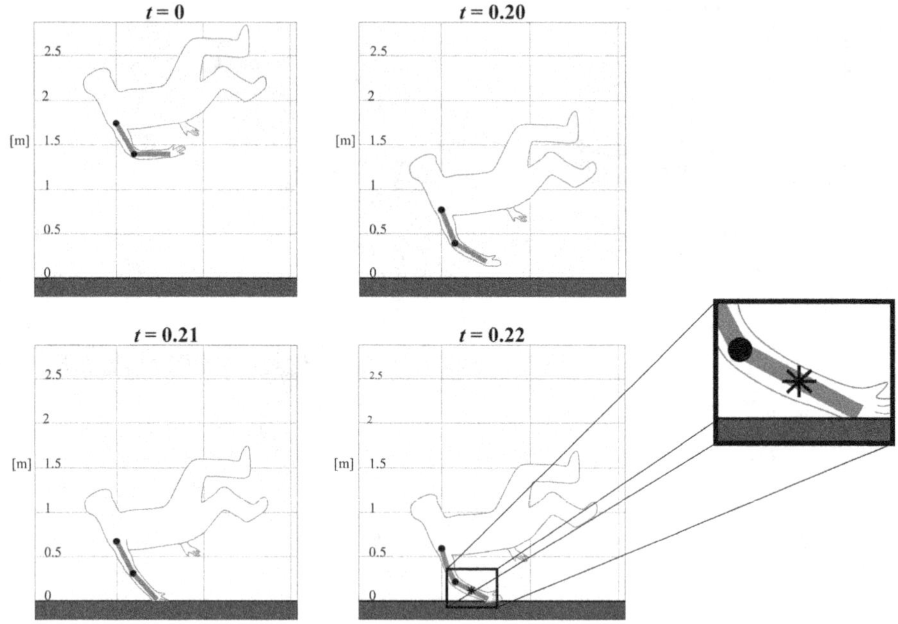

Fig. 3. Stick-figures of humerus and ulna in a simulated drop.

of reference formulation. In flexible multi-body dynamics, this formulation has proven itself for the consideration of small deformation.

The developed approach was used to investigate a person's free fall landing on the ulna, which was a hypothetical case without any reference data. A simulation of a real injury situation is in progress.

5 Conclusion

The developed computational approach allows the investigation of injuries with a musculoskeletal simulation model considering the structural behaviour of bones.

In future applications, this approach might also be used to study the mechanical behaviour of prostheses using musculoskeletal simulation models.

References

1. Heinrich, D., van den Bogert, A.J., Nachbauer, W.: Relationship between jump landing kinematics and peak ACL force during a jump in downhill skiing: a simulation study. Scand. J. Med. Sci. Sports **24**(3), e180–e187 (2014)
2. Shabana, A.A.: Dynamics of Multibody Systems. Cambridge University Press, Cambridge (2013)
3. Eberle, R., Kaps, P., Oberguggenberger, M.: A multibody simulation study of alpine ski vibrations caused by random slope roughness. J. Sound Vib. **446**, 225–237 (2019)

4. Begeman, P.C., Pratima, K.: Bending strength of the human cadaveric forearm due to lateral loads. In: Proceedings of 43rd Stapp Car Crash Conference SAE Technical Paper (No. 99SC24). Warrendale (1999)
5. Eberle, R., Kaps, P., Filippi-Oberegger, U., Heinrich, D., Mössner, M., Schindelwig, K., Niederkofler, A., Nachbauer, W.: Damped vibrations of alpine skis on inelastic foundations. In: Science and Skiing V, pp. 142–151. Meyer & Meyer Verlag, Maidenhead (2012)
6. Wanner, G., Hairer, E.: Solving Ordinary Differential Equations II. Springer, Berlin (1996)
7. McLean, S.G., Su, A., van den Bogert, A.J.: Development and validation of a 3-D model to predict knee joint loading during dynamic movement. J. Biomech. Eng.-T ASME **125**(6), 864–874 (2003)
8. van den Bogert, A.J., Blana, D., Heinrich, D.: Implicit methods for efficient musculoskeletal simulation and optimal control. Procedia Iutam **2**, 297–316 (2011)
9. Erdemir, A., McLean, S., Herzog, W., van den Bogert, A.J.: Model-based estimation of muscle forces exerted during movements. Clin. Biomech. **22**(2), 131–154 (2007)

Biomechanical Assessment of the Iris in Relation to Angle-Closure Glaucoma: A Multi-scale Computational Approach

Vineet S. Thomas[1], Sam D. Salinas[1], Anup D. Pant[2], Syril K. Dorairaj[3], and Rouzbeh Amini[1](✉)

[1] Department of Biomedical Engineering, The University of Akron, Akron, OH 44304, USA
ramini@uakron.edu

[2] Department of Engineering, East Carolina University, Greenville, NC 27858, USA

[3] Department of Ophthalmology, Mayo Clinic, Jacksonville, FL 32224, USA

Abstract. The abnormalities in the iris shape and deformation could cause closure of the space between the iris and cornea. Such closure may lead to development of certain types of glaucoma, a mysterious disease causing irreversible blindness. As such, the mechanical response of the iris and its deformation have been studied extensively in the context of glaucoma. The collagen fibrils of the iris stroma provide support while undergoing continuous large mechanical deformation. The relationship between micrometer-scale and macro-scale mechanical environment, however, remains unknown. We have used a multiscale computational framework, linking the volume-averaged stress in micrometer-scale representative volume elements to a macro-scale finite-element continuum. We fitted the multiscale model response to experimental data obtained from uniaxial tension tests of intact irides. We hope to employ our model in pathophysiological states of the iris to understand how the microscale deformation may differ in glaucomatous eyes as compared to the healthy controls.

Keywords: Multiscale model · Angle-closure glaucoma · Iris

1 Introduction

The iris controls the pupil diameter in response to surrounding changes in the ambient light. It is located posterior to the cornea and anterior to the lens. The iris consists of a passive stromal region and two active smooth muscles: the *sphincter iridis* and the *dilator pupillae*. The constriction and relaxation of iris smooth muscles controls the pupil diameter as dictated via the autonomic nervous system [1]. In addition to the activity of the iris smooth muscles, the iris configuration is affected by the aqueous humor flow [2–6]. The aqueous humor is a transparent fluid secreted from the ciliary body into the posterior chamber of

G. A. Ateshian et al. (Eds.): CMBBE 2019, LNCVB 36, pp. 470–482, 2020.
https://doi.org/10.1007/978-3-030-43195-2_38

the eye and flows around the lens through the iris-lens gap continuously into the anterior chamber. It provides the cornea and lens, avascular ocular tissues, with oxygen and nutrients. It drains into the venous circulation system mostly via trabecular meshwork located between the cornea and iris and further through the Schlemm's canal.

Resistance to the aqueous humor flow generates the intra ocular pressure (IOP). The normal IOP of approximately 15 mmHg is essential for optical functions and proper shape of the eye [7–9]. An increase in IOP could put individuals at higher risk of glaucoma [10]. Glaucoma is a group of diseases, in which damage to the optic nerve of the eye may lead to irreversible blindness [11]. Abnormality in the iris and its interaction with aqueous humor flow contribute towards glaucomatous IOP increases [2,6,12]. In primary angle closure glaucoma (PACG), in particular, the blockage of the drainage canal due to anterior apposition of the iris results in elevated IOP [13].

Due to its importance in the pathophysiology of glaucoma, the iris biomechanics has been studied extensively [2–6,14–19]. Recently, it has been observed that patients suffering from PACG have significantly stiffer irides when compared to those in healthy individuals [20,21]. A major component of the iris stroma is collagen fibers that provide continuous support while the iris undergoes large mechanical deformations [22]. As such, a computational model capable of predicting changes in the microstructure associated to the macroscale deformations is necessary to study iris biomechanics. Structural components of the eye have been previously studied by various investigators using computational methods [23–30]. The goal of this project was to develop a microstructurally faithful computational model of the iris to link the macroscale responses to its extracellular matrix (ECM) microstructural architecture. This multi-scale model will potentially help us understand how regulation of ECM collagen fiber synthesis in iris stroma relates to the iris stiffness in healthy and glaucomatous eyes.

2 Methods and Materials

Macroscale mechanical responses of the iris was obtained from previously quantified uniaxial data for an intact iris [15] (Fig. 1a). Undeformed microstructural architecture of the Iris tissue was characterized experimentally in this study. In doing so, fresh porcine eyes were obtained from local slaughterhouse (3-D Meats, Dalton, Ohio) and submerged immediately in a cold isotonic phosphate buffer saline solution. After the eye samples arrived at the lab, an incision was made on the sclera at a latitude that lies approximately behind the lens of the eye. The portion of the eye consisting of iris along with the cornea was then fixed in 2.5% gluteraldehyde for three hours. Thereafter, the iris was isolated and the pigmented layer was scrapped off carefully. The sample was then prepared to be mounted on the small angle light scattering (SALS) equipment. Details about SALS method are provided in our previous publications [31,32]. The specimen was dehydrated in 100% glycerine for an hour and then sandwiched between two glass-slides. A light source was passed through the sample placed in the sample

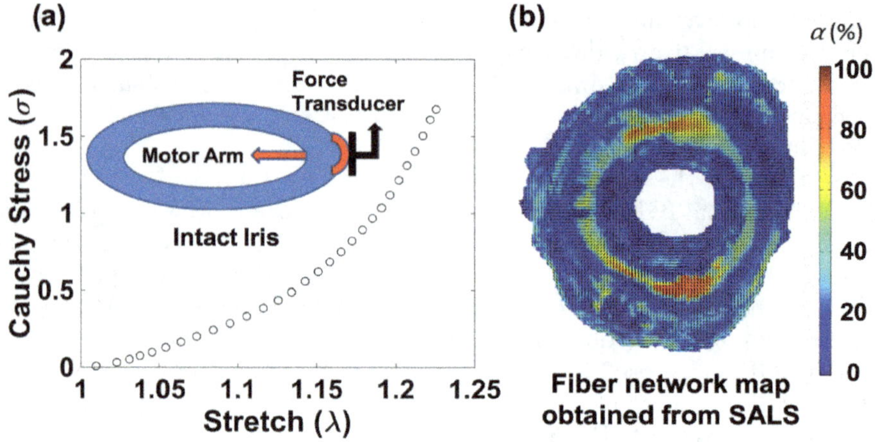

Fig. 1. (a) Schematic diagram showing the uniaxial test setup and uniaxial stretch response of an intact iris [15]. (b) Contour map of the fiber network parameters obtained from SALS.

holder to generate a scattering pattern on a mylar sheet. The scattering pattern was digitized using a digital camera installed behind the screen which was post-processed to obtain the contour map of anisotropy index α and preferred fiber direction μ [31] (Fig. 1b). These microstructure network parameters (α, μ) were computed using a second-order orientation tensor H (Eq. 1) [33,34]. Eigen analysis was performed to obtain the eigenvalues (λ_1, λ_2) and eigenvectors (V_1, V_2) of H. The eigen values and vector components were used to compute anisotropy index α (Eq. 2) and the mean fiber direction μ (Eq. 3).

$$H = \oint r \otimes r \; R(\theta) d\theta \tag{1}$$

$$\alpha = 100(1 - \frac{\lambda_1}{\lambda_2}) \quad where \quad \lambda_1 \leq \lambda_2 \tag{2}$$

$$\mu = tan^{-1}(\frac{V_2.j}{V_2.i}) \tag{3}$$

As described previously [31,35], the symbol $\otimes$ represents the dyadic multiplication in Eq. 1. In addition, r is a unit vector in the direction of θ and $R(\theta)$ represents the fiber network distribution density function. Higher values of anisotropy index α indicate more well-oriented fiber network. A library of three dimensional representative volume elements (RVEs) were generated based on the experimentally-quantified fiber network distribution. Random points were generated between -1 & 1 in a three dimensional space. Three unique points were identified to form a triangle whose edges were treated as fibers (using the Delaunay triangulation algorithm). A unit domain $(1 \times 1 \times 1)$ was clipped from the central region. The clipped domain, known as representative volume element,

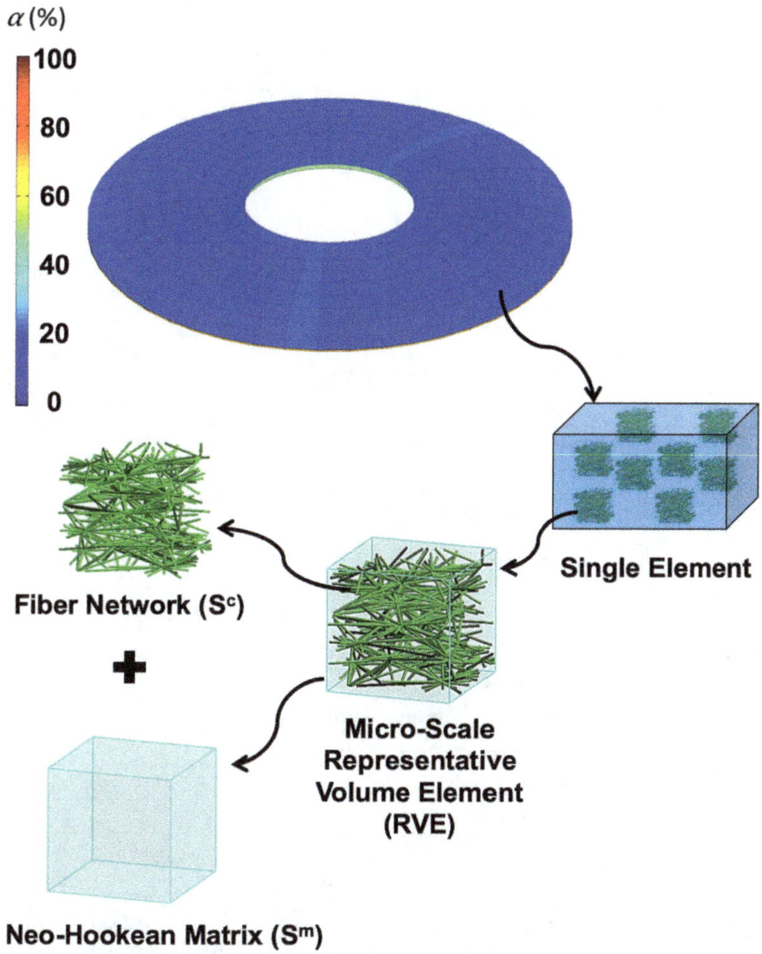

Fig. 2. Three dimensional model of an undeformed intact iris and schematic outline of the multiscale finite element model arrangement.

represents the fibrous portion of the ECM. An average number of 450 fibers was maintained while generating RVEs for all elements, based on the microstructural parameters (α, μ). Detailed steps to generate RVEs are provided in our previous publication [35].

A three dimensional finite element model of iris was generated based on published data [15]. The model was meshed with 1504 8-node brick elements (Fig. 2). Based on the fiber network distribution maps obtained from SALS, three different networks with α of 20% - radially aligned, 50% - aligned circumferentially and 70% - radially aligned, were assigned to the stroma, spinchter, and dilator regions, respectively. Details about the computational framework of our model are also provided in our previous publication [35] as well as in

Table 1. List of model parameters and their descriptions

Model parameter	Description
ψ	Fiber volume fraction
γ	RVE edge length
L	Dimensionless length of the fibers
A_f	Fiber cross sectional area
V	RVE volume
x_i	Position of the boundary cross-link
f_j	Force generated at the fiber level
ε_G	Fiber Green strain
ϵ_G^c	Green strain at threshold fiber stretch
E_f	Fiber modulus
λ_f	Fiber longitudinal stretch ratio
λ_c	Fiber critical stretch threshold
β	Exponential parameter representing the non-linearity of the force response
G	Shear modulus
ν	Poisson's ratio
J	Determinant of the deformation gradient tensor F ($F_{ij} = \frac{\partial x_i}{\partial X_j}$)
B_{ij}	Left Cauchy–Green deformation tensor ($B_{ij} = F_{ik}F_{jk}$)
S_{ij}^c	Volume averaged Cauchy stress from fibrous matrix
S_{ij}^m	Stress contribution from non-fibrous matrix
s_{ij}^c	Microscopic stress tensor from fibrous matrix
u_k	Displacement of the RVE boundary
n_k	Unit normal vector

those of others [36–41]. Briefly, the generated RVE's are arranged at eight Gauss points for every element in the model. When boundary conditions are applied to the model, induced macroscale deformations are mapped to microscale RVE boundaries. Each assigned RVE fiber network realigns and deforms in response to RVE boundary deformation and generates a macroscopic volume-averaged Cauchy stress S'_{ij} in terms of microscopic stress tensor s'_{ij} [42]:

$$S'_{ij} = \frac{1}{V'} \int_V s'_{ij} dV \tag{4}$$

Assuming microscopic equilibrium (i.e. $s'_{ij,i} = 0$) and applying Green's theorem, Eq. 4 in its discrete form is given by Eq. 5, where the prime denotes dimensional quantity.

$$S'_{ij} = \frac{1}{V'} \sum_{boundary\ \ crosslinks} x'_i f_j \tag{5}$$

Dimensionless undeformed RVEs are converted to dimensional quantities (i.e., $V' = \gamma V$ and $x'_i = \gamma x_i$) and scaled to represent tissue level physical space. More details about model parameters are given in Table 1. The fiber volume is computed as $\gamma L A_f$, in which A_f is the cross-sectional area of the fiber, γ represents the RVE edge length at the tissue level and γL is the total fiber length at the tissue level. The volume fraction ψ is given by:

$$\psi = \frac{LA_f}{\gamma^2} \tag{6}$$

which can be rewritten as

$$\gamma = \sqrt{\frac{LA_f}{\psi}} \tag{7}$$

Substituting the scaled quantities in Eq. 5, the macroscopic RVE volume averaged Cauchy stress tensor S^c_{ij} is computed as in Chandran and Barocas [36]. The force f in the fibers (modeled as nonlinear springs) are governed by a nonlinear constitutive equation from Billiar and Sacks (Eq. 8) [43] for fiber stretch λ_f less than critical stretch value of λ_c. The force in the fibers are constrained to vary linearly with fiber stretch λ_f beyond the critical stretch value of λ_c:

$$f = \begin{cases} \dfrac{E_f A_f}{\beta}(e^{\beta\varepsilon_G} - 1), & \lambda_f < \lambda_c \\ \dfrac{E_f A_f}{\beta}(e^{\beta\varepsilon^c_G} - 1) + E_f A_f \lambda_c e^{\beta\varepsilon^c_G}(\lambda_f - \lambda_c), & \lambda_f > \lambda_c \end{cases} \tag{8}$$

where $\varepsilon_G = \frac{1}{2}(\lambda_f^2 - 1)$ is the Green strain of the fiber and ε^c_G denotes the strain value corresponding to the critical stretch value λ_c.

For the non-fibrous matrix, the stress contribution S^m_{ij} is computed using a neo-Hookean model from Bonet and Wood [44]:

$$S^m_{ij} = \frac{G}{J}(B_{ij} - \delta_{ij}) + \frac{2G\nu}{J(1-2\nu)}(lnJ)\delta_{ij} \tag{9}$$

Leading to

$$S^c_{ij,i} + S^m_{ij,i} = \frac{1}{V}\oint_{\partial V}(s^c_{ij} - S^c_{ij})u_{k,i}n_k dS \tag{10}$$

A balance between the stress generated (both fibrous part of the matrix and non-fibrous components of the tissue) and the macroscale deformation is obtained in an iterative manner [39]. The macroscopic deformation field is mapped to the RVE boundaries to solve the macroscopic boundary value problem with the applied boundary conditions. The force in the fibers are computed for each RVE generating volume-averaged Cauchy stress for the ECM fibers and solving the macroscopic stress balance iteratively. The nonlinear macroscopic finite element (FE) problem and microscopic force balance were iterated using the Newton-Raphson method until the solution converged. More details about this modeling approach are available in the literature [35–41].

The collagen fiber diameter and fiber volume fraction for our simulations were assumed to be 300 nm [22] and 91% [45], respectively. Uniaxial boundary conditions were applied to the intact iris finite element model similar to the experiment described by Whitcomb et al. [15]. A portion of the outer circumference was fixed and 25% strain was simulated by stretching a portion of the inner circumference uniaxially. A trial set of fitting parameters (shear modulus G, fiber modulus E_f, fiber critical stretch threshold λ_c, and parameter β to control the nonlinearity of the response) was chosen to simulate the uniaxial stretch. The quality of fit between the model and the experimental data was evaluated by computing the coefficient of determination (R^2) [46]. The R^2 value was defined using Eq. 11:

$$R^2 = 1 - \frac{\sum_{i=1}^{n}(y_i - \hat{y_i})^2}{\sum_{i=1}^{n}(y_i - \overline{y})^2} \tag{11}$$

where y_i refers to the experimental data, $\overline{y}$ is the mean of y_i and $\hat{y_i}$ refers to the value obtained from the simulation. The trial set of parameters was revised and the model was simulated until the quality of fit R^2 approached 1.

An average of ∼450 fibers per RVE accounts to more than 32 million degrees of freedom for our FE model with 1504 elements. The code was parallelized using the message passing interface (MPI) technique to speed-up the solving process and submitted to Ohio Supercomputer Center (OSC) to run on its Ruby Cluster (which uses 200 2.6-GHz Intel Xeon E5-2670 processors) [47]. A total of 8 h (clock time) was required for each simulation.

3 Results

The first column of Fig. 3 shows the anterior view of the deformed configuration of an intact iris FE model at different stages of uniaxial stretching simulation. The multiscale model fitted the uniaxial test data with an overall R^2 value of 0.99 for an intact porcine iris sample (Fig. 4). The model fit parameters were the following: $G = 1.5$ kPa, $E_f = 50$ kPa, $\beta = 1.5$ and $\lambda_c = 1.05$.

To demonstrate changes in the microstructure response of the ECM to macro-scale uniaxial stretch as predicted by our multiscale model, a representative network was chosen from the model, shown in the second column of Fig. 3. The color contour of true strain in the fiber shows that the initial strain for all fibers in the undeformed configuration is zero. The fiber networks were found to have translated and rotated based on the RVE boundary deformation, and they realigned primarily in the direction of the applied stretch. The radially aligned network showed higher anisotropy index values after deformation (Fig. 3, third column), indicating that these networks became more aligned.

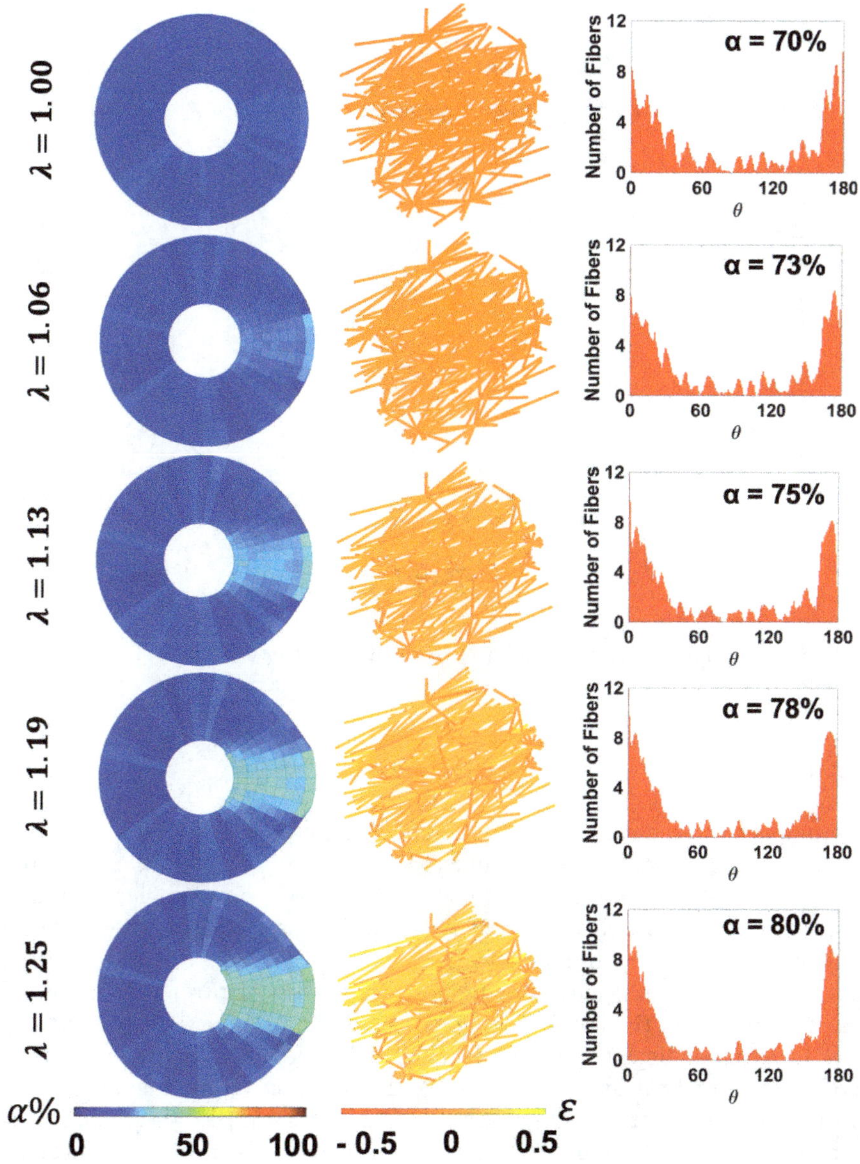

Fig. 3. Column 1: anterior view of the deformed configuration of an intact iris under uniaxial stretching from initial to the final step. Column 2: changes in the microstructure at different stages of macroscale deformation. Column 3: the fiber distribution histograms of a selected network under uniaxial stretch.

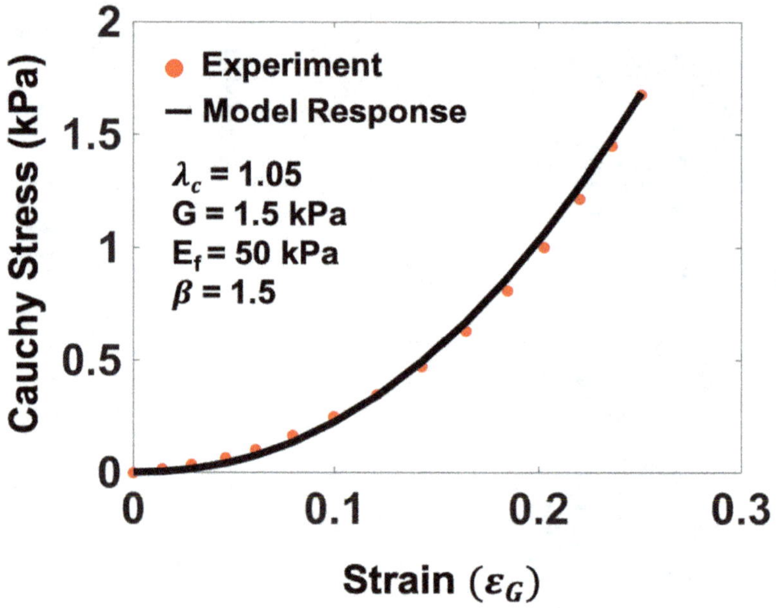

Fig. 4. Simulated multiscale model response after fitting the model parameters to uniaxial stretch.

4 Discussion

Iris responds to the changes in the surrounding light intensity adjusting the pupil diameter by activity of its smooth muscles. The underlying ECM fibers provide support while the iris undergoes deformations under physiological conditions, whether it is due to the activities of its smooth muscle or because of its interactions with the aqueous humor flow. Previous studies have reported differences in iris stiffness and structural components in healthy controls as compared to diseased eyes. In particular, it has been found that patients with a history of PACG had higher density of type-I collagen in the stromal tissue when compared to those of healthy volunteers [48]. Further, ex-vivo measurements have confirmed that irides of PACG patients are also mechanically stiffer than those of healthy controls [21]. In many soft tissues, collagen synthesis or its disintegration is heavily controlled by mechanical strains [49,50]. As such, biomechanical differences in healthy and glaucomatous irides has motivated us to develop models capable of predicting the macroscale responses of irides due to changes in their microenvironment and vice versa. Such multi-scale finite element models provide unprecedented information about biomechanics and mechanobiology of the iris that cannot be obtained from other alternative methods.

In this study, a multiscale finite element modeling approach was implemented to predict the response of an intact iris tissue at multiple length scales. Although an uniaxial extension test cannot be directly associated to physiological function of the iris, data were used to demonstrate and validate the model responses. The iris consists of circumferentially aligned sphincter and radially oriented dilator muscles. However, the current version of the model does not include any active component to represent smooth muscles (i.e., the sphincter and dilator muscles). Comparing the contour of the final deformed configuration with the undeformed configuration (Fig. 3: column 1), we observed more prominent changes in the anisotropy index with deformation. Similar trend was observed by examining individual RVEs in the model. For instance, the anisotropy index increased from 70% to 80% for the chosen RVE (Fig. 3, third column). For an applied macroscale strain of 25%, we observed a maximum of ~50% strain in the fibers. The two strains on different length scales were noticeably different. It is therefore important to quantify ECM-level strains to understand the mechanobiology of the iris tissue. Deviation from homeostatic strains caused either by pathological conditions or surgical interventions could alter the mechanical integrity of iris via damage or remodeling responses.

Cellular configuration linked with ECM fiber architecture may vary nonlinearly in response to tissue level stresses. Cell nuclei deform and realign in association with the surrounding ECM fiber network in response to the mechanical loading. Cell level deformations are influenced through complex micro- mechanical interactions that occur within physiological conditions, imparting altered stress levels on the nucleus due to changes in the cell cytoskeleton, which affects an array of cell functions. Gene expression and protein synthesis have been associated directly to certain optimal change in nuclear shape [51]. Our model can be a tool to investigate nuclear deformation in stromal cells, which is an important factor in mechanical regulation of genome and gene expression. While recent studies show that ECM type-I collagen content increases in PACG irides, it is unclear how regulation of collagen synthesis in iris stroma relates to the iris stiffness. Although there is considerable understanding of relationship between mechanical stimulation on cellular processes, identifying specific mechanisms responsible for vital physiological phenomenon remains poorly understood.

Our modeling approach is not without limitations. We broadly categorized the tissue sample into a fibrous component representing the collagen and a non-fibrous component representing the rest of the ECM and cells, which are all passive components. As a next step towards model development, we will introduce an active contribution representing the smooth muscles in addition to the passive contributions (Eq. 10). Such modification will enable us to simulate a more realistic physiological response of irides during miosis and mydriasis.

5 Conclusions

Our multi-scale modeling approach can provide insight into complex microstructural interactions in the iris which can aid in identifying changes in the iris ECM architecture and fiber strains due to an altered mechanical loading environment. It has the potential to predict remodeling responses in pathophysiological conditions that affect the function of iris, similar to remodeling response models developed for other biological materials.

Acknowledgement. Computations were facilitated by a supercomputing resource grant from the Ohio Supercomputer Center (Columbus, OH). Funding for this study was provided in part by a grant from BrightFocus Foundation (G2018177).

References

1. Thompson, H.S., Hart, W.M.: The pupil. In: Adler's Physiology of the Eye, 9th Edn., pp. 412–441. Mosby-Year Book (1992)
2. Heys, J.J., Barocas, V.H., Taravella, M.J.: Modeling passive mechanical interaction between aqueous humor and iris. J. Biomech. Eng. **123**(6), 540–547 (2001)
3. Huang, E.C., Barocas, V.H.: Active iris mechanics and pupillary block: steady-state analysis and comparison with anatomical risk factors. Ann. Biomed. Eng. **32**(9), 1276–1285 (2004)
4. Amini, R., Barocas, V.H.: Reverse pupillary block slows iris contour recovery from corneoscleral indentation. J. Biomech. Eng. **132**(7), 071010 (2010)
5. Amini, R., Jouzdani, S., Barocas, V.H.: Increased iris–lens contact following spontaneous blinking: mathematical modeling. J. Biomech. **45**(13), 2293–2296 (2012)
6. Jouzdani, S., Amini, R., Barocas, V.H.: Contribution of different anatomical and physiologic factors to iris contour and anterior chamber angle changes during pupil dilation: theoretical analysis. Invest. Ophthalmol. Vis. Sci. **54**(4), 2977–2984 (2013)
7. Carel, R.S., Korczyn, A.D., Rock, M., Goya, I.: Association between ocular pressure and certain health parameters. Ophthalmology **91**(4), 311–314 (1984)
8. Gsellman, L., Amini, R.: Patients with intravitreal gas bubbles at risk of high intraocular pressure without exceeding elevation of surgery: theoretical analysis. Invest. Ophthalmol. Vis. Sci. **57**(7), 3340–3347 (2016)
9. Rashidi, N., Thomas, V.S., Amini, R.: Theoretical assessment of the risk of ocular hypotony in patients with intravitreal gas bubbles who travel through subsea tunnels. Transl. Vis. Sci. Technol. **8**(1), 4 (2019)
10. Sommer, A.: Glaucoma risk factors observed in the baltimore eye survey. Curr. Opin. Ophthalmol. **7**(2), 93–98 (1996)
11. Tham, Y.-C., Li, X., Wong, T.Y., Quigley, H.A., Aung, T., Cheng, C.-Y.: Global prevalence of glaucoma and projections of glaucoma burden through 2040: a systematic review and meta-analysis. Ophthalmology **121**(11), 2081–2090 (2014)
12. Amini, R., Whitcomb, J.E., Al-Qaisi, M.K., Akkin, T., Jouzdani, S., Dorairaj, S., Prata, T., Illitchev, E., Liebmann, J.M., Ritch, R., et al.: The posterior location of the dilator muscle induces anterior iris bowing during dilation, even in the absence of pupillary block. Invest. Ophthalmol. Vis. Sci. **53**(3), 1188–1194 (2012)
13. Lowe, R.F.: Aetiology of the anatomical basis for primary angle-closure glaucoma. Biometrical comparisons between normal eyes and eyes with primary angle-closure glaucoma. Br. J. Ophthalmol. **54**(3), 161 (1970)
14. Heys, J., Barocas, V.H.: Mechanical characterization of the bovine iris. J. Biomech. **32**(9), 999–1003 (1999)
15. Whitcomb, J.E., Barnett, V.A., Olsen, T.W., Barocas, V.H.: Ex vivo porcine iris stiffening due to drug stimulation. Exp. Eye Res. **89**(4), 456–461 (2009)
16. Heys, J.J., Barocas, V.H.: Computational evaluation of the role of accommodation in pigmentary glaucoma. Invest. Ophthalmol. Vis. Sci. **43**(3), 700–708 (2002)
17. Whitcomb, J.E., Amini, R., Simha, N.K., Barocas, V.H.: Anterior–posterior asymmetry in iris mechanics measured by indentation. Exp. Eye Res. **93**(4), 475–481 (2011)

18. Pant, A.D., Dorairaj, S.K., Amini, R.: Appropriate objective functions for quantifying iris mechanical properties using inverse finite element modeling. J. Biomech. Eng. **140**(7), 074502 (2018)
19. Pant, A.D., Amini, R.: Iris biomechanics. Biomech. Eye 281 (2018)
20. Pant, A.D., Gogte, P., Pathak-Ray, V., Dorairaj, S.K., Amini, R.: Increased iris stiffness in patients with a history of angle-closure glaucoma: an image-based inverse modeling analysis. Invest. Ophthalmol. Vis. Sci. **59**(10), 4134–4142 (2018)
21. Narayanaswamy, A.K., Hoon, N.M., Nongpiur, M.E., Htoon, H.M., Thomas, A., Sangtam, T., Lim, C.T., Wong, T.T., Aung, T.: Young's modulus determination of normal and glaucomatous human iris. Invest. Ophthalmol. Vis. Sci. **60**(7), 2690–2695 (2019)
22. Marshall, G.E., Konstas, A.G., Lee, W.R.: Collagens in ocular tissues. Br. J. Ophthalmol. **77**(8), 515 (1993)
23. Schwaner, S.A., Kight, A.M., Perry, R.N., Pazos, M., Yang, H., Johnson, E.C., Morrison, J.C., Burgoyne, C.F., Ross Ethier, C.: A methodology for individual-specific modeling of rat optic nerve head biomechanics in glaucoma. J. Biomech. Eng. **140**(8), 084501 (2018)
24. Sigal, I.A., Flanagan, J.G., Ethier, C.R.: Factors influencing optic nerve head biomechanics. Invest. Ophthalmol. Vis. Sci. **46**(11), 4189–4199 (2005)
25. Amini, R., Barocas, V.H.: Anterior chamber angle opening during corneoscleral indentation: the mechanism of whole eye globe deformation and the importance of the limbus. Invest. Ophthalmol. Vis. Sci. **50**(11), 5288–5294 (2009)
26. Voorhees, A.P., Jan, N.-J., Sigal, I.A.: Effects of collagen microstructure and material properties on the deformation of the neural tissues of the lamina cribrosa. Acta Biomater. **58**, 278–290 (2017)
27. Ayyalasomayajula, A., Park, R.I., Simon, B.R., Vande Geest, J.P.: A porohyperelastic finite element model of the eye: the influence of stiffness and permeability on intraocular pressure and optic nerve head biomechanics. Comput. Methods Biomech. Biomed. Eng. **19**(6), 591–602 (2016)
28. Downs, J.C., Roberts, M.D., Burgoyne, C.F., Hart, R.T.: Multiscale finite element modeling of the lamina cribrosa microarchitecture in the eye. In: 2009 Annual International Conference of the IEEE Engineering in Medicine and Biology Society, pp. 4277–4280. IEEE (2009)
29. Notghi, B., Bhardwaj, R., Bailoor, S., Thompson, K.A., Weaver, A.A., Stitzel, J.D., Nguyen, T.D.: Biomechanical evaluations of ocular injury risk for blast loading. J. Biomech. Eng. **139**(8), 081010 (2017)
30. Grytz, R., El Hamdaoui, M.: Multi-scale modeling of vision-guided remodeling and age-dependent growth of the tree shrew sclera during eye development and lens-induced myopia. J. Elast. **129**(1–2), 171–195 (2017)
31. Pant, A.D., Thomas, V.S., Black, A.L., Verba, T., Lesicko, J.G., Amini, R.: Pressure-induced microstructural changes in porcine tricuspid valve leaflets. Acta Biomaterialia **67**, 248–258 (2018)
32. Amini, R., Voycheck, C.A., Debski, R.E.: A method for predicting collagen fiber realignment in non-planar tissue surfaces as applied to glenohumeral capsule during clinically relevant deformation. J. Biomech. Eng. **136**(3), 031003 (2014)
33. Advani, S.G., Tucker III, C.L.: The use of tensors to describe and predict fiber orientation in short fiber composites. J. Rheol. **31**(8), 751–784 (1987)
34. Sander, E.A., Barocas, V.H.: Comparison of 2D fiber network orientation measurement methods. J. Biomed. Mater. Res. Part A **88**(2), 322–331 (2009)

35. Thomas, V.S., Lai, V., Amini, R.: A computational multi-scale approach to investigate mechanically-induced changes in tricuspid valve anterior leaflet microstructure. Acta Biomaterialia (2019). https://doi.org/10.1016/j.actbio.2019.05.074
36. Chandran, P.L., Barocas, V.H.: Deterministic material-based averaging theory model of collagen gel micromechanics. J. Biomech. Eng. **129**(2), 137–147 (2007)
37. Hadi, M.F., Sander, E.A., Ruberti, J.W., Barocas, V.H.: Simulated remodeling of loaded collagen networks via strain-dependent enzymatic degradation and constant-rate fiber growth. Mech. Mater. **44**, 72–82 (2012)
38. Stylianopoulos, T., Barocas, V.H.: Volume-averaging theory for the study of the mechanics of collagen networks. Comput. Methods Appl. Mech. Eng. **196**(31–32), 2981–2990 (2007)
39. Hadi, M.F., Sander, E.A., Barocas, V.H.: Multiscale model predicts tissue-level failure from collagen fiber-level damage. J. Biomech. Eng. **134**(9), 091005 (2012)
40. Quindlen, J.C., Lai, V.K., Barocas, V.H.: Multiscale mechanical model of the Pacinian corpuscle shows depth and anisotropy contribute to the receptor's characteristic response to indentation. PLoS Comput. Biol. **11**(9), e1004370 (2015)
41. Zhang, S., Zarei, V., Winkelstein, B.A., Barocas, V.H.: Multiscale mechanics of the cervical facet capsular ligament, with particular emphasis on anomalous fiber realignment prior to tissue failure. Biomech. Modeling Mechanobiol. **17**(1), 133–145 (2018)
42. Nemat-Nasser, S., Hori, M.: Micromechanics: Overall Properties of Heterogeneous Elastic Solids. North-Holland, Amsterdam (1999)
43. Billiar, K.L., Sacks, M.S.: Biaxial mechanical properties of the native and glutaraldehyde-treated aortic valve cusp: part ii: a structural constitutive model. J. Biomech. Eng. **122**(4), 327–335 (2000)
44. Bonet, J., Wood, R.D.: Nonlinear Continuum Mechanics for Finite Element Analysis. Cambridge University Press, Cambridge (1997)
45. Lam, M.W.Y., Baranoski, G.V.G.: A predictive light transport model for the human iris. In: Computer Graphics Forum, vol. 25, pp. 359–368. Wiley Online Library (2006)
46. Neter, J., Kutner, M.H., Nachtsheim, C.J., Wasserman, W.: Applied Linear Statistical Models, vol. 4. Irwin Chicago (1996)
47. Ohio Supercomputer Center (1987). http://osc.edu/ark:/19495/f5s1ph73
48. He, M., Lu, Y., Liu, X., Ye, T., Foster, P.J.: Histologic changes of the iris in the development of angle closure in Chinese eyes. J. Glaucoma **17**(5), 386–392 (2008)
49. Robitaille, M.C., Zareian, R., DiMarzio, C.A., Wan, K.T., Ruberti, J.W.: Small-angle light scattering to detect strain-directed collagen degradation in native tissue. Interface Focus **1**(5), 767–776 (2011)
50. Zareian, R., Church, K.P., Saeidi, N., Flynn, B.P., Beale, J.W., Ruberti, J.W.: Probing collagen/enzyme mechanochemistry in native tissue with dynamic, enzyme-induced creep. Langmuir **26**(12), 9917–9926 (2010)
51. Thomas, C.H., Collier, J.H., Sfeir, C.S., Healy, K.E.: Engineering gene expression and protein synthesis by modulation of nuclear shape. Proc. Natl. Acad. Sci. **99**(4), 1972–1977 (2002)

Representative Knee Kinematic Patterns Identification Using Within-Subject Variability Analysis

Mariem Abid[1,2(✉)], Youssef Ouakrim[1,2], Pascal-André Vendittoli[3], Nicola Hagemeister[2], and Neila Mezghani[1,2]

[1] Centre de Recherche LICEF, TELUQ University, Montreal, Canada
mariem.abid@etsmtl.ca
[2] Laboratoire LIO, Centre de Recherche du CHUM, Montreal, Canada
[3] Centre de recherche Hôpital Maisonneuve-Rosemont, Montreal, Canada

Abstract. The purpose of this study is to perform data pre-processing to identify, for a subject, robust representative knee kinematic patterns from a set of three-dimensional (3D) curves, recorded during gait. In this study, we identify a representative gait of a given subject using the within-subject variability evaluation for outliers' removal and reliable curves' selection. The proposed pre-processing method encompasses steps of gait events detection, normalization, outlier detection and cycles' selection. In order to measure the reliability of the subjects' curves before and after pre-processing, we computed the intra-class correlation (ICC) estimates and their 95% confidence intervals for knee kinematics of a multicentric dataset of 226 asymptomatic subjects and knee osteoarthrosis (OA). The experimental results show that the proposed pre-processing method allows to identify representative knee kinematic patterns.

Keywords: Knee kinematics · Representative patterns · Stride-to-stride variability · Reliability · Pre-processing

1 Introduction

In biomechanical gait analysis performed using motion analysis systems, data are curves of measurements made over a gait cycle (for example, the knee joint angles in degrees, expressed in different planes during a gait cycle). Knee pathologies classification using knee joint kinematic data is a challenging endeavour due to the complexity of such data [25]. The classification system flow consists generally of three parts that performs various tasks such as pre-processing, feature extraction and selection, and classification, while facing three complex characteristics of kinematic data namely temporal dependence, high-dimensionality, and high variability [5]. The two latter have been the focus of most studies. Indeed, dimensionality reduction is generally achieved during the feature extraction and selection task via local and global feature extraction methods [1]. Local features,

G. A. Ateshian et al. (Eds.): CMBBE 2019, LNCVB 36, pp. 483–494, 2020.
https://doi.org/10.1007/978-3-030-43195-2_39

most often considered, are characteristic points on the data curves, such as peaks, ranges, and minimum values. Global features account for the time-series of the data by considering the shape of kinematic curves as a whole or its representation. The classification is then performed based on knee kinematic features using statistical and machine learning methods [1], showing the complexity of the kinematic data [25]. However, withstanding of this complexity, a good pre-processing is manifestly as important as good feature extraction and classification. Variability in knee joint measurements arises from different sources [7]: between-subjects (i.e., inter-individual), within-subject (i.e., intra-individual), instrumentation, and experimenter. In this study, we consider within-subject variability, defined as the fluctuation in gait parameters from one stride to the next. Indeed, due to this variability, knee kinematics measurements do not generate a single curve, but a family of curves that differ from each other in magnitudes and timings, and are possibly affected by outliers that have a different shape from the rest of the curves. The objective of this study is to determine a representative pattern of a subject's kinematic curves by averaging these after addressing variability in order to improve the rigor and objectivity of kinematic data. In light of this variability, three types of problem are encountered. The first one concerns the robust estimation of variability among a family of curves. The second problem is how to determine whether a curve is an outlier. The last problem concerns the reliability of gait curves. We demonstrate how robust spread estimation [18] would be advantageous in conjunction with an appropriate choice of outlier curves removal and cycles selection methods for augmenting the curves' reliability. The proposed methodology is represented schematically in Fig. 1. Following data collection, we performed pre-processing steps of gait events detection, normalization, outlier detection and cycles' election. This allows us to determine the most representative shape by averaging the pre-processed kinematic curves.

2 Related Work

In this section, we present related work at each step of the pre-processing flowchart depicted in Fig. 1. Indeed, the choice of an appropriate method in each step of kinematic data pre-processing, while significantly affecting performance, is not always obvious, and requires a combination of experience and trial-and-error [6]. Any gait events, outliers detection and cycle selection methods developed must therefore be both accurate and reliable.

Several methods of determining gait events from knee kinematic curves have been shown in the literature. Kinetics-based gait event detection methods have been widely used. That is, a force plate is conventionally used to identify gait events instants, whereby a vertical ground reaction force (GRF) threshold of 2% of the patients' body weight is defined [4,10,17]. Velocity-based Algorithms was also used to determine when gait events occur with relatively successful results. For example, using the foot velocity algorithm proposed by O'Connor et al. [27], the beginning and the final of each cycle are determined using the vertical velocity signals, derived from heel markers placed on each foot [19].

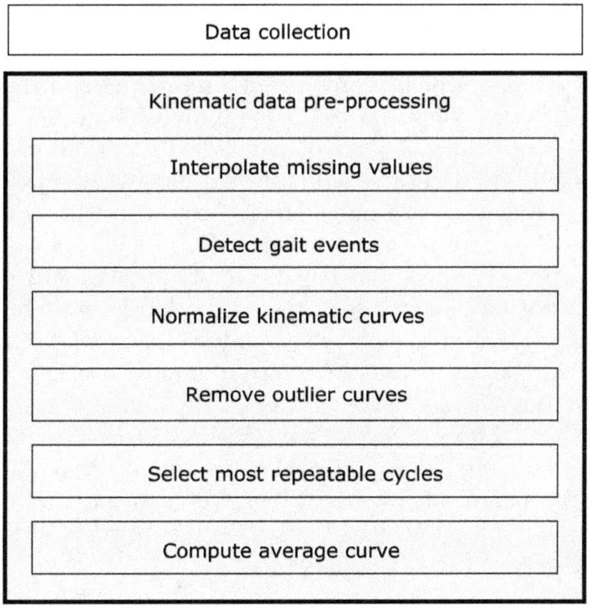

Fig. 1. The flowchart describes the detailed pre-processing steps of knee kinematic data.

When instrumentation is not available to determine gait events timing, some studies rely on kinematics-based gait event detection methods [13, 31].

Within a sample of single-cycle gait curves, there is both phase and amplitude variation. The literature refers to lateral displacements in curve features as phase variation, as opposed to amplitude variation in curve height. Typically, when we describe variability in gait curves, we refer to amplitude variability. Mechmeche *et al.* [22] proposed to deal with phase variability in knee kinematic curves using curve registration. A popular approach to estimating curve variability is to peg prediction bands around a group of curves. However, the presence of a small fraction of outliers can unduly inflate our estimates of gait variability and subsequent analysis. Johnson [14] defines an outlier to be 'an observation in a data set which appears to be inconsistent with the remainder of that set of data'. One common way of estimating curve variability in knee kinematics is the calculation of the standard deviation band, and then to mark as a potential outlier any curve that is more than two standard deviations away from the mean [17]. The problem with this method is that the mean of the data can be greatly affected by outliers. Other related work for detecting outlying observations consists of the construction of a functional boxplot based on the concept of band-depth, and then to mark as a potential outlier any curve that is outside the maximum non-outlying envelope obtained by inflating the inter-quartile range (IQR) 1.5 times [22]. One of the main limitations of the band-depth computation is its computational complexity.

Two similarity indices that attempt to assess the within-subject repeatability of knee kinematic data were considered in the literature. The coefficient of multiple correlation (CMC), which represents the root square of the adjusted coefficient of multiple determination as reported by Kadaba et al. [15], was used to identify the most repeatable 15 curves [9,24]. The limitations of the use of CMC to assess within-subject repeatability in kinematic gait data have been discussed [28], notably the influence of the range of motion of the joint. Duhameh et al. [8] compute an intra-class correlation coefficient (ICC) to determine which representative curves to select. For each participant, a minimum of four and a maximum of 10 gait cycles were obtained for each knee angle. However, in his work the computation of the ICC is based on a one-way, random, linear model, and doesn't take into account the correlation between the repeated measurements.

Traditionally, the average curve of all observations was used as a representative gait cycle [26]. Two gait cycles were included to represent the variability of the individual gait instead of using the average of some gait cycles [19]. A representative data curve was also determined by a variational method to characterize a subject [3]. Other related works average over the most repeatable 15 cycles among all observations [10,23].

3 Methods

This section describes in more detail the proposed method at each individual step, as depicted in Fig. 1.

3.1 Data Collection

This study has been approved by ethics committee of the Centre de Recherche du Centre Hospitalier de l'Université de Montréal (CRCHUM), the École de Technologie Supérieure (ÉTS), and the Hôpital Maisonneuve-Rosemont (HMR), Montreal, Canada. All subjects provided an informed consent before participation. We used three-dimensional (3D) knee kinematics of two cohort of subjects. The first group included 81 asymptomatic (AS) subjects (37 males and 44 females), with mean age of 32.7 ± 10 years, and mean body mass index (BMI) $24.5 \pm 4.1\,\text{kg/m}^2$. The second group included 145 knee osteoarthritis (OA) patients (59 males and 86 females), with mean age of 62.8 ± 10.1 years, and mean body mass index (BMI) $031.7 \pm 7.4\,\text{kg/m}^2$.

3D knee kinematics are acquired with the KneeKG^{TM} system (Emovi Inc. Canada) during a comfortable walk on a treadmill. The KneeKG is a noninvasive system consisting of a harness that is placed on the participant's knee, an infrared camera, and a computer equipped with the Knee3D^{TM} software suite (Emovi, Inc.) [20]. The accuracy [29], reproducibility [12], repeatability [11], and reliability [17] of the system have been assessed. Each participant undergoes a series of successive gait trials during a given session. In each trial, the motion trajectories in the sagittal, frontal, and transverse planes are recorded.

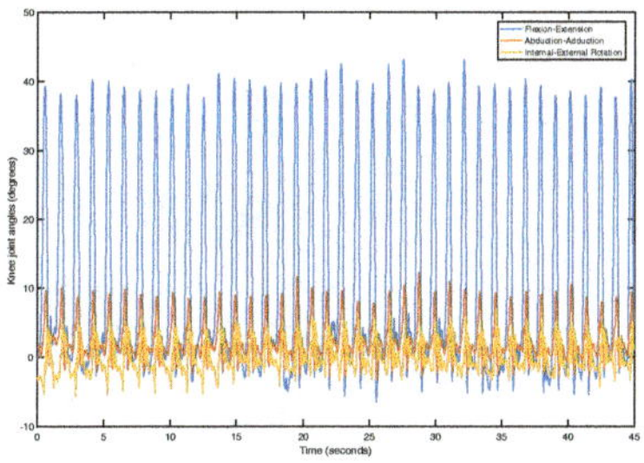

(a)

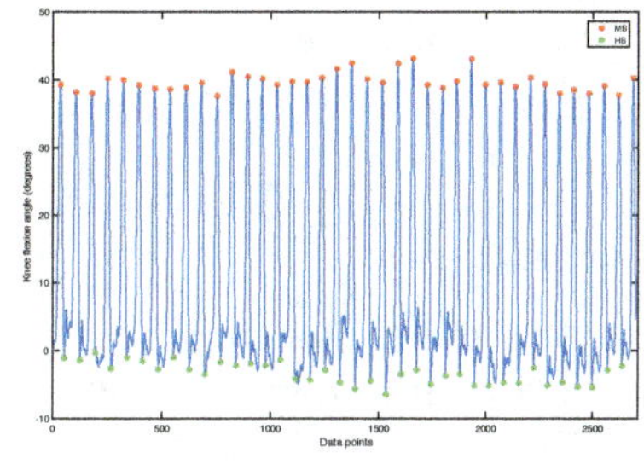

(b)

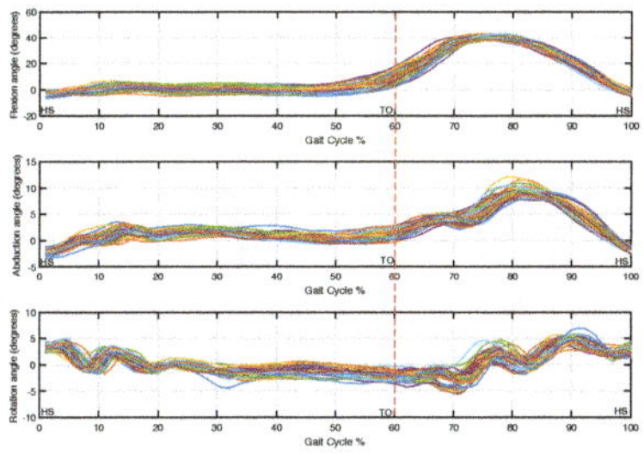

(c)

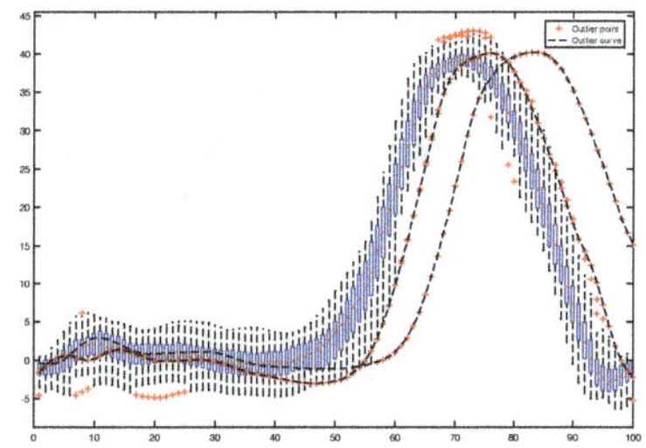

(d)

Fig. 2. (a) Plot of measured knee joint angles during a 45 s walking trial on the treadmill for one AS subject. (b) The detection of the HS and MS instants based on the knee flexion extrema of one AS subject. (c) Depicted are gait cycles for one AS subject, beginning at HS and ending at the next HS, for knee joint angles in all three planes. (d) Boxplot (in blue) of knee flexion angles, for one AS subject, highlighting outliers (in red).

These data are filtered using a non-parametric time series analysis called Singular Spectrum Analysis (SSA) [2], and transformed into 3D knee joint angles using a knee joint coordinate system method [11]. A set of data is then created for each participant containing the 3D knee joint angles in the sagittal, frontal, and transverse planes respectively: flexion-extension, abduction-adduction, and internal-external rotation. Only valid trials are selected for further processing, based on the experimenter's judgment. Conclusively, for each participant, the resulting data involve a set of three kinematic curves in all three planes representing time-varying angle values, i.e., time-series (Fig. 2a).

3.2 Kinematic Data Pre-processing

Once data collection is completed, the raw kinematic data need first to be processed in order to find robust representative patterns for each participant. The proposed method encompasses steps of missing data interpolation, gait events detection, normalization, outlier detection, cycles selection, and averaging.

In a first step, each curve is interpolated using the cubic spline interpolation method, guaranteeing a smooth curve, to fill missing data between kinematic data
points. A detailed description of the remaining pre-processing steps is given in the following:

Gait Events Detection. The raw kinematic curves of each participant are then divided into distinct gait cycles. The gait cycle is the time between two successive heel contacts of the same foot, and consists of a stance and swing phase. Analysis of knee kinematics relies on the accurate determination of the timing of key gait events such as heel strike (HS), the time at which the heel first hits the walking surface, toe-off (TO), the time at which the foot leaves the walking surface, and mid-swing (MS), the maximum knee flexion. In the context of kinematics-based gait event detection methods, our approach center around the location of local maxima values in the sagittal plane curve, since the data from the sagittal plane are more reproducible than those from the other two planes [30]. MS points correspond to local maxima. HS points are the first local minima after the local maxima and are specified as the start of each gait cycle (Fig. 2b). By convention, HS initiates the cycle, TO initiates the swing phase, and stance is followed by swing phases (Fig. 2c).

Normalization. Knowing gait events allows for normalization of the resulting knee kinematic curves over percentages of the gait cycle rather than absolute time. A typical gait cycle normalized in time is represented on a linear 1–100% scale, therefore giving 100 sample points. That is, HS is generally taken as the starting point of a complete gait cycle (0%), TO occurs at about 60–62%, denoting the initiation of the swing phase, and the end of the cycle (100%) occurs with the next HS which will be the HS of the next cycle. Thus, for each knee angle, the superposed normalized cycles (about 30 to 40 cycles depending on the person's stride) constitute the observations to describe with representative patterns characterizing the given participant (Fig. 2c).

Outlier Removal. These observations correspond to a family of curves each one slightly different from the other, due to within-subject variability from stride-to-stride. In this work, we first propose to use cross-validation to quantify the true achieved coverage probability for a robust estimate of the spread of the sample of gait curves [18]. We argued that bootstrap prediction bands provide inadequate coverage probability toward boxplot, when applied to the knee angle curves of AS and knee OA subjects employed in this study. Afterward, the variability among the group of curves, as estimated by bloxplot, is minimized by the subsequent removal of outlying curves in all three planes of motion. In case of gait curves, the outlier is not a single point, but an entire curve (Fig. 2d). In order to affirm that a curve is an outlier, we apply the Chebyshev's theorem stating the percentage of data points falling outside the sample values that are a factor k of the IQR below the 25th percentile or above the 75th percentile.

Most Repeatable Cycles Selection. After estimating within-subject variability, we evaluate the similarity of curves to decide which curves can be selected as characterizing the subject. In other words, we intend to determine whether these curves are repeatable, i.e. sufficiently similar to consider that their mean estimates the true, unknown curve [8]. In this study, a cross-validation methodology was applied to the set of observations. The idea is to remove one curve from the original data set, then calculate the Root Mean Square Error (RMSE) on the remaining curves. The number of RMSE calculated is equal to the number of curves in the data set. The curve resulting in the highest RMSE is removed. A set of 15 curves for each knee angle is selected that demonstrate the best repeatability [4].

Averaging. In this study, each subject is represented by a single gait curve, which is the mean of the 15 most repeatable cycles, as proposed in [10,23].

3.3 Pattern Validation Using Within-Subject Analyzes

The representative patterns have been validated using the within-subject variation. We computed the intra-class correlation (ICC) estimates and their 95% confident intervals based on a single measurement, absolute agreement, 2-way mixed-effects model [21], using Matlab. ICC values less than 0.5 are indicative of poor repeatability, values between 0.5 and 0.75 indicate moderate repeatability, values between 0.75 and 0.9 indicate good repeatability, and values greater than 0.90 indicate excellent repeatability [16].

4 Results

The proposed pre-processing steps have been performed separately on each population databases (OA and AS), for ease of analysis and visualization. Figure 3 shows the time-series kinematic signals in the sagittal plane, frontal plane, and transverse plane for each population.

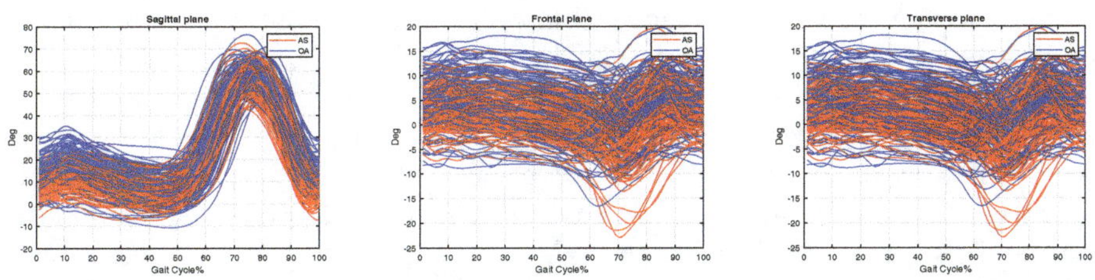

(a) (b) (c)

Fig. 3. Plots of pre-processed knee kinematic time series for AS population (in red) and OA population (in blue), in all three planes.

For each subject, the gait curve reliability was gauged by fulfilling the following steps: (1) summarize the variability within the time-normalized curves of the subject, which is further reduced by the subsequent removal of outlying curves; (2) identify a subset of representative curves for the subject; (3) compute the ICC estimates and their 95% confidence intervals for knee kinematics of a multi-centric dataset of 226 knee OA and AS subjects (presented in Sect. 3.1), in order to measure the reliability of the subjects' curves before and after pre-processing. To summarize the variability, we have adopted a displaying of the distribution via boxplot. Based on the true achieved coverage estimation, boxplot is a robust measurement of variability in knee angle curves of AS and knee OA subjects employed in this study. In Fig. 2d, the curves of this subject are graphically superimposed in order to visually assess reliability. We observe that two curves seem to be different from the others. Using interquartile range, we detect with no doubt that these curves can be considered as outliers. Figure 4 shows, for each plane, the frequency distribution of the ICC computed on the dataset before and after processing, for all the subjects. The graph shows that the subset of the 15 curves selected to represent the gait of the subject are perfectly reliable ($ICC \geq 0.7$). The experimental results show that the proposed pre-processing methodology allows to identify representative Knee Kinematic Patterns.

5 Discussion and Conclusion

We presented pre-processing steps for summarizing the knee kinematic curves of asymptomatic and osteoarthritis subjects. Our analysis takes into consideration the within-subject variability. The proposed methods make it possible to solve two main problems encountered in clinical practice: the removal of outliers and the selection of reliable curves to represent the gait of a given subject. The robust estimation of variability in a family of gait curves is itself a non-trivial challenge. For this issue, we supported the use of bloxplot which provide adequate coverage to the kinematic curves employed in this study. We demonstrated that the variability among a subject's family of curves, as estimated by boxplot, can be minimized by the removal of outlying curves and further reduced by the subsequent selection of the most repeatable cycles as representative of a subject's gait. We believe that gait curves classification relies heavily on the output of the pre-processing step. We hypothesize that the spread estimator and outlier detection methods, suggested during the pre-processing step of robustly representing the gait of a given subject, might be also useful for between-subjects (i.e., inter-individual) variability reduction to further characterize a population group and perform classification tests. We intend also to continue this research by defining the gait cycle as the period between successive TOs, meaning that the swing phase is followed by the stance phase. In this study, We considered subjects with knee osteoarthritis in addition to asymptomatic subjects, in order to validate the proposed pre-processing method on data that show greater variability. We point out that reducing variability has been used to obtain representative patterns. However, it is worth mentioning that the within-subject variability from stride-to-stride carries important information, and is an important predictor for various neurological (such as cerebral palsy) and age-related

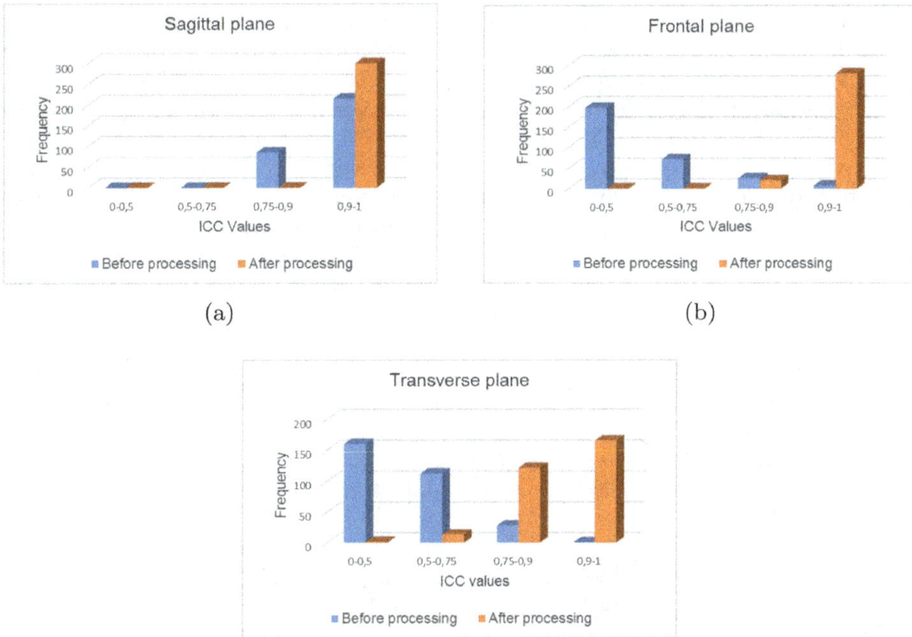

(a) (b)

(c)

Fig. 4. ICC Values of knee kinematics for all subjects in all three planes, before processing (in blue), and after processing (in orange). High values of ICC indicate high similarity between the kinematic curves of a subject.

diseases, which lead to inflated stride-to-stride variability during gait. In these contexts, pre-processing techniques should be performed prudently for addressing variability issues.

Acknowledgment. The authors would like to thank Gerald Parent for his technical assistance regarding the automated internal tool developed in the Imaging and Orthopaedics Research Laboratory (LIO), which provides a framework for biomechanical data analysis. The authors would also to thank Julien Clément for making the data collected from Hôpital Maisonneuve-Rosemont (HMR) available.

References

1. Abid, M., Mezghani, N., Mitiche, A.: Knee joint biomechanical gait data classification for knee pathology assessment: a literature review. Appl. Bionics Biomech. **2019**(7472039), 14 (2019). https://doi.org/10.1155/2019/7472039
2. Aissaoui, R., Husse, S., Mecheri, H., Parent, G., de Guise, J.A.: Automatic filtering techniques for three-dimensional kinematics data using 3D motion capture system. In: 2006 IEEE International Symposium on Industrial Electronics, vol. 1, pp. 614–619 (2006). https://doi.org/10.1109/ISIE.2006.295530

3. Ben Nouma, B., Mezghani, N., Mitiche, A., Ouakrim, Y.: A variational method to determine the most representative shape of a set of curves and its application to knee kinematic data for pathology classification. In: Proceedings of the 2nd Mediterranean Conference on Pattern Recognition and Artificial Intelligence, New York, NY, USA, pp. 22–26 (2018). https://doi.org/10.1145/3177148.3180095
4. Boivin, K.: Développement d'une approche d'évaluation clinique de la cinématique tridimensionnelle du genou durant la marche pour des patients gonarthrosiques. Ph.D. thesis, École Polytechnique de Montréal (2010). https://publications.polymtl.ca/317/
5. Chau, T.: A review of analytical techniques for gait data. Part 1: fuzzy, statistical and fractal methods. Gait Posture **13**, 49–66 (2001). https://doi.org/10.1016/S0966-6362(00)00094-1
6. Chau, T.: A review of analytical techniques for gait data. Part 2: neural network and wavelet methods. Gait Posture **13**, 102–120 (2001). https://doi.org/10.1016/S0966-6362(00)00095-3
7. Chau, T., Young, S., Redekop, S.: Managing variability in the summary and comparison of gait data. J. Neuroeng. Rehabil. **2**(1), 22 (2005). https://doi.org/10.1186/1743-0003-2-22
8. Duhamel, A., Bourriez, J., Devos, P., Krystkowiak, P., Destée, A., Derambure, P., Defebvre, L.: Statistical tools for clinical gait analysis. Gait Posture **20**(2), 204–212 (2004). https://doi.org/10.1016/j.gaitpost.2003.09.010
9. Fuentes-Dupré, A.: Apport d'une évaluation biomécanique 3D du genou dans la prise en charge orthopédique de patients ayant une rupture du ligament croisé antérieur. Ph.D. thesis, Université de Montréal (Faculté de médecine) (2010). http://hdl.handle.net/1866/4764
10. Gaudreault, N., Hagemeister, N., Poitras, S., de Guise, J.A.: Comparison of knee gait kinematics of workers exposed to knee straining posture to those of non-knee straining workers. Gait Posture **38**(2), 187–191 (2013). https://doi.org/10.1016/j.gaitpost.2012.11.004
11. Hagemeister, N., Parent, G., de Putte, M.V., St-Onge, N., Duval, N., de Guise, J.: A reproducible method for studying three-dimensional knee kinematics. J. Biomech. **38**(9), 1926–1931 (2005). https://doi.org/10.1016/j.jbiomech.2005.05.013
12. Hagemeister, N., Yahia, L., Duval, N., de Guise, J.: In vivo reproducibility of a new non-invasive diagnostic tool for three-dimensional knee evaluation. Knee **6**(3), 175–181 (1999). https://doi.org/10.1016/S0968-0160(99)00004-6
13. Hreljac, A., Marshall, R.N.: Algorithms to determine event timing during normal walking using kinematic data. J. Biomech. **33**(6), 783–786 (2000). https://doi.org/10.1016/S0021-9290(00)00014-2
14. Johnson, R.A.: Applied Multivariate Statistical Analysis. https://doi.org/10.1007/978-3-642-17229-8
15. Kadaba, M.P., Ramakrishnan, H.K., Wootten, M.E., Gainey, J., Gorton, G., Cochran, G.V.B.: Repeatability of kinematic, kinetic, and electromyographic data in normal adult gait. J. Orthop. Res. **7**(6), 849–860 (1989). https://doi.org/10.1007/978-1-4471-5451-8_101
16. Koo, T.K., Li, M.Y.: A guideline of selecting and reporting intraclass correlation coefficients for reliability research. J. Chiropr. Med. **15**(2), 155–163 (2016). https://doi.org/10.1016/j.jcm.2016.02.012

17. Labbe, D.R., Hagemeister, N., Tremblay, M., de Guise, J.A.: Reliability of a method for analyzing three-dimensional knee kinematics during gait. Gait Posture **28**(1), 170–174 (2008). https://doi.org/10.1016/j.gaitpost.2007.11.002
18. Lenhoff, M.W., Santner, T.J., Otis, J.C., Peterson, M.G.: Bootstrap prediction and confidence bands: a superior statistical method for analysis of gait data. Gait Posture **9**, 10–17 (1999). https://doi.org/10.1016/S0966-6362(98)00043-5
19. Leporace, G., Batista, L.A., Muniz, A.M., Zeitoune, G., Luciano, T., Metsavaht, L., Nadal, J.: Classification of gait kinematics of anterior cruciate ligament reconstructed subjects using principal component analysis and regressions modelling. In: 2012 Annual International Conference of the IEEE Engineering in Medicine and Biology Society, pp. 6514–6517 (2012). https://doi.org/10.1109/EMBC.2012.6347486
20. Lustig, S., Magnussen, R.A., Cheze, L., Neyret, P.: The KneeKG system: a review of the literature. Knee Surg. Sports Traumatol. Arthrosc. **20**(4), 633–638 (2012). https://doi.org/10.1007/s00167-011-1867-4
21. Mcgraw, K., Wong, S.: Forming inferences about some intraclass correlation coefficients. Psychol. Methods **1**, 30–46 (1996). https://doi.org/10.1037/1082-989X.1.1.30
22. Mechmeche, I., Mitiche, A., Ouakrim, Y., de Guise, J.A., Mezghani, N.: Data correction to determine a representative pattern of a set of 3D knee kinematic measurements. In: 2016 38th Annual International Conference of the IEEE Engineering in Medicine and Biology Society (EMBC), pp. 884–887 (2016). https://doi.org/10.1109/EMBC.2016.7590842
23. Mezghani, N., Dunbar, M., Ouakrim, Y., Fuentes, A., Mitiche, A., Whynot, S., Richardson, G.: Biomechanical signal classification of surgical and non-surgical candidates for knee arthroplasty. In: 2016 International Symposium on Signal, Image, Video and Communications (ISIVC), pp. 287–290 (2016). https://doi.org/10.1109/ISIVC.2016.7894002
24. Mezghani, N., Gaudreault, N., Mitiche, A., Ayoubian, L., Ouakrim, Y., Hagemeister, N., de Guise, J.A.: Kinematic gait analysis of workers exposed to knee straining postures by Bayes decision rule. Artif. Intell. **4**(2), 106–111 (2015). https://doi.org/10.5430/air.v4n2p106
25. Mezghani, N., Mechmeche, I., Mitiche, A., Ouakrim, Y., de Guise, J.A.: An analysis of 3D knee kinematic data complexity in knee osteoarthritis and asymptomatic controls. PLoS ONE **13**(10), 1–14 (2018). https://doi.org/10.1371/journal.pone.0202348
26. Mezghani, N., Ouakrim, Y., Fuentes, A., Hagemeister, N., Aissaoui, R., Pelletier, M., de Guise, J.: Knee osteoarthritis severity assessment using knee kinematic data classification. Osteoarthr. Cartil. **20**, S97 (2012). https://doi.org/10.1016/j.joca.2012.02.102
27. O'Connor, C.M., Thorpe, S.K., O'Malley, M.J., Vaughan, C.L.: Automatic detection of gait events using kinematic data. Gait Posture **25**(3), 469–474 (2007). https://doi.org/10.1016/j.gaitpost.2006.05.016
28. Røislien, J., Skare, Ø., Opheim, A., Rennie, L.: Evaluating the properties of the coefficient of multiple correlation (CMC) for kinematic gait data. J. Biomech. **45**(11), 2014–2018 (2012). https://doi.org/10.1016/j.jbiomech.2012.05.014

29. Sati, M., de Guise, J., Larouche, S., Drouin, G.: Improving in vivo knee kinematic measurements: application to prosthetic ligament analysis. Knee **3**(4), 179–190 (1996). https://doi.org/10.1016/S0968-0160(96)00209-8
30. Yu, B., Kienbacher, T., Growney, E.S., Johnson, M.E., An, K.N.: Reproducibility of the kinematics and kinetics of the lower extremity during normal stair-climbing. J. Orthop. Res. **15**(3), 348–352 (1997). https://doi.org/10.1002/jor.1100150306
31. Zeni, J.A., Richards, J.G., Higginson, J.S.: Two simple methods for determining gait events during treadmill and overground walking using kinematic data. Gait posture **27**(4), 710–714 (2008). https://doi.org/10.1016/j.gaitpost.2007.07.007

Towards Particle Tracking Velocimetry of Cell Flow in Developing Tissue Using Deep Neural Network

Yukitaka Ishimoto(✉) and Tomoaki Watanabe

Department of Mechanical Engineering, Akita Prefectural University, Yurihonjo, Akita 015-0055, Japan
ishimoto@akita-pu.ac.jp

Abstract. In medical and biological research, measurements of cellular dynamics have been paid much attention in recent years. Among such quantifications, particle tracking velocimetry (PTV) of cells is one of the major tools to elucidate the dynamics. For this purpose, it is critical to establish precise detection of individual cell positions at different timepoints. However, in live imaging of confluent cell system, cells are densely packed and touching images of nuclei prevent naive PTV analysis. In this work, we focus on precise detection of cell nucleus positions particularly in a very confluent situation and construct the detection algorithm with deep neural network (DNN). We have tested our system in the case of fly pupal wing epithelium and found the centers of the spots within nuclei with high probability. An implementation of this spot detection to PTV and an extension of DNN to recurrent neural network model (RNN) will also be discussed.

Keywords: Cellular dynamics · Tissue mechanics · Particle tracking velocimetry · Machine learning · Deep neural network

1 Introduction

In biology and regenerative therapy, or in a broader sense, medical research, measurements of cellular dynamics in living body have been paid much attention in recent years, because they provide deeper insight on tissue mechanics and engineering. Among measurements of such dynamics, those of cell motility or cell flow play a key role for elucidation of tissue/organ morphogenesis. To perform the measurements, or more specifically to track individual cells, it is critical to establish precise detection of individual cell positions and correspondence between them at different timepoints in a time series of tissue images. These demands are reflected in recent developments in the field, for example in the intelligent image-activated cell sorter [1]. Recognition and sorting of cells are becoming tractable through such emerging technologies, yet a general platform for measurements on cell positions and cell tracking have been anticipated.

G. A. Ateshian et al. (Eds.): CMBBE 2019, LNCVB 36, pp. 495–504, 2020.
https://doi.org/10.1007/978-3-030-43195-2_40

Difficulties of such measurements lie chiefly on cell heterogeneity and autonomy upon which one may not naively assume linearity of their movements and may assume cell-specific events such as cell divisions and deaths. In addition, in confluent cell system like epithelium, cells are densely packed and even cell nuclei are touching in acquired images, whose resolution is inevitably poor in the case of live imaging.

In this work, we aim at establishing a method of particle tracking velocimetry (PTV) for four-dimensional cell flow by using deep neural network model (DNN) of machine learning. We focus on precise detection of cell nucleus positions particularly in a very confluent situation, which would enable us to quantify cell flow not only of the fly case but of a broader range of morphologically changing tissues. That is, as an example of intriguing cellular dynamics, we consider morphogenesis of fly wing in its pupal stage, when cells nontrivially exhibit inhomogeneous and autonomous behaviors. Nontriviality means that, for instance, Navier-Stokes equation cannot be directly applied onto the cell flow, nor can conventional algorithms of flow visualization other than PTV with manual tracking of cells. An implementation of this spot detection to PTV and an extension of DNN to recurrent neural network model (RNN) will also be discussed in the end.

2 Method and Model

In what follows, we consider morphogenesis of fruit fly wing in its pupal stage. Precisely speaking, a time series of three-dimensional images of epithelial nuclei, *i.e.*, a 4D image, were taken during about 13–16 h after pupation (AP) by live-imaging technique with confocal laser microscope (a 3D reconstructed snapshot is shown in Fig. 1), whose interval was set to be 5 min. Some important conditions of the image acquisition are that, in live-imaging, image quality should be sacrificed for faster process, the laser power should be low enough to avoid

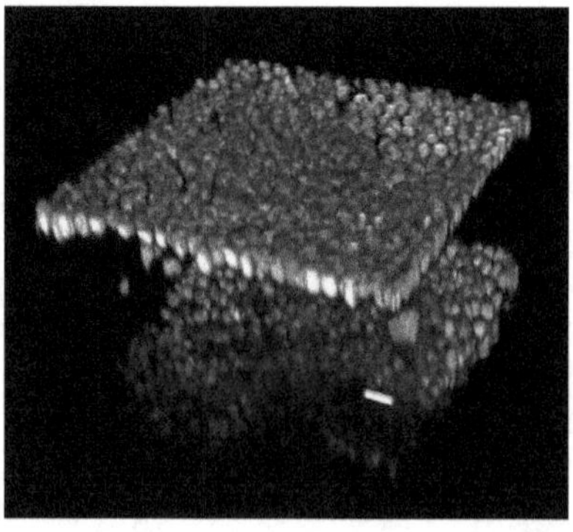

Fig. 1. A 3D reconstructed snapshot of cell nuclei in a fruit fly pupal wing during development (scale bar: 20 μm, $t = 0$, by the courtesy of Dr. O. Shimmi's group).

excessive damage on cells, and a long enough rest should be taken between consecutive measurements to clear the fluorescence raised by the previously injected laser. The resulting images (by the courtesy of Dr. O. Shimmi's group) have relatively low resolutions in space – 0.457 µm in the lateral directions and 1.041 µm in depth, so that the images of nuclei are frequently touching to each other. Besides, shapes of nuclei in developing tissues are usually changing due to a high rate of proliferation and rapid turnovers of nuclear membranes.

The above restrictions imply that the cross correlation method of particle imaging velocimetry would not work and the ordinary automated machinery of particle tracking velocimetry (PTV) would not work in a naive way. Namely, neither Gaussian fitting nor ellipsoidal fitting of nucleus image would work with less ambiguity, and touching objects would prevent even precise counting of nuclei in the images. For this very reason, we construct and implement a deep neural network model of machine learning as a part of spot detection algorithm and intend to detect positions of nuclei with higher accuracy.

In this section, we describe our method of spot detection with machine learning and explain the relevant neural network models.

2.1 Method

In order to make our machinery work properly, it is important to pre-process the images for lower computational cost, and to choose an efficient neural network model and a learning method. As the pre-processing, we first cut off dark noisy pixels and performed the 3D median filter with the smallest kernel to eliminate salt-and-pepper noise. Then, we performed the 3D mean filter with the smallest kernel to fill possible gaps taken away by the first cut off. For the learning, we adopted the supervised learning by manually creating the labels of three-dimensional positions of nuclei. We define the positions by their centers of mass (COMs), that is, intensity-weighted averages of the positions. To increase the number of the supervising dataset, we segmented the original three-dimensional images so that each fragment may contain, at least, one whole picture of a nucleus, and applied data augmentation by shifting and flipping the segmented images three-dimensionally. Thus prepared datasets were input to a set of neural network models whose cost function is set to be a simple sum of squared differences between inferred positions and the calculated labels. In addition to the supervised learning, we used the mini-batch training to avoid over- and under-fittings. The learnt model will be ready to use for spot detection in a cropped image.

2.2 Deep Neural Network Models

Since we deal with images, we chose deep convolutional neural network (CNN) as the basic design among deep neural network models (DNN). As is well known, the AlexNet [2] of CNN made DNN popular in 2012 by winning the competition of image recognition, ImageNet Large Scale Visual Recognition Competition (ILSVRC). The AlexNet consists of five convolutional layers and three fully connected layers with the ReLU function as activation and the max pooling as an overlapping layer, ending with the softmax function to classify the images. As its pre-process, their primary data was down sampled by cropping the image, and data augmentation and dropout layer were used to avoid overfitting. Training was done on multiple GPUs. It is now known that the convolutional layers of AlexNet had bigger kernel sizes and strides than necessary. Later, the second major update on CNN was accomplished by ResNet [3] which also won ILSVRC in 2015 and is now a de facto standard among CNNs. ResNet introduces the so called "residual module" which consists of two convolutional layers with batch normalization and the addition of the input in the end. The last addition helps optimization of the variables on earlier layers so that the whole network can be optimized better than AlexNet. It should be mentioned that many of the latest versions of ResNet include pre-activation [4].

Along the same line, we have constructed six types of CNNs – single convolutional layer, double convolutional layers, triple convolutional layers, single residual module, single pre-activated residual module, and double pre-activated residual layers. We set the smaller kernel sizes and strides than those of AlexNet and the dropout layer was also considered. A main difference between ours and the standard CNNs is that others are well established for classification of images by integer labeling whereas ours is to find the positions by numbers with decimal point. So, we put a projection layer that produces three decimal numbers from a fully connected layer. During the training, we take as the loss function the sum of squared differences between the labels of three-dimensional positions and the calculated numbers. The Adams optimizer is used to minimize the loss function. It should be noted here that, in ResNet, layers usually end up with the global average pooling (GAP layer) which would correspond to the projection layer in our case since the projection is another way of averaging the layer.

In the next section, we will present the results of two superior models to the rest. Python 2.7 with the NumPy package and the GPU version of TensorFlow (version 0.1.14) were used to implement the models and to perform the learning. For the TensorFlow visualization, the TensorBoard was used.

3 Results

From the first six timepoints, we calculated 650 COMs of the nuclei as the labels of the input dataset and extracted corresponding small three-dimensional images of $23 \times 23 \times 20$ pixels out of the original image of $262 \times 262 \times 84$ pixels with some random shifts of a few pixel widths. After the data augmentation of shifting and flipping, we have prepared for the 3563 input dataset, 3000 of which have been used for learning and the rest of which for evaluation of the learnt models. An example of an original image and COMs are shown in Fig. 2.

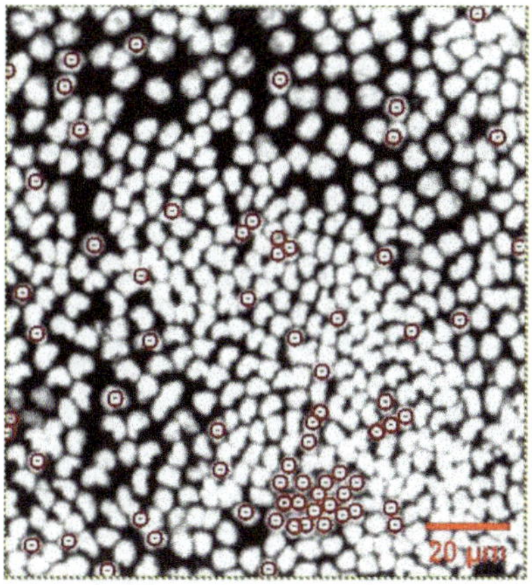

Fig. 2. An example of the cell nuclei image and the calculated centers of mass (circles) (Max Intensity Projection of 20 slices at $t = 2$, by the courtesy of Dr. O. Shimmi's group).

As for the mini-batch training, we have tested several patterns with the 3000 sets on the CNN model with a single convolutional layer, and found that the size of the batch, $BATCH = 10$, and the number of training, $NUM\_TRAIN = 300$, produced a better result and a declining trend in the loss function. So, we used this combination throughout the whole tests.

Among six tested models, CNN with two convolutional layers and that with two pre-activated residual modules produced the best results. The explicit diagrammatic expressions of these two models are shown in Figs. 3 and 4. During the learning of 3000 dataset, the loss functions declined slowly (Fig. 5). The results suggest that the double convolution model swiftly learnt the data and approached to a certain value, while the other residual module model succeeded to obtain an optimized value in its earliest steps and struggled to minimize it while learning. The evaluation of the models was done with 563 dataset and the projections of the resulting errors onto the x-y plane (lateral directions of the tissue) and the x-z plane (z being the depth dimension) are shown in Fig. 6. The errors in the z direction are relatively larger than those of the other lateral directions due to the z resolution of the original 4D data. By converting the errors in pixels, one can find that they are comparative in all directions. From the values of the loss functions, the average errors of the model predictions are estimated as $1.599\,\mu$m for the double convolution model, and $1.634\,\mu$m for the double pre-activated residual module model.

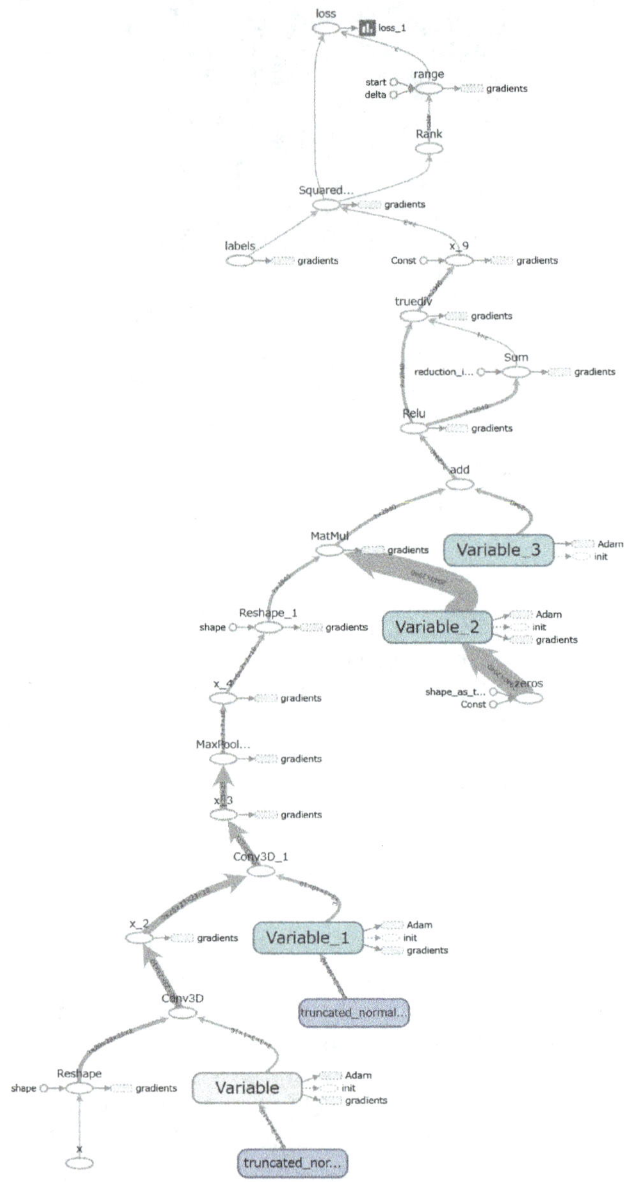

Fig. 3. Graph representation of the CNN model with two convolutional layers which produced the best result (visualized by TensorBoard). X stands for the input data, Conv3D is the three-dimensional convolutional layer, and Variable's are the model parameters to optimize.

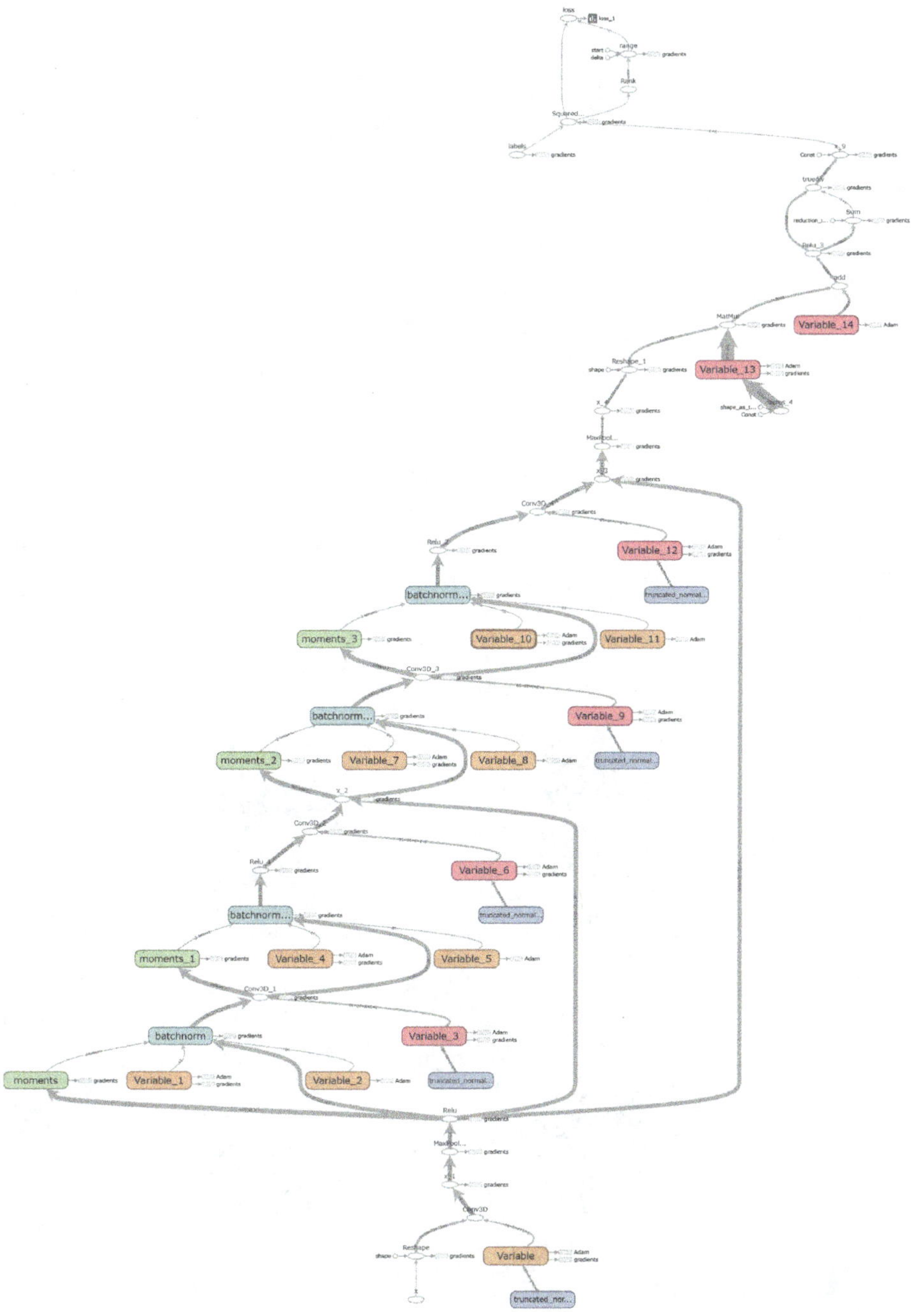

Fig. 4. Graph representation of the CNN model with two pre-activated residual modules which produced the second best result (visualized by TensorBoard). *X* stands for the input data, Conv3D is the three-dimensional convolutional layer, batchnorm is the batch normalization layer, and Variable's are the model parameters to optimize.

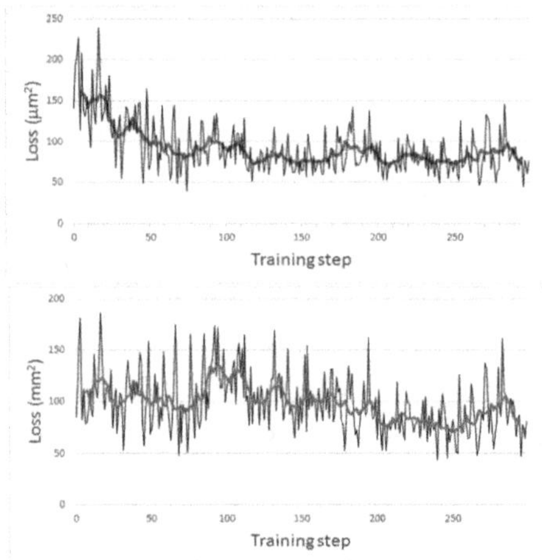

Fig. 5. Plots of the loss functions of the two models while training. The loss of the two convolutional layers (UPPER), and that of the two pre-activated residual modules (LOWER). In both plots, the thin lines represent bare values of the functions, while the thick lines do their trends within 11 steps.

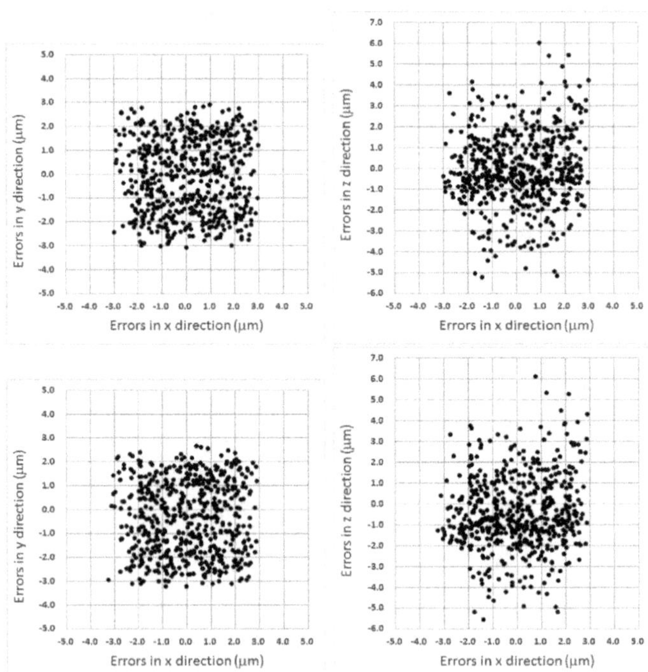

Fig. 6. The errors of the models during evaluation. The result of the two convolutional layers on the x-y plane (TOP LEFT), that on the x-z plane (TOP RIGHT), the result of the two pre-activated residual modules on the x-y plane (BOTTOM LEFT), and that on the x-z plane (BOTTOM RIGHT).

4 Concluding Remarks

We have constructed the inference model of the cell positions from 4D live-imaging data by using deep neural network model (DNN) of machine learning. The data we have used is of a very confluent tissue, so that the ordinary PTV method could not be applied in a naive way. We have created the 3563 cell data from 650 bare positions with their COMs as labels and trained the models. Among the tested six, the two models accomplished the best results: the average errors are 1.599 µm for the double convolution model, and 1.634 µm for the double pre-activated residual module model. Compared to the average size of 650 nuclei, which is about 4.65 µm, most of the inferred positions are within the three-dimensional nucleus images with high probability as can be seen in Fig. 6. This result can be regarded sufficiently well, and the models will be applicable not only to the confluent cellular system but a broader range of morphologically changing tissues.

There are two points to remark on further developments of the models. First, it should be noted that the loss functions declined rather slowly (Fig. 5) and we have not confirmed their asymptotic minimums during the training. This implies that the more data are used for training, the less errors can be expected. It would be valuable to create more data and find the models' capability. Besides, the depth of the DNN models usually works effectively when the size of data becomes larger. So, with more data, other untested deep models may produce better results than ours in this work.

In this work, we aim at establishing a PTV method for four-dimensional cell flow, but we have only constructed a machinery to detect the cellular positions. To make it a part of PTV, one should implement our machinery onto the PTV method. One of the simplest ways is to put the inferred positions to the existing PTV codes and find its results. Another yet complicated way is to place our tested models as a submodule on a node of a family of DNN models called recurrent neural network models (RNNs), where a time series of data can be learnt and the model can be optimized in a real-time fashion. It may also be possible for thus constructed RNNs to predict the cellular positions or trajectories after a certain learning period. To construct such, various RNN models should be constructed and tested which is beyond the scope of this manuscript.

Acknowledgements. The authors are grateful to Dr. O. Shimmi and his group for providing the precious imaging data and stimulating discussions during YI's stay in Helsinki, and to Dr. P. Marcq for his valuable discussions and hospitality during YI's stay in Paris. The authors would also like to acknowledge Prof. H. Madokoro for his comments on the machine learning techniques. This work was supported by JSPS KAKENHI (Grants-in-Aid for Scientific Research) Grant Number 17K00410 and 17KK0007.

References

1. Nitta, N., et al.: Intelligent image-activated cell sorting. Cell **175**, 266–276.e13 (2018)
2. Krizhevsky, A., Sutskever, I., Hinton, G.E.: ImageNet classification with deep convolutional neural networks. In: Proceedings of NIPS (2012)
3. He, K., Zhang, X., Ren, S., Sun, J.: Deep residual learning for image recognition. In: Proceedings of CVPR (2016)
4. Han, D., Kim, J., Kim, J.: Deep pyramidal residual networks. In: Proceedings of CVPR (2017)

Nanoindentation of Subchondral Bone During Osteoarthritis Pathological Process Using Atomic Force Microscopy

Lisa Manitta[1,2,3(✉)], Clemence Fayolle[1], Lucile Olive[1], and Jean-Philippe Berteau[1,4,5]

[1] Department of Physical Therapy, City University of New York - College of Staten Island, New York, USA
[2] Department of Biology, City University of New York - College of Staten Island, New York, USA
[3] Macaulay Honors College, City University of New York, New York, NY, USA
Lisa.Manitta@Macaulay.cuny.edu
[4] New York Center for Biomedical Engineering, City University of New York – City College of New York, New York, USA
[5] Nanoscience Initiative, Advanced Science Research Center, City University of New York, New York, USA

Abstract. Osteoarthritis (OA) is characterized by the alteration of connective tissue in the joints [1]. The current theory believes that changes to the subchondral bone, such as increased stiffness actually precede and contribute to cartilage loss in the early stages of OA [2], however, subchondral bone changes are not yet fully understood [3]. The goal of this project is to understand the effects of osteoarthritis on the mechanical properties of the subchondral bone femur of mice. The hypothesis is that at the nanoscale the subchondral bone (SB) Elastic Modulus is lower in early OA mice than in control. To test our hypothesis, we investigated the SB of C57Bl6 mice femurs from an early OA group and a control group (n = 6 and n = 6, respectively). We performed eight indentations at three locations on both the medial and lateral condyles for each femur. Our results show that the average Elastic Modulus in early OA mice is 3.57E+09 Pa, while that of the control group mice is 1.09E+10 Pa. This indicates a higher average Elastic Modulus in the control group mice than in the early OA mice, which may indicate an initial decrease in Elastic Modulus during the OA pathological process.

Keywords: Elastic modulus · Young's modulus · Stiffness · Hardness · Nanoindentation · Atomic Force Microscope · Subchondral bone · Femur · Osteoarthritis · Murine

1 Introduction

Osteoarthritis is a degenerative joint disease that affects around 9.3 million Americans over the age of forty-five and can have a lifetime cost between $209,800 and $228,600 per patient [4]. Osteoarthritis is characterized by the alteration of connective tissue in the joints [1]. A joint consists of both cartilage and subchondral bone. The cartilage

G. A. Ateshian et al. (Eds.): CMBBE 2019, LNCVB 36, pp. 505–517, 2020.
https://doi.org/10.1007/978-3-030-43195-2_41

serves to protect bones during movement by acting as a cushion, and the subchondral bone provides support for the cartilage and also acts to absorb some of the shock during loading [5]. In osteoarthritis the cartilage is broken down, and the subchondral bone undergoes mechanical changes such as increased stiffness and bone hardening [2]. The degradation of cartilage decreases the amount of protection provided to the subchondral bone, thus contributing to pain associated with osteoarthritis.

The progression of OA has been depicted as a vicious cycle involving both biological and mechanical changes (Fig. 1) [1]. The starting point to this vicious cycle is not yet known [1]. Joint changes that contribute to pain such as cartilage break down and increased sensitization of the bone due to vascularization during remodeling cause patients to reduce their physical activity, thus altering loading at the joint [1]. Altering the mechanical load at the joint induces further bone remodeling triggering the cycle to start over [1].

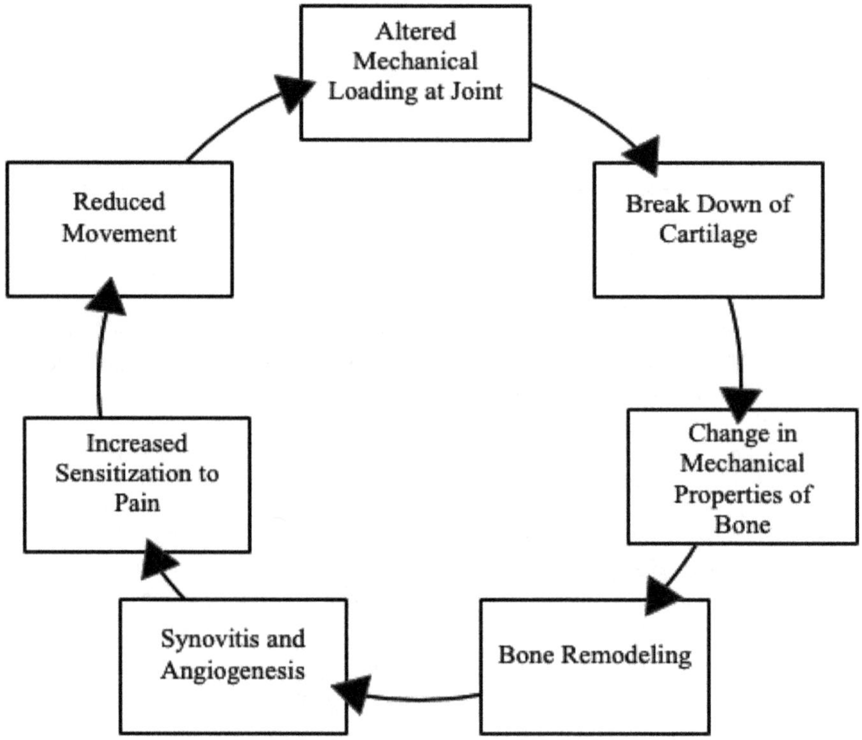

Fig. 1. Osteoarthritis Pathological Progression adapted from Wieland et al. [1]

It is believed that changes to the mechanical properties of subchondral bone occur during remodeling, such as bone hardening [2]. Thus, there has been an increased focus on the subchondral bone rather than cartilage when studying osteoarthritis progression [2]. The current theory believes that changes to the subchondral bone, such as increased stiffness actually precede and contribute to cartilage loss in the early stages of OA [2].

It has been shown through a rat model of OA that changes in gene expression of the subchondral bone precede cartilage degeneration and alters the activity of aggrecan and metalloproteinases by chondrocytes, contributing to the degeneration of cartilage [3]. Current evidence indicates that the subchondral bone plays a much more important role in the early stages of OA than previously thought, however, subchondral bone changes in later stages of OA progression are not yet fully understood [3].

Early stage changes in subchondral bone are currently of great interest because there is currently no treatment for osteoarthritis that provides a cure, and often times OA is detected late in its progression [6]. Thus, a biomarker for the progression of osteoarthritis is needed to make early detection possible, and the subchondral bone is a likely candidate.

The aim of this study is to gain an understanding of the effects of osteoarthritis on the mechanical properties of the subchondral bone femur of mice at the nano-scale. To do so, cantilever-based-nanoindentation experiments were performed on wild-type and osteoarthritic murine subchondral bone femurs. A timeline of the experimental setup is depicted below (Fig. 2).

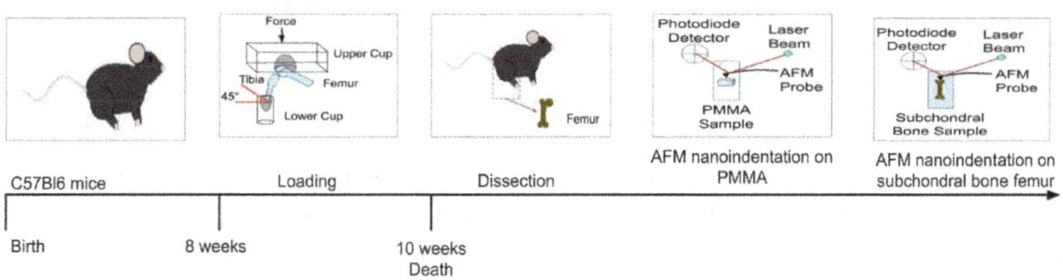

Fig. 2. Timeline of experimental setup

2 Materials and Methods

2.1 Murine OA Model

Osteoarthritis was induced in C57Bl6 mice by performing loading experiments on the mice eight weeks after birth using the ADMET eXpert 4000 machine (ADMET, Inc., Norwood, MA, USA). The protocol for these experiments followed Blandine Poulet's repetitive loading model [7]. The mice were loaded three times per week for two weeks. A 9 N force was applied for 40 cycles on the right joint of the mice with ADMET eXpert 4000 machine (ADMET, Inc., Norwood, MA, USA). The mice were placed on their back and their tibia was placed between two customized cups with a rotation of 45°. This rotation increases the shear force on the joint. The force applied on their knee was tracked with the software MTESTQuattro® (ADMET, Inc., Norwood, MA, USA). This procedure has been shown to induce joint changes similar to those that occur during the OA pathology, such as cartilage lesions and subchondral bone thickening [7]. Before starting the loading experiments the mice were anesthetized with isoflurane at a concentration of 1.5 to 5% with an oxygen rate of 1 liter per minute. At ten weeks after birth the mice were sacrificed and dissected and their femurs were harvested.

2.2 Analysis of Bone Biomechanics

Atomic Force Microscope Principle. The Atomic Force Microscope (AFM) is a multifunctional tool that can be used for topographic imaging of samples or for force measurement. The AFM consists of a cantilever with a very small sharp tip that acts as a probe. A laser is projected onto the cantilever and deflected into a photodiode. Thus, as the cantilever bends, the deflection of the laser is altered, and this change in deflection is measured by the photodiode [8]. This project employs the AFM technique known as cantilever based nanoindentation in order to measure the Elastic Modulus, or stiffness, of murine subchondral bone femur samples. During the loading phase the tip is lowered into the sample resulting in an increasing amount of force being applied to the sample, and is followed by unloading which involves the tip being raised and decreasing the amount of force applied to the sample. The resulting output is a force-distance curve. The unloading portion of the force distance curve is used to determine the Elastic Modulus using the Oliver-Pharr Method [9]. During the loading and unloading stages the cantilever bends as the tip indents into the sample and then raises away from the sample. This bending is detected by the change in deflection of the laser, which is measured in volts. The deflection in volts can be converted into nanometers using the inverse optical lever sensitivity (InvOLS) using the equation:

$$D_{nm} = D_V * InvOLS \tag{1}$$

where Dnm is the deflection in nanometers and DV is the deflection in Volts. The InvOLS is determined by performing an indentation on an infinitely hard surface, such as mica, and measuring the slope of the resulting force-distance curve. The unit of the value for InvOLS will be nm/V. The deflection in nanometers can then be used to determine the amount of force applied in Newtons (P) by using Hook's Law:

$$P = D_{nm} * 10^{-9} * k \tag{2}$$

where k is the spring constant in units N/m.

Data Acquisition. *Instruments.* The AFM used was the AFM MPF-3D (Asylum Research, Oxford Instruments, Abingdon, Oxfordshire, UK) at the Imaging Facility, CSI, CUNY, and its associated software is Igor Pro (WaveMetrics, Inc., Portland, OR, USA). A DT-NCHR (Nanosensors, NanoWorld AG, Matterhorn, Switzerland) probe was used for the nanoindentations, because a hard tip is necessary to perform these experiments. This probe type is a diamond coated probe and characterized by a tip radius between 100 and 200 nm, a resonance frequency of 400 kHz, and a spring constant of 80 N/m.

Sample Preparation. Murine femurs were embedded in PMMA using Technovit 9100 (Kulzer Technique International, Hanau, Germany). The medial and lateral condyles were cut from the rest of the bone using the IsoMet Slow Speed Saw (Buehler, Lake Bluff, IL, USA). The cut samples were then polished to expose the subchondral bone using the MetaServ Grinder and Polisher (Buehler, Lake Bluff, IL, USA). The samples were then glued onto glass microscope slides.

Calibration. Prior to beginning the experiments, the InvOLS of the cantilever must be found. This value is determined by performing an indentation on an infinitely hard material such as mica, and the slope of the force-displacement curve is equal to the InvOLS in nm/V [10]. Then the machine is calibrated by performing nanoindentations on a sample of a known Elastic Modulus, PMMA. The expected range for the Elastic Modulus of PMMA is from 2.24–3.8 GPa [11].

Acquisitions. The regions of interest were both the medial and lateral condyles of the subchondral bone (Fig. 3). Eight indentations were performed on three different locations on both the medial and lateral condyles. Data was collected from six osteoarthritic bones and six wild-type bones.

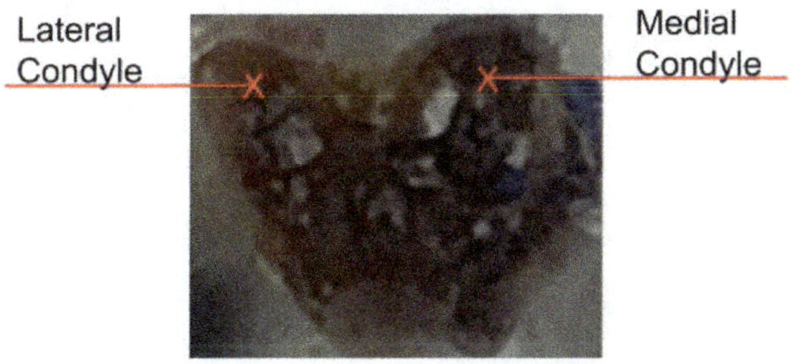

Fig. 3. Region of Interest on Murine Femur embedded in PMMA

Oliver and Pharr Method. The data obtained was analyzed using a Matlab (MathWorks, Inc., Natick, MA, USA) code based on the Oliver and Pharr Method. The Oliver and Pharr Method is a method used to calculate the Elastic Modulus of a material using force-displacement curves produced using nanoindentation experiments. The Elastic Modulus is the ratio of stress to strain, and stress is the ratio of the force applied to the cross-sectional area to which the force is applied [12]. A high elastic modulus indicates a stiff material, while a lower elastic modulus indicates a less stiff material. The Oliver and Pharr Method has become one of the most commonly used methods for determining the Elastic Modulus using nanoindentation [13]. This method assumes that both elastic and plastic deformation occurs during indentations, but only the elastic deformation reverses following the unloading phase of the indentation [14, 15]. There are three quantities that must be extracted from the force-displacement curve and these are the maximum applied load, P_{max}, the maximum indentation depth, h_{max}, and the elastic unloading stiffness, S, which is equal to the slope of the unloading curve [14, 15]. The unloading curve is non-linear and is fitted to a linear model using the following power law relation [15]:

$$P = \alpha(h - h_f)^m \tag{3}$$

where P is the load applied on the cantilever (N), h is the vertical displacement of the tip, h_f is the final displacement of the tip after unloading (nm), and α and m are fitting constants dependent on the indenter geometry [15]. The indentation depth, h_c, is then calculated using the equation [15]:

$$h_c = h_{max} - \varepsilon * \frac{P_{max}}{S} \quad (4)$$

where ε is a constant dependent on the geometry of the tip [15]. The standard value of ε is 0.75, which was used in this project [14, 15]. The indentation depth is less than the maximum indentation due to the sink in effect that results from elastic deformation during the loading phase [15].

During nanoindentation the elastic modulus is often found as the reduced, or effective, elastic modulus, E_r, which takes into account contributions from both the sample and the tip [15]. The reduced elastic modulus can be calculated using two equations [15]:

$$E_r = \frac{1}{\beta} * \frac{\sqrt{\pi}}{2} * \frac{S}{\sqrt{(\mathrm{Ac}\ (\mathrm{h}))}} \quad (5)$$

Or

$$\frac{1}{\mathrm{E_r}} = \frac{\left(1 - v_{sample}^2\right)}{E_{sample}} + \frac{\left(1 - v_{tip}^2\right)}{E_{tip}} \quad (6)$$

In Eq. (5) β is a parameter dependent on the geometry of the tip, and A_c is the area of contact between the sample and the tip [15]. In Eq. (6) v_{sample} is the Poisson's ratio of the sample, v_{tip} is the Poisson's ratio of the tip, E_{sample} is the elastic modulus of the sample, and E_{tip} is the elastic modulus of the tip. However, the elastic modulus of the tip is significantly higher than that of the sample, thus Eq. (6) can be simplified to [10]:

$$\frac{1}{E_r} = \frac{\left(1 - v_{sample}^2\right)}{E_{sample}} \quad (7)$$

The elastic modulus of the sample can then be found by combining Eqs. 5 and 7, resulting in [10]:

$$E_{sample} = \frac{1}{\beta} * \frac{\sqrt{\pi}}{2} \left(1 - v_{sample}^2\right) \frac{S_c}{\sqrt{(A_c(h))}} \quad (8)$$

Influence of Tip Geometry on Measured Elastic Modulus. The elastic modulus is dependent on the area of contact between the tip and the sample, thus a reliable method for calculating this area is crucial for accurately determining the elastic modulus of the sample. The method used in this study to measure the area of contact was scanning electron microscopy (SEM) of the tip. An SEM uses an electron gun that directs a beam of electrons towards the sample, and a detector will then detect the backscattered electrons producing a topographic image of the sample. The SEM is useful because it enables us to obtain images at a magnification greater than 100,000 times while maintaining high resolution. The tip was held in the sample chamber of the SEM using a 3D printed sample holder, which was carbon taped onto a pin mount. The custom sample holder had a slit for the tip that allowed the tip to come in contact with the carbon tape. This is necessary because the tip needs to be in contact with a conductive surface to reduce charging on the sample, which would result in an unclear image. When the image is obtained, it is analyzed using ImageJ (NIH, Bethesda, MD, USA). Due to limitations of manufacturing such a small tip, a radius between 100 and 200 nm, the tip does not come to a point and instead has a rounded end [13]. Thus, our lab has come up with a model to describe the geometry of the tip which was adapted from the work of Calabri et al. [13]. The method designed by Calabri et al. [13] assumes an axisymmetric sphere-cone hybrid model of the tip (Fig. 4) [13]. This assumes that the tip corner angle, θ, is the same on either side of the tip.

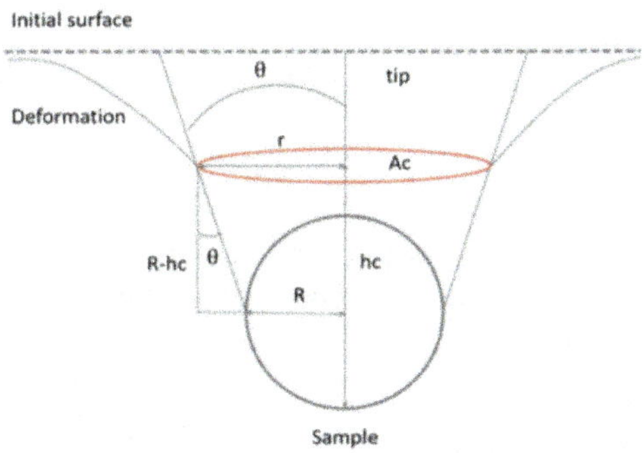

Fig. 4. Axisymmetric Model of Tip adapted from Calabri et al. [13]

The area of contact for this model is calculated using Eqs. 8 and 9 [13].

$$r = \tan\theta\,(h_c - R) + R \tag{9}$$

$$A_c = \pi r^2 = \pi(tan\theta(h_c - R) + R)^2 \tag{10}$$

Upon SEM imaging of the tip it was found that this is not the case. Thus, this model was adapted to assume that the geometry of the tip is a non-axisymmetric sphere-cone hybrid. When the indentation depth is less than the radius of the sphere portion of the tip, it can be assumed that the tip has a spherical shape (Fig. 5).

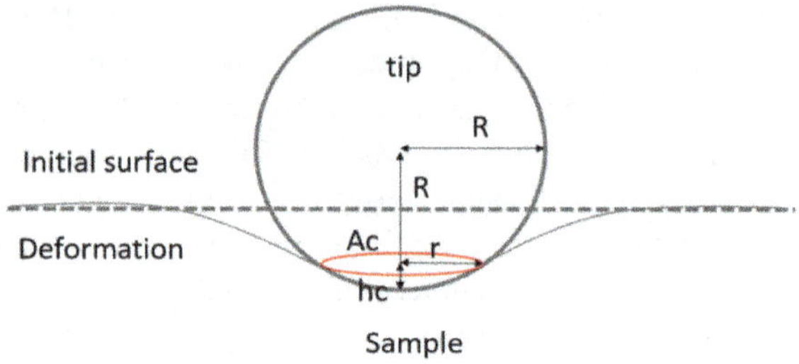

Fig. 5. Sphere model of the tip

Using this model, the following equations are used to arrive at the area of contact, A_c:

$$r = \sqrt{R^2 - (R - h_c)^2} \tag{11}$$

$$r = \sqrt{2Rh_c - h_c^2} \tag{12}$$

$$A_c = \pi r^2 = \pi\left(2Rh_c - h_c^2\right) \tag{13}$$

When the indentation depth is greater than the radius of the tip, this model can no longer be used, so a non-axisymmetric sphere-cone hybrid (Fig. 6) was employed to calculate the area of contact.

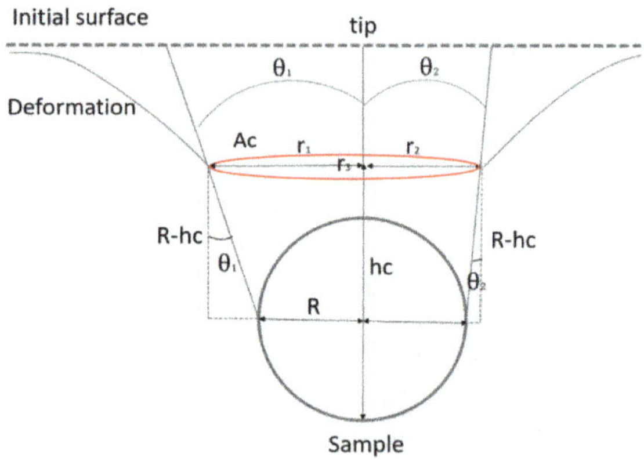

Fig. 6. Non-axisymmetric hybrid model of tip geometry adapted from Calabri et al. [13]

The equations used to calculate the elastic modulus using this model are:

$$r_1 = tan\theta_1(h_c - R) + R \tag{14}$$

$$r_2 = tan\theta_2(h_c - R) + R \tag{15}$$

$$r_3 = \frac{r_1}{(r_1 + r_2)} * r_2 + \frac{r_2}{(r_1 + r_2)} * r_1. \tag{16}$$

$$A_c = \frac{\pi}{2}(r_1 * r_3 + r_2 * r_3) \tag{17}$$

The parameters necessary for these calculations are obtained from the SEM image using the ImageJ (NIH, Bethesda, MD, USA) software. The SEM image is converted into a binary image to which the sphere-cone hybrid model is applied. The radius can be measured by converting the scale on ImageJ (NIH, Bethesda, MD, USA) from pixels into nm using the scale bar on the SEM image. The process of ImageJ (NIH, Bethesda, MD, USA) analysis is depicted below (Fig. 7).

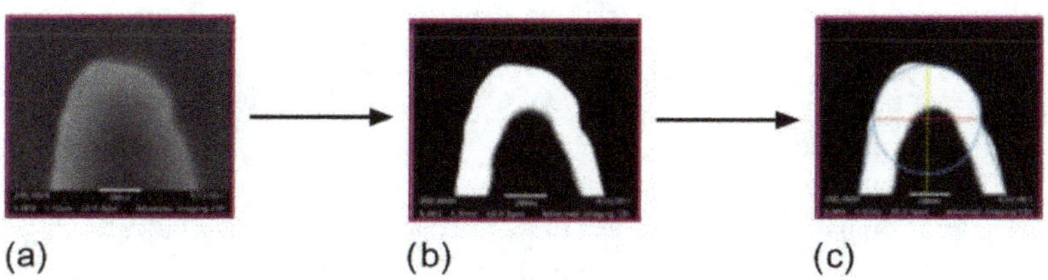

Fig. 7. (a) SEM image of tip. (b) SEM image converted to binary. (c) Sphere-Cone hybrid model applied to image.

Statistical Analysis. The results obtained from this study were analyzed using IBM's SPSS® software (IBM, Armonk, NY, USA). The first step in the analysis was to test for a normal distribution using a Shapiro-Wilke's test. A paired t-test was then used to test for a significant difference between paired groups that demonstrate a normal distribution. The Mann-Whitney test was used to test for a significant difference between non-normally distributed and unpaired groups. The Wilcoxon matched pairs test was used for comparing paired groups not demonstrating a normal distribution. Finally, an unpaired t-test was used to test for a significant difference between unpaired groups demonstrating a normal distribution. During the statistical analysis, we compared the results from the OA group, both the medial and lateral condyles, to the results of the control group, both medial and lateral condyles. Next, we tested for a difference between the elastic moduli of OA medial condyles compared to OA lateral condyles. Then, we tested for a difference between the control medial condyles compared to the control lateral condyles. Finally, we compared the OA medial condyle to the control medial condyle and the OA lateral condyle to the control lateral condyle. The power and sample size calculating software PSS (OriginLab, Northampton, MA, USA) was used to determine the power of the results and the likelihood of type I and type II errors present in the results.

3 Results

Prior to testing bone samples, PMMA was tested to ensure that our method produces reliable results. First, experiments using the axisymmetric tip model proposed by Calabri et al. [13] (Fig. 4) were performed [13]. This model assumes that the two tip corner angles are the same, however, the tips used in this study were shown to have a non-axisymmetric geometry (Fig. 6). Meaning that the two tip corner angles were not equal to each other. In order to analyze the data using the axisymmetric model, the data was analyzed twice. One time using θ_1 (Fig. 6) to calculate the area of contact between the tip and the sample, and one time using θ_2 (Fig. 6) to calculate the area of contact. The results are shown in Table 1.

Table 1. PMMA elastic modulus using axisymmetric model

	E (GPa) for θ_1	E (GPa) for θ_2
Indentation 1	1.562	1.525
Indentation 2	3.160	3.012
Indentation 3	3.060	2.862
Indentation 4	2.036	1.836
Indentation 5	2.860	2.654
Indentation 6	3.290	3.038
Indentation 7	3.490	3.275
Indentation 8	5.614	5.509
Average	3.134	2.964
Standard deviation	1.199	1.197

The results from both values for tip corner angle are within the expected range for PMMA. However, upon statistical analysis a significant difference was found between these two sets of results. Thus, this model was adapted into the non-axisymmetric sphere-cone hybrid model in order to factor both tip corner angles into the calculation of contact area. This model was tested by performing eight indentations at five different locations on PMMA. The results are shown in Tables 2 and 3.

Table 2. Average elastic modulus of PMMA

Sample	Average elastic modulus (GPa)
PMMA 1	2.21 ± 0.942
PMMA 2	4.65 ± 5.51
PMMA 3	3.56 ± 2.43
PMMA 4	1.49 ± 0.882
PMMA 5	4.31 ± 4.58
Mean of the means	3.24
Standard deviation	1.36

Nanoindentations have been performed on six osteoarthritic and six healthy mice. The preliminary data indicates a higher average Elastic Modulus in the control group mice than in the early OA mice, which may indicate an initial decrease in Elastic Modulus during the OA pathological process. Upon statistical analysis, a significant difference was found between the elastic moduli of the early OA group and the control group. In both groups the elastic modulus of the medial condyle was shown to be higher than that of the lateral condyle, however no significant difference was found after statistical analysis. Further tests will be performed in order to increase the power of these results.

Table 3. Average elastic modulus values reported in GPa

	Elastic modulus in GPa for early OA group		Elastic modulus in GPa for control group	
Sample number	Medial condyle	Lateral condyle	Medial condyle	Lateral condyle
1	2.51 ± 1.28	2.75 ± 1	19.9 ± 29.6	2.46 ± 1.42
2	2.60 ± 2.28	0.466 ± 6.04	3.47 ± 4.56	5.76 ± 8.09
3	0.740 ± 5.72	0.323 ± 2.19	2.53 ± 4.36	5.19 ± 3.33
4	3.02 ± 2.74	2.07 ± 1.74	2.19 ± 3.55	9.45 ± 15.6
5	9.15 ± 5.41	9.29 ± 1.63	29.6 ± 17.5	34.2 ± 30.1
6	6.52 ± 6.24	3.39 ± 5.04	10.6 ± 4.7	5.87 ± 2.41
Mean of the means	4.09	3.05	11.4	10.5
Standard deviation	3.12	3.29	11.2	11.8

4 Discussion

We have adapted the model of Calabri et al. [13], so that it more accurately represents the geometry of the tips used in this study. The results show that the non-axisymmetric sphere-cone hybrid model for calculating the area of contact between the tip and sample produces reliable results when compared to literature. The average elastic modulus of PMMA was 3.24 ± 1.36 GPa, which is in the range found in literature of 2.24–3.8 GPa [11]. Testing on bone samples has begun, and preliminary data shows an initial decline in elastic modulus of the subchondral bone femur during the OA pathology. The results of this study show that the average Elastic Modulus in early OA mice is 3.57 GPa, while that of the control group mice is 10.9 GPa. These results fall into the range of 2–30 GPa for healthy bone found in literature [16].

5 Conclusion

The results found in this study show that the non-axisymmetric sphere-cone hybrid model has been able to produce reliable results while taking into account the geometry of the tips used. The preliminary findings on bone samples indicate that the elastic modulus of the subchondral bone femur decreases during the initial stages of osteoarthritis in mice. This supports the hypothesis, that the subchondral bone initially decreases in hardness during the pathological process of OA. It is hoped that the discovery of these mechanical changes that occur to the subchondral bone femur in mice with OA, could contribute to the development of a biomarker for earlier diagnosis of OA.

Ethical Approval. The use of mice in this study was approved by the College of Staten Island Institutional Animal Care and Use Committee (CSI IACUC).

References

1. Wieland, H.A., Michaelis, M., Kirschbaum, B.J., Rudolphi, K.A.: Osteoarthritis - an untreatable disease? Nat. Rev. Drug Discov. (2005). https://doi.org/10.1038/nrd1693
2. Amini, M., Nazemi, S.M., Lanovaz, J.L., Kontulainen, S., Masri, B.A., Wilson, D.R., Szyszkowski, W., Johnston, J.D.: Individual and combined effects of OA-related subchondral bone alterations on proximal tibial surface stiffness: a parametric finite element modeling study. Med. Eng. Phys. (2015). https://doi.org/10.1016/j.medengphy.2015.05.011
3. Fang, H., Huang, L., Welch, I., Norley, C., Holdsworth, D.W., Beier, F., Cai, D.: Early changes of articular cartilage and subchondral bone in the DMM mouse model of osteoarthritis. Sci. Rep. **8**(1), 2855 (2018). https://doi.org/10.1038/s41598-018-21184-5
4. Losina, E., Paltiel, A.D., Weinstein, A.M., Yelin, E., Hunter, D.J., Chen, S.P., Klara, K., Suter, L.S., Solomon, D.H., Burbine, S.A., Walensky, R.P., Katz, J.N.: Lifetime medical costs of knee osteoarthritis management in the United States: impact of extending indications for total knee arthroplasty. Arthritis Care Res. **67**(2), 203–215 (2015). https://doi.org/10.1002/acr.22412
5. Madry, H., van Dijk, C.N., Mueller-Gerbl, M.: The basic science of the subchondral bone. Knee Surg. Sports Traumatol. Arthrosc. **18**(4), 419–433 (2010). https://doi.org/10.1007/s00167-010-1054-z
6. Williams, F.M., Spector, T.D.: Biomarkers in osteoarthritis. Arthritis Res. Therapy **10**(1), 101 (2008). https://doi.org/10.1186/ar2344
7. Poulet, B.: Non-invasive loading model of murine osteoarthritis. Curr. Rheumatol. Rep. **18** (7), 40 (2016). https://doi.org/10.1007/s11926-016-0590-z
8. Rugar, D., Hansma, P.: Atomic Force Microscopy. American Institute of Physics (1990). http://www.if.ufrj.br/~tclp/estadosolido/phystoday23(90).pdf

9. Thurner, P.J.: Atomic force microscopy and indentation force measurement of bone. Wiley Interdisc. Rev.: Nanomed. Nanobiotechnol. **1**(6), 624–649 (2009). https://doi.org/10.1002/wnan.56
10. Andriotis, O.G., Manuyakorn, W., Zekonyte, J., Katsamenis, O.L., Fabri, S., Howarth, P.H., Davies, D.E., Thurner, P.J.: Nanomechanical assessment of human and murine collagen fibrils via atomic force microscopy cantilever-based nanoindentation. J. Mech. Behav. Biomed. Mater. **39**, 9–26 (2014). https://doi.org/10.1016/J.JMBBM.2014.06.015
11. Han, Y., Elliott, J.: Molecular dynamics simulations of the elastic properties of polymer/carbon nanotube composites. Comput. Mater. Sci. **39**(2), 315–323 (2007). https://doi.org/10.1016/J.COMMATSCI.2006.06.011
12. Stolz, M., Raiteri, R., Daniels, A.U., Van Landingham, M.R., Baschong, W., Aebi, U.: Dynamic elastic modulus of porcine articular cartilage determined at two different levels of tissue organization by indentation-type atomic force microscopy. Biophys. J. **86**(5), 3269–3283 (2004). https://doi.org/10.1016/S0006-3495(04)74375-1
13. Calabri, L., Pugno, N., Menozzi, C., Valeri, S.: AFM nanoindentation: tip shape and tip radius of curvature effect on the hardness measurement. J. Phys.: Condens. Matter **20**(47), 474208 (2008). https://doi.org/10.1088/0953-8984/20/47/474208
14. Oliver, W.C., Pharr, G.M.: An improved technique for determining hardness and elastic modulus using load and displacement sensing indentation experiments. J. Mater. Res. **7**(6), 1564–1583 (1992). https://doi.org/10.1557/JMR.1992.1564
15. Oliver, W.C., Pharr, G.M.: Measurement of hardness and elastic modulus by instrumented indentation: advances in understanding and refinements to methodology (2004). www.mrs.org/publications/jmr/policy.html
16. Tai, K., Dao, M., Suresh, S., Palazoglu, A., Ortiz, C.: Nanoscale heterogeneity promotes energy dissipation in bone. Nat. Mater. **6**(6), 454–462 (2007). https://doi.org/10.1038/nmat1911

Computational Modeling of Blood Flow with Rare Cell in a Microbifurcation

Iveta Jančigová(✉)

Cell in Fluid: Biomedical Modelling and Computation Group,
Department of Software Technologies, University of Žilina, Žilina, Slovakia
iveta.jancigova@fri.uniza.sk
https://www.pyoif.eu/

Abstract. The study of microfluidic blood flow in bifurcations is important for understanding the blood behavior in capillary network, where the important gas exchange happens. This computational investigation looks at rare cells suspended in 3D red blood cell flow in symmetric and asymmetric bifurcations. We observe how the presence of a large rare cell, which flows through the wider daughter branch, impacts the red blood cell distribution in both branches. The model also allows us to quantitatively examine the fluid forces acting on the red blood cells and compare them in situations when a rare cell is and is not present.

Keywords: Microfluidic bifurcations · Red blood cell · Computational modeling · Rare cell

1 Introduction

Investigation of vascular networks that contain bifurcations and have topological similarity to microvasculature in vivo is important both for understanding the biological behavior and for design of microfluidic devices. In general, a bifurcation manifests in one of two ways: the classical or reverse partitioning of the flow. In the classical, the fluid entering a side branch tends to receive disproportionally fewer cells, as the region of low hematocrit near the vessel wall is 'skimmed'. This phenomenon is known as plasma skimming. However, under certain conditions a reverse effect is observed. The behavior is so complex that it can vary in time between these two types either due to a non-uniform upstream RBC distribution or due to a temporary increase in cell concentration near bifurcations, [1].

Several computational approaches have been taken to consider 2D [2] or 3D [3] red blood cell flow in bifurcations. Such works typically do not take into account that there are other cells suspended in the blood plasma, some of them comparable in size to the diameter of the bifurcating branches, e.g. white blood cells or circulating tumor cells. These cells occur at much lower frequency than red blood cells, however, when they do, they have a significant effect on the local flow.

G. A. Ateshian et al. (Eds.): CMBBE 2019, LNCVB 36, pp. 518–525, 2020.
https://doi.org/10.1007/978-3-030-43195-2_42

The diverging bifurcations are responsible for non-uniform partitioning of red blood cells within the network [4]. While there are quantitative data on the RBC velocity, line density and flux in the daughter branches and it is known that the RBCs have the tendency to enter the daughter branch with higher flow rate (Zweifach-Fung effect), in some cases an inversion of this effect was observed [5].

The converging bifurcations are helpful in controlling the concentration of cells at a given position across the width of a channel and thus also need to be considered in the process of design of microfluidic devices. [6] experimentally investigated the collective spreading of RBCs in a straight microchannel after such bifurcation and observed that the spreading increased considerably as the hematocrit increased, mainly due to cell-cell interactions while the flow rate had a small effect.

In this paper we present a general framework for investigating 3D blood flow in microbifurcations. It takes into account the RBC deformability and cell-cell interactions which are two of the key physical determinants in the RBC distribution in the microcirculation. It also allows easy inclusion of other types of cells and quantification of their impact on the flow, which we demonstrate by considering a larger spherical stiffer cell that may represent a while blood cell or a circulating tumor cell.

2 Computational Model

The computational model has two main parts. One is the fluid, which represents the blood plasma, for which we use the lattice-Boltzmann method with a regular cubical grid. The second component are the elastic cells that are modeled as moving objects represented by triangular meshes. The elastic behavior is achieved using five elastic moduli: stretching, bending, local area conservation, global area conservation and global volume conservation. The fluid and cells interact and affect each other through dissipative coupling. The detailed derivation and description of the model can be found in [7]. Here we provide a brief summary of the individual elastic moduli.

Stretching Modulus. The nonlinear stretching force at point A on edge AB is defined as

$$\mathbf{F}_s(A) = k_s \kappa(l_{AB}/l_{AB0}) \Delta l_{AB} \mathbf{p}_{AB}, \tag{1}$$

where k_s is the stretching coefficient, $\mathbf{p}_{AB}$ is a unit vector pointing from A to B, κ represents the neo-Hookean nonlinearity, l_{AB0} is the relaxed length of the edge AB and $\Delta l_{AB} = l_{AB} - l_{AB0}$ is the prolongation of this edge. Opposite force is applied at mesh point B.

Bending Modulus. Bending forces are defined for position vectors A, B, C and D of two triangles ABC and ABD that share a common edge AB:

$$\begin{aligned}
\mathbf{F}_b(A) &= -k_b \Delta\theta \left(\frac{\mathbf{N}_C}{|\mathbf{N}_C|^2} \frac{(A-B, C-B)}{|B-A|} + \frac{\mathbf{N}_D}{|\mathbf{N}_D|^2} \frac{(A-B, D-B)}{|B-A|} \right), \\
\mathbf{F}_b(B) &= -k_b \Delta\theta \left(\frac{\mathbf{N}_C}{|\mathbf{N}_C|^2} \frac{(A-B, A-C)}{|B-A|} + \frac{\mathbf{N}_D}{|\mathbf{N}_D|^2} \frac{(A-B, A-D)}{|B-A|} \right), \\
\mathbf{F}_b(C) &= k_b \Delta\theta |B-A| \frac{\mathbf{N}_C}{|\mathbf{N}_C|^2}, \\
\mathbf{F}_b(D) &= k_b \Delta\theta |B-A| \frac{\mathbf{N}_D}{|\mathbf{N}_D|^2},
\end{aligned} \tag{2}$$

where k_b is the bending coefficient, $\Delta\theta$ is the difference between the current angle θ (between triangles ABC and ABD) and θ_0, the angle between these triangles in relaxed state. The vector $\mathbf{N}_C = (A-C) \times (B-C)$ is the normal vector to triangle ABC and $\mathbf{N}_D = (B-D) \times (A-D)$ is the normal vector to triangle ABD.

Local Area Modulus. The local area force applied at vertex A of triangle ABC with area S_{ABC} and centroid T is

$$\mathbf{F}_{al}(A) = k_{al} \frac{1}{3} \frac{\Delta S_{ABC}}{t_a^2 + t_b^2 + t_c^2} \mathbf{AT}, \tag{3}$$

where k_{al} is the local area coefficient, ΔS_{ABC} is the difference between current S_{ABC} and area S_{ABC0} of the triangle in the relaxed state and t_a, t_b, t_c are the distances from points A, B, C to centroid T. Analogous forces are applied at vertices B and C.

Global Area Modulus. The global area modulus keeps the total surface area of the cell fairly constant. It is similar to the local area in that we have the proportional distribution according to the distance of vertices from centroid t_a, t_b, t_c. In addition to that, we have a weight that takes into account the area of the triangle with respect to the total surface area of the cell:

$$\mathbf{F}_{ag}(A) = k_{ag} \Delta S_{cell} \frac{S_{ABC}}{S_{cell0}} \frac{\mathbf{AT}}{t_a^2 + t_b^2 + t_c^2}, \tag{4}$$

where k_{ag} is the global area coefficient, ΔS_{cell} is the difference between the current S_{cell} and area S_{cell0} in relaxed state. This formula represents the global area contribution due to the triangle ABC with area S_{ABC} and vector $\mathbf{AT}$. Since vertex A belongs also to other triangles, it receives more contributions of this kind.

Volume Modulus. The volume modulus keeps the volume of the cell fairly constant. The force applied at vertex A of triangle ABC is:

$$\mathbf{F}_v(A) = -k_v \frac{1}{3} \frac{\Delta V_{cell}}{V_{cell0}} S_{ABC} \mathbf{n}_{ABC}, \tag{5}$$

where k_v is the volume coefficient, ΔV_{cell} is the difference between the current volume V_{cell} and volume V_{cell0} in relaxed state. The vector $\mathbf{n}_{ABC}$ is the unit normal vector to the plane ABC. The direction of the force is along the triangle's normal pointing inside the cell.

2.1 Implementation

The model is implemented in the open-source computational package ESPResSo 4.0 [8] as a PyOIF module [9]. It has an easy-to-use python scripting interface, can be used for other applications or extended as needed.

There are two types of cells in this computational study: the elastic red blood cells and a stiffer larger spherical cell. The elastic parameters k_s, k_b, k_{al}, k_{ag} and k_v of all red blood cells were the same and were set to values provided by the calibration performed in [10]. The values of k_s and k_b of the rare cell were approximately 1.5 times larger than those of RBCs and the value of k_{al} for the rare cell was about 4 times larger than the corresponding RBC value. The values of global coefficients of the rare cell were the same as for RBCs. These larger values were chosen empirically to approximate the fact that both white blood cells and circulating tumor cells are stiffer than the red blood cells.

3 Simulation Setup

3.1 Geometry

The simulated domain was a periodic channel with bifurcation into two daughter branches that later converge together to satisfy the periodic condition, see Fig. 1. The width of the primary channel was $w_p = 30\,\mu\text{m}$. For the two daughter branches (denoted upper and lower), we considered a symmetric case (C: $w_u = w_l = 15\,\mu\text{m}$) and two asymmetric cases. In one of them the ratio of sizes of daughter branches was 1:2 (A: $w_u = 10\,\mu m, w_l = 20\,\mu\text{m}$) and in the other 2:3 (B: $w_u = 12\,\mu\text{m}$, $w_l = 18\,\mu\text{m}$).

3.2 Cells and Fluid

The red blood cells had the typical biconcave discoid shape with radius 3.91 μm. They were represented by meshes with 374 nodes. We ran two kinds of simulations for each case: with 40 cells ($Ht \approx 4\%$) and with 60 cells ($Ht \approx 6\%$). Each of these was repeated 10 times with a different random initial seeding of cells. The results are aggregated into six cases by geometry and hematocrit: A40, A60, B40, B60, C40, C60.

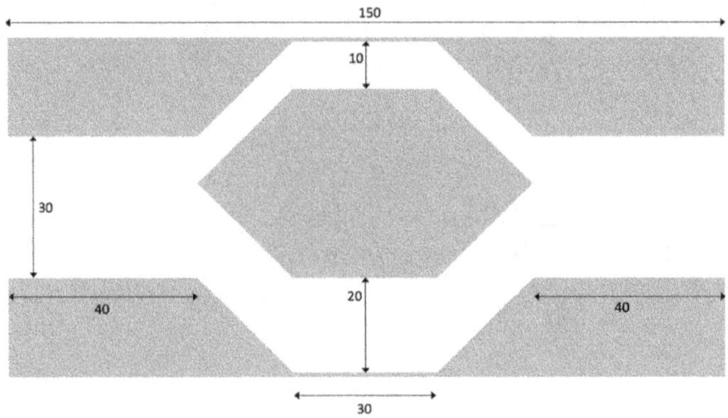

Fig. 1. Bifurcating channel, geometry A. All dimensions are in micrometers. The height of the channel was $h = 20\,\mu m$.

The rare cell was spherical, with radius $7.5\,\mu$m and 642 mesh nodes. In all cases, it was seeded at the center of the primary channel.

The kinematic viscosity of the fluid was $1.5 \cdot 10^{-6}\,\text{m}^2/\text{s}$, the density $1 \cdot 10^3\,\text{kg/m}^3$. The stabilized fluid velocity at the center of the primary channel was $1.3 \cdot 10^{-3}\,\text{m/s}$ similar to the value reported in [3].

4 Results

We look at the RBC characteristics in the daughter branches in the following scenarios:
Situation 1 - RBCs in narrow branch, rare cell inside the wider branch
Situation 2 - RBCs in wider branch, rare cell inside the wider branch
Situation 3 - RBCs in narrow branch, rare cell in parent channel
Situation 4 - RBCs in wider branch, rare cell in parent channel
In all asymmetric cases (A, B), the rare cell entered the wider branch. In symmetric cases (C), we kept the same notation and report the branch, which the rare cell entered, as *wider* (even though both are the same size). In cases A40, B40, C40, there were about 800 occurrences of RBC entering a daughter branch in each. In cases A60, B60, C60, there were about 1200 each.

4.1 Red Blood Cell Distributions

As can be seen in Table 1 and as is expected, more red blood cells enter the wider daughter branch than the narrow daughter branch. This is true for both geometries A and B and both values of hematocrit. For the symmetric geometry C, we see a fairly even split. Except for the case A40, in all other asymmetric cases we see greater disparity between narrow and wider branch while the rare cell is inside the wider branch (Situations 1 and 2) than while the rare cell is in the parent channel either before or after the bifurcation (Situations 3 and 4).

Table 1. Percentage of RBCs entering the branches

Case	Situation 1	Situation 2	Situation 3	Situation 4
A40	29.1	70.9	28.6	71.4
B40	35.0	65.0	38.5	61.5
C40	52.3	47.7	50.1	49.9
A60	24.9	75.1	26.3	73.7
B60	33.9	66.1	35.1	64.9
C60	50.2	49.8	49.7	50.3

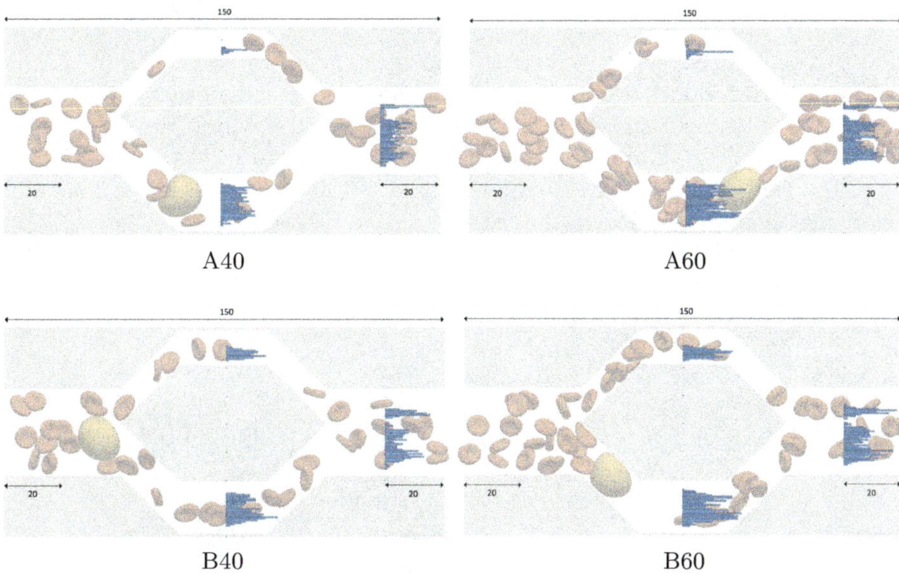

Fig. 2. Distribution of red blood cells in daughter branches and after the confluence. The histograms indicate y-coordinates of RBC centers as the cells crossed the $x = 75\,\mu\mathrm{m}$ and $x = 130\,\mu\mathrm{m}$ positions. The ratio of sizes of daughter branches is 1:2 in case A and 2:3 in case B. In the two cases on the left, there were 40 red blood cells ($Ht \approx 4\%$) and in the two cases on the right, there were 60 red blood cells ($Ht \approx 6\%$). The indicated dimensions are in micrometers.

When looking at red blood cells in the asymmetric bifurcations, we observe that they tend to travel closer to the wall towards the wider branch, see Fig. 2. The histograms of y-coordinates of cell centers shown at $x = 75\,\mu\mathrm{m}$ (middle of the daughter branch) indicate that the cells are the closer to the *right* wall (with respect to the direction of the flow). The double peak distribution in the wider branch is due to the presence of the spherical rare cell. While it travels through the branch, the red blood cells *try to squeeze past it* on the sides.

The histograms at $x = 130\,\mu\text{m}$ (middle of the parent branch after confluence) indicate that the short distance from the confluence of the two daughter branches is not sufficient for mixing the flow, which still exhibits two distinct sets of cells (based on their trajectory through the bifurcations).

4.2 Forces Acting on Cells

We calculated an approximation of fluid force acting on a single red blood cell as a difference between the mesh node velocity and fluid velocity at the mesh node position, summed over all mesh nodes. The mean fluid force acting on red blood cells was calculated by first obtaining the mean approximate fluid force on each RBC while it traveled through the daughter branch and then averaged over all RBCs that traveled the same branch in the given case.

The averaged values are presented in Table 2. Counterintuitively, we see that (except the A40 case) the fluid forces are larger in the wider branch daughter branches (Situations 2 and 4), regardless where the rare cell is. This might be due to the fact that the rare cell partially blocks the wider passage and the fluid is carrying the red blood cells against it.

The second observation is the disparity in acting forces between the channels is typically larger while the rare cell is inside the wider daughter branch (Situations 1 and 2) compared to when it is in the parent branch (Situations 3 and 4).

Table 2. Approximation of average fluid force acting on red blood cells ($\cdot 10^{-3}$ m/s)

Case	Situation 1	Situation 2	Situation 3	Situation 4
A40	7.65	7.41	6.27	8.03
B40	4.51	7.29	4.41	9.07
C40	4.38	10.95	4.88	6.77
A60	6.07	7.46	5.77	8.07
B60	5.60	8.08	5.07	9.05
C60	5.35	9.88	5.46	6.55

5 Conclusion

In this work we demonstrated the the computational model is suitable for investigation of blood flow characteristics in various geometries. It enables quantitative investigation of the impact of a larger immersed cell on the blood flow and can bring insight into the micro-flow behavior and red blood cell stress in micro-bifurcations. In addition to the cell distribution among branches and approximate fluid forces acting on cells, also the cell velocities, rotations, cell distributions in channel cross-sections and other characteristics can be examined in a similar fashion. The knowledge obtained in this way is useful in understanding collective behaviors of RBCs with rare cell present in a microbifurcations. In the future it can also be extended with model of rare cell clusters.

Acknowledgments. This work was supported by the Slovak Research and Development Agency (APVV-15-0751) and by the Ministry of Education, Science, Research and Sport of the Slovak Republic (VEGA 1/0643/17).

References

1. Balogh, P., Bagchi, P.: Analysis of red blood cell partitioning at bifurcations in simulated microvascular networks. Phys. Fluids **30**, 051902 (2018)
2. Wang, T., Rongin, U., Xing, Z.: A micro-scale simulation of red blood cell passage through symmetric and asymmetric bifurcated vessels. Nat. Sci. Rep. **6**, 20262 (2016)
3. Ye, T., Peng, L., Li, Y.: Three-dimensional motion and deformation of a red blood cell in bifurcated microvessels. J. Appl. Phys. **123**, 064701 (2018)
4. Medhi, B., Agrawal, V., Singh, A.: Experimental investigation of particle migration in suspension flow through bifurcating microchannels. Am. Inst. Chem. Eng. J. **64**(6), 2293–2307 (2018)
5. Clavica, F., Homsy, A., Jeandupeux, L., Obrist, D.: Red blood cell phase separation in symmetric and asymmetric microchannel networks: effect of capillary dilation and inflow velocity. Nat. Sci. Rep. **6**, 36763 (2016)
6. Chuang, C., Kikuchi, K., Ueno, H., et al.: Collective spreading of red blood cells flowing in a microchannel. J. Biomech. **69**, 64–69 (2018)
7. Cimrák, I., Jančigová, I.: Computational Blood Cell Mechanics: Road Towards Models and Biomedical Applications, 1st edn. CRC Press, Taylor and Francis Group, Boca Raton (2018)
8. Weik, F., Weeber, R., Szuttor, K., et al.: ESPResSo 4.0 - an extensible software package for simulating soft matter bystems. Eur. Phys. J. Spec. Top. **227**(14), 1789–1816 (2019)
9. PyOIF: Computational tool for modelling of multi-cell flows in complex geometries. http://www.pyoif.eu
10. Jančigová, I., Kovalčíková, K., Bohiniková, A., Cimrák, I.: From red blood cell membrane mechanics to a validated mesh-based model, preprint (2019)

Chondrocyte and Pericellular Matrix Deformation and Strain in the Growth Plate Cartilage Reserve Zone Under Compressive Loading

Masumeh Kazemi(✉) and John L. Williams

University of Memphis, Memphis, USA
{mkzmmghd, jlwllml7}@memphis.edu

Abstract. Long bones grow by a process in which chondrocytes within growth plate cartilage near the bone ends synthesize and mineralize a cartilaginous matrix that serves as a template for bone cells. The growth plate consists of several distinct zones, including a reserve zone (RZ) that lies between the epiphyseal subchondral bone plate and the proliferative zone (PZ). Mechanical loading of the growth plate modulates chondrocyte activity and bone growth, but the role of RZ in relation to this is unclear. To explore this, an axisymmetric, large deformation model was developed. In this model, chondrocytes were embedded at four different depths within the RZ between the SB and PZ. Growth plate cartilage was partitioned into sections to represent the RZ and the proliferative/hypertrophic zones and zone of provisional calcification (PC). Chondrocytes were surrounded by a layer of pericellular matrix (PCM). By including or excluding the PCM, we could examine the influence of the PCM on stress-strain distributions within and around the chondrocytes. The volume-averaged height, width and principal tensile strains in the cells and PCM varied with the cell depth within the RZ. The presence of the PCM resulted a 10% decrease in cellular hydrostatic pressure and in a 20% increase in the cellular maximum principal strains, except near the SB plate border where cellular maximum principal strains were amplified by 45%. This suggests that the PCM is a component of the cell's mechanosensory mechanism and acts to reduce intra-cellular pressure while amplifying cellular strains.

Keywords: Growth plate cartilage · Pericellular matrix · Chondrocyte · Mechanobiology

1 Introduction

The growth plate or physis consists of a thin layer of hyaline cartilage sandwiched between epiphyseal and metaphyseal bone; it is responsible for most of the longitudinal bone growth by an endochondral ossification mechanism. Growth plate cartilage consists of three main regions: a highly cellular region of proliferative and hypertrophic chondrocytes located in tubular structures aligned with the bone long axis and a region

G. A. Ateshian et al. (Eds.): CMBBE 2019, LNCVB 36, pp. 526–538, 2020.
https://doi.org/10.1007/978-3-030-43195-2_43

that is more sparsely populated by chondrocytes called the resting zone, or reserve zone, which is located adjacent to the epiphyseal bone [1]. Blood vessels pass through the subchondral bone plate and reserve zone ending at the beginning of the proliferative zone [2, 3], but the RZ itself has no blood supply. In each location throughout the growth plate every chondrocyte is surrounded by an extracellular matrix (ECM). Generally, chondrocytes are responsible for making, maintaining and repairing extracellular matrix in response to mechanical loading. It has been shown that mechanical loading modulates chondrocyte metabolic activities [4]. However, we have a limited understanding of how biomechanical signals are sensed by the cell and what the relevant molecular pathways are. Chondrocyte behavior is partly regulated by the stress-strain state in and around the cell [4–6]. The morphology and mechanical properties of the growth plate have been measured at the cell and tissue level by confocal microscopy/fluorescent labeling techniques [7, 8] and atomic force microscopy (AFM) [9, 10]. These studies have revealed differences in the structural and mechanical properties between the zones of the cartilage [9]. It has been shown that the chondrocytes are surrounded by a narrow region of pericellular matrix (PCM) [11, 12].

Chondrocytes and their surrounding PCM and territorial matrix were described by Benninghof as the basic functional units of cartilage and named chondrons using the analogy of osteons in bone [11, 12, 15]. The PCM has been characterized by type IV collagen [13, 14]. Experimental and theoretical studies have reported the mechanical properties of the PCM and extracellular matrix (ECM) within articular cartilage by various techniques, such as micropipette aspiration and atomic force microscopy (AFM) [11, 12, 16, 17] and showed that the PCM mechanical and microstructural properties significantly influence the biomechanical environment of chondrocytes [11, 18, 19]. In contrast to the nonuniformity and mechanical anisotropy of cartilage ECM, the PCM has zonal uniformity and constant mechanical properties at the microscale [20, 21]. There is a large difference between the elastic moduli of chondrocytes, PCM and ECM [4, 7, 22]. Therefore, the stress-strain fields within the cells are expected to differ from the surrounding matrix. Although several biomechanical studies have reported microscale models to study the stress-strain state around the chondrocytes and PCM within articular cartilage ECM, relatively few studies have been published on the deformation and strain of chondrocytes in different zones of growth plate cartilage [8, 23, 24] and to our knowledge none have examined the chondrocytes within the growth plate reserve zone.

The overall goal of this study was to explore the role of the PCM in modulating chondrocyte deformation and stress-strain state as a function of depth within the reserve zone under normal physiological loads. Specifically, we sought to answer the following three questions: (1) How do deformation, strain and stress within the chondrocyte and PCM vary through the depth of the reserve zone? (2) Does the PCM protect reserve zone chondrocytes from excessive stress? (3) Does the PCM amplify or reduce strains in the reserve zone chondrocytes?

2 Method

2.1 Model Description

An axisymmetric linearly elastic microscale model was developed in which circular chondrocytes were embedded within the RZ extracellular matrix (ECM) at four depths between the subchondral bone plate and the proliferative zone. The PCM wall thickness was set to be half of the chondrocyte radius (cell radius = 10 μm, PCM thickness = 5 μm) [25, 26]. Chondrocytes make up only about 10% of the reserve zone volume [7]. Therefore, to simplify the model, we simulated four cells in the RZ in a finite element microscale model using ABAQUS/CAE 2019 (SIMULIA, USA). A chondrocyte was placed in four different locations of the RZ, starting with a location near the subchondral bone (SB) interface and moving along the symmetry axis to a location near the proliferative zone (PZ) interface. Homogeneous isotropic linearly elastic materials (Table 1) were used for all components. The cartilage was unconfined along the perichondrial periphery. This static elastic analysis represents a state during fast loading (short duration relative to the stress relaxation time), such as heal strike in gait, during which the cartilage does not have time to undergo relaxation and creep and the influence of the fluid component is negligible.

Table 1. Material properties and dimensions

	Young's modulus (MPa)	Poisson's ratio	Thickness (mm)	Radius (mm)
Chondrocyte	0.002 [27]	0.4999 [27]	–	0.01 [26]
PCM	0.265 [17]	0.45 [17]	0.005 [26]	–
RZ (ECM)	0.98 [27]	0.47 [27]	0.94 [28]	–
PZ/HZ	0.49 [27]	0.47 [27]	0.4 [28]	–
PC	300 [29]	0.2 [30]	0.16*	–
SB	1100 [31]	0.3 [31]	0.5*	–

*Based on our histology of a 20-day piglet distal ulnar growth plate. The harvesting of tissue was exempt from IACUC review as it was obtained from euthanized animals in another University of Memphis IACUC approved study.

To account for large deformation, the effect of geometrical nonlinearity was considered by turning on NLGEOM. The idealized geometry of our model is shown in Fig. 1, in which each cell is surrounded by a PCM and both are embedded in the ECM of the reserve zone (RZ) along the symmetry axis of the axisymmetric model. The growth plate layer was partitioned into two sections to represent the reserve zone (RZ) and the proliferative/hypertrophic zones (PZ/HZ). The bottom of the model was fixed at the junction between the zone of provisional calcification (PC) and the metaphyseal cancellous bone; and the nodes at the subchondral bone plate surface were constrained to prevent translation in the horizontal direction and a vertical displacement was prescribed equal to 15% of the overall growth plate thickness (height) representing normal physiological loading along the bone long axis (vertical axis). Axisymmetric quadrilateral bilinear, hybrid with constant pressure elements (CAX4H) were used for all parts to avoid hour-glassing and volumetric locking during analysis, which is caused by zero energy deformation modes because nearly incompressible material properties have been assigned for all growth plate regions. A mesh convergence study for the microscale model was done to optimize the mesh. The microscale model was run with all four cells and with a single cell at each depth which confirmed that there were no cell-cell interactions in the stress-strain field.

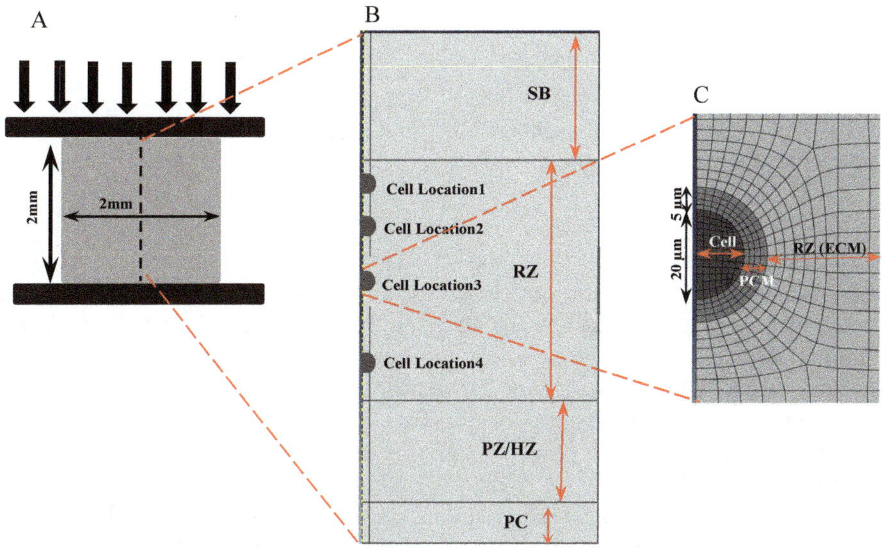

Fig. 1. Overview of the modeling approach. (A) Cross-section of a cylindrical plug of growth plate cartilage, subjected to a displacement of 15% of the overall cartilage thickness. (B) Idealized axisymmetric model consisting of subchondral bone plate (SB), reserve zone (RZ), proliferative/hypertrophic zone (PZ/HZ), provisional calcification (PC). (C) Chondrocyte cell wrapped with a narrow region of pericellular matrix (PCM) embedded in extracellular matrix (ECM).

To evaluate the effect of the PCM on the stress-strain distribution in the chondrocytes, ECM material properties were assigned for the elements in the PCM region to compare it with the model in which chondrocyte, PCM and ECM material properties were defined to be different for each individual location. In addition, a 'macroscale' model without cells and PCM was analyzed to evaluate the volume-averaged stresses and strains in the same four cell/PCM locations within the reserve zone by assigning material properties of the

ECM to the regions otherwise occupied by the cells and PCM. This allowed for comparisons of stresses and strains over the same cell and PCM regions of the mesh in the microscale cell model as the macroscale cell model devoid of cells and PCM.

2.2 Stress-Strain Measurement

To analyze the interaction between cell and matrix and study the local stress-strain state of the cell, hydrostatic stresses and maximum tensile principal strains of the cells were calculated as the main key mechanobiological factors. The volume averaged hydrostatic compressive stress (volumetric or dilatational stress) and principal tensile strain (related to octahedral shear stress) of the cells and PCM were calculated according to Eqs. (1) and (2), respectively.

$$\text{Average cellular stress} = \sum\nolimits_{i=1}^{n} \sigma\text{iEvoli} \Big/ \sum\nolimits_{i=1}^{n} \text{Evoli} \tag{1}$$

where i represent the element number, N is total number of the elements for the cell, Evol_i refers to the element volume and σ_i is hydrostatic stress for the ith element.

$$\text{Average cellular strain} = \sum\nolimits_{i=1}^{n} \varepsilon\text{i}/\text{n} \tag{2}$$

where i is the element number, ε_i represents the maximum tensile principal strain for each element within the cell and n is the total number of elements for each chondrocyte.

In this paper, the maximum principal strain was used as a substitute for octahedral shear stress since the maximum principal (tensile) strains in the cells and PCM of the reserve zone were found to vary by location in a very similar manner to the octahedral shear stresses. The maximum principal strain results were extracted in terms of logarithmic strains.

Cell height and width engineering strains were calculated based on the change in the horizontal diameter and vertical radius of the cell, respectively before and after compression. PCM height and width strains were also defined as the change in the vertical and horizontal wall thickness, respectively, of the PCM divided by the original wall thickness. This was also done for the reserve zone macroscale model which had no cells or PCM, by using the same mesh used for the microscale model in which cells and PCM were present, but in which the reserve zone was homogenous (assigned Young's modulus and Poisson's ratio of the ECM).

3 Results

The results revealed a highly non-homogenous and depth-dependent strain field within and around the cells for the model that included the PCM and the model that excluded the PCM. Figure 2 shows the contour plots of maximum principal tensile strain (logarithmic strain) distribution within and around the cell at different locations within the growth plate reserve zone (RZ). The magnitudes of the cell-averaged hydrostatic stresses and maximum principal tensile strains change throughout the depth of RZ as we move our view point from close to the epiphyseal subchondral bone (cell location 1) to near the PZ/HZ interface (cell location 4) (Fig. 2).

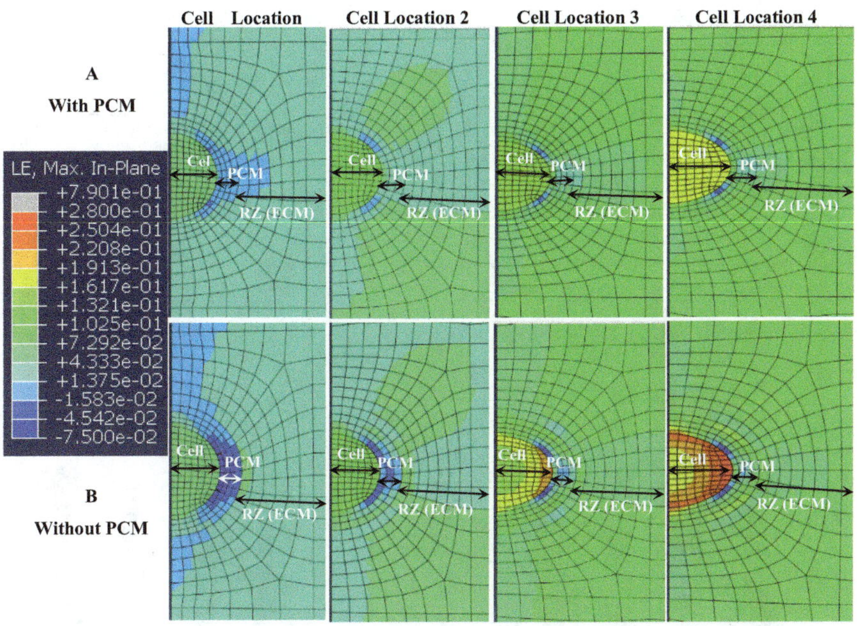

Fig. 2. Distribution of maximum principal strains around and within the cell for models (A) with PCM and (B) without PCM (when material properties of ECM are assigned to the PCM, i.e. PCM = ECM), for different locations of the cell within the reserve zone, moving from subchondral bone to proliferative zone, subjected to 15% compression and using CAX4H elements.

3.1 Influence of Cell Location in the Reserve Zone on Chondrocyte Stress and Strain

When the PCM was included in the model, the cell-averaged maximum principal tensile strains within the chondrocytes increased in magnitude from a value of 6% strain near the epiphyseal subchondral bone (SB) to 22% near the proliferative zone (PZ) interface, representing a nearly 4-fold increase (Fig. 3A). Cellular height and width strains followed a similar trend (Fig. 3C). However, the opposite pattern was observed for the cell hydrostatic compressive stresses which decreased from a value 0.12 near the SB to 0.09 MPa near the PZ, an approximate location-dependent decline of 26%. Figure 2 represents how the presence of the PCM changes the stress-strain state around and within the cell.

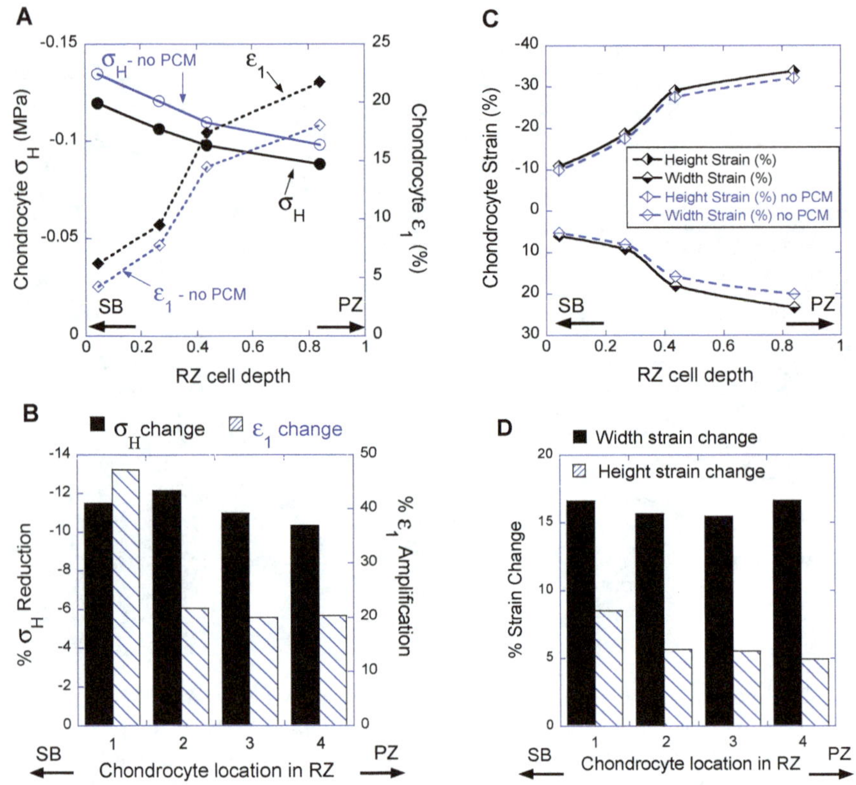

Fig. 3. (A) Cell-averaged hydrostatic stress and maximum principal tensile strain for cells in different locations within the reserve zone moving from subchondral bone (SB) to proliferative zone (PZ) subjected to 15% compression of the growth plate cartilage for two models (with/without PCM). (B) Changes in hydrostatic stress and maximum principal strain due to the presence of the PCM, (C) Chondrocyte height and width strains in locations 1–4, (D) Changes in width and height strains due to the presence of the PCM.

The height and width strains of the matrix within the cell locations in the macroscale model in which there were no cells within the matrix, are compared with the microscale model in which cells and PCM were present, in Fig. 4.

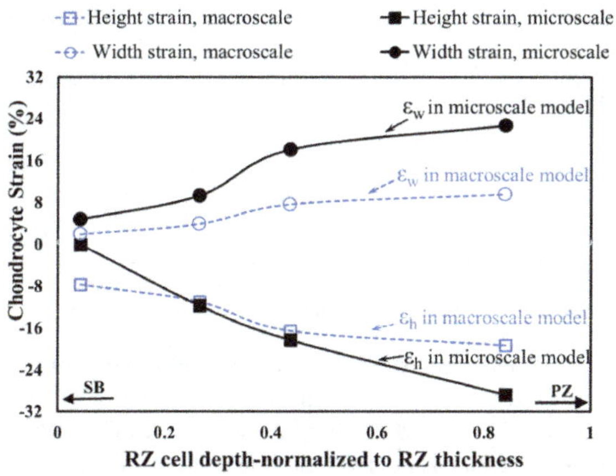

Fig. 4. Chondrocyte region height and width strains within each cell location moving from subchondral bone (SB) to proliferative zone (PZ) for: the microscale model in which cells and PCM are present; and the macroscale model in which there are no cells or PCM ('Chondrocyte Strain' in this case refers the ECM strain in the macroscale model location of the cell).

3.2 Influence of the Pericellular Matrix on Chondrocyte Stress and Strain

The presence of the pericellular matrix (PCM) resulted in an approximate 20% increase in the cell-averaged maximum principal strains in reserve zone locations 2, 3 and 4 and in a 10–12% decrease in cell-averaged hydrostatic pressure. The influence of the PCM on strains was greatest for the cell located near the SB border where the presence of the PCM amplified the cell-averaged maximum principal strains by 45% (Fig. 3B). Likewise, cellular height strain of the cell closet to the SB was most influenced by the presence of the PCM (Fig. 3D), while the influence of the PCM on cell width strains appeared not to vary much with cell locations (Fig. 3D).

The height and width strains of the PCM followed similar trends to those for cell height and width (Fig. 5). Heights of both cells and PCM decreased under the applied compression, while their widths increased.

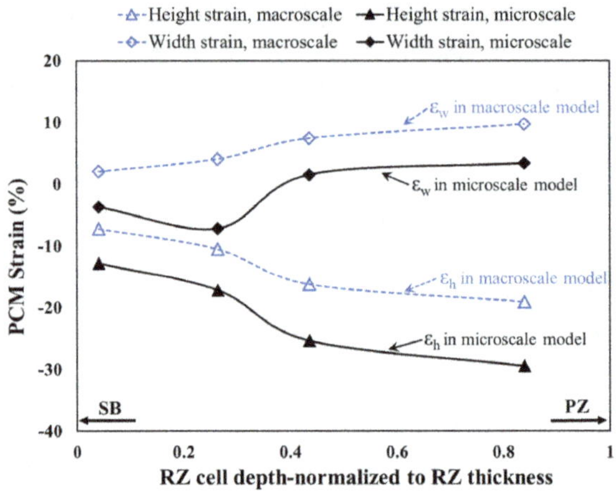

Fig. 5. PCM region height and width strains within each cell location moving from subchondral bone (SB) to proliferative zone (PZ) for: the microscale model in which cells and PCM are present; and the macroscale model in which there are no cells or PCM ('PCM Strain' in this case refers the ECM strain in the macroscale model location of the PCM).

4 Discussion

We developed a computational model to simulate chondrocyte deformation and study chondrocyte–matrix interactions as a function of cell depth within reserve zone of growth plate cartilage under physiological compression. The main objectives of the current study were to characterize the depth-dependent chondrocyte stress-strain distribution within the reserve zone and the influence of the microenvironment of the chondrocytes. Our findings indicate that the pericellular matrix (PCM) amplifies the cell's internal strains and the cell's height and width strains, while slightly reducing the cell's internal hydrostatic stress. The results showed that the state of stress and strain in the reserve zone chondrocytes are significantly depth dependent. The cell-averaged maximum principal strain increased from a minimum near the SB interface and reached maximum near the PZ interface. Cell height and width strains were also smallest at the SB border and greatest at the border with the PZ. This may play a role in preparing the reserve zone chondrocyte for cell alignment in the proliferative zone column of cells (chondron) and for the cell's function as a daughter cell. In the this study, the presence of PCM with a Young modulus value between that of the chondrocyte and ECM, decreased the cell-averaged hydrostatic stress, while it had the opposite effect on the cellular strains and acted as a strain amplifier, which is similar to the results reported for articular cartilage [4, 11].

In particular, the cell-averaged maximum principal strains, cell height, and cell width strains all increased with cell location from the SB towards the PZ border where they reached maximal values. The addition of the PCM in the model further increased the maximum principal strains, width, and height strains near the PZ by about 20%, 5%, and 16.6%, respectively. The most notable influence of the PCM in amplifying cellular strains occurred for the cell located near the subchondral bone, where cellular strains were at a minimum. This further suggests that the PCM may act as a mechano-transducer and serve as a transitional zone between cell and ECM. Our results revealed that PCM plays a more dominant role in regulating the micro-mechanical strain environment of chondrocytes near the interface with the much 'stiffer' subchondral bone.

Experimental and computational studies on PCM biological function and its chemical and mechanical properties, suggest that the PCM chemical composition, and physical and mechanical properties are directly proportional to one another [19]. The PCM can function as a thin barrier between the cell and ECM to filter molecules and regulate the biochemical environment of articular cartilage chondrocytes, and is hypothesized to have a prominent role in cartilage degeneration and osteoarthritis [12, 32]. Computational studies of micromechanical environment of chondrocytes in different zones of articular cartilage imply that the PCM plays a dominant role in modifying the cellular stress-strain state; and any change in PCM properties with aging or cartilage abnormality can significantly affect the biophysical/biomechanical environment of the chondrocytes [18, 33]. Computational modeling of articular cartilage has given rise to the idea that the PCM can act as a mechano-transducer between cells and ECM [4, 11].

Generally, chondrocytes are responsible for making, maintaining and repairing extracellular matrix in response to mechanical loading. Despite the cited foregoing studies on cell-matrix interactions, it is not known how biomechanical signals are transmitted to the cell and transduced–exact molecular mechanisms and pathways are still unknown. One proposed general mechanism by which chondrocytes may sense mechanical signal variations, is through the alteration of cell shape and volume [6]. A confocal microscopy study revealed that the micromechanical environment in articular cartilage is depth and zone dependent and causes an inhomogeneous and anisotropic deformation of the cells [6]. Other studies have proposed that primary cilia on the surface of the cell can act as mechanosensory organelles, which transduce mechanical forces into biological signals [34, 35]. The PCM may function in conjunction with cilia to provide a feasible signal transducer [4, 11]. In articular cartilage, the PCM reduces the local stress of the cell, providing a protective role for the PCM while amplifying the cellular strain [4, 11]. The current study on growth plate cartilage reserve zone suggests a similar role for the PCM in the growth plate reserve zone. Understanding the sequence of events by which cells and ECM can communicate and convert the biomechanical and biochemical signals or signal transduction is still a challenge to investigators. Further evaluation on how PCM components contribute to chondrocyte mechano-transduction in the reserve zone of growth plate cartilage may elucidate factors contributing growth abnormalities and disturbances and enhance our understanding of regulatory mechanical factors that modulate bone growth.

In the current model we have chosen to represent the reserve zone with a thickness slightly more than double the height or thickness of the proliferative and hypertrophic zones combined, based on our histological studies of twenty-day old mixed breed piglet distal ulnar growth plates. The relative thickness of each zone varies by age and location and may influence the reserve zone chondrocyte stress and strain magnitudes. We also chose to model a flat growth plate and need to explore the influence of mammillary processes on reserve zone cell stress and strain and the possible role that reserve zone cells might play in mammillary processes development.

Although macroscale models have been reported of the three zones of the growth plate for various loading conditions [24, 27, 36–39], to date there has been no model that includes chondrocytes in the reserve zone or of the interaction between reserve zone cells, PCM and ECM. Further computational studies of cell-PCM-ECM interaction in growth plate cartilage may generate new insights into the role played by the micromechanical environment around chondrocytes in the mechano-transduction mechanisms of bone and cartilage growth.

References

1. Abad, V., Meyers, J.L., Weise, M., Gafni, R.L., Barnes, K.M., Nilsson, O., Bacher, J.D., Baron, J.: The role of the resting zone in growth plate chondrogenesis. Endocrinology **143**(4), 1851–1857 (2002)
2. Brighton, C.T.: The growth plate. Orthop. Clin. North Am. **15**(4), 571–595 (1984)
3. DeCamp, C.E.: The epiphyseal plate: physiology, anatomy, and trauma. Compendium (Yardley, PA) **31**(8), 1–11 (2009)
4. Guilak, F., Mow, V.C.: The mechanical environment of the chondrocyte: a biphasic finite element model of cell–matrix interactions in articular cartilage. J. Biomech. **33**(12), 1663–1673 (2000)
5. Guilak, F.: Compression-induced changes in the shape and volume of the chondrocyte nucleus. J. Biomech. **28**(12), 1529–1541 (1995)
6. Guilak, F., Ratcliffe, A., Mow, V.C.: Chondrocyte deformation and local tissue strain in articular cartilage: a confocal microscopy study. J. Orthop. Res. **13**(3), 410–421 (1995)
7. Amini, S., Veilleux, D., Villemure, I.: Three-dimensional in situ zonal morphology of viable growth plate chondrocytes: a confocal microscopy study. J. Orthop. Res. **29**(5), 710–717 (2011)
8. Amini, S., Veilleux, D., Villemure, I.: Tissue and cellular morphological changes in growth plate explants under compression. J. Biomech. **43**(13), 2582–2588 (2010)
9. Radhakrishnan, P., Lewis, N.T., Mao, J.J.: Zone-specific micromechanical properties of the extracellular matrices of growth plate cartilage. Ann. Biomed. Eng. **32**(2), 284–291 (2004)
10. Campbell, S.E., Ferguson, V.L., Hurley, D.C.: Nanomechanical mapping of the osteochondral interface with contact resonance force microscopy and nanoindentation. Acta Biomater. **8**(12), 4389–4396 (2012)
11. Alexopoulos, L.G., Setton, L.A., Guilak, F.: The biomechanical role of the chondrocyte pericellular matrix in articular cartilage. Acta Biomater. **1**(3), 317–325 (2005)
12. Zhang, Z.: Chondrons and the pericellular matrix of chondrocytes. Tissue Eng. Part B: Rev. **21**(3), 267–277 (2014)
13. Alini, M., Matsui, Y., Dodge, G.R., Poole, A.R.: The extracellular matrix of cartilage in the growth plate before and during calcification: changes in composition and degradation of type II collagen. Calcif. Tissue Int. **50**(4), 327–335 (1992)
14. Alexopoulos, L.G., Haider, M.A., Vail, T.P., Guilak, F.: Alterations in the mechanical properties of the human chondrocyte pericellular matrix with osteoarthritis. J. Biomech. Eng. **125**(3), 323–333 (2003)

15. Poole, C.A., Glant, T.T., Schofield, J.R.: Chondrons from articular cartilage (IV). Immunolocalization of proteoglycan epitopes in isolated canine tibial chondrons. J. Histochem. Cytochem. **39**(9), 1175–1187 (1991)
16. Ferguson, V.L., Bushby, A.J., Boyde, A.: Nanomechanical properties and mineral concentration in articular calcified cartilage and subchondral bone. J. Anat. **203**(2), 191–202 (2003)
17. Allen, D.M., Mao, J.J.: Heterogeneous nanostructural and nanoelastic properties of pericellular and interterritorial matrices of chondrocytes by atomic force microscopy. J. Struct. Biol. **145**(3), 196–204 (2004)
18. Guilak, F., Alexopoulos, L.G., Haider, M.A., Ting-Beall, H.P., Setton, L.A.: Zonal uniformity in mechanical properties of the chondrocyte pericellular matrix: micropipette aspiration of canine chondrons isolated by cartilage homogenization. Ann. Biomed. Eng. **33** (10), 1312–1318 (2005)
19. Wilusz, R., Sanchez-Adams, E.J., Guilak, F.: The structure and function of the pericellular matrix of articular cartilage. Matrix Biol. **39**, 25–32 (2014)
20. McLeod, M.A., Wilusz, R.E., Guilak, F.: Depth-dependent anisotropy of the micromechanical properties of the extracellular and pericellular matrices of articular cartilage evaluated via atomic force microscopy. J. Biomech. **46**(3), 586–592 (2013)
21. Sanchez-Adams, J., Leddy, H.A., McNulty, A.L., O'Conor, C.J., Guilak, F.: The mechanobiology of articular cartilage: bearing the burden of osteoarthritis. Curr. Rheumatol. Rep. **16**(10), 451–460 (2014)
22. Alexopoulos, L.G., Williams, G.M., Upton, M.L., Setton, L.A., Guilak, F.: Osteoarthritic changes in the biphasic mechanical properties of the chondrocyte pericellular matrix in articular cartilage. J. Biomech. **38**(3), 509–517 (2005)
23. Sergerie, K., Lacoursière, M.O., Lévesque, M., Villemure, I.: Mechanical properties of the porcine growth plate and its three zones from unconfined compression tests. J. Biomech. **42**(4), 510–516 (2009)
24. Piszczatowski, S.: Geometrical aspects of growth plate modelling using Carter's and Stokes's approaches. Acta Bioeng. Biomech. **14**(1), 93–106 (2012)
25. Maggiano, C.: Confocal laser scanning microscopy as a tool for the investigation of tetracycline fluorescence in archaeological human bone. Master thesis, University of Florida (2005)
26. Korhonen, R.K., Julkunen, P., Wilson, W., Herzog, W.: Importance of collagen orientation and depth-dependent fixed charge densities of cartilage on mechanical behavior of chondrocytes. J. Biomech. Eng. **130**(2), 1–11 (2008)
27. Gao, J., Roan, E., Williams, J.L.: Regional variations in growth plate chondrocyte deformation as predicted by three-dimensional multi-scale simulations. PLoS One **10**(4), 1–18 (2015)
28. Vendra, B.B., Roan, E., Williams, J.L.: Chondron curvature mapping in growth plate cartilage under compressive loading. J. Mech. Behav. Biomed. Mater. **84**, 168–177 (2018)
29. Mente, P.L., Lewis, J.L.: Elastic modulus of calcified cartilage is an order of magnitude less than that of subchondral bone. J. Orthop. Res. **12**(5), 637–647 (1994)
30. Carter, D.R., Mikić, B., Padian, K.: Epigenetic mechanical factors in the evolution of long bone epiphyses. Zool. J. Linn. Soc. **123**(2), 163–178 (1998)
31. Wei, H.W., Sun, S.S., Jao, S.E., Yeh, C., Cheng, C.: The influence of mechanical properties of subchondral plate, femoral head and neck on dynamic stress distribution of the articular cartilage. Med. Eng. Phys. **27**(4), 295–304 (2005)
32. Leddy, H.A., Christensen, S.E., Guilak, F.: Microscale diffusion properties of the cartilage pericellular matrix measured using 3D scanning microphotolysis. J. Biomech. Eng. **130**(6), 1–20 (2008)

33. Guilak, F., Jones, W.R., Ting-Beall, H.P., Lee, G.M.: The deformation behavior and mechanical properties of chondrocytes in articular cartilage. Osteoarthritis Cartilage **7**(1), 59–70 (1999)
34. Shao, Y.Y., Wang, L., Welter, J.F., Ballock, R.T.: Primary cilia modulate IHH signal transduction in response to hydrostatic loading of growth plate chondrocytes. Bone **50**(1), 79–84 (2012)
35. Seeger-Nukpezah, T., Golemis, E.A.: The extracellular matrix and ciliary signaling. Curr. Opin. Cell Biol. **24**(5), 652–661 (2012)
36. Narváez-Tovar, C.A., Garzón-Alvarado, D.A.: Computational modeling of the mechanical modulation of the growth plate by sustained loading. Theor. Biol. Med. Model. **9**(1), 9–41 (2012)
37. Gao, J., Williams, J.L., Roan, E.: On the state of stress in the growth plate under physiologic compressive loading. Open J. Biophys. **4**(1), 13–21 (2014)
38. Farzaneh, S., Paseta, O., Gómez-Benito, M.J.: Multi-scale finite element model of growth plate damage during the development of slipped capital femoral epiphysis. Biomech. Model. Mechanobiol. **14**(2), 371–385 (2015)
39. Gao, J., Williams, J.L., Roan, E.: Multiscale modeling of growth plate cartilage mechanobiology. Biomech. Model. Mechanobiol. **16**(2), 667–679 (2017)

Finite Element Simulations for Investigating the Cause of Catastrophic Wear and/or Failure of Polyethylene Acetabular Cup Liner in Hip Prosthesis

Changhee Cho[1(✉)], Toshiharu Mori[2], and Makoto Kawasaki[3]

[1] Department of Mechanical Systems Engineering, Faculty of Environmental Engineering, The University of Kitakyushu, Kitakyushu, Japan
cho@kitakyu-u.ac.jp

[2] Department of Orthopaedic Surgery, Shin-Kokura Hospital, Kitakyushu, Japan

[3] Department of Orthopaedic Surgery, University of Occupational and Environmental Health, Kitakyushu, Japan

Abstract. Wear and/or failure of ultra-high molecular weight polyethylene (UHMWPE) component after total joint replacement are major factors restricting the clinical longevity of artificial joints. In order to minimize the wear and failure of the UHMWPE and to improve the clinical longevity of artificial joints, it is necessary to clarify the factors influencing the wear and failure mechanism of the UHMWPE. The generations of catastrophic wear and/or failure of the UHMWPE acetabular cup liner are frequently observed in retrieved hip prostheses. The primary purpose of this study was to investigate the cause of catastrophic wear and/or failure of the UHMWPE liner in hip prosthesis. The authors focused on change in mechanical state of the UHMWPE liner due to contact with metallic component as a factor influencing the wear and failure mechanism of the UHMWPE liner. Contact analyses between the metallic components and the UHMWPE liner by using the finite element method (FEM) were performed to investigate the mechanical state of the UHMWPE liner. It was found that high contact stresses, which exceed the yield stress of UHMWPE, and considerable plastic deformations occurred in the rim of the UHMWPE liner. It was also found that high stress concentrations occurred near screw holes in the acetabular cup and notches in the UHMWPE liner. This study confirmed that change in mechanical state due to contact with metallic component is the cause of catastrophic wear and/or failure of the UHMWPE liner.

Keywords: Hip prosthesis · UHMWPE acetabular cup liner · Wear · Failure · Contact analysis · Finite element method

1 Introduction

Ultra-high molecular weight polyethylene (UHMWPE) has been used as a major polymeric material for weight-bearing surfaces in total joint replacement. However, the wear and failure of UHMWPE component in the human body after total joint

G. A. Ateshian et al. (Eds.): CMBBE 2019, LNCVB 36, pp. 539–549, 2020.
https://doi.org/10.1007/978-3-030-43195-2_44

replacement cause serious clinical and biomechanical reactions, such as osteolysis and eventual loosening of artificial joints [1]. Therefore, the wear and failure of the UHMWPE component are now recognized as the major factors restricting the clinical longevity of artificial joints. In order to minimize the wear and failure of the UHMWPE component and to improve the clinical longevity of artificial joints, it is necessary to clarify the factors influencing the wear and failure mechanism of the UHMWPE component. The generations of catastrophic wear and/or failure of the UHMWPE acetabular cup liner are frequently observed in retrieved hip prostheses as shown in Figs. 1(a) to (d) [2–4]. The primary purpose of this analytical study was to investigate the causes of catastrophic wear and/or failure of the UHMWPE acetabular cup liners in retrieved hip prostheses shown in Fig. 1.

In this study, we focused on change in mechanical state, such as contact stress and plastic strain of the UHMWPE acetabular cup liners due to contact with metallic components of the hip prosthesis as a factor influencing the wear and/or failure mechanism of the UHMWPE acetabular cup liner. The elasto-plastic contact analyses between the metallic components and the UHMWPE acetabular cup liner by using the finite element method (FEM) were performed in order to investigate the mechanical state of the retrieved UHMWPE acetabular cup liner.

2 Materials and Methods

In this study, the elasto-plastic contact analyses based on geometric measurements of several retrieved UHMWPE acetabular cup liners showing catastrophic wear and/or failure were performed by using the FEM, in order to investigate the change in mechanical state, such as contact stress and plastic strain due to contact with metallic components and the causes of the catastrophic wear and/or failure of the retrieved UHMWPE acetabular cup liners.

The retrieved UHMWPE acetabular cup liners of the hip prosthesis shown in Fig. 1 were used to produce the FEM models of the hip prosthesis. The *in vivo* duration of the retrieved UHMWPE liner shown in Fig. 1(a) was 157 months. In this retrieved case, excessive abrasive wear of the contact surface was observed in and around the rim of the UHMWPE liner. The *in vivo* duration of the retrieved UHMWPE liner shown in Fig. 1(b) was 158 months. In this retrieved case, partial breakage failure of the rim was observed. Figures 1(c) and (d) [3, 4] show the main contact surface (rubbing surface) and the back surface of an identical UHMWPE acetabular cup liner. The *in vivo* duration of this retrieved liner was 290 months. This retrieved case demonstrated the most catastrophic wear and failure of the UHMWPE liner in the retrieved cases in this study. Excessive wear, plastic deformation and failure were observed in the rim of the liner as shown in Fig. 1(c). Furthermore, through-hole due to excessive wear of the liner was also observed as shown in Fig. 1(d).

The metallic components of the hip prosthesis shown in Fig. 2 [4] were also used to produce the FEM models of the hip prosthesis. Figure 2(a) shows the metallic femoral head and stem neck of the hip prosthesis. Figure 2(b) shows the metallic acetabular cup of the hip prosthesis with screw holes for fixation to the pelvis. The geometric measurements of the retrieved liners and metallic components were performed.

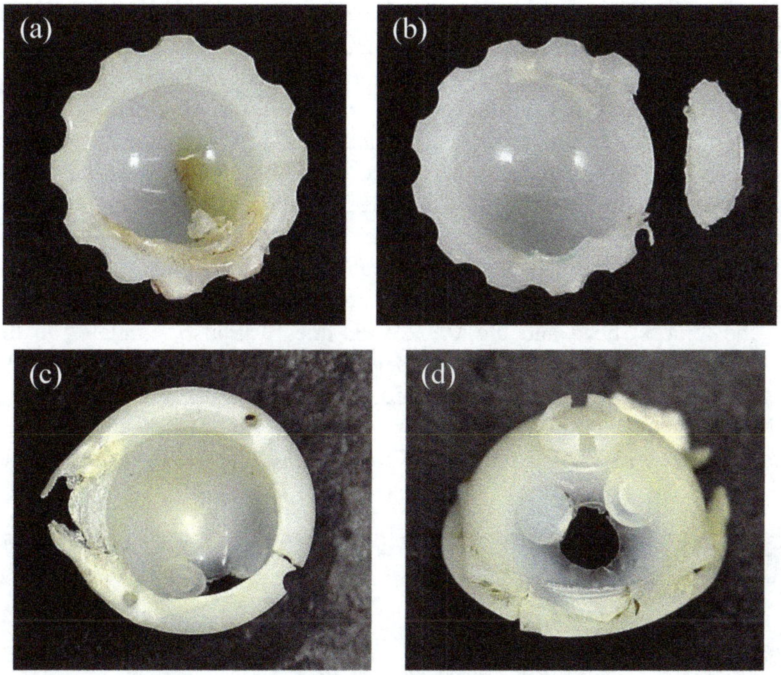

Fig. 1. Retrieved UHMWPE acetabular cup liners showing catastrophic wear and/or failure used to produce the FEM models of the hip prosthesis: (a) 157 months *in vivo*, (b) 158 months *in vivo*, (c) and (d) 290 months *in vivo*.

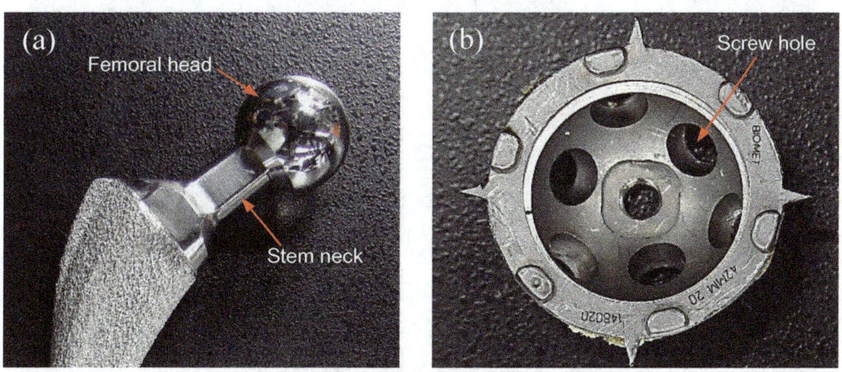

Fig. 2. Metallic components of the hip prosthesis used to produce the FEM models of the hip prosthesis: (a) metallic femoral head and stem neck, (b) metallic acetabular cup with screw holes for cementless fixation to the pelvis.

On the basis of the geometrical measurements from the retrieved UHMWPE acetabular cup liners and the metallic components of the hip prosthesis, various 2D and 3D FEM models of the hip prosthesis were produced as shown in Fig. 3 [4, 5]. The UHMWPE acetabular cup liner shown in Fig. 1 was assumed to be an elasto-plastic body that has the Poisson's ratio equal to 0.45. The contacting metallic components shown in Fig. 2 were assumed to be rigid bodies. The coefficient of friction between the contact surfaces was also assumed to be 0.1 [6]. The contact between the stem neck and the UHMWPE liner was simulated as a rigid body deforming a soft elasto-plastic body by applying to the femoral head the rotational displacement (rotation angle), which represents the flexion angle of the hip prosthesis. The contact between the femoral head and the UHMWPE liner was also simulated in the same manner. Only the static contact condition between the femoral head and the UHMWPE liner in the standing position was considered, the femoral head was simply pressed into the UHMWPE liner with a constant normal load of 1 kN. The degrees of freedom of the acetabular cup with three screw holes shown in Fig. 3(d) were entirely constrained.

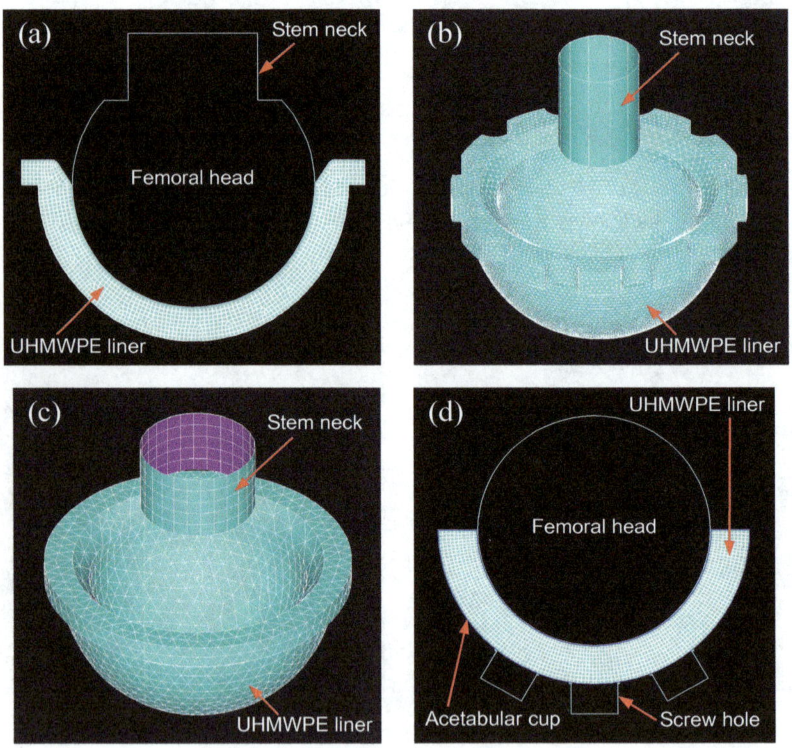

Fig. 3. FEM models of the hip prosthesis used in the contact analyses: (a)–(c) FEM models for the contact analysis between the stem neck and the UHMWPE liner, (d) FEM model for the contact analysis between the femoral head and the UHMWPE liner and the acetabular cup.

As for the material model of the UHMWPE used in the FEM contact analyses performed in this study, an elasto-plastic material model of the UHMWPE established in a previous study [7] was used. The linear elastic modulus, the yield stress and the tensile strength of the UHMWPE were 498, 13.3 and 167.6 MPa respectively. All the FEM contact analyses in this study were performed by using the commercial finite element analysis software ANSYS (ANSYS, Inc., USA).

3 Results

The results of the elasto-plastic contact analysis between the metallic component and the UHMWPE acetabular cup liner by using the FEM are shown in Figs. 4, 5, 6 and 7.

Figures 4(a) and (b) [5] show the contour plots of the contact stress (von Mises equivalent stress) with deformed shape (plastic deformation) of the retrieved UHMWPE acetabular cup liner shown in Fig. 1(a) for the contact of the stem neck with the rotational displacement (rotation angle) of 62.5° (flexion angle of 125°) [Fig. 4(a)], and for the reverse rotation of the femoral head after the neck-liner contact [Fig. 4(b)]. It was found that extremely high contact stresses, which far exceed the yield stress of UHMWPE (13.3 MPa), and considerable plastic deformations occurred in and around the rim of the UHMWPE acetabular cup liner as shown in Figs. 4(a) and (b). It was also found that locally high contact stresses, which far exceed the yield stress of UHMWPE, occurred on the liner surface near the rim due to contact with the edge of the reversing femoral head as shown in Fig. 4(b).

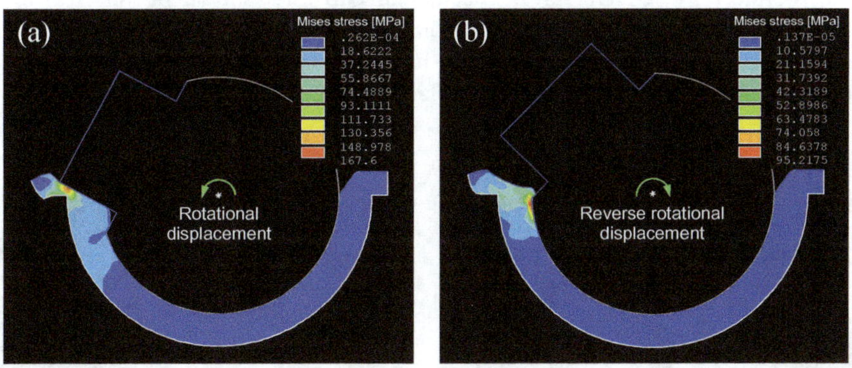

Fig. 4. Contour plots of the contact stress with deformed shape of the retrieved UHMWPE acetabular cup liner shown in Fig. 1(a): (a) for the contact of the stem neck with the rotation angle of 62.5°, (b) for the reverse rotation of the femoral head.

Figure 5 shows the contour plots of the contact stress (von Mises equivalent stress) with deformed shape (plastic deformation) of the retrieved UHMWPE acetabular cup liner shown in Fig. 1(b) for the contact of the stem neck with the rotational displacement (rotation angle) of 65° (flexion angle of 130°). Figures 5(a) and (b) show the main contact surface with the stem neck and the back surface of the liner respectively. It was found that locally high contact stresses, which far exceed the yield stress of UHMWPE, due to stress concentration at notch (groove) occurred in the rim of the UHMWPE acetabular cup liner as shown in Figs. 5(a) and (b).

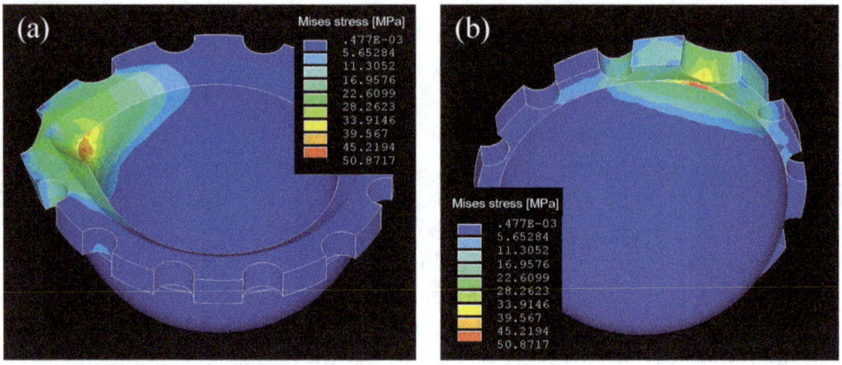

Fig. 5. Contour plots of the contact stress with deformed shape of the retrieved UHMWPE acetabular cup liner shown in Fig. 1(b) for the contact of the stem neck with the rotation angle of 65°: (a) main contact surface with the stem neck, (b) back surface of the liner.

Figures 6(a) and (b) [3, 4] show the contour plots of the contact stress (von Mises equivalent stress) and the equivalent plastic strain with deformed shape (plastic deformation) of the retrieved UHMWPE acetabular cup liner shown in Fig. 1(c) for the contact of the stem neck with the rotational displacement (rotation angle) of 62.5° (flexion angle of 125°). It was found that high contact stresses, which far exceed the yield stress of UHMWPE, and considerable plastic strains occurred in and around the rim of the UHMWPE acetabular cup liner as shown in Figs. 6(a) and (b). The magnitude

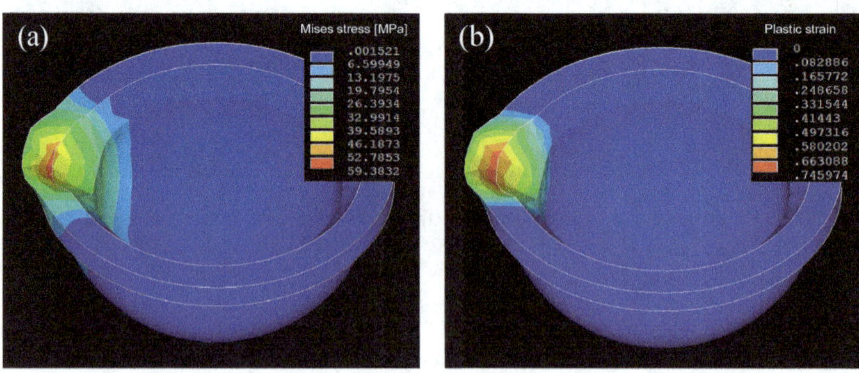

Fig. 6. Contour plots of the contact stress and the plastic strain with deformed shape of the retrieved UHMWPE acetabular cup liner shown in Fig. 1(c) for the contact of the stem neck with the rotation angle of 62.5°: (a) contact stress, (b) plastic strain.

of plastic deformation which occurred in the rim of the liner at this rotation angle shown in Fig. 6(b) corresponded approximately to the magnitude of plastic deformation that was measured in the rim of the retrieved UHMWPE liner shown in Fig. 1(c).

Figures 7(a) and (b) [4] show the contour plots of the contact stress (von Mises equivalent stress) and the equivalent plastic strain with deformed shape (plastic deformation) of the retrieved UHMWPE acetabular cup liner shown in Fig. 1(d) for the contact of the femoral head with a constant normal load of 1 kN (approximately 2.7 times the patient's body weight of 37.6 kg).

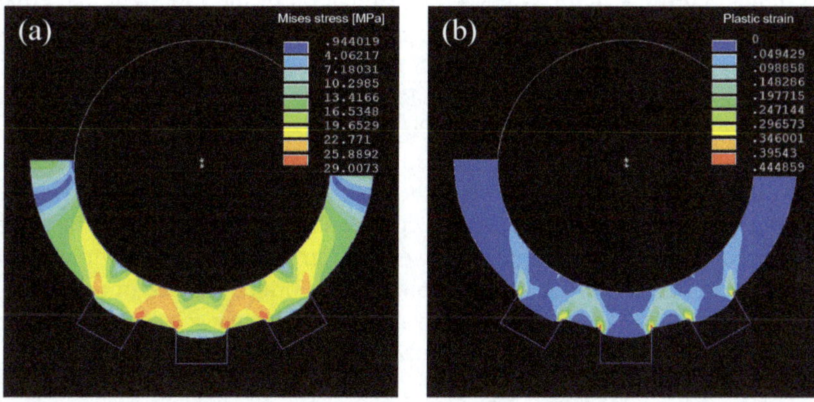

Fig. 7. Contour plots of the contact stress and the plastic strain with deformed shape of the retrieved UHMWPE acetabular cup liner shown in Fig. 1(d) for the contact of the femoral head with a constant normal load of 1 kN: (a) contact stress, (b) plastic strain.

It was found that plastic deformations into the screw holes in the acetabular cup occurred on the back surface of the UHMWPE liner. Polyethylene protrusions were generated on the back surface of the liner by these plastic deformations. The magnitude of maximum plastic deformation, that is maximum height of the protrusion, which occurred on the back surface of the UHMWPE liner was 0.90 mm. This numerical value corresponded approximately to the maximum height of the protrusion of 0.89 mm that was measured on the back surface of the retrieved UHMWPE liner shown in Fig. 1(d). It was also found that locally high contact stresses, which far exceed the yield stress of UHMWPE, and considerable plastic strains due to high stress concentrations near the sharp corners of the protrusions were generated throughout the overall thickness between the undersurface and the top surface of the UHMWPE liner as shown in Figs. 7(a) and (b). In addition, overlapping sites of the locally high contact stresses between the polyethylene protrusions were observed in the regions which the plastic deformations into the screw holes occurred.

4 Discussion

As the results of the FEM contact analyses performed in this study, it was found that high contact stresses, which far exceed the yield stress of UHMWPE, and considerable plastic strains caused by the contact with metallic components occurred in the retrieved UHMWPE acetabular cup liners, as shown in Figs. 4, 5, 6 and 7.

The contact between the metal stem neck and the rim of the UHMWPE acetabular cup liner in the hip prosthesis shown in Figs. 4, 5 and 6 is called neck-liner impingement. The neck-liner impingement is induced by the range of motion (ROM) limitation of the hip prosthesis. The limitation of the ROM can be influenced by the patient's activities and the geometric shape of the hip prosthesis components. All of the analysis results shown in Figs. 4, 5 and 6 are results of contact analyses performed with the flexion angle beyond the ROM of each retrieved hip prosthesis.

In the case of the retrieved UHMWPE acetabular cup liner shown in Fig. 1(a), considerable inward plastic deformation into the gap between the femoral head and the stem neck due to neck-liner impingement occurred as shown in Fig. 4(a). This inward plastic deformation can cause locally high contact stresses on the liner surface near the rim due to contact with the edge of the reversing femoral head as shown in Fig. 4(b). This phenomenon can finally lead to scraping the deformed surface of the UHMWPE liner. The cause for the excessive abrasive wear (scraping wear) observed in and around the rim of the retrieved UHMWPE liner shown in Fig. 1(a) seems to be this phenomenon.

In the case of the retrieved UHMWPE acetabular cup liner shown in Fig. 1(b), a notch (groove) for fixation to the acetabular cup with a metal locking ring exists around the rim of the UHMWPE liner. Locally high contact stresses occurred at the notch (groove) due to stress concentration as shown in Figs. 5(a) and (b). Sharp notches in components result in high stress concentration which reduces fatigue strength of the components. This phenomenon is known as the notch effect. The notch effect can lead to early fatigue failure of the components. The cause for the partial breakage failure of the rim of the UHMWPE liner observed in this case shown in Fig. 1(b) seems to be this notch effect that is associated with the stress concentration at the notch (groove) for the metal locking ring.

In the case of the retrieved UHMWPE acetabular cup liner shown in Fig. 1(c), considerable outward plastic deformation due to neck-liner impingement occurred on the rim of the liner, and high contact stresses, which far exceed the yield stress of UHMWPE, also occurred on the rim edge of the liner as shown in Figs. 6(a) and (b). The causes for the severe wear and failure observed on the rim of the retrieved UHMWPE liner shown in Fig. 1(c) seem to be these outward plastic deformation and locally high contact stresses due to neck-liner impingement.

In retrieved hip prostheses with screw holes for cementless fixation to the pelvis, in the metallic acetabular cup like an example shown in Fig. 2(b), plastic deformations of the polyethylene into the screw holes are frequently observed on the back surface (undersurface) of the retrieved UHMWPE acetabular cup liner as shown in Fig. 1(d) [2]. This phenomenon, that is plastic deformations of the polyethylene into the screw holes due to mechanical force or pressure, is called cold flow [8]. Polyethylene protrusions generated by the cold flow were located above the screw holes in the acetabular cup. Also, in most retrieved UHMWPE liners, the polyethylene protrusions on the back surface were located on the reverse side of the severely worn and plastically deformed contact surface of the UHMWPE liner.

In the case of the retrieved UHMWPE acetabular cup liner shown in Fig. 1(d), high stress concentrations occurred near the sharp corners of the polyethylene protrusions (near the edges of screw holes). Furthermore, overlapping of the contact stresses occurred between the polyethylene protrusions as shown in Fig. 7(a). These changes in mechanical state of the UHMWPE liner are caused by the generation of cold flow into the screw holes. The results of the contact analysis for the case which has no screw holes, that is no cold flow occurs, are shown in Fig. 8 for the purpose of comparison with the results of the case with screw holes. The values of the contact stresses and the plastic strains were much lower than the case with screw holes shown in Fig. 7. The results of these FEM analyses suggest that the generation of cold flow into the screw holes on the back surface of the UHMWPE liner is closely related to the generations of the excessive wear and through-hole in the UHMWPE liner observed in this retrieved case shown in Fig. 1(d). The results also suggest that the cold flow generated due to the existence of the screw holes in the acetabular cup of the hip prosthesis reduces the wear resistance of the UHMWPE liner. It would appear that the cold flow into the screw holes contributes to reduction in thickness of the UHMWPE liner and increase of internal stresses and plastic strains in and around the regions of cold flow, and thus acceleration and increase of wear and/or failure of the UHMWPE liner.

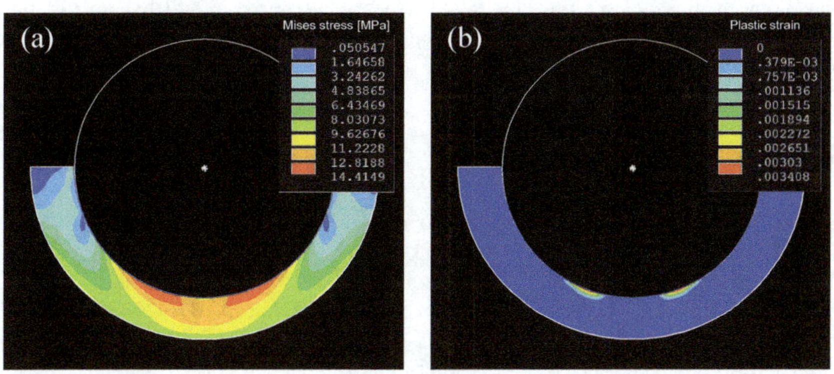

Fig. 8. Contour plots of the contact stress and the plastic strain of the UHMWPE acetabular cup liner of a hip prosthesis with no screw holes in the acetabular cup for the contact of the femoral head with a constant normal load of 1 kN: (a) contact stress, (b) plastic strain.

Considering the causes of the catastrophic wear and/or failure of the retrieved UHMWPE acetabular cup liners shown in Fig. 1 from the results of the FEM contact analyses performed in this study, it would appear that the change in mechanical state, such as contact stress and plastic strain due to contact with metallic components of the hip prosthesis contributes to structural weakening of the UHMWPE acetabular cup liner and increase of internal stresses and plastic strains in and around the contact regions, hence the catastrophic wear and/or failure of the UHMWPE liner.

5 Conclusion

In this study, FEM simulations of the contact between the UHMWPE acetabular cup liner and the metallic component of the hip prosthesis were performed in order to investigate the causes of the catastrophic wear and/or failure of the retrieved UHMWPE acetabular cup liners. The results of this study confirmed that change in mechanical state due to contact with the metallic component is the cause of the catastrophic wear and/or failure of the UHMWPE liner in the hip prosthesis.

The findings from this analytical study suggest that the generations of the neck-liner impingement, notch effect and cold flow have significant influences on the generation of wear and/or failure of the UHMWPE acetabular cup liner. Therefore, it is necessary to improve resistance to the neck-liner impingement, notch effect and cold flow in order to decrease wear and failure of the UHMWPE acetabular cup liner and to increase clinical longevity of the hip prosthesis.

Acknowledgements. This work was supported by JSPS KAKENHI Grant Number JP17K01391. The authors gratefully acknowledge receipt of the retrieved hip prosthesis components from the Department of Orthopaedic Surgery, University of Occupational and Environmental Health, Japan. Figures 1, 2, 3, 4, 6 and 7 were adapted from the Japanese Journal of Clinical Biomechanics, vol. 37, pp. 243–249 (2016), vol. 39, pp. 21–30 (2018), vol. 40, pp. 159–165 (2019) by Cho *et al.*, permission was granted by the Council of the Japanese Society for Clinical Biomechanics.

References

1. Ingham, E., Fisher, J.: Biological reactions to wear debris in total joint replacement. Proc. Inst. Mech. Eng. Part H J. Eng. Med. **214**, 21–37 (2000)
2. Cho, C., Mori, T., Kawasaki, M.: Observation of wear characteristics on retrieved polyethylene liners of hip prostheses. Jpn. J. Clin. Biomech. **37**, 243–249 (2016). (in Japanese)
3. Cho, C., Mori, T., Kawasaki, M.: Neck-liner impingement and cold flow induce catastrophic wear and failure of polyethylene acetabular cup liner in total hip arthroplasty. In: Session Book of the 8th World Congress of Biomechanics, no. 186 (2018)
4. Cho, C., Mori, T., Kawasaki, M.: Finite element analysis of contact state of a retrieved polyethylene acetabular cup liner showing excessive wear. Jpn. J. Clin. Biomech. **40**, 159–165 (2019). (in Japanese)
5. Cho, C., Mori, T., Kawasaki, M.: Influence of chamfer angle and size of polyethylene liner rim on the impingement failure in hip prosthesis. Jpn. J. Clin. Biomech. **39**, 21–30 (2018). (in Japanese)
6. Fisher, J., Dowson, D.: Tribology of total artificial joints. Proc. Inst. Mech. Eng. Part H: J. Eng. Med. **205**, 73–79 (1991)

7. Cho, C., Murakami, T., Sawae, Y., Sakai, N., Miura, H., Kawano, T., Iwamoto, Y.: Elasto-plastic contact analysis of an ultra-high molecular weight polyethylene tibial component based on geometrical measurement from a retrieved knee prosthesis. Proc. Inst. Mech. Eng. Part H: J. Eng. Med. **218**, 251–259 (2004)
8. Cho, C., Murakami, T., Sawae, Y., Miura, H., Kawano, T., Nagamine, R., Urabe, K., Matsuda, S., Iwamoto, Y.: Evaluation of wear in retrieved knee prostheses. J. Jpn. Soc. Clin. Biomech. **22**, 169–173 (2001). (in Japanese)

Constitutive Modelling of Knitted Abdominal Implants in Numerical Simulations of Repaired Hernia Mechanics

Agnieszka Tomaszewska[1(✉)], Daniil Reznikov[1], Czesław Szymczak[2], and Izabela Lubowiecka[1]

[1] Faculty of Civil and Environmental Engineering, Gdansk University of Technology, Narutowicza 11/12, 80-233 Gdansk, Poland
{atomas,lubow}@pg.edu.pl, danrezni.pg@gmail.com

[2] Faculty of Ocean Engineering and Ship Technology, Gdansk University of Technology, Narutowicza 11/12, 80-233 Gdansk, Poland
szymcze@pg.edu.pl

Abstract. The paper presents a numerical approach to describe mechanical behavior of anisotropic textile material, which is a selected abdominal prosthesis. Two constitutive nonlinear concepts are compared. In the first one the material is considered composed from two families of threads (dense net model) and in the second one the material is homogeneous but anisotropic (as proposed by Gassel, Ogden, Holzapfel). Parameters of both models are identified based on experimental tensile tests (uni-axial and bi-axial, simple and cyclic). The constitutive relations are applied in numerical membrane model of the prosthesis applied in the abdominal wall. Its mechanical responses to the pressure loading has been compared, also to deflection experimentally observed in physical model of the operated hernia of the same geometry. The authors find that both constitutive models properly describe the implant's mechanics, but further studies are needed to possibly approach the outcome of hyperelastic anisotropic model to experimental results obtained for synthetic knit mesh.

Keywords: Hernia repair · Abdominal prosthesis · Mechanics · FEM modelling · Experiment

1 Introduction

Abdominal prostheses are applied to prevent hernia occurrence in post-operational scar or to reconstruct abdominal wall in a case of hernia so that its structural function is restored. As typical in the human body reparation, hernia management deals with searching for the best solutions [1]. In a number of cases the operation is followed by permanent pain or even by the sickness recurrence [2, 3]. The hernia recurrence is observed when the implant fixation device is overloaded and the prosthesis is disconnected with the abdominal tissue. The load bearing capacity of selected tacks and sutures has been described e.g. in [4]. Proper hernia management depends on an accurate to a given case selection of the implant and its fixation. Many medical papers discuss that problem, also other than medical studies are undertaken to understand the

G. A. Ateshian et al. (Eds.): CMBBE 2019, LNCVB 36, pp. 550–559, 2020.
https://doi.org/10.1007/978-3-030-43195-2_45

biology and mechanics of operated abdominal hernia. They cover experiments on animals, in which tissue-mesh incorporation is observed [5, 6], *ex-vivo* experiments on operated hernia models with the use of animal tissue [7–9] experiments on living human abdominal wall to recognize its mechanical properties [10–12], mechanical tests of abdominal prostheses to observe their behavior and to identify their mathematical models [13–15] and finally, many numerical studies and simulations [16–19].

In the present paper numerical modelling of the prosthesis implanted in the abdominal wall is considered. The study refers to DynaMesh®-IPOM mesh. Finite Elements Method (FEM) is applied. The study is focused on constitutive modelling of the prosthesis. Two different concepts are compared. In the first one the mesh is modelled as a woven textile comprising two families of threads with non-linear stress-strain relation. Dense net material model is applied here [20]. In the second one the material is modelled with the use of homogeneous hyperelastic anisotropic material model, as proposed in [21]. That model is defined rather for tissues but the authors were tempted to analyze its suitability for modeling the implant as its knit wear structure can be treated as fibrous. The model was already applied in similar sense as it is described in [22].

In both cases the model is loaded by 'intra-abdominal' pressure, the deflection is calculated and compared to experimentally measured on corresponding physical model (experimental results are described in [23]).

2 Materials and Methods

2.1 The Implant

DynaMesh®-IPOM (FEG Textiltechnik GmbH, Aachen, Germany) is selected. It is a synthetic knit mesh, in which polypropylene (PP) filaments (12%, placed on parietal side) are interlinked with polyvinylidene fluoride (PVDF) threads (88%, placed on visceral side).

2.2 Constitutive Models

Dense Net Constitutive Model. This model is dedicated for woven materials [24, 25]. It has been applied in static and dynamic analysis performed for designing structures built with the use of textile material, e.g. Forest Opera in Sopot (Poland). In the authors team it serves for modelling textile, reticular or knitted implants. In this concept woven material is treated as a continuum without explicit reference to its discrete microstructure. Two directions $\xi \in (1, 2)$ in the structure plane are distinguished and it is assumed that cross-sectional membrane forces T in the two directions ξ depend solely on the uniaxial strains in these directions $(\varepsilon_1, \varepsilon_2)$. Thus, the following constitutive equation is postulated:

$$\mathbf{T}_\xi = \begin{Bmatrix} T_1 \\ T_2 \end{Bmatrix} = \begin{bmatrix} F_1 & 0 \\ 0 & F_2 \end{bmatrix} \begin{Bmatrix} \varepsilon_1 \\ \varepsilon_2 \end{Bmatrix}, \tag{1}$$

where F_1 and F_2 denote the material's tension stiffness in the two selected directions (1, 2). The details of the model can be found e.g., in [20] where it was defined and, in [16] where it was applied in implant modelling, in [23] where some mechanical analyses with this model were undertaken. In this study this stiffness is identified based on uni-axial tension tests that is possible due to the model specifics as mentioned before.

Hyperelastic Anisotropic Model. The anisotropic hyperelastic model is described using the Gasser-Ogden-Holzapfel (GOH) model [26]. The strain energy density function (SEDF) for this model is expressed as:

$$\Psi = C_{10}(I_1 - 3) + \sum\nolimits_{i=4,6} \frac{k_1}{2k_2}\left(e^{k_2(\kappa I_1 + (1-3\kappa)I_i - 1)^2} - 1\right), \tag{2}$$

where C_{10} and k_1 are stress – like parameters, k_2 – is a dimensionless parameter and κ describes the dispersion of the fibers. The SEDF contains two parts. The first term describes an isotropic behavior of the material (the influence of the matrix material) and the second term describes an anisotropic behavior of the material (the contribution of collagen fibers). I_1 is the first invariant of the Cauchy-Green tensor $\mathbf{C} = \mathbf{F}^T\mathbf{F}$.

$$I_1 = tr(\mathbf{C}) \tag{3}$$

The terms I_4 and I_6 are two pseudo – invariants of $\mathbf{C}$. They describe the properties of the fiber family

$$I_4 = \mathbf{a_0} \cdot \mathbf{C}\mathbf{a_0}, \quad I_6 = \mathbf{g_0} \cdot \mathbf{C}\mathbf{g_0}. \tag{4}$$

$$\mathbf{a_0} = \begin{bmatrix} \cos(\alpha) \\ \sin(\alpha) \\ 0 \end{bmatrix}, \ \mathbf{g_0} = \begin{bmatrix} \cos(\alpha) \\ -\sin(\alpha) \\ 0 \end{bmatrix} \tag{5}$$

where $\mathbf{a_0}$ and $\mathbf{g_0}$ are the unit vectors which describe the directions of fibers in the undeformed configuration [27]. The second Piolla – Kirchoff stress can be calculated as:

$$\mathbf{S} = -p\mathbf{C}^{-1} + 2\frac{\partial \Psi(\mathbf{C})}{\partial \mathbf{C}} \tag{6}$$

where p is the Lagrangian multiplier. Parameters of the model are identified based on biaxial tensile tests of DynaMesh-IPOM samples.

2.3 Description of the Experiments

Simple Tension Tests. Rectangular samples cut in two orthogonal directions of the material have been prepared and subjected to tests. The directions specified are parallel (direction 1) and perpendicular (direction 2) to the mesh knitting pattern. The samples of the width of 30 mm have been subjected to failure tension tests and to cyclic loading experiments, with various force ranges, between 0.5 and 2–20 N. Zwick Roel Z020 machine with video extensometer has been utilized. The details of the experiments are presented in [13].

Biaxial Tension Tests. Square sample of DynaMesh-IPOM has been prepared. Its edges are parallel to the knitting pattern of the mesh. It has been placed on Biax Zwick Roel machine using specially constructed rakes. The square field of the material, with side dimension of 50 mm has been subjected to biaxial tension tests. From uniaxial tests it is known that the mesh reveals orthotropic properties – ratio of elastic moduli determined for two orthogonal directions is approximately 4.5. Thus, the following various force ratios have been applied in the tests: 1:1, 1:2, 1:1.5. Bigger force has been applied in the stiffer direction of the mesh. Maximal force applied equals 12 N. The experimental set up is shown in Fig. 1. To identify Cauchy-Green deformation tensor 2-cameras Digital Image Correlation system has been used. The system tracks positions of four markers placed on the sample (see Fig. 1).

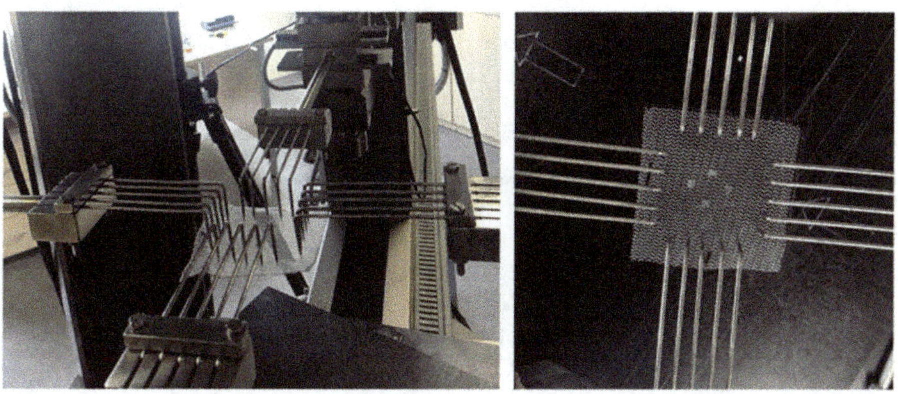

Fig. 1. Set up in biaxial tests

2.4 Numerical Models and Simulation

The models geometry responses to physical model of operated ventral hernia built of a porcine abdominal wall and DynaMesh-IPOM, which has been subjected to cyclic pressure loading (simulation of post-operational cough). The details of the experimental setup and the results are described in the paper [23].

In the numerical modelling the implant is represented by membrane finite elements with four nodes and three translational degrees of freedom in each node. The structure is circular, with a diameter of 13.5 cm. It is supported in 19 points evenly distributed on the circumference. In the central circular region, with a diameter of 7 cm, hernia orifice is supposed so this region is built only of the membrane (prosthesis). The ring around that hernia orifice is the overlap of the implant and the abdominal wall. It is modelled by a membrane (prosthesis) supported by elastic foundation (abdominal wall). Such set up is sufficient for the action simulated, which is pressure loading. The stiffness of the elastic foundation is 2.7 MPa, as identified in earlier study [16]. The pressure is applied as in the experiment, linearly growing from 0 to 7.75 kPa within 4 s.

Two FEM models created in commercial software are compared here. The first one, *M1* model, built in the MSC. Marc is described in details in [23]. It is supplemented by linear springs placed radially in the supporting points in the model plane. The springs mimic the abdominal wall elasticity, their stiffness coefficient is 1500 N/m. Dense net material model is applied in this case. Dynamic analysis is performed here with damping coefficients as described in [16]. The second, *M2* model, is made in Abaqus. Hyperelastic anisotropic material model is applied here. Nonlinear static analysis is performed with an increment size 0.05. Due to the numerical instabilities that may occur in membranes analysis both models demand initial tension, as discussed in [28]. In the model *M1* the initial stress is applied directly to the elements while in the model *M2* it is achieved by initial displacements of the model supports.

3 Results

3.1 Parameters of the Constitutive Models

The models are identified with the use of Marquardt–Levenberg variant of the least squares method. Compatibility of the hyperelastic anisotropic model with the experimental data is shown in Fig. 2. The data come from fifth in a row test of biaxial tension, so the sample is in the preconditioned state. The applied force ratio in two directions is 1:1.5. Force value of 12 N is applied in the stiffer direction of the material. The fitting accuracy is acceptable, as the correlation coefficients are 0.9982 and 0.9838 for two curves considered. The parameters of the model are placed in Table 1.

Table 1. Parameters of hyperelastic anisotropic model of DynaMesh-IPOM

Parameter	C_{10} [MPa]	k_1 [MPa]	k_2 [-]	κ [-]	α [rad]
Value	1.3005	2.8813	50.3756	0.0188	0.5170

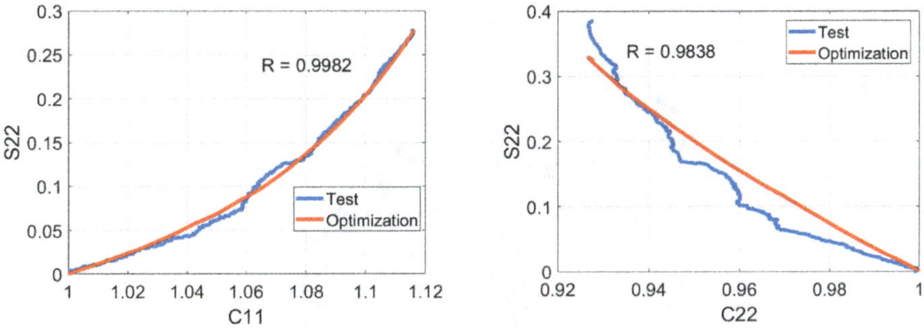

Fig. 2. Results of the hyperelastic model identification

Based on the uni-axial tests the DynaMesh-IPOM stiffness function necessary in dense net model in the two directions has been specified by determining elastic modulus values for each loading path. Baseline (based on failure tension tests) and preconditioned (based on cyclic loading tests) states of the material have been described. Identification details are described in [23]. Here the elastic moduli values, which form piecewise constant stiffness functions in the two distinguished directions of the prosthesis in the preconditioned state are considered. They are shown in Table 2.

Table 2. Parameters of stiffness functions in dense net model

Strain range, direction 1	Elastic modulus value [N/m]	Strain range, direction 2	Elastic modulus value [N/m]
0.00–0.10	594.00	0.00–0.06	1678.00
0.10–0.20	824.00	0.006–0.13	2650.00
0.20–0.25	1130.00	0.13–0.18	3850.00
0.25–0.35	1603.00	0.18–0.22	5700.00
0.35–0.45	2520.00	0.22–0.28	10650.00
0.45–0.55	4000.00		
0.55–0.65	6000.00		

3.2 Numerical Simulations Results

Maximum principal stress distribution calculated in the models *M1* and *M2* are presented in Fig. 3. The maximum value of the reaction force is obtained in the direction x in *M1* model (the values is 1.42 N) and in the direction y in *M2* model (with the value of 3.01 N). The deflection value experimentally observed equals 17 mm (as described in [23]). The value calculated in *M1* model is 16 mm and in the *M2* model it is 8 mm.

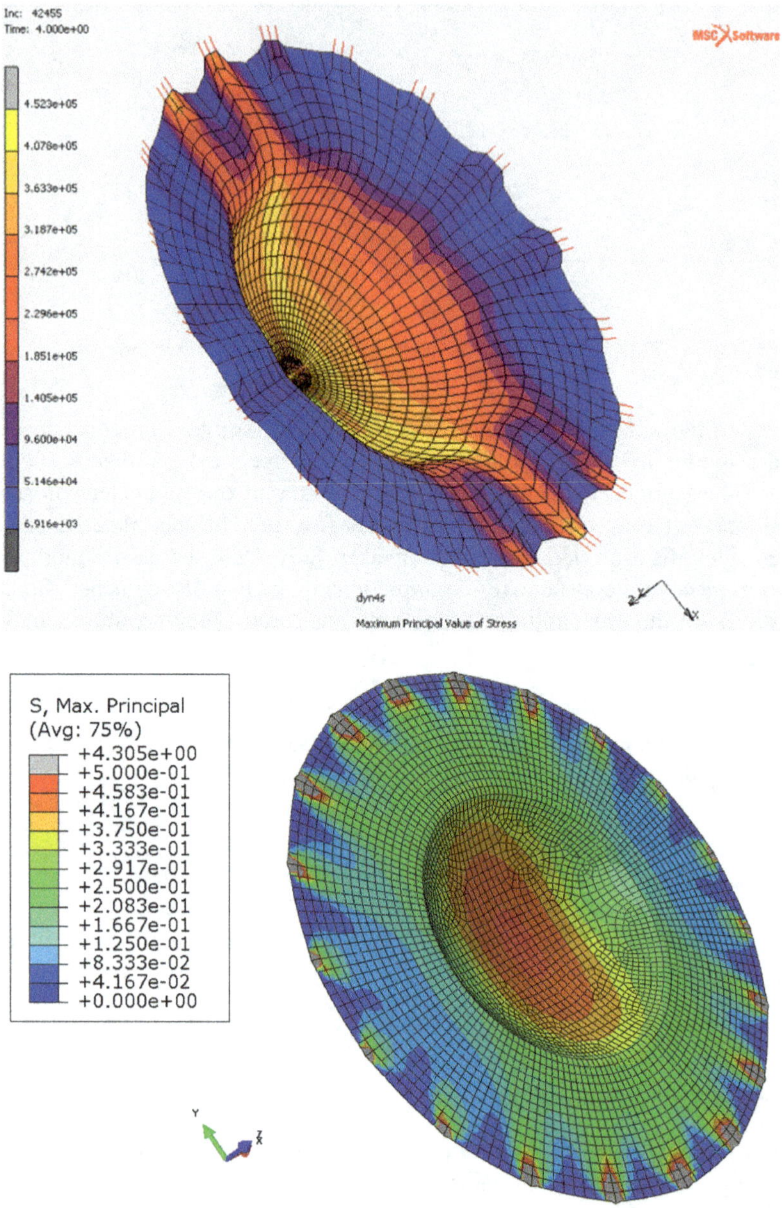

Fig. 3. Maximum principal stress calculated in *M1* model in [Pa] (the upper one) and in *M2* model in [MPa] (the bottom one)

4 Discussion and Conclusion

The two numerical models reveal similar response to the pressure loading. Zone of increased stress is observed in the direction of bigger stiffness of the material. Maximum stress value in both models equals 4.5 MPa approximately. However, the deflection calculated in the *M2* model is twice smaller than in the *M1* model and the maximum reaction in *M2* model is twice bigger than in *M1* model. Such relation between deflection and reaction in supporting points is typical in membrane and cable models (see e.g. [29]). The results obtained may suggest that an application of GOH constitutive relation makes the model more stiff than an application of dense net material model. But both numerical models have been created differently. The *M1* model has been validated to the experiment on hernia model, which is described in [23] and the deflection calculated in it responses to the physically measured. The *M2* model has been built based on parameters of *M1* model, including boundary conditions. The difference between outcome of the two models suggest that *M2* model should be validated to the experiment separately.

The aim of the study was to compare the effectiveness of two different constitutive concepts in application to a selected abdominal prosthesis, which is an anisotropic textile material. The authors have a well-established experience in the dense net model application in the cases of this kind. However, other groups apply hyperelastic anisotropic model to mimic mechanical behavior of such meshes (see e.g. [22]). In general both models reveal similar response to the load applied. However, by comparison of the results concerning deflection obtained in the two models one may hold a preliminary opinion that dense net material model describes the prosthesis behavior better than the homogeneous one. Further research aiming at obtaining bigger similarity between *M1* and *M2* models are needed, e.g. stiffness functions for dense net model should be identified from biaxial tests, the same as used in GOH model, *M2* model should be validated separately to the experiments to determine boundary conditions.

Acknowledgments. This work has been partially supported by the National Science Centre (Poland) [grant No. UMO-2017/27/B/ST8/02518]. Calculations have been carried out at the Academic Computer Centre in Gdansk.

References

1. Pawlak, M., Bury, K., Śmietański, M.: The management of abdominal wall hernias - in search of consensus. Videosurg. Other Miniinvasive Tech./Kwart. Pod patronatem Sekc. Wideochirurgii TChP oraz Sekc. Chir. Bariatrycznej TChP **10**(1), 49–56 (2015)
2. Sauerland, S., Walgenbach, M., Habermalz, B., Cm, S., Miserez, M.: Laparoscopic versus open surgical techniques for ventral or incisional hernia repair (review). Cochrane Libr. **3**, 1–62 (2011)
3. Bansal, V.K., Misra, M.C., Kumar, S., Rao, Y.K., Singhal, P., Goswami, A., Guleria, S., Arora, M.K., Chabra, A.: A prospective randomized study comparing suture mesh fixation versus tacker mesh fixation for laparoscopic repair of incisional and ventral hernias. Surg. Endosc. **25**(5), 1431–1438 (2011)

4. Tomaszewska, A., Lubowiecka, I., Szymczak, C., Smietański, M., Meronk, B., Kłosowski, P., Bury, K.: Physical and mathematical modelling of implant-fascia system in order to improve laparoscopic repair of ventral hernia. Clin. Biomech. **28**(7), 743–751 (2013). (Bristol, Avon)
5. Anurov, M.V., Titkova, S.M., Oettinger, P.: Biomechanical compatibility of surgical mesh and fascia being reinforced: dependence of experimental hernia defect repair results on anisotropic surgical mesh positioning. Hernia **16**(2), 199–210 (2012)
6. Gomez-Gil, V., Rodriguez, M., Garcia-Moreno, N.F., Perez-Kohler, B., Pascual, G.: Evaluation of synthetic reticular hybrid meshes designed for intraperitoneal abdominal wall repair: preclinical and in vitro behavior. PLoS ONE **14**(2), 1–26 (2019)
7. Podwojewski, F., Otténio, M., Beillas, P., Guérin, G., Turquier, F., Mitton, D.: Mechanical response of human abdominal walls ex vivo: effect of an incisional hernia and a mesh repair. J. Mech. Behav. Biomed. Mater. **38**, 126–133 (2014)
8. Röhrnbauer, B., Ozog, Y., Egger, J., Werbrouck, E., Deprest, J., Mazza, E.: Combined biaxial and uniaxial mechanical characterization of prosthetic meshes in a rabbit model. J. Biomech. **46**(10), 1626–1632 (2013)
9. Kallinowski, F., Gutjahr, D., Vollmer, M., Harder, F., Nessel, R.: Increasing hernia size requires higher GRIP values for a biomechanically stable ventral hernia repair. Ann. Med. Surg. **42**, 1–6 (2019)
10. Song, C., Alijani, A., Frank, T., Hanna, G., Cuschieri, A.: Elasticity of the living abdominal wall in laparoscopic surgery. J. Biomech. **39**(3), 587–591 (2006)
11. Todros, S., de Cesare, N., Pianigiani, S., Concheri, G., Savio, G., Natali, A.N., Pavan, P.G.: 3D surface imaging of abdominal wall muscular contraction. Comput. Methods Programs Biomed. **175**, 103–109 (2019)
12. Szymczak, C., Lubowiecka, I., Tomaszewska, A., Smietański, M.: Investigation of abdomen surface deformation due to life excitation: implications for implant selection and orientation in laparoscopic ventral hernia repair. Clin. Biomech. **27**(2), 105–110 (2012). (Bristol, Avon)
13. Tomaszewska, A.: Mechanical behavior of knit synthetic mesh used in hernia surgery. Acta Bioeng. Biomech. **18**(1), 77–86 (2016)
14. Deeken, C.R., Thompson, D.M., Castile, R.M., Lake, S.P.: Biaxial analysis of synthetic scaffolds for hernia repair demonstrates variability in mechanical anisotropy, non-linearity and hysteresis. J. Mech. Behav. Biomed. Mater. **38**, 6–16 (2014)
15. Röhrnbauer, B., Mazza, E.: Uniaxial and biaxial mechanical characterization of a prosthetic mesh at different length scales. J. Mech. Behav. Biomed. Mater. **29**, 7–19 (2014)
16. Lubowiecka, I.: Mathematical modelling of implant in an operated hernia for estimation of the repair persistence. Comput. Methods Biomech. Biomed. Eng. **18**(4), 438–445 (2015)
17. Lubowiecka, I., Szepietowska, K., Szymczak, C., Tomaszewska, A.: A preliminary study on the optimal choice of an implant. J. Theor. Appl. Mech. **53**(2), 411–421 (2016)
18. Pavan, P.G., Todros, S., Pachera, P., Pianigiani, S., Arturo, N.: The effects of the muscular contraction on the abdominal biomechanics: a numerical investigation. Comput. Methods Biomech. Biomed. Eng. **22**(2), 139–148 (2019)
19. Simón-Allué, R., Hernández-Gascón, B., Lèoty, L., Bellón, J.M., Peña, E., Calvo, B.: Prostheses size dependency of the mechanical response of the herniated human abdomen. Hernia **20**(6), 839–848 (2016)
20. Branicki, C., Kłosowski, P.: Statical analysis of hanging textile membranes in nonlinear approach. Arch. Civ. Eng. **XXIX**(3), 189–219 (1983)
21. Holzapfel, G.A., Gasser, T.C., Ogden, R.W.: A new constitutive framework for arterial wall mechanics and a comparative study of material models. J. Elast. **61**, 1–48 (2000)

22. Hernández-Gascón, B., Peña, E., Grasa, J., Pascual, G., Bellón, J.M., Calvo, B.: Mechanical response of the herniated human abdomen to the placement of different prostheses. J. Biomech. Eng. **135**(5), 051004 (2013)
23. Tomaszewska, A., Lubowiecka, I., Szymczak, C.: Mechanics of mesh implanted into abdominal wall under repetitive load. Experimental and numerical study. J. Biomed. Mater. Res. Part B Appl. Biomater. **107**(5), 1400–1409 (2019)
24. Ambroziak, A., Klosowski, P.: Review of constitutive models for technical woven fabrics in finite element analysis. AATCC Rev. **11**(3), 58–67 (2011)
25. Klosowski, P., Zerdzicki, K., Woznica, K.: Identification of Bodner-Partom model parameters for technical fabrics. Comput. Struct. **187**(187), 114–121 (2017)
26. Gasser, T.C., Ogden, R.W., Holzapfel, G.A.: Hyperplastic modeling of arterial layers with distributed collagen fiber orientations. J. R. Soc. Interface **3**, 15–35 (2006)
27. Holzapfel, G.: Nonlinear Solid Mechanics. A Continuum Approach for Engineering. Wiley, New York (2000)
28. Lubowiecka, I., Szymczak, C., Tomaszewska, A., Śmietański, M.: A FEM membrane model of human fascia – synthetic implant system in a case of a stiff ventral hernia orifice. In: Shell Structures. Theory and Applications, vol. 2, pp. 311–314 (2010)
29. Szymczak, C., Lubowiecka, I., Tomaszewska, A., Śmietański, M.: Modeling of the fascia-mesh system and sensitivity analysis of a junction force after a laparoscopic ventral hernia repair. J. Theor. Appl. Mech. **48**(4), 933–950 (2010)

A Novel Image Reconstruction Method in Three Dimensions

Zhe Zhang[1(✉)], Shiwei Zhou[1], Yi Min Xie[1,2], and Qing Li[3]

[1] Centre for Innovative Structures and Materials, School of Engineering, RMIT University, GPO Box 2476, Melbourne 3001, Australia
s3479043@student.rmit.edu.au

[2] XIE Archi-Structure Design (Shanghai) Co., Ltd., Shanghai 200433, China

[3] School of Aerospace, Mechanical and Mechatronic Engineering, The University of Sydney, Sydney, NSW 2006, Australia

Abstract. The reconstruction of three-dimensional biological structures from magnetic resonance image and computed tomography data remains challenging because of the limitations of existing numerical techniques and substantial computer resources required. The work renders the structure as the zero-level contour of a level set function, which converges to the model when an objective functional, the sum of a fitting energy term used to extract the local intensity and a diffusion term acting as a regularization contributor, is minimized. In addition, a reaction-diffusion method developed to replace the original time-difference algorithm by finite element analysis. Numerical examples illustrate that correspondingly clear and smooth structures of a woodpecker's skull can be obtained. Compared with the model reconstructed in a commercial tool, Mimics, our results presents substantially clearer and smoother interfaces without any isolated part.

Keywords: Biological structures · Image segmentation · Reaction-diffusion equation · Level set method · Reconstruction

1 Introduction

It is highly desirable to build effective and reliable three-dimensional (3D) models from tomographic images collected from computed tomography (CT) and magnetic resonance (MR) because it will provide practicians with comprehensive and intuitive structural information in clinic. The 3D reconstruction largely relies on image segmentation techniques [1, 2] to separate the raw data into non-overlapped geometric regions in line with their intensity information. However, there are still some difficulties, such as inhomogeneity of tissues, artefacts of images, a lack of computer resources, and irregularities of noise hinder the development of reconstruction of complex biological structures.

Recently, the global fitting energy in Chan-Vese model, which is a popular piecewise smoothed model and applies the specific energy functional as sum of the contour perimeter and fitting term has been replaced by a term of local data fitting energy, aiming to measure the image intensities on both sides of the contour. Its

G. A. Ateshian et al. (Eds.): CMBBE 2019, LNCVB 36, pp. 560–574, 2020.
https://doi.org/10.1007/978-3-030-43195-2_46

minimization allows the local intensity to participate in directing the motion of the boundary [18]. In addition, a regularization term can be included into the energy functional to increase the robustness of updating algorithm for the conventional Hamilton-Jacobi equation. Nevertheless, there are still two issues associated with this method: (1) small time step is compulsory when updating the governing equation with time-difference method; and (2) the level set function should be re-initialized frequently to remain the regularity [19].

In this study, we develop the reaction-diffusion approach to expediting the updating process of Hamilton-Jacobi equation, which involves both a reaction process to express local interaction between intensities and a diffusion process to allow the intensities homogenized in a spatial context [20]. The original idea of the reaction-diffusion method is to reveal the propagation of a gene population [21], and then it has been successfully introduced to structural topology optimization [22] to make the objective functional get convergent within fewer iteration steps. Another attractive character of reaction-diffusion method is the dispensability of the re-initialization process because the smoothness of the level set function is enhanced mathematically by a diffusion term in the energy functional. The regularity of the contour can be fully retained by a diffusion energy, thus the perimeter length term can be removed from the reaction-diffusion model. Furthermore, the finite element method used in renewing the Hamilton-Jacobi equation is able to allow a much larger time step than time difference method. Finally, we use this reaction-diffusion based level set method to reconstruct the skull of a woodpecker from the raw data, and hence a well-constructed structure is obtained within few iteration steps. Compared with the model reconstructed in a commercial tool, Mimics [23], our results presents substantially clearer and smoother interfaces without any isolated part.

The rest of paper is organized as follows. In Sect. 2, we describe the image reconstruction methodology. Section 3 presents some numerical implementations and examples. Conclusion is drawn in Sect. 4.

2 Image Reconstruction Methodology

1. Level set method

Level set method distinguishes the individual parts from the raw data by tracing the zero-level contour of a high-dimensional level set functional [13]. Mathematically, it is represented as:

$$\begin{cases} \varphi(x) > 0 & \forall x \in \Omega_1 \\ \varphi(x) = 0 & \forall x \in \Gamma \\ \varphi(x) < 0 & \forall x \in \Omega_2 \end{cases} \quad (1)$$

where the interface is represented as Γ, which commonly separates the solid region Ω_1 and void region Ω_2 in domain Ω, and x denotes the spatial variable. The evolution of the boundaries with respect to the fictitious time t is tracked by the well-known Hamilton-Jacobi equation, formulated as:

$$\frac{\partial \varphi(x,t)}{\partial t} - V_N(x,t)|\nabla \varphi(x,t)| = 0 \quad \text{in } \Omega \tag{2}$$

where V_N is the velocity along the normal direction, which is normally determined as the negative variational of an objective functional J with respect to the level set function as:

$$V_N = -\frac{\partial J(\varphi(x,t))}{\partial \varphi(x,t)} \tag{3}$$

where the negative sign allows the functional being minimized. By introducing the time difference method, Eq. (2) becomes:

$$\varphi^{n+1} = \varphi^n + \Delta t V_N |\nabla \varphi^n| = 0 \tag{4}$$

where the finite time-step Δt needs to satisfy the Courant-Friedrichs-Lewy (CFL) stability condition [24]:

$$V^n_{\max} \cdot \Delta t \leq min\left(h_x, h_y, h_z\right) \tag{5}$$

Thereinto, h_x, h_y and h_z denote the grid space in x, y and z directions, respectively.

Obviously, the size-dependent time step could result in very slow convergence for the raw data with extremely high resolution in the level-set based image separation. Conversely, large time increment may cause the level set function unstable and diverge largely from the value of signed distance quickly in the evolution. Another reason of low efficiency in updating the level set function is the compulsory re-initialization process in each or every a few steps to guarantee the level set function remaining as a distance function from the interface and therefore avoiding the ill-conditioning (e.g. small gradient of level set function) when numerically locating the interface [25].

2. Reaction-diffusion equation

We herein propose to adopt the reaction-diffusion equation for improving the efficiency of updating the level set function. Since the Hamilton-Jacobi equation is solved using the finite element analysis in this algorithm, the constraint to the time step resulted from CFL condition is released, and thus the re-initialization process becomes unnecessary. Without specific indication, a general objective function J is employed without showing its physical meanings in the following derivative. In the next Section, this function will be replaced by a sum of two energy terms used for image segmentation.

The optimization problem is mathematically expressed as:

$$\inf_{\varphi} \quad F(\varphi) = J(\varphi) + \frac{1}{2}\int_{\Omega} \tau |\nabla \varphi|^2 d\Omega + \lambda\left(\int_{\Omega} d\Omega - V_0\right) \tag{6}$$

in which λ represents the Lagrangian multiplier for the volume constraint V_0. The second term in Eq. (6) represents the interface energy, which is widely used in image processing and other physical model like the Cahn-Hilliard equation [26]. For the sake of controlling the regularization effect, it is weighted by a small positive factor $\tau > 0$. This energy term serves as number of roles, such as a regulator to the optimization problem, a smoother of the level set function, and a stabilizer of the algorithm. Due to the competences between regularity and the smoothness, the re-initialization process becomes unnecessary [22].

In accordance with the Karush-Kuhn-Tucker conditions [27], by introducing a fictitious time t and assuming that the level set function φ is an implicit function of time t, the structural changes in domain Ω are naturally and flexibly implemented. In the level set-based optimization method, the object functions are updated by solving time evolutional equation as Eq. (2). Here, a standard algorithm to minimize function F is to find the steady state solution of the gradient follow equation which assumes that the variation of the level set function φ with respect to time t is equal to $\partial F/\partial\varphi$ which is the Gâeaux derivative [28] of the function F, as:

$$\frac{\partial\varphi}{\partial t} = -\frac{\partial F}{\partial\varphi} \tag{7}$$

Substituting Eq. (6) into Eq. (7), the following equation is obtained:

$$\frac{\partial\varphi}{\partial t} = -\left(\frac{\partial J(\varphi)}{\partial\varphi} - \tau\Delta\varphi\right) \tag{8}$$

where Δ is the Laplace operator. Based upon the definition that the boundary is composed of Dirichlet boundary on the non-design boundary and Neumann boundary on the rest, the time evolutionary equation related to boundary conditions is expressed as:

$$\begin{cases} \frac{\partial\varphi}{\partial t} = -\frac{\partial F}{\partial\varphi} = -\left(C\frac{\partial J(\varphi)}{\partial\varphi} - \tau\Delta\varphi\right) & \text{in } \Omega \\ \frac{\partial\varphi}{\partial n} = 0 & \text{on } \partial\Omega\backslash\partial\Gamma_N \\ \varphi = 1 & \text{on } \partial\Gamma_N \end{cases} \tag{9}$$

In the reaction-diffusion method, $\tau\Delta\varphi$ is considered to be the diffusion term other than the regularization term in the time-difference algorithm, while the derivative of the cost function $dJ(\varphi)/d\varphi$ acts as a reaction term to account for the change in the curve (2D) or surface (3D) which is weighted by a factor C for normalization, defined as [22]:

$$C = \frac{c\int_\Omega d\Omega_1}{\int_\Omega \left|\frac{\partial J(\varphi)}{\partial\varphi}\right| d\Omega_1} \tag{10}$$

Rearrange Eq. (9) in a weak form for implementing discretization by finite element method, as follows:

$$\begin{cases} \int_\Omega \frac{\varphi(t+\Delta t)}{\Delta t}\varphi d\Omega + \int_\Omega \nabla^T \varphi(t+\Delta t)(\tau \nabla \varphi) d\Omega \\ = \int_\Omega (C\frac{\partial J(\varphi)}{\partial \varphi} + \frac{\varphi(t)}{\Delta t})\varphi d\Omega & \text{for} \, \forall \varphi \in \widetilde{\varphi} \\ \varphi = 1 & \text{on } \partial \Gamma_N \end{cases} \tag{11}$$

where $\widetilde{\varphi}$ is the Sobolev functional space of level set function.

For the time-related discretization, the implicit scheme is applied and the domain Ω is discretized based on finite element method. Thus, the discretized evolution equation is formulated as:

$$\begin{cases} (\frac{1}{\Delta t}\mathbf{T}_1 + \tau \mathbf{T}_2)\zeta(t+\Delta t) = \frac{\varphi(t)}{\Delta t} + C\frac{\partial J(\varphi)}{\partial \varphi} & \text{in } \Omega \\ \varphi = 1 & \text{on } \partial \Gamma_N \end{cases} \tag{12}$$

where $\zeta(t)$ is the nodal value vector of the level set function at time t, $\mathbf{T}_1$ and $\mathbf{T}_2$ can be described as follows:

$$\begin{cases} \mathbf{T}_1 = \bigcup_{j=i}^{e} \int_{V_e} \mathbf{N}^T \mathbf{N} dV_e \\ \mathbf{T}_2 = \bigcup_{j=i}^{e} \int_{V_e} \nabla \mathbf{N}^T \nabla \mathbf{N} dV_e \end{cases} \tag{13}$$

where e is the number of element. j is the number of the elements and $\mathbf{N}$ is the interpolation function of the level set function.

Through application of the reaction-diffusion equation to update the level set function, the compulsory re-initialization process in the level set method and the typical weaknesses from time difference method can be avoided. Therefore, introducing reaction-diffusion equation into the level set method is able to establish a more efficient, powerful and re-initialization-free algorithm for 3D reconstruction.

3 Numerical Implementation and Examples

1. Numerical implementation

Compared with current popular energy formulation in dealing with image segmentation schemes, here we only apply a data fitting energy [18] and a diffusion term which replaces the sum of two necessary terms in others, including an arc length term and a regularization term.

$$F(\varphi, f_1, f_2) = \sum_{i=1}^{2} \lambda_i \int (\int K_\sigma(x-y)|\mathbf{I}(y) - f_i(x)|^2 M_i^\varepsilon(\varphi(y))dy)dx + \tau \int (|\nabla \varphi(x)|)^2 dx \tag{14}$$

where $M_1(\varphi) = H(\varphi)$ and $M_2(\varphi) = 1 - H(\varphi)$, $\mathbf{I}$ denotes an image, λ_1 and λ_2 are positive constants, and $f_1(x)$ and $f_2(x)$ are the two values that approximate the weighted averages of the image intensities in Ω_{in} and Ω_{out} which are inside and outside a close contour in the design domain Ω, respectively. The intensities $\mathbf{I}(y)$ effectively involved in the above fitting energy are in a local region, whose size can be controlled by the Gaussian kernel K_σ which is shown as follows:

$$K_\sigma(\mathbf{I}) = \frac{1}{(2\pi)^{n/2}\sigma^n} e^{-|\mathbf{I}|^2/2\sigma^2} \tag{15}$$

where the scale parameter $\sigma > 0$.

The Heaviside function used in the above energy functional is defined by

$$H(x) = \frac{1}{2}\left[1 + \frac{2}{\pi}\arctan\left(\frac{x}{\varepsilon}\right)\right] \tag{16}$$

The derivative of H is

$$\delta(x) = H'(x) = \frac{1}{\pi}\frac{\varepsilon}{\varepsilon^2 + x^2} \tag{17}$$

The functional $F(\varphi, f_1, f_2)$ is the minimized postulate of a fixed level set function φ. By applying the calculus of variations, the functions $f_1(x)$ and $f_2(x)$ need to satisfy the Euler-Lagrange equations, as follows:

$$\int K_\sigma(x-y)|\mathbf{I}(y) - f_i(x)|^2 M_i^\varepsilon(\varphi(y))dy = 0 \quad i = 1, 2 \tag{18}$$

From Eq. (18), we obtain

$$f_i(x) = \frac{K_\sigma(x) * [M_i^\varepsilon(\varphi(x))\mathbf{I}(x)]}{K_\sigma(x) * M_i^\varepsilon(\varphi(x))} \quad i = 1, 2 \tag{19}$$

where the symbol * represents a convolution operator. Keeping f_1 and f_2 be fixed, the derivative of energy functional $F(\varphi, f_1, f_2)$ with respect to t is:

$$\frac{\partial\varphi}{\partial t} = -\delta(\varphi)(\lambda_1 c_1 - \lambda_2 c_2) + \tau\nabla^2\varphi \tag{20}$$

where δ is Delta function and c_1 and c_2 is given by:

$$c_i(x) = \int K_\sigma(x-y)|\mathbf{I}(x) - f_i(y)|^2 dy \quad i = 1, 2 \tag{21}$$

Next, the updating process is carried out by using the reaction-diffusion scheme with respect to Eq. (12). The approach in this section will be implemented through the following numerical examples.

2. Numerical example

In this section, we reconstruct three numerical models for a woodpecker from the raw CT data to assess the validity and utility of our proposed method. In all examples, the value of ε in Eq. (16) is set to be 0.1 and parameter $K_\sigma = 1.0$ and the value of λ_1 and λ_2 in Eq. (20) are set to be $\lambda_1 = 1.0$ and $\lambda_2 = 20$, respectively.

Due to huge amount of invalid information in CT images, the intensity distribution (see Fig. 1) shows that the useful data related to the bone of a woodpecker is mainly located in the region where intensity is greater than 500. Thus, in these given case scenarios, only the part where conforms to intensity $T > 500$ is considered.

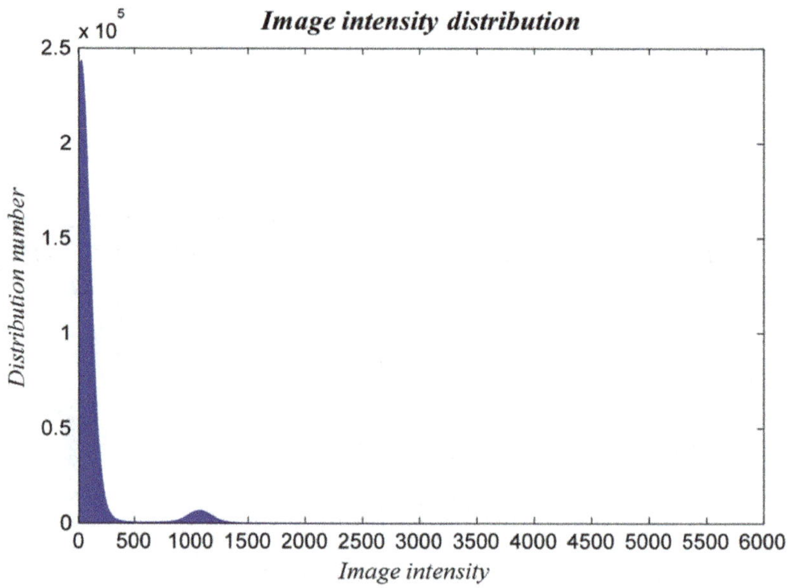

Fig. 1. Image intensity distribution

A. Case 1

In the first case, the diffusion parameter τ is set to be $\tau = 5.0 \times 10^{-6}$, parameter $\sigma = 6.0$ and the time increment $\Delta t = 0.01$. Figure 2 illustrates the segmentation process using the proposed method, the value of m stands for the number of iteration steps. As shown in the 500th step, a smooth and clear model for a woodpecker's skull is obtained by using the proposed method.

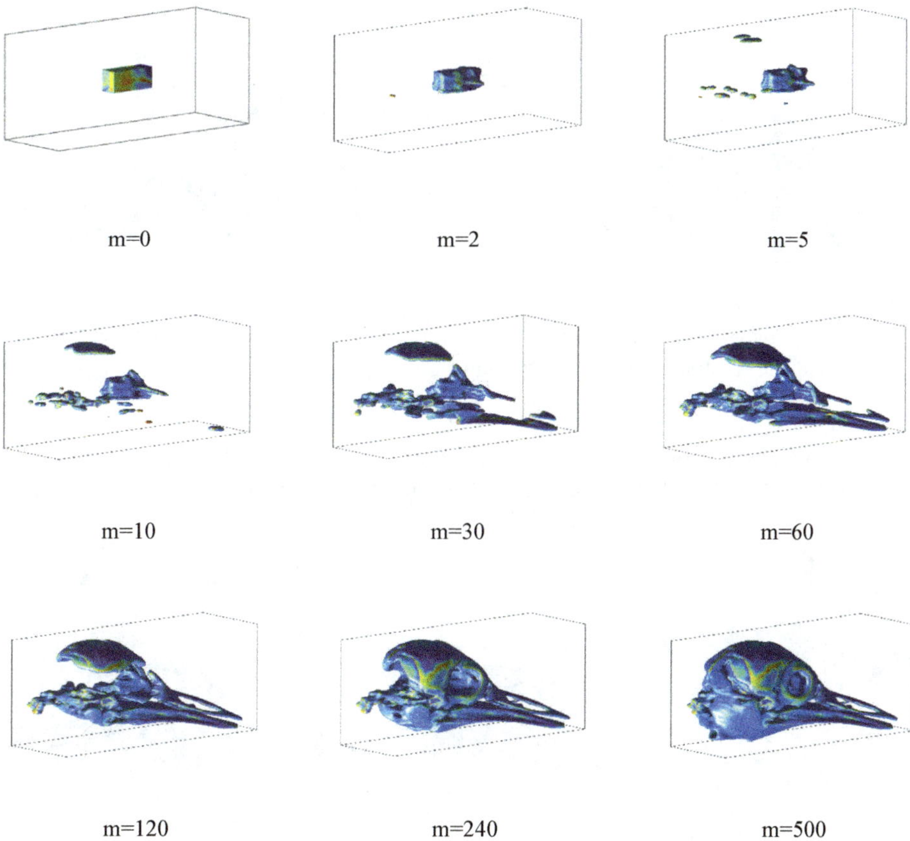

Fig. 2. Configurational transformation in case 1

As mentioned before, the major novelty of this new method lies in applying the reaction-diffusion equation to update of the level set function. To demonstrate the superiority of the proposed method to traditional algorithms, an example obtained from time difference method was considered, which needs to add an extra perimeter length term into the energy functional for maintaining the smoothness of level set contour. Herein, the arc length parameter in perimeter length term $\mu = 1.0$ and all other sets are not changed for reasonable comparison.

The two sub-figures in Fig. 3 specifically show that the surface obtained by the proposed method (Fig. 3(a)) is clearer, smoother and more accurate than that in the conventional time difference method (Fig. 3(b)). In the next stage of our research, we will investigate the impact-resistance of the woodpecker's skull. Thus, clarity, smooth and accuracy of the model are primary considerations. Some preliminary studies have shown that the asperous result in Fig. 3(b) is not suitable for carrying out impact simulation.

(a) (b)

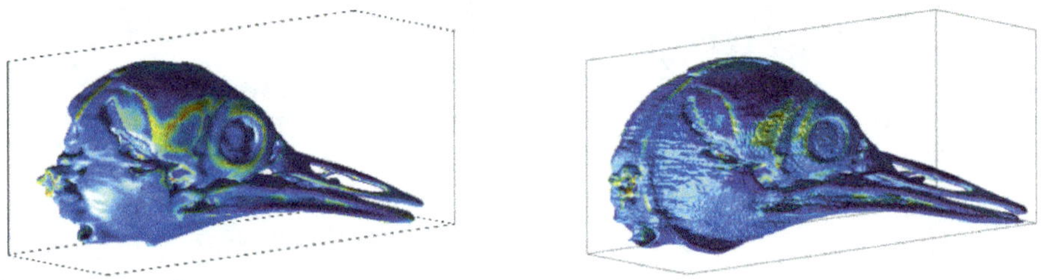

Fig. 3. Comparison of the configurational performance between (a) the reaction diffusion algorithm and (b) the time difference algorithm

B. Case 2

In the second case, diffusion parameter τ is set to be $\tau = 5.0 \times 10^{-5}$, parameter $\sigma = 3.0$ and time step $\Delta t = 0.001$. Instead of variable values with regard to Eq. (10), parameter C is assigned to be the fixed value of $C = 0.1$. The segmentation process in this case is displayed in Fig. 4.

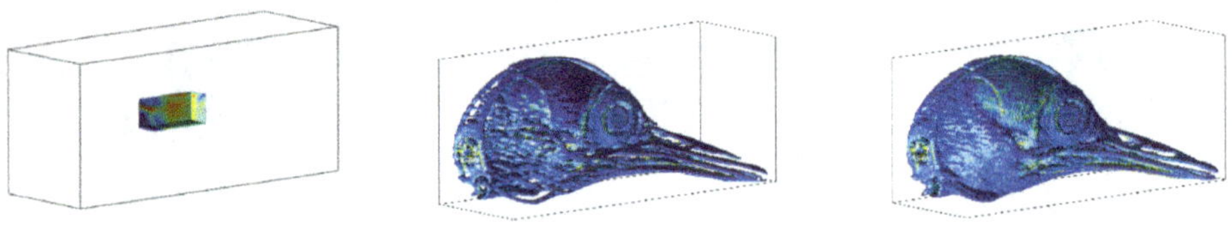

m=0 m=1 m=10

Fig. 4. Configurational transformation in case 2

Further algorithmic comparison between the proposed method and time difference method are shown in Figs. 5 and 6, which includes the performance of configurations and numerical analysis. Note that the parameter μ in the length term for the conventional model is set to be $\mu = 1.0$ by remaining the other parameters unchanged.

(a) (b)

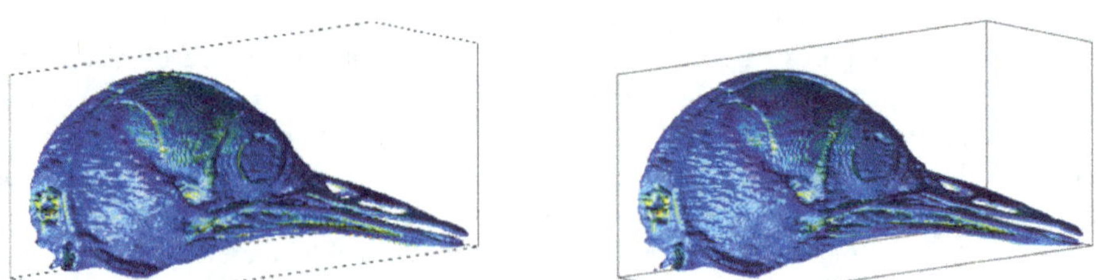

Fig. 5. Comparison of the configurational performance between (a) the proposed method and (b) the time difference method

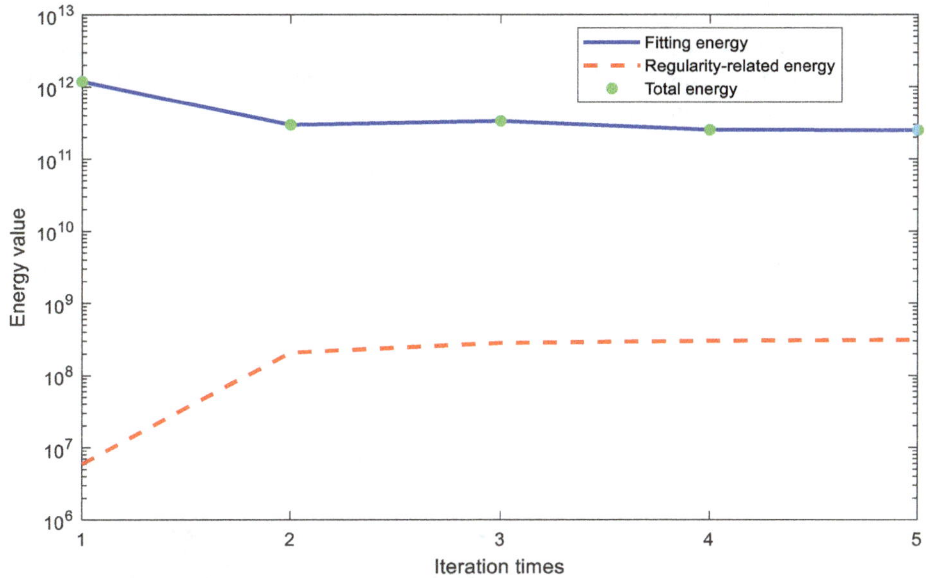

Fig. 6. The comparison of energy variation between the proposed method and time difference method, including (a) total energy, (b) fitting energy, (c) regularity-related energy, and (d) perimeter length energy

Differing from the comparison in Fig. 3, both the results in Fig. 5(a) and (b) are fairly similar. It is difficult to distinguish which provides better configuration by eyes. Thus, an in-depth comparison between the models with the reaction-diffusion equation and time difference method is needed.

In these four line charts in Fig. 6, the blue line presents the energy variation in the model using the proposed method, while the red line shows the transformation using the conventional finite difference method. In order to check the convergence of energy $J(\phi, f_1, f_2)$, we use the following criterion: if $(J_{n+1} - J_n)/J_n \leqslant 10^{-3}$, the nth iteration step attains convergence. Herein, $J_n(\phi, f_1, f_2)$ represents the energy value in the nth iteration step. It is evident that both models are able to converge quickly after 2 iterations. It means that the proposed model can not only generate more accurate results, but also achieve a goal of immediate image segmentation by adjusting a certain parameter.

Note that Sections A and B confirmed that the proposed method has potential to reconstruct complex three-dimensional raw data into biological structures with well-defined boundaries. In comparison with the conventional time difference method, the proposed method exhibits certain superiority in 3D reconstruction.

C. Parameter investigations

Note that the scale parameter σ plays an important role in determining the width of the Gaussian kernel. The alteration of σ leads to the changes in K_σ, f_1 and f_2. In addition to σ, diffusion parameter τ that is used to control the effect of regularization also plays a significant role in running the proposed method. As two of the significant parameters, the effects of σ and τ are investigated further in this section. Before discussing the effects of parameters, a case scenario is presented in Fig. 7. Except that the diffusion parameter τ is changed to $\tau = 7.0 \times 10^{-6}$ and scale parameter σ is set to $\sigma = 3.0$, the other parameters are remained the same as case 1.

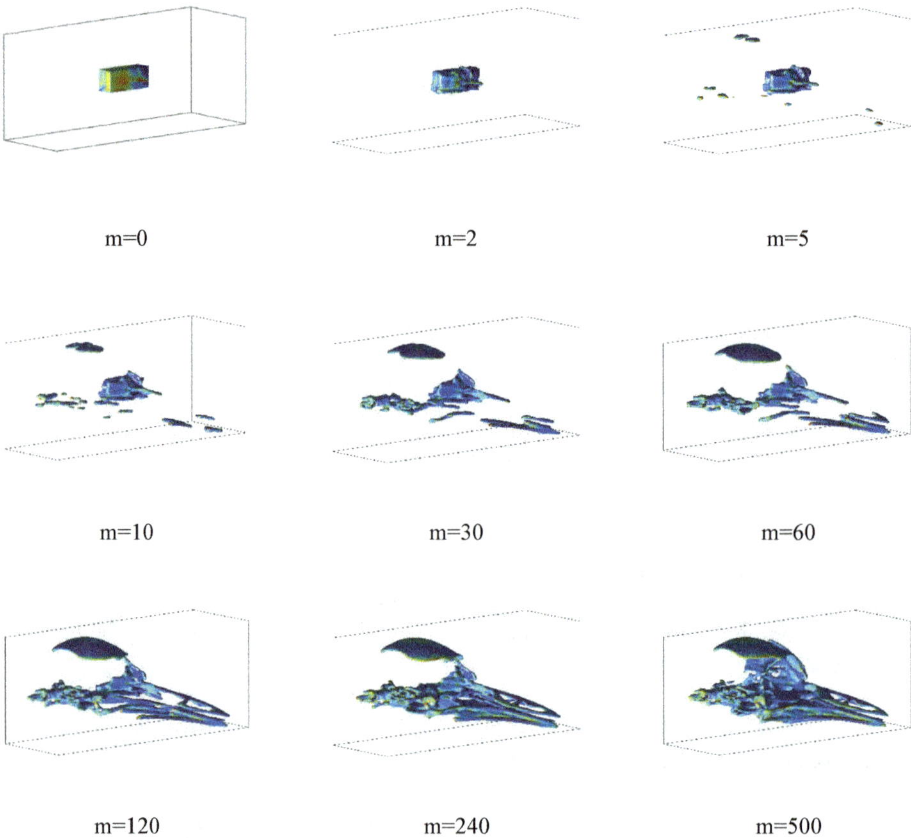

Fig. 7. Configurational transformation in case 3

Compared with the result of case 1 shown in Fig. 2, while the configurations in the both cases are of almost the same level of smoothness, huge differences occur in the other aspects. Therefore, it proves that the change of τ and σ play significant roles in the transformation of model.

Figure 8 compares the performances of the obtained configurations under different values of σ. In terms of clarity and smoothness, the surfaces on all results are nearly same, but increasing σ increases the volume of final configurations. Nevertheless, the complexity of calculation is also raised numerically. It means that more computational resource and cost are needed to complete the entire modelling process. Therefore, taking the efficiency into considerations, setting σ as big as possible does not mean the best choice.

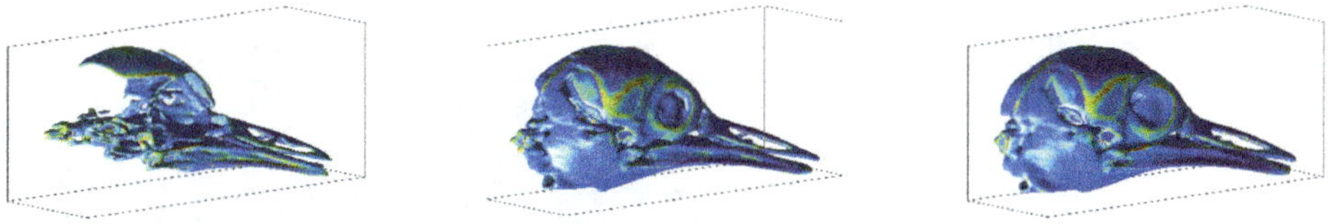

(a) σ=3.0 (b) σ=6.0 (c) σ=9.0

Fig. 8. Comparison of the configurational performance when σ is equal to (*a*) 3.0, (*b*) 6.0 and (*c*) 9.0

Next, the effects from regularization parameter τ are analysed as shown in Fig. 9 in which the results at the 500th iteration are compared for the different values of τ. Owing to the functions of controlling the effects of the regularization and adjusting length of level set contours, the rise of parameter τ leads to length shortening in the configurations (see Fig. 8). It is estimated that the case (*b*) and (*c*) in Figs. 8 require more iteration steps to achieve a similar configuration to case (*a*). Namely, the change of parameter τ makes the segmentation process longer. To some degree, an appropriate reduction in τ is beneficial to improving the efficiency. Nevertheless, specifying τ at a smaller value could potentially make the proposed method unable to run.

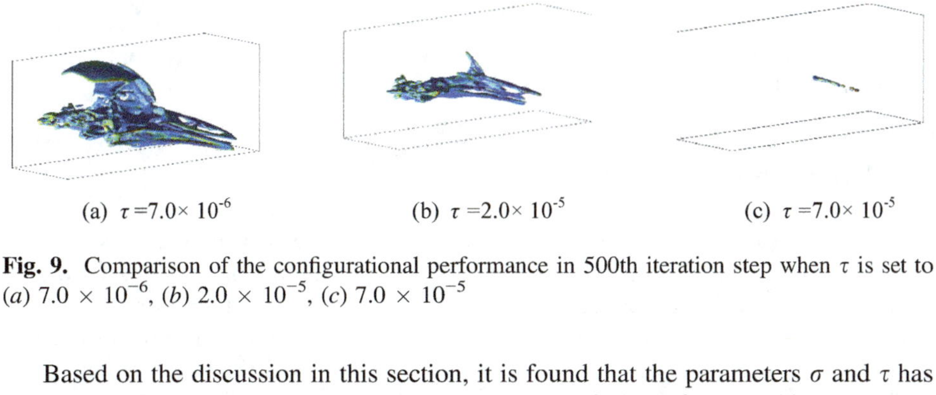

(a) τ=7.0× 10^{-6} (b) τ =2.0× 10^{-5} (c) τ =7.0× 10^{-5}

Fig. 9. Comparison of the configurational performance in 500th iteration step when τ is set to (*a*) 7.0×10^{-6}, (*b*) 2.0×10^{-5}, (*c*) 7.0×10^{-5}

Based on the discussion in this section, it is found that the parameters σ and τ has significant effects on the reconstruction results. Prescription of reasonable parameters would be helpful for dealing with different 3D structures.

D. Reconstruction in Mimics

At present, several commercial codes are able to conduct image processing, such as Mimics, Simpleware [29] and 3D-doctor [30]. As one of the most popular reconstruction programs, Mimics uses 2D cross-sectional images, e.g. obtained from computer tomography and magnetic resonance image, to reconstruct 3D models. In order to make a comparison between both the results generated from Mimics and the proposed method, the woodpecker's CT data is imported into Mimics for 3D modelling.

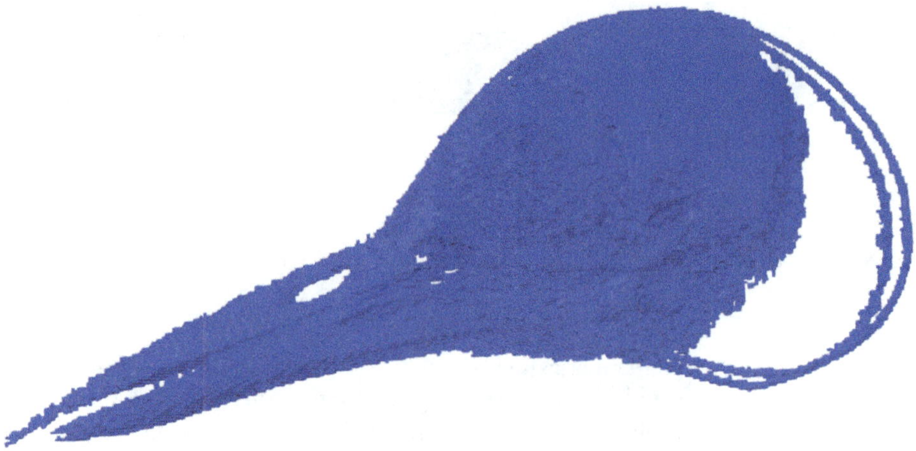

Fig. 10. Reconstruction result in Mimics

The modelling in Mimics is relying on defining a certain range of grey scale for identifying the eligible sections in each CT image. These distinguished parts are used to make up a new 3D model. However, due to proximate greyscales for skull, spine and weasand, it is difficult to segment an accurate scope for reconstructing the woodpecker's skull. Further, as presented in Fig. 10, the marginal area in the materialized model is fuzzy with a huge number of isolated regions. Thus, the simulation of a woodpecker's impacting process is hard to be conducted by directly using Mimics results unless further tunning is taken.

4 Conclusions

This study proposed a novel level set method by incorporating the reaction-diffusion equation, and applied it to reconstruct the three-dimensional structures from the high resolution raw data. We defined the energy functional with a fitting energy, which extracted the initial contour toward desired boundaries by utilizing local intensity and a diffusion term to regularize the level set function and smooth the interface contour.

The proposed method is capable of reconstruction images with high inhomogeneity of intensity, and provide desirable reconstructed model for complicated scanning images with blurred boundaries. With the diffusion term in the reaction-diffusion equation, the regularity of the level set function and the smoothness of the contour can be intrinsically maintained to ensure accurate computation, and thus the expensive re-initialization process and perimeter length term which are compulsory in conventional level set algorithms can be eliminated.

In addition, the finite element method replaces the time difference method to allow a greater time step when updating the level set function. For this reason, the efficiency to a certain degree is improved. The numerical example showcased the desired performance of the proposed method for CT image based reconstruction of a woodpecker's skull. Compared with the model reconstructed using commercial code Mimics, the proposed method is able to provide a better result with clearer and smoother interfaces without isolated parts.

References

1. Haralick, R.M., Shapiro, L.G.: Image segmentation techniques. Comput. Vis. Graph. Image Process. **29**(1), 100–132 (1985)
2. Marion, A.: Introduction to Image Processing. Springer, New York (2013)
3. Hojjatoleslami, S., Kittler, J.: Region growing: a new approach. IEEE Trans. Image Process. **7**(7), 1079–1084 (1998)
4. Adams, R., Bischof, L.: Seeded region growing. IEEE Trans. Pattern Anal. Mach. Intell. **16** (6), 641–647 (1994)
5. Level Otsu, N.: A threshold selection method from gray-level histogram. IEEE Trans. Syst. Man Cybern. **9**(1), 62–66 (1979)
6. Sezgin, M., Sankur, B.: Survey over image thresholding techniques and quantitative performance evaluation. J. Electron. Imaging **13**(1), 146–166 (2004)
7. Chan, T., Vese, L.: An active contour model without edges. In: Scale-Space Theories in Computer Vision, pp. 141–151 (1999)
8. Chan, T.F., Vese, L.A.: Active contours without edges. IEEE Trans. Image Process. **10**(2), 266–277 (2001)
9. Roberts, L.: Machine perception of three-dimensional solids. In: Tippett, J. (ed.) Optical and Electro-Optical Information Processing, pp. 159–197. MIT Press, Cambridge (1965)
10. Shrivakshan, G., Chandrasekar, C.: A comparison of various edge detection techniques used in image processing. IJCSI Int. J. Comput. Sci. Issues **9**(5), 272–276 (2012)
11. Canny, J.: A computational approach to edge detection. In: Readings in Computer Vision, pp. 184–203. Elsevier (1987)
12. Kass, M., Witkin, A., Terzopoulos, D.: Snakes: active contour models, p. 268
13. Osher, S., Sethian, J.A.: Fronts propagating with curvature-dependent speed: algorithms based on Hamilton-Jacobi formulations. J. Comput. Phys. **79**(1), 12–49 (1988)
14. Vese, L.A., Chan, T.F.: A multiphase level set framework for image segmentation using the Mumford and Shah model. Int. J. Comput. Vision **50**(3), 271–293 (2002)
15. Li, C., Kao, C.-Y., Gore, J.C., Ding, Z.: Implicit active contours driven by local binary fitting energy, pp. 1–7
16. Tian, Y., Zhou, M., Wu, Z., Wang, X.: A region-based active contour model for image segmentation, pp. 376–380
17. Li, B., Jia, F., Cao, Y.P., Feng, X.Q., Gao, H.J.: Surface wrinkling patterns on a core-shell soft sphere. Phys. Rev. Lett. **106**(23), 234301 (2011)
18. Li, C., Kao, C.-Y., Gore, J.C., Ding, Z.: Minimization of region-scalable fitting energy for image segmentation. IEEE Trans. Image Process. **17**(10), 1940–1949 (2008)
19. Li, C., Xu, C., Gui, C., Fox, M.D.: Distance regularized level set evolution and its application to image segmentation. IEEE Trans. Image Process. **19**(12), 3243–3254 (2010)
20. Choi, J.S., Yamada, T., Izui, K., Nishiwaki, S., Yoo, J.: Topology optimization using a reaction–diffusion equation. Comput. Methods Appl. Mech. Eng. **200**(29–32), 2407–2420 (2011)
21. Fisher, R.A.: The wave of advance of advantageous genes. Ann. Hum. Genet. **7**(4), 355–369 (1937)

22. Yamada, T., Izui, K., Nishiwaki, S., Takezawa, A.: A topology optimization method based on the level set method incorporating a fictitious interface energy. Comput. Methods Appl. Mech. Eng. **199**(45–48), 2876–2891 (2010)
23. Materialise: Mimics. http://www.materialise.com/en/medical/software/mimics. Accessed 19 June 2019
24. Chaudhury, K.N., Ramakrishnan, K.: Stability and convergence of the level set method in computer vision. Pattern Recogn. Lett. **28**(7), 884–893 (2007)
25. Li, C., Xu, C., Gui, C., Fox, M.D.: Level set evolution without re-initialization: a new variational formulation, pp. 430–436
26. Cahn, J.W., Hilliard, J.E.: Free energy of a nonuniform system. I. Interfacial free energy. J. Chem. Phys. **28**(2), 258–267 (1958)
27. Murat, F.: Optimality conditions and homogenization. In: Marino, A., et al. (eds.) Nonlinear Variational Problems, pp. 1–8. Pitman Publishing Program, Boston (1985)
28. Preiss, D.: Gâteaux differentiable functions are somewhere Fréchet differentiable. Rendiconti del Circolo Matematico di Palermo **33**(1), 122–133 (1984)
29. Synopsys: Simpleware. https://www.synopsys.com/simpleware.html. Accessed 19 June 2019
30. 3D-Doctor. http://www.ablesw.com/3d-doctor/. Accessed 19 June 2019
31. Oasys: LS-DYNA. https://www.oasys-software.com/dyna/software/ls-dyna/. Accessed 19 June 2019

CFD Analysis of Flow Around a Serrated Feather

Tetsuhiro Tsukiji(✉) and Hiroki Takase

Sophia University, Tokyo 1028554, Japan
t-tukiji@sophia.ac.jp

Abstract. It is known generally that nocturnal owls fly quietly with low flight sound compared with other birds. The nocturnal owls have serrated feathers of the shape of fine saw at the leading edge of the feather. It is called serrations. It is a unique feather which cannot be seen with other birds and the serrations seem to have the effect to reduce the flight sound. The effect is called the silencing effect. Therefore, it is interesting to investigate the effect of the serrations on the flow properties around the feather and the acoustic effect. From those background, the present research is focused on the investigation of the influence of the serrations on the flow properties such as the velocity distribution, the vorticity distribution and the acoustic effect by analyzing the flow field around the feathers using CFD technique.

Keywords: Owls · Serrated feather · CFD (Computational Fluid Dynamics) analysis · Fluid Mechanics

1 Introduction

It is said that the flying sound of owls is less than that of other birds. The sound of the owls is thought to be low because the owls have serrations at the leading edge of the feathers. The serrations are the serrated feathers of the shape of fine saw. The serrations which have minute teeth at regular intervals shown in Fig. 1. The serrations cannot be seen with other birds and those seems to have the effect to reduce the sound level during owls flying.

In the previous researches, some applications for mechanical device and fluid machine using the design of the serrations are done to reduce the sound level. The sound level of the axil flow fan was tried to be reduced using the serrations located at the edges of the blades [1]. As the basic research to investigate the serration effect, the effect of serrations on the mixing layer past a thin splitter plate was investigated using CFD [2]. The aerodynamic characteristics of the airfoil with serrations were measured and the flow fields around the airfoil was visualized in a smoke tunnel [3].

However the effect of the serrations on the flow properties such as vortex structure, lift and drag force and velocity distributions and the sound level around the feathers with serrations at the leading edge is not clear.

G. A. Ateshian et al. (Eds.): CMBBE 2019, LNCVB 36, pp. 575–580, 2020.
https://doi.org/10.1007/978-3-030-43195-2_47

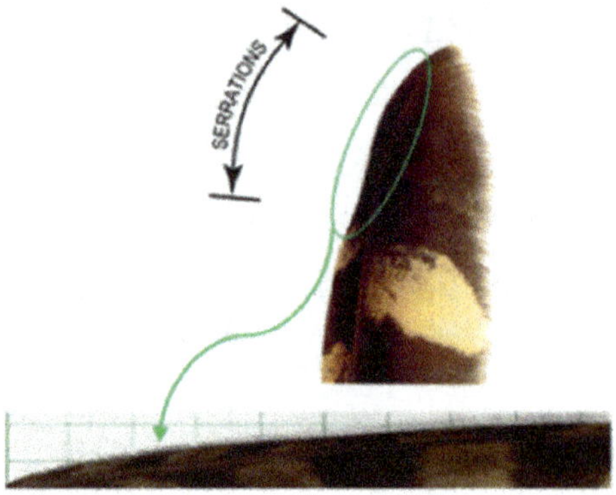

Fig. 1. Serrations

From those background, the present research is focused on the investigation of the influence of the serrations on the flow properties such as the velocity distribution, the vorticity distribution and the acoustic effect by analyzing the flow field around the feathers using CFD technique.

2 Feather and Calculation Model

2.1 Feather Model

The model of the feather designed by 3D CAD base on the real feather of the owl is shown in Fig. 2 and the normal feather and the serrated feather for CFD model are shown in Fig. 3. The width of the feathers is 5 mm.

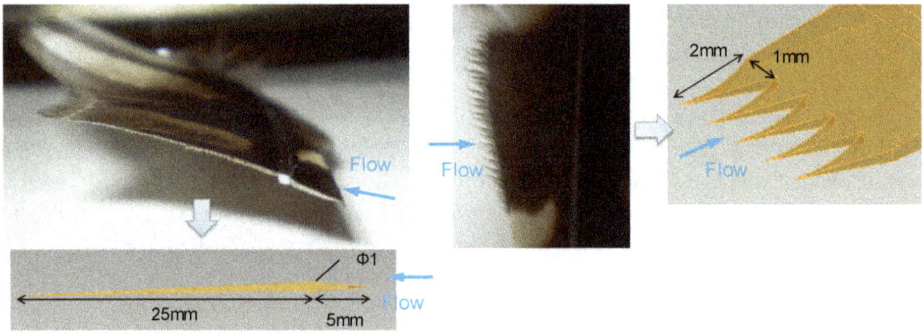

Fig. 2. Model of the feather

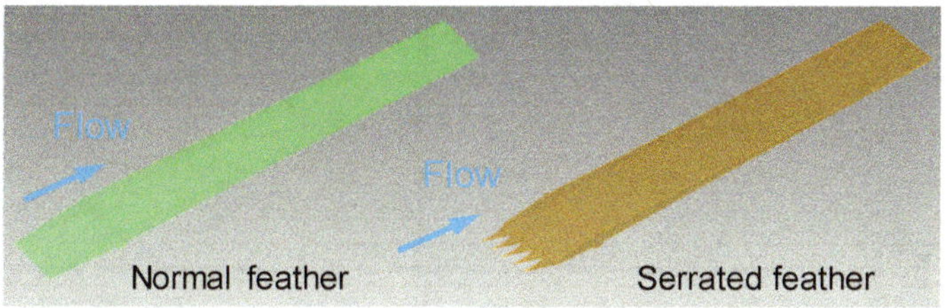

Fig. 3. Normal feather and serrated feather

2.2 CFD Method

In this calculation, we used the Large Eddy Simulation as a CFD (Computational Fluid Mechanics) technique. Standard Smagorinsky model is used as sub-grid scale models. Ffowcs Williams and Hawkings equation was used for the evaluation of the sound [4]. In the formulation of Ffowcs Williams and Hawkings, the form improved from the generality of acoustic analogy of Lighthill equation is adopted [5]. Solver is Fluent ver. 6.3 CFD package. Flow velocity is equal to owl's cruising speed 9 m/s. Reynolds number based on the flow velocity and the chord length is 17880 in all cases. The time step is set to 0.0001 s for analyzing complicated unsteady flow field. Both serrated and normal feather are simulated till 0.15 s. The attack angle is 10°. The feather models are supposed to be inflexible. The numerical solution using the $k - \varepsilon$ standard model is used as the initial values of the Large Eddy Simulation.

3 Calculated Results and Discussion

The computational grid used in this study is shown in Fig. 4. It is composed of tetrahedral elements. The calculating zone consists of 2,000,000 cells.

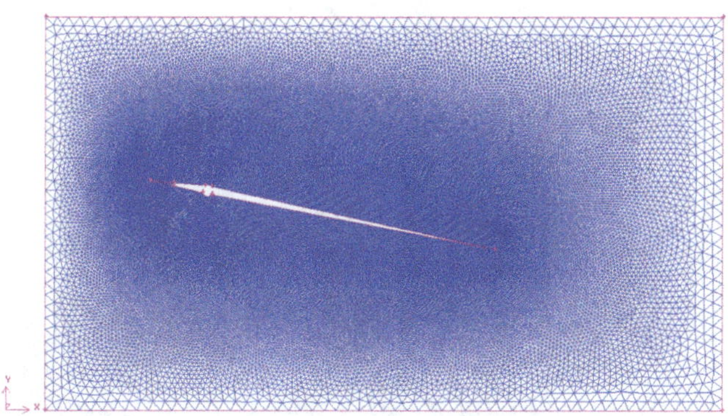

Fig. 4. Grids around the feather

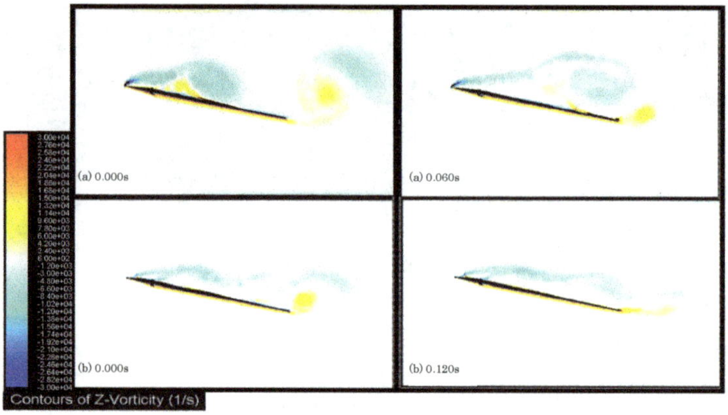

Fig. 5. Comparison of the flow pattern

The contours of the three-dimensional vorticity distributions around the feather are shown in Fig. 5(a) and (b). The results for normal feather without serrations are shown in Fig. 5(a) and those with serrations are shown in Fig. 5(b). Flow separation from the leading edge is seen for feather without serrations. However no separation until the vortices reaches around the center of the chord of the airfoil is seen on serrated feather. It turned out that the flow separation controlled on the upper side of an airfoil is performed by the serrations on the leading edge. Figure 6 shows that the serrations generate longitudinal vortices, and the turbulent transition is moved downstream on the upper surface of the airfoil to delay the separation point by the longitudinal vorticities. The sound pressure level (SPL) along frequency at angle of attack 10° is calculated. The results for the back receiver of the feather are shown in Fig. 7 and those of the top receiver are shown in Fig. 8. From both results the SPL of owl feather is lower than

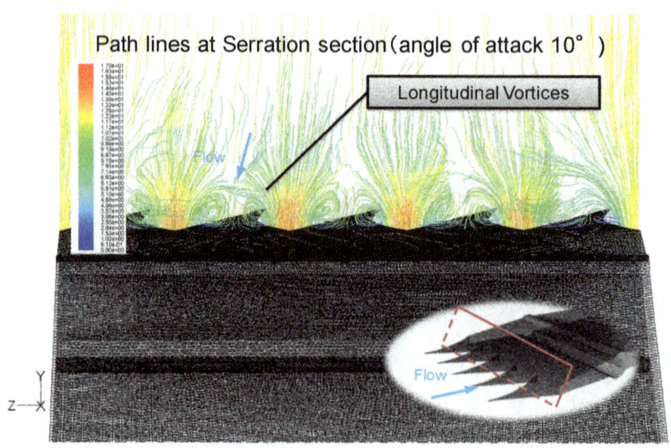

Fig. 6. Longitudinal vortices

that of normal feather through the wide range of frequency. So the silencing effect seems to be caused by the longitudinal vortices generated in the case of the serrated feather.

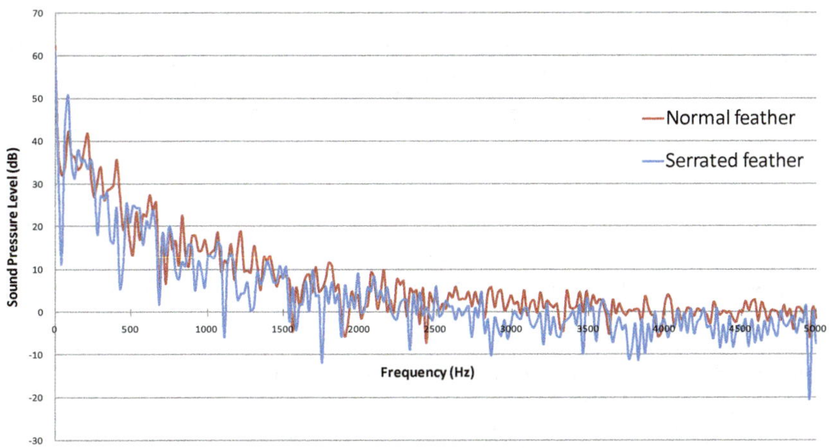

Fig. 7. SPL for back receiver (10° attack angle)

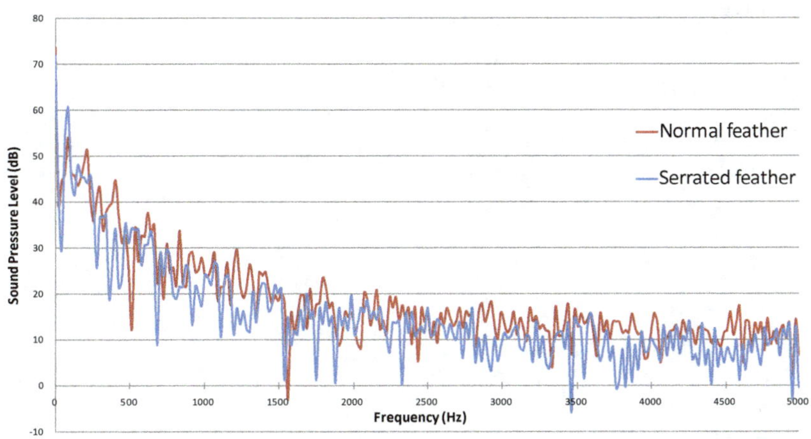

Fig. 8. SPL for top receiver (10° attack angle)

4 Conclusions

In this study, the flow field around the feathers was simulated by CFD to investigate the effect of serrations at the leading edge of owl feather on the flow structure and the SPL around the feathers. The main conclusions are as follows.

1. The separation points of the feather with the serrations move downstream because of the longitudinal vortices generated at the leading edge at angle of attack 10°. Sound pressure level was also reduced by serrations.
2. The SPL of the serrated feather is lower than that of normal feather.

References

1. Longhouse, R.E.: Vortex shedding noise of low tip speed, axial flow fans. J. Sound Vib. **53** (1), 25–46 (1977)
2. Babucke, A., Kloker, M., Rist, U.: Direct numerical simulation of the mixing layer past serrated nozzle ends. In: Seventh IUTAM Symposium on Laminar-Turbulent Transition, pp. 93–98. Springer (2009)
3. Ito, S.: Aerodynamic influence of leading-edge serrations on an airfoil in a low Reynolds number. J. Biomech. Sci. Eng. **4**(1), 117–123 (2009)
4. Ffowcs Williams, J.E., Hawkings, D.L.: Sound generation by turbulence and surfaces in arbitrary motion. Philos. Trans. R. Soc. A **246**, 321–342 (1969)
5. Lighthill, M.J.: On sound generated aerodynamically I. General theory. Proc. R. Soc. A **211**, 564–587 (1952)

Fluid-Structure Interaction Modeling of Genetically Engineered Micro-calcification at the Luminal Surface of the Aorta in Mice

Ian Kelly, Olga Savinova, and Dorinamaria Carka(✉)

New York Institute of Technology, Old Westbury, NY 11568, USA
dcarka@nyit.edu

Abstract. It is hypothesized that atherosclerosis can be influenced or promoted by early signs of arterial stiffening caused by a polydisperse micron distribution of calcified lesions on the endothelial cell layer. Optical confocal scans (Sensofar S Neox) of open luminal surfaces from mice group with overexpressed tissue-nonspecific alkaline phosphatase in endothelial cells (eTNAP), were treated to optimize resolution and used as geometrical import on a 3D fluid-structure interaction (FSI) finite element model built using COMSOL Multiphysics Software. To investigate shear stress distribution on the endothelium due to the presence of calcified nodules, the aorta was modeled as an idealized cylindrical shell, rigidly attached on the rest of the vessel wall. The calcifications were modeled as linear elastic materials with five times the Young's modulus of the endothelium, whereas the healthy part of the endothelium was modeled as a hyper elastic material representative of the nonlinear elastic behavior of biological tissue. Due to the small Reynolds number, a fully formed parabolic laminar flow was considered with mean velocity of 17 cm/s. The distribution of the wall shear stress (WSS) magnitude and the disturbed direction of the tangent components of the viscous stress at the fluid-structure interface show a reduction of magnitude by 50% around the calcified nodules. Surface plots of the Mises stress on radial and axial planes, demonstrate a highly non-uniform distribution of deviatoric stress levels and a considerable grading in shear stress transfer from the fluid-structure interface through the thickness of the endothelium.

Keywords: Aortic micro-calcification · Fluid-structure interaction · WSS distribution

1 Introduction

Atherosclerosis, the full hardening of arteries, constitutes the main culprit of cardiovascular disease and mortality. There are evidence that early signs of coronary calcium are linked and independently contribute to the acceleration of atherosclerosis and poor cardiovascular outcomes [1–5]. The exact biological mechanisms, however, are yet not fully understood, posing serious limitations on the development of therapies that target vascular calcification [6]. The Savinova group has demonstrated that transgenic upregulation of tissue-nonspecific alkaline phosphatase (TNAP) expression in endothelial cells (EC) leads to pathological micro-calcification in the intima and

G. A. Ateshian et al. (Eds.): CMBBE 2019, LNCVB 36, pp. 581–591, 2020.
https://doi.org/10.1007/978-3-030-43195-2_48

internal elastic lamina in animal models [7, 8]. Studies in the last few years in a sample of primary cardiovascular prevention patients, show that increased blood alkaline phosphatase activity (ALP) can be associated to prognosis of human coronary artery disease [9, 10].

Of greater importance is the fact that pre-existing calcification, linked to TNAP activity overexpression, can independently accelerate atherosclerosis in coronary and in mesenteric arteries in a mouse model [7, 8]. Such micro calcification can progress to a full-thickness medial calcification fitting the classical definition of Monckeberg sclerosis [11], obscuring the detection and isolation of the initial event of the pathogenesis of a particular lesion, since intimal and medial calcification often co-exist in the same vascular site in human samples. Although calcification can be detected in the elastic lamina by regular histological methods, more sensitive methods could perhaps detect earlier micro-calcifications and link their presence to increased blood ALP levels, yielding inhibition of TNAP activity as a therapeutic approach of early vascular calcification.

Experimental studies indicate that the onset of atherosclerotic plaque formation and development is triggered by interaction between the aortic wall strain, subsequent aortic stenosis and blood hemodynamics. Anatomical observation of dissected apolipoprotein E (apoE)-null mice aorta show high plaque spatial density on the aortic arch and Doppler ultrasound measurements link the plaque formation rate to reduced mean blood flow velocities and to regions of reduced mean relative wall shear stress (rWSS) [12].

Computational fluid dynamics (CFD) models using microcomputed tomography (micro-CT), in addition to ultrasound data in mice [13, 14] and human patient specific models [15, 16] have contributed great insight in identifying the role of WSS spatial differentials in both magnitude and direction and not only absolute values of WSS [17, 18], as the main parameter affecting the size and distribution of plaque formation. This inhomogeneous plaque formation, in turn triggers mechano-sensitive biochemical and biological events of importance [19]. Traditionally, low values of WSS or oscillatory WSS regions are correlated to larger plaque size formation and high values to aneurysm growth and rupture [20, 21], however values of WSS of pathogenic character are very different in mice and human aortas limiting the knowledge transfer between the two models. The fact that the spatial distribution of calcification in the mouse aorta is similar to plaque localization found in human aorta [13] provides the incent for detailed computational studies on mouse models, that pose less computational complexity [16], towards the understanding of early cardiovascular mechanosensitive pathogenic mechanisms.

The elastic effects of the vessel wall studied in Fluid-structure interaction (FSI) models mainly concerned with aneurysm growth and rupture show the limitations of the rigid wall assumption of the CFD models, with the latter overestimating WSS values and neglecting pressure differences and stresses on the wall. Extensive review of the effect of the elasticity and the deformation of the entire vessel flow can be found in and references therein [18, 20, 22, 23].

In this work, although the full elastic effect of the entire vessel is not considered, we employ a 3D fluid-structure interaction (FSI) finite element model to investigate the localized stress differences on the endothelium layer due to early signs of arterial calcification and the resulting stiffening. A realistic optical scan of the luminal surface of the aorta from a mouse, showing polydisperse calcification lesions on the micron scale are used as the geometrical reference of the fluid-structure interface boundary. It is of importance to map nonuniform redistribution of the shear stresses and WSS due to micron scale plaque on the endothelium as this effect could further be linked and promote plaque formation by triggering stress sensitive low-density lipoprotein (LDL) uptake in the endothelial cells [8].

2 Materials and Methods

2.1 Geometry Data Acquisition of TNAP Mouse Abdominal Aorta

Specimen and Surface Topology Acquisition

The specimen of aortic tissue was acquired from an archived mouse model from a previously published study [8], affected by TNAP-induced calcification and atherosclerosis (WHC-eTNAP + placebo group, described in detail in [8]). Segment of formalin-fixed thoracic aorta was cut axially and mounted luminal side up onto a drop of ClearMount™ Mounting Solution (Electron Microscopy Sciences) that resulted in sample immobilization and dehydration for 24 h prior to scanning.

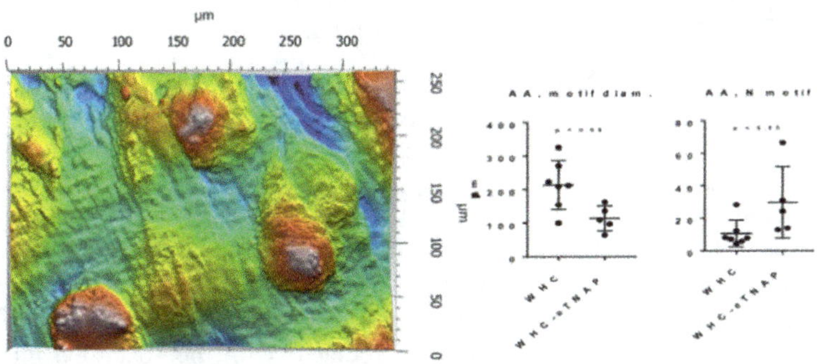

Fig. 1. Luminal surface of a representative segment of flat-mounted AA, scanned using Sensofar S Neox optical confocal microscope at 200 nm vertical resolution (left) and number and size of topological motifs between the TNAP and control groups (right).

The luminal surfaces were scanned using Sensofar S Neox optical confocal microscope at 1 μ vertical resolution using a 50X objective. The surface topographies per sample, each measuring 500 × 700 μ, were analyzed for ISO 25178-2 areal roughness parameters using SensoMap. Parameters were compared by t-test (n = 5 eTNAP, n = 6–7 control) and significant differences in the incidence and morphology of the plaques in the eTNAP group compared to control mice was observed. Specifically, the number of motifs was increased in the abdominal aorta of the eTNAP mice compared to controls (2.8 fold; $p = 0.06$), however the average area of the motifs was significantly larger in the controls compared to the eTNAP group, Fig. 1. The maximum height of the lesions increased by 74% in the abdominal aorta of the eTNAP compared to controls (34.0 ± 7.4 µm $p < 0.0099$).

Reconstruction of the Aortic Surface with Non-uniform Rational Basis Spline (NURBS)
Surface profile data of a representative scanned segment of the TNAP aorta scanned at 200 nm vertical resolution (using a 50X objective) and sampled on a rectangular grid of approximately 0.7 mm width and 3 mm length along blood flow direction were converted to a triangular mesh using the SensoMAP software, Fig. 2. The mapped mesh points were then converted and used directly as the control points for the construction of a NURBS surface using the nurbs_toolbox from MATLAB as described in [24].

Fig. 2. Triangular mesh (left) and corresponding rendered surface image (right) of flat-mounted TNAP aorta scan extracted from SensoMAP software.

To ensure smoothness for the finite element mesh stability the control point sampled grid was reduced from the initial triangular mesh points to 200×50 along the length and width respectively. Subsequently, two band-pass filters were applied to the control point grid. Initially a low pass filter was used to eliminate low spatial frequency artificial components due to height errors associated with the specimen mount (winkles). A high pass filter was applied to eliminate high frequency spatial components associated with noise on the calcified area measurements. The height data were mapped using a sigmoid function with the maximum slope corresponding to the arithmetic mean height of the TNAP surface data. The sigmoid mapping reduces roughness and texture on the surface of the calcifications and natural wrinkles on the adjacent healthy endothelium layer where there are no calcifications. Finally in order to approximate a closed channel geometry for the aorta, the reduced and filtered control sample points were mapped to a cylindrical surface representing the aorta at peak diameter of 1.2 mm [14] that covered a 55° slice of the aorta at the peak diameter, Fig. 3.

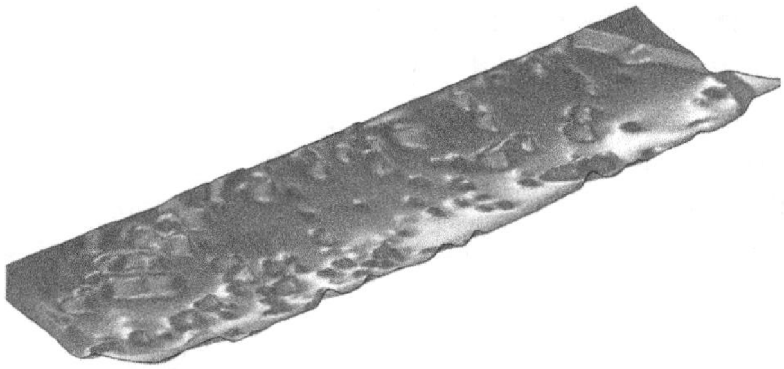

Fig. 3. NURBS surface generated by reduced and bandpass filtered control point grid, mapped on a cylindrical 55° slice at 1.2 mm peak diameter.

3 FSI Model and Results

A fully coupled FSI model was constructed based on the realistic geometries to test this endothelial roughness hypothesis. To investigate the shear stress distribution on the endothelium due to the presence of calcified nodules we represented the aorta as an idealized cylindrical shell of constant radius of 0.6 mm, of which we modeled a periodic slice. The NURBS surface was imported in COMSOL Multiphysics and a cylindrical slice of 55° domain with a 0.6 mm radius and 2 mm length was created and used to partition the NURBS surface in order to isolate protruding calcified lesions and trim the edges of the imported and filtered geometry. It is assumed that the isolated calcified lesions with heights ranging from 5–40 μ are imbedded on top of a smooth and uniform 3.8 μ thick cylindrical layer corresponding to the healthy endothelium cell layer. In this pilot study, the entire vessel wall is not modeled and the endothelium is considered to be rigidly attached on the rest of the vessel wall so displacements on the outer boundary are fixed and no radial stretching is considered. It is to be noted that the even the lesions with the maximum height of 40 μ create a localized maximum stenosis of 3%, so no considerable effect on blood flow resistance according to Poiseuille's equation is anticipated in the unbranched portion of the simulated aorta.

The calcified lesions were modeled as linear elastic materials with five times the Young's modulus of the endothelium layer. The healthy part of the endothelium layer was modeled as a hyper-elastic (Neo-Hookean) material with a Young's modulus of 2 kPa [25] and a Poisson's ratio of 0.36 [26]. The difference in the material models and stiffness properties between the calcified nodules was a first step in the analysis of the very complex heterogeneous properties of the calcified endothelium. Due to small Reynolds number, a fully formed parabolic laminar flow with mean velocity 17 cm/s was considered and applied incrementally as the inlet boundary condition on the fluid domain and zero pressure on the outlet of the modeled segment. Since the considered vessel diameter was larger than 1 mm, the effect of the red blood cell region in the flow was neglected, and the blood was modeled as a Newtonian fluid with a constant viscosity of 0.00282 Pa · s. Periodic flow and cyclic symmetry boundary conditions were applied in the hoop direction of the vessel to simulate a closed channel flow. The

Navier-Stokes equations were solved in the fluid domain and Newton's equation on the calcified lesions and endothelium domains.

Results

Fully formed parabolic laminar flow with mean velocity 17 cm/s was established and a surface plot of the velocity profile and the distribution of the von Mises stress is shown in Fig. 4 with blood flow direction along increasing vessel axis. Values of Mises stress on the fluid-structure interface boundary vary from 5 Pa on smooth regions and on regions with lesions with mean diameter less than 50 μ and height less than 10 μ to 20 Pa on regions of congregated lesions with spatial extension in the order of 200 μ and more with a mean height in the order of 20 μ. For distributed intermediate lesion size in the order of 100 μ and maximum height of 30–40 μ the values of Misses stress rise to 40 Pa. The pressure distribution along the fluid and solid domains is also shown in Fig. 4. The simulated length of the aorta was approximately 2 times the diameter and the pressure drop on the aortic segment modeled is on the order of 23 Pa in accordance $\Delta p = 32\,\mu VL/D^2 = 23\,\mathrm{Pa}$ (0.16 mm Hg)^2, where μ, V, L and D are the dynamic viscosity, mean velocity, length and vessel diameter respectively.

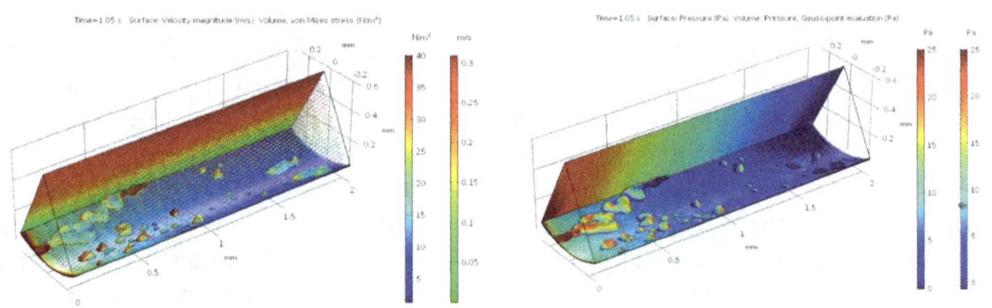

Fig. 4. Fluid-structure interaction model results: Surface and arrow plot of fully formed parabolic laminar flow with mean velocity 17 cm/s and volume plot of von Mises stress distribution (left) and corresponding surface and volume pressure distribution on the fluid and solid domain (right).

Volume distribution and contour plots of the Mises stress on radial and axial planes on the heterogeneous endothelium layer demonstrate a highly non-uniform distribution of the deviatoric stress levels as shown in Fig. 5. It is observed that values of the deviatoric stress greatly vary in different proportions around the calcified lesions. In the axial direction along the blood flow a reduction in value of 20–30% is observed around lesions with diameters of 50 μ and less than 10 μ height, whereas the reduction of the mises stress escalates to up to 75% in regions around the larger diameter lesions of intermediate height. Significant changes are also observed on radial planes, perpendicular to blood flow direction. Symmetry in von Mises values is present in this direction around regions of standalone lesions, however a reduction of 30% is observed when lesions of different size are at a radial distance of 100 μ or less. In Fig. 5 a considerable grading in deviatoric stress transfer from the fluid-structure interface through the thickness of the endothelium and calcified lesions can be observed even

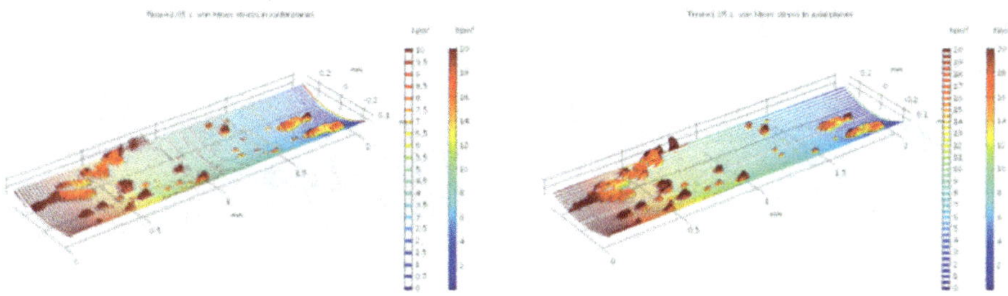

Fig. 5. Volume plot of non-uniform Mises stress distribution along radial (left) and axial (right) planes

when a thin material layer of 3.8 μ was included in the calculations. In sub-endothelial regions below larger calcification lesions the Mises stress values are of approximately the same order or slightly higher compared to the values of stress at the fluid interface boundary, however a reduction of up to 50% in stress distribution is observed on the endothelium below the smallest regions, pointing towards the fact that very early signs of microcalcification can significantly change the mechano-transduction properties of the endothelial cells.

The distribution of the WSS magnitude and the disturbed direction of the tangent components of the viscous stress at the fluid-structure interface are shown in Fig. 6 and further demonstrate the hypothesis that the presence of the micro-calcifications introduces oscillatory disturbances in the direction of the WSS vector in addition to a reduction in the absolute value. Around standalone small lesions a reduction in the order of 30–50% is observed in the absolute value of WSS when compared to the uniform value of 2.7 Pa smooth regions of the vessel, with the largest reduction observed behind the lesions with respect to the blood flow direction. Regions of congregated lesion distributions and lesions with highly irregular boundaries present more complexity on the WSS value distribution, with the magnitude of WSS being reduced to zero. Similar behavior between lesion distribution and oscillatory response is also observed in the tangent viscous stress direction, also seen in Fig. 6. Oscillations in viscous stress direction although not significant, still exist even around the smallest featured standalone lesion of less than 10 μ diameter and become increasingly prominent on regions of interacting lesions of 50–100 μ diameter and around larger lesions with irregular boundaries. The effect of the microcalcification on the hemodynamics is also shown in Fig. 6. Disturbances in the vorticity field are observed around intermediate diameter standalone lesions with maximum height and on regions of axially congregated small features with a radial distance distribution in the order of 100 μ or less.

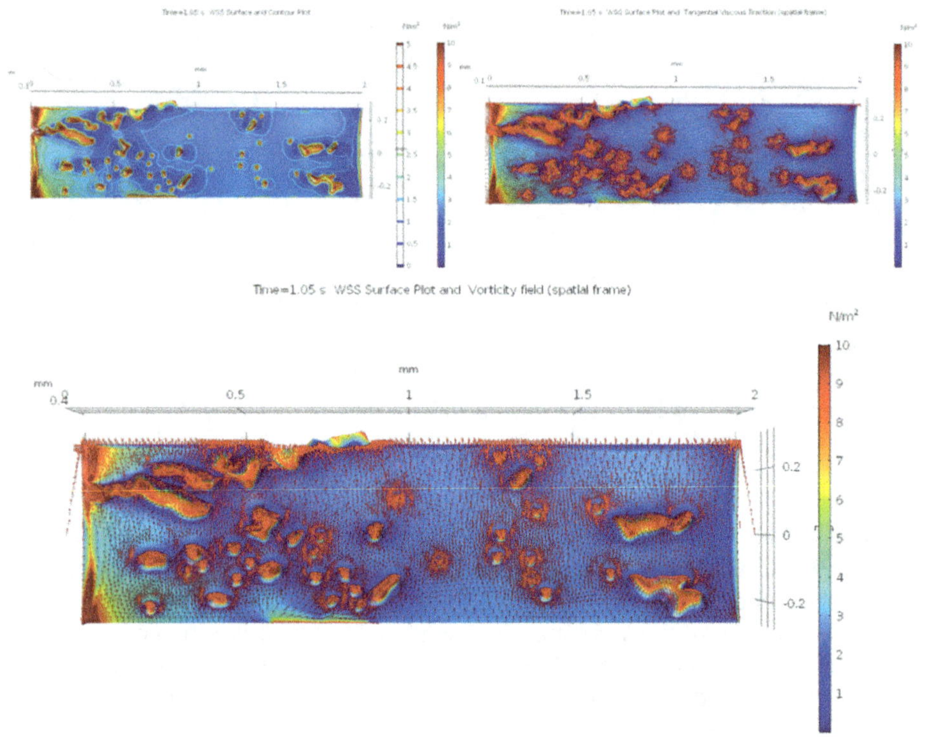

Fig. 6. WSS magnitude contour and surface plot on the fluid-structure interface (top-left), surface plot of WSS magnitude and arrow plot of the tangent component of the viscous shear stress vector (top-right) and WSS magnitude and vorticity field (bottom).

4 Discussion

It is well established that coronary artery calcium (CAC) score is an independently predictor of poor cardiovascular outcomes [1–5] however pathophysiologic mechanisms of arterial calcification and the role of calcium in initiation and progression of early atherosclerotic lesions are yet not fully understood, posing limitations on the development of treatment strategies targeting early stages of vascular calcification. In this study, a simplified 3D FSI finite element model was employed to investigate any effects of micron scale calcium lesion distribution on WSS spatial differentials and shear stress values on the ED layer of mice aorta with transgenic upregulation of TNAP.

It was found that under conditions of fully developed laminar flow, values of the deviatoric stress greatly vary in different proportions around the calcified lesions. Along blood flow direction a reduction in deviatoric stress value of 20–30% is observed around lesions with diameters of up to 50 μ and height less than 10 μ, whereas the reduction of the mises stress escalates to up to 75% in regions around the larger diameter lesions of intermediate height of 20 μ. A reduction in the order of

30–50% is observed in the absolute value of WSS when compared to the uniform value on smooth regions of the vessel around standalone small lesions, with the largest reduction observed behind the lesions with respect to the blood flow direction. Oscillations in WSS appear at the smallest featured standalone lesion of less than 10 μ diameter and become increasingly prominent on regions of interacting lesions of 50–100 μ diameter by approximately 40–50° when compared to uniform flow direction on the smooth parts of vessel.

Although the model predictions of deviatoric stress and WSS redistribution and percentage reduction, point towards a significant effect of early micron scale calcium lesions and offers a general classification of the polydispersity of the lesions with respect to their effect on mechanical parameters which can be tied to biologically relevant mechanisms of calcification progression, the present model has serious limitations for the exact values to be considered further. First the images were taken from a dried tissue after significant, unknown shrinkage had occurred and histological data of the specific mouse showed that significant calcification exists deeper in the intima and extends in the median layer. Here we considered and modeled the surface roughness optical data, limited on the ED layer attached to a rigid vessel, ignoring residual stresses and even redistribution of the roughness geometry due to deformation. The Neo-Hookean material model used for the ED while might approximate the dried tissue cannot capture the complexity of the actual aortic vessel [27, 28] and more elaborate models should be used [29–34] in conjunction with microtomography for the full 3D caption of the microcalcification in order to draw conclusion for the effect of micro-calcification on mechanics of the full vessel.

Acknowledgments. "We thank Dr. James Scire for help with image data processing and Anton Mararenko for help with Sensofar Scans. This work was supported in part by the NIH grant 1R56HL131547-01A1".

Compliance with Ethical Standards. "Animal studies were approved by the Institutional Animal Care and Use Committees (IACUC) of Sanford Research (Sioux Falls, SD) and the New York Institute of Technology College of Osteopathic Medicine (Old Westbury, NY) and complied with the National Institutes of Health guidelines for humane treatment of laboratory animals."

References

1. Polonsky, T.S.: Coronary artery calcium score and risk classification for coronary heart disease prediction. JAMA **303**(16), 1610 (2010)
2. Folsom, A.R.: Coronary artery calcification compared with carotid intima-media thickness in prediction of cardiovascular disease incidence: the Multi-Ethnic Study of Atherosclerosis (MESA). Arch. Intern. Med. **168**(12), 1333 (2008)
3. Taylor, A.J., Bindeman, J., Feuerstein, I., Cao, F., Brazaitis, M., O'Malley, P.G.: Coronary calcium independently predicts incident premature coronary heart disease over measured cardiovascular risk factors. J. Am. Coll. Cardiol. **46**(5), 807–814 (2005)

4. Greenland, P., LaBree, L., Azen, S.P., Doherty, T.M., Detrano, R.C.: Coronary artery calcium score combined with Framingham score for risk prediction in asymptomatic individuals. JAMA **291**(2), 210 (2004)
5. Vliegenthart, R., et al.: Coronary calcification improves cardiovascular risk prediction in the elderly. Circulation **112**(4), 572–577 (2005)
6. Wanner, C., Amann, K., Shoji, T.: The heart and vascular system in dialysis. Lancet **388** (10041), 276–284 (2016)
7. Savinov, A.Y., Salehi, M., Yadav, M.C., Radichev, I., Millán, J.L., Savinova, O.V.: Transgenic overexpression of tissue-nonspecific alkaline phosphatase (TNAP) in vascular endothelium results in generalized arterial calcification. J. Am. Heart Assoc. **4**(12), e002499 (2015)
8. Romanelli, F., et al.: Overexpression of tissue-nonspecific alkaline phosphatase (TNAP) in endothelial cells accelerates coronary artery disease in a mouse model of familial hypercholesterolemia. PLoS ONE **12**(10), e0186426 (2017)
9. Panh, L., et al.: Association between serum alkaline phosphatase and coronary artery calcification in a sample of primary cardiovascular prevention patients. Atherosclerosis **260**, 81–86 (2017)
10. Ndrepepa, G., et al.: Alkaline phosphatase and prognosis in patients with coronary artery disease. Eur. J. Clin. Invest. **47**(5), 378–387 (2017)
11. Micheletti, R.G., Fishbein, G.A., Currier, J.S., Fishbein, M.C.: Mönckeberg sclerosis revisited: a clarification of the histologic definition of Mönckeberg sclerosis. Arch. Pathol. Lab. Med. **132**(1), 43–47 (2008)
12. Tomita, H., Hagaman, J., Friedman, M.H., Maeda, N.: Relationship between hemodynamics and atherosclerosis in aortic arches of apolipoprotein E-null mice on 129S6/SvEvTac and C57BL/6J genetic backgrounds. Atherosclerosis **220**(1), 78–85 (2012)
13. Suo, J., Ferrara, D.E., Sorescu, D., Guldberg, R.E., Taylor, W.R., Giddens, D.P.: Hemodynamic shear stresses in mouse aortas: Implications for atherogenesis. Arterioscler. Thromb. Vasc. Biol. **27**(2), 346–351 (2007)
14. Huo, Y., Guo, X., Kassab, G.S.: The flow field along the entire length of mouse aorta and primary branches. Ann. Biomed. Eng. **36**(5), 685–699 (2008)
15. Huo, Y., Wischgoll, T., Kassab, G.S.: Flow patterns in three-dimensional porcine epicardial coronary arterial tree. AJP Hear. Circ. Physiol. **293**(5), H2959–H2970 (2007)
16. Zhong, L., Zhang, J.M., Su, B., Tan, R.S., Allen, J.C., Kassab, G.S.: Application of patient-specific computational fluid dynamics in coronary and intra-cardiac flow simulations: challenges and opportunities. Front. Physiol. **9**, 742 (2018)
17. Kulcsar, Z., Ugron, A., Marosfoi, M., Berentei, Z., Paal, G., Szikora, I.: Hemodynamics of cerebral aneurysm initiation: the role of wall shear stress and spatial wall shear stress gradient. Am. J. Neuroradiol. **32**(3), 587–594 (2011)
18. Liu, J., Shar, J.A., Sucosky, P.: Wall shear stress directional abnormalities in BAV aortas: toward a new hemodynamic predictor of aortopathy? Front. Physiol. **9**, 1–9 (2018)
19. Lu, D., Kassab, G.S.: Role of shear stress and stretch in vascular mechanobiology. J. R. Soc. Interface **8**(63), 1379–1385 (2011)
20. Hsu, M.-C., Bazilevs, Y.: Blood vessel tissue prestress modeling for vascular fluid-structure interaction simulation. Finite Elem. Anal. Des. **47**, 593–599 (2011)
21. Cheng, C., Tempel, D., Van Haperen, R., Van Der Baan, A.: Atherosclerotic lesion size and vulnerability are determined by patterns of fluid shear stress. Circulation **113**, 2744–2753 (2006)
22. Yahya, M.: Three dimensional finite-element modeling of blood flow in elastic vessels: effects of arterial geometry and elasticity on aneurysm growth and rupture. Ryerson University (2010)

23. Trachet, B., et al.: An animal-specific FSI model of the abdominal aorta in anesthetized mice. Ann. Biomed. Eng. **43**(6), 1298–1309 (2015)
24. Hart, R.T.: Generation of lofted NURBS curves for 3D model generation with COMSOL Multiphysics® (2013)
25. Kataoka, N., et al.: Measurements of endothelial cell-to-cell and cell-to-substrate gaps and micromechanical properties of endothelial cells during monocyte adhesion. Proc. Natl. Acad. Sci. U. S. A. **99**(24), 15638–15643 (2002)
26. Trickey, W.R., Baaijens, F.P.T., Laursen, T.A., Alexopoulos, L.G., Guilak, F.: Determination of the Poisson's ratio of the cell: recovery properties of chondrocytes after release from complete micropipette aspiration. J. Biomech. **39**(1), 78–87 (2006)
27. Humphrey, J.D.: Cardiovascular Solid Mechanics: Cells, Tissues. And Organs. Springer, New York (2002)
28. Holzapfel, G.A., Ogden, W.R.: Constitutive modelling of arteries. Proc. R. Soc. A. **466**, 1551–1597 (2010)
29. Schmid, H., et al.: Impact of transmural heterogeneities on arterial adaptation: application to aneurysm formation. Biomech. Model. Mechanobiol. **9**(3), 295–315 (2010)
30. Humphrey, J.D., Holzapfel, G.A.: Mechanics, mechanobiology, and modeling of human abdominal aorta and aneurysms. J. Biomech. **45**(5), 805–814 (2012)
31. Holzapfel, G.A., Ogden, R.W.: Modelling the layer-specific three-dimensional residual stresses in arteries, with an application to the human aorta. J. R. Soc. Interface **7**(46), 787–799 (2010)
32. Ateshian, G.A., Albro, M.B., Maas, S., Weiss, J.A.: Finite element implementation of mechanochemical phenomena in neutral deformable porous media under finite deformation. J. Biomech. Eng. **133**(8), 081005 (2011)
33. Thon, M.P., et al.: A multiphysics approach for modeling early atherosclerosis. Biomech. Model. Mechanobiol. **17**(3), 617–644 (2018)
34. Xiong, L., Chui, C.K., Fu, Y., Teo, C.L., Li, Y.: Modeling of human artery tissue with probabilistic approach. Comput. Biol. Med. **59**, 152–159 (2015)

Experimental Evaluation of Pad Degradation of Helmet for American Football and Its Application to Numerical Design

Yoshiki Umaba[1], Atsushi Sakuma[2(✉)], and Yuelin Zhang[3]

[1] Graduate School, Kyoto Institute of Technology, Matsugasaki, Sakyo-ku, Kyoto 606-8585, Japan

[2] Division of Textile Science, Kyoto Institute of Technology, Matsugasaki, Sakyo-ku, Kyoto 606-8585, Japan

sakuma@kit.ac.jp

[3] Faculty of Science and Technology, Sophia University, 7-1 Kioi-cho, Chiyoda-ku, Tokyo 102-8554, Japan

Abstract. American football is a sport that requires high safety precaution for players because it entails very fast and powerful movements. The players' helmets are one of the most important means of ensuring their safety, and so the mechanical design of these helmets is extremely difficult for the manufacturer. Therefore, in this report, the mechanical behavior of the pad material is measured experimentally and then it is formulated by theoretical simulation for design of the helmet numerically. Then, an analysis of the collision problem of the helmet is carried out using the characteristics of the pad material obtained in these mechanical tests. Numerical analyses are carried out by changing the material characteristics and movement of physical model while keeping the structure of the helmet and head constant. LS-Dyna is used in the numerical simulation because of its availability in crash analysis. In the analysis results, it was first confirmed that the optimum characteristics of the pad material can be defined by the mechanics of plateau stress. With these results, the concept of a usage limit due to deterioration of the pad material is discussed in addition to the suggestion of helmet specifications according to the physique and exercise capacity of the football players.

Keywords: Helmet · Shock absorption · Foam material · Constitutive equation · Elasticity · Plateau stress · Softening rate · Indentation test · FEM

1 Introduction

Helmets are one of the most effective means of reducing impacts on the human head, and as such are required to have a highly technical design of dynamics. The American football helmet is a product that is particularly required to reduce the effects of impact to the head, and it has attracted great deal of many researches. As a method of realizing the buffering effect of a helmet, it is the common practice to use materials with high shock absorption capabilities, and foam material is frequently used because of its high performance in this regard. In detailed investigations of the mechanical properties of

G. A. Ateshian et al. (Eds.): CMBBE 2019, LNCVB 36, pp. 592–596, 2020.
https://doi.org/10.1007/978-3-030-43195-2_49

foam materials, it is standard to use the uniaxial compression test method [1]. However, the ball indentation test procedure [2] has also been found applicable in a variety of studies [3–11]. In particular, the foam material used for helmets has a head-like shape, so the indentation test is very advantageous in evaluating the mechanical properties of helmet products manufactured with this material.

Therefore, this paper presents a study on the optimization of the mechanical properties of the foam materials used in American football helmets by using the indentation test method that can quantitatively evaluate the mechanical characteristics of actual products with various shapes. With regard to the American football helmet, the foam material used for shock absorption in the forehead of the football helmet is evaluated as the object of the optimization problem. To reduce the analysis load of the optimization problem, numerical analyses using the finite element method (FEM) are used for the evaluation.

2 Experimental Evaluation of Helmet Pad

2.1 Pad Foam of American Football Helmets

There are various types of helmets used in American football, but the model shown in Fig. 1(a), which is one of the most popular products in the sport, was chosen as the object of this study. This model is the Speed Helmet, produced by Riddell Sports Group, Inc. Its front pad of foam material is shown in Fig. 1(b).

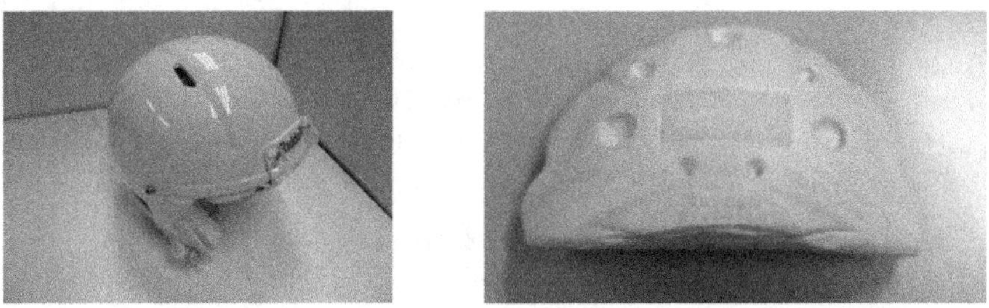

(a) (b)

Fig. 1. The helmet used as a specimen (a) and the pad part of its forehead (b).

To evaluate the mechanical characteristics of this pad nondestructively, indentation test measurements were carried out as shown in Fig. 2(a). The indentation test apparatus used is that of YAWASA, a softness evaluation system, which is manufactured by Tech Gihan Co., Ltd. of Japan. Here, the MSES-05 series (rated load 5N) is adopted to evaluate the mechanical properties of the pad foam.

The indentation depth δ-reaction forces F curve obtained in this test is shown in Fig. 2(b). Here, the solid red line indicates the experimental result obtained by the indentation test, and the black solid line indicates the ideal curve of the Hertzian elastic contact theory, which can be formulated as follows:

$$F = \frac{4}{3}\frac{E}{1-\upsilon^2}\left(\frac{\phi}{2}\right)^{\frac{1}{2}}\delta^{\frac{3}{2}} \tag{1}$$

Here, E, v, and ϕ represent the Young's modulus, Poisson's ratio, and diameter of ball indenter, respectively.

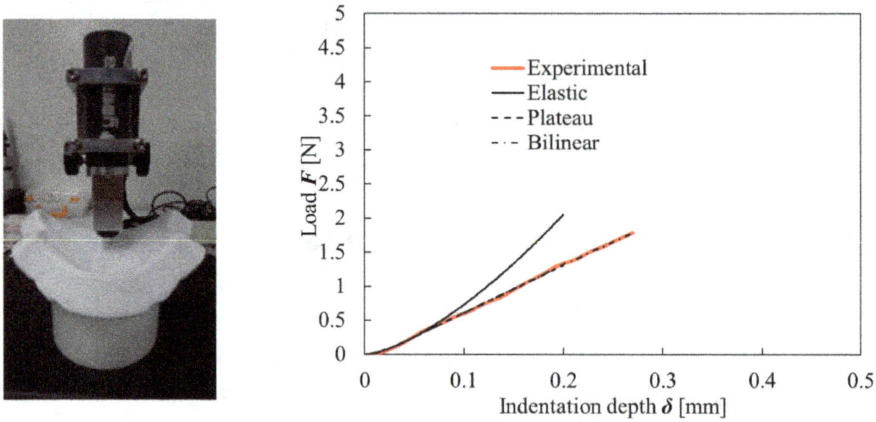

(a) (b)

Fig. 2. Indentation test device (a) and the experimental results (b)

As shown in Fig. 2(b), it is difficult to evaluate the deformation characteristics of a foam pad part with the Hertzian contact theory. Therefore, Eq. (2), first presented by Nagai *et al.* [2], is also applied to evaluate the deformation of the foam material.

$$F = (1-\zeta)\frac{4}{3}\frac{E}{1-v^2}\left(\frac{\phi}{2}\right)^{\frac{1}{2}}\delta^{\frac{3}{2}} + \zeta\left\{\frac{\pi}{2}\phi\sigma_{\mathrm{p}}\delta - \frac{\pi^3\phi^2\sigma_{\mathrm{p}}^3\left(1-\upsilon^2\right)^2}{48E^2}\right\} \tag{2}$$

Here, ζ and σp indicates the softening rate and plateau stress, respectively. The plateau stress σp, which is equivalent to the elastic limit, is a standard property by which to analyze the softening rate ζ, which is 0 for an ideal elastic body and 1 for a body without further stress increase. In the results shown in Fig. 2(b), the evaluated softening rate is x = 0.998 for the foam of the helmet pad. An ideal plateau body has a softening rate of $\zeta = 1$, which is in good agreement with the evaluated results. This means that the foam pad material has a large capacity to absorb shock behavior.

3 Numerical Analysis

3.1 Head and Helmet Model

An analysis model including the head is considered for mechanically designing the collision characteristics of the helmet with high precision. Figure 3(a) shows the FE model for the collision analysis. In this figure, a green sphere in the center represents the model of a human head, and the red spherical shell around the head model represents the outer shell of the helmet. The light blue trapezoid between these two models is the foam material of the helmet's forehead pad. These three parts move to right-hand side of this figure at the same initial velocity, and collide with a rigid wall.

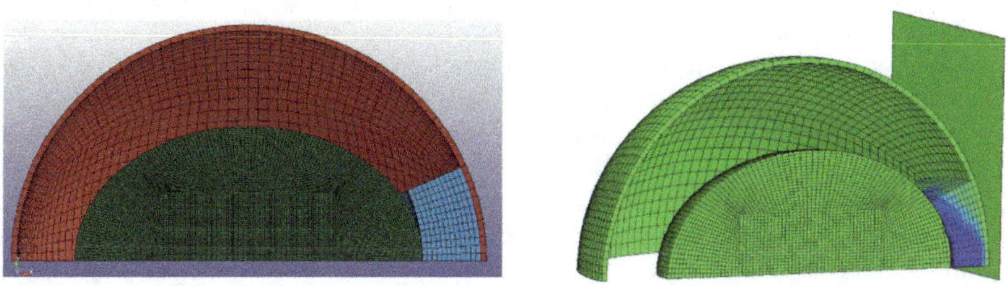

(a) (b)

Fig. 3. FE analysis model of helmet collision problem (a) and an analysis result (b)

3.2 Collision Analysis Model of Head and Helmet

Figure 3(b) shows an example of the FE analysis of the helmet collision. In this analysis, the parts of head and helmet shell are modeled with rigid properties for simplification of the problem and improvement of its analysis accuracy. By using the analytical model shown here, it is possible to investigate the influence of the mechanical properties of foam material on the helmet-collision problem with high sensitivity.

4 Conclusion

To establish the technology for precisely and mechanically designing American football helmets, which require an extremely high buffering performance; material testing methods for helmet pad parts, and a helmet design technology using FE modeling of these results were presented. These results can promote the development of superior pad materials for American football helmets.

References

1. Azusawa, N., Sakuma, A., Nagaki, S.: Constitutive equation of low-density porous materials incorporating structure transformations induced by cellular crush. Trans. Jpn. Soc. Mech. Eng. Ser. C **76**(761), 96–101 (2010)
2. Nagai, N., Takayama, E., Sakuma, A.: In-elasticity evaluation and its design technique for soft materials by using ball indentation test. In: Proceedings of the ASME International Mechanical Engineering Congress and Exposition, ASME, Tampa, USA (2017). https://doi.org/10.1115/IMECE2017-70386
3. Tani, M., Sakuma, A., Shinomiya, M.: Evaluation of thickness and Young's modulus of soft materials by using spherical indentation testing. Trans. Jpn. Soc. Mech. Eng. Ser. A **75**(755), 901–908 (2009)
4. Tani, M., Sakuma, A.: Applicability evaluation of Young's modulus measurement using equivalent indentation strain in spherical indentation testing for soft materials. Trans. Jpn. Soc. Mech. Eng. Ser. A **76**(761), 102–108 (2010)
5. Sakuma, A.: Softness measurement technics of human skin by indentation devices imitating palpation. In: Proceedings of the ASME International Mechanical Engineering Congress and Exposition, ASME, San Diego, USA (2013). https://doi.org/10.1115/IMECE2013-66258
6. Sakuma, A., Sango, Y.: Viscoelasticity measurement of softness by indentation devices for evaluation of human skin. In: Proceedings of the ASME International Mechanical Engineering Congress and Exposition, Montreal, Canada (2014). https://doi.org/10.1115/IMECE201438261
7. Zhimeng, L., Sakuma, A.: Error reduction and performance improvement of palpation for human soft tissues based on 3D indentation system. In: Proceedings of the ASME International Mechanical Engineering Congress and Exposition, Montreal, Canada (2014). https://doi.org/10.1115/IMECE201439736
8. Sakuma, A.: Handy-device development of softness measurement by indentation imitating palpation for evaluation of elasticity and visco-elasiticity. In: Proceedings of the ASME International Mechanical Engineering Congress and Exposition, Houston, USA (2015). https://doi.org/10.1115/IMECE2015-51589
9. Sakuma, A., Igarashi, K.: Behavior evaluation of deformation, damage and fracture of biological soft tissue by using indentation test. In: Proceedings of the ASME International Mechanical Engineering Congress and Exposition, Phoenix, USA (2016). https://doi.org/10.1115/IMECE2016-66672
10. Sakuma, A. Shirai, Y.: Palpation-like devices of ball indentation for softness evaluation of biological tissue. In: Proceedings of the ASME International Design Engineering Technical Conferences and Computers and Information in Engineering Conference, Cleveland, USA (2017). https://doi.org/10.1115/DETC2017-67010
11. Shirai, Y., Zhimeng, L., Sakuma, A.: Position identification of inclusions in soft objects by indentation testing. In: Proceedings of the ASME International Mechanical Engineering Congress and Exposition, Tampa, USA (2017). https://doi.org/10.1115/IMECE2017-70546

Creation of Categorical Mandible Atlas to Benefit Non-Rigid Registration

Heather Borgard[1(✉)], Amir H. Abdi[1], Eitan Prisman[1,2], and Sidney Fels[1]

[1] University of British Columbia, Vancouver, Canada
hbrgrd@gmail.com

[2] Vancouver General Hospital, Vancouver, Canada

Abstract. The adoption of statistical shape modeling into the realm of medical technologies has been explored with applications ranging from medical image analysis to registration techniques. Here, we have created 6 category-specific mandibular mesh templates, i.e. atlases, to enable subject-specific deformable registration methods. Our approach considers the characteristics of individuals including gender and dentition, compiling specific atlases based on these categories. Our objective is to improve deformable registration techniques through these category-dependent templates, which have not previously been investigated for the human mandible. We have evaluated non-rigid transformation techniques both with and without the use of the mandibular atlases and validated the results by comparison metrics between the surface of the resulting meshes and the patient model by Hausdorff distances and volumetric overlap scores. Our results showed no statistically significant difference between the average maximum Hausdorff 95 distance for cases using templates that matched the category of the test mandible when compared to templates that directly did not match $F(2, 72) = 0.64$, $p = 0.53$. The volumetric overlap scores offered more promising results where the matched group had a statistically significant greater mean than the unmatched group $F(2, 33) = 12$, $p = 0.00012$. Furthermore, the overlap percentage for matched cases ($M = 74.08$, $SD = 8.91$) was higher than unmatched cases ($M = 59.42$, $SD = 6.05$), $t(22) = 4.72$, $p < .001$. Together these results indicate potential in the use of categorical templates for improving the non-rigid registration.

Keywords: Morphometrics · Medical atlases · Non-rigid registration

1 Introduction

Surgical planning using computer assisted technology offers the potential to improve and predict patient outcomes, decrease operative time, and effectively train surgeons for various cases. Along with these benefits comes a need for real-time simulations, with models that capture the unique anatomy of an individual patient as well as defected areas. Rather than building simulations for each specific case, creating more robust and efficient subject-specific modeling has the

G. A. Ateshian et al. (Eds.): CMBBE 2019, LNCVB 36, pp. 597–607, 2020.
https://doi.org/10.1007/978-3-030-43195-2_50

potential to limit the simulation development time and to use computer assisted simulations intraoperatively. Subject-specific modeling of this nature typically involves some type of model registration technique, the process of transforming one object to another object's configuration through either rigid or non-rigid transformations. Rigid-based transformations capture translational and rotational movements preserving the original shape and size of the translated object, while deformable or nonrigid-based transformations include affine transforms, shearing, and scaling which allow the shape to morph.

The field of non-rigid registration is especially useful for medical applications because it allows the shape matching of features, sizes, and other complex variations. While non-rigid registration is often intricate due to the process of determining the deformation field, statistical shape modeling (SSM) offers a systematic method for investigating transformations through the analysis of a dataset's variation. SSM can be used to create mean shape templates which are also referred to as atlases. Atlases function to provide a foundation for complex topographies and to capture the variation properties across a wide range of individuals. Previous studies have developed atlases from large datasets that are not categorically specific in order to benefit automatic segmentation, develop surgical planning guides, and create dense models where data is otherwise sparse or missing [1–3,15].

The main purpose of this work is to evaluate if the use of categorically-dependent mandibular atlases improve deformable registration by resulting in a closer correspondence of the template to the final target specimen. The eventual application for this work is to improve mandibular reconstructive surgery planning through subject-specific modeling of the main structures associated with mastication, with resulting simulations that can easily be registered for each unique patient case.

1.1 Characteristics

In this section, we highlight the two classifiers of gender and denture profile that contribute to changes in the morphology of the human mandible. There have been several investigations into these variations through both static measurements of the mandible and from the field of forensic medicine. The following information is not provided to test the reported findings on mandible morphological variations, but rather to signify the evidence of these morphological differences between both genders and denture profiles.

Chrcanovic et al. investigated mandibles from 80 Brazilian modern humans, finding morphological differences between in dentae individuals and endentulous individuals among males and females [4]. Their results showed that dental status could significantly alter the morphology of the mandible, and that it had a greater influence of the mandible anatomy than gender differences.

Another study from the Journal of Clinical and Diagnostic Research [5] analyzed 250 undamaged human adult mandibles from the South Indian population and found statistically significant differences between male and female mandibles for parameters of bigonial breadth, bicondylar breadth, and mandibular length.

In the field of forensic odontology, it has been reported that maxillary and mandibular canines and mandibular second molars can provide accurate gender determination in 77% of cases [6]. Additionally, morphological features of the mandibular gonial angle, contour of the lower border of the mandible, shape of mental region, and ramus dimensions all have been found to differentiate significantly between genders. Bizygomatic breadth was the most dimorphic variable with 80% diagnostic accuracy [6].

1.2 Statistical Shape Modeling

A similar work to our template generation approach is one proposed by Raith et al. where they used principal component analysis (PCA) generated mandibular templates categorized according to defect classifications [3]. The six defected mandible models were composed of a combination of the first 30 principal components and found to be sufficiently well representative of the missing geometry with potential applications for clinical practice.

When describing morphology variations across patients' anatomy, clinical value is an important consideration. To best address this issue we have incorporated the following insurances into our work: working with a physician to define key landmarks and features as well as end applications, using a source that is comprised entirely from patient data, and carefully defining exclusion criteria as outlined in the data acquisition section of this paper.

We utilize PCA to describe a large dataset of mandibles by components, an approach that has evolved from geometric morphometrics. A general Procrustes analysis (GPA) was performed on all mandibles in the dataset to exclude all information such as position, orientation, and scaling from the data so that only the shape information remains.

1.3 Nonrigid Registration

A similar research to our current registration approach [2] used patient-specific atlases and non-rigid registration to optimize corrective surgeries for asymmetric patients. This study created custom patient-specific atlases to eliminate surgical planning subjectivity and to increase reproducible treatment planning for types of surgeries where significant data is missing. Our approach categorizes multiple mandible templates based on groups of individual characteristics for the potential of more clinically reliable surgical simulations.

Nonrigid or deformable registration includes affine transformations of shearing and scaling applications to an object. This transforms 3D data sets into same coordinate system as other sets through morphing and articulating deformations and is preferred for clinical applications where individual features are important for attachment sites, simulation, identifying landmarks, and medical image segmentation [7].

With the Iterative Closest Point (ICP) method, a target mesh is kept fixed, as the source mesh is transformed to match the target configuration. This process is capable through correspondence of points across each mesh, evaluation of their spacial proximity, and a optimization function to bring these points within as close as possible spacial proximity to one another.

Given two datasets points:

$$A = a_1, ..., a_n \tag{1}$$

$$B = b_1, ..., b_n \tag{2}$$

Find the translation t and rotation R that minimizes:

$$\sum_{i=1}^{n} ||R(a_i) + t - b_i||^2 \tag{3}$$

2 Methods

2.1 Data Acquisition

The total data set was comprised of 80 mandibular meshes segmented from CT imaging. Gender information was only available for 48 of the mandibles. All meshes were broken down into 6 template categories (37 with teeth, 37 without teeth, 11 females with teeth, 11 females without teeth, 11 males with teeth, 11 males without teeth) and a remaining 6 reserved for testing each of the respective categories. We also created a universal template made from all the mandibles for comparison.

One subset of the data was provided by Vancouver General Hospital, Canada and consisted of 48 segmented mandibles that have been differentiated by gender and dentition (12 males with teeth, 12 females with teeth, 12 males without teeth, and 12 females without teeth). The study was approved by the UBC Clinical Research Ethics Board, certificate number (H17-02922). Additional data sets were used to create a larger analysis, however these did not provide information regarding gender classification. These were 7 segmented mandibles published from Wallner et al. [8] and 25 mandibles from The Cancer Imaging Archive (TCIA) [9]. TCIA test set contains open source CT scans, some of which have been segmented into bony structures [10].

While all mandible mesh data was segmented prior, exclusion criteria included any deficits, poor segmentation (where much of the mesh was incomplete), and metal fillings or other artifacts that led to a noisy mesh. In addition, we excluded mandibles where denture profile could not be determined (those that were segmented to not include teeth with a flat aveolar process).

2.2 Template Creation

We identified 33 distinct landmarks on the human mandible that represented features present on all mandibles in the study. These landmarks were discussed with an an Otolaryngology surgeon and most matched the attachment sites for muscles as well as fossas for vessels and nerves. Preliminary studies determined that 33 landmarks captured adequate information to perform a analysis of variation. We verified the amount of landmarks by calculating pairwise distances for increasingly dense sets of landmarks. To counteract intra-observer error associated with manual landmark selection, each landmark was selected 3 times and averaged on the mandible's surface.

Description of Landmark	Left		Right
Anterior point of condylar process	1		23
Medial point of condylar process	2		22
Posterior point of condylar process	3		21
Lateral point of condylar process	4		20
Deepest point of mandibular notch	5		19
Maxima point of coronoid process	6		18
Anterior end of coronoid process	7		17
Mandibular foramen	8		16
Retromolar fossa	9		15
Midpoint of oblique line on anterior surface	10		14
Midpoint of mylohyoid groove on posterior surface	11		13
Deepest point on posterior surface of ramus	24		26
Mental tubercle	25		29
Angle of mandible	33		28
Anterior end of ramus	32		27
Genial tubercle on the sagittal plane		12	
Mental protuberance on the sagittal plane		30	
Maxima point of alveolar crest on the sagittal plane		31	

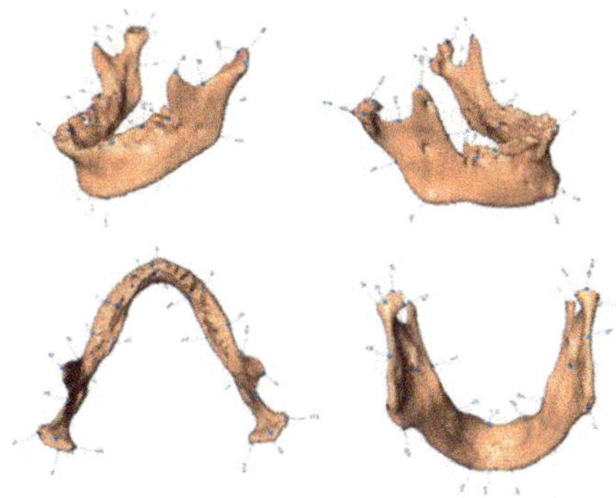

Fig. 1. Description of each landmark with 4 mandibles

After performing a GPA to bring all landmarks into the same configuration space, we used PCA as the basis for morphing these templates. All variation analysis was done using the Geomorph package in R [11]. We combined the 1st 10 principal components which captured around 70% of the total variation. Thin Plate Spline was used to warp a mandible in the category to the mean shape.

2.3 Registration Testing

We employed the ICP algorithm to register all templates to test meshes. We used two different software programs, to reduce any biases for using a specific system to register the templates. The first, a biomechanical toolkit ArtiSynth, has been used in a wide range of applications across the biomechanical modeling and simulation [12]. In this program, we placed the source mesh (template mandible) into a deformable FEM grid and used the ICP method to warp the grid and thereby the inner mesh to the target rigid body of the test mandible. The second software we utilized was the commercial software, 3D Slicer that has been used for registering both images and surfaces [13]. Within Slicer, all mesh warping was performed using the surface affine registration found in the registration toolkit.

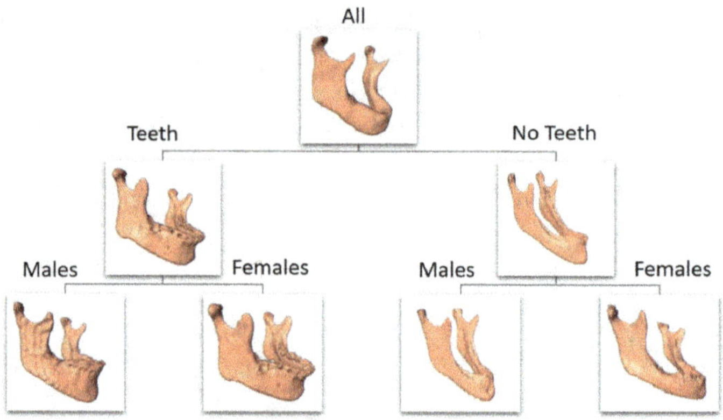

Fig. 2. Atlas tree depicting all templates

This also has a foundation based on the ICP algorithm, and includes a set range of allowable adjustments for sample size, landmarks, and number of iterations. These parameters were chosen according to standard values used for deforming bony structures in literature and were adjusted according to our unique mesh properties.

Our 6 test mandibles consisted of those not included in the template, with each fitting a specific category of template (see Fig. 3). We began with registering each template to those with characteristics that directly matched the template and then registration of templates (matched) to those that did not match (unmatched). We also registered the universal template to all test meshes.

2.4 Evaluation Methods

To test the effectiveness of the registration and to evaluate if templates improved the process, we compared the final deformation result (warped template) with the target mesh (test mandible). We used two verified methods that have been implemented in previous studies for comparing surface and 3D mesh data: Hausdorff at 95 percentile (HD95) and volumetric overlap percentage [13,14]. The combination of these analyses provide meaningful information in order to evaluate the resulting deformations compared to the target specimens.

Hausdorff 95. While Hausdorff distance is an analysis of the maximum distance between points of one set to points of another set, this distance alone can result in outliers. For example, two meshes may be closely similar to one another, but even one single point with a greater distance than threshold will always result in a higher Hausdorff distance for the whole mesh. To account for these errors, we employed 95 Hausdorff which only takes into account 95% of the points in the set, excluding any noisy outlier information.

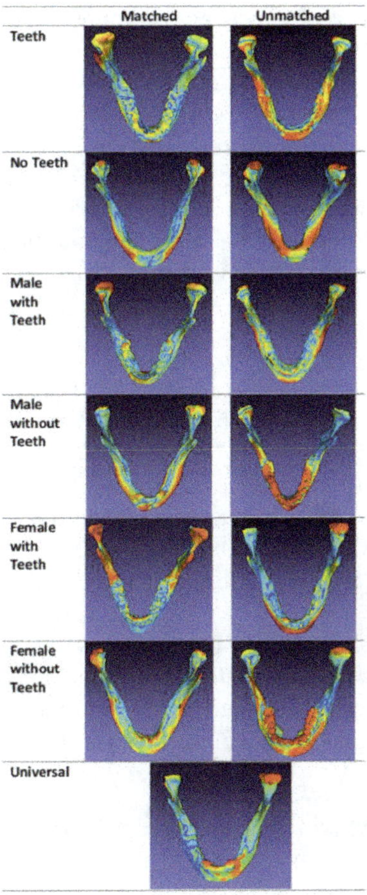

Fig. 3. Quality map of all test cases where red indicates maximum variation and blue represents minimum variation

The equation is listed as follows,

$$h(A, B) = max[min[d(a, b)]] \tag{4}$$

$$max = a \epsilon A \tag{5}$$

$$min = b \epsilon B \tag{6}$$

where A and B represent two diverse datasets, a and b represent points of those datasets respectively, and d represents Euclidean distance.

Mesh Overlap Percentage. While HD95 provides a surface comparison, an overlap percentage approach provided a volumetric evaluation of the 3D meshes by measuring the percentage of total mesh that overlaps the volume of another mesh. This method is also more visually demonstrative as meshes with a high overlap percentage will typically appear more visual similar. We evaluated this metric by performing a Boolean operation that removes the non-intersecting (non-overlapping) volume of the target mesh from the source mesh and calculated this remaining volume as a percentage of whole volume. A Two-way Anova was performed on both Hausdorff distance and surface overlap data to determine statistical difference between the means and variance of the data sets. T-tests were performed to find differences between the matched and unmatched groups, as well as the matched and universal groups.

3 Results

Regarding the average Hausdorff 95 distance, We found there was no significant statistical difference between the three groups $F(2,72) = 0.64$, $p = 0.53$. For overlap percentage, approximately equal variance was found across all three groups. The matched group had a statistically significant greater mean than the unmatched group $F(2,33) = 12$, $p = 0.00012$. The overlap percentage for matched cases ($M = 74.08$, $SD = 8.91$) was higher than unmatched cases ($M = 59.42$, $SD = 6.05$), $t(22) = 4.72$, $p < .001$. There was no statistical significant difference found between the matched ($M = 74.08$, $SD = 8.91$) and universal ($M = 71.92$, $SD = 8.48$) groups $t(22) = 0.61$, $p > .001$. A larger dataset and more analysis of shape matching is needed to conclusively determine if the templates had a significant effect on the result of deformable registration.

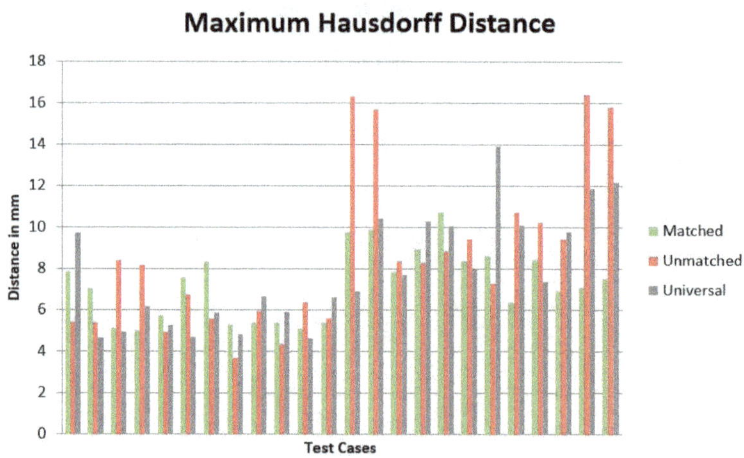

Fig. 4. Maximum Hausdorff distance for all test cases between matched, unmatched, and universal templates

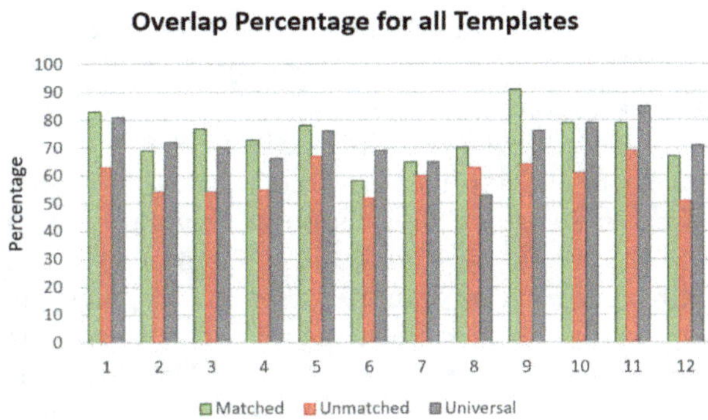

Fig. 5. Overlap percentage for all test cases between matched, unmatched, and universal templates

4 Discussion

At this time, we cannot fully accept our hypothesis that using this characteristic matching template improves deformable registration. However, the results of this study show promise that using some type of atlas can improve the registration process and potentially optimize computer assisted surgeries. More data and testing is needed to fully examine this potential and to determine which characteristics or factors are most beneficial in atlas creation.

Some of the challenges we found during the course of this project included the registration method itself. We used ICP as a test algorithm for the experiment of the templates, but other techniques may be better suited with using categorical templates than others depending on the shape and properties of the anatomy. We chose to employ the ICP method because of its prevalence in registration literature, its stability, and its advantage of being available in various software packages [16]. The overall purpose of this study was not to evaluate the effectiveness of a particular registration algorithm, but instead determine if characteristic templates would improve a type of deformable registration.

One issue we found during the registration test for both software was that critical regions of interest, namely the condyles of each mesh (test and template), did not match well both visually and physically. This is problematic for clinical applications, where the location of these regions is critical for surgical cutting guides, attachment sites (joints and muscles), and the process of reconstructive surgery overall. Pairing ICP registration with landmarks over regions of interest may offer a better solution. While the verification methods of evaluated surface matching gave an indication of how closely the source matched the target overall, none provided information on specific region matching. Specific region matching is worthwhile to examine in future work, and the incorporation of landmark registration or region of interest (ROI) registration may help with this problem.

We only included the first 10 components of the PCA in our study, with a total of 70 percent of variance captured by these components. The issue of determining the appropriate number of component to capture variation throughout a large set of data is both widely debated and dependent on that specific dataset [17]. Another challenge in this study was landmark validation. While we attempted to prevent any error by recording landmarks multiple times between different methods, using an automatic 3D landmark selection algorithm would the most optimal choice to reduce these errors. In addition, preliminary experimentation was done to evaluate both the number and location of landmarks, but these parameters would benefit from further examination. For one, the most clinically relevant landmark sites may not be the most effective for the creation of categorical atlases. Investigations into computation of segmentation and landmark selection through computer algorithms rather than clinical anatomy show potential for more stable registration outcomes [18]. We chose the characteristics of denture profile and gender because this information was available from our datasets, and previous research suggests morphological variation by these factors.

Future work includes creating templates based on additional characteristics such as age and ethnicity as well as for defect classifications to investigate if these templates will provide better registration results. There are a variety of complex processes that affect these classifiers on an individual basis including metabolic changes, hormones, mechanics, nutrition, diet, and lifestyle. In addition, these variables have unique and complicated changes over time.

This method is a proposed step in the overall process of developing and utilizing categorical atlases for improving registration as well the process of creating patient-specific simulations in the long run. Our next steps include investigation of additional anatomy registration, other classifications for template creation, and more nonrigid registration methods.

Acknowledgments. We would like to thank Antonio Sánchez for his support in the development of the ArtiSynth registration framework utilized in this work.

References

1. Virzi, A., Marret, J., Muller, C.O., Berteloot, L., Boddaert, N., Sarnacki, S., Bloch, I.: A new method based on template registration and deformable models for pelvic bones semi-automatic segmentation in pediatric MRI. In: 2017 IEEE 14th International Symposium on Biomedical Imaging (ISBI 2017) (2017). https://doi.org/10.1109/isbi.2017.7950529
2. AlHadidi, A., Cevidanes, L.H., Cook, R., Festy, F., Tyndall, D., Paniagua, B.: The use of a custom made atlas as a template for corrective surgeries of asymmetric patients. In: Proceedings of SPIE 8317, Medical Imaging 2012: Biomedical Applications in Molecular, Structural, and Functional Imaging, 83171Y, 5 April 2012. https://doi.org/10.1117/12.911048
3. Raith, S., Wolff, S., Steiner, T., Modabber, A., Weber, M., Hölzle, F., Fischer, H.: Planning of mandibular reconstructions based on statistical shape models. Int. J. Comput. Assist. Radiol. Surg. **12**(1), 99–112 (2016). https://doi.org/10.1007/s11548-016-1451-y
4. Chrcanovic, B.R., Abreu, M.H.N.G., Custódio, A.L.N.: Surg. Radiol. Anat. **33**, 203 (2011). https://doi.org/10.1007/s00276-010-0731-4

5. Cutright, B., Quillopa, N., Schubert, W.: An anthropometric analysis of the key foramina for maxillofacial surgery. J. Oral Maxillofac. Surg. **61**, 354–357 (2003). https://doi.org/10.1053/joms.2003.50070
6. Rai, B., Kaur, J.: Sex determination methods in forensic odontology. In: Evidence-Based Forensic Dentistry. Springer, Heidelberg (2013). https://doi.org/10.1007/978-3-642-28994-1_5
7. Tam, G.K., Cheng, Z., Lai, Y., Langbein, F.C., Liu, Y., Marshall, D., Martin, R.R., Sun, X.F., Rosin, P.L.: Registration of 3D point clouds and meshes: a survey from rigid to nonrigid. IEEE Trans. Visual Comput. Graphics **19**(7), 1199–1217 (2013). https://doi.org/10.1109/tvcg.2012.310
8. Wallner, J., Egger, J.: Mandibular CT dataset collection, November 2018
9. Nikolov, S., Blackwell, S., Mendes, R., De Fauw, J.-r., Meyer, C., Hughes, C., Askham, H., Romera-Paredes, B., Karthikesalingam, A., Chu, C., Carnell, D., Boon, C., D'Souza, D., Moinuddin, S.A., Sullivan, K.: DeepMind Radiographer Consortium, Montgomery, H., Rees, G., Sharma, R., Suleyman, M., Back, T., Ledsam, J.R., Ronneberger, O.: Deep learning to achieve clinically applicable segmentation of head and neck anatomy for radiotherapy. arXiv e-prints (2018)
10. Bosch, W.R., Straube, W.L., Matthews, J.W., Purdy, J.A.: Head-neck cetuximab-the cancer imaging archive (2015)
11. R Core Team: R: a language and environment for statistical computing. R Foundation for Statistical Computing, Vienna, Austria (2013). http://www.R-project.org/
12. Lloyd, J.E., Stavness, I., Fels, S.: ArtiSynth: a fast interactive biomechanical modeling toolkit combining multibody and finite element simulation. In: Payan, Y. (ed.) Soft Tissue Biomechanical Modeling for Computer Assisted Surgery, pp. 355–394. Springer, Heidelberg (2012)
13. Fedorov, A., Beichel, R., Kalpathy-Cramer, J., Finet, J., Fillion-Robin, J.-C., Pujol, S., Bauer, C., Jennings, D., Fennessy, F.M., Sonka, M., Buatti, J., Aylward, S.R., Miller, J.V., Pieper, S., Kikinis, R.: 3D slicer as an image computing platform for the quantitative imaging network. Magn. Reson. Imaging **30**(9), 1323–1341 (2012). PMID: 22770690. PMCID: PMC3466397
14. Kieselmann, J.P., Kamerling, C.P., Burgos, N., Menten, M.J., Fuller, C.D., Nill, S., Cardoso, M.J., Oelfke, U.: Geometric and dosimetric evaluations of atlas-based segmentation methods of MR images in the head and neck region. Phys. Med. Biol. **63**(14), 145007 (2018). https://doi.org/10.1088/1361-6560/aacb65
15. Zachow, S., Lamecker, H., Elsholtz, B., Stiller, M.: Reconstruction of mandibular dysplasia using a statistical 3D shape model. Int. Congr. Ser. **1281**, 1238–1243 (2005). https://doi.org/10.1016/j.ics.2005.03.339
16. Amberg, B., Romdhani, S., Vetter, T.: Optimal step nonrigid ICP algorithms for surface registration. In: 2007 IEEE Conference on Computer Vision and Pattern Recognition (2007). https://doi.org/10.1109/cvpr.2007.383165
17. Timmerman, M.E., Kiers, H.A.: Three-mode principal components analysis: choosing the numbers of components and sensitivity to local optima. Br. J. Math. Stat. Psychol. **53**(1), 1–16 (2000). https://doi.org/10.1348/000711000159132
18. Ibragimov, B., Vrtovec, T.: Landmark-based statistical shape representations. In: Statistical Shape and Deformation Analysis, pp. 89–113 (2017). https://doi.org/10.1016/b978-0-12-810493-4.00005-5

3D Constitutive Model of the Rat Large Intestine: Estimation of the Material Parameters of the Single Layers

F. Bini[1(✉)], M. Desideri[1], A. Pica[1], S. Novelli[1,2], and F. Marinozzi[1]

[1] Department of Mechanical and Aerospace Engineering, Sapienza University of Rome, Eudossiana 18, 00184 Rome, Italy
fabiano.bini@uniroma1.it

[2] Institute for Liver and Digestive Health, University College London (UCL), Gower Street, WC1E6BT London, UK

Abstract. Several functions of the large intestine depend on its morphology and biomechanical properties. Since some classes of soft tissues have non-linear mechanical behavior, they may be modeled as hyperelastic materials applying the strain energy function. Moreover, as the arterial walls, the colonic walls are composed by collagen fibers characterized by an anisotropic behavior. It is known a mechanical model of artery that considered its walls composed of two cylindrical layers reinforced with fibers of collagen suitably oriented. Afterwards, it has been proposed a structure-based mathematical model for mechanical passive behavior of the rat colon, fitted to data obtained from inflation/extension tests that doesn't subdivide the walls into layers. However, the wall of the rat large intestine is composed by four distinct layers, i.e. mucosa, submucosa, muscle layer and serosa. Thus, the aim of this paper is to identify a method for estimating the parameters of a computational model that considers each layer of the colonic walls. We use the Nelder-Mead nonlinear regression technique for minimizing the residual sum of squares between experimental data reported in literature and the outcomes of the proposed model. The estimated material parameters (k_1, k_2, c) are used to develop a 3D finite element model. Furthermore, we computed the components of the Cauchy stress over the colonic wall across each layer for different values of internal pressure and axial stretch.

Keywords: Colon constitutive model · Hyperelastic material · 3D multilayer FEM

1 Introduction

Computational models play an ever-increasing role in advancing the understanding of the complex behavior of biological tissues in physiological and pathological conditions. Mathematical and computational modelling techniques are versatile and powerful tools applied to predict biological and biomechanical properties of hard [1, 2] and soft

G. A. Ateshian et al. (Eds.): CMBBE 2019, LNCVB 36, pp. 608–623, 2020.
https://doi.org/10.1007/978-3-030-43195-2_51

[3–11] tissues. A significant amount of experimental data [12–20] is used to parametrize, validate and refine the accuracy of the estimation method of model parameters.

The development of computational models for soft tissues poses many challenges due to their complex structure [21]. In order to better mimic their non-linear mechanical properties, these structures can be studied as hyperelastic materials, adopting the classical Mooney-Rivlin and Neo-Hookean models [21] or developing new specific models. For instance, in the 1970, Veronda and Westmann [22] developed a model validated upon tensile uniaxial tests from skin of cats and, in 1987, Humphrey and Yin [5] proposed a model to reproduce the passive behavior of myocardium tissue. Martins et al. [6] modified this model in order to create a suitable model for the skeletal muscles.

Furthermore, soft tissues like arterial and colonic walls, are composed of collagen fibers, whose hierarchical organization leads to an anisotropic behavior [3, 4]. The anisotropy was taken into account by Holzapfel et al. [4], that created a mechanical model of artery considering its walls composed of two cylindrical layers reinforced with collagen fibers suitably oriented. In order to reproduce the purely elastic passive response of the layered structure of the intestine, Ciarletta el al. [7] developed a model based on the hyperelastic theory. Carniel et al. [8] presented a constitutive model of intestine based on multi-layered conformation of the colonic wall. However, both models were fitted to data on porcine intestine obtained only from multiple uniaxial tensile tests. Hence, they didn't consider the tubular geometric conformation of the large intestine. In order to take into account also the contribution of the geometric features, Sokolis et al. [3, 12] proposed a structure-based constitutive model for mechanical passive behavior of rat colon, fitted to data obtained from inflation\extension test. Lately, Patel et al. [9] estimated parameters for a constitutive model of the passive behavior of the swine colon from an inflation\extension tests. The main difference between the mentioned models of the colon [3, 9, 12] and the model presented by Holzapfel et al. [4] for the arteries is that they consider the colonic wall as a single layer composed by collagen fibers with different orientations. Nevertheless, the walls of the large intestine of the pig and the rat are composed by four distinct layers, i.e. mucosa, submucosa, muscle layer and serosa [4, 9].

Therefore, the aim of this study is to develop a method for estimating the parameters of a mathematical model that consider each layer of the colonic walls. Starting from the experimental data obtained by Sokolis and Sassani [3], we use the Nelder-Mead [23] nonlinear regression technique for minimizing the residual sum of squares between the experimental data and the outcomes of the proposed mathematical model. Subsequently, the estimated parameters are used to develop a finite element model and compute the principal Cauchy stress through the wall for each layer.

2 Materials and Methods

We consider the large intestine wall composed of four cylindrical layers reinforced with collagen fibers. Each layer is assumed to be a nonlinear, homogeneous, hyperelastic and orthotropic material subjected to finite deformation with the incompressible material assumption. The strain energy function for incompressible material can be express as:

$$W = \overline{W}(\mathbf{F}) - p(J - 1) \tag{1}$$

The first term describes the isochoric deformation and the second considers the isovolumetric deformation. In particular, p is the Lagrange multiplier and has a pressure-like dimension. The models presented in literature propose different form for the isochoric term. As Sokolis and Sassani [3], we decompose the isochoric term of the strain energy function into an isotropic part (W_{iso}) and an anisotropic part (W_{anis}):

$$\overline{W} = \overline{W_{iso}} + \overline{W_{anis}} \tag{2}$$

The isotropic part describes the non-collagenous matrix and a Neo-Hookean model can be used, whereas the anisotropic part considers the direction of the collagen fibers. Thus, the isochoric term of the strain energy function is expressed for each layer of the large intestine as follows:

$$\overline{W}_{muc} = c_{muc} \cdot (I_1 - 3) + \frac{k_{1muc}}{4k_{2muc}} \cdot \left[e^{k_{2muc} \cdot (\lambda_r^2 - 1)^2} - 1\right], R_i \leq R \leq R_i + h_{muc} \tag{3}$$

$$\overline{W}_{sub} = c_{sub} \cdot (I_1 - 3) + \frac{k_{1sub}}{2k_{2sub}} \cdot \left[e^{k_{2sub} \cdot (\lambda_d^2 - 1)^2} - 1\right], \quad R_i + h_{muc} < R \leq R_i + h_{muc} + h_{sub} \tag{4}$$

$$\lambda_d = \sqrt{\lambda_z^2 \cos(a_0)^2 + \lambda_\theta^2 \sin(a_0)^2} \tag{5}$$

$$\overline{W}_{mus} = c_{mus} \cdot (I_1 - 3) + \frac{k_{1mus\_c}}{4k_{2mus\_c}} \cdot \left[e^{k_{2mus\_c} \cdot (\lambda_\theta^2 - 1)^2} - 1\right] + \frac{k_{1mus\_l}}{4k_{2mus\_l}} \cdot \left[e^{k_{2mus\_l} \cdot (\lambda_z^2 - 1)^2} - 1\right], R_i + h_{muc} + h_{sub} < R \leq R_i + h_{muc} + h_{sub} + h_{mus}. \tag{6}$$

$$\overline{W}_{si} = c_{si} \cdot (I_1 - 3) + \frac{k_{1si}}{4k_{2si}} \cdot \left[e^{k_{2si} \cdot (\lambda_z^2 - 1)^2} - 1\right], R_i + h_{muc} + h_{sub} + h_{mus} < R \leq R_0 \tag{7}$$

where the subscripts muc, sub, mus and si indicate the layer competing to mucosa, submucosa, muscle layer and serosa, respectively. The letter h represents the thickness of the different layers, R is the radius of the cylinder, R_i is the radius of the internal surface of the cylinder and R_0 the radius of the external surface. The material parameters k_1, c, μ have stress-like dimension, k_2 is dimensionless. For the muscle layer we considered separately the contribution of the circular muscle cells by means of the material parameters k_{1mus_c}, k_{2mus_c}, and the contribution of the longitudinal muscle cells, by means of the material parameters k_{1mus_l}, k_{2mus_l}. The term I_1 is the first invariant of the right Cauchy-Green strain tensor, defined as follows:

$$I_1 = \lambda_r^2 + \lambda_\theta^2 + \lambda_z^2 \tag{8}$$

where $\lambda_r, \lambda_\theta, \lambda_z$ are the principal stretches in radial, circumferential and axial direction, respectively.

The direction of the fibers (a_0) in different layers is determined from the histological observation performed by Sokolis and Sassani [3]. Frequently, it is assumed that collagen fibers are active only in elongation [3]. Hence, we impose the following condition:

$$\text{if } \lambda_i < 1 \rightarrow \overline{W}_{anis} = 0 \tag{9}$$

Moreover, as mentioned by Gao and Gregersen [13], the residual stress in the walls should be considered. In [13], the authors proposed a method to identify the residual Green strain at the internal (E_i) and external (E_o) surface of the Wistar rat colon from morphological observations. The corresponding equations for E_i and E_o are expressed as follows:

$$E_i = \frac{1}{2}\frac{(C_{i-n}^2 C_{i-z}^2)}{C_{i-z}^2} \tag{10}$$

$$E_o = \frac{1}{2}\frac{(C_{o-n}^2 C_{o-z}^2)}{C_{o-z}^2} \tag{11}$$

where, C_{i-n} and C_{o-n} are the internal and external circumference of the cylinder section at no-load state, whereas C_{i-z} and C_{o-z} are the internal and external circumference of the cylinder section at zero-stress state (see Fig. 1).

Choung and Fung [24] presented an analysis for the problem of a residually stressed thin cylinder subjected to internal pressure and axial stress. The zero-stress configuration is assumed as the reference configuration and it is obtained applying a radial cut along the length of the specimen in no load configuration (see Fig. 1). The deformation χ_{nl} transports the tissue from zero stress configuration to no load configuration [9, 10].

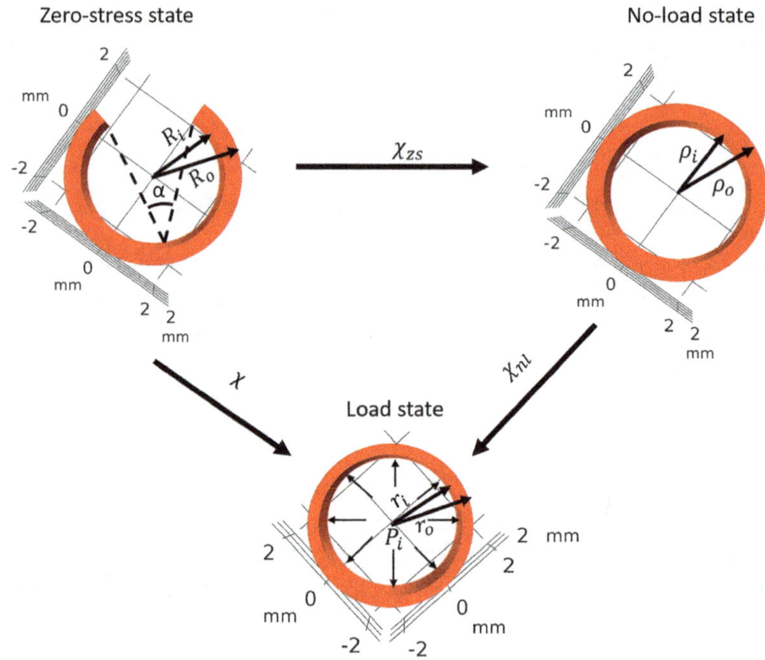

Fig. 1. Kinematics of a thin cylinder residually stressed subjected to internal pressure and axial force

The principal stretches of the associated deformation tensor are expressed as:

$$\mathbf{F}_{\mathbf{zs}}^{\mathbf{nl}} = \begin{bmatrix} \frac{d\rho}{dR} & 0 & 0 \\ 0 & k \cdot \frac{\rho}{R} & 0 \\ 0 & 0 & 1 \end{bmatrix} \tag{12}$$

In particular, ρ is the radius of the cylinder at no load state, R is the radius at zero stress state and $k = \frac{\pi}{\pi - \alpha}$ where α is the opening angle after the cut. Applying an internal pressure and an axial stretch, the tissue is transported to the loaded state. The deformation tensor is expressed as:

$$\mathbf{F}_{\mathbf{nl}}^{\mathbf{l}} = \begin{bmatrix} \frac{dr}{d\rho} & 0 & 0 \\ 0 & \frac{r}{\rho} & 0 \\ 0 & 0 & \lambda_z \end{bmatrix} \tag{13}$$

In this state r represents the radius and $\lambda_z = \frac{L_l}{L_{nl}}$ the ratio between the length of cylinder at load state and at no load state (axial stretch). It is possible to find a deformation that describes the passage from zero stress state to load state $\chi = \chi_{zs} \circ \chi_{nl}$ and its associated deformation tensor **F**:

$$\mathbf{F} = \mathbf{F}^{nl}_{zs}\mathbf{F}^{l}_{nl} = \begin{bmatrix} \frac{dr}{dR} & 0 & 0 \\ 0 & k\frac{r}{R} & 0 \\ 0 & 0 & \lambda_z \end{bmatrix} = \begin{bmatrix} \lambda_r & 0 & 0 \\ 0 & \lambda_\theta & 0 \\ 0 & 0 & \lambda_z \end{bmatrix} \tag{14}$$

Due to incompressibility assumption we can write the radial stretch as a function of the axial and circumferential stretch:

$$\lambda_r = \frac{dr}{dR} = \frac{1}{\lambda_\theta \lambda_z} = \frac{R}{k \cdot \lambda_z \cdot r} \tag{15}$$

Data reported by Sokolis and Sassani [3] show the values of the outer radius of the specimen for different values of the internal pressure and axial stretch. The authors presented also the values of the axial force imposed to stretches. All data are obtained varying the pressure between 0 and 15 mmHg in increment of 0.025 mmHg for three different value of axial stretch, namely 1.1, 1.2 and 1.3 [3]. We have estimated the unknown material parameters matching the analytical values of the axial force F_z and the internal pressure P_i with the Sokolis et al. [3] measures by minimizing of the residual sum of squares [3]:

$$RSS = \sum_i \sum_k \frac{\left(P_{exp_{i,k}} - P_{an_{i,k}}\right)^2}{P^2_{exp_{i,k}}} + \sum_i \sum_k \frac{\left(F_{exp_{i,k}} - F_{an_{i,k}}\right)^2}{F^2_{exp_{i,k}}} \tag{16}$$

The minimization is performed by means of Nelder-Mead nonlinear regression technique [23] and the quality of fitting are checked with the root-mean error computed by the following equation [3]:

$$\varepsilon = \sqrt{\chi^2} = \sqrt{\frac{RSS}{n_{exp} - n_{par}}} \tag{17}$$

where n_{exp} is the number of the experimental data and n_{par} is the number of the unknown parameters. To compute the analytical values, it can be started from the equilibrium equation in absence of the body force and at the static condition [9]:

$$\nabla \cdot \boldsymbol{\sigma} = 0 \rightarrow \frac{\partial \sigma_r}{\partial r} + \frac{\sigma_r - \sigma_\vartheta}{r} = 0 \tag{18}$$

where $\boldsymbol{\sigma}$ is the Cauchy stress tensor given by the following equations [9]:

$$\boldsymbol{\sigma} = \frac{\partial \overline{W}(\mathbf{F})}{\partial \mathbf{F}} \cdot \mathbf{F}^T - p\mathbf{I} \tag{19}$$

$$\sigma_r = \lambda_r \cdot \frac{\partial \overline{W}(\lambda_r, \lambda_\theta, \lambda_z)}{\partial \lambda_r} - p \tag{20}$$

$$\sigma_{\vartheta} = \lambda_{\theta} \cdot \frac{\partial \overline{W}(\lambda_r, \lambda_\theta, \lambda_z)}{\partial \lambda_\theta} - p \tag{21}$$

$$\sigma_z = \lambda_z \cdot \frac{\partial \overline{W}(\lambda_r, \lambda_\theta, \lambda_z)}{\partial \lambda_z} - p \tag{22}$$

The terms σ_ϑ, σ_r, σ_z are the component of the Cauchy stress tensor along the radial (r), circumferential (θ) and axial (z) direction.

Appling a pressure P_i on the internal surface of the cylinder and integrating the Eq. (18) with the boundary conditions $\sigma_{ri} = -P_I \mid \sigma_{ro} = 0$, an analytical formulation of P_i is obtained:

$$P_i = \int_{ri}^{ro} \frac{\sigma_\vartheta - \sigma_r}{r} dr \tag{23}$$

The analytical form of the axial force F_z imposed to obtain the necessary stretch to maintain the tissue at a fixed length is achieved from the equilibrium between the sum of the axial forces and the axial stress upon the cross section of the cylinder [9]:

$$F_z + P_i \pi r^2 = 2\pi \cdot \int_{ri}^{ro} \sigma_z r dr \rightarrow F_z = \pi \cdot \int_{r_i}^{r_o} (2\sigma_z - \sigma_\theta - \sigma_r) \cdot r dr \tag{24}$$

Since the external radius in loaded state r_o is achieved from literature data [3] and the internal (R_i) and external (R_o) radius are known in the reference configuration, the internal radius in loaded state r_i is computed from Eq. (15), splitting the variables and integrating between the internal and external radius:

$$r \cdot dr = \frac{R}{k \cdot \lambda_z} \cdot dR \tag{25}$$

$$\int_{r_i}^{r_o} r \cdot dr = \int_{R_i}^{R_o} \frac{R}{k \cdot \lambda_z} \cdot dR \tag{26}$$

$$r_o^2 - r_i^2 = \frac{1}{k \cdot \lambda_z} \cdot (R_o^2 - R_i^2) \tag{27}$$

$$r_i = \sqrt{r_o^2 - (R_o^2 - R_i^2) \cdot \frac{1}{k \cdot \lambda_z}} \tag{28}$$

We highlight that, differently than the method presented in [3], we compute the value of P_i and F_z for each layer and we consider the P_{an} and F_{an} presented in the Eq. (13) as the sum of the P_i and F_z of the single layers:

$$P_{an} = \int_{r_i}^{r_i + h_{muc}} \frac{\sigma_{\theta muc} - \sigma_{rmuc}}{r} dr + \int_{r_{i+h_{muc}}}^{r_i + h_{muc} + h_{sub}} \frac{\sigma_{\theta sub} - \sigma_{rsub}}{r} dr + \ldots \quad (29)$$

$$F_{an} = \pi \cdot \int_{r_i}^{r_i + h_{muc}} (2\sigma_{zmuc} - \sigma_{\theta muc} - \sigma_{rmuc}) \cdot r dr + \ldots \quad (30)$$

where h_{muc} is the thickness of the mucosa layer, h_{sub} the thickness of the submucosa layer and $\sigma_{zmuc}, \sigma_{\theta muc}, \sigma_{rmuc}$ are the component of the Cauchy stress tensor along the axial (z), radial (r) and circumferential (θ) direction calculated on the mucosa layer.

In a second step, we implemented a finite element model (FEM) of the rat descending colon as a cylinder with four layers that represent the mucosa, submucosa, muscle layer and serosa, respectively (see Fig. 2). Hence, to each layer we assigned the corresponding estimated material parameters. In fact, our aim is to reproduce the experiment of Sokolis and Sassani [3] and to compute the components of the Cauchy stress over the colonic wall across each layer for different values of internal pressure and axial stretch.

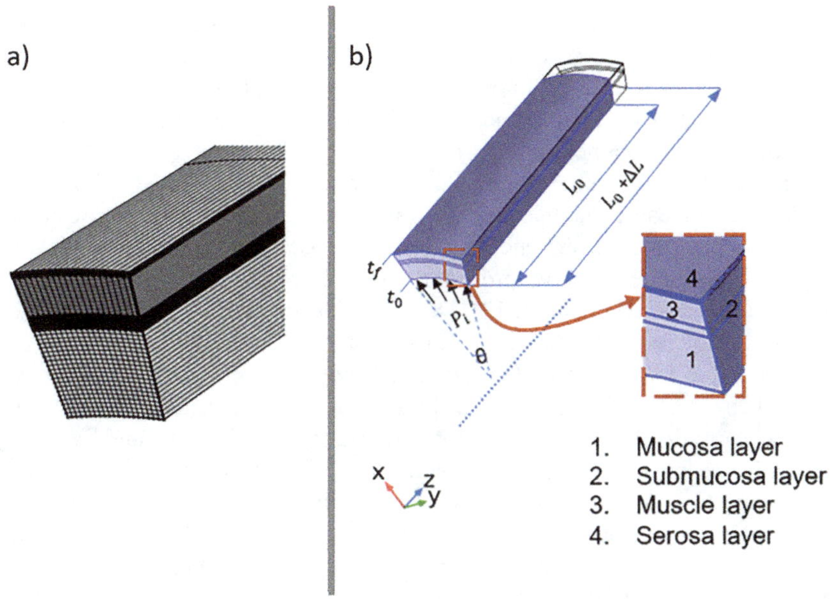

Fig. 2. (a) A 3D-meshed section of the cylinder used for the FEM simulation. (b) The layered cylinder section pressurized and axially stretched where L_0 is the length at no load state and $L_0 + \Delta L$ is the length at loaded state

Table 1. Comparison between the parameters obtained by Holzapfel et al. [4] for a carotid artery and the parameters estimated with our method

	c (kPa)	k_1 (kPa)	k_2	a_0 (deg)	ε	Reference
Media	3	2.3632	0.8393	29	–	*Holzapfel* [4]
	2.96	2.3725	0.8390	29	0.16	*Present study*
Adventitia	0.3	0.5620	0.7112	62	–	*Holzapfel* [4]
	0.36	0.5606	0.7107	62	0.16	*Present study*

Due to this reason, we imposed three different value for the axial stretch, and we applied a pressure P_i on the internal surface from 0 kPa to 2 kPa (see Fig. 2b). Subsequently, we performed the FEM simulation to compute the solution of the mechanical problem under the assumption previously exposed that on the entire cylinder we have considered a different form of strain energy function for each layer (Eqs. 3–7).

3 Results

Before applying the proposed method to estimate the unknown material parameter of the descending rat colon, we validated it on a carotid artery since its material parameters were available from the studies of Holzapfel et al. [4, 14]. However, we had to reduce the number of layers and consider different fibers orientation (Table 1).

The condition of goodness on fit [4, 9], i.e. $\varepsilon < 0.2$, was respected (Table 1) and the obtained values are similar to the values achieved by Holzapfel et al. [4]. Thus, we considered the procedure suitable to our scope and we have applied it to estimate the unknown parameters of the descending colon of a Wistar rat. Nevertheless, we decided to perform two different fitting procedure. In the first procedure (see Fig. 3a, b), contrary to Sokolis and Sassani [3, 12], we didn't consider the residual stress (Table 2).

Therefore, we choose the unloaded state as the reference condition, i.e. $\alpha = 0$, and we used the value for the internal and external radius reported in a previous study of Sokolis and co-workers [12], i.e. $R_i = 2.09$ mm and $R_o = 2.54$ mm. The minimization process was carried out on 90 experimental data taken from Sokolis and Sassani [3] in order to estimate the set of 15 material parameters. Whilst the orientation angle of the collagen fibres (a_0) into submucosa layer was estimated, we didn't consider the orientation angles of the fibres within the circumferential muscle layer, the longitudinal muscle layer and the serosa layer as unknown parameters and we used the value reported by Sokolis and Sassani [3].

However, several Authors have supposed the presence of the residual stress inside tubular structures as colon and arteries in the human body [4, 6, 9–13, 15, 25, 26]. Hence, in the second fitting procedure (see Fig. 3c, d) we selected the zero-stress state as reference condition (Table 3). The external and internal radius at this state, i.e. $R_i = 3.76$ mm and $R_o = 4.21$ mm was calculated by means of the Eqs. 10, 11. The values of the residual strains, the opening angle ($\alpha = 77°$), the internal radius and external radius at no-load state required in Eqs. 10, 11 were individuated in the study of Sokolis et al. [12].

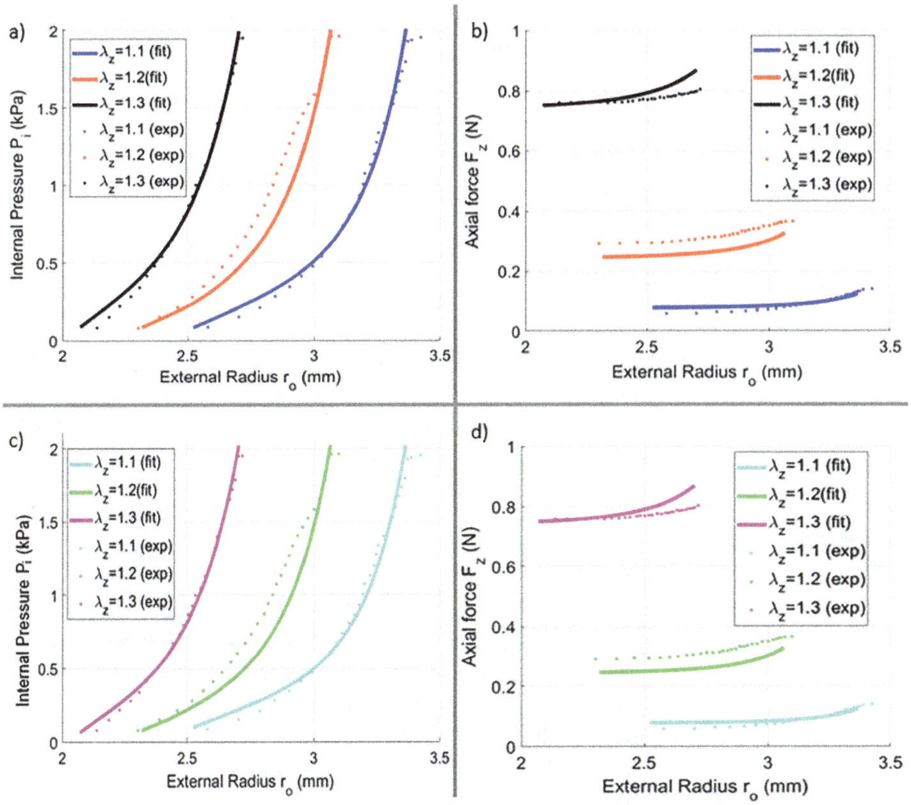

Fig. 3. Fitting of the experimental data (exp) by Eqs. 29 (right) and 30 (left) with the estimated material parameters with (a, b) and without (c, d) residual stress

The minimization process was performed as reported for the previous procedure and the fibres orientation angle (a_0) was estimated only for the fibres into the submucosa layer. Subsequently, we implemented the FEM model with the estimated material parameters. First, we generated the geometry at the no-load state as a layered cylinder with length L_0 of 4 cm, an internal radius R_i of 2.09 mm, an external radius R_o of 2.45 mm according to the morphological observations of Gao and Gregersen [13] and Sokolis and Sassani [3]. The thickness of each layer is indicated in Table 2. Consequently, we defined the strain-energy function for each layer according to Eqs. 3–7. Then, we imposed an axial stretch and internal pressure. A stationary study was performed by parametric sweep for different values of axial stretch, namely 1.1, 1.2 and 1.3 and internal pressure (from 0 kPa to 2 kPa) with 80000 mesh elements and 1110267 degrees of freedom (see Fig. 2a).

Table 2. Material parameters and the wall thickness for each layer and the root-mean error ε calculated without the residual stress for the descending colon of a Wistar rat

	h (mm)	c (kPa)	k_1 (kPa)	k_2	a_0 (deg)	ε
Mucosa	0.271	0.661	1.760	0.869	–	0.21
Submucosa	0.043	1.468	23.540	11.076	38.7	
Circumferential muscle layer	0.118	0.4218	4.855	0.001	0	
Longitudinal muscle layer			162.490	2.104	90	
Serosa	0.016	0.384	2.637	2.613	90	

Therefore the solution of the equation system allows to determine the pressure-outer radius, force-outer radius plots, the Cauchy stress and the strain over the entire wall, for each layer at any value of the internal pressure and axial stretch within the evaluated range (see Figs. 4 and 5).

4 Discussion and Conclusion

In the current study, we proposed a method to estimate the material parameters of each layer that compose the colonic wall, i.e. mucosa, submucosa, muscle and serosa layers. Two approaches were considered to achieve the constitutive parameters: the first algorithm neglects the residual stress, while the second procedure takes it into account. The estimated parameters from both methods allow to reproduce the nonlinear behavior of the colon (see Fig. 3). As reported by Sokolis et al. [3, 12], Gao and Gregersen [13] and Patel et al. [9], we can observe a high compliance at lower pressure values, in particular when we impose an axial stretch of 1.1, which represents the in situ condition [3]. It can also be noticed that the stiffness increases as rapidly as higher is the value of the pressure and the axial stretch. The axial force required to maintain the cylinder at fixed length, is almost constant until a specific value of external radius is reached, while afterwards it rises rapidly. This behavior seems confirming the previous observation and it is related to its storage capacity of the fecal content [3, 9]. Moreover, the increasing of the stiffness reveals the protection mechanism of the tissue that should avoid over distention and rupture [3]. As Patel et al. [9] observed for the swine large intestine, the submucosa and the muscle layer provide the mechanical support at the colon structure. In particular, in line with previous studies [3, 9], the axial load is provided by submucosa efficiently as circumferential load (see Figs. 4b and 5b). Therefore, the mechanical contribution of the mucosa and serosa layer is negligible.

Table 3. Material parameters and the wall thickness for each layer and the root-mean error ε calculated with the residual stress for the descending colon of a Wistar rat

	h (mm)	c (kPa)	k_1 (kPa)	k_2	a_0 (deg)	ε
Mucosa	0.267	1.040	8.390	69.150	–	0.23
Submucosa	0.044	0.099	16.453	10.789	38	
Circumferential muscle layer	0.121	0.004	3.600	0.001	0	
Longitudinal muscle layer			160.210	2.085	90	
Serosa	0.017	0.503	49.259	2.129	90	

Moreover, in addition to the quantities analyzed in the previous studies [3, 9], we were able to quantify the components of Cauchy stress tensor for each layer at different values of internal pressure and axial stretch (see Figs. 4 and 5). Furthermore, the computed values of strain, specifically the circumferential strain, are consistent with the previous work of Sokolis et al. [12] (see Figs. 4 and 5).

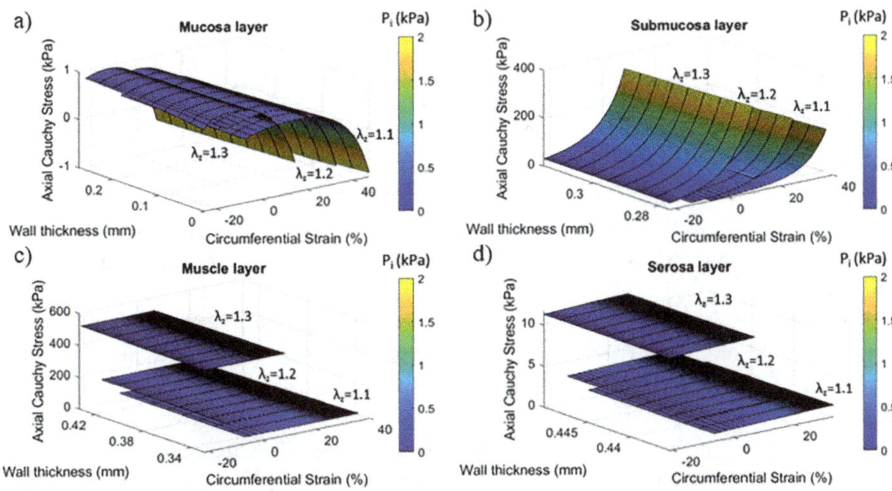

Fig. 4. Axial component of the Cauchy stress tensor over the wall thickness and circumferential strain for each layer at different values of internal pressure and axial stretch

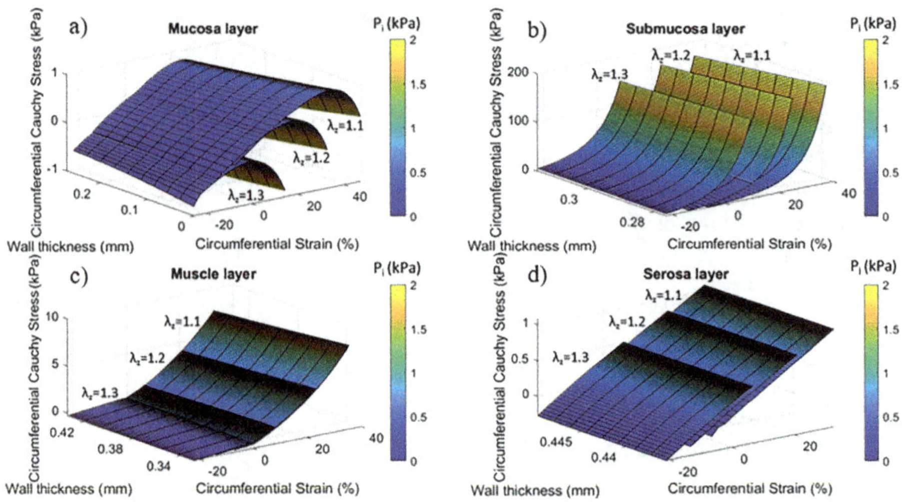

Fig. 5. Circumferential component of the Cauchy stress tensor over the wall thickness and circumferential strain for each layer at different values of internal pressure and axial stretch

The fibers of elastin characterize the linear behavior of the colon tissue. However, its content within the different layers is lower than the content of collagen. This may be confirmed by the lower values of c_{10} compared with the k_1 values [9] for each layer in both estimation procedure. We also found reasonable values for the orientation angle (a_0) of the cross-ply fibers into submucosa layer compared with the value of Sokolis and Sassani [3], but in disagreement with the values estimated by Patel et al. [9] for swine colon.

Although the presented predictive model appears suitable to reproduce the passive biomechanical behavior of the Wistar rat descending colon, several assumptions have been made. First, we considered a simplified colon geometry as a layered cylinder where each layer was considered as a homogeneous material. Moreover, the orientation angle of the collagen fibers into the submucosa was not directly measured but it was estimated, considering negligible the dispersion of the fibers as proposed by Gasser et al. [11]. We also make the hypothesis that the colon wall is an incompressible material, in accordance with previous studies [3, 8, 9]. However, the effect of this assumption on the estimation of material parameters has not been analyzed on the colon tissue [9]. With the aim of a more accurate representation of physiological conditions, a further aspect of improvement of the model should concern taking into account the compressibility behavior of the colon tissue.

The workflow presented in this study leads to an estimation of the material parameters competing to each layer of the colonic wall (Tables 2 and 3). Subsequently, we achieved by means of the finite element method the prediction of the components of the Cauchy stress tensor and the circumferential strain through the wall thickness of the different layers (see Figs. 4 and 5). To the best of our knowledge, it is the first study that analyzes the mechanical behavior of every layer that composes the colon structure.

In fact, the previous studies [3, 7–9, 12] reported the overall mechanical response of the colonic wall, without considering separately its layers.

Nevertheless, in the context of the specified limitations, the model could provide new insights into biomechanical response of the colon tissue. An accurate tissue characterization could be of interest in the improvement of mechanical properties of soft tissue substitutes, especially if combined with biosensors based on ZnO nanomaterials [27], or in the development of 3D printed colon models for surgical planning [28, 29].

References

1. Bini, F., Pica, A., Marinozzi, A., Marinozzi, F.: A 3D model of the effect of tortuosity and constrictivity on the diffusion in mineralized collagen fibril. Sci. Rep. **9**(1) (2019). https://doi.org/10.1038/s41598-019-39297-w. Article number 2658
2. Bini, F., Pica, A., Marinozzi, A., Marinozzi, F.: 3D diffusion model within the collagen apatite porosity: an insight to the nanostructure of human trabecular bone. PLoS ONE **12** (12), e0189041 (2017). https://doi.org/10.1371/journal.pone.0189041
3. Sokolis, D.P., Sassani, S.G.: Microstructure-based constitutive modeling for the large intestine validated by histological observations. J. Mech. Behav. Biomed. Mater. **21**, 149–166 (2013). https://doi.org/10.1016/j.jmbbm.2013.02.016
4. Holzapfel, G.A., Gasser, T.C., Ogden, R.W.: A new constitutive framework for arterial wall mechanics and a comparative study of material models. J. Elast. Phys. Sci. Solids **61**(1), 1–48 (2000). https://doi.org/10.1023/A:1010835316564
5. Humphrey, J.D., Yin, F.C.P.: On constitutive relations and finite deformations of passive cardiac tissue: I. A pseudostrain-energy function. J. Biomech. Eng. **109**(4), 298–304 (1987). https://doi.org/10.1115/1.3138684
6. Martins, J.A.C., Pires, E.B., Salvado, R., Dinis, P.B.: A numerical model of passive and active behavior of skeletal muscles. Comput. Methods Appl. Mech. Eng. **151**(3), 419–433 (1998). https://doi.org/10.1016/S0045-7825(97)00162-X
7. Ciarletta, P., Dario, P., Tendick, F., Micera, S.: Hyperelastic model of anisotropic fiber reinforcements within intestinal walls for applications in medical robotics. Int. J. Robot. Res. **28**(10), 1279–1288 (2009). https://doi.org/10.1177/0278364909101190
8. Carniel, E.L., Gramigna, V., Fontanella, C.G., Frigo, A., Stefanini, C., Rubini, A., Natali, A. N.: Characterization of the anisotropic mechanical behaviour of colonic tissues: experimental activity and constitutive formulation. Exp. Physiol. **99**(5), 759–771 (2014). https://doi.org/10.1113/expphysiol.2013.076091
9. Patel, B., Chen, H., Ahuja, A., Krieger, J.F., Noblet, J., Chambers, S., Kassab, G.S.: Constitutive modeling of the passive inflation-extension behavior of the swine colon. J. Mech. Behav. Biomed. Mater. **77**, 176–186 (2018). https://doi.org/10.1016/j.jmbbm.2017.08.031
10. Duong, M.T.: Hyperelastic modeling and soft-tissue growth integrated with the smoothed finite element method-SFEM. Dissertation, RWTH Aachen University (2014)
11. Gasser, T.C., Ogden, R.W., Holzapfel, G.A.: Hyperelastic modelling of arterial layers with distributed collagen fibre orientations. J. R. Soc. Interface **3**(6), 15–35 (2006). https://doi.org/10.1098/rsif.2005.0073
12. Sokolis, D.P., Orfanidis, I.K., Peroulis, M.: Biomechanical testing and material characterization for the rat large intestine: regional dependence of material parameters. Physiol. Meas. **32**(12), 1969–1982 (2011). https://doi.org/10.1088/0967-3334/32/12/007

13. Gao, C., Gregersen, H.: Biomechanical and morphological properties in rat large intestine. J. Biomech. **33**(9), 1089–1097 (2000). https://doi.org/10.1016/S0021-9290(00)00067-1
14. Holzapfel, H., Ogden, R.W.: Biomechanics of Soft Tissue in Cardiovascular Systems. CISM Courses and Lectures, vol. 441. Springer, Wien (2003). https://doi.org/10.1007/978-3-7091-2736-0
15. Liao, D., Zhao, J., Gregersen, H.: 3D mechanical properties of the partially obstructed guinea pig small intestine. J. Biomech. **43**(11), 2079–2086 (2010). https://doi.org/10.1016/j.jbiomech.2010.04.008
16. Marinozzi, F., Bini, F., Quintino, A., Corcione, M., Marinozzi, A.: Experimental study of diffusion coefficients of water through the collagen–apatite porosity in human trabecular bone tissue. BioMed. Res. Int. **2014** (2014). https://doi.org/10.1155/2014/796519. Article ID 796519, 8 pages
17. Marinozzi, F., Bini, F., Marinozzi, A., Zuppante, F., De Paolis, A., Pecci, R., Bedini, R.: Technique for bone volume measurement from human femur head samples by classification of micro-CT image histograms. Annali dell'Istituto superiore di sanità **49**(3), 300–305 (2013). https://doi.org/10.4415/ANN_13_03_11
18. Iosa, M., Morone, G., Bini, F., Fusco, A., Paolucci, S., Marinozzi, F.: The connection between anthropometry and gait harmony unveiled through the lens of the golden ratio. Neurosci. Lett. **612**, 138–144 (2016). https://doi.org/10.1016/j.neulet.2015.12.023
19. Serrao, M., Chini, G., Iosa, M., Casali, C., Morone, G., Conte, C., Bini, F., Marinozzi, F., Coppola, G., Pierelli, F., Draicchio, F., Ranavolo, A.: Harmony as a convergence attractor that minimizes the energy expenditure and variability in physiological gait and the loss of harmony in cerebellar ataxia. Clin. Biomech. **48**, 15–23 (2017). https://doi.org/10.1016/j.clinbiomech.2017.07.001
20. Marinozzi, F., Bini, F., Marinozzi, A.: Evidence of entropic elasticity of human bone trabeculae at low strains. J. Biomech. **44**(5), 988–991 (2011). https://doi.org/10.1016/j.jbiomech.2010.11.030
21. Martins, P.A.L.S., Jorge, R.M.N., Ferreira, A.J.M.: A comparative study of several material models for prediction of hyperelastic properties: application to silicone-rubber and soft tissues. Strain **42**(3), 135–147 (2006). https://doi.org/10.1111/j.1475-1305.2006.00257.x
22. Veronda, D.R., Westmann, R.A.: Mechanical characterization of skin—finite deformations. J. Biomech. **3**(1), 111–124 (1970). https://doi.org/10.1016/0021-9290(70)90055-2
23. Lagarias, C.J., Reeds, J.A., Wright, M.H., Wright, P.E.: Convergence properties of the Nelder–Mead simplex method in low dimensions. SIAM J. Optim. **9**(1), 112–147 (1998). https://doi.org/10.1137/S1052623496303470
24. Chuong, C.J., Fung, Y.C.: Residual stress in arteries. Front. Biomech. **108**, 117–129 (1986). https://doi.org/10.1007/978-1-4612-4866-8_9
25. Merodio, J., Ogden, R.W.: Extension, inflation and torsion of a residually stressed circular cylindrical tube. Continuum Mech. Thermodyn. **28**(1–2), 157–174 (2016). https://doi.org/10.1007/s00161-015-0411-z
26. Pierce, D.M., Fastl, T.E., Rodriguez-Vila, B., Verbrugghe, P., Fourneau, I., Maleus, G., Herjgers, P., Gomez, E.J., Holzapfel, G.A.: A method for incorporating three-dimensional residual stretches/stresses into patient-specific finite element simulations of arteries. J. Mech. Behav. Biomed. Mater. **47**, 147–164 (2015). https://doi.org/10.1016/j.jmbbm.2015.03.024
27. Araneo, R., Rinaldi, A., Notargiacomo, A., Bini, F., Marinozzi, F., Pea, M., Lovat, G., Celozzi, S.: Effect of the scaling of the mechanical properties on the performances of ZnO piezo-semiconductive nanowires. In: Nanoforum 2013, AIP Conference Proceedings, vol. 1603, pp. 14–22. (2014). https://doi.org/10.1063/1.4883037

28. Guachi, R., Bini, F., Bici, M., Campana, F., Marinozzi, F.: Finite element model set-up of colorectal tissue for analyzing surgical scenarios. In: Tavares, J., Natal Jorge, R. (eds.) VipIMAGE2017. ECCOMAS2017. Lecture Notes in Computational Vision and Biomechanics, vol. 27, pp. 599–609. Springer, Cham (2018). https://doi.org/10.1007/978-3-319-68195-5_65
29. Bici, M., Cardini, V., Eugeni, M., Guachi, R., Bini, F., Campana, F., Marinozzi, F., Gaudenzi, P.: Digital design of medical replicas via desktop systems: shape evaluation of colon parts. J. Healthc. Eng. **2018** (2018). https://doi.otg/10.1155/2018/3272596. Article 3272596, 10 pages

Differences Between Static and Dynamical Optimization Methods in Musculoskeletal Modeling Estimations to Study Elite Athletes

Rodrigo Mateus(✉), Filipa João(✉), and António P. Veloso(✉)

Faculdade de Motricidade Humana,
University of Lisbon, Cruz Quebrada, Portugal
rodrigo.mateus@tecnico.ulisboa.pt,
{filipajoao,apveloso}@fmh.ulisboa.pt

Abstract. Abrupt deceleration is a common practice in several sports, where sudden changes of direction are needed to perform at the highest level. The aim of this study is the comparison between static and dynamic optimization methods for muscle force estimation using a musculoskeletal modelling approach. Six elite male indoor elite athletes participated in this work. Musculoskeletal models, consisting of 12 segments, 23 degrees of freedom and 92 musculotendon actuators was used. Kinematic and kinetic data were collected at 300 Hz using 8 infrared cameras (Qualisys) and 2 force plates (Kistler). Muscle forces were attained through OpenSim. Similarities between force estimations using static optimization (SO) and computed muscle control (CMC) were quantified using a correlation coefficient. Stronger correlations occurred along the muscles *Vasti* (0.712 ± 0.292), *Gluteus Maximus* (0.619 ± 0.277), *Soleus* (0.755 ± 0.255) and *Erector Spinae* (0.855 ± 0.150). These muscles only span one joint. Moderate to weak correlations arise when comparing both methods in biarticular muscles, such as the hamstrings (0.422 ± 0.475), *Rectus Femoris* (0.356 ± 0.404), *Gastrocnemius* (0.325 ± 0.387), with the *Tibialis Anterior* (-0.211 ± 0.321) showing a weak negative correlation. muscles synergies are in agreement with the joint moments and measured kinematic data. Both SO and CMC predicted similar results in terms of force profile and magnitudes during an abrupt A/P deceleration task, albeit caution must be taken when biarticular muscles, such as the hamstrings or gastrocnemius, are concerned, so CMC might be the better choice.

Keywords: A/P deceleration · Musculoskeletal modelling · Optimization techniques

1 Introduction

The magnitudes of the forces generated by the muscles and acting upon joints and ligaments during tasks involving deceleration and direction changes, while considering the amount of times an athlete needs to decelerate and change directions during the course of its career, are hard to fathom. Hence, being cognizant of the internal forces acting on the human body during such tasks is key for understanding the functionalities of muscles, to prevent injuries and improving the athlete's performance through specific training programs.

G. A. Ateshian et al. (Eds.): CMBBE 2019, LNCVB 36, pp. 624–631, 2020.
https://doi.org/10.1007/978-3-030-43195-2_52

These forces can be obtained in vivo [1, 2], however ethical issues hinder the viability of such methods. An alternative to this is the implementation of musculoskeletal models to obtain these forces [3–5].

They are estimated using optimization techniques, which calculate the solution that minimizes a cost function related to a physiological criterion whilst depicting the equations of motion for a selection of kinematic and kinetic data.

These optimization techniques are divided into static and dynamic techniques [6, 7]. The first operates via solving the inverse dynamics problem, followed by attempting to sort out the joint moments into muscle forces at each instant, individual forces, which are obtained through the minimization of a cost function, normally the sum of the squared muscle activations. Several works performed simulations using this optimization method [3, 6, 8–13]. However, it carries some important drawbacks, mainly related not only to the dependency between the results legitimacy and the certainty of the motion data measured, but also to the difficulty of inserting muscle physiology in the management of the optimization problem [14]. On the other hand, in dynamic optimization techniques, output variables such as muscle forces are explained by a set of differential equations, which correlate to the physiological properties of the system [15], and the solution is achieved by solving the optimization problem per unit of distance for a full cycle of the motion. Computed Muscle Control [16, 17] is a dynamic optimization method that uses a static optimization (SO) step added to feedforward and feedback controls to move the model into the wanted kinematics. Due to the employment of static optimization and feedback procedures, this algorithm is able to resolve the dynamics optimization problem much more efficiently.

Both of these methods are available in OpenSim [18], which is an open – source software that allows the joint implementation of a musculoskeletal model and experimental data. Anderson and Pandy [6] compared these optimization methods for normal simulated human gait data, revealing that either approach is appropriate in this case, albeit that a dynamic optimization technique may be recommended in certain conditions. Similarly, Lin et al. [19] measured three different methods – SO, CMC and Neuromusculoskeletal Tracking (NMT) –, up against each other for gait and running, by comparing muscle forces estimations. This study concluded that factors, such as the dynamics of muscle activation or time – dependence of the performance criterion, do not alter the estimations of muscle forces when walking or running is concerned. Nonetheless, for more ballistic tasks, SO might not be as reliable. The drawbacks inherent to this work mainly reside on the population size, since only one subject was analyzed. A previous study comparing SO and CMC for a ballistic task – single legged hop –, deemed both methods as valid to be implemented in musculoskeletal modelling [20].

Therefore, the motivation for this study stems from the need to understand which technique for muscle estimation is ideal in tasks performed by elite athletes at maximum capacity. The aims of this study are to estimate muscle forces using both a static and dynamic optimization techniques, and to quantitatively compare them.

2 Methods

In this work, six elite male team sports injury free athletes consented to participate in this study (22 ± 4 years, 183 ± 8 cm, 79 ± 14 kg). The procedures used in this study were approved by the Ethics Committee of the Catalonian Sports Council. All subjects gave their written informed consent after an explanation of the experimental procedures and before commencement of the study.

Kinematic and kinetic data were collected using 8 infrared cameras (Qualisys) working at a frequency of 300 Hz and 2 force plates (Kistler). The Inverse kinematics (IK) problem was solved as a global optimization problem. Muscle forces and contributions were attained through OpenSim [18].

A 12 segment, 23 degrees of freedom musculoskeletal model was used to create the simulation. Each lower extremity had five degrees-of-freedom. The hip was modeled as a ball-and-socket joint (3 degrees of freedom), the knee was modeled as a custom joint with 1 degree of freedom, and the ankle was modeled as a revolute joint (1 degree of freedom) [21]. Lumbar motion was modeled as a ball-and-socket joint (3 degrees of freedom) [22]. The lower extremity and back joints were actuated by 92 musculotendon actuators [21, 22].

The musculoskeletal model was manually scaled to match each subject's anthropometry. A residual reduction algorithm (RRA) step was used to minimize errors related to kinematic inconsistencies and modelling assumptions. Positional errors were kept within a range of acceptable values (<5 cm for translational and <5° for angular degrees of freedom). After the first passage, RRA tracking weights related to the degrees of freedom for smaller errors between experimental data and the adjusted kinematics were lowered. Several RRA iterations were performed until the residual forces and torques converged. Joint moments obtained from RRA were compared to inverse dynamics (ID). Optimizer derivative step size and convergence tolerance were set to 10–4 and 10–5, respectively. The subtalar and metatarsophalangeal joints were locked in a neutral position. Muscle forces were estimated using a Static Optimization and Computed Muscle Control optimization technique. To compare both methods for muscle force estimation, a Shapiro–Wilk test was employed to test both curves for normality. Posteriorly, a correlation coefficient was estimated to quantify how similar both force profiles are.

3 Results

During this movement, the right hip remains constantly flexed and abducted. For the majority of the task, the right hip stays in a state of external rotation. The motions of the hip are accompanied by a posterior tilt along the sagittal plane (with varying angles between subjects) and a posterior pelvis rotation to the contralateral side along the transverse. As for the pelvis kinematics along the coronal plane, it moves towards a downwards pelvic rotation, with varying angles between subjects. In order to dissipate as much as possible the impact forces occurring during this task, the knee joints stays flexed and the ankle joint in dorsiflexion. The lumbar joint stays extended, tilted to the side of the supporting leg and in contralateral lumbar rotation.

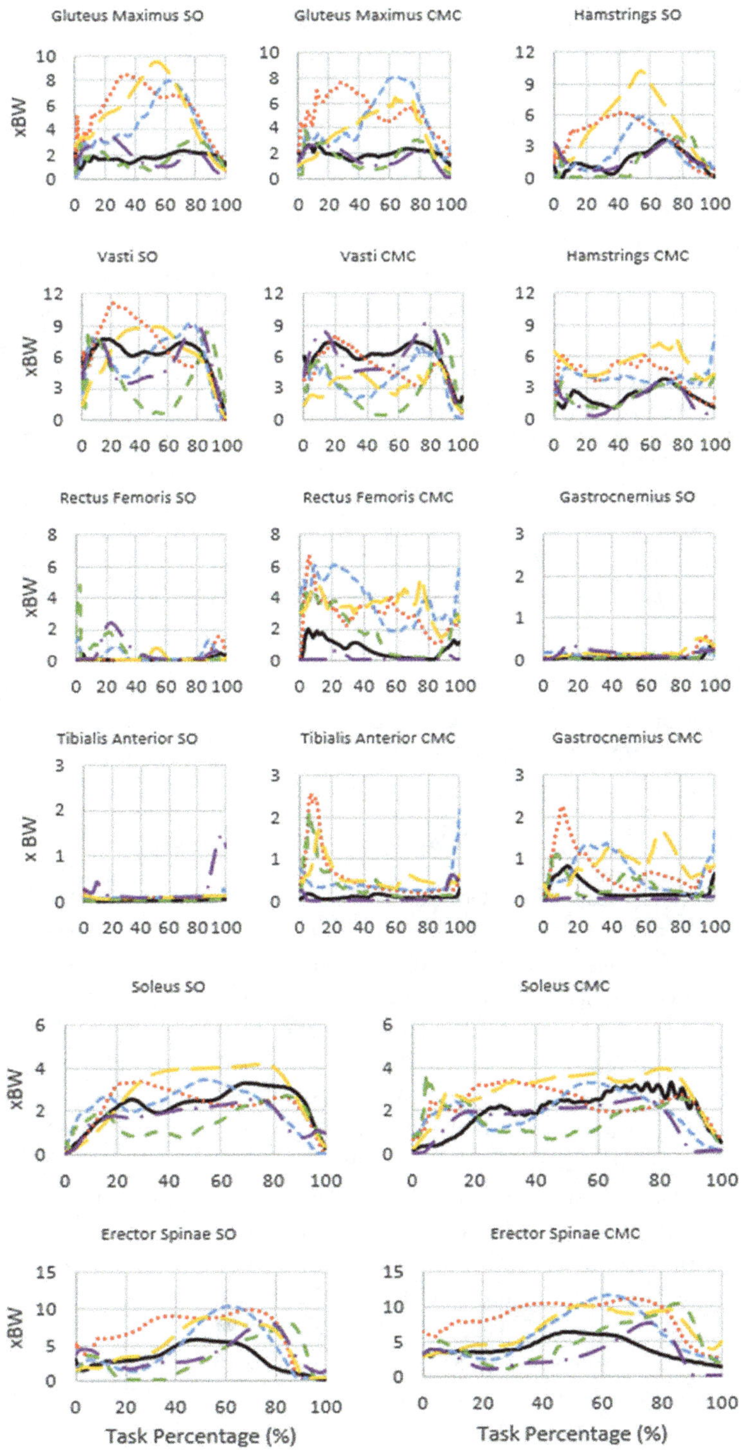

Fig. 1. Resulting muscle forces obtained for all the subjects in this work from SO and CMC.

The net internal joint moments obtained from inverse dynamics are caused by the action of internal forces, which results in moments across a joint axis, and balance out the moments created by forces acting externally to the rigid body, creating moments across the same joint axis. Hence, it is expected that the curves of the joint moments along each of the available anatomical planes at a particular joint contradict the general behaviour of the correspondent kinematic data.

By evaluating the results of the muscle forces presented in Fig. 1 below, one can see that the muscle which exert the most force are the *vasti*, *gluteus maximus*, hamstrings and *erector spinae*. This was expected, since the *gluteus maximus* is the main contributor to the extensor moment at the hip joint, the *vasti* is the protagonist in the entensor moment observed at the knee joint, working synergistically with the hamstrings along this same joint, and the *erector spinae* are responsible for the lumbar extensor moment present during this task.

By observing the values present in Table 1, stronger correlations occur along the muscles *vasti* (0.712 ± 0.292), *gluteus maximus* (0.619 ± 0.277), *soleus* (0.755 ± 0.255) and *erector spinae* (0.855 ± 0.150). These muscles only span one joint. Moderate to weak correlations arise when comparing both methods in biarticular muscles, such as the hamstrings (0.422 ± 0.475), *rectus femoris* (0.356 ± 0.404), *gastrocnemius* (0.325 ± 0.387), with the *tibialis anterior* (−0.211 ± 0.321) showing a weak negative correlation.

Table 1. Correlation coefficient between SO and CMC estimation of muscle forces and RMS magnitude differences in terms of bodyweight (BW).

Correlation coefficient						
Muscles	Subject					
	1	2	3	4	5	6
Gluteus maximus	0.356	0.749	0.892	0.265	0.532	0.921
Vasti	0.775	0.862	0.905	0.677	0.144	0.908
Hamstrings	0.846	0.583	−0.352	0.32	0.2	0.937
Erector spinae	0.934	0.975	0.95	0.681	0.645	0.944
Soleus	0.865	0.899	0.889	0.292	0.623	0.963
Rectus femoris	0.465	−0.432	0.327	0.686	0.521	0.567
Gastrocnemius	0.267	0.394	0.518	−0.271	0.153	0.888
Tibialis anterior	−0.017	−0.494	−0.234	0.292	0.623	0.963
RMS magnitude differences in x BW						
Gluteus maximus	0.472	1.296	0.651	1.096	2.683	0.474
Vasti	0.931	2.596	2.44	1.974	3.879	0.906
Hamstrings	0.661	1.286	2.649	1.552	2.931	0.485
Erector spinae	0.724	1.898	1.265	2.387	2.843	1.337
Soleus	0.652	0.507	0.612	0.84	0.82	0.306
Rectus femoris	0.917	3.016	3.885	2.168	3.574	0.775
Gastrocnemius	0.316	0.769	0.664	0.49	0.826	0.118
Tibialis anterior	0.096	0.761	0.478	0.816	0.642	0.217

4 Discussion

The results obtained for both SO and CMC showed similar profiles and magnitudes regarding muscles, such as the *gluteus maximus*, *vasti*, *erector spinae* and *soleus*, whilst for the rest of the muscles presented, which are the rectus femoris and hamstrings, slight differences are depicted, which can be explained by two different reasons. Firstly, regarding biarticular muscles, like the *rectus femoris* and *gastrocnemius*, CMC estimates higher forces than SO, possibly due to the planar knee model implemented in this simulations, which only requires the attainment of muscle forces along the sagittal plane. In addition to this, the fact that SO does not take into account muscle activation dynamics may be a significant drawback for explosive tasks such as the one being studied in this work. Secondly, SO only needs to provide a set of muscle forces that will satisfy the net joint moments, kinematic properties and ground reaction forces of the task, so it will tilt towards the muscles with higher maximum isometric force to perform to get the end results. Hence, muscles such as the *tibialis anterior* may suffer from this and result in higher muscle force estimate using CMC, as it was observed.

The large values for the force estimates taken for the muscles is representative of the explosiveness inherent to the task being analyzed. *Gluteus maximus*, in this task, works as a stabilizer by supporting both the HAT segment and pelvis upon the femur head during the entirety of the task and by eccentrically controlling the forward bending motion of the HAT segment. Co – contraction is observed around the knee joint, as both the *vasti* and hamstrings work together to stabilize the knee joint. *Vasti* also provides support by absorbing the impact of the braking portion of the task, by offsetting the action of the knee flexion activity from the hamstrings and by aiding in the final stage of the movement.

Supporting these inferences, studies showed similar findings for single – leg triple hop test [3], jumping [5] and running [23]. *Rectus femoris* operates largely as a hip flexor during this task, leaving the vasti to be the main knee extensor during this task. Concerning the soleus, one of the main plantarflexors of the foot, works, along with the vasti, to absorb the impact of the braking phase, mainly by preventing the anterior translation of the tibia.

With respect to the *erector spinae*, they play a major role in lumbar rotation. Consequently, they also participated in hip rotation as well as hip flexion. They also provided support in keeping the back of the athletes straight throughout the whole task. All athletes showed an ipsilateral side lumbar bending during the movement.

According to Lin et al. [19], SO is the more robust and efficient of the two methods for muscle force estimation, however using SO for high-velocity tasks that require large amounts of force might not be the best of choices. Therefore, for these types of tasks, CMC might be the better choice, as it takes into account the dynamics of the movement, as well as activation dynamics.

5 Conclusions

To begin with, the musculoskeletal model used in this work is a valid option to portray reliable results in this work. Also, performing multiple passages of RRA prior to the other steps in this work revealed to be key to minimize the effect of kinematic inconsistencies and modelling assumptions.

Furthermore, muscles synergies are in agreement with the joint moments and measured kinematic data. Both SO and CMC predicted similar results in terms of force profile and magnitudes during an abrupt A/P deceleration task, albeit caution must be taken when biarticular muscles, such as the hamstrings or gastrocnemius, are concerned. However, work including electromyography (EMG) data should be done to have an extra validation step. To this end, CMC might be the better choice to analyze these types of tasks.

Acknowledgments. This work was supported by CIPER-FCT (PEST-OE/SAU/UI447/2014 & UID/DTP/00447/2013).

References

1. Bergmann, G., Deuretzbacher, G., Heller, M., Graichen, F., Rohlmann, A.: Hip contact and gait patterns from routine activities. J. Biomech. **34**, 859–871 (2001)
2. Renström, P., Arms, S.W., Stanwyck, T.S., Johnson, R.J., Pope, M.H.: Strain within the anterior cruciate ligament during hamstring and quadriceps activity. Am. J. Sports Med. **14** (1), 83–87 (1986)
3. Alvim, F.C., Lucareli, P.R.G., Menegaldo, L.L.: Predicting muscle forces during the propulsion phase of single leg triple hop test. Gait Posture **59**, 298–303 (2018)
4. Pandy, M.G., Zajac, F.E., Sim, E., Levine, W.S.: An optimal control model for maximum-height human jumping. J. Biomech. **23**(12), 1185–1198 (1990)
5. Pandy, M.G., Zajac, F.E.: Optimal muscular coordination strategies for jumping. J. Biomech. **24**(1), 1–10 (1991)
6. Anderson, F.C., Pandy, M.G.: Static and dynamic optimization solutions for gait are practically equivalent. J. Biomech. **34**(2), 153–161 (2001)
7. Erdemir, A., McLean, S., Herzog, W., van den Bogert, A.J.: Model-based estimation of muscle forces exerted during movements. Clin. Biomech. **22**(2), 131–154 (2007)
8. Crowninshield, R.D., Johnstonn, R.C., Andrews, J.G., Brand, R.A.: A biomechanical investigation of the human hip. J. Biomech. **11**, 75–85 (1978)
9. Glitsch, U., Baumann, W.: The three-dimensional determination of internal loads in the lower extremity. J. Biomech. **30**(11–12), 1123–1131 (1997)
10. Lin, Y.C., Walter, J.P., Banks, S.A., Pandy, M.G., Fregly, B.J.: Simultaneous prediction of muscle and contact forces in the knee during gait. J. Biomech. **43**(5), 945–952 (2010)
11. Edwards, W.B., Gillette, J.C., Thomas, J.M., Derrick, T.R.: Internal femoral forces and moments during running: implications for stress fracture development. Clin. Biomech. **23** (10), 1269–1278 (2008)
12. Schache, A.G., Kim, H.J., Morgan, D.L., Pandy, M.G.: Hamstring muscle forces prior to and immediately following an acute sprinting-related muscle strain injury. Gait Posture **32**(1), 136–140 (2010)

13. Crowninshield, R.D., Brand, R.A.: A physiologically based criterion of muscle force prediction in locomotion. J. Biomech. **14**(11), 793–801 (1981)
14. Davy, D.T., Audu, M.L.: A dynamic optimization technique for predicting muscle forces in the swing phase of gait. J. Biomech. **20**(2), 187–201 (1987)
15. Hatze, H.: The complete optimization of a human motion. Math. Biosci. **28**(1–2), 99–135 (1976)
16. Thelen, D.G., Anderson, F.C., Delp, S.L.: Generating dynamic simulations of movement using computed muscle control. J. Biomech. **36**(3), 321–328 (2003)
17. Thelen, D.G., Anderson, F.C.: Using computed muscle control to generate forward dynamic simulations of human walking from experimental data. J. Biomech. **39**(6), 1107–1115 (2006)
18. Delp, S.L., et al.: OpenSim: open-source software to create and analyze dynamic simulations of movement. IEEE Trans. Biomed. Eng. **54**(11), 1940–1950 (2007)
19. Lin, Y.-C., Dorn, T.W., Schache, A.G., Pandy, M.G.: Comparison of different methods for estimating muscle forces in human movement. Proc. Inst. Mech. Eng. Part H J. Eng. Med. **226**(2), 103–112 (2011)
20. Mokhtarzadeh, H., et al.: A comparison of optimisation methods and knee joint degrees of freedom on muscle force predictions during single-leg hop landings. J. Biomech. **47**(12), 2863–2868 (2014)
21. Delp, S.L.L., Loan, J.P.P., Hoy, M.G.G., Zajac, F.E., Topp, E.L., Rosen, J.M.: An interactive graphics-based model of the lower extremity to study orthopedic surgical procedures. IEEE Trans. Biomed. Eng. **37**, 757–767 (1990)
22. Anderson, F.C., Pandy, M.G.: A dynamic optimization solution for vertical jumping in three dimensions. Comput. Methods Biomech. Biomed. Eng. **2**(3), 201–231 (1999)
23. Hamner, S.R., Delp, S.L.: Muscle contributions to fore-aft and vertical body mass center accelerations over a range of running speeds. J. Biomech. **46**(4), 780–787 (2013)

Dynamical Rheological Properties of *In-Silico* Epithelial Tissue by Vertex Models

Takuya Toyoshima and Yukitaka Ishimoto(✉)

Akita Prefectural University, 84-4 Ebinokuchi,
Yurihonjo, Akita 015-0055, Japan
ishimoto@akita-pu.ac.jp

Abstract. Many biological systems including humans are made up of various cells. In addition, cells of a specific type gather together to form tissues, and organs are organized as an aggregate of multiple tissues. One of the tissues constituting organs is epithelial tissue. Epithelial tissue is composed of epithelial cells, and almost all the surfaces inside and outside the body is covered with epithelial tissue, many of which are monolayered. Mathematical models expressing morphogenesis of epithelial tissue have been developed by many researchers. One of the models, the cell vertex model [1] discusses tissue growth through polygonal approximation of cells, but it does not take into account the curvature of cell boundaries. Ishimoto *et al.*, developed the bubbly vertex model [2] introducing the curvature into the model and studied the morphology of the epithelial tissue.

In this study, we focus on two mathematical models of epithelial tissue: the cell vertex model and the bubbly vertex model. We analyze the rheological properties of the models by simulation and aim at finding a correspondence with actual viscoelastic epithelial tissues.

Keywords: Vertex model · Bubbly vertex model · Epithelial tissue · Rheology · Viscoelasticity

1 Introduction

A number of researchers have developed mathematical models that express the morphology of epithelial tissue, which we focuses on in this research: the cell vertex model (VM) [1, 3, 4], the cell-centered model [5], the cellular Potts model [6, 7] and so on. Among them, vertex model discussed tissue growth by polygonal approximation of cells, but it does not take into account the curvature of cell boundaries. In 2014, Ishimoto *et al.* developed the bubbly vertex model (BVM) [2] introducing the curvature into this model and studied the morphology of the epithelial tissue.

Morphogenesis of epithelial tissues is nothing but the formation of multicellular mechanical structures by intercellular communication and intracellular activities. Computer-aided elucidation of such formation mechanisms has been awaited for further applications. However, correspondence between existing simulation models and epithelial tissues has not been established due to the complexity of even the basic physical properties of the tissues [8]. In addition, the properties may vary depending on morphogenetic stages and the affiliated organs.

G. A. Ateshian et al. (Eds.): CMBBE 2019, LNCVB 36, pp. 632–638, 2020.
https://doi.org/10.1007/978-3-030-43195-2_53

In this study, we will quantify the rheological properties of the two models (VM and BVM) by simulation, aiming at establishing a correspondence with various epithelial tissues with viscoelasticity.

2 Rheology and Complex Modulus

In this study, we use some concepts of rheology to analyze the viscoelastic properties of epithelial tissue. Since epithelial tissue is a collection of epithelial cells, and the cells are composed of many proteins, epithelial tissue can be considered as a highly self-assembled macromolecular substance. Polymeric substances are said to be typical viscoelastic bodies. Since a viscoelastic body is a substance having both viscosity and elasticity, in the mechanical behavior of the tissue viscous fluid and ideal elastic responses would be combined in a complex manner.

One of the simplest viscoelastic models is the Maxwell model. The system can be represented by a dashpot and a spring arranged in series (Fig. 1).

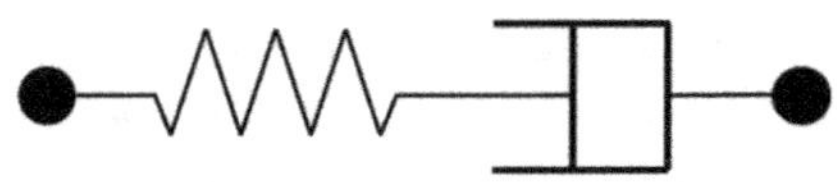

Fig. 1. Simple Maxwell model.

To quantify the mechanical properties of this linear viscoelastic model, the relationship between stress and strain is used when strain $\epsilon(t)$ changes periodically as in Eq. (1).

$$\epsilon(t) = \epsilon_0 \cdot e^{i\omega t}, \tag{1}$$

where ϵ_0 is the strain amplitude and ω is the angular velocity. Since, in general, there is found a phase lag between the strain $\epsilon(t)$ and the acting stress $\tau(t)$, the stress $\tau(t)$ is represented with the phase difference δ by

$$\tau(t) = \tau_0 \cdot e^{i(\omega t + \delta)}, \tag{2}$$

where τ_0 is the stress amplitude. Setting $|G| = \tau_0/\epsilon_0$, $G' = |G|\cos\delta$ and $G'' = |G|\sin\delta$ for simplicity, Eq. (2) becomes

$$\tau(t) = |G| \cdot \epsilon_0 \cdot e^{i\omega t} \cdot e^{i\delta} = |G| \cdot e^{i\delta} \cdot \epsilon. \tag{3}$$

Here, $|G|$ is called the absolute dynamic elastic modulus, $|G'|$ is the storage modulus, $|G''|$ is the loss modulus, and G^* is expressed by the sum of G' and G'' as

$$G^* = G' + iG'', \tag{4}$$

where G^* is the complex modulus. In what follows, we will quantify the complex modulus of the two models. Note that we usually do not intend to do shear test of the tissues in the lateral direction. So, we do not consider the shear modulus but the definitions of the moduli given above in what follows.

3 Numerical Analysis of *In-Silico* Epithelial Tissues by Mathematical Models

3.1 Vertex Model

In the VM, cells in tissue are approximated by polygons, and these polygons are expressed by vertices, and edges connecting them. Here, the vertices represent multicellular connections and the edges represent cell boundaries. In this model, the edges are defined by straight lines, and in most cases, it is assumed that each vertex is connected to only three edges. Also, the energy in the network is described as

$$E(\{\boldsymbol{x}_i\}) = \sum\nolimits_{\langle i,j\rangle} \Lambda_{ij} l_{ij} + \sum\nolimits_{\alpha} \frac{\Gamma_\alpha}{2} L_\alpha^2 + \sum\nolimits_{\alpha} \frac{\kappa_\alpha}{2} \left(A_\alpha - A_\alpha^0\right)^2 \tag{5}$$

using the potential function [3], where i and j are adjacent vertices, and α labels a cell. $\boldsymbol{x}_i$ is the 2D position of the ith vertex, *i.e.*, the junction of the cell boundaries. Λ_{ij} is a constant, Γ_α and κ_α are nonnegative constants, l_{ij} is the cell edge length, and L_α is the cell perimeter. A_α is the area of cell α and A_α^0 is its preferred area. The area "constraint" of the third term is introduced as a "mechanism to make the cell size uniform" in [4].

The equation of motion at the vertex position is given by

$$\eta_i \frac{d\boldsymbol{x}_i}{dt} = -\frac{\partial E(\{\boldsymbol{x}_i\})}{\partial \boldsymbol{x}_i}. \tag{6}$$

The left hand side of Eq. (6) is the friction force, and η_i represent the friction coefficient. The right hand side is the potential force, and the inertia term is ignored in this equation.

3.2 Bubbly Vertex Model

In the VM, the cell boundary is expressed by linear approximation, but one can introduce curvature to it as in the BVM [2]. In the BVM, each cell has a unique hydrostatic pressure, and each cell boundary has a unique curvature resulting from the mechanical properties of the boundary and the surrounding pressure. The general form of the potential energy function of this model is defined in [2] as

$$\begin{aligned} E(\{\boldsymbol{x}_i\}, \{R_{ij}\}) = {} & \sum\nolimits_{\langle i,j\rangle} \Lambda_{\alpha\beta} l_{ij} + \sum\nolimits_{i,j} \frac{\Gamma_{\alpha\beta}^{(l)}}{2} \left(l_{ij}\right)^2 + \sum\nolimits_{\alpha} \frac{\Gamma_\alpha^{(L)}}{2} L_\alpha^2 + \sum\nolimits_{\alpha} \Big\{ \kappa_\alpha^{(1)} A_\alpha + \\ & \frac{\kappa_\alpha^{(2)}}{2} \left(A_\alpha - A_\alpha^{(0)}\right)^2 \Big\} - \sum\nolimits_{\alpha} \int_{A_\alpha^{(0)}}^{A_\alpha} dS_\alpha \Pi_\alpha(S_\alpha) - P_{outer}\left(\sum\nolimits_{\alpha} A_\alpha\right), \end{aligned} \tag{7}$$

where $\{R_{ij}\}$ is the curvature radius of the cell boundaries, α and β are the cell number. The pair $\langle i,j \rangle$ represents the cell boundary connecting vertices i and j. $\Lambda_{\alpha\beta}$ and $\Gamma^{(l)}_{\alpha\beta}$ are constants dependent on the cells α and β on both sides of the cell boundary, and $\Gamma^{(L)}_{\alpha}$ is a nonnegative constant for cell α. l_{ij} is the cell boundary length. $\kappa^{(1,2)}_{\alpha}$ is a nonnegative number depending on the cell height of epithelial tissue [4]. dS_α is the surface element of the cell α for the integration of the osmotic pressure Π_α, and P_{outer} is the pressure in the outer region of the tissue when the tissue boundary exists.

The equation of motion of the BVM is given by

$$\eta_i \frac{d(\boldsymbol{x}_i - \boldsymbol{X}_i)}{dt} = -\frac{\partial E(\{\boldsymbol{x}_i\})}{\partial \boldsymbol{x}_i}. \tag{8}$$

The left hand side of Eq. (8) is the friction, and the right hand side is the potential force. The variable $\boldsymbol{X}_i$ is introduced as a referential position to the vertex i. The noise term is suppressed for convenience, and the inertia term is neglected as in the VM.

3.3 Analysis Method and Initial Conditions

Most cell shapes in epithelial tissue show hexagonal shapes in a mechanically stable state, as Staple *et al.* analyzed the behavior of cell shapes in tissue based on the relationship between Λ and Γ in Eq. (6) [3]. In our study, in order to analyze the rheological properties of the tissue in which the hexagon seems to be in the mechanical stable state, analysis is performed by changing parameters $\Lambda_{\alpha\beta}$, $\Gamma^{(l)}_{\alpha\beta}$ and $\Gamma^{(L)}_{\alpha}$ within the parameter range for the stable hexagonal lattice in [3].

As a method of evaluating the simulation, complex modulus is used to quantify the rheological properties of the epithelial tissue, and its mechanical indications is to be analyzed (Fig. 2).

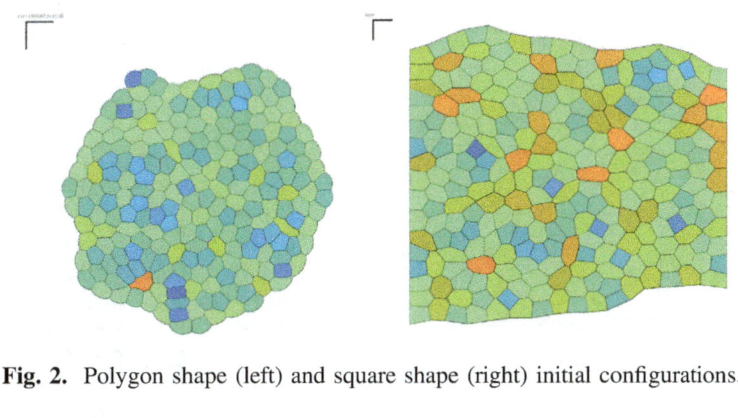

Fig. 2. Polygon shape (left) and square shape (right) initial configurations.

We use two types initial configurations for a tissue shape (polygon and square) for different *in-silico* experiments, and the polygonally shaped tissue is composed of

217cells, and the square-shaped tissue is composed of 250 cells. Here, each cell is randomly arranged first by Voronoi division, and the tissue is relaxed mechanically enough to be the equilibrated initial configuration.

4 Results and Discussion

In this section, we present our simulation results of viscoelastic measurement of the polygonal stress test in the VM and BVM, and the results of the square stretch test in the VM. In these simulations, the polygonal tissue is isotopically stressed from the outside of the tissue, and in the square tissue, the left side of the tissue is fixed and the stress is applied to the right side so that a prescribed strain is achieved.

Figure 3 shows the storage and loss moduli of the polygonal stress test in the VM and BVM cases. Here, there are 5 data points, and a spline curve interpolation is applied. The complex moduli are converted from the complex compliances.

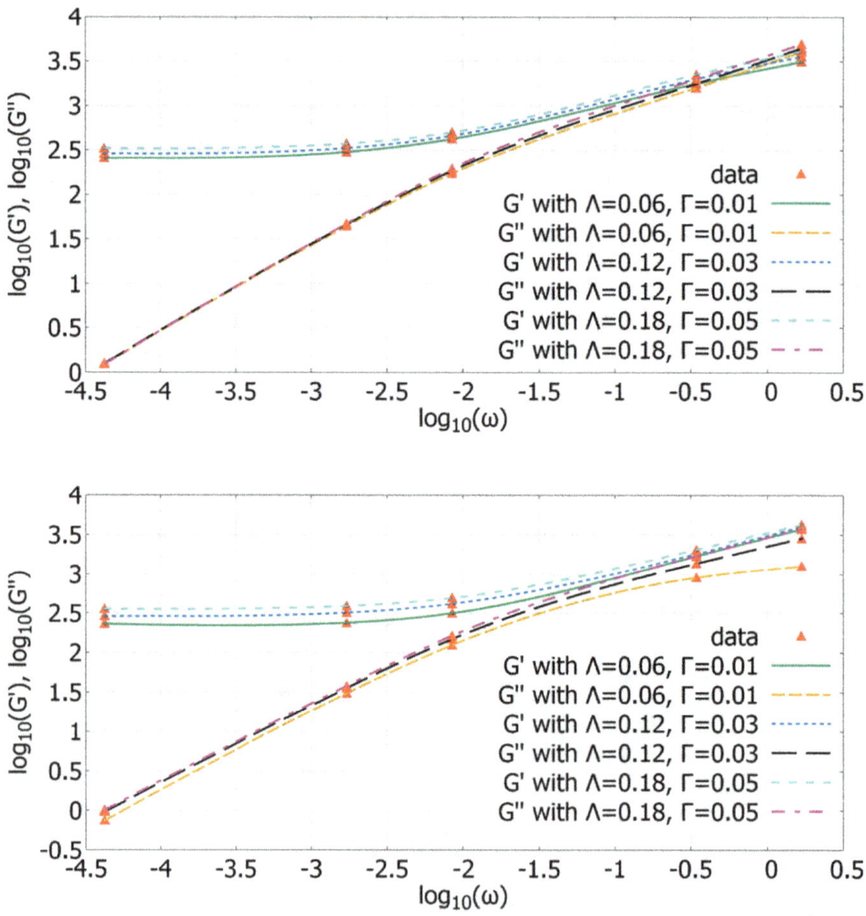

Fig. 3. Viscoelastic properties of the VM (top) and the BVM (bottom) in the polygonal tissue.

In Fig. 3, the tissue behaves in a solid-like behavior in the range where G' is larger than G'', and the liquid-like behavior in the range where G'' is larger than G'. In the VM, G' and G'' intersect in the vicinity of $\log_{10} \omega = -0.5$ in each parameter, but in the BVM, it can be seen that both moduli do not intersect in the frequency range for this simulation. From this result, the BVM is considered to exhibit complex viscoelasticity in a wider area compared to the VM.

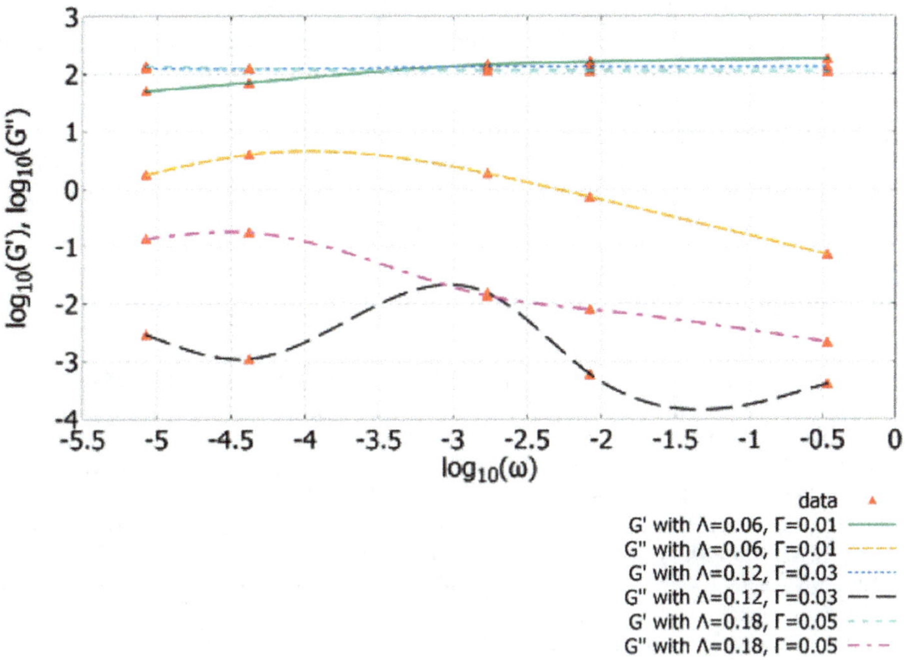

Fig. 4. Viscoelastic properties of the VM in the square tissue.

Figure 4 shows the simulation results of the VM in a square tissue. Here, there are 5 data points, and the same curve interpolation is applied. The viscoelastic properties of the square tissue by the stretch test are largely different from the properties plotted in Fig. 3. A reason why they are largely different is because, in the latter test, a tissue is adhered to a substrate such as PDMS and vibrated together. The different friction works between the substrate and the tissue.

Among the tested parameter values in the stretch test, at $\Lambda = 0.12$ and $\Gamma = 0.03$, the loss modulus has a different response from the others. With regard to storage modulus, there is no significant change, so the test might investigate only the viscosity component of the tissue. To deepen the discussion, it would be interesting to see whether this phenomenon is unique to this set of values and, if so, the set characterizes the tissue.

5 Conclusion

We analyzed the viscoelastic properties of polygonal and square epithelial tissues using the cell vertex model and the bubbly vertex model. In this study, simulation parameters were set for the tissue to be hexagonally stable state, and the viscoelasticity of the epithelial tissue was confirmed from the rheological analysis by *in-silico* experiments. It would be better to continue to analyze the viscoelastic properties with more parameter values in the future.

Although the initial configurations used in these simulations are constructed from random Voronoi division, it would be more implicative to use initial configurations constructed from experimental images.

Acknowledgments. We would like to thank Prof. H. Honda, Dr. P. Marcq, and Dr. M. Merkel for their discussions and comments. We would like to thank all of the members of our laboratory for their continuous support and stimulating discussions. This work was supported by JSPS Kakenhi (Grant-in-Aid for Scientific Research) Grant Number 17K00410.

References

1. Honda, H.: Geometrical models for cells in tissues. Int. Rev. Cytol. **81**, 191–248 (1983)
2. Ishimoto, Y., Morishita, Y.: Bubbly vertex dynamics: a dynamical and geometrical model for epithelial tissues with curved cell shapes. Phys. Rev. E **90**, 052711 (2014)
3. Staple, D.B., Farhadifar, R., Röper, J.C., Aigouy, B., Eaton, S., Jülicher, F.: Mechanics and remodelling of cell packing in epithelia. Eur. Phys. J. E **33**(2), 117–127 (2010)
4. Nagai, T., Honda, H.: A dynamic cell model for the formation of epithelial tissues. Philos. Mag. B **81**, 699 (2001)
5. Salazar-Ciudad, I., Jernvall, J.: A computational model of teeth and the developmental origins of morphological variation. Nature **464**, 583 (2010)
6. Graner, F., Glazier, J.A.: Simulation of biological cell sorting using a two-dimensional extended potts model. Phys. Rev. Lett. **69**, 2013 (1992)
7. Glazier, J.A., Graner, F.: Simulation of the differential adhesion driven rearrangement of biological cells. Phys. Rev. E **47**, 2128 (1993)
8. Fujii, Y., Ochi, Y., Tuchiya, M., Kajita, M., Fujita, Y., Ishimoto, Y., Okajima, T.: Spontaneous spatial correlation of elastic modulus in jammed epithelial monolayers observed by AFM. Biophys. J. **116**, 1152–1158 (2019)

Male Sphincter-Urinary System Behavior Under a Valsalva Maneuver

António André[1(✉)], Sérgio Pinto[1], and Pedro Martins[2]

[1] INEGI, Porto, Portugal
{aandre,spinto}@inegi.up.pt
[2] FEUP and INEGI, Porto, Portugal
palsm@fe.up.pt

Abstract. Male incontinence is a real problem for an aging adult population and it can be described as the incapacity to hold urine.

With this paper, the authors intend to explore a methodology to simulate the male urinary system using the Finite Element Method (FEM). In this approach we consider Valsalva manoeuvre, a procedure that increases the intrathoracic pressure, thus affecting the abdominal pressure.

This preliminary work aims to prepare a basis for understanding the physiological changes in the male urinary system with an abdominal pressure increase, such as a sneeze and its impact on the male urinary continence mechanisms, whose failure often leads to incontinence.

Keywords: Valsalva manoeuvre · Urinary incontinence · Male urinary system · Finite Element Model (FEM)

1 Introduction

The male sphincter-urinary complex has unique characteristics that influence its normal physiology and pathophysiology [1]. Urinary incontinence (UI) is counted among the most common medical conditions, as supported by the 6 million men (estimated) affected just in the United States of America [2].

This medical problem has a huge impact on the quality of life and overall patients' health, which can result in higher rates of depression and social isolation [2].

An abdominal pressure increase is commonly associated to urine leakage for both men an women. When this phenomenon occurs after the Valsalva manoeuvre it is often called the Valsalva voiding [3]. Valsalva manoeuvre is a procedure that increases the intrathoracic pressure, causing physiological changes in the human body [4]. In the absence of bladder contraction, when vesical pressure exceeds the urethral pressure, it can cause urine leakage [5] for those affected by UI. This occurs, for example, when someone sneezes, coughs or jumps.

This preliminary work aims to prepare a basis for understanding the physiological changes in the male urinary system with an abdominal pressure increase,

G. A. Ateshian et al. (Eds.): CMBBE 2019, LNCVB 36, pp. 639–644, 2020.
https://doi.org/10.1007/978-3-030-43195-2_54

such as a sneeze and its impact on the male urinary continence mechanisms, whose failure often leads to incontinence. UI can affect the daily life of men thus leading to severe psychological and social problems. Ultimately, our objective is to provide a significant understanding of incontinence mechanics, that helps to minimize the social problems caused by UI.

2 FEM Model Development

The anatomical models used for the male urinary system were reconstructed based on typical human male anatomical features obtained from three different anatomy atlas [6–8]. The organs considered in this study were the bladder and the urethra, since they are crucial for the urinary continence mechanisms. Our main objective was to investigate the interaction of an increasing external pressure with the bladder. The outcome of such interaction will determine the urine flow through the urethra - in particular the occurrence of UI, characterized by a failure of the continence mechanisms.

It was difficult to obtain relevant information on the mechanical properties of tissues and structures from the male urinary system. In fact the work from Barzegari et al. [5] was the main source for the mechanical properties used in this work. This led us to consider a reduced FEM model of the male urinary system composed by the urethra and the bladder. The initial geometry considered, represents the anatomy of urethra and bladder for the typical human male.

To assemble the FEM model required the preparation of a three-dimensional geometrical mesh. Two additional steps had to be carried out. The work-flow for mesh preparation includes FEMAP [9] followed by ABAQUS [10].

FEMAP was used to correct the mesh geometry of each component, which means deleting, creating or remeshing nodes and elements of the anatomical model. ABAQUS [10], was used to assemble the finite element model, with boundary conditions, loads, displacements and mechanical properties. The procedure carried out in ABAQUS will be detailed in next sections. The resulting model can be observed in Fig. 1.

2.1 Mechanical Properties

Every type of material, including human tissue, skin, bones or organs, can be modelled through a constitutive model with adequate mechanical properties. This characterization should be intrinsically connected to the mechanical behaviour of the studied materials. However, in the consulted literature we were not able to find adequate mechanical properties for the tissues (organs) considered. The reference properties were for a linear model. In this work we compromise by considering that tissues behave linearly, while in reality their mechanical behaviour is nonlinear (hyperelastic). This allowed us to use the mechanical properties from literature [5].

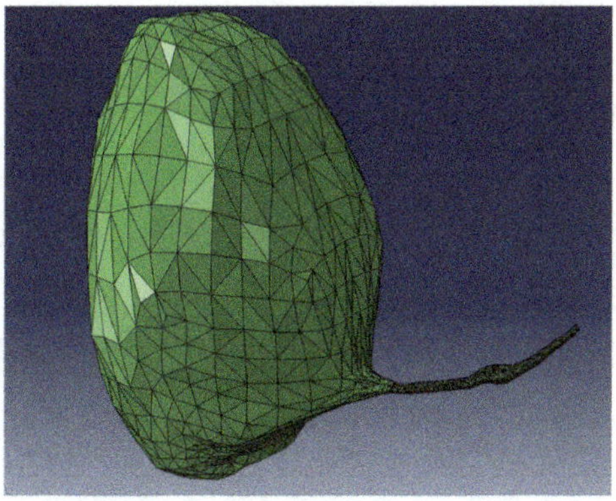

Fig. 1. The assembled model, including the bladder and urethra.

To prepare the FEM model, the mechanical properties must be attributed to all components (organs) considered in the simulation. This will allow to simulate the behaviour of the model developed under the conditions of a Valsalva manoeuvre, i.e., corresponding to a sneeze.

Taking into account our choices for material behaviour and properties, the model was defined with a Young Modulus (E) of 0.05 MPa, a Poisson ratio (ν) of 0.45 and a density of $1030\,kg/m^3$ [5]. Urine was assumed to be present inside the whole model (both bladder and the urethra). This aims to replicate as truly as possible a real-life situation. The urine density was approximated to water density, $1\,g/cm^3$ [5].

3 Simulation

The goal of the FEM simulation was to study the interaction between the bladder walls and the urethra, with urine. Such interaction takes place under an increasing pressure due to external factors, such as a sneeze.

The male urinary system model was prepared for simulation following the steps detailed on the previous sections. This Section will present the FEM simulation procedure used in ABAQUS. The procedure consists in defining the Step, Interaction, Loads, Mesh and finally creating the Job. All of this process was done using a SPH (smoothed particle hydrodynamics) simulation [11].

Regarding the Step, since the objective is to simulate what happens to the urine inside the bladder when pressure outside increases, a "Dynamic, Explicit" procedure was adopted. A time period of 0.2 s was applied, as well as non-linear effects of large deformations and displacements activated (Nlgeom).

The global model was defined considering the bladder with urine inside, therefore the bladder-urine interaction was considered in the simulation. A contact instance was created to make sure the global model kept together during the simulation.

To approximate the circumstances typically found in the human body, and to properly define the problem mathematically, the boundary conditions were defined in ABAQUS. The applied external loads must be established as well, thus simulating the external pressure caused by a sneeze. This value was set to 7 kPa [5].

Figure 2, show that both the bladder and the end of urethra are clamped.

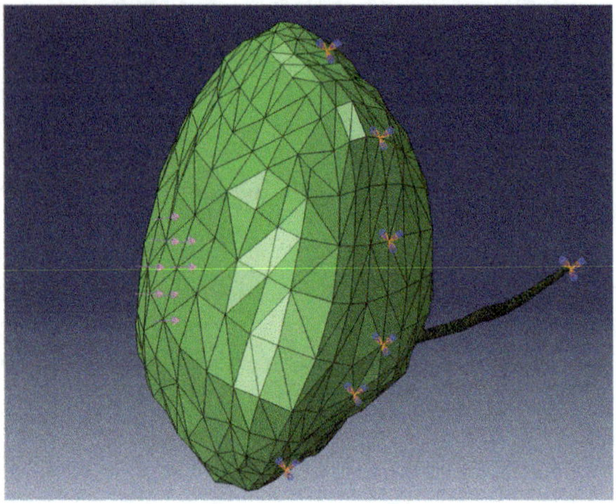

Fig. 2. Boundary conditions and loads applied.

The next step is to define the Mesh, and in order to accomplish that, the model representing the bladder was assembled using S3R elements (3 node triangular element), because it was assumed as a shell, and the urine by C3D4 elements (tetrahedron element with 4 nodes). Figure 3 illustrates the elements used in both.

After these procedures, the Job can be created and the FE simulation can be executed.

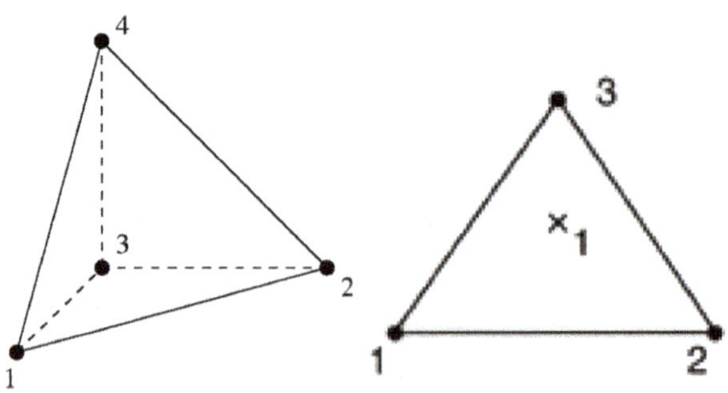

Fig. 3. C3D4 element (left) and S3R element (right) [10].

4 Results and Discussion

The displacements suffered by the organs (bladder and urethra), when an external pressure is applied to the bladder, can be seen in Fig. 4. The results are consistent with the pressure application, in terms of displacement direction and intensity. These preliminary results agree with literature [5] despite the simplicity of the model.

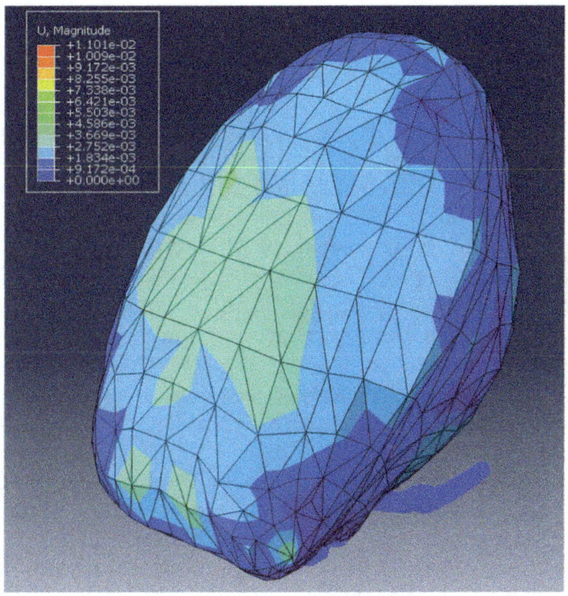

Fig. 4. Results of the simulation.

This study constitutes a first effort to model the male urinary system using FEM. While a realistic geometry was considered, there are significant limitations to this work. The mechanical behaviour of the tissues considered is typically nonlinear, except if affected by carcinoma and/or a few other pathologies. This means that hyperelastic constitutive models should have been considered, instead of a linear model. The mechanical properties should be estimated adequately, according to the target model.

This step towards a (more) realistic simulation of the male urinary system is significant, but there are other improvements such as the inclusion of the urinary sphincter. This muscle structure has an active functional role (of preventing urine leaks) in the continence mechanism, and therefore must be included in future iterations.

The prostate should be include as well, since its normal anatomical position contributes to the continence mechanism. There is ample evidence of prostate's contribution to continence, with incontinence being a common unfortunate outcome of radical prostatectomy.

Acknowledgements. The authors gratefully acknowledge funding from FCT, Portugal, under grant SFRH/BPD/111846/2015 and projects UROSPHINX - Project 16842 (COMPETE2020), and MImBI - PTDC/EME-APL/29875/2017 financed through FEDER and FCT.

References

1. Martins, P., et al.: A virtual reality approach to the study of the male urinary system. In: 8th World Congress of Biomechanics, July 2018
2. Irwin, G.M.: Urinary incontinence. Prim. Care Clin. Off. Prat. **46**, 233–242 (2019)
3. Ullrich, N.F.E., Comiter, C.V.: The male sling for stress urinary incontinence: urodynamic and subjective assessment. J. Urol. **172**(1), 204–206 (2004)
4. Low, P.A.: Valsalva maneuver. Encycl. Neurol. Sci. **2**, 591–592 (2014)
5. Barzegari, M., Vahidi, B., Safarinejad, M.R.: A clinical and finite element study of stress urinary incontinence in women using fluid-structure interactions, August 2017
6. Netter, F.H.: Atlas of Human Anatomy. Elsevier, Amsterdam (2006)
7. Seeley, R.R., Stephens, T.D., Tate, P.: Anatomy & Physiology, 6th edn. McGraw-Hill, New York (2003)
8. Visible Body: Human Anatomy Atlas. Argosy Publishing, Inc. 2007–2013
9. Siemens. https://www.plm.automation.siemens.com/global/pt/products/simcenter/femap.html
10. Dassault Systems. https://www.3ds.com/products-services/simulia/products/abaqus/
11. Hibbit and Karlsson & Sorensen: Abaqus Theory Manual Version 5.5 (1995)

Virtual Reality Tools Applied to the Male Urinary System: Visualization and Interaction

Sérgio Pinto[1(✉)], António André[1], and Pedro Martins[2]

[1] INEGI, Porto, Portugal
spinto@inegi.up.pt
[2] FEUP/INEGI, Porto, Portugal

Abstract. The male urinary system (MUS) can be affected by pathologies which disrupt the patient's quality of life. The consequences of the common treatments are not fully studied and the procedures are not a guarantee for the complete recovery of the physiological functions. In this context, imageology is used for planning or training, but may not provide the best data for medical experts. As an improved approach, a virtual reality tool is suggested, allowing the interaction with 3D models portraying the MUS, in both healthy and pathological states. This constitutes a limitless way of exploring scientific and medical data, for both training and monitoring clinical cases. The models used in this application were reconstructed using anatomy atlas and imported into Unity, responsible for creating the virtual reality scenario. The Oculus Rift device is used for the visualization and Leap Motion for detection of the user's hands, creating an intuitive and interactive system. Some functionalities were implemented using C# scripts to animate the simulations and control the position of the models by dragging them across the 3D space, as well as interact with them using a menu with several buttons, attached to the user's left palm. The models are stored in color coded cubes, which can be used to open and close the desired simulations. The environment was constructed to promote a familiar feel, thus reducing the "disorganization" associated with large amounts of information.

Keywords: Virtual reality · Leap Motion · Male urinary system · Biomechanics

1 Introduction

The human body is composed by multiple systems which interact and that, together, contribute to the normal functioning of the whole and the welfare of the person. However these systems may be affected by pathologies which may disrupt the health state, either at tissue or organ levels. This may compromise the integrity of the patient's condition and negatively impact its daily life.

One of those systems is the male urinary system (MUS), which may be disturbed by pathologies such as urinary incontinence, benign prostatic hyperplasia

G. A. Ateshian et al. (Eds.): CMBBE 2019, LNCVB 36, pp. 645–654, 2020.
https://doi.org/10.1007/978-3-030-43195-2_55

or oncological diseases [1–3]. These may have great impacts in the normal physiology and lead to more severe problems.

The common treatments for these pathologies may sometimes include drastic procedures, such as the removal of entire organs (for example, the prostatectomy), and the outcomes of such proceedings are not fully understood yet. Moreover, the diagnosis tools more commonly used by the medical community nowadays consist in outdated imaging techniques, frequently in two dimensions, and with limited physical space, which does not allow the observation of multiple cases at the same time. This method may not be beneficial and may mislead the medical professionals [4,5]. This implies that further studies in this area are needed, and the health professionals may be needing new ways to understand and study clinical cases [4–10].

In order to provide the medical community, researchers and health educators with an alternative way to study the MUS, a new tool is proposed, using virtual reality (VR) techniques. In this approach, the user is immersed in a virtual environment, where he is able to visualize and interact with anatomical models (both healthy and pathological) using his own hands, manipulating them across the three-dimensional space, performing gestures and triggering some functionalities in real-time, with the referred limitations.

In the developed application, some clinical cases can be found, using pathological data and the mechanical properties of the tissues found in the literature, with the objective of mimicking the real-life scenarios. This resulted in a new tool that grants the user the ability to interact with realistic cases, which may be patient-specific, and that may help the training and teaching of medical professionals.

2 Methodology

Using medical atlas and anatomy books, the anatomical models for the MUS were reconstructed considering the typical features of the male body [11,12]. In the current state of the application, only the bladder, prostate and a portion of the urethra were considered for the finite element simulations. However, other structures may be added at any time with relative ease.

To grant these organs with behaviors that resemble those of the actual human body, the mechanical properties found in the literature were introduced into the simulation software (Abaqus). These simulations are continuously being improved and expanded by the work team and the methods used can be found in previous works [13–15].

Whenever a selected case is simulated and some results are obtained, it is divided into segments spaced at a constant temporal pace, each of these being a different frame of that simulation. To introduce these frames into the virtual environment, each of them needs to be extracted and converted to a compatible format.

This process starts with exporting the frame from Abaqus as a VRML file (Virtual Reality Modeling Language), which is able to maintain the properties of the different colors resultant of the simulation process, and that retains the physical meaning regarding the stresses or the displacement of the tissues, for example.

The next step is to use the software Blender, where some unnecessary data (automatically introduced in the previous step) is removed from each frame. When only the model portraying the organ remains, each frame is exported into a new format called Filmbox (FBX), directly compatible with the VR software, Unity.

After the application of the described process (summed up in Fig. 1) to each of the frames of the simulation, they are aggregated inside Unity into sets (corresponding to the appropriate simulated case), called *gameObjects*, as shown in Fig. 2. This provides a simple way to apply functionalities to all of the frames in a single simulation, which was important since an array of C# scripts were developed to animate each simulation independently.

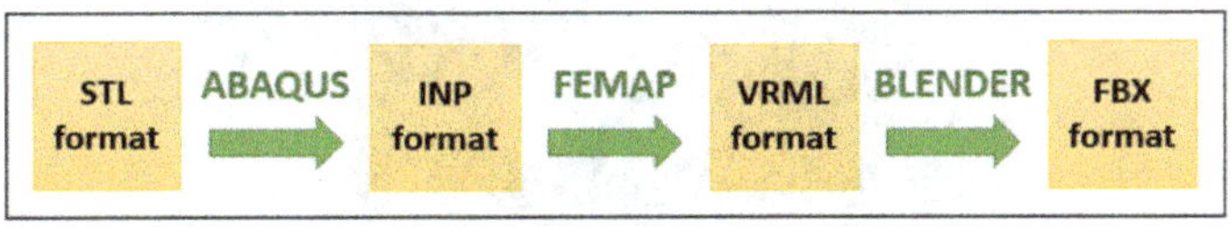

Fig. 1. Conversion process applied to every frame of each simulation.

For this purpose, a script was associated with each *gameObject* containing a set of frames, responsible for determining the total number of frames in the hierarchy below, and finally activating the next and deactivating the current one, in a sequential order, resulting in an animation of the original simulation, now inside Unity.

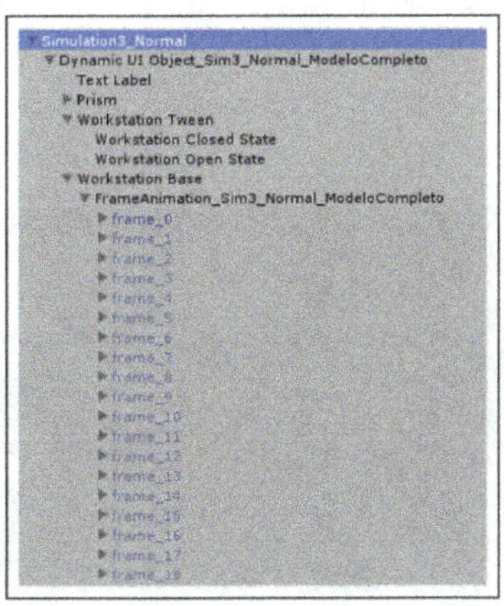

Fig. 2. Hierarchical distribution of the gameObjects.

As a result of this process, some animated simulations are already introduced in the VR software, so the next step is the inclusion of the VR devices. The Oculus Rift is an head-mounted display capable of showing a virtual environment to the user in an immersive way, while at the same time tracking its position. The Leap Motion device provides the ability to detect each segment of the user's hands and fingers, and includes some ready-to-use packages and resources, aiding the development and integration of interactive tools in VR. This setup (Fig. 3) has a straightforward implementation into the most recent versions of Unity.

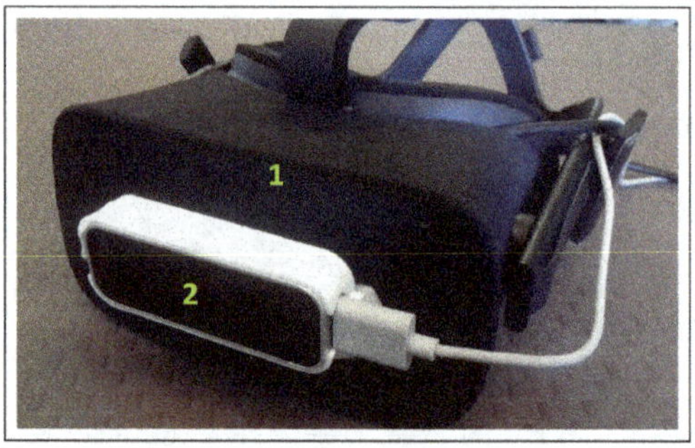

Fig. 3. Hardware setup used. 1 - Oculus Rift; 2 - Leap Motion.

To keep the simulations organized at a reference point in the virtual environment, some of the resources provided by Leap Motion were used, such as the "anchor" behavior. This feature allows an object (in this case a colored cube) to have a corresponding place where only that specific cube can be stored (Fig. 4).

Whenever the cube is in its anchor or is being grabbed by the virtual hands of the user, the simulations are closed inside the cube. If the user then throws the cube in front of him or releases it away from the anchor radius, the simulation inside the cube expands and appears, and can be observed and manipulated by the user.

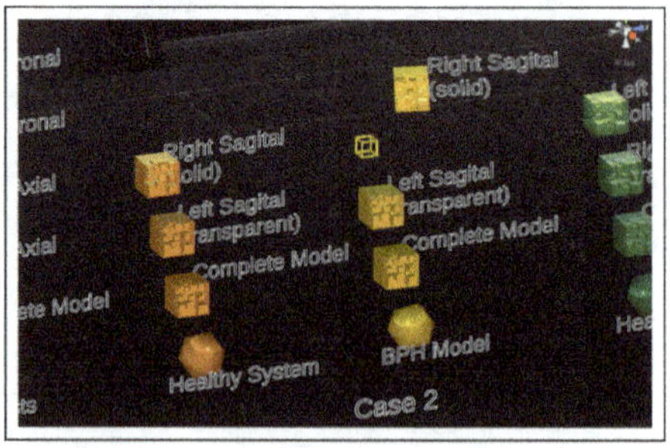

Fig. 4. Anchor behavior for each cube containing the simulations.

The opposite behavior is also implemented, so when the user grabs the cube again, the simulation is closed, and if that cube is released near its anchor, it can be pulled to its place automatically, keeping the environment clean and organized, which is an important feature, since there are many clinical cases, and one of the goals of the application is to allow the most intuitive use possible by medical professionals.

Another main objective of the application is the immersion and the familiar feel that the virtual environment conveys to the user. To achieve this, a real-life look alike room was constructed. This feature allowed the introduction of some furniture, walls, floor and ceiling, with realistic textures, which provides some visual references to the user, reducing the motion sickness that is sometimes caused when the brain feels tricked by deceitful visual queues. In order to reduce even more the sterile feeling from a virtual environment, some plant "life" and daily-based objects were introduced in the scene (Fig. 5).

Fig. 5. Constructed virtual environment.

The furniture also provided a support to display useful information (such as the description of the clinical cases and tips on how to use the implemented functionalities). The table was chosen as the place where the different cubes containing the simulations are stored, providing an easily accessible area within user reach.

To further improve the organization, the cubes belonging to the same clinical case were compacted into a single object, which folds and unfolds depending on the proximity of the user's hands.

Two features were implemented to interact with the anatomical models. The first allows the user to move and rotate each model in the three-dimensional space, simply by grabbing the model itself. This is possible since every model is surrounded by an invisible collision box that is able to intersect and interact with the virtual hands, allowing multiple perspectives and arrangements of the clinical cases. The second functionality consists in a graphical interface (menu), containing some buttons and sliders with associated C# scripts; these are triggered when pressed by the user. This menu can be found attached to the user's left palm, giving him the freedom to observe the simulations while simultaneously using this feature.

At the current state of the application, the menu consists in three collapsible panels (Fig. 6). One of these panels contains some information on the implemented scripts, and is called "Info Panel". The other two contain the referred buttons and sliders. The first of these, called "Rotation Control Panel", is an alternative way to rotate the models around each axis, with more precision. The second panel, labeled "Frame Control Panel", allows the user to stop or resume the animation at any given time, or navigate through all the frames (Fig. 7).

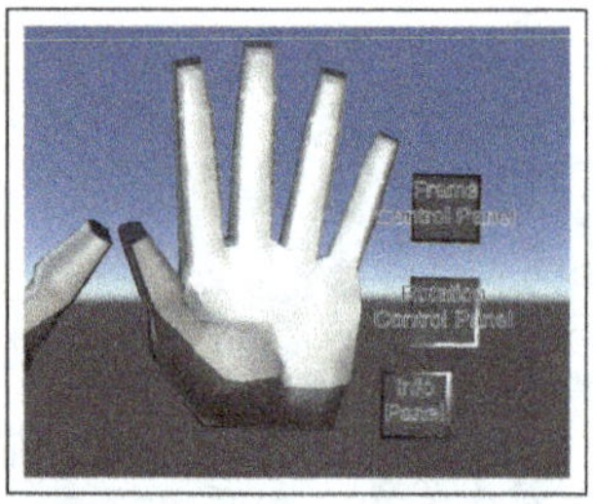

Fig. 6. Collapsed menu, attached to the left palm of the user.

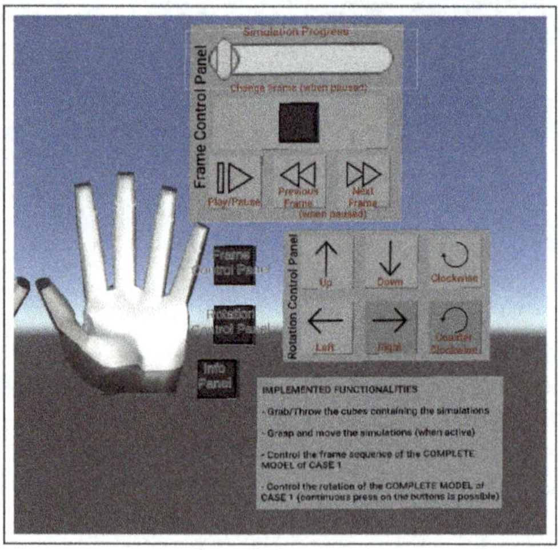

Fig. 7. Panels in the menu, with implemented functions triggered with buttons and sliders.

3 Results

The current version of the VR application includes multiple simulations, on several clinical cases found in literature, namely: cancerous prostatic tissue, benign prostatic hyperplasia, among others. All of these models are able to be compared with healthy cases, and observed side by side, at the same time (Fig. 8).

The pipeline used to add additional simulations is already established, so new clinical cases can be easily introduced to the application, and it can even be extended to other scientific domains, which gives the project the ability to be fully customized.

The main functionalities are implemented and the VR application has been tested in medical and engineering environments, with mostly positive feedback on its usability for medical training and teaching.

The resulting tool is a very intuitive and easy-to-use application, with high potential for positive outcomes in the medical educational system [4,7–10]; as a way to improve the storage environment of clinical cases; or to evaluate the evolution of a patient through time [5].

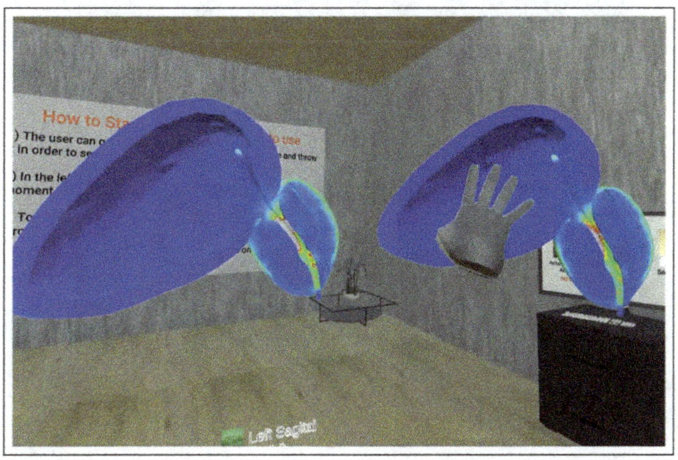

Fig. 8. Comparing multiple clinical cases, side by side.

4 Discussion

The described approach using VR in the study of the pathologies and physiology of the MUS may prove to be an upgrade to the current methods which are being used in the assessment of clinical cases, as well as in the teaching of new health professionals [6]. In the current stage of the application, the user is able to choose any model, change to his preferred perspective (Fig. 9), and even enter inside the tissues or analyse the interior surfaces of the MUS organs (Fig. 10).

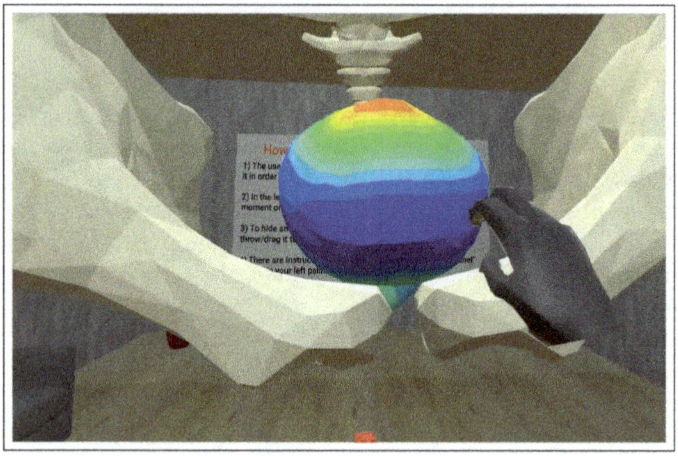

Fig. 9. Manipulation of the position of a model.

Fig. 10. Entering inside the tissues and observing the interior surfaces.

There is the possibility of (arbitrary) expansion of the amount of clinical cases, since the virtual environment can be fully customized to accommodate new cases while maintaining the organization of the space.

The proposed tool can be adapted to suit multiple scenarios, making it a versatile solution able to cover multiple domains where VR would have a positive impact, creating beneficial outcomes and complementing the current methods.

Compared to the current techniques, this application has several advantages in the diagnosis and study of pathologies in the MUS, such as the ability interact directly with anatomical models using the user's own hands and comparing multiple cases at once. This high level of interaction is achieved without the need of integrating remote controls or physical two-dimensional images, which deviates from other approaches [9,16–19].

The positive feedback received, despite it being in preliminary versions, may be taken as a positive indicator for the implementation of this solution with useful outcomes to the medical community and the healthcare system [5,8]. As for future developments, augmented reality is being looked as a next step for coming works.

Acknowledgments. The authors gratefully acknowledge funding from FCT, Portugal, under grant SFRH/BPD/111846/2015 and projects UROSPHINX - Project 16842 (COMPETE2020), and MImBI - PTDC/EME-APL/29875/2017 financed through FEDER and FCT.

References

1. Gacci, M., et al.: Best practice in the management of storage symptoms in male lower urinary tract symptoms: a review of the evidence base. Ther. Adv. Urol. **10**(2), 79–92 (2018)
2. Sinha, S., et al.: Exploration of male urethral sphincter complex using diffusion tensor imaging (DTI)-based fiber-tracking. J. Magn. Reson. Imaging **48**, 1002–1011 (2018)
3. Lim, K.B.: Epidemiology of clinical benign prostatic hyperplasia. Asian J. Urol. **4**(3), 148–151 (2017). https://doi.org/10.1016/j.ajur.2017.06.004
4. Shirk, J.D., Kwan, L., Saigal, C.: The use of 3-dimensional, virtual reality models for surgical planning of robotic partial nephrectomy. Urology (2019). https://doi.org/10.1016/j.urology.2018.12.026
5. Hamacher, A., et al.: Virtual reality and simulation for progressive treatments in urology. Int. Neurourol. J. **22**(3), 151–160 (2018)
6. Hoogenes, J., et al.: A randomized comparison of 2 robotic virtual reality simulators and evaluation of trainees' skills transfer to a simulated robotic urethrovesical anastomosis task. Urology **111**, 110–115 (2018)
7. Parkhomenko, E., et al.: Pilot assessment of immersive virtual-reality renal models as an educational and preoperative planning tool for percutaneous nephrolithotomy. J. Endourol. **33**, 283–288 (2019)
8. Schulz, G.B., et al.: Validation of a high-end virtual reality simulator for training transurethral resection of bladder tumors. J. Surg. Educ. **76**, 568–577 (2019). https://doi.org/10.1016/j.jsurg.2018.08.001
9. Rai, A., et al.: Patient-specific virtual simulation - a state of the art approach to teach renal tumor localization. Urology **120**, 42–48 (2018)
10. Neumann, E., et al.: Transurethral resection of bladder tumors: next-generation virtual reality training for surgeons. Eur. Urol. Focus (2019). https://doi.org/10.1016/j.euf.2018.04.011
11. Netter, F.H.: Atlas of Human Anatomy. Saunders/Elvesier, Philadelphia (2006)
12. Seeley, R.R., Stephens, T.D., Tate, P.: Anatomy & Physiology, 6th edn. McGraw-Hill, New York (2003)
13. Pinto, S., et al.: Application of virtual reality techniques to a birth simulation. In: 4th Experiment@International Conference (exp.at 2017) (2017)
14. Martins, P., et al.: A virtual reality approach to the study of the male urinary system. In: 8th World Congress of Biomechanics, July 2018
15. André, A., Pinto, S., Martins, P.: Male sphincter-urinary system: model development and fem simulation. In: 8° Congresso Nacional de Biomecânica (2019)

16. Silva, J.N.A., et al.: Emerging applications of virtual reality in cardio-vascular medicine. JACC Basic Transl. Sci. **3**(3), 420–430 (2018)
17. Chamunyonga, C., Burbery, J., Rutledge, P.: Utilising the virtual environment for radiotherapy training system to support undergraduate teaching of IMRT, VMAT, DCAT treatment planning, and QA concepts. J. Medical Imaging Radiat. Sci. **49**(1), 31–38 (2018)
18. Boudissa, M., et al.: Computer-assisted surgery in acetabular fractures: virtual reduction of acetabular fracture using the first patient-specific biomechanical model simulator. Orthop. Traumatol. Surg. Res. **104**, 359–362 (2018)
19. van Oosterom, M.N., et al.: Computer-assisted surgery: virtual- and augmented-reality displays for navigation during urological interventions. Curr. Opin. Urol. **28**(2), 205–213 (2018)

Author Index

G. A. Ateshian et al. (Eds.): CMBBE 2019, LNCVB 36, pp. 655–657, 2020.
https://doi.org/10.1007/978-3-030-43195-2

CPSIA information can be obtained
at www.ICGtesting.com
Printed in the USA
LVHW082118120420
653141LV00002B/52

9 783030 431945